PREFACE TO THE FIR

This book is based on a one year course I held at the University of Heidelberg and on a series of lectures I gave at the "Autumn College on Techniques in Many-Body Problems" at Lahore, Pakistan, in november of 1987. These lectures have been published in the proceedings to this school by World Scientific (Rothe, 1989). I was later encouraged by the editors of World Scientific to expand on the material presented at the autumn college. This I have done in this book.

The purpose of my lectures at Lahore was to introduce lattice gauge theories to young physicists who may not have the opportunity to attend a course on this subject at their home universities. I had therefore kept the discussion as elementary as possible, including only enough thechnical details to enable the reader to follow the published literature on this subject. In this book I have expanded substantially on the material presented at Lahore, and have included a number of technical details which I felt would be very helpful to those readers who may want to carry out research in this branch of elementary particle physics. I did, however, arrange the material in such a way that those physicists who are mainly interested in getting a bird eyes view on the subject can safely skip the technical parts, without the danger of getting lost at a later stage. This concerns, in particular, the discussion in sections 4 and 5 of chapter 4 on lattice fermions, and the weak coupling expansion in lattice quantum chromodynamics (QCD), chapter 14. I have included this material for the readers convenience, since it is not discussed in such detail in the literature. I also decided to include a chapter on the path integral formalism, since the entire book is based on the path integral approach to quantization, and I do not assume that everybody is familiar with this formalism. Those readers that have never come in touch with the path integral formulation of quantum field theory may find this chapter a bit technical. However, the results we derive, of which we will make ample use in this book, are very simple, and are easily understood by everybody.

This book is mainly addressed to graduate students interested in particle physics. But it is also of interest to physicists actively engaged in research in the field of lattice gauge theories, and who may want to get a more general view on this subject. It assumes that the reader has a fair background in quantum field theory. A moderate knowledge of the continuum formulation of quantum chromodynamics would certainly be very helpful. Also physicists working in statistical mechanics may profit from reading this book, since the lattice formulation of field theories resembles closely that of complex statistical mechanical systems.

The book is divided in two parts. In the first part, comprising chapters 1 to 16, I discuss the zero temperature formulation of field theories on a space-time lattice, and in particular QCD. They are the lattice analogues of the usual continuum field theories discussed in standard text books. The second part, consisting of chapters 17 to 20, deals with finite temperature field theory. The emphasis will be on QCD, but I shall use a scalar field theory to introduce the reader to a number of new concepts which play an important role in finite temperature QCD.

Since the main goal of this book is to stimulate the readers interest in this fascinating branch of elementary particle physics, I have taken an optimistic standpoint, selecting some results of Monte Carlo calculations which illustrate the phenomena in a particularily dramatic way. I did not attempt to present a critical analysis of the results, and have left it to the reader to confer the original literature. Nor did I attempt to give a complete list of references, which the reader can find in the numerous proceedings to lattice conferences. More detailed discussions of most of the topics presented in this book can be found in the proceedings to various schools. An introduction to lattice gauge theories can also be found in the monograph by M. Creutz: *Quarks, Gluons and Lattices*, published by Cambridge University Press (1983).

Hopefully this book will stimulate some of the readers to carry out some research in the field of lattice gauge theories. If so, I have achieved the purpose it has been written for.

I like to take this opportunity to thank a number of colleages for their constructive criticisms and for having read several chapters of this book. In particular I am grateful to A. Actor, I. Bender, D. Gromes, F. Karsch, K.H. Mütter, I.O. Stamatescu and W. Wetzel. I am especially grateful to W. Wetzel for having checked a number of formulae, and for his extensive technical help in getting the manuscript into its final form. I also want to express my gratitude to Mrs. U. Einecke, and Mrs. M. Steiert for having typed so patiently the manuscript in TEX. Finally, I am particularily thankful to my family, whose continued support has made this book possible. In particular my children had to dispense of their father for many (!) hours.

CONTENTS

CHAPTER 1

INTRODUCTION

It is generally accepted that quantum field theory is the appropriate framework for describing the strong, electromagnetic and weak interactions between elementary particles. As for the electromagnetic interactions, it has been known for a long time that they are described by a quantum gauge field theory. But that the principle of gauge invariance also plays a fundamental role in the construction of a theory for the strong and weak interactions has been recognized only much later. The unification of the weak and electromagnetic interactions by Glashow, Salam and Weinberg was a major breakthrough in our understanding of elementary particle physics. For the first time one had been able to construct a renormalizable quantum field theory describing simultaneously the weak and electromagnetic interactions of hadrons and leptons. The "electro-weak" theory of Glashow, Salam and Weinberg is based on a non-abelian $SU(2) \times U(1)$ gauge symmetry, which is broken down spontaneously to the U(1) symmetry of the electromagnetic interactions. This breaking manifests itself in the fact that, in contrast to the massless photon, the particles mediating the weak interactions, i.e., the W^+, W^- and Z^0 vector bosons, become massive. In fact they are very massive, which reflects the fact that the weak interactions are very short ranged. The detection of these particles constituted one of the most beautiful tests of the Glashow-Salam-Weinberg theory.

The fundamental fermions to which the vector bosons couple are the quarks and leptons. The quarks, which are the fundamental building blocks of hadronic matter, come in different "flavours". There are the "up", "down", "strange", "charmed", "bottom" and "top" quarks. The weak interactions can induce transitions between different quark flavours. For example, a "u" quark can convert into a "d" quark by the emission of a virtual W^+ boson. The existence of the quarks has been confirmed (indirectly) by experiment. None of them have been detected as free particles. They are permanently confined within the hadrons which are built from the different flavoured quarks and antiquarks. The forces which are responsible for the confinement of the quarks are the strong interactions. Theoretical considerations have shown, that the "up", "down", etc., quarks should come in three "colours". The strong interactions are flavour blind, but sensitive to colour. For this reason one calls the theory of strong interactions *Quantum Chromodynamics*, or in short, QCD. It is a gauge theory based on the unbroken non-abelian SU(3)-colour group (Fritzsch and Gell-Mann, 1972; Fritzsch,

Gell-Mann and Leutwyler, 1973). The number "3" reflects the number of colours carried by the quarks. Since there are eight generators of SU(3), there are eight massless "gluons" carrying a colour charge which mediate the strong interactions between the fundamental constituents of matter. By the emission or absorption of a gluon, a quark can change its colour.

QCD is an asymptotically free theory (t'Hooft, 1972; Politzer, 1973; Gross and Wilczek, 1973). Asymptotic freedom tells us that the forces between quarks become weak for small quark separations. Because of this asymptotic freedom property it was possible for the first time to carry out quantitative perturbative calculations of observables in strong interaction physics which are sensitive to the short distance structure of QCD.* In particular it allowed one to study the Bjorken scaling violations observed in deep inelastic lepton nucleon scattering at SLAC. QCD is the only theory we know that can account for these scaling violations.

The asymptotic freedom property of QCD is intimatley connected with the fact that it is based on a non-abelian gauge group. As a consequence of this non-abelian structure the coloured gluons, which mediate the interactions between quarks, can couple to themselves. These self couplings, one believes, are responsible for quark confinement. Since the coupling strength becomes small for small separations of the quarks, one can speculate that the forces may become strong for large separations. This could explain why these fundamental constituents of matter have never been seen free in nature, and why only colour neutral hadrons are observed. A confirmation that QCD accounts for quark confinement can however only come from a non-perturbative treatment of this theory, since confinement is a consequence of the dynamics at large distances where perturbation theory breaks down.

Until 1974 all predictions of QCD were restricted to the perturbative regime. The breakthrough came with the lattice formulations of QCD by Kenneth Wilson (1974), which opened the way to the study of non-perturbative phenomena using numerical methods. By now lattice gauge theories have become a branch of particle physics in its own right, and their intimate connection to statistical mechanics make them of interest to elementary particle physicists as well as to physicists working in the latter mentioned field. Hence also those readers who are not acquainted with quantum field theory, but are working in statistical mechanics, can profit from a study of lattice gauge theories. Conversely, elementary particle physicists have profited enormously from the computational methods used in statistical mechanics, such as the high temperature expansion, cluster expansion, mean field

---

* for an early review see Politzer (1974).

approximation, renormalization group methods, and numerical methods.

Once the lattice formulation of QCD had been proposed by Wilson, the first question that physicists were interested in answering, was whether QCD is able to account for quark confinement. Wilson had shown that within the strong coupling approximation QCD confines quarks. As we shall see, however, this is not a justified approximation when studying the continuum limit. Numerical simulations however confirm that QCD indeed accounts for quark confinement.

There are of course many other questions that one would like to answer: does QCD account for the observed hadron spectrum? It has always been a dream of elementary particle physicists to explain why hadrons are as heavy as they are. Are there other particles predicted by QCD which have not been observed experimentally? Because of the self-couplings of the gluons, one expects that the spectrum of the Hamiltonian also contains states which are built mainly from "glue". Does QCD account for the spontaneous breakdown of chiral symmetry? It is believed that the (light) pion is the Goldstone Boson associated with a spontaneous breakdown of chiral symmetry. How do the strong interactions manifest themselves in weak decays? Can they explain the $\Delta I = 1/2$ rule in weak non-leptonic processes? How does hadronic matter behave at very high temperatures and/or high densities? Does QCD predict a phase transition to a quark gluon plasma at sufficiently high temperatures, as is expected from general theoretical considerations? This would be relevant, for example, for the understanding of the early stages of the universe.

An answer to the above mentioned questions requires a non-perturbative treatment of QCD. The lattice formulation provides the only possible framework at present to study QCD non-perturbatively.

The material in this book has been organized as follows. In the following chapter we first discuss in some detail the path integral formalism in quantum mechanics, and the path integral representation of Green functions in field theory. This formalism provides the basic framework for the lattice formulation of field theories. If the reader is well acquainted with the path integral method, he can skip all the sections of this chapter, except the last. In chapters 3 and 4 we then consider the lattice formulation of the free scalar field and the free Dirac field. While this formulation is straight-forward for the case of the scalar field, this is not the case for the Dirac field. There are several proposals that have been made in the literature for placing fermions on a space-time lattice. Of these we shall discuss in detail the Wilson and the Kogut-Susskind fermions, which have been widely used in numerical simulations, and introduce the reader to Ginsparg-

Wilson fermions, which have become of interest in more recent times, but whose implementation in numerical simulations is very time consuming. In chapters 5 and 6 we then introduce abelian and non-abelian gauge fields on the lattice, and discuss the lattice formulation of QED and QCD.

Having established the basic theoretical framework, we then present in chapter 7 a very important observable: The Wilson loop, which plays a fundamental role for studying the confinement problem. This observable will be used in chapter 8 to calculate the static potential between two charges in some simple solvable models. The purpose of that chapter is to verify in some explicit calculations that the interpretation of the Wilson loop given in chapter 7, which may have left the reader with some uneasy feelings, is correct. In chapter 9 we then discuss the continuuum limit of QCD and show that this limit, which is realized at a critical point of the theory where correlations lengths diverge, corresponds to vanishing bare coupling constant. Close to the critical point the behaviour of observables as a function of the coupling constant can be determined from the renormalization group equation. Knowledge of this behaviour will be crucial for establishing whether one is extracting continuum physics in numerical simulations.

Chapter 10 is devoted to the discussion of the Michael lattice action and energy sum rules, which relate the static quark-antiquark potential to the action and energy stored in the chromoelectric and magnetic fields of a $q\bar{q}$-pair. These sum rules are relevant for studying the energy distribution in the flux tube connecting a quark and antiquark at large separations.

Chapters 11 to 15 are devoted to various approximation schemes. Of these, the weak coupling expansion of correlation functions in lattice QCD is the most technical one. In order not to confront the reader immediately with the most complicated case, we have divided our presentation of the weak coupling expansion into three chapters. The first one deals with a simple scalar field theory and merely demonstrates the basic structure of Feynman lattice integrals. It also includes a discussion of an important theorem proved by Reisz, which is the lattice version of the well known power counting theorem for continuum Feynman integrals. In the following chapter we then increase the degree of difficulty by considering the case of lattice quantum electrodynamics (QED). Here several new concepts will be discussed, which are characteristic of a gauge theory. Readers having a fair background in the perturbative treatment of continuum QED will be able to follow easily the presentation. As an instructive application of lattice perturbation theory, we include in this chapter a 1-loop computation of the renormalization constant for the axial vector current with Wilson fermions, departing from a lattice regularized Ward identity. Also included is a discussion of the ABJ-anomaly within the framework of Ginsparg-Wilson fermions. The next chapter then treates the case of QCD, which from the conceptional point of view is quite similar to the case of QED, but is technically far more involved. The Feynman rules are applied to the computation of the ABJ anomaly which is shown to be independent of the form of the lattice regularized action.

At this point we leave the analytic "terrain" and discuss in chapter 16 various algorithms that have been used in the literature to calculate observables numerically. All algorithms are based on the concept of a Markov process. We will keep the discussion very general, and only show in the last two section of this chapter, how such algorithms are implemented in an actual calculation. Chapter 17 first summarizes some earlier numerical results obtained in the pioneering days. Because of the ever increasing computer power the numerical data becomes always more refined, and we leave it to the reader to confer the numerous proceedings for more recent results. We have however also included in this chapter some important newer developments which concern the vacuum structure of QCD and the dynamics of quark confinement.

The remaining part of the book is devoted to the study of field theories at finite temperature. It has been expected for some time that QCD undergoes a phase transition to a quark-gluon plasma, where quarks and gluons are deconfined. In chapter 18 we consider some simple bosonic and fermionic models, and discuss in detail the path-integral representation for the thermodynamical partition function. In particular we will construct such a representation for a simple fermionic system which is exact for arbitrary time step, and point out some subtle points which are not discussed in the literature. Chapter 19 is devoted to finite temperature perturbation theory in the continuum and on the lattice. The basic steps leading to the finite-temperature Feynman rules are first exemplified for a scalar field theory in the continuum. We then extend our discussion to the case of QED and QCD in the continuum as well as on the lattice and discuss in detail the temporal structure of the free propagator for naive and Wilson fermions. The Feynman rules are then applied to calculate the screening mass in QED and QCD in one-loop order, off and on the lattice. These computations will at the same time illustrate the power of frequency summation formulae, whose derivation has been relegated, in part, to two appendices.

Chapter 20 is devoted to non-perturbative aspects of QCD at finite temperature. The lattice formulation of this theory is the appropriate framework for studying the deconfinement and chiral phase transitions, and deviations of thermodynamical observables from the predictions of perturbation theory at temperatures well above the phase transition. In this chapter we discuss how thermodynamical observables are computed on the lattice, and introduce an order parameter (the Wilson line or Polyakov loop) which characterizes the phases of the pure gauge theory. This order parameter plays a central role in a later section, where we present some early Monte Carlo data which gave strong support for the existence of a deconfinement phase transition. The theoretical concepts introduced in this chapter are then implemented in a simple lattice model which also serves to illustrate the power of the character expansion, a technique which is used to study $SU(N)$ gauge theories for strong coupling. The remaining part of this chapter is devoted to the high temperature phase of QCD which, as already mentioned, is expected to be that of a quark gluon plasma.

The material covered in this book should enable the reader to follow the extensive literature on this fascinating subject. What the reader will not have learned, is how much work is involved in carrying out numerical simulations. A few paragraphs in a publication will in general summarize the results obtained by several physicists over many months of very hard work. The reader will only become aware of this by speaking to physicists working in this field, or if he is involved himself in numerical calculations. Although much progress has been made in inventing new methods for calculating observables on a space time lattice, some time will still pass before one has sufficiently accurate data available to ascertain that QCD is the correct theory of strong interactions.

CHAPTER 2

THE PATH INTEGRAL APPROACH TO QUANTIZATION

Since its introduction by Feynman (1948), the path integral (PI) method has become a very important tool for elementary particle physicists. Many of the modern developments in theoretical elementary particle physics are based on this method. One of these developments is the lattice formulation of quantum field theories which, as we have mentioned in the introduction, opened the gateway to a non-perturbative study of theories like QCD. Since the path integral representation of Green functions in field theory plays a fundamental role in this book, we have included a chapter on the path integral method in order to make this monograph self-contained. In the literature it is customary to derive the PI-representation of Green functions in Minkowski space. But for the lattice formulation of field theories, we shall need the corresponding representation for Green functions continued to imaginary time. Usually a rule is given for making the transition from the real-time to the imaginary-time formulation. This rule is not self-evident. Since we shall make use of it on several occasions, we will verify the rule for the case of bosonic Green functions, by deriving directly their path integral representation for imaginary time. What concerns the fermionic Green functions, we will not derive the PI-representation from scratch, but shall present strong arguments in favour of it.

In the following section, we first discuss the case of non-relativistic quantum mechanics.* The results we shall obtain will be relevant in section 2, where we derive the PI-representation of bosonic Green functions which are of interest to the lattice formulation of quantum field theories involving Bose-fields. In section 3 we then discuss the transfer matrix for bosonic systems. Green functions of fermionic operators are considered in section 4.

As we shall see, the PI-representation of Green functions is only formally defined for systems whose degrees of freedom are labeled by a continuous variable, as is the case in field theory. One is therefore forced to regularize the path integral expressions. In section 5 we discuss this problem on a qualitative level, and motivate the introduction of a space-time lattice. This, as we shall comment

* For a comprehensive discussion of the PI-method in quantum mechanics in the real-time formulation, the reader should confer the book by Feynman and Hibbs (1965).

on, corresponds in perturbation theory to a particular choice of regularization of Feynman integrals.

2.1 The Path Integral Method in Quantum Mechanics

In the Hilbert space formulation of quantum mechanics, the states of the system are described by vectors in a Hilbert space, and observables are represented by hermitean operators acting in this space. The time evolution of the quantum mechanical system is given by the Schrödinger equation, or equivalently by*

$$|\psi(t)>= e^{-iH(t-t_0)}|\psi(t_0)>, \tag{2.1}$$

where H is the Hamiltonian. Thus if we know the state of the system at time t_0, (2.1) determines the state at a later time t. Let $q = \{q_\alpha\}$ denote collectively the coordinate degrees of freedom of the system and $|q>$ the simultaneous eigenstates of the corresponding operators $\{Q_\alpha\}$, i.e.

$$Q_\alpha|q>= q_\alpha|q>, \quad \alpha = 1,\ldots,n.$$

Then (2.1) implies the following equation for the wave function $\psi(q,t) =< q|\psi(t)>$

$$\psi(q',t') = \int dq G(q',t';q,t)\psi(q,t),$$

where

$$G(q',t';q,t) =< q'|e^{-iH(t'-t)}|q> \tag{2.2}$$

is the Green function describing the propagation of the state $|\psi(t)>$, and where the integration measure is given by

$$dq = \prod_{\alpha=1}^{n} dq_\alpha.$$

A very important property of the Green function (2.2) is that it satisfies the following composition law

$$G(t',q';q,t) = \int dq'' G(t',q';q'',t'')G(q'',t'';q,t). \tag{2.3}$$

This relation follows immediately by writing $\exp(-iH(t'-t)) = \exp(-iH(t'-t''))\exp(-iH(t''-t))$ in (2.2) and introducing a complete set of intermediate states

* We set $\hbar = 1$ throughout this book.

$|q''>$ between the two exponentials. Using the property (2.3), Feynman derived a path integral representation for the matrix element (2.2), which exhibits in a very transparent way the connection between the classical and quantum theory. In classical physics the time evolution of the system is given by the Lagrange equations of motion which follow from the principle of least action. To quantize the system, one then constructs the Hamiltonian, and writes the equation of motion in terms of Poisson brackets. This provides the starting point for the canonical quantization of the theory. By proceeding in this way, one has moved far away from the original action principle. The path integral representation of Feynman reestablishes the connection with the classical action principle. In the following we derive this representation for the Green function (2.2) continued to imaginary time, $t \to -i\tau, t' \to -i\tau'$, since we shall need it in the following section.

Consider the matrix element

$$< q', t'|q, t >=< q'|e^{-iH(t'-t)}|q >, \tag{2.4}$$

where

$$|q, t >= e^{iHt}|q >$$

are eigenstates of the Heisenberg operators

$$Q_\alpha(t) = e^{iHt}Q_\alpha e^{-iHt}, \tag{2.5}$$

i.e.

$$Q_\alpha(t)|q, t >= q_\alpha|q, t > .$$

Inserting a complete set of energy eigenstates to the right and left of the exponential in (2.4), we have that

$$< q', t'|q, t >= \sum_n e^{-iE_n(t'-t)}\psi_n(q')\psi_n^*(q),$$

where $\psi_n(q) =< q|E_n >$ is the eigenfunction of H with energy E_n. The sum over n extends over the discrete as well as the continuous spectrum of the Hamiltonian. This expression can now be continued to imaginary time. Making the replacements $t \to -i\tau, t' \to -i\tau'$, we arrive at an expression which is dominated by the ground state in the limit $\tau' - \tau \to \infty$:

$$< q', t'|q, t >_{\substack{t=-i\tau \\ t'=-i\tau'}} = \sum_n e^{-E_n(\tau'-\tau)}\psi_n(q')\psi_n^*(q) .$$

The right-hand side is just the matrix element $< q'|\exp(-H(\tau'-\tau))|q >$. Hence, as expected from (2.4), the Green function continued to imaginary times is given by

$$< q', t'|q, t >_{\substack{t\to -i\tau \\ t'\to -i\tau'}} =< q'|e^{-H(\tau'-\tau)}|q > . \tag{2.6}$$

To arrive at a path integral representation for the right-hand side of (2.6), we split the time interval* $[\tau, \tau']$ into N infinitesimal segments of length $\epsilon = (\tau' - \tau)/N$. Let $\tau_1, \tau_2, ..., \tau_{N-1}$ denote the intermediate times, i.e. $\tau < \tau_1 < \tau_2 < ... < \tau'$. Then the imaginary-time Green function can be obtained by a sequence of infinitesimal time steps as follows,

$$\begin{aligned} &< q'|e^{-H(\tau'-\tau)}|q >=< q'|e^{-H(\tau'-\tau_{N-1})}e^{-H(\tau_{N-1}-\tau_{N-2})}...e^{-H(\tau_1-\tau)}|q > \\ &= \int \prod_{\ell=1}^{N-1} dq^{(\ell)} < q'|e^{-H\epsilon}|q^{(N-1)} >< q^{(N-1)}|e^{-H\epsilon}|q^{(N-2)} > ... < q^{(1)}|e^{-H\epsilon}|q >, \end{aligned} \tag{2.7}$$

where

$$dq^{(\ell)} = \prod_{\alpha} dq_{\alpha}^{(\ell)}.$$

Here $|q^{(\ell)} >$ denote the complete set of eigenstates which have been introduced in the ℓ'th intermediate time step.

In order to evaluate the matrix elements in (2.7), we must now specify the structure of the Hamiltonian. Let us assume it to be of the form

$$H = \frac{1}{2}\sum_{\alpha=1}^{n} P_{\alpha}^2 + V(Q), \tag{2.8}$$

where P_{α} are the momenta canonically conjugate to Q_{α}. Making use of the Baker–Campbell–Hausdorff formula,

$$e^A e^B = e^{A+B+\frac{1}{2}[A,B]+...},$$

we conclude that $\exp(-H\epsilon)$ can be approximated for small ϵ by

$$e^{-H\epsilon} \approx e^{-\epsilon\frac{1}{2}\sum_{\alpha} P_{\alpha}^2} e^{-\epsilon V(Q)}.$$

It follows that

$$< q^{(\ell+1)}|e^{-H\epsilon}|q^{(\ell)} >\approx< q^{(\ell+1)}|e^{-\frac{\epsilon}{2}\sum_{\alpha} P_{\alpha}^2}|q^{(\ell)} > e^{-\epsilon V(q^{(\ell)})}.$$

---

* We shall henceforth refer to τ as "time".

To evaluate the remaining matrix element, we introduce a complete set of momentum eigenstates to the right and left of $\exp\left(-\frac{\epsilon}{2}\sum_\alpha P_\alpha^2\right)$. With

$$< q|p >= \prod_{\alpha=1}^{n} \frac{1}{\sqrt{2\pi}} e^{ip_\alpha q_\alpha},$$

we have that

$$< q^{(\ell+1)}|e^{-H\epsilon}|q^{(\ell)} >$$
$$\approx e^{-\epsilon V(q^{(\ell)})} \int \prod_{\beta=1}^{n} \frac{dp_\beta^{(\ell)}}{2\pi} \prod_{\alpha=1}^{n} \exp\left\{-\epsilon\left[\frac{1}{2}p_\alpha^{(\ell)2} - ip_\alpha^{(\ell)}\left(\frac{q_\alpha^{(\ell+1)} - q_\alpha^{(\ell)}}{\epsilon}\right)\right]\right\}$$

Substituting this expression into (2.7) we arrive at the following approximate path integral representation in phase space, valid for small ϵ:

$$< q'|e^{-H(\tau'-\tau)}|q >\approx \int DqDp e^{ip_\alpha^{(\ell)}\left(q_\alpha^{(\ell+1)}-q_\alpha^{(\ell)}\right)} e^{-\epsilon H(q^{(\ell)},p^{(\ell)})} , \tag{2.9a}$$

where

$$q^{(0)} = q \quad , q^{(N)} = q' , \tag{2.9b}$$

$$DqDp = \prod_{\beta=1}^{n} \prod_{\ell=1}^{N-1} dq_\beta^{(\ell)} \prod_{\ell=0}^{N-1} \frac{dp_\beta^{(\ell)}}{2\pi} , \tag{2.9c}$$

and

$$H(q^{(\ell)}, p^{(\ell)}) = \sum_{\alpha=1}^{n} \frac{1}{2} p_\alpha^{(\ell)^2} + V(q^{(\ell)}) . \tag{2.9d}$$

Notice that the number of momentum integrations exceeds that of the coordinates.

Actually, as the reader can readily verify, the above formula holds just as well for any Hamiltonian of the form $H(Q,P) = T(P) + V(Q)$, with $T(P)$ a polynomial in the canconical momenta. For the case where $T(P)$ has the quadratic form given in (2.8), we can also obtain a configuration space path integral representation, by carrying out the Gaussian integration over the momenta. The following expression is valid for infinitessimal time slices,

$$< q'|e^{-H(\tau'-\tau)}|q >\approx \int \prod_{\ell'=1}^{N-1} \prod_{\alpha=1}^{n} \frac{dq_\alpha^{(\ell')}}{\sqrt{2\pi\epsilon}} e^{-\sum_{\ell=0}^{N-1} \epsilon L_E(q^{(\ell)},\dot{q}^{(\ell)})}, \tag{2.10a}$$

where

$$L_E(q^{(\ell)}, \dot{q}^{(\ell)}) = \sum_\alpha \frac{1}{2} \dot{q}_\alpha^{(\ell)^2} + V(q^{(\ell)}), \tag{2.10b}$$

$$\dot{q}_\alpha^{(\ell)} \equiv \frac{q_\alpha^{(\ell+1)} - q_\alpha^{(\ell)}}{\epsilon} , \tag{2.10c}$$

and $q^{(0)} \equiv q$, $q^{(N)} \equiv q'$. The subscript "E" on L_E is to remind us that we are studying the Green function in the "euclidean" formulation.* Let us interpret the right-hand side of (2.10a). Consider an arbitrary path in q-space connecting the space-time points (q, τ) and (q', τ'), consisting of straight line segments in every infinitesimal time interval. Let $q^{(\ell)}$ denote the set of coordinates of the system at time τ_ℓ (see fig. (2-1)). To emphasize this correspondence let us set

$$q_\alpha^{(\ell)} = q_\alpha(\tau_\ell) ,$$

where $\tau_0 \equiv \tau$, $\tau_N \equiv \tau'$. Then (2.10c) is the "euclidean velocity" in the time interval $[\tau_\ell, \tau_{\ell+1}]$ of a "particle" moving in an n-dimensional configuration space, and L_E is the discretized version of the classical Lagrangean in the euclidean formulation (notice the "plus" sign between the kinetic term and the potential). The action associated with the path depicted schematically in fig. (2-1) is given by

$$S_E[q] = \sum_{\ell=0}^{N-1} \epsilon \left[\sum_\alpha \frac{1}{2} \left(\dot{q}_\alpha(\tau_\ell)\right)^2 + V\left(q(\tau_\ell)\right) \right] , \tag{2.11}$$

This is the expression appearing in the argument of the exponential in (2.10a). We therefore arrive at the following prescription for calculating the Green function for imaginary time:

i) Divide the interval $[\tau, \tau']$ into infinitesimal segments of length $\epsilon = (\tau' - \tau)/N$.
ii) Consider all possible paths starting at q at time τ and ending at q' at time τ'. Approximate these paths by straight-line segments as shown in fig. (2-1), and calculate the action (2.11) for each path.
iii) Weigh each path with $\exp(-S_E[q])$ and sum these exponentials over all paths, by integrating over all possible values of the coordinates at intermediate times.
iv) Multiply the resulting expression with $(1/\sqrt{2\pi\epsilon})^{nN}$, where n is the number of coordinate degrees of freedom and take the limit $\epsilon \to 0, N \to \infty$, keeping the product $N\epsilon = (\tau' - \tau)$ fixed.

---

* In the following chapters, where we will study the PI–representation of field theories in detail, the transition to imaginary time corresponds to formulating the theories in euclidean space–time. We shall therefore refer in the following to the imaginary–time formulation as the euclidean formulation.

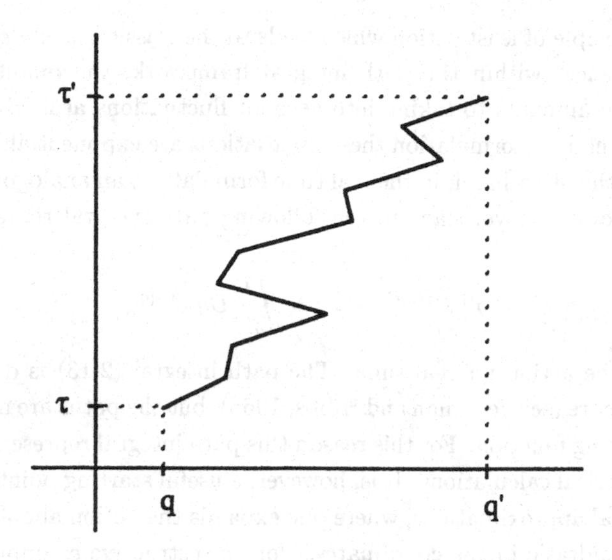

Fig. 2-1 Path connecting the space-time points (q,τ) and (q',τ') contributing to the integral (2.11a)

The result of steps i) to iv) we formally denote by

$$(q',\tau'|q,\tau) = \int_q^{q'} Dq\, e^{-S_E[q]}, \tag{2.12a}$$

where

$$S_E[q] = \int_\tau^{\tau'} d\tau'' L_E(q(\tau''),\dot{q}(\tau'')), \tag{2.12b}$$

and where, for later convenience, we have introduced the short-hand notation

$$(q',\tau'|q,\tau) \equiv < q'|e^{-H(\tau'-\tau)}|q>, \tag{2.12c}$$

in analogy to (2.4). This is the path integral expression we wanted to obtain. Notice that because the paths are weighted with $\exp(-S_E)$, important contributions to (2.12a) are expected to come from those paths for which $S_E[q]$ takes values close to the minimum, where

$$\delta S_E[q] = 0.$$

This is the principle of least action which leads to the classical *euclidean* equations of motion. Hence, within the path integral framework, the quantization of a classical system amounts to taking into account fluctuations around the classical path. In the euclidean formulation these fluctuations are exponentially suppressed if $S_E \geq 0$. On the other hand, in the real time formulation, an analogous procedure to the one followed above, leads to the following path integral representation of (2.4):

$$< q'|e^{-iH(t'-t)}|q >= \int_q^{q'} Dq \; e^{iS[q]}, \tag{2.13}$$

where $S[q]$ is the action for real time. The path integral (2.13) is defined in the same way as before (see Feynman and Hibbs, 1965), but the paths are now weighted with an oscillating function. For this reason this path integral representation is not suited for numerical calculations. It is, however, a useful starting point for carrying out semiclassical approximations, where one expands the action about a minimum up to terms quadratic in the coordinates. For an instructive example the reader may consult the paper by Bender et al. (1978), where the energy spectrum and eigenfunctions are calculated in the WKB approximation for a one-dimensional periodic potential.

An exact evaluation of the path integral (2.12) or (2.13) is only possible in a few cases. The standard example in the real time formulation is the harmonic oscillator. It is discussed in detail in the book by Feynman and Hibbs (1965). The Coulomb potential already provides a quite non-trivial example (Duru and Kleinert, 1979). It therefore may appear that the path integral method is of little practical use. This is true for quantum mechanics, where more efficient methods are available to calculate scattering amplitudes, bound state energies and eigenfunctions. But in field theory, we only know how to compute Green functions in perturbation theory (except for some simple models which can be solved exactly). It is here where physicists first became very interested in the path integral method, since it allowed one to derive the Feynman rules for gauge theories like QCD in a very straightforward way. This is, however, only one of the merits of the method. As we have already pointed out, many of the modern developments in theoretical elementary particle physics are based on the path integral formalism. In the following section we extend the above discussion to bosonic Green functions of interest in field theory.

2.2 Path Integral Representation of Bosonic Green Functions in Field Theory

In quantum mechanics all physical information about the quantum system is contained in the Green function (2.2). In field theory, on the other hand, this information is stored in an infinite set of vacuum expectation values of time-ordered products of Heisenberg field operators. The simplest such operator is the real scalar field $\phi(x) \equiv \phi(\vec{x}, t)$. Its time evolution is given by

$$\phi(\vec{x}, t) = e^{iHt}\phi(\vec{x}, 0)e^{-iHt},$$

where H is the Hamiltonian of the system. The coordinates $\vec{x}$ label the infinite number of coordinate degrees of freedom of the system. They play the role of the discrete index "α" labeling the Heisenberg operators $Q_\alpha(t)$ defined in (2.5). The Green functions of the scalar field are defined by

$$G(x_1, x_2, ..., x_\ell) =< \Omega|T\left(\phi(x_1)\phi(x_2)...\phi(x_\ell)\right)|\Omega >, \tag{2.14}$$

where $x_i = (\vec{x}_i, t_i)$, and $|\Omega >$ denotes the ground state (vacuum) of the system whose dynamics is determined by the Hamiltonian H. The time-ordering operation "T" orders the operators from left to right according to descending time. The analogue of (2.14) in our quantum mechanical example is evidently given by

$$G_{\alpha_1\alpha_2...\alpha_\ell}(t_1, t_1, ..., t_\ell) =< E_0|T(Q_{\alpha_1}(t_1)Q_{\alpha_2}(t_2)...Q_{\alpha_\ell}(t_\ell))|E_0 > . \tag{2.15}$$

Let us assume that we have ordered the operators in (2.15) according to descending time from left to right; then

$$G_{\alpha_a\alpha_2...\alpha_\ell}(t_1, t_1, ..., t_\ell) =< E_0|Q_{\alpha_1}(t_1)Q_{\alpha_2}(t_2)...Q_{\alpha_\ell}(t_\ell)|E_0 >, \quad (t_1 > t_2 > ... > t_\ell). \tag{2.16}$$

We are interested in a path integral representation of (2.16) continued to imaginary times, $t_i \to -i\tau_i$. It is this representation which we shall need to formulate bosonic field theories on a lattice. The transition to imaginary times is made by replacing the operators $Q_{\alpha_i}(t_i)$ by

$$\hat{Q}_{\alpha_i}(\tau_i) = e^{H\tau_i}Q_{\alpha_i}e^{-H\tau_i}. \tag{2.17}$$

This corresponds to setting $t = -i\tau$ in the right-hand side of (2.5). The euclidean version of (2.16) is therefore given by

$$< E_0|\hat{Q}_{\alpha_1}(\tau_1)\hat{Q}_{\alpha_2}(\tau_2)...\hat{Q}_{\alpha_\ell}(\tau_\ell)|E_0 > . \tag{2.18}$$

To derive a path integral representation for this ground state expectation value we proceed in two steps. We first show that (2.18) can be extracted from the matrix elements

$$\begin{aligned}(q',\tau'|\hat{Q}_{\alpha_1}(\tau_1)\hat{Q}_{\alpha_2}(\tau_2)...\hat{Q}_{\alpha_\ell}(\tau_\ell)|q,\tau)\\ \equiv < q'|e^{-H\tau'}\hat{Q}_{\alpha_1}(\tau_1)\hat{Q}_{\alpha_2}(\tau_2)...\hat{Q}_{\alpha_\ell}(\tau_\ell)e^{H\tau}|q>,\end{aligned} \tag{2.19}$$

by studying this expression for large positive and negative values of τ' and τ. We then demonstrate that (2.19) has a path integral representation which is an (almost) obvious generalization of (2.12a).

We begin with the first mentioned step. Inserting a complete set of energy eigenstates to the left and right of the operators $\exp(H\tau)$ and $\exp(-H\tau')$ in (2.19), we have that

$$\begin{aligned}&(q',\tau'|\hat{Q}_{\alpha_1}(\tau_1)...\hat{Q}_{\alpha_\ell}(\tau_\ell)|q,\tau)\\ &=\sum_{\kappa,\kappa'} e^{-E_{\kappa'}\tau'}e^{E_\kappa\tau}\psi_{\kappa'}(q')\psi^*_\kappa(q) < E_{\kappa'}|\hat{Q}_{\alpha_1}(\tau_1)...\hat{Q}_{\alpha_\ell}(\tau_\ell)|E_\kappa > .\end{aligned} \tag{2.20}$$

Assuming that there exists an energy gap between the ground state and first excited state, we therefore find that

$$\begin{aligned}&(q',\tau'|\hat{Q}_{\alpha_1}(\tau_1)...\hat{Q}_{\alpha_\ell}(\tau_\ell)|q,\tau)\\ &\underset{\substack{\tau'\to\infty\\ \tau\to-\infty}}{\longrightarrow} e^{-E_0(\tau'-\tau)}\psi_0(q')\psi^*_0(q) < E_0|\hat{Q}_{\alpha_1}(\tau_1)...\hat{Q}_{\alpha_\ell}(\tau_\ell)|E_0 > .\end{aligned} \tag{2.21a}$$

Furthermore, replacing the $\hat{Q}_{\alpha_i}(\tau_i)$'s in this expression by the unit operator, we have that

$$(q',\tau'|q,\tau) \underset{\substack{\tau'\to\infty\\ \tau\to-\infty}}{\longrightarrow} e^{-E_0(\tau'-\tau)}\psi_0(q')\psi^*_0(q). \tag{2.21b}$$

From (2.21a,b) we are led to the following important statement:

$$\frac{(q',\tau'|\hat{Q}_{\alpha_1}(\tau_1)...\hat{Q}_{\alpha_\ell}(\tau_\ell)|q,\tau)}{(q',\tau'|q,\tau)} \underset{\substack{\tau'\to\infty\\ \tau\to-\infty}}{\longrightarrow} < E_0|\hat{Q}_{\alpha_1}(\tau_1)...\hat{Q}_{\alpha_\ell}(\tau_\ell)|E_0 > . \tag{2.22}$$

Notice that according to (2.22) we are free to choose for $q=\{q_\alpha\}$ and $q'=\{q'_\alpha\}$ any values as long as $< q|E_0 >$ and $< q'|E_0 >$ are different from zero! In other words, the ground state wave function must have non-vanishing support at q and q'. Since (2.22) actually holds for arbitrary time, $\tau_1,...,\tau_\ell$, it follows that a corresponding expression holds for the time-ordered product of the operators, i.e.

$$\frac{(q',\tau'|T(\hat{Q}_{\alpha_1}(\tau_1)...\hat{Q}_{\alpha_\ell}(\tau_\ell))|q,\tau)}{(q',\tau'|q,\tau)} \underset{\substack{\tau'\to\infty\\ \tau\to-\infty}}{\longrightarrow} < E_0|T(\hat{Q}_{\alpha_1}(\tau_1)...\hat{Q}_{\alpha_\ell}(\tau_\ell)|E_0 > . \tag{2.23}$$

This completes the first step of our program. We now proceed with the second step, and construct the path integral representation for the numerator appearing on the left-hand side of (2.23). The PI–representation for the denominator has already been obtained in the previous section. Our starting point, however, is not this numerator, but the matrix element (2.19), with $\tau_1 > \tau_2 > ... > \tau_\ell$. Let us first write out the time dependence explicitly, by making use of the definition (2.17):

$$
\begin{aligned}
&(q',\tau'|\hat{Q}_{\alpha_1}(\tau_1)\hat{Q}_{\alpha_2}(\tau_2)...\hat{Q}_{\alpha_\ell}(\tau_\ell)|q,\tau) \\
&=< q'|e^{-H(\tau'-\tau_1)}\hat{Q}_{\alpha_1}e^{-H(\tau_1-\tau_2)}\hat{Q}_{\alpha_2}e^{-H(\tau_2-\tau_3)}...e^{-H(\tau_{\ell-1}-\tau_\ell)}\hat{Q}_{\alpha_\ell}e^{-H(\tau_\ell-\tau)}|q> .
\end{aligned}
$$

We next insert a complete set of eigenstates of $\{\hat{Q}_\alpha\}$ to the right and left of each of the operators $\hat{Q}_{\alpha_i}$. These operators are diagonal in this representation. Let us denote the integration variables associated with $\hat{Q}_{\alpha_i}$ collectively by $q^{(i)}$. Then

$$
\begin{aligned}
&(q',\tau'|\hat{Q}_{\alpha_1}(\tau_1)...\hat{Q}_{\alpha_\ell}(\tau_\ell)|q,\tau) \\
&= \int \prod_{i=1}^{l} dq^{(i)} (q',\tau'|q^{(1)},\tau_1)q^{(1)}_{\alpha_1}(q^{(1)},\tau_1|q^{(2)},\tau_2)q^{(2)}_{\alpha_2}...q^{(\ell)}_{\alpha_\ell}(q^{(\ell)},\tau_\ell|q,\tau).
\end{aligned}
$$

Inserting for $(q^{(i)},\tau_i|q^{(j)},\tau_j)$, etc., the path integral expressions analogous to (2.12a) one finds that

$$
\begin{aligned}
&(q',\tau'|\hat{Q}_{\alpha_1}(\tau_1)...\hat{Q}_{\alpha_\ell}(\tau_\ell)|q,\tau) \\
&= \int_q^{q'} Dq q_{\alpha_1}(\tau_1)...q_{\alpha_\ell}(\tau_\ell) e^{-\int_\tau^{\tau'} d\tau'' L_E(q(\tau''),\dot{q}(\tau''))}, \quad (\tau' > \tau_1 > ... > \tau_\ell > \tau),
\end{aligned}
\tag{2.24}
$$

where the path integral is calculated as follows:

i) Split the interval $[\tau,\tau']$ into N infinitesimal time intervals of length $\epsilon = (\tau'-\tau)/N$.

ii) Consider all paths starting at q at time τ and ending at q' at time τ'. Approximate these paths by straight line segments in each infinitesimal time interval.

iii) Weigh each path with $\exp(-S_E[q])$, where $S_E[q]$ is the action defined in (2.11), and with the product of the coordinates $q_{\alpha_1},...,q_{\alpha_\ell}$ at times $\tau_1,...,\tau_\ell$, respectively. Sum the contributions over all paths by integrating over all possible values of the coordinates at intermediate times.

iv) Multiply the resulting expression with $(1/\sqrt{2\pi\epsilon})^{nN}$, where n is the number of degrees of freedom of the system, and take the limit $\epsilon \to 0, N \to \infty$, keeping the product $N\epsilon = \tau' - \tau$ fixed.

In deriving the above expression we have assumed that $\tau' > \tau_1 > \tau_2 > ... > \tau_\ell > \tau$. Instead of (2.24) we can therefore also write

$$\begin{aligned}(q',\tau'|T\left(\hat{Q}_{\alpha_1}(\tau_1)...\hat{Q}_{\alpha_\ell}(\tau_\ell)\right)|q,\tau) \\ = \int_q^{q'} Dq q_{\alpha_1}(\tau_1)...q_{\alpha_\ell}(\tau_\ell)e^{-\int_\tau^{\tau'} d\tau'' L(q(\tau''),\dot{q}(\tau''))}.\end{aligned} \tag{2.25}$$

But because of the definition of the T-product, we can now write the product of the operators $\hat{Q}_{\alpha_i}(\tau_i)$ in any order we wish. This symmetry under the exchange of any two operators is reflected in the path integral, since the q_α's are ordinary commuting variables.

We are now ready to write down the path integral expression for the right-hand side of (2.23). Inserting the expressions (2.25) and (2.12a) into the left-hand side of (2.23), and taking the indicated limit, we find that

$$< E_0|T\left(\hat{Q}_{\alpha_1}(\tau_1)...\hat{Q}_{\alpha_\ell}(\tau_\ell)\right)|E_0 >= \frac{\int Dq q_{\alpha_1}(\tau_1)...q_{\alpha_\ell}(\tau_\ell)e^{-S_E[q]}}{\int Dq e^{-S_E[q]}}, \tag{2.26a}$$

where (2.12b) is now replaced by

$$S_E[q] = \int_{-\infty}^{\infty} d\tau L_E(q(\tau),\dot{q}(\tau)). \tag{2.26b}$$

S_E is the euclidean action associated with the path $q(\tau)$. The integrals in (2.26a) are carried out over all paths starting and ending at arbitrary points at times $\tau = -\infty$ and $\tau = +\infty$, respectively. This is true as long as the ground state wave function has non-vanishing support at q and q'. In practical calculations the size of $< q|E_0 >$ and $< q'|E_0 >$ is important. The reason is that in most cases of interest we cannot evaluate the path integral analytically, but must recur to numerical methods. This forces one to calculate the multiple integrals on finite time lattices. It is then essential that the contributions to the sum in (2.20) coming from higher energy states are suppressed as much as possible. When (2.26a) is calculated numerically one usually imposes periodic boundary conditions, i.e., $q = q'$, and allows q to take arbitrary values. This choice of boundary conditions turns out to be very convenient.

We now make some further comments about the path integral expression (2.26a). The evaluation of the right-hand side demands that we first calculate the multiple integrals on a time lattice with finite lattice spacing ϵ, and then take the limit $\epsilon \to 0$. Unfortunately, one can only carry out this program for path integrals of the Gaussian type. As an example consider the following integral,

$$I_{\alpha_1 \dots \alpha_\ell} = \int \prod_{i=1}^{N} dq_i q_{\alpha_1} q_{\alpha_2} \dots q_{\alpha_\ell} e^{-\frac{1}{2} \sum_{n,m} q_n M_{nm} q_m} , \tag{2.27}$$

where M is a real, positive definite symmetric matrix, and where the sum extends over $n, m = 1, \dots, N$. This integral can be calculated as follows. Introduce the generating functional

$$Z_0[J] = \int \prod_{i=1}^{N} dq_i e^{-\frac{1}{2} \sum_{nm} q_n M_{nm} q_m + \sum_n J_n q_n} . \tag{2.28}$$

Then (2.27) is evidently given by

$$I_{\alpha_1 \dots \alpha_\ell} = \left(\frac{\partial^\ell Z_0[J]}{\partial J_{\alpha_1} \partial J_{\alpha_2} \dots \partial J_{\alpha_\ell}} \right)_{J=0} . \tag{2.29}$$

We therefore need to calculate the integral (2.28). This can be easily done by performing an orthogonal transformation on the coordinates $\{q_\alpha\}$ which diagonalizes the matrix M. One then finds

$$Z_0[J] = \frac{(2\pi)^{N/2}}{\sqrt{\det M}} e^{\frac{1}{2} \sum_{n,m} J_n M^{-1}_{nm} J_m} , \tag{2.30}$$

where M^{-1} is the inverse of the matrix M, and $\det M$ is the determinant of M. This expression is very useful for carrying out a perturbative expansion of Green functions in theories where the potential is a polynomial in the coordinates and can be treated as a small perturbation. Thus suppose we want to calculate the integral

$$K_{\alpha_1 \dots \alpha_\ell} = \int \prod_{i=1}^{N} dq_i q_{\alpha_1} q_{\alpha_2} \dots q_{\alpha_\ell} e^{-S[q]} , \tag{2.31a}$$

where

$$S[q] = \frac{1}{2} \sum_{n,m} q_n M_{nm} q_m + S_I[q] , \tag{2.31b}$$

with $S_I[q]$ a polynomial in the coordinates $\{q_n\}$. The integral (2.31a) is given by (2.29), but with the generating functional (2.28) replaced by

$$Z[J] = \int \prod_{i=1}^{N} dq_i e^{-S[q] + \sum_n J_n q_n} .$$

Expanding $\exp(-S_I[q])$, we have that

$$Z[J] = \sum_{k=0}^{\infty} \frac{(-1)^k}{k!} \int \prod_{i=1}^{N} dq_i (S_I[q])^k e^{-\frac{1}{2}\sum_{n,m} q_n M_{nm} q_m + \sum_n J_n q_n}.$$

This expression can also be written in the form

$$Z[J] = \sum_{\kappa=0}^{\infty} \frac{(-1)^k}{k!} \left(S_I[\frac{\partial}{\partial J}] \right)^k Z_0[J],$$

where $S_I[\partial/\partial J]$ is obtained from $S_I[q]$ by making the replacements $q_n \rightarrow \partial/\partial J_n$ ($n = 1, \dots, N$) in the argument of S_I, and where $Z_0[J]$ is the generating functional (2.30). The above formula allows us to compute the generating functional $Z[J]$ of the interacting theory in every order of S_I. This is the first comment we wanted to make. The second comment concerns our earlier claim, that the path integral representation of Green functions opens the possibility of studying field theories non–perturbatively. The reason for this is the following. Consider the right-hand side of (2.26a). If the action is bounded from below, then this expression has the form of a statistical ensemble average, with a Boltzmann distribution given by $\exp(-S_E[q])$. This allows us to use well-known statistical methods to calculate Green functions in theories with a large number of degrees of freedom. The entire book is based on this simple observation. Because of this similarity with statistical mechanics we shall speak of the euclidean Green functions as *correlation functions*, and write (2.26) in the form

$$< q_{\alpha_1}(\tau_1) \dots q_{\alpha_\ell}(\tau_\ell) > = \frac{1}{Z} \int Dq q_{\alpha_1}(\tau_1) \dots q_{\alpha_\ell}(\tau_\ell) e^{-S_E[q]}, \tag{2.32a}$$

where

$$Z = \int Dq e^{-S_E[q]}. \tag{2.32b}$$

We want to point out, however, that the right-hand side of (2.32a) should not be confused with a canonical ensemble average in classical statistical mechanics. Nevertheless, we shall refer to (2.32b) as the partition function.

For reasons mentioned at the beginning of this chapter, we have concentrated our attention on Green functions continued to imaginary times. The derivation of the path integral representation for the real–time Green functions (2.15) can be found in the review article by Abers and Lee (1973), and in most modern text books on field theory. We only quote here the result:

$$< E_0 | T \left(Q_{\alpha_1}(t_1) \dots Q_{\alpha_\ell}(t_\ell) \right) | E_0 > = \frac{\int Dq \; q_{\alpha_1}(t_1) \dots q_{\alpha_\ell}(t_\ell) e^{iS[q]}}{\int Dq \; e^{iS[q]}}. \tag{2.33}$$

Here $S[q]$ is the action whose variation leads to the equations of motion for real times. In our quantum mechanical example $S[q]$ is given by

$$S[q] = \int_{-\infty}^{\infty} dt \left[\frac{1}{2} \sum_{\alpha} \left(\frac{dq_\alpha}{dt} \right)^2 - V(q(t)) \right] . \tag{2.34}$$

The quantity appearing within brackets is the Lagrangean describing the dynamics of the classical system. On the other hand, we have seen that the "euclidean" action $S_E[q]$ has the form

$$S_E[q] = \int_{-\infty}^{\infty} d\tau \left[\frac{1}{2} \sum_{\alpha} \left(\frac{dq_\alpha}{d\tau} \right)^2 + V(q(\tau)) \right] , \tag{2.35}$$

whose variation leads to the "euclidean" equations of motion. By comparing (2.35) and (2.34), we see that $S_E[q]$ can be obtained from $S[q]$ by the following formal rule: Consider the action $S[q]$. Replace t by $-i\tau$ wherever t appears explicitly, and $q_\alpha(t)$ by $q_\alpha(\tau)$, where the coordinates are treated in both cases as real valued functions of their arguments. Then

$$iS[q] \underset{\text{"}t \to -i\tau\text{"}}{\longrightarrow} -S_E[q],$$

where "$t \to -i\tau$" stands for the above formal prescription. Of course we have only proved this rule for systems, where the kinetic part of the Lagrangean is quadratic in the velocities. For fermionic systems this is not the case, but the prescription is still correct. Since we are usually given the action of the system for real times, the above rule is useful for determining the form of the euclidean action that enters the path integral expression (2.32a).

As we have demonstrated in this section, euclidean path integrals involve the integration over real valued coordinates on an euclidean time lattice. A similar statement holds for the PI–representation of Green functions defined for real times. The difference between the two representation, merely resides in the structure of the action. Thus in the real–time formulation the paths $q(t)$ are weighted with a phase, while the corresponding weight in the euclidean formulation can be interpreted as a "Boltzmann factor", if the action is a real valued functional of the coordinates, bounded from below. This was the main point we wanted to demonstrate by deriving directly the path integral representation for Green functions continued to imaginary times.

2.3 The Transfer Matrix

Consider the partition function (2.32b). For a system whose dynamics is dictated by the Hamiltonian (2.8), Z has the following explicit form on a finite, periodic, euclidean time–lattice:*

$$Z = \int \prod_{\ell'=0}^{N-1} \prod_{\beta} \frac{dq_\beta^{(\ell')}}{\sqrt{2\pi\epsilon}} e^{-\sum_{\ell=0}^{N-1} \epsilon \left[\sum_{\alpha=1}^{n} \frac{1}{2} \left(\frac{q_\alpha^{(\ell+1)} - q_\alpha^{(\ell)}}{\epsilon} \right)^2 + V(q^{(\ell)}) \right]} . \tag{2.36}$$

Let us write this expression in the form

$$Z = \int \prod_{\ell'=0}^{N-1} dq^{(\ell')} \prod_{\ell=0}^{N-1} T_{q^{(\ell+1)} q^{(\ell)}} , \tag{2.37a}$$

where $dq^{(\ell)} \equiv \prod_\beta dq_\beta^{(\ell)}$, and

$$T_{q^{(\ell+1)} q^{(\ell)}} = \left(\frac{1}{2\pi\epsilon} \right)^{n/2} e^{-\epsilon \left[\sum_\alpha \frac{1}{2} \left(\frac{q_\alpha^{(\ell+1)} - q_\alpha^{(\ell)}}{\epsilon} \right)^2 + V(q^{(\ell)}) \right]} . \tag{2.37b}$$

From (2.9) we see that

$$T_{q^{(\ell+1)} q^{(\ell)}} = < q^{(\ell+1)} | e^{-H\epsilon} | q^{(\ell)} > . \tag{2.38}$$

The matrix defined by (2.38) is the so–called transfer matrix. It describes the evolution of the system in an infinitesimal timestep ϵ. Actually, the more fundamental definition of the partition function is given by (2.37a) , with the transfer matrix defined in (2.38). This is evident from our discussion in the previous two sections, where the matrix elements of $\exp(-\epsilon H)$ played a fundamental role.

Suppose now that we were given the transfer matrix. Can we extract from it the Hamiltonian (2.8)? Indeed, this can be done by reversing the steps which led us from $< q^{(\ell+1)} | \exp(-H\epsilon) | q^{(\ell)} >$ to (2.9).* We now give the details. The states $|q^{(k)} >$ are simultaneous eigenstates of the coordinate operators Q_α:

$$Q_\alpha | q^{(k)} > = q_\alpha^{(k)} | q^{(k)} > .$$

Let us introduce the momentum operators P_α canonically conjugate to Q_α, which satisfy the commutation relations

$$[Q_\alpha, P_\beta] = i\delta_{\alpha\beta} .$$

---

* i.e., $q^{(0)}$ and $q^{(N)}$ are identified.
* see e.g., Creutz (1977)

Then $\exp(-i\xi \cdot P)$, with $\xi \cdot P \equiv \sum_\alpha \xi_\alpha P_\alpha$, generates finite translations by $\xi = (\xi_1, \ldots, \xi_n)$:

$$e^{-i\xi \cdot P}|q>=|q+\xi> .$$

Since

$$<q'|q>=\prod_\alpha \delta(q'_\alpha - q_\alpha),$$

we conclude that

$$<q^{(\ell+1)}|e^{-i\xi \cdot P}|q^{(\ell)}>=\prod_\alpha \delta(q_\alpha^{(\ell+1)} - q_\alpha^{(\ell)} - \xi_\alpha). \tag{2.39}$$

Now the matrix element (2.38) can also be written in the form

$$T_{q^{(\ell+1)}q^{(\ell)}} = \int d\xi \prod_\alpha \left\{ \delta(q_\alpha^{(\ell+1)} - q_\alpha^{(\ell)} - \xi_\alpha) e^{-\frac{1}{2\epsilon}\xi_\alpha^2} \right\} e^{-\epsilon V(q^{(\ell)})},$$

where

$$d\xi = \prod_{\alpha=1}^{n} \frac{d\xi_\alpha}{\sqrt{2\pi\epsilon}}.$$

By making use of the relation (2.39), this expression becomes

$$T_{q^{(\ell+1)}q^{(\ell)}} =<q^{(\ell+1)}| \left[\int d\xi e^{-\frac{1}{2\epsilon}\sum_\alpha (\xi_\alpha^2 + 2i\epsilon\xi_\alpha P_\alpha)} \right] e^{-\epsilon V(Q)}|q^{(\ell)}> .$$

Performing the Gaussian integral we therefore find that

$$T_{q^{(\ell+1)}q^{(\ell)}} =<q^{(\ell+1)}|e^{-\epsilon[\sum_\alpha \frac{1}{2}P_\alpha^2 + V(Q)]}|q^{(\ell)}> .$$

By comparing this expression with (2.38), we conclude that the Hamiltonian is given by (2.8).

The above described procedure for constructing the Hamiltonian, given the transfer matrix, will be relevant later on, when we discuss the lattice Hamiltonian of a gauge theory. In the lattice formulation of field theories we are given the partition function. By writing the partition function in the form (2.37a), the identification (2.38) will allow us to deduce the lattice Hamiltonian.

2.4 Path Integral Representation of Fermionic Green Functions

So far we have considered quantum mechanical systems involving only bosonic degrees of freedom. But the fundamental matter fields in nature are believed to carry spin 1/2. In contrast to the bosonic case these fields anticommute in the limit $\hbar \to 0$, and hence become elements of a Grassmann algebra in this limit. We therefore expect that the path integral representation of Green functions built from fermion fields will involve the integration over anticommuting (Grassmann) variables.* We hence begin this section with a discussion of how one differentiates and integrates functions of Grassmann variables. The integration rules are then applied to calculate specific integrals, which will play an important role throughout this book. The results we shall obtain will give us a strong hint regarding the path integral representation of fermionic Green functions in theories of interest for elementary particle physics. We begin our discussion with some basic definitions.

Grassmann Algebra

The elements $\eta_1, ..., \eta_N$ are said to be the generators of a Grassmann algebra, if they anticommute among each other, i.e. if

$$\{\eta_i, \eta_j\} = \eta_i \eta_j + \eta_j \eta_i = 0, \quad i, j = 1, ..., N. \tag{2.40}$$

From here it follows that

$$\eta_i^2 = 0. \tag{2.41}$$

A general element of a Grassmann algebra is defined as a power series in the η_i's. Because of (2.41), however, this power series has only a finite number of terms:

$$f(\eta) = f_0 + \sum_i f_i \eta_i + \sum_{i \neq j} f_{ij} \eta_i \eta_j + ... + f_{12...N} \eta_1 \eta_2 ... \eta_N. \tag{2.42}$$

As an example consider the function

$$g(\eta) = e^{-\sum_{i,j=1}^{N} \eta_i A_{ij} \eta_j}.$$

It is defined by the usual power series expansion of the exponential. Since the terms appearing in the sum - being quadratic in the Grassmann variables - commute among each other, we can also write $g(\eta)$ as follows

$$g(\eta) = \prod_{i,j} e^{-\eta_i A_{ij} \eta_j},$$

* For a comprehensive discussion of the functional formalism for fermions the reader may consult the book by Berezin (1966).

or, making use of (2.41),

$$g(\eta) = \prod_{\substack{i,j=1 \\ i\neq j}}^{N} (1 - \eta_i A_{ij} \eta_j).$$

Next we consider the following function of a set of $2N$-Grassmann variables which we denote by $\eta_1, .., \eta_N, \bar{\eta}_1, ..., \bar{\eta}_N$:

$$h(\eta, \bar{\eta}) = e^{-\sum_{ij} \bar{\eta}_i A_{ij} \eta_j}.$$

Proceeding as above, we now have that

$$h(\eta, \bar{\eta}) = \prod_{i,j=1}^{N} (1 - \bar{\eta}_i A_{ij} \eta_j).$$

Notice that in contrast to previous cases, this expression also involves diagonal elements of A_{ij}.

Integration over Grassmann variables

We now state the Grassmann rules for calculating integrals of the form

$$\int \prod_{i=1}^{N} d\eta_i f(\eta),$$

where $f(\eta)$ is a function whose general structure is given by (2.42). Since a given Grassmann variable can at most appear to the first power in $f(\eta)$, the following rules suffice to calculate an arbitrary integral [Berezin (1966)]:

$$\int d\eta_i = 0, \qquad \int d\eta_i \eta_i = 1. \tag{2.43a}$$

When computing multiple integrals one must further take into account that the integration measures $\{d\eta_i\}$ also anticommute among themselves, as well as with all η_j's

$$\{d\eta_i, d\eta_j\} = \{d\eta_i, \eta_j\} = 0, \quad \forall i, j. \tag{2.43b}$$

These integration rules look indeed very strange. But, as we shall see soon, they are the appropriate ones to allow us to obtain a PI representation of fermionic

Green functions. As an example let us apply these rules to calculate the following integral:

$$I[A] = \int \prod_{\ell=1}^{N} d\bar{\eta}_\ell d\eta_\ell e^{-\sum_{i,j=1}^{N} \bar{\eta}_i A_{ij} \eta_j} . \tag{2.44}$$

We could have also denoted the Grassmann variables by $\eta_1, \ldots, \eta_{2N}$, by setting $\eta_{N+i} = \bar{\eta}_i$. But for reasons which will become clear later, we prefer the above notation. To evaluate (2.44), we first write the integrand in the form

$$e^{-\sum_{i,j} \bar{\eta}_i A_{ij} \eta_j} = \prod_{i=1}^{N} e^{-\bar{\eta}_i \sum_{j=1}^{N} A_{ij} \eta_j} .$$

Since $\bar{\eta}_i^2 = 0$, only the first two terms in the expansion of the exponential will contribute. Hence

$$e^{-\sum_{i,j} \bar{\eta}_i A_{ij} \eta_j} = (1 - \bar{\eta}_1 A_{1i_1} \eta_{i_1})(1 - \bar{\eta}_2 A_{2i_2} \eta_{i_2}) \ldots (1 - \bar{\eta}_N A_{Ni_N} \eta_{i_N}), \tag{2.45}$$

where a summation over repeated indices $i_\ell (\ell = 1, \ldots, N)$ is understood. Now because of the Grassmann integration rules (2.43a), the integrand of (2.44) must involve the product of all the Grassmann variables. We therefore only need to consider the term

$$K(\eta, \bar{\eta}) = \sum_{i_1, \ldots, i_N} \eta_{i_1} \bar{\eta}_1 \eta_{i_2} \bar{\eta}_2 \ldots \eta_{i_N} \bar{\eta}_N A_{1i_1} A_{2i_2} \ldots A_{Ni_N} , \tag{2.46}$$

where we have set $\bar{\eta}_k \eta_{i_k} = -\eta_{i_k} \bar{\eta}_k$ to eliminate the minus signs appearing in (2.45). The summation clearly includes only those terms for which all the indices $i_1, \ldots, i_N$ are different. Now, the product of Grassmann variables in (2.46) is antisymmetric under the exchange of any pair of indices i_ℓ and $i_{\ell'}$. Hence we can write expression (2.46) in the form

$$K(\eta, \bar{\eta}) = \eta_1 \bar{\eta}_1 \eta_2 \bar{\eta}_2 \ldots \eta_N \bar{\eta}_N \sum_{i_1 \ldots i_N} \epsilon_{i_1 i_2 \ldots i_N} A_{1i_1} A_{2i_2} \ldots A_{Ni_N} ,$$

where $\epsilon_{i_1 i_2 \ldots i_N}$ is the ϵ–tensor in N–dimensions. Recalling the standard formula for the determinant of a matrix A, we therefore find that

$$K(\eta, \bar{\eta}) = (\det A) \eta_1 \bar{\eta}_1 \eta_2 \bar{\eta}_2 \ldots \eta_N \bar{\eta}_N .$$

We now replace the exponential in (2.44) by this expression and obtain

$$I[A] = \left[\prod_{i=1}^{N} \int d\bar{\eta}_i d\eta_i \eta_i \bar{\eta}_i \right] \det A = \det A .$$

Let us summarize our result for later convenience:

$$\begin{gathered} \int D(\bar{\eta}\eta) e^{-\sum_{i,j=1}^{N} \bar{\eta}_i A_{ij} \eta_j} = \det A, \\ D(\bar{\eta}\eta) = \prod_{\ell=1}^{N} d\bar{\eta}_\ell d\eta_\ell . \end{gathered} \tag{2.47}$$

There is another important formula we shall need, which is the analog of (2.30). It will allow us to calculate integrals of the type

$$I_{i_1 \ldots i_\ell i'_1 \ldots i'_\ell}[A] = \int D(\bar{\eta}\eta) \eta_{i_1} \ldots \eta_{i_\ell} \bar{\eta}_{i'_1} \ldots \bar{\eta}_{i'_\ell} e^{-\sum_{i,j=1}^{N} \bar{\eta}_i A_{ij} \eta_j} . \tag{2.48}$$

Consider the following generating functional

$$Z[\rho, \bar{\rho}] = \int D(\bar{\eta}\eta) e^{-\sum_{i,j} \bar{\eta}_i A_{ij} \eta_j + \sum_i (\bar{\eta}_i \rho_i + \bar{\rho}_i \eta_i)} , \tag{2.49}$$

where all indices are understood to run from 1 to N, and where the "sources" $\{\rho_i\}$ and $\{\bar{\rho}_i\}$ are now also anticommuting elements of the Grassmann algebra generated by $\{\eta_i, \bar{\eta}_i, \rho_i, \bar{\rho}_i\}$. To evaluate (2.49) we first rewrite the integral as follows:

$$Z[\rho, \bar{\rho}] = \left[\int D(\bar{\eta}\eta) e^{-\sum_{i,j} \bar{\eta}'_i A_{ij} \eta'_j} \right] e^{\sum_{i,j} \bar{\rho}_i A^{-1}_{ij} \rho_i}$$

where

$$\begin{aligned} \eta'_i &= \eta_i - \sum_k A^{-1}_{ik} \rho_k , \\ \bar{\eta}'_i &= \bar{\eta}_i - \sum_k \bar{\rho}_k A^{-1}_{ki} , \end{aligned}$$

and A^{-1} is the inverse of the matrix A. Making use of the invariance of the integration measure under the above transformation * and of (2.47), we find that

$$Z[\rho, \bar{\rho}] = \det A e^{\sum_{i,j} \bar{\rho}_i A^{-1}_{ij} \rho_j} . \tag{2.50}$$

Notice that in contrast to the bosonic case, this generating functional is proportional to $\det A$ [instead of $(\det A)^{-1/2}$; see (2.30)].

---

* This is ensured by the Grassmann integration rules.

Differentiation of Grassmann Variables

We now complete our discussion on Grassmann variables by introducing the concept of a partial derivative on the space of functions defined by (2.42). Suppose we want to differentiate $f(\eta)$ with respect to η_i. Then the rules are the following:

i) If $f(\eta)$ does not depend on η_i, then $\partial_{\eta_i} f(\eta) = 0$

ii) If $f(\eta)$ depends on η_i, then the left derivative $\partial/\partial\eta_i$ is performed by first bringing the variable η_i (which never appears twice in a product!) all the way to the left, using the anticommutation relations (2.40), and then applying the rule

$$\frac{\partial}{\partial\eta_i}\eta_i = 1.$$

Correspondingly, we obtain the right derivative $\overleftarrow{\partial}/\partial\eta_i$ by bringing the variable η_i all the way to the right and then applying the rule

$$\eta_i \frac{\overleftarrow{\partial}}{\partial\eta_i} = 1.$$

Thus for example

$$\frac{\partial}{\partial\eta_i}\eta_j\eta_i = -\eta_j \quad (i \neq j),$$

or

$$\bar{\eta}_i\eta_j \frac{\overleftarrow{\partial}}{\partial\bar{\eta}_i} = -\eta_j .$$

Notice that, because of the peculiar definition of Grassmann integration, we have that

$$\int d\eta_i f(\eta) = \frac{\partial}{\partial\eta_i} f(\eta).$$

Hence integration over η_i is equivalent to partial differentiation with respect to this variable! Another property, which can be easily proved, is that

$$\left\{\frac{\partial}{\partial\eta_i}, \frac{\partial}{\partial\eta_j}\right\} f(\eta) = 0.$$

Let us apply these rules to some cases of interest. Consider the function

$$E(\bar{\rho}) = e^{\sum_j \bar{\rho}_j\eta_j},$$

where $\{\eta_i, \bar{\rho}_i\}$ are the generators of a Grassmann algebra. If they were ordinary c–numbers then we would have that

$$\frac{\partial}{\partial\bar{\rho}_i}E(\bar{\rho}) = \eta_i E(\bar{\rho})$$

This result is in fact correct. To see this let us write $E(\bar{\rho})$ in the form

$$E(\bar{\rho}) = \prod_j (1 + \bar{\rho}_j \eta_j).$$

Applying the rules of Grassmann differentiation, we have that

$$\frac{\partial}{\partial \bar{\rho}_i} E(\bar{\rho}) = \eta_i \prod_{j \neq i} (1 + \bar{\rho}_j \eta_j).$$

But because of the appearance of the factor η_i we are now free to include the extra term $1 + \bar{\rho}_i \eta_i$ in the above product. Hence we arrive at the above-mentioned naive result. It should, however, be noted, that the order of the Grassmann variables in $\sum_i \bar{\rho}_i \eta_i$ was important. By reversing this order we get a minus sign, and the rule is not the usual one! By a similar argument one finds that

$$e^{\sum_j \bar{\eta}_j \rho_j} \frac{\overleftarrow{\partial}}{\partial \rho_i} = \bar{\eta}_i e^{\sum_j \bar{\eta}_j \rho_j}.$$

Let us now return to the generating functional defined in (2.49). Proceeding as above one can easily show that

$$I_{i_1 \dots i_\ell ; i'_1 \dots i'_\ell}[A] = \left[\frac{\partial}{\partial \bar{\rho}_{i_1}} \cdots \frac{\partial}{\partial \bar{\rho}_{i_\ell}} Z[\rho, \bar{\rho}] \frac{\overleftarrow{\partial}}{\partial \rho_{i'_1}} \cdots \frac{\overleftarrow{\partial}}{\partial \rho_{i'_\ell}} \right]_{\rho = \bar{\rho} = 0}, \tag{2.51}$$

where the left-hand side has been defined in (2.48). By making use of the explicit expression for $Z[\rho, \bar{\rho}]$ given in (2.50), one can calculate the right-hand side of (2.51). Since we shall need this expression in later chapters, we will derive it here. To this effect we first rewrite (2.50) as follows

$$\begin{aligned} Z[\rho, \bar{\rho}] &= \det A \prod_i e^{\bar{\rho}_i \sum_j A^{-1}_{ij} \rho_j} = (\det A)(1 + \bar{\rho}_{i_1} A^{-1}_{i_1 k_1} \rho_{k_1}) \\ &\quad \cdot (1 + \bar{\rho}_{i_2} A^{-1}_{i_2 k_2} \rho_{k_2}) \dots (1 + \bar{\rho}_{i_\ell} A^{-1}_{i_\ell k_\ell} \rho_{k_\ell}) [\dots], \end{aligned} \tag{2.52}$$

where the indices $k_1, \dots, k_\ell$ are summed, and [...] stands for the remaining factors not involving the variables $\bar{\rho}_{i_1}, \dots, \bar{\rho}_{i_\ell}$. The only terms which contribute to the left derivatives in (2.51) are those involving the product $\bar{\rho}_{i_1} \dots \bar{\rho}_{i_\ell}$. Furthermore since we will eventually set all "sources" ρ_i and $\bar{\rho}_i$ equal to zero, we can replace [...] by 1.

The contribution in (2.52), which is relevant when computing (2.51), is therefore given by

$$\tilde{Z}[\rho,\bar{\rho}] = \det A \sum_{\{k_i\}'} \bar{\rho}_{i_1} A^{-1}_{i_1 k_1} \rho_{k_1} ... \bar{\rho}_{i_\ell} A^{-1}_{i_\ell k_\ell} \rho_{k_\ell},$$

where all k_i's are different, and the "prime" on $\{k_i\}'$ indicates that the k_i's take only values in the set $(i'_1, i'_2, ..., i'_\ell)$, labeling the right derivatives in (2.51). Thus we can write the above expression in the form

$$\tilde{Z}[\rho,\bar{\rho}] = \det A \sum_P A^{-1}_{i_1 i'_{P_1}} A^{-1}_{i_2 i'_{P_2}} ... A^{-1}_{i_\ell i'_{P_\ell}} \bar{\rho}_{i_1} \rho_{i'_{P_1}} \bar{\rho}_{i_2} \rho_{i'_{P_2}} ... \bar{\rho}_{i_\ell} \rho_{i'_{P_\ell}}, \tag{2.53a}$$

where the sum extends over all permutations

$$P : \begin{pmatrix} i'_1 & i'_2 & ... & i'_\ell \\ i'_{P_1} & i'_{P_2} & ... & i'_{P_\ell} \end{pmatrix}. \tag{2.53b}$$

Each of the products of Grassman variables appearing in the sum (2.53a) can be put into the form

$$F^{i'_1 i'_2 ... i'_\ell}_{i_1 i_2 ..., i_\ell} \equiv \bar{\rho}_{i_1} \rho_{i'_1} \bar{\rho}_{i_2} \rho_{i'_2} ... \bar{\rho}_{i_\ell} \rho_{i'_\ell}$$

by using the anticommutation rules for Grassmann variables. It follows that

$$\tilde{Z}[\rho,\bar{\rho}] = (\det A) \left[\sum_P (-1)^{\sigma_P} A^{-1}_{i_1 i'_{P_1}} ... A^{-1}_{i_\ell i'_{P_\ell}} \right] F^{i'_1 ... i'_\ell}_{i_1 ... i_\ell}(\rho, \bar{\rho}),$$

where $(-1)^{\sigma_P}$is the signum of the permutation (2.53b). We now apply the left and right derivatives indicated in (2.51) to the above expression and obtain the following important result:

$$\begin{aligned} &\int D(\bar{\eta}\eta) \eta_{i_1} ... \eta_{i_\ell} \bar{\eta}_{i'_1} ... \bar{\eta}_{i'_\ell} e^{-\sum_{i,j} \bar{\eta}_i A_{ij} \eta_j} \\ &\quad = \xi_\ell (\det A) \sum_P (-1)^{\sigma_P} A^{-1}_{i_1 i'_{P_1}} .. A^{-1}_{i_\ell i'_{P_\ell}}, \end{aligned} \tag{2.54}$$

where $\xi_\ell = (-1)^{\ell(\ell-1)/2}$. As a particular case of (2.54) we have that

$$\int D(\bar{\eta}\eta) \eta_i \bar{\eta}_j e^{-\sum_{i,j} \bar{\eta}_i A_{ij} \eta_j} = (\det A) A^{-1}_{ij}. \tag{2.55}$$

Let us define the two-point correlation function

$$< \eta_i \bar{\eta}_j >= \frac{\int D(\bar{\eta}\eta) \eta_i \bar{\eta}_j e^{-\sum_{i,j} \bar{\eta}_i A_{ij} \eta_j}}{\int D(\bar{\eta}\eta) e^{-\sum_{i,j} \bar{\eta}_i A_{ij} \eta_j}}. \tag{2.56}$$

Then it follows from (2.47) and (2.55) that

$$\eta_i \bar{\eta}_j \equiv < \eta_i \bar{\eta}_j >= A^{-1}_{ij}. \tag{2.57}$$

We shall refer to (2.57) as a *contraction*. The generalization of (2.57) to arbitrary "correlation" functions,

$$\begin{aligned} &< \eta_{i_1} ... \eta_{i_\ell} \bar{\eta}_{i'_1} ... \bar{\eta}_{i'_\ell} > \\ &= \frac{\int D(\bar{\eta}\eta) \eta_{i_1} ... \eta_{i_\ell} \bar{\eta}_{i'_1} ... \bar{\eta}_{i'_\ell} e^{-\sum_{i,j} \bar{\eta}_i A_{ij} \eta_j}}{\int D(\bar{\eta}\eta) e^{-\sum_{i,j} \bar{\eta}_i A_{ij} \eta_j}}, \end{aligned} \tag{2.58}$$

follows from (2.54) and (2.47). One finds that

$$\begin{aligned} < \eta_{i_1} \eta_{i_2} ... \eta_{i_\ell} \bar{\eta}_{i'_1} \bar{\eta}_{i'_2} .. \bar{\eta}_{i'_\ell} > &= \eta_{i_1} \eta_{i_2} ... \eta_{i_\ell} \bar{\eta}_{i'_1} \bar{\eta}_{i'_2} .. \bar{\eta}_{i'_\ell} \\ &+ \eta_{i_1} \eta_{i_2} ... \eta_{i_\ell} \bar{\eta}_{i'_1} \bar{\eta}_{i'_2} .. \bar{\eta}_{i'_\ell} + \eta_{i_1} \eta_{i_2} ... \eta_{i_\ell} \bar{\eta}_{i'_1} \bar{\eta}_{i'_2} .. \bar{\eta}_{i'_\ell} + ... \end{aligned} \tag{2.59}$$

where the right-hand side stands for the sum of all possible pairwise contractions (2.57) of the Grassmann variables, multiplied by a phase $(-1)^p$, where p is the number of transpositions required to place the contracted variables next to each other in the form $\eta\bar{\eta}$.

This completes our discussion of Grassmann variables. Let us now answer the question what all these exercises with Grassmann variables have to do with the path integral representation of fermionic Green functions. To this effect let us consider the simplest type of relativistic field theory involving only fermionic fields: the free Dirac field. The corresponding action in Minkowski space is given by*

$$S_F[\psi, \bar{\psi}] = \int d^4x \bar{\psi}(x)(i\gamma^\mu \partial_\mu - M)\psi(x),$$

where γ^μ are the Dirac γ-matrices, and where the Lorentz index μ is summed. Let us write this action in the form

$$S_F = \sum_{\alpha,\beta} \int d^4x d^4y \bar{\psi}_\alpha(x) K_{\alpha\beta}(x,y) \psi_\beta(y),$$

where

$$K_{\alpha\beta}(x,y) = (i\gamma^\mu \partial_\mu - M)_{\alpha\beta} \delta^{(4)}(x-y).$$

---

* We assume the reader is familiar with the quantization of the free Dirac field.

The two-point function (fermion propagator) is related to the inverse of the matrix K as follows:

$$\langle\Omega|T(\Psi_\alpha(x)\bar{\Psi}_\beta(y))|\Omega\rangle = iK^{-1}_{\alpha\beta}(x,y),$$

where the time-ordering operation "T" orders the operators from left to right according to descending time, treating the operators Ψ_α and $\bar{\Psi}_\beta$ as elements of a Grassmann algebra. But we have just learned above how to compute the inverse of a matrix by means of Grassmann integrals. Thus a naive application of the formulae (2.56) and (2.57) leads to the following PI-expression for $iK^{-1}_{\alpha\beta}(x,y)$:

$$iK^{-1}_{\alpha\beta}(x,y) = \frac{\int D(\bar{\psi}\psi)\psi_\alpha(x)\bar{\psi}_\beta(y)e^{iS_F[\psi,\bar{\psi}]}}{\int D(\bar{\psi}\psi)e^{iS_F[\psi,\bar{\psi}]}},$$

where the measure is formally defined by

$$D(\bar{\psi}\psi) = \prod_{\alpha,x} d\bar{\psi}_\alpha(x)d\psi_\alpha(x),$$

and where ψ and $\bar{\psi}$ are Grassmann-valued fields. The above observation suggests that the PI-representation of Green functions involving an equal number of Dirac fields of type Ψ and $\bar{\Psi}$ is given in Minkowski space by *

$$\begin{aligned}&\langle\Omega|T(\Psi_{\alpha_1}(x_1)...\Psi_{\alpha_\ell}(x_\ell)\bar{\Psi}_{\beta_1}(y_1)...\bar{\Psi}_{\beta_\ell}(y_\ell))|\Omega\rangle\\ &=\frac{\int D(\bar{\psi}\psi)\psi_{\alpha_1}(x_1)...\psi_{\alpha_\ell}(x_\ell)\bar{\psi}_{\beta_1}(y_1)...\bar{\psi}_{\beta_\ell}(y_\ell))e^{iS_F[\psi,\bar{\psi}]}}{\int D(\bar{\psi}\psi)e^{iS_F[\psi,\bar{\psi}]}}.\end{aligned} \tag{2.60}$$

This is certainly true for the free Dirac field, as follows from a naive application of formula (2.59), which is nothing but Wick's theorem. But it is also true for theories like QED or QCD, where the fermionic contribution to the action is again a bilinear function in the fields ψ and $\bar{\psi}$. In these theories, this fermionic contribution also depends on a collection of bosonic variables (the gauge potentials) and the path integral (2.60) gives the Green function evaluated in external gauge fields. When quantum fluctuations of these fields are taken into account, the appearance of the determinant of the matrix A in (2.55) (rather than $1/\sqrt{\det A}$, which is characteristic of the bosonic case) will play a crucial role. This will become clear later on, when we discuss these theories in detail.

---

* All other functions vanish because of the Grassmann integration rules. For a derivation of the path integral expression for fermions from fundamental principles, see Berezin (1966)

As the reader will have noticed, we have not discussed the path integral representation of fermionic Green functions continued to imaginary times. We shall do this in chapter 4, using the rule derived in section 2 for Green functions involving bosonic variables. In the case of the free Dirac theory, the correctness of this rule can be checked explicitly by comparing the results obtained by the PI-method with those derived using conventional canonical Hilbert space methods. In field theories with interactions like QED or QCD, such a comparison can be made in perturbation theory. The non-perturbative definition of the correlation functions in these theories is assumed to be given by the PI expressions.

We close this section with a remark. In contrast to the bosonic case, we cannot calculate numerically "ensemble averages" of products of Grassmann variables using statistical methods. Nevertheless, we will still be able to study theories like QED or QCD numerically. The reason is that, as we have just mentioned, the fermionic contributions to the action in these theories is bilinear in the fields ψ and $\bar{\psi}$. This allows one to perform the Grassmann integrals and to recast the path integral expression for the euclidean correlation functions in the form of a statistical mechanical ensemble average, with a new effective action. This action depends in a non-local way on the bosonic fields to which the fermions are coupled. It is this non-locality that makes numerical computations of correlation functions involving fermions very time-consuming.

2.5 Discretizing Space-Time. The Lattice as a Regulator of a Quantum Field Theory

As we have pointed out repeatedly in the previous sections, the path integral expressions for Green functions have only a well-defined meaning for systems with a denumerable number of degrees of freedom. In field theory, however, where one is dealing with an infinite number of degrees of freedom, labeled by the coordinates $\vec{x}$ and, in general, by some additional discrete indices, the multiple integrals are only formally defined. To give the path integrals a precise meaning, we will therefore have to discretize not only time, but also space; i.e., we will be forced to introduce a space-time lattice. Eventually we will have to remove again this lattice structure. This is a quite non-trivial task. Those readers acquainted with the renormalization program in continuum perturbation theory know that the renormalization of Green functions first requires the regularization of the corresponding Feynman integrals in momentum space. These integrals will then depend on one or more parameters which are introduced in the regularization process (momentum cut-off, Pauli-Villars masses, dimensional regularization parameter). Since the effect of any regularization procedure is to render the momentum integrations in Feynman integrals ultraviolet finite, let us loosely say that the first step in the renormalization program consists in the introduction of a momentum cutoff. If the original Feynman integrals are divergent, then the regularized integrals will be strongly dependent on the cutoff. The second step in the renormalization program now consists in defining renormalized Green functions, which approach a finite limit as the cutoff is removed. This demands that the bare parameters of the theory become cutoff dependent. This dependence is determined by imposing a set of renormalization conditions, which merely state that such quantities as the physical coupling strength measured at some momentum transfer, and particle masses are to be held fixed as the cutoff is removed.

The above described renormalization program is carried out on the level of Feynman integrals in momentum space. In the lattice approach this program can be formulated without reference to perturbation theory. The first step (regularization) consists in introducing a space-time lattice at the level of the path integral. This regularization merely corresponds to defining what we mean by a path integral. The second step of the renormalization program then corresponds to removing the lattice structure. This amounts to studying the continuum limit. It is therefore not surprising that the bare parameters of the theory will have to be tuned to the lattice spacing in a very definite way depending in general on the dynamics, if physical observables are to become insensitive to the underlying lattice structure. Thus if the reader had some uneasy feelings about the way the infinities are removed in conventional perturbation theory, he will probably feel much better after having read this book. In this connection we also want to mention that within the perturbative framework the introduction of a space-time lattice corresponds to a particular way of regularizing Feynman integrals. As we shall see, this regularization does not amount to the naive introduction of a momentum cutoff. Although the momentum space integrals will indeed be cut off at a momentum of the order of the inverse lattice spacing, the integrands of Feynman integrals will not have the usual structure, but are modified in a non-trivial way. This is one of the reasons why lattice weak coupling perturbation theory is so difficult. The other reason is that in the lattice formulation of gauge theories, new interaction vertices pop up, which have no analogue in the continuum formulation.

The appearance of a momentum cutoff in the lattice formulation is not surprising. Consider a function $f(x)$ of a single continuous variable. If its absolute value is square integrable, then $f(x)$ has the following Fourier representation:

$$f(x) = \int_{-\infty}^{\infty} \frac{dk}{2\pi} \tilde{f}(k) e^{ikx} . \tag{2.61}$$

On the other hand, if x is restricted to a multiple of a "lattice spacing" a, i.e. $x = na$ with n an integer, then $f(na)$ can be Fourier-decomposed as follows:

$$f(na) = \int_{-\pi/a}^{\pi/a} \frac{dk}{2\pi} \tilde{f}_a(k) e^{ikna} , \tag{2.62}$$

where $\tilde{f}_a(-\pi/a) = \tilde{f}_a(\pi/a)$. Hence the "momentum" integration is now restricted to the so-called Brillouin zone (BZ) $[-\pi/a, \pi/a]$. $\tilde{f}_a(k)$ can be represented by a Fourier series. The coefficient of $\exp(-ikna)$ is given by (2.62) multiplied by a:

$$\tilde{f}_a(k) = a \sum_{n=-\infty}^{\infty} f(na) e^{-inka} . \tag{2.63}$$

The right-hand side is just the discretized version of the expression for $\tilde{f}(k)$ obtained by inverting (2.61). By setting $f(na) = 1/2\pi$ in (2.63), we obtain a Fourier series representation of the δ-function in the BZ,

$$\delta_P(k) = \frac{a}{2\pi} \sum_n e^{-inka} , \tag{2.64}$$

where the subscript P stands for "periodic". It emphasizes the fact that $\delta_P(k)$ has non-vanishing support at $k = 0$ modulo $2n\pi$. The Dirac δ-function, $\delta(x - y)$, of course becomes the Kronecker-δ (multiplied by $1/a$) on the x-lattice:

$$\delta_{nm} = a \int_{-\pi/a}^{\pi/a} \frac{dp}{2\pi} e^{ip(n-m)a} . \tag{2.65}$$

The above formulae are trivially extended to functions depending on an arbitrary number of variables. In particular, in four space-time dimensions, all four components of momenta will be restricted to the interval $[-\pi/a, \pi/a]$. Thus the introduction of a lattice provides a momentum cutoff of the order of the inverse lattice spacing.

We are now ready to embarque on the main task of this book, i.e. the formulation of field theories on a space-time lattice. As a warm-up, we begin in the following chapter with a very simple field theory: the free scalar field. Although the lattice formulation will be trivial in this case, we will nevertheless learn a number of important facts by studying it in detail.

CHAPTER 3

THE FREE SCALAR FIELD ON THE LATTICE

Consider the classical field equation

$$(\Box + M^2)\phi(x) = 0, \tag{3.1}$$

where ϕ is a real field, $\Box$ is the d'Alembert operator, and x stands for the space-time vector with components $x_\mu(\mu = 0, 1, 2, 3)$. This equation of motion follows from an action principle, $\delta S = 0$, where

$$S = -\frac{1}{2}\int d^4x\phi(x)(\Box + M^2)\phi(x) \tag{3.2}$$

is the action associated with the Lagrangian density

$$\mathcal{L} = \frac{1}{2}\partial^\mu\phi\partial_\mu\phi - \frac{1}{2}M^2\phi^2 .$$

In the quantum theory the coordinates $q_{\vec{x}}(t) \equiv \phi(x)$ and momenta $p_{\vec{x}}(t) \equiv \dot{\phi}(x)$ become operators, $\Phi(x)$ and $\dot{\Phi}(x)$, satisfying canonical commutation relations. The information about the quantum theory is contained in the Green functions

$$G(x, y \cdots) = \langle\Omega|T(\Phi(x)\Phi(y)\cdots)|\Omega\rangle, \tag{3.3}$$

where $|\Omega\rangle$ stands for the ground state of the system (physical vacuum) and T denotes the time-ordered product of the operators $\Phi(x)$. These Green functions have a path integral representation which can be formally obtained from (2.33) by making the replacements $Q_\alpha(t) \to \Phi(\vec{x}, t)$ and $q_\alpha(t) \to \phi(\vec{x}, t)$:

$$G(x, y, \cdots) = \frac{\int D\phi\phi(x)\phi(y)\cdots e^{iS[\phi]}}{\int D\phi e^{iS}}. \tag{3.4}$$

Here $\int D\phi$ denotes the sum over all possible field configurations $\phi(x)$. The effects arising from quantum fluctuations are contained in those contributions to the integral (3.4) coming from field configurations which are not solutions to the classical equation of motion (3.1) and hence do not lead to a stationary action. Now for reasons mentioned in chapter 2 we are interested in the analytic continuation of (3.4) to imaginary times, $x^0 \to -ix_4, y^0 \to -iy_4$, etc. Let, from now on, x and y denote the euclidean four vectors with components x_μ and y_μ $(\mu = 1, ..., 4)$.

It then follows from our discussion in chapter 2 that the Green functions (3.3) continued to imaginary times have the following path integral representation

$$\langle\phi(x)\phi(y)...\rangle = \frac{\int D\phi(\phi(x)\phi(y)...)e^{-S_E[\phi]}}{\int D\phi e^{-S_E[\phi]}}, \tag{3.5}$$

where we have made use of the notation for the euclidean Green functions, introduced in chapter 2. The euclidean action $S_E[\phi]$ appearing in (3.5) is obtained from (3.2) by i) making the replacement $x_0 \to -ix_4$ where-ever x^0 appears explicitly, ii) substituting for $\phi(\vec{x},t)$ the (real valued!) field $\phi(x) = \phi(\vec{x},x_4)$,* and iii) multiplying the resultant expression by $-i$. This leads to the following expression for $S_E[\phi]$,

$$S_E[\phi] = \frac{1}{2}\int d^4x\phi(x)(-\Box + M^2)\phi(x), \tag{3.6a}$$

where $\Box$ denotes from now on the 4-dimensional Laplacean

$$\Box = \sum_{\mu=1}^{4} \partial_\mu\partial_\mu. \tag{3.6b}$$

In passing to the imaginary time formulation, the Green functions take the form of correlation functions of a statistical mechanical system defined by the partition function

$$Z = \int D\phi e^{-S[\phi]},$$

where the integration measure $D\phi$ is *formally* defined by

$$D\phi = \prod_{\vec{x},x_4} d\phi(\vec{x},x_4).$$

So far the path integral (3.5) has not been given a precise mathematical meaning. We do this now by introducing a space-time lattice with lattice spacing a. Every point on the lattice is then specified by four integers which we denote collectively by $n \equiv (n_1,n_2,n_3,n_4)$. By convention the last component will denote euclidean time. The transition from the continuum to the lattice is then effected by making

---

* We are a bit sloppy in our notation: $\phi(\vec{x},x_4)$ is not obtained from $\phi(\vec{x},t)$ by substituting x_4 for t, but denotes a real field which is a function of the euclidean variables $x_\mu(\mu = 1,...,4)$.

the following substitutions:

$$
\begin{aligned}
&x_\mu \to n_\mu a\ ,\\
&\phi(x) \to \phi(na)\ ,\\
&\int d^4x \to a^4 \sum_n\ ,\\
&\Box\phi(x) \to \frac{1}{a^2}\hat{\Box}\phi(na),\\
&D\phi \to \prod_n d\phi(na)\ ,
\end{aligned} \tag{3.7a}
$$

where the action of the dimensionless lattice Laplacean $\hat{\Box}$ is defined by

$$
\hat{\Box}\phi(na) = \sum_\mu \left(\phi(na+\hat{\mu}a) + \phi(na-\hat{\mu}a) - 2\phi(na)\right). \tag{3.7b}
$$

Here $\hat{\mu} \equiv \hat{e}_\mu$, where $\hat{e}_\mu$ is a unit vector pointing along the μ-direction.

We next want to obtain a path integral expression involving dimensionless variables only. To this effect we scale the mass parameter M and the field ϕ according to their "canonical" dimension. As seen from (3.6a) ϕ has the dimension of inverse length (the same as M). Hence we define the dimensionless quantities $\hat{M}$ and $\hat{\phi}_n$ by

$$
\begin{aligned}
\hat{\phi}_n &= a\phi(na),\\
\hat{M} &= aM.
\end{aligned} \tag{3.8}
$$

With (3.7) and (3.8) expression (3.5) translates into

$$
\langle \hat{\phi}_n\hat{\phi}_m\cdots\rangle = \frac{\int \prod_\ell d\hat{\phi}_\ell \hat{\phi}_n\hat{\phi}_m\cdots\ e^{-S_E[\hat{\phi}]}}{\int \prod_\ell d\hat{\phi}_\ell e^{-S_E[\hat{\phi}]}}, \tag{3.9a}
$$

where

$$
S_E = -\frac{1}{2}\sum_{n,\hat{\mu}} \hat{\phi}_n\hat{\phi}_{n+\mu} + \frac{1}{2}(8+\hat{M}^2)\sum_n \hat{\phi}_n\hat{\phi}_n, \tag{3.9b}
$$

and where the sum over μ extends over all positive and negative directions. Notice that the lattice spacing no longer appears in these expressions! This is not particular to the free field theory considered here, for it is merely a consequence of the fact that, measured in units of $\hbar$, the action is dimensionless, which is true for any theory.

It is important to realize that the form of the lattice action (3.9b) is not unique, and that we have merely chosen the simplest one. Thus, a priori, the only requirement that *any* lattice action should fulfil is that it reproduces the correct *classical* expression in the naive continuum limit*. Indeed, the scalar field is the only case in which a simple prescription of the above type gives the correct lattice action describing the quantum theory. Already the free Dirac theory will require a more careful treatment!

Let us now consider the integral (3.9a) in more detail. Its structure is analogous to that encountered in the statistical mechanics of a spin system with nearest neighbour interactions. In the present case, however, the theory is easily solved since the variables $\hat{\phi}_n$ are allowed to take on any real value. To carry out the integral (3.9a) we rewrite the action (3.9b) in the form

$$S_E = \frac{1}{2}\sum_{n,m} \hat{\phi}_n K_{nm} \hat{\phi}_m, \tag{3.10a}$$

where K_{nm} is given by

$$K_{nm} = -\sum_{\mu>0}[\delta_{n+\hat{\mu},m} + \delta_{n-\hat{\mu},m} - 2\delta_{nm}] + \hat{M}^2\delta_{nm}. \tag{3.10b}$$

Consider the generating functional

$$Z_0[J] = \int \prod_\ell d\hat{\phi}_\ell \; e^{-S_E[\hat{\phi}]+\sum_n \hat{J}_n\hat{\phi}_n}. \tag{3.11}$$

It can be easily calculated, since the (multiple) integral is of the Gaussian type. Apart from an overall constant, which we shall always drop since it plays no role when computing ensemble averages, we have that (cf. e. g. (2.28) and (2.30)

$$Z_0[J] = \frac{1}{\sqrt{det\ K}} e^{\frac{1}{2}\sum_{n,m} J_n K^{-1}_{nm} J_m} \quad . \tag{3.12}$$

Here K^{-1} is the inverse of the matrix (3.10b), and $det\ K$ is the determinant of K. By differentiating (3.12) with respect to the sources we obtain any desired correlation function. For our purpose it suffices to consider the 2-point function. From what we have learned in chapter 2, we get

$$\langle \hat{\phi}_n \hat{\phi}_m \rangle = K^{-1}_{nm} \quad . \tag{3.13}$$

The inverse matrix K^{-1} is determined from the equation

$$\sum_\ell K_{n\ell} K^{-1}_{\ell m} = \delta_{nm}, \tag{3.14}$$

---

* i.e. scaling the variables with a appropriately, and letting $a \to 0$.

and is easily computed by working in momentum space, where δ_{nm} is given by

$$\delta_{nm} = \int_{-\pi}^{\pi} \frac{d^4\hat{k}}{(2\pi)^4} e^{i\hat{k}\cdot(n-m)}. \tag{3.15}$$

We have introduced the "hat" on $\hat{k} = (\hat{k}_1, ..., \hat{k}_4)$ to emphasize that these variables are dimensionless. Making use of the Fourier representation (3.15), one finds that (3.10b) is given by

$$K_{nm} = \int_{-\pi}^{\pi} \frac{d^4\hat{k}}{(2\pi)^4} \tilde{K}(\hat{k}) e^{i\hat{k}\cdot(n-m)}, \tag{3.16a}$$

where

$$\tilde{K}(\hat{k}) = 4\sum_{\mu=1}^{4} \sin^2 \frac{\hat{k}_\mu}{2} + \hat{M}. \tag{3.16b}$$

Notice that the integration in (3.16a) is restricted to the Brillouin zone (BZ), $-\pi \leq \hat{k}_\mu \leq \pi$. The inverse matrix (3.13) is now easily determined from (3.14) by making the ansatz

$$K^{-1}_{nm} = \int_{-\pi}^{\pi} \frac{d^4\hat{k}}{(2\pi)^4} G(\hat{k}) e^{i\hat{k}\cdot(n-m)} \quad ,$$

and performing the sum over ℓ using the expression (2.64) with ($a = 1$) for the periodic delta function:

$$K^{-1}_{nm} = \langle\hat{\phi}_n\hat{\phi}_m\rangle = \int_{-\pi}^{\pi} \frac{d^4\hat{k}}{(2\pi)^4} \frac{e^{i\hat{k}\cdot(n-m)}}{4\sum_\mu \sin^2 \frac{\hat{k}_\mu}{2} + \hat{M}^2} \quad . \tag{3.17}$$

The right-hand side of this expression depends on the lattice sites n and m, and on the dimensionless mass parameter $\hat{M}$. To make this explicit let us define

$$G(n, m; \hat{M}) = < \hat{\phi}_n\hat{\phi}_m > .$$

Suppose we were given (3.17) and were asked to study its continuum limit in order to extract the physical two-point correlation function, $\langle\phi(x)\phi(y)\rangle$. The obvious thing to try would be to introduce the lattice spacing by rescaling $\hat{\phi}_n$ and $\hat{M}$ according to (3.8), and to take the limit $a \to 0$, holding M, ϕ, $x = na$ and $y = ma$ fixed. Hence one must know which quantities are to be held fixed as one removes the lattice structure! For example, in an interacting theory the mass parameter M would in general be unphysical and cannot be held fixed as we let the lattice spacing go to zero. In the present simple case, however, the naive procedure just described gives the correct continuum limit; i.e., we claim that the right-hand side of

$$\langle \phi(x)\phi(y)\rangle = \lim_{a\to 0}\frac{1}{a^2}G(\frac{x}{a},\frac{y}{a};Ma) \tag{3.18}$$

approaches a finite limit, and reproduces the well-known result for the scalar two-point function. For this to be the case, $G(x/a, y/a; Ma)$ must clearly vanish in the limit $a \to 0$. From (3.17) one finds after a trivial change of integration variables that

$$G\left(\frac{x}{a},\frac{y}{a};Ma\right) = a^2\int_{-\pi/a}^{\pi/a}\frac{d^4k}{(2\pi)^4}\frac{e^{ik\cdot(x-y)}}{\sum_\mu \tilde{k}_\mu^2 + M^2} \tag{3.19a}$$

where $\tilde{k}_\mu$ is given by

$$\tilde{k}_\mu = \frac{2}{a}\sin\frac{k_\mu a}{2} \ . \tag{3.19b}$$

Since the integration in (3.19a) is restricted to the interval $[-\frac{\pi}{a}, \frac{\pi}{a}]$, the integral will be dominated by momenta which are small compared to the inverse lattice spacing; hence we may set $\tilde{k}_\mu \to k_\mu$. Taking the limit $a \to 0$ we arrive at the well-known result:

$$\langle \phi(x)\phi(y)\rangle = \int_{-\infty}^{\infty}\frac{d^4k}{(2\pi)^4}\frac{e^{ik\cdot(x-y)}}{k^2+M^2} \ . \tag{3.20}$$

The above discussion has made explicit use of the lattice spacing. But suppose we were not in the position of performing the continuum limit analytically, but must rely on a numerical calculation of the path integral (3.9) where the lattice spacing does not appear. What does it mean to study the continuum limit in such a case? The idea is of course to make the lattice finer and finer with physics remaining the same as we approach the continuum limit. Consider for example a physical correlation length ξ. Decreasing the lattice spacing means increasing the correlation length $\hat{\xi}$ measured in lattice units. But $\hat{\xi}$ may be controlled by the parameters on which the theory depends! Thus, doubling $\hat{\xi}$ by choosing these parameters appropriately, we have cut down the lattice spacing by a factor of $1/2$. Now what controles the correlation length $\hat{\xi}$ in our example is the dimensionless parameter $\hat{M}$. In the continuum limit the correlation function (3.20) decays exponentially for large $|x-y|$ with a correlation length given by the inverse mass. Hence the corresponding correlation length measured in lattice units, i.e., $\hat{\xi} = \frac{1}{Ma}$, diverges as $a \to 0$! Thus from the point of view of a statistical mechanical system described by the partition function (3.11) with $J = 0$, the continuum limit is realized for $\hat{M} \to 0$ at a critical point of the theory! It is therefore evident that in any practical numerical calculation carried out on a finite-size lattice we can never actually go to the continuum limit. How do we then decide whether we are extracting continuum physics? In principle the answer is very simple: We must ensure that our lattice is fine enough (i.e. $\hat{M}$ small enough in the present case) so that physical quantities become insensitive to the lattice structure. But quantities like $\langle\phi(x)\phi(y)\rangle$ are dimensioned, and we can only calculate dimensionless objects! So we must consider dimensionless ratios of physical quantities. In our case the simplest quantity is $\langle\phi(x)\phi(y)\rangle/M^2$; hence we must study the lattice ratio $\langle\hat{\phi}_n\hat{\phi}_m\rangle/\hat{M}^2$ for small values of $\hat{M}$, keeping $\hat{M}|n-m| \approx M|x-y|$ fixed. If for sufficiently small $\hat{M}$ this ratio becomes independent of $\hat{M}$, then our lattice is fine enough and we are extracting continuum physics.

Admittedly the case of a free scalar field was a very simple one. Nevertheless it has served to elucidate several ideas that go into a lattice formulation. In an interacting theory the story will be certainly more complicated. But the main messages of this chapter will remain.

CHAPTER 4

FERMIONS ON THE LATTICE

In the preceding chapter we have shown that the lattice formulation of the free scalar field theory poses no problems. The correct continuum limit was reached by simply scaling all dimensionless variables appropriately with the lattice spacing a, and taking the limit $a \to 0$ holding physical quantities fixed. The purpose of these lectures, however, is to arrive at a lattice formulation of QCD which describes the interaction of quarks and gluons. Hence we must learn how to deal with fermions and gauge fields on a lattice. While there is a clear-cut and elegant way of introducing gauge fields on a space-time lattice, the situation regarding fermions is not so clear. As we shall see, the difficulties arise already on the level of the free Dirac field. From the psychological point of view it would therefore be preferable to discuss that part of lattice gauge theories first which one believes to be well understood, and to introduce lattice fermions at a later stage, since they are endowed with special problems not encountered for bosonic fields. On the other hand, having discussed the scalar field, it is only natural to attempt a similar naive formulation for the other kind of matter field. Thus it is interesting that in the case of fermions the lattice forces us to deviate from the naive type of prescription adopted in the previous chapter, in order to avoid the so-called fermion "doubling" problem. Several proposals have been made in the literature to get around this problem, and we shall discuss the two most popular ones.

4.1 The Doubling Problem

We begin by pointing out the difficulties one encounters when latticizing the free Dirac field.

Consider first the Dirac equation in Minkowski space

$$(i\gamma^\mu \partial_\mu - M)\psi(x) = 0,$$

where γ^μ are 4×4 Dirac matrices satisfying the following anticommutation relations

$$\{\gamma^\mu, \gamma^\nu\} = 2g^{\mu\nu},$$

and ψ is a 4-component field, whose components we shall label by a Greek index (α, β, etc.). The equation of motion for ψ and $\bar{\psi}(\equiv \psi^\dagger \gamma^0)$ follow from the independent variation of the action

$$S_F[\psi, \bar{\psi}] = \int d^4x \bar{\psi}(x)(i\gamma^\mu \partial_\mu - M)\psi(x)$$

with respect to the fields ψ and $\bar{\psi}$. In the quantum theory, ψ and $\psi^\dagger$ become operators, Ψ and $\Psi^\dagger$ satisfying the following canonical equal-time commutation relations

$$\{\Psi_\alpha(\vec{x},t), \Psi^\dagger_\beta(\vec{y},t)\} = \delta_{\alpha\beta}\delta^{(3)}(\vec{x}-\vec{y}).$$

The path integral representation of the Green function *

$$\langle\Omega|T(\Psi_{\alpha_1}(x_1)...\Psi_{\alpha_\ell}(x_\ell)\bar{\Psi}_{\beta_1}(y_1)...\bar{\Psi}_{\beta_\ell}(y_\ell))|\Omega\rangle$$

is given by (2.60), i.e.

$$\langle\Omega|T(\Psi_\alpha(x)...\bar{\Psi}_\beta(y)...)|\Omega\rangle = \frac{\int D\bar{\psi}D\psi\psi_\alpha(x)...\bar{\psi}_\beta(y)...e^{iS_F}}{\int D\bar{\psi}D\psi e^{iS_F}}.$$

The corresponding representation of the Green functions continued to imaginary times is obtained by replacing $iS_F[\psi,\bar{\psi}]$ by $-S_F^{(eucl.)}[\psi,\bar{\psi}]$, where $S_F^{(eucl.)}$ is the euclidean action, and identifying x, y etc. with the euclidean four-vectors. We denote the euclidean Green functions by $\langle\psi_\alpha(x)...\bar{\psi}_\beta(y)...\rangle$. Then

$$\langle\psi_\alpha(x)...\bar{\psi}_\beta(y)...\rangle = \frac{\int D\bar{\psi}D\psi(\psi_\alpha(x)...\bar{\psi}_\beta(y)...)e^{-S_F^{(eucl.)}[\psi,\bar{\psi}]}}{\int D\bar{\psi}D\psi e^{-S_F^{(eucl.)}[\psi,\bar{\psi}]}}. \tag{4.1}$$

The euclidean action can be obtained from S_F by the prescription discussed in section 2 of the previous chapter. But since in euclidean space the Lorentz group is replaced by the rotation group in four dimensions, it is convenient to express the action in terms of a new set of γ-matrices $\gamma_\mu^E(\mu = 1, ..., 4)$, satisfying the algebra

$$\{\gamma_\mu^E, \gamma_\nu^E\} = 2\delta_{\mu\nu}.$$

With the hermitean choice $\gamma_4^E = \gamma^0, \gamma_i^E = -i\gamma^i$, the euclidean action then takes the form

$$S_F^{(eucl.)} = \int d^4x\bar{\psi}(x)(\gamma_\mu^E\partial_\mu + M)\psi(x). \tag{4.2}$$

Since from now on we shall be interested only in the euclidean formulation, we shall drop any labels reminding us of this.

So far the path integrals in (4.1) are only formally defined, since x, y etc. are continuous variables. So let us introduce a space-time lattice. The fields ψ and $\bar{\psi}$

---

* Here x_i, y_i $(i = 1, ..., \ell)$ denote four vectors in Minkowski space.

then live on the lattice sites na, where a is the lattice spacing, and the integration measure is given by

$$D\bar{\psi}D\psi = \prod_{\alpha,n} d\bar{\psi}_\alpha(na) \prod_{\beta,m} d\psi_\beta(ma).$$

Next we rewrite (4.1) in terms of dimensionless lattice variables, by scaling M, ψ and $\bar{\psi}$ with a according to their canonical dimensions. This is achieved by making the replacements

$$\begin{aligned} M &\rightarrow \frac{1}{a}\hat{M}, \\ \psi_\alpha(x) &\rightarrow \frac{1}{a^{3/2}}\hat{\psi}_\alpha(n), \\ \bar{\psi}_\alpha(x) &\rightarrow \frac{1}{a^{3/2}}\hat{\bar{\psi}}_\alpha(n), \\ \partial_\mu\psi_\alpha(x) &\rightarrow \frac{1}{a^{5/2}}\hat{\partial}_\mu\hat{\psi}_\alpha(n), \end{aligned} \tag{4.3a}$$

where $\hat{\partial}_\mu$ is the antihermitean lattice derivative defined by

$$\hat{\partial}_\mu\hat{\psi}_\alpha(n) = \frac{1}{2}[\hat{\psi}_\alpha(n+\hat{\mu}) - \hat{\psi}_\alpha(n-\hat{\mu})] \quad . \tag{4.3b}$$

Then the lattice version of (4.2) reads

$$S_F = \sum_{\substack{n,m \\ \alpha,\beta}} \hat{\bar{\psi}}_\alpha(n) K_{\alpha\beta}(n,m) \hat{\psi}_\beta(m), \tag{4.4a}$$

where

$$K_{\alpha\beta}(n,m) = \sum_\mu \frac{1}{2}(\gamma_\mu)_{\alpha\beta}[\delta_{m,n+\hat{\mu}} - \delta_{m,n-\hat{\mu}}] + \hat{M}\delta_{mn}\delta_{\alpha\beta}. \tag{4.4b}$$

With this action the lattice correlation functions are given by the following path integral expression

$$\langle\hat{\psi}_\alpha(n)\cdots\hat{\bar{\psi}}_\beta(m)\cdots\rangle = \frac{\int D\hat{\bar{\psi}}D\hat{\psi}\ \hat{\psi}_\alpha(n)\cdots\hat{\bar{\psi}}_\beta(m)\cdots e^{-S_F}}{\int D\hat{\bar{\psi}}D\hat{\psi}\ e^{-S_F}}, \tag{4.5a}$$

where the integration measure is defined by

$$D\hat{\bar{\psi}}D\hat{\psi} = \prod_{n,\alpha} d\hat{\bar{\psi}}_\alpha(n) \prod_{m,\beta} d\hat{\psi}_\beta(m). \tag{4.5b}$$

The correlation functions (4.5a) can be obtained from the generating functional

$$Z[\eta,\bar{\eta}] = \int D\hat{\bar{\psi}} D\hat{\psi} e^{-S_F+\sum_{n,\alpha}[\bar{\eta}_\alpha(n)\hat{\psi}_\alpha(n)+\hat{\bar{\psi}}_\alpha(n)\eta_\alpha(n)]} \tag{4.6}$$

by carrying out the appropriate differentiations with respect to the Grassmann-valued sources $\eta_\alpha(n)$ and $\bar{\eta}_\alpha(n)$ (see chapter 2). The integral (4.6) can be performed (cf. eqs. (2.49) and (2.50)) and we obtain

$$Z[\eta,\bar{\eta}] = det\ K e^{\sum_{n,m,\alpha,\beta}\bar{\eta}_\alpha(n)K^{-1}_{\alpha\beta}(n,m)\eta_\beta(m)}.$$

Hence the two-point function is given by

$$\langle\hat{\psi}_\alpha(n)\hat{\bar{\psi}}_\beta(m)\rangle = K^{-1}_{\alpha\beta}(n,m).$$

So far everything is quite analogous to the scalar case considered in the previous chapter. In particular we want to emphasize that the lattice action (4.4) was obtained by proceeding in the most naive way possible. Such a prescription was shown to work in the case of the free scalar field. Hence there is no a priori reason why it should fail to do so in the present case. But the fact is that it fails! To see this let us compute the physical correlation function $\langle\psi_\alpha(x)\bar{\psi}_\beta(y)\rangle$ by carrying out the continuum limit in a manner analogous to the scalar case (cf. eq. (3.18)), i.e.

$$\langle\psi_\alpha(x)\bar{\psi}_\beta(y)\rangle = \lim_{a\to 0}\frac{1}{a^3}G_{\alpha\beta}\left(\frac{x}{a},\frac{y}{a};Ma\right),$$

where $G_{\alpha\beta}(n,m,\hat{M}) \equiv K^{-1}_{\alpha\beta}(n,m)$. The factor $1/a^3$ arises from scaling the fields according to (4.3a). The inverse matrix $K^{-1}_{\alpha\beta}(n,m)$, defined by

$$\sum_{\lambda,\ell}K^{-1}_{\alpha\lambda}(n,\ell)K_{\lambda\beta}(\ell,m) = \delta_{\alpha\beta}\delta_{nm},$$

can be easily calculated by proceeding as in the previous chapter, and one obtains

$$\langle\psi_\alpha(x)\bar{\psi}_\beta(y)\rangle = \lim_{a\to 0}\int_{-\pi/a}^{\pi/a}\frac{d^4p}{(2\pi)^4}\frac{[-i\sum\gamma_\mu\tilde{\bar{p}}_\mu + M]_{\alpha\beta}}{\sum_\mu\tilde{\bar{p}}^2_\mu + M^2}e^{ip(x-y)}, \tag{4.7a}$$

where $\tilde{\bar{p}}_\mu$ is given by

$$\tilde{\bar{p}}_\mu = \frac{1}{a}\sin(p_\mu a). \tag{4.7b}$$

For $\tilde{\bar{p}}_\mu \to p_\mu$ the above integral would reduce to the well-known 2-point function in the limit $a \to 0$. Recall that in the scalar case, we had encountered a similar situation, but with an all important difference! The argument of the sine-function in eq. (3.19b) is only half of that in (4.7b)! This makes a big difference and is the origin of the so-called "fermion doubling" problem. While in the case of the scalar field we could argue that $\tilde{k}_\mu$ in (3.19a) can be replaced by k_μ in the continuum limit, such a replacement cannot be made in the present case. The reason for this is most clearly seen by looking at fig. (4-1) where we have plotted $\tilde{\bar{p}}_\mu$ as a function of p_μ, for p_μ within the Brillouin zone. The straight line corresponds to $\tilde{\bar{p}}_\mu = p_\mu$. Within half of the BZ the situation is analogous to that encountered in the scalar case: near the continuum limit, the deviation from the straight line behaviour occurs only for large momenta where p_μ *and* $\tilde{\bar{p}}_\mu$ are both of order $1/a$.

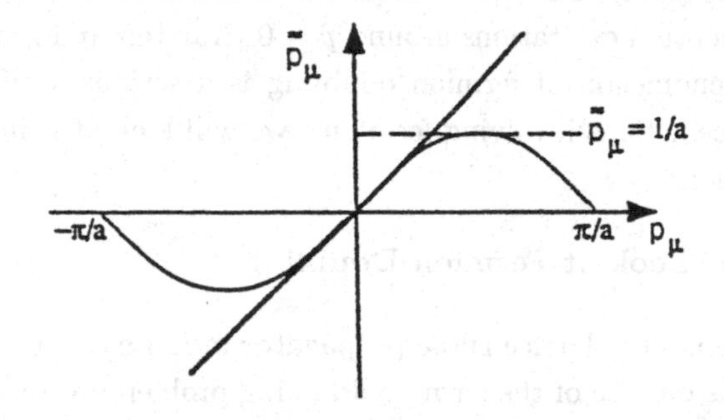

Fig. 4-1 Plot of $\sin(p_\mu a)/a$ versus p_μ in the Brillouin zone. The straight line corresponds to $\tilde{\bar{p}}_\mu = p_\mu$. The continuum limit is determined by the momenta in the neighbourhood of $p_\mu = 0$ and $p_\mu = \pm\pi/a$.

What destroys the correct continuum limit in the fermionic case are the zeros of the sine-function in (4.7b) at the edges of the BZ. Thus there exist sixteen regions of integrations in (4.7a), where $\tilde{\bar{p}}_\mu$ takes a finite value in the limit $a \to 0$. Of these, fifteen regions involve high momentum excitations of the order of π/a (and -π/a), which give rise to a momentum distribution function having the form resembling that of a single particle propagator. Hence in the continuum limit, the Green function (4.7a) receives contributions from sixteen fermion-like excitations in momentum space, of which fifteen are pure lattice artefacts having no continuum analog. In d space-time dimensions the number would be 2^d; i.e. it doubles for each additional dimension.

The "doubler" contributions, arising from momentum excitations near the corners of the Brillouin zone, are in fact essential for avoiding an apparent clash with a well known result in continuum QED. For vanishing fermion mass the QED action is invariant under the global chiral transformation

$$\psi \to e^{i\theta\gamma_5}\psi \ ; \ \ \bar{\psi} \to \bar{\psi}e^{i\theta\gamma_5} \ , \tag{4.8}$$

where θ is a parameter, and $\gamma_5 = \gamma_1\gamma_2\gamma_3\gamma_4$ is a hermitean matrix which anticommutes with γ_μ ($\mu = 1, 2, 3, 4$). Naively this implies the existence of a conserved axial vector current. But because of quantum fluctuations this current has actually an anomalous divergence [Adler (1969); Bell and Jackiw (1969)]. In a lattice regularized theory, on the other hand, such a symmetry implies that this current is strictly conserved for any lattice spacing. The way the lattice resolves this apparent puzzle, consists in generating extra excitations ($\to$ doublers) that have no analog in the continuum, and which cancel the anomaly of the continuum theory arising from momentum excitations around $\hat{p} = 0$ [Karsten and Smit (1981)].

Since the phenomenon of fermion doubling is a serious stumbling block in constructing lattice actions involving fermions, we will look at it in more detail in the following section.

4.2 A Closer Look at Fermion Doubling

Before we discuss the lattice Dirac propagator in more detail, it is instructive to demonstrate the essence of the fermion doubling problem in some simple examples. In particular we want to show that the origin of fermion doubling lies in the use of the symmetric form for the lattice derivative.

i) Example 1

Consider the following eigenvalue equation

$$-i\frac{d}{dx}f(x) = \lambda f(x) \ .$$

The solution is given by $f_\lambda(x) = f_\lambda(0)e^{i\lambda x}$. Next consider the discretized version of this equation, where the derivative is replaced by the right "lattice" derivative. Let $f(n)$ be the value of $f(x)$ at the lattice site $x = na$, where a is the lattice spacing. Then

$$-i[f((n+1)) - f(n)] = \hat{\lambda}f(n) \ ,$$

where $\hat{\lambda}$ is the eigenvalue measured in units of the lattice spacing, i.e. $\hat{\lambda} = \lambda a$. The equation can be solved immediately by iteration:

$$f_{\hat{\lambda}}(n) = e^{n\ln(1+i\hat{\lambda})}f_{\hat{\lambda}}(0)$$

In the continuum limit, which is obtained by setting $n = \frac{x}{a}$, $\hat{\lambda} = \lambda a$, and taking the limit $a \to 0$ with λ fixed, we recover the above solution.

Let us now consider a discretization which respects the hermiticity of the operator $i\frac{\partial}{\partial x}$. This requires the use of the symmetric lattice derivative.

$$-\frac{i}{2}[f((n+1)) - f((n-1))] = \hat{\lambda} f(n) \ . \tag{4.9}$$

Thus our estimate of the derivative now involves twice the lattice spacing! As a consequence one finds that for each eigenvalue $\hat{\lambda}$ there exist two solutions to the eigenvalue equation (4.8). Not both of the solutions can possess a continuum limit, since the continuum eigenfunctions are non-degenerate. Indeed, equation (4.8) can be solved with the Ansatz

$$f_{\hat{\lambda}}(n) = Ce^{i\hat{p}n} \ ,$$

where $\hat{p}$ satisfies the equation

$$\sin\hat{p} = \hat{\lambda} \ .$$

For a given positive (negative) eigenvalue $\hat{\lambda}$ this equation possesses two solutions for $\hat{p}$: one lying in the range $0 < \hat{p} < \frac{\pi}{2}$ $(-\frac{\pi}{2} < \hat{p} < 0)$, and the other in the interval $\frac{\pi}{2} < \hat{p} < \pi$ $(-\pi < \hat{p} < -\frac{\pi}{2})$. The corresponding eigenfunctions are given by

$$\hat{f}^{(1)}_{\hat{\lambda}}(n) = Ae^{i(arcsin\hat{\lambda})n} \ , \tag{4.10a}$$

$$\hat{f}^{(2)}_{\hat{\lambda}}(n) = B(-1)^n e^{-i(arcsin\hat{\lambda})n} \ , \tag{4.10b}$$

where

$$-\frac{\pi}{2} < \arcsin\hat{\lambda} < \frac{\pi}{2} \ . \tag{4.10c}$$

The solution (4.10a) possesses a continuum limit which is realized by setting $n = x/a$, $\hat{\lambda} = \lambda a$, and taking the limit $a \to 0$ with λ fixed. In this limit we recover the solution to the original continuum eigenvalue equation. On the other hand (4.10b) does not possess such a limit because of the factor $(-1)^n$ which alternates in sign as one proceeds from one lattice site to the next. Notice that the origin of the "doubler" solution (4.10b) is a consequence of having used the symmmetric form for the lattice derivative. One might be tempted to merely ignore this solution. In this case, however, our eigenfunctions would no longer constitute a complete set.

ii) Example 2

Consider the Green function for the differential operator $\frac{d}{dt} + M$:

$$[\frac{d}{dt} + M]G(t,t') = \delta(t-t') \ . \tag{4.11}$$

The general solution is given by

$$G(t,t') = Ae^{-M(t-t')} + \theta(t-t')e^{-M(t-t')} \ . \tag{4.12}$$

Next consider the dimensionless discretized version of (4.11), where the derivative is replaced by the symmetric lattice derivative:

$$\sum_{n'}[\frac{1}{2}(\delta_{n+1,n'} - \delta_{n-1,n'}) + \hat{M}\delta_{nn'}]\hat{G}(n',m) = \delta_{nm} \ . \tag{4.13}$$

Here $\hat{M}$ is the "mass" M measured in lattice units, i.e., $\hat{M} = Ma$. The most general solution to the homogeneous equation reads:

$$\hat{G}^{(0)}_{\hat{M}}(n,m) = Ae^{-(n-m)arsinh\hat{M}} + B(-1)^{n-m}e^{(n-m)arsinh\hat{M}} \ . \tag{4.14}$$

Thus, when discretized, the homogeneous equation has an aditional solution which, because of the factor $(-1)^{n-m}$, possesses no continuum limit. This limit is realized for $\hat{M} \to 0$, $n, m \to \infty$ with $\hat{M}n = Mt$ and $\hat{M}m = Mt'$ fixed. The homogeneous solution to (4.11) is then seen to correspond to the first term appearing on the rhs of (4.14).

A particular solution to the inhomogeneous equation (4.13) can be obtained by making the Fourier Ansatz

$$\hat{G}^{(part)}_{\hat{M}}(n,m) = \int_{-\pi}^{\pi} \frac{d\hat{p}}{2\pi}\hat{G}(\hat{p})e^{i\hat{p}(n-m)} \ .$$

Introducing this expression into (4.13) we obtain

$$\hat{G}^{(part)}_{\hat{M}}(n,m) = \int_{-\pi}^{\pi} \frac{d\hat{p}}{2\pi}\frac{e^{i\hat{p}(n-m)}}{i\sin\hat{p} + \hat{M}} \ . \tag{4.15}$$

The integral can be easily evaluated by introducing the variable $z = e^{i\hat{p}}$. Then

$$\hat{G}^{(part)}_{\hat{M}}(n,m) = \frac{1}{\pi i}\int_C dz \frac{z^{n-m}}{z^2 + 2\hat{M}z - 1} \ ,$$

where integration is carried in the counterclockwise sense along a unit circle in the complex z-plane centered at $z = 0$. For $n - m \geq 0$ the integral is determined by the residue of the pole at $z = -\hat{M} + \sqrt{1 + \hat{M}^2}$. On the other hand, for $n - m < 0$

we can distort the integration contour to infinity, taking into account the pole at $z = -\hat{M} - \sqrt{1+\hat{M}^2}$, located outside the unit circle. One then finds that

$$\hat{G}^{(part)}_{\hat{M}}(n,m) = \theta(n-m)\frac{e^{-(n-m)arsinh\hat{M}}}{\sqrt{(1+\hat{M}^2)}} + (-1)^{n-m}\theta(m-n)\frac{e^{(n-m)arsinh\hat{M}}}{\sqrt{(1+\hat{M}^2)}} \tag{4.16}$$

where $\theta(0) = \frac{1}{2}$. Note again, that only the first term on the rhs possesses a continuum limit. For lattice spacings small compared to $\frac{1}{M}$, i.e. for small $\hat{M}$, and for fixed $t = na$ and $t' = ma$, we have that

$$\hat{G}^{(part)}_{\hat{M}}(\frac{t}{a}, \frac{t'}{a}) \approx \theta(t-t')e^{-M(t-t')} + (-1)^{\frac{t-t'}{a}}\theta(t'-t)e^{M(t-t')} . \tag{4.17}$$

While the second term is not defined for $a \to 0$, the first term reproduces the inhomogeneous solution to (4.11), given by the second term in (4.12).

The expression (4.17) could also have been obtained as follows, without actually carrying out the integrals. Let us set $\hat{p} = pa$, $\hat{M} = Ma$, and $t = na$, $t' = ma$ in (4.15). Then

$$\hat{G}^{(part)}_{\hat{M}}(\frac{t}{a}, \frac{t'}{a}) = \int_{-\frac{\pi}{a}}^{\frac{\pi}{a}} \frac{dp}{2\pi} \frac{e^{ip(t-t')}}{\frac{i}{a}\sin pa + M} . \tag{4.18}$$

For $a \to 0$ the relevant contributions to the integral come from momenta for which $\sin pa \approx O(a)$, i.e. from i) finite (dimensioned) momenta p, and ii) momenta close to the corners of the Brillouine zone. Let us therefore decompose the integral (4.18) as follows:

$$\hat{G}^{(part)}_{\hat{M}}(\frac{t}{a}, \frac{t'}{a}) = \int_{-\frac{\pi}{2a}}^{\frac{\pi}{2a}} \frac{dp}{2\pi} \frac{e^{ip(t-t')}}{\frac{i}{a}\sin pa + M} + \int_{\frac{\pi}{2a}}^{\frac{\pi}{a}} \frac{dp}{2\pi} \frac{e^{ip(t-t')}}{\frac{i}{a}\sin pa + M}$$
$$+ \int_{-\frac{\pi}{a}}^{-\frac{\pi}{2a}} \frac{dp}{2\pi} \frac{e^{ip(t-t')}}{\frac{i}{a}\sin pa + M} .$$

Making the change of variables $p = \frac{\pi}{a} + p'$ and $p = -\frac{\pi}{a} + p'$ in the last two integrals, respectively, one finds that

$$\hat{G}^{(part)}_{\hat{M}}(\frac{t}{a}, \frac{t'}{a}) = \int_{-\frac{\pi}{2a}}^{\frac{\pi}{2a}} \frac{dp}{2\pi} \frac{e^{ip(t-t')}}{\frac{i}{a}\sin pa + M} + (-1)^{\frac{t-t'}{a}} \int_{-\frac{\pi}{2a}}^{\frac{\pi}{2a}} \frac{dp}{2\pi} \frac{e^{ip(t-t')}}{-\frac{i}{a}\sin pa + M} .$$

The integrations now extend over only one half of the Brillouin zone. For $a \to 0$ the two integrals are therefore dominated by finite momenta p, for which $\sin pa$ is of $\mathcal{O}(a)$. Hence for $a \to 0$ we can replace $\sin pa$ by pa and obtain

$$\hat{G}^{(part)}(\frac{t}{a}, \frac{t'}{a}) \underset{a\to 0}{\longrightarrow} \int_{-\infty}^{\infty} \frac{dp}{2\pi} \frac{e^{ip(t-t')}}{ip + M} - (-1)^{\frac{t-t'}{a}} \int_{\infty}^{\infty} \frac{dp}{2\pi} \frac{e^{ip(t-t')}}{ip - M} ,$$

which can be readily integrated to yield (4.17).

The simple examples we discussed show how the discretization of an equation can lead to a doubling of solutions. The "doubler"-like solutions manifestated themselves in the appearance of a phase factor which changes sign as one proceeds from one "lattice" site to the next. This is not only peculiar to the above examples. In fact that doubler contributions to the Dirac propagator (4.7a) are expected to manifest themselves in the same way. This we will show below.

iii) Fermion Propagator

In the case of the fermion propagator one is confronted with matrix valued integrals over four dimensional momentum space. The characteristic structure of the "doublers" contributions to the Green function can nevertheless be easily exhibited by proceeding in the way we have just described. Our starting point is the expression (4.7a). To exhibit the effect of fermion doubling we decompose each of the momentum integrations into the following two regions

$$i)\quad |p_\mu| < \frac{\pi}{2a}$$

and

$$ii)\quad \frac{\pi}{2a} < |p_\mu| < \frac{\pi}{a}\ . \tag{4.19}$$

After changes of variables similar to those we made before, the reader can easily convince himself that the two-point function (4.7a) can be written in the form

$$\langle\psi_\alpha(x)\bar{\psi}_\beta(y)\rangle = \sum_{\bar{p}} e^{i\bar{p}\cdot(n-m)} \int_{-\frac{\pi}{2a}}^{\frac{\pi}{2a}} \frac{d^4p}{(2\pi)^4} \frac{[-i\sum\delta_{\bar{p}_\mu}\gamma_\mu\tilde{\tilde{p}}_\mu + M]_{\alpha\beta}}{\sum_\mu \tilde{\tilde{p}}_\mu^2 + M^2} e^{ip\cdot(x-y)}, \tag{4.20a}$$

where $x = na$, $y = ma$, and

$$\delta_{\bar{p}_\mu} = e^{i\bar{p}_\mu}\ . \tag{4.20b}$$

The sum in (4.20a) runs over all possible sets of four-momenta $\bar{p}$ (measured in lattice units), labeling the 16 corners of the hypercube in the first quadrand in momentum space: $(0,0,0,0)$, $(\pi,0,0,0)\cdots$, $(\pi,\pi,0,0)\cdots$, $(\pi,\pi,\pi,0)\cdots$, (π,π,π,π). The dots stands for all possible permutations of the components. Notice that all the integrations in (4.20a) extend over the reduced Brillouine zone $[-\frac{\pi}{2a},\frac{\pi}{2a}]$. Of the sixteen terms in the sum, only the term corresponding to $\bar{p} = (0,0,0,0)$ yields the familiar result in the continuum limit. While the remaining 15 integrals also possess a continuum limit, the phase factor $\exp[i\bar{p}\cdot(n-m)]$ does not. The structure of these fifteen integrals is the same as that corresponding to $\bar{p} = (0,0,0,0)$, except that each term is the Dirac propagator in a different representation of the gamma-matrices. Thus in each of the integrals the sign of the gamma matrix γ_μ is reversed if $\bar{p}_\mu = \pi$. The new set of gamma matrices are related to the standard set by a similarity transformation. Let us denote the matrices which induce these transformations by $\mathcal{T}_{\bar{p}}$:

$$\mathcal{T}_{\bar{p}}\gamma_\mu\mathcal{T}_{\bar{p}}^{-1} = \delta_{\bar{p}_\mu}\gamma_\mu \, . \tag{4.21}$$

They have a different structure depending on the number of non-vanishing components of $\bar{p}$. Let $\tilde{\mathcal{T}}_\mu$, $\tilde{\mathcal{T}}_{\mu\nu}$, $\tilde{\mathcal{T}}_{\mu\nu\lambda}$, $\tilde{\mathcal{T}}_{\mu\nu\lambda\rho}$ (all indices distinct) stand for the matrices $\mathcal{T}_{\bar{p}}$ which induce a reversal of signs of those γ-matrices corresponding to the set of subscripts. Their explicit form is given by $\tilde{\mathcal{T}}_\mu = \gamma_\mu\gamma_5$, $\tilde{\mathcal{T}}_{\mu\nu} = \gamma_\mu\gamma_\nu$, $\tilde{\mathcal{T}}_{\mu\nu\lambda} = \gamma_\mu\gamma_\nu\gamma_\lambda\gamma_5$, and $\tilde{\mathcal{T}}_{\mu\nu\lambda\rho} = \gamma_5$, where γ_5 is the hermitean matrix: $\gamma_5 = \gamma_1\gamma_2\gamma_3\gamma_4$. By the same reasoning as that given in example ii), the propagator is found to take the following form close to the continuum limit:

$$\begin{aligned} S_F(x-y) &\approx \sum_{\bar{p}} e^{i\bar{p}\cdot\frac{x-y}{a}}\mathcal{T}_{\bar{p}}\Big[\int_{-\infty}^{\infty}\frac{d^4p}{(2\pi)^4}\frac{-i\sum\gamma_\mu p_\mu + M}{\sum_\mu p_\mu^2 + M^2}e^{ip\cdot(x-y)}\Big]\mathcal{T}_{\bar{p}}^{-1} \\ &= \sum_{\bar{p}} V_{\bar{p}}(x)S_F^{(0)}(x-y)V_{\bar{p}}^{-1}(y) \, , \end{aligned} \tag{4.22a}$$

where

$$V_{\bar{p}}(z) = e^{i\bar{p}\cdot\frac{z}{a}}\mathcal{T}_{\bar{p}} \, , \tag{4.22b}$$

and

$$S_F^{(0)}(x-y) = \int_{-\infty}^{\infty}\frac{d^4p}{(2\pi)^4}\Big[\frac{-i\sum\gamma_\mu p_\mu + M}{\sum_\mu p_\mu^2 + M^2}\Big]e^{ip\cdot(x-y)} \tag{4.22c}$$

is the continuum Feynman propagator.

The above structure of the correlation function is intimately connected with a symmetry of the lattice action. Indeed one readily verfies that the gamma matrices satisfy the following relation,

$$V_{\bar{p}}(z)\gamma_\mu V_{\bar{p}}^{-1}(z \pm \hat{\mu}a) = \gamma_\mu \, .$$

As a consequence the lattice action

$$S = \frac{1}{2}\sum_x a^4\bar{\psi}(x)\gamma_\mu[\psi(x+\hat{\mu}a) - \psi(x-\hat{\mu}a)] + M\sum_x a^4\bar{\psi}(x)\psi(x)$$

is invariant under the transformation

$$\psi(x) \to V_{\bar{p}}(x)\psi(x) \, ,$$

$$\bar{\psi}(x) \to \bar{\psi}(x)V_{\bar{p}}^{-1}(x) .$$

In momentum space the action of the operator $V_{\bar{p}}$ corresponds to a shift in the momenta by $\bar{p}$. Corresponding to the sixteen edges of the BZ there exist sixteen such symmetry transformations. The fermion doubling phenomenon is a consequence of the existence of these symmetry transformations.

Finally, let us take a look at how the doublers manifest themselves in the naive lattice fermion two-point function for a massless field with definite chirality (as is the case for the neutrino). This field is an eigenstate of the projection operator $P_L = \frac{1}{2}(1-\gamma_5)$. The two point function is given by

$$< \psi_L(x)\bar{\psi}_L(y) >= P_L < \psi(x)\bar{\psi}(y) > P_R$$

where $\psi_L(n) = P_L\psi(n)$, and $P_R = \frac{1}{2}(1+\gamma_5)$. It therefore has the form

$$< \psi_L(x)\bar{\psi}_L(y) >= P_L \int \frac{d^4p}{(2\pi)^4}\frac{-i\gamma_\mu p_\mu}{p^2}e^{ip\cdot(x-y)} ,$$

where we have made use of the fact that γ_5 anticommutes with the γ-matrices γ_μ, and that $P_L^2 = 1$. On the other hand the corresponding lattice version is given by

$$< \psi_L(x)\bar{\psi}_L(y) >_{latt}= \sum_{\bar{p}} e^{i\bar{p}\cdot\frac{x-y}{a}} P_L \int_{\frac{-\pi}{2a}}^{\frac{\pi}{2a}} \frac{d^4p}{(2\pi)^4}\mathcal{T}_{\bar{p}}\Big[\frac{-i\sum\gamma_\mu\tilde{\bar{p}}_\mu}{\sum_\mu \tilde{\bar{p}}_\mu^2}\Big]\mathcal{T}_{\bar{p}}^{-1}e^{ip\cdot(x-y)} . \tag{4.23}$$

Now one can easily verify that

$$\mathcal{T}_{\bar{p}}(1-\gamma_5)\mathcal{T}_{\bar{p}}^{-1} = (1-\epsilon_{\bar{p}}\gamma_5) , \tag{4.24a}$$

where

$$\epsilon_{\bar{p}} = \prod_\mu \delta_{\bar{p}_\mu} . \tag{4.24b}$$

Hence we can also write (4.23) in the form

$$< \psi_L(x)\bar{\psi}_L(y) >= \sum_{\bar{p}} e^{i\bar{p}\cdot\frac{x-y}{a}}\mathcal{T}_{\bar{p}}\Big[\int_{\frac{-\pi}{2a}}^{\frac{\pi}{2a}} \frac{d^4p}{(2\pi)^4}\frac{1}{2}(1-\epsilon_{\bar{p}}\gamma_5)\frac{-i\sum\gamma_\mu\tilde{\bar{p}}_\mu}{\sum_\mu \tilde{\bar{p}}_\mu^2}e^{ip\cdot(x-y)}\Big]\mathcal{T}_{\bar{p}}^{-1} . \tag{4.25}$$

Since there are eight momenta $\bar{p}$ for which $\epsilon_{\bar{p}} = 1$, and eight momenta $\bar{p}$ for which $\epsilon_{\bar{p}} = -1$, it follows that the sum involves sixteen integrals which in the continuum limit have the form of eight left handed and eight right handed correlators in different representations of the γ-matrices. Fifteen of these integrals are however multiplied by phase factors characteristic for the doubler contributions.

As we have already stressed, the origin of the doubling problem lies in the use of the (antihermitean) symmetric form for the lattice derivative (4.3b). Thus while our lattice scale is a, our estimate of the derivative involves twice the lattice spacing. By using the right derivative

$$\hat{\partial}_\mu^R \hat{\psi}(n) = \hat{\psi}(n+\hat{\mu}) - \hat{\psi}(n) , \tag{4.26a}$$

or left derivative

$$\hat{\partial}_\mu^L \hat{\psi}(n) = \hat{\psi}(n) - \hat{\psi}(n-\hat{\mu}) , \tag{4.26b}$$

our estimate of the derivative would involve a distance which is just the lattice spacing and the doubling problem would not occur. In this case the hermitean conjugate of $\hat{\partial}_\mu^L$ would be $-\hat{\partial}_\mu^R$. A detailed analysis however shows that in the presence of interactions the use of the left or right derivative gives rise, for example, to non-covariant contributions to the fermion self energy and vertex function in QED which render the theory non-renormalizable. *

That the doubling phenomenon must occur in a lattice regularization which respects the usual hermiticity, locality and translational invariance requirements, follows from a theorem by Nielsen and Ninomiya [Nielson (1981)] which states that, under the above assumptions, one cannot solve the fermion doubling problem without breaking chiral symmetry for vanishing fermion mass.** This suggests that one may get rid of the doubling problem at the expense of breaking chiral symmetry explicitly on the lattice. A proposal in this direction was made originally by Wilson (1975), and is one of the two most popular schemes dealing with the doubling problem.

---

* A very detailed study of the precise form of these non-covariant contributions has been carried out by Sadooghi and Rothe (1996). Actions using one side lattice differences have been considered in the literature before, where by a suitable averaging procedure the correct continuum behaviour of the quantum theory is restored. The reader may confer the references cited in the above mentioned paper.

** For a simple derivation of the theorem, based on the Poincare-Hopf index theorem, we refer the reader to the book by Itzykson and Drouffe (1989).

4.3 Wilson Fermions

As we have emphasized in the last chapter, there are many different lattice actions which have the same naive continuum limit, and we have merely chosen the simplest one. We now exploit this ambiguity to modify the action (4.4) in such a way that the zeros of the denominator in (4.7a) at the edges of the BZ are lifted by an amount proportional to the inverse lattice spacing. This appears to be a perfectly legitimate procedure. What will, however, be particular to it is that, while one usually constructs the lattice action in accordance with the symmetries of the classical theory, one is forced to break explicitly the chiral symmetry which the original theory possesses for vanishing fermion mass. This is the price one has to pay to eliminate the fermion doubling problem and to ensure the correct continuum limit.

Let us now modify the action (4.4) by a term which vanishes in the naive continuum limit and is not invariant under chiral transformations. As we shall see below a second derivative term is a good candidate. Thus consider the action

$$S_F^{(W)} = S_F - \frac{r}{2}\sum_n \hat{\bar{\psi}}(n)\hat{\Box}\hat{\psi}(n), \tag{4.27}$$

where r is the Wilson parameter and $\hat{\Box}$ is the four-dimensional lattice Laplacean defined by (3.7b) with $a = 1$. Setting $\hat{\psi} = a^{3/2}\psi$ and $\hat{\Box} = a^2\Box$, we see that the additional term in (4.27) vanishes linearly with a in the naive continuum limit. Inserting for $\hat{\Box}\hat{\psi}(n)$ the expression analogous to (3.7b), the Wilson action can be written in the form

$$S_F^{(W)} = \sum_{n,m} \hat{\bar{\psi}}_\alpha(n) K_{\alpha\beta}^{(W)}(n,m)\hat{\psi}_\beta(m), \tag{4.28a}$$

where

$$\begin{aligned} K_{\alpha\beta}^{(W)}(n,m) =& (\hat{M} + 4r)\delta_{nm}\delta_{\alpha\beta} \\ & -\frac{1}{2}\sum_\mu [(r-\gamma_\mu)_{\alpha\beta}\delta_{m,n+\hat{\mu}} + (r+\gamma_\mu)_{\alpha\beta}\delta_{m,n-\hat{\mu}}]. \end{aligned} \tag{4.28b}$$

Notice that for $r \neq 0$ this expression breaks chiral symmetry even for $\hat{M} \to 0$.

The action (4.28) leads to the following two-point function of the continuum theory

$$\langle \psi_\alpha(x)\bar{\psi}_\beta(y)\rangle = \lim_{a\to 0}\int_{-\pi/a}^{\pi/a} \frac{d^4p}{(2\pi)^4} \frac{[-i\gamma_\mu \tilde{\tilde{p}}_\mu + M(p)]_{\alpha\beta}}{\sum_\mu \tilde{\tilde{p}}_\mu^2 + M(p)^2} e^{ip\cdot(x-y)}, \tag{4.29a}$$

where $\tilde{\bar{p}}_\mu$ is given by (4.7b) and

$$M(p) = M + \frac{2r}{a}\sum_\mu \sin^2(p_\mu a/2) \quad . \tag{4.29b}$$

From (4.29b) we see that for any fixed value of p_μ, $M(p)$ approaches M for $a \to 0$. Near the corners of the BZ, however, $M(p)$ diverges as we let the lattice spacing go to zero. This eliminates the fermion doubling problem, but at the expense that the chiral symmetry of the original action (4.4) for $M = 0$ has been broken. This makes this scheme less attractive for studying such questions as spontaneous chiral symmetry breaking in QCD (which requires a fine tuning of the parameter $\hat{M}$). So let us turn to the discussion of an alternative scheme for putting fermions on the lattice, known in the literature as the staggered fermion formulation (Kogut and Susskind (1975); Susskind, 1977; Banks et al. 1977). In contradistinction to Wilson fermions one then speaks of Susskind, Kogut-Susskind, or staggered fermions.

4.4 Staggered Fermions

As we have seen above, the fermion doubling problem owes its existence to the fact that the function (4.7b) vanishes at the corners of the Brillouin zone. This suggests the possibility of eliminating the unwanted fermion modes by reducing the BZ, i.e. by doubling the effective lattice spacing. This could in principle be accomplished if a) we were able to distribute the fermionic degrees of freedom over the lattice in such a way that the effective lattice spacing for each type of Grassmann variable is twice the fundamental lattice spacing, and b) if in the naive continuum limit the action reduces to the desired continuum form. Hence let us first take a look at the number of degrees of freedom required to double the effective lattice spacing. To this end consider a d-dimensional space-time lattice and subdivide it into elementary d-dimensional hypercubes of unit length. At each site within a given hypercube place a different degree of freedom, and repeat this structure periodically throughout the lattice. Then the effective lattice spacing has been doubled for each degree of freedom. In fig. (4-2) we show the case of a 2-dimensional lattice. Since there are 2^d sites within a hypercube, but only $2^{d/2}$ components of a Dirac field (in even space-time dimensions) we need $2^{d/2}$ different Dirac fields to reduce the BZ by a factor of $1/2$. In four space-time dimensions such a prescription may therefore be appropriate for describing 2^2 different "flavoured" (i.e. "up", "down", etc.) quarks. The corresponding Dirac fields we denote by ψ^f_α, where f denotes the flavour and α the spinor index. The concrete realization of the above program is, however, not as simple as it sounds. Thus the sites of an elementary hypercube will be occupied by certain linear combinations of the fields ψ^f_α chosen in such a way that the lattice action reduces in the naive continuum limit to a sum of free fermion actions, one for each of the quark flavours:

$$S_f^{(stag)} \to \int d^4x \sum_{\alpha,\beta,f} \bar{\psi}^f_\alpha(x)(\gamma_\mu \partial_\mu + M)_{\alpha\beta}\psi^f_\beta(x) \quad . \tag{4.30}$$

In a staggered-fermion formulation much of the work goes into the construction of the quark fields from the different one-component fields populating the lattice sites within an elementary hypercube, and into the study of the lattice symmetries.* In the following we briefly describe the main steps involved in arriving at a staggered fermion formulation, so that the reader will be acquainted with some expressions occurring frequently in the literature. Our presentation will essentially follow the work of Kluberg-Stern, Morel, Napoly and Petersson (1983). The technical details are relegated to the next section.

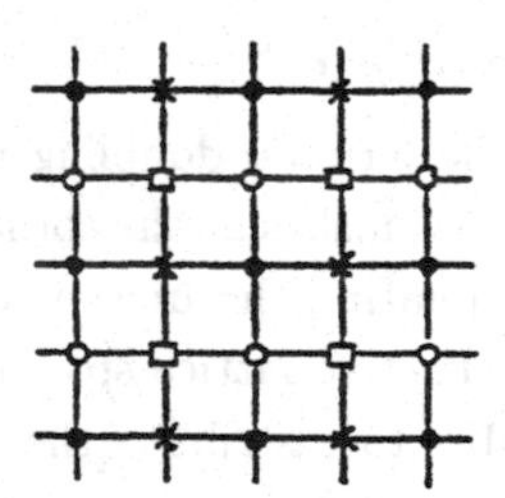

Fig. 4-2 Distributing 2^d degrees of freedom on a two dimensional (d=2) lattice.

Consider the naive action (4.4) for a free Dirac field on the lattice:

$$S = \frac{1}{2}\sum_{n,\mu}[\hat{\bar{\psi}}(n)\gamma_\mu\hat{\psi}(n+\hat{\mu}) - \hat{\bar{\psi}}(n)\gamma_\mu\hat{\psi}(n-\hat{\mu})] + \hat{M}\sum_n \hat{\bar{\psi}}(n)\hat{\psi}(n). \tag{4.31}$$

By making a local change of variables

$$\begin{aligned} \hat{\psi}(n) &= T(n)\chi(n), \\ \hat{\bar{\psi}}(n) &= \bar{\chi}(n)T^\dagger(n), \end{aligned} \tag{4.32}$$

* See e.g., Kluberg-Stern et al. (1983), Golterman (1986a), Jolicoeur, Morel, and Petersson (1986), Kilcup and Sharpe (1987).

where $T(n)$ are unitary $2^{d/2} \times 2^{d/2}$ matrices, one can "spin diagonalize" this expression by choosing the matrices $T(n)$ in such a way that

$$T^{\dagger}(n)\gamma_{\mu}T(n+\hat{\mu}) = \eta_{\mu}(n)\mathbb{1}, \tag{4.33}$$

where η_{μ} are c-numbers (notice that different space-time points are involved!), and $\mathbb{1}$ is the unit matrix. The matrices

$$T(n) = \gamma_1^{n_1}\gamma_2^{n_2}\cdots\gamma_d^{n_d} \tag{4.34}$$

satisfy (4.33) with the following phases $\eta_{\mu}(n)$:

$$\eta_{\mu}(n) = (-1)^{n_1+n_2+\cdots+n_{\mu-1}}, \ \eta_1(n) = 1. \tag{4.35}$$

Written in terms of the fields $\chi(n)$ and $\bar{\chi}(n)$ the action (4.31) reads

$$S = \sum_{\substack{n,\alpha \\ \mu}} \eta_{\mu}(n)\bar{\chi}_{\alpha}(n)\hat{\partial}_{\mu}\chi_{\alpha}(n) + \hat{M}\sum_{n,\alpha}\bar{\chi}_{\alpha}(n)\chi_{\alpha}(n),$$

where $\hat{\partial}_{\mu}$ is the lattice derivative defined in (4.3b). So far, of course, we have merely rewritten (4.31). Now comes a crucial step. Since we have got rid of the Dirac matrix γ_{μ} we can in principle let α run over any number of possible values, $\alpha = 1, 2, \cdots, k$. The minimal choice is $k = 1$, so that we shall omit this index from now on. The corresponding action

$$S_F^{(stag.)} = \frac{1}{2}\sum_{n,\mu}\eta_{\mu}(n)[\bar{\chi}(n)\chi(n+\hat{\mu}) - \bar{\chi}(n)\chi(n-\hat{\mu})] + \hat{M}\sum_{n}\bar{\chi}(n)\chi(n) \tag{4.36}$$

now involves only one degree of freedom per lattice site, and the only remnants of the original Dirac structure are the phases $\eta_{\mu}(n)$. The expression (4.36) is the action of the staggered fermion formulation in the absence of interactions. For it to be of physical relevance one must still show that in the continuum limit it may be written in the form (4.30) where the flavoured Dirac field components ψ_{α}^{f} are given as linear combinations of the one component fields living at the lattice sites within an elementary hypercube. Hence for finite lattice spacing the space-time coordinates of the fields ψ_{α}^{f} will be those labeling the position of the particular hypercube considered. Let us look at the "reconstruction" problem in somewhat more detail.

Consider a hypercube with origin at $\hat{x}_{\mu} = 2N_{\mu}$ (N_{μ} integers); then the lattice coordinates at the 2^d sites within this hypercube are given by

$$\hat{r}_{\mu} = 2N_{\mu} + \rho_{\mu},$$

where $\rho_\mu = 0$ or 1. This suggests the following relabeling of the fields $\chi(n)$

$$\chi_\rho(N) \equiv \chi(2N + \rho), \tag{4.37}$$

and similarly for $\bar{\chi}$. Notice that $N = (N_1, \cdots, N_d)$ now labels the space-time points of a lattice with lattice spacing $2a$, and that the (multi) index ρ labels the 2^d components of the new field χ. From these components one then constructs $2^{d/2}$ flavoured Dirac fields $\psi^f (f = 1, \cdots, 2^{d/2})$ with components ψ^f_α $(\alpha = 1, \cdots, 2^{d/2})$, by taking appropriate linear combinations:

$$\hat{\psi}^f_\alpha(N) = \mathcal{N}_0 \sum_\rho (T_\rho)_{\alpha f} \chi_\rho(N). \tag{4.38a}$$

Here

$$T_\rho = \gamma_1^{\rho_1} \gamma_2^{\rho_2} \cdots \gamma_d^{\rho_d}. \tag{4.38b}$$

By choosing the normalization constant $\mathcal{N}_0$ in (4.38a) appropriately, the action (4.36) then takes the following form in terms of the fields $\bar{\hat{\psi}}^f$ and $\hat{\psi}^f$:

$$S_F^{(stag)} = \sum_f \sum_N \bar{\hat{\psi}}^f(N)(\gamma_\mu \hat{\partial}_\mu + \hat{M})\hat{\psi}^f(N) + \cdots, \tag{4.39}$$

where $\hat{\partial}_\mu$ is now the lattice derivative on the new (blocked) lattice, and where the "dots" stand for terms which vanish in the naive continuum limit.* For finite lattice spacing, however, these terms are no longer invariant under the full chiral group. Nevertheless for $\hat{M} = 0$, (4.39) preserves a continuous $U(1) \times U(1)$ symmetry which is a remnant of the original chiral symmetry group. Because of this, one can use the staggered fermion formulation to study the spontaneous breakdown of this remaining lattice symmetry, and the associated Goldstone phenomenon (zero mass excitation, accompanying the spontaneous breakdown). This is a major advantage of staggered fermions over the Wilson fermions discussed before. Of course the staggered formulation has its drawbacks. Thus for example it can only be the lattice version of a theory with $2^{d/2}$ degenerate quark flavours, whereas there is no restriction on the flavour number in the Wilson formulation.

We close this section with a remark. Making use of the relation (4.38), one can calculate correlation functions of flavoured quark fields, by taking appropriate

---

* See the following section.

linear combinations of the χ-correlation functions. The latter are given by the following path integral expression

$$\langle \chi_{\rho_1}(N_1) \cdots \bar{\chi}_{\rho_\ell}(N_\ell) \rangle = \frac{\int D\bar{\chi} D\chi \, \chi_{\rho_1}(N_1) \cdots \bar{\chi}_{\rho_\ell}(N_\ell) e^{-S_F^{(stag)}}}{\int D\bar{\chi} D\chi \, e^{-S_F^{(stag)}}}, \tag{4.40a}$$

where

$$D\bar{\chi} D\chi = \prod_{\rho, N} d\bar{\chi}_\rho(N) \prod_{\rho', N'} d\chi_{\rho'}(N'). \tag{4.40b}$$

This all sounds rather simple. Nevertheless, the details can become rather involved. Thus one must make sure that the composite fields that one constructs to study the properties of hadronic matter carry the correct quantum numbers in the continuum limit. In particular one wants to know what are the flavour, spin, and parity contents of the states excited by these operators. These are the problems which demand a lot of effort. We shall not discuss them here. The interested reader may consult for example the papers by Morel and Rodrigues (1984), Golterman and Smit (1985), Golterman (1986b).

4.5 Technical Details of the Staggered Fermion Formulation

Having sketched the main ideas which go into the staggered formulation, we now present some mathematical details. Should the reader not be interested in the details for the moment, he may skip this and the following section and go on to the next chapter without any danger of getting lost at a later stage.

Consider the action (4.36). As we have already pointed out, the coordinates n_μ of any lattice site may be written in the form $n_\mu = 2N_\mu + \rho_\mu$, where $2N_\mu$ are the coordinates labeling the hypercube to which the site belongs, and ρ is a vector whose components are either one or zero. Since it follows from the definition (4.35) that $\eta_\mu(2N + \rho) = \eta_\mu(\rho)$, we can rewrite the action in the form

$$\begin{aligned} S_F^{(stag.)} = \frac{1}{2} \sum_{N,\rho,\mu} \eta_\mu(\rho) \bar{\chi}(2N + \rho)[\chi(2N + \rho + \hat{\mu}) - \chi(2N + \rho - \hat{\mu})] \\ + \hat{M} \sum_{N,\rho} \bar{\chi}(2N + \rho) \chi(2N + \rho). \end{aligned} \tag{4.41}$$

We next express (4.41) in terms of the 2^d–component field defined in (4.37). To this effect we must remember that the components of ρ are restricted to the values one or zero. Hence we must exercise some care in rewriting the difference $\chi(2N + \rho + \hat{\mu}) - \chi(2N + \rho - \hat{\mu})$ appearing in the expression (4.41). Consider for example

$\chi(2N+\rho+\hat{\mu})$. If $\rho+\hat{\mu}$ is a vector of type "ρ", i.e. if the components of $\rho+\hat{\mu}$ are either one or zero, then $2N+\rho+\hat{\mu}$ labels a site within the hypercube with base at $\hat{r}=2N$. Hence in this case $\chi(2N+\rho+\hat{\mu})$ can be identified with $\chi_{\rho+\hat{\mu}}(N)$. On the other hand if $\rho+\hat{\mu}$ is not a vector of type ρ, then $\rho-\hat{\mu}$ is such a vector, and $\chi(2N+\rho+\hat{\mu})=\chi_{\rho-\hat{\mu}}(N+\hat{\mu})$. These conclusions can be summarized by the equation

$$\chi(2N+\rho+\hat{\mu})=\sum_{\rho'}[\delta_{\rho+\hat{\mu},\rho'}\chi_{\rho'}(N)+\delta_{\rho-\hat{\mu},\rho'}\chi_{\rho'}(N+\hat{\mu})], \tag{4.42a}$$

where ρ' is understood to be a vector whose components are either one or zero.* In a similar way one obtains that

$$\chi(2N+\rho-\hat{\mu})=\sum_{\rho'}[\delta_{\rho-\hat{\mu},\rho'}\chi_{\rho'}(N)+\delta_{\rho+\hat{\mu},\rho'}\chi_{\rho'}(N-\hat{\mu})]. \tag{4.42b}$$

Inserting the expressions (4.42a,b) into (4.41), we obtain

$$\begin{aligned} S_F^{(stag.)}=\frac{1}{2}\sum_{N,\rho,\rho',\mu}&\eta_\mu(\rho)\bar{\chi}_\rho(N)[\delta_{\rho+\hat{\mu},\rho'}\hat{\partial}^L_\mu\chi_{\rho'}(N)\\ &+\delta_{\rho-\hat{\mu},\rho'}\hat{\partial}^R_\mu\chi_{\rho'}(N)]+\hat{M}\sum_{N,\rho}\bar{\chi}_\rho(N)\chi_\rho(N), \end{aligned} \tag{4.23a}$$

where $\hat{\partial}^L_\mu$ and $\hat{\partial}^R_\mu$ are the left and right (block) derivatives analogous to (4.26a,b), defined by

$$\begin{aligned} \hat{\partial}^R_\mu\chi(N)=&\chi(N+\hat{\mu})-\chi(N),\\ \hat{\partial}^L_\mu\chi(N)=&\chi(N)-\chi(N-\hat{\mu}). \end{aligned} \tag{4.43b}$$

Before proceeding, let us digress for a moment and calculate the 2–point correlation function $\langle\chi_\rho(N)\bar{\chi}_{\rho'}(N')\rangle$ from the path integral expression (4.40). A quick way of proceeding is to rewrite (4.43a) in momentum space by introducing the Fourier transforms**

$$\begin{aligned} \chi_\rho(N)&=\int_{-\pi}^{\pi}\frac{d^4\hat{p}}{(2\pi)^4}\tilde{\chi}_\rho(\hat{p})e^{i\hat{p}\cdot N},\\ \bar{\chi}_\rho(N)&=\int_{-\pi}^{\pi}\frac{d^4\hat{p}}{(2\pi)^4}\tilde{\bar{\chi}}_\rho(\hat{p})e^{-i\hat{p}\cdot N}, \end{aligned} \tag{4.44}$$

* It will be understood from now on, that vectors denoted by the symbol rho (i.e. ρ,ρ' etc.) have components restricted to these values.

** Here, and in the following we shall restrict ourselves to four space–time dimensions.

into (4.43a), and performing the sum over N using eq. (2.64). A simple calculation yields

$$S_F^{(stag.)} = \sum_{\rho,\rho'} \int_{-\pi}^{\pi} \frac{d^4\hat{p}}{(2\pi)^4} \bar{\tilde{\chi}}_\rho(\hat{p}) K_{\rho\rho'}(\hat{p}) \tilde{\chi}_{\rho'}(\hat{p}), \tag{4.45a}$$

where

$$K_{\rho\rho'}(\hat{p}) = \sum_\mu i\Gamma^\mu_{\rho\rho'}(\hat{p}) \sin\frac{\hat{p}_\mu}{2} + \hat{M}\delta_{\rho\rho'}, \tag{4.45b}$$

$$\Gamma^\mu_{\rho\rho'}(\hat{p}) = e^{i\hat{p}\cdot(\rho-\rho')/2}\Gamma^\mu_{\rho\rho'}, \tag{4.45c}$$

and

$$\Gamma^\mu_{\rho\rho'} = [\delta_{\rho+\hat{\mu},\rho'} + \delta_{\rho-\hat{\mu},\rho'}]\eta_\mu(\rho). \tag{4.45d}$$

Hence the Fourier representation of the $\chi\bar{\chi}-$ correlation function is given by

$$\langle \chi_\rho(N)\bar{\chi}_{\rho'}(N')\rangle = \int_{-\pi}^{\pi} \frac{d^4\hat{p}}{(2\pi)^4} K^{-1}_{\rho\rho'}(\hat{p}) e^{i\hat{p}\cdot(N-N')}. \tag{4.46}$$

The inverse of the matrix (4.45b) can be easily calculated by making use of the relation

$$\{\Gamma^\mu, \Gamma^\nu\} = 2\delta_{\mu\nu}1\!\!1, \tag{4.47}$$

where $1\!\!1$ is the unit matrix.* This relation follows from the definition (4.45d) and can be proved by making use of the following properties of the phases (4.35):

$$\begin{aligned}\eta_\mu(\rho \pm \hat{\mu}) &= \eta_\mu(\rho), \\ \eta_\mu(\rho)\eta_\nu(\rho + \hat{\mu}) &= -\eta_\nu(\rho)\eta_\mu(\rho + \hat{\nu}), \qquad (\mu \neq \nu).\end{aligned}$$

From (4.47) and the definition of $\Gamma^\mu_{\rho\rho'}(\hat{p})$ given in (4.45c), it follows that for given $\hat{p}$, these matrices also satisfy anticommutation relations analogous to (4.47). Hence the inverse of the matrix (4.45b) can be written down immediately:

$$K^{-1}(\hat{p}) = \frac{-i\sum_\mu \Gamma^\mu(\hat{p}) \sin\frac{\hat{p}_\mu}{2} + \hat{M}}{\sum_\mu \sin^2\frac{\hat{p}_\mu}{2} + \hat{M}^2}.$$

Because of the appearance of the factor 1/2 in the argument of the sine function in the denominator, the integral (4.46) will be dominated for $\hat{M} \to 0$ (i.e. in the continuum limit) by the momenta in the immediate neighbourhood of $\hat{p} = 0$. Hence no doubling problem of the type discussed before arises here. The same

* The Γ^μ's therefore satisfy the same algebra as the Dirac matrices γ^μ.

conclusion will then also hold for the quark correlation functions, since they can be constructed from (4.46) by taking appropriate linear combinations.

After this intermezzo, let us return now to expression (4.43a), and express the left and right derivatives (4.43b) in terms of the symmetric first and second block derivatives, defined by

$$\hat{\partial}_\mu \chi_\rho(N) = \frac{1}{2}(\chi_\rho(N+\hat{\mu}) - \chi_\rho(N-\hat{\mu})),$$
$$\hat{\Box}_\mu \chi_\rho(N) = \chi_\rho(N+\hat{\mu}) + \chi_\rho(N-\hat{\mu}) - 2\chi_\rho(N).$$

In terms of these derivatives, the action (4.43a) takes the form

$$S_F = \frac{1}{2}\sum_{N,\rho,\rho'} \bar{\chi}_\rho(N)\left[\sum_\mu \left(\Gamma^\mu_{\rho\rho'}\hat{\partial}_\mu + \frac{1}{2}\Gamma^{5\mu}_{\rho\rho'}\hat{\Box}_\mu\right) + 2\hat{M}\delta_{\rho\rho'}\right]\chi_{\rho'}(N), \tag{4.48a}$$

where $\Gamma^\mu_{\rho\rho'}$ has been defined in (4.45d), and

$$\Gamma^{5\mu}_{\rho\rho'} = (\delta_{\rho-\hat{\mu},\rho'} - \delta_{\rho+\hat{\mu},\rho'})\eta_\mu(\rho). \tag{4.48b}$$

The matrices Γ^μ and $\Gamma^{5\mu}$ satisfy the same anticommutator algebra as the direct products $\gamma_\mu \otimes 1\!\!1$ and $\gamma_5 \otimes \gamma_\mu\gamma_5$, respectively, where $\gamma_5 = \gamma_1\gamma_2\gamma_3\gamma_4$; i.e., in addition to (4.47) we have that

$$\{\Gamma^\mu, \Gamma^{5\nu}\} = 0,$$
$$\{\Gamma^{5\mu}, \Gamma^{5\nu}\} = -2\delta_{\mu\nu}1\!\!1.$$

This suggests that Γ^μ and $\Gamma^{5\mu}$ are unitarily equivalent to the above mentioned direct products. If the second matrix in these products is interpreted to operate in flavour space, while the first matrix acts in Dirac space, then $(\gamma_\mu \otimes 1\!\!1)\partial_\mu$ would be the matrix version of the kinetic term in (4.30). This suggests that the fields ψ^f_α and $\bar{\psi}^f_\alpha$ are related to χ_ρ and $\bar{\chi}_\rho$ by the unitary transformation which brings Γ^μ and $\Gamma^{5\mu}$ into the form $\gamma_\mu \otimes 1\!\!1$ and $\gamma_5 \otimes \gamma_\mu\gamma_5$, respectively. We now construct this unitary transformation. Because of the way the χ-fields had been introduced originally (cf. eqs. (4.32) and (4.34)) and the appearance of the phase $\eta_\mu(\rho)$ (rather than $\eta_\mu(n)$) in the definitions of Γ^μ and $\Gamma^{5\mu}$, it is not surprising that the transformation will involve the matrix T_ρ, defined by

$$T_\rho = \gamma_1^{\rho_1}\gamma_2^{\rho_2}\gamma_3^{\rho_3}\gamma_4^{\rho_4}. \tag{4.49}$$

Thus consider the following sixteen component fields

$$\hat{\psi}_{\alpha\beta}(N) = \mathcal{N}_0 \sum_\rho U_{\alpha\beta,\rho}\chi_\rho(N), \tag{4.50a}$$

$$\bar{\hat{\psi}}_{\alpha\beta}(N) = \mathcal{N}_0 \sum_\rho \bar{\chi}_\rho(N)U^\dagger_{\rho,\alpha\beta}, \tag{4.50b}$$

where $\mathcal{N}_0$ is a normalization constant to be determined later, and where

$$U_{\alpha\beta,\rho} = \frac{1}{2}(T_\rho)_{\alpha\beta}. \qquad (4.50c)$$

Since in four space–time dimensions α and β take the values one to four, and since $\rho = (\rho_1, ..., \rho_4)$ runs over the sixteen sites within a hypercube, we see that U is a 16×16 square matrix whose rows are labeled by the double index $\alpha\beta$. Equations (4.50a,b) can be readily inverted by noting that because the trace of any product of distinct γ–matrices vanishes, T_ρ satisfies the following orthogonality relation

$$Tr(T_\rho^\dagger T_{\rho'}) = 4\delta_{\rho\rho'}.$$

For the matrix U defined in (4.50c) this relation reads

$$(U^\dagger U)_{\rho\rho'} = \delta_{\rho\rho'},$$

where $(U^\dagger)_{\rho,\alpha\beta} = U^*_{\alpha\beta,\rho} = \frac{1}{2}(T_\rho)^*_{\alpha\beta}$. We hence obtain

$$\chi_\rho(N) = \frac{1}{\mathcal{N}_0}\sum_{\alpha,\beta} U^\dagger_{\rho,\alpha\beta}\hat{\psi}_{\alpha\beta}(N)$$

$$\bar{\chi}_\rho(N) = \frac{1}{\mathcal{N}_0}\sum_{\alpha,\beta} \bar{\hat{\psi}}_{\alpha\beta}(N)U_{\alpha\beta,\rho}.$$

Introducing these expressions into (4.48a), and making use of*

$$(UU^\dagger)_{\alpha\beta,\alpha'\beta'} = \delta_{\alpha\alpha'}\delta_{\beta\beta'},$$

we arrive at the following expression for the staggered fermion action in terms of the "quark" fields $\hat{\psi}$ and $\bar{\hat{\psi}}$:

$$S^{(stag.)} = \frac{1}{2\mathcal{N}_0^2}\sum_{N,\alpha,\beta,\alpha',\beta'} \bar{\hat{\psi}}_{\alpha\beta}(N)\Big\{\sum_\mu [\Lambda^\mu_{\alpha\beta,\alpha'\beta'}\hat{\partial}_\mu + \frac{1}{2}\Lambda^{5\mu}_{\alpha\beta,\alpha'\beta'}\hat{\Box}_\mu] + 2\hat{M}\delta_{\alpha\alpha'}\delta_{\beta\beta'}\Big\}\hat{\psi}_{\alpha'\beta'}(N), \qquad (4.51a)$$

---

* This relation can be verified by using the following representation for the γ-matrices in (4.49)

$$\gamma_i = \begin{pmatrix} 0 & -i\sigma_i \\ i\sigma_i & 0 \end{pmatrix}, \quad \gamma_4 = \begin{pmatrix} 1\!\!1 & 0 \\ 0 & -1\!\!1 \end{pmatrix},$$

where σ_i are the Pauli matrices, and $1\!\!1$ the 2×2 unit matrix.

where

$$\Lambda^{\mu}_{\alpha\beta,\alpha'\beta'} = \sum_{\rho,\rho'} U_{\alpha\beta,\rho} \Gamma^{\mu}_{\rho\rho'} U^{\dagger}_{\rho',\alpha'\beta'}, \tag{4.51b}$$

and

$$\Lambda^{5\mu}_{\alpha\beta,\alpha'\beta'} = \sum_{\rho,\rho'} U_{\alpha\beta,\rho} \Gamma^{\mu}_{5\rho\rho'} U^{\dagger}_{\rho',\alpha'\beta'}. \tag{4.51c}$$

Making use of the explicit representation of the γ-matrices given in the previous footnote, one finds that

$$\Lambda^{\mu}_{\alpha\beta,\alpha'\beta'} = (\gamma_\mu)_{\alpha\alpha'} \delta_{\beta\beta'},$$
$$\Lambda^{5\mu}_{\alpha\beta,\alpha'\beta'} = (\gamma_5)_{\alpha\alpha'} (t_\mu t_5)_{\beta\beta'},$$

where $t_\mu = \gamma^*_\mu$ and $t_5 = \gamma_5$. Since in the naive continuum limit only the first and third term in (4.51a) contribute to the action, it follows by comparing this expression with (4.39) that α and β should be identified with the Dirac and flavour quark–degrees of freedom, respectively. In (4.38a) we had made this fact explicit by writing the field components in the form ψ^f_α, rather than $\hat{\psi}_{\alpha\beta}$. Accordingly, Λ^μ and $\tilde{\Lambda}^\mu$ can be written as the following direct products,

$$\Lambda^\mu = \gamma_\mu \otimes \mathbb{1},$$
$$\Lambda^{5\mu} = \gamma_5 \otimes t_\mu t_5,$$

where the first matrix appearing on the right hand side acts in Dirac space, while the second matrix acts in flavour space.

Finally, we must determine the normalization constant $\mathcal{N}_0$ in (4.51a). To this effect we study the naive continuum limit of this expression by introducing the dimensioned fields $\psi_{\alpha\beta}$ and block derivatives ∂_μ in the standard way; i.e., $\psi = b^{-3/2}\hat{\psi}$ and $\partial_\mu = \frac{1}{b}\hat{\partial}_\mu$. Here b is the lattice spacing of the blocked lattice. By choosing $\mathcal{N}_0 = 1/\sqrt{2}$ the action (4.51a) then takes the form

$$\begin{aligned} S^{(stag.)} = \sum_{x,\rho,\mu} b^4 \bar{\psi}(x)[(\gamma_\mu \otimes \mathbb{1})\partial_\mu + \\ + \frac{1}{2} b(\gamma_5 \otimes t_\mu t_5)\Box_\mu]\psi(x) + 2M \sum_x \bar{\psi}(x) \mathbb{1} \otimes \mathbb{1} \psi(x) \end{aligned} \tag{4.52}$$

where $M = \frac{1}{b}\hat{M}$ and where the sum over x $(= Nb)$ runs over all hypercubes of the blocked lattice. In the naive continuum limit $(b \to 0)$ this expression reduces to the form

$$S^{(stag.)} \underset{b\to 0}{\longrightarrow} \sum_f \int d^4x \bar{\psi}^f(x) (\sum_\mu \gamma_\mu \partial_\mu + M_0) \psi^f(x)$$

where $M_0 = 2M$, and where $\psi_{\alpha\beta}$ has been replaced by ψ^f_α. Notice that the appearance of the mass $2M$ instead of M in eq. (4.52) is not surprising since we have scaled $\hat{M}$ with the lattice spacing of the blocked lattice. Thus $M_0 = \frac{1}{a}\hat{M}$ is the dimensioned mass parameter of the original fine lattice.

Consider now the two point correlation function of the quark fields, $\langle \psi^f_\alpha(N)\psi^{f'}_{\alpha'}(N')\rangle$, where we have introduced the more suggestive notation ψ^f_α mentioned above. This two–point function can either be computed from the $\chi - \bar{\chi}$ correlation functions (4.46) by taking appropriate linear combinations, as dictated by the eqs. (4.50a,b), i.e.

$$\langle \hat{\psi}^f_\alpha(N)\hat{\bar{\psi}}^{f'}_{\alpha'}(N')\rangle = \frac{1}{2}\sum_{\rho,\rho'} U_{\alpha f,\rho}\langle \chi_\rho(N)\bar{\chi}_{\rho'}(N')\rangle U^\dagger_{\rho',\alpha' f'} \quad ,$$

and multiplying the expression with b^{-3}, or by inverting the matrix operator appearing within square brackets in eq. (4.52). This inversion is easiest done by going to momentum space. Introducing the Fourier decompositions

$$\psi(Nb) = \int_{-\pi/b}^{\pi/b} \frac{d^4p}{(2\pi)^4}\tilde{\psi}(p)e^{ip\cdot Nb}$$

$$\bar{\psi}(Nb) = \int_{-\pi/b}^{\pi/b} \frac{d^4p}{(2\pi)^4}\tilde{\bar{\psi}}(p)e^{-ip\cdot Nb}$$

into (4.52), and making use of the relation

$$\sum_N b^4 e^{i(p-p')\cdot Nb} = (2\pi)^4\delta^{(4)}_P(p-p'),$$

where δ_P is the periodic delta function, one finds that

$$\begin{aligned} S^{(stag.)} = \int_{-\pi/b}^{\pi/b}\frac{d^4p}{(2\pi)^4}\tilde{\bar{\psi}}(p)\Big\{\sum_\mu\Big[(\gamma_\mu\otimes 1\!\!1)\frac{i}{b}\sin(p_\mu b) \\ -\frac{1}{b}(1-\cos(p_\mu b))\gamma_5\otimes t_\mu t_5\Big] + M_0\Big\}\tilde{\psi}(p). \end{aligned} \tag{4.53}$$

The propagator in momentum space is given by the inverse of the matrix appearing within curly brackets:

$$S(p) = \frac{\sum_\mu\left[-i(\gamma_\mu\otimes 1\!\!1)\frac{1}{b}\sin p_\mu b + \frac{2}{b}(\gamma_5\otimes t_\mu t_5)\sin^2\frac{p_\mu b}{2}\right] + M_0 1\!\!1\otimes 1\!\!1}{\sum_\mu \frac{4}{b^2}\sin^2\frac{p_\mu b}{2} + M_0^2}. \tag{4.54}$$

Notice that the denominator is the same as for the scalar field discussed in chapter 3. In the naive continuum limit ($b \to 0$), the above expression reduces to

$$S(p) \xrightarrow[b\to 0]{} \frac{-i\sum_\mu (\gamma_\mu \otimes 1\!\!1)p_\mu + M_0 1\!\!1 \otimes 1\!\!1}{p^2 + M_0^2},$$

which is the correct fermion propagator describing four degenerate flavoured Dirac particles. Because the denominator in (4.54) has the same structure as for the scalar field, the two-point function $\langle \psi^f_\alpha(x)\bar{\psi}^{f'}_{\alpha'}(x)\rangle$ obtained by Fourier transforming this expression and taking the limit $b \to 0$, is given by

$$\langle \psi^f_\alpha(N)\bar{\psi}^{f'}_\beta(N')\rangle = \int_{-\infty}^{\infty} \frac{d^4p}{(2\pi)^4} S_{\alpha\beta}(p)\delta_{ff'} e^{ip\cdot(N-N')b},$$

where

$$S(p) = \frac{-i\gamma\cdot p + M_0}{p^2 + M_0^2}.$$

Finally let us compare the expression (4.52) with the lattice action for four flavoured Wilson fermions. Clearly, the generalization of eq. (4.27) (with S_F defined in (4.4)) to the case of four quark flavours is

$$S_F^{(W)} = \sum_n a^4 \bar{\psi}(n) \left\{ \sum_\mu \left[(\gamma_\mu \otimes 1\!\!1)\partial_\mu - \frac{ar}{2}(1\!\!1 \otimes 1\!\!1)\Box_\mu \right] + M 1\!\!1 \otimes 1\!\!1 \right\} \psi(n). \quad (4.55)$$

Thus the only difference between the Wilson and the staggered fermion actions, consists in the matrix structure of the second derivative term. This term does not contribute in the naive continuum limit, but is responsible for lifting the fermion degeneracy. So why not choose the simpler (Wilson) version, which has the merit that the number of quark flavours is not restricted? We only give a brief answer to this question without going into details: In the naive continuum limit both actions (4.52) and (4.55) possess a $U(4) \times U(4)$ (chiral) symmetry for $M = 0$. This symmetry is broken in both cases by the second term. But whereas the axial symmetry (involving the generator γ_5 in Dirac space) is completely lost for Wilson fermions, the staggered fermion action preserves a non–trivial piece of the full chiral symmetry, whose generator is $\gamma_5 \otimes t_5$. Under this abelian subgroup the fields transform as follows

$$\psi(N) \to e^{i\alpha(\gamma_5\otimes t_5)}\psi(N)$$
$$\bar{\psi}(N) \to \bar{\psi}(N)e^{i\alpha(\gamma_5\otimes t_5)}$$

where α is an arbitrary parameter which does not depend on N. For this reason the staggered fermion formulation is more adequate for studying spontaneous chiral symmetry breaking and the associated Goldstone phenomenon in theories like QCD with massless quarks.

4.6 Staggered Fermions in Momentum Space

In the previous section the transition from a one–component field to a 2^d–component field χ_ρ was carried out in configuration space. We now want to construct an alternative lattice action by working in momentum space. As we shall see, this action differs from (4.52) in several interesting respects. For some papers which are of relevance to this section confer Sharatchandra, Thun and Weisz (1981), Van den Doel and Smit (1983), Kluberg-Stern et al. (1983) and Golterman and Smit (1984).

Our starting point is again the action (4.36). Inserting for $\chi(n)$ and $\bar{\chi}(n)$ the Fourier decomposition analogous to (4.44), and writing the phase $\eta_\mu(n)$ as

$$\eta_\mu(n) = e^{in\cdot\delta^{(\mu)}},$$

where

$$\delta^{(1)} = (0,0,0,0), \quad \delta^{(2)} = (\pi,0,0,0),$$
$$\delta^{(3)} = (\pi,\pi,0,0), \quad \delta^{(4)} = (\pi,\pi,\pi,0),$$

the action takes the following form in momentum space

$$S^{(stag.)} = \int_{-\pi}^{\pi} \frac{d^4\hat{p}}{(2\pi)^4} \int_{-\pi}^{\pi} \frac{d^4\hat{p}'}{(2\pi)^4} \hat{\bar{\chi}}(\hat{p}')M(\hat{p}',\hat{p})\hat{\chi}(\hat{p}). \tag{4.56a}$$

Here

$$M(\hat{p},\hat{p}') = (2\pi)^4 \left\{ \sum_\mu \delta_P^{(4)}(\hat{p}-\hat{p}'+\delta^{(\mu)}) i \sin\hat{p}_\mu + \hat{M}\delta_P^{(4)}(\hat{p}-\hat{p}') \right\}. \tag{4.56b}$$

and $\delta_P(\hat{k})$ is the periodic delta function. For the following discussion it is convenient to extend the fields $\hat{\chi}(\hat{p})$ and $\hat{\bar{\chi}}(\hat{p})$ periodically with period 2π. Let us denote the corresponding fields by $\hat{\phi}(\hat{p})$ and $\hat{\bar{\phi}}(\hat{p})$, respectively. Because (4.56b) is itself 2π-periodic, we can then shift the integration intervals $[-\pi,\pi]$ by $\pi/2$. In fig. (4-3) we show the new integration region for the case of two space-time dimensions.

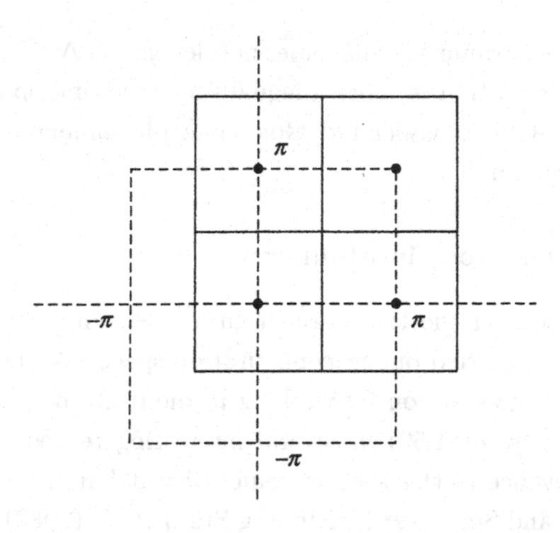

Fig. 4-3 The four quadrants of the shifted integration region (solid lines).

Next, let us divide the new Brillouin zone, BZ', into sixteen domains, whose centers are the corners of the first quadrant in the original Brillouin zone. For two dimensions these domains are shown in fig. (4-3). Because of the appearance of $\sin\hat{p}_\mu$ rather than $\sin\hat{p}_\mu/2$ in (4.56b), the $\chi-\bar{\chi}$ correlation function will receive, in the continuum limit, contributions from momenta in the immediate neighbourhood of the above-mentioned corners. The momenta in the sixteen integration regions can be parametrized as follows.

$$\hat{p}_\mu = k_\mu + \pi_\mu^{(A)} \quad (A = 1, ..., 16),$$

where

$$-\pi/2 \le k_\mu \le \pi/2,$$

and $\pi^{(A)}$ are constant vectors, defined by

$$\pi^{(1)} = (0,0,0,0), \quad \pi^{(2)} = (\pi,0,0,0), \ldots, \quad \pi^{(16)} = (\pi,\pi,\pi,\pi).$$

The integral (4.56a) may then be written in the form

$$S^{(stag.)} = \sum_{A,B} \int_{-\pi/2}^{\pi/2} \frac{d^4\hat{k}}{(2\pi)^4} \frac{d^4\hat{k}'}{(2\pi)^4} \bar{\hat{\phi}}_A(\hat{k}') M_{AB}(\hat{k}',\hat{k}) \hat{\phi}_B(\hat{k}), \tag{4.57a}$$

where

$$\begin{aligned} M_{AB}(\hat{k}',\hat{k}) = (2\pi)^4 \{ &\sum_\mu \delta_P^{(4)}(\hat{k} - \hat{k}' + \pi^{(B)} - \pi^{(A)} + \delta^{(\mu)}) e^{i\pi_\mu^{(B)}} i \sin\hat{k}_\mu \\ &+ \hat{M}\delta_P^{(4)}(\hat{k} - \hat{k}' + \pi^{(B)} - \pi^{(A)}) \}, \end{aligned} \tag{4.57b}$$

and where we have introduced the 16–component (dimensionless) fields $\hat{\phi}_A(\hat{k})$ as follows

$$\hat{\phi}_A(\hat{k}) \equiv \hat{\phi}(\hat{k} + \pi^{(A)}). \tag{4.57c}$$

This is the analog of (4.37) in the momentum space approach. Notice that the integral (4.57a) now extends only over half the BZ. Because $\hat{k}$ and $\hat{k}'$ are restricted to the interval $[-\pi/2, \pi/2]$, the periodic δ–functions appearing in (4.57b) can be written in the form

$$\delta_P^{(4)}(\hat{k} - \hat{k}' + \pi^{(B)} - \pi^{(A)}) = \delta^{(4)}(\hat{k} - \hat{k}')\delta_{AB}$$
$$\delta_P^{(4)}(\hat{k} - \hat{k}' + \pi^{(B)} - \pi^{(A)} + \delta^{(\mu)}) = \eta^\mu_{AB}\delta^{(4)}(\hat{k} - \hat{k}')$$

where $\delta(\hat{k} - \hat{k}')$ has only support at $k = \hat{k}'$, and where

$$\eta^\mu_{AB} = \prod_{\nu=1}^{4} \frac{1}{2}(e^{i(\pi_\nu^{(B)} - \pi_\nu^{(A)} + \delta_\nu^{(\mu)})} + 1).$$

This factor just expresses the fact that the 4–dimensional periodic δ–function only has support in the integration domain given in (4.57a) if $\pi_\nu^{(B)} - \pi_\nu^{(A)} + \delta_\nu^{(\mu)}$ equals zero or 2π for every component ν. Inserting these expressions into (4.57b), and performing the integration over $\hat{k}'$ in (4.57a), one obtains

$$S_F^{(stag.)} = \sum_{A,B} \int_{-\pi/2}^{\pi/2} \frac{d^4k}{(2\pi)^4} \bar{\hat{\phi}}_A(\hat{k}) K_{AB}(\hat{k}) \hat{\phi}_B(\hat{k}), \tag{4.58a}$$

where

$$K_{AB}(\hat{k}) = i \sum_\mu \hat{\Gamma}^\mu_{AB} \sin \hat{k}_\mu + \delta_{AB}\hat{M}, \tag{4.58b}$$

and

$$\hat{\Gamma}^\mu_{AB} = e^{i\pi_\mu^{(B)}} \eta^\mu_{AB}. \tag{4.58c}$$

These matrices satisfy the following anticommutation relations.

$$\{\hat{\Gamma}^\mu, \hat{\Gamma}^\nu\} = 2\delta_{\mu\nu}1\!\!1.$$

Furthermore it can be shown that the $\hat{\Gamma}_\mu$'s are unitarily equivalent to $(\gamma_\mu \otimes 1\!\!1)$. Hence we can write (4.58a) in the form

$$S_F^{(stag.)} = \int_{-\pi}^{\pi} \frac{d^4\hat{p}}{(2\pi)^4} \bar{\hat{Q}}(\hat{p}) \left[\sum_\mu (\gamma_\mu \otimes 1\!\!1) 2i \sin \frac{\hat{p}_\mu}{2} + 2\hat{M} 1\!\!1 \otimes 1\!\!1 \right] \hat{Q}(\hat{p}) \tag{4.59}$$

where we have rescaled the range of integration to the interval $[-\pi, \pi]$. This is, at first sight, a surprising result, for the action is invariant for $\hat{M} = 0$ under the transformations of the full $U(4) \otimes U(4)$ chiral group, and in particular under

$$\hat{Q} \to e^{i\alpha_B(1\!\!1 \otimes T_B)}\hat{Q}$$
$$\hat{\bar{Q}} \to \hat{\bar{Q}}e^{-i\alpha_B(1\!\!1 \otimes T_B)}$$

and

$$\hat{Q} \to e^{i\beta_B(\gamma_5 \otimes T_B)}\hat{Q},$$
$$\hat{\bar{Q}} \to \hat{\bar{Q}}e^{i\beta_B(\gamma_5 \otimes T_B)},$$

for each generator T_B in the N_f-dimensional flavour space. This $U(4) \otimes U(4)$ symmetry has however been gained at the expense of giving up the locality of the action in configuration space. Indeed, the action (4.59) is a non-local function of the fields $\hat{Q}(n)$ and $\hat{\bar{Q}}(n)$, obtained by inverting the Fourier series

$$\hat{Q}(\hat{p}) = \sum_n \hat{Q}(n)e^{-i\hat{p}\cdot n}$$

$$\hat{\bar{Q}}(\hat{p}) = \sum_n \hat{\bar{Q}}(n)e^{i\hat{p}\cdot n}.$$

Substituting these expressions into (4.59), one finds that

$$S_F^{(stag)} = \sum_{n,m} \hat{\bar{Q}}(n)\Big[\sum_\mu (\gamma_\mu \otimes 1\!\!1)(\Delta_\mu)_{nm} + 2\hat{M}1\!\!1 \otimes 1\!\!1\Big]\hat{Q}(m), \tag{4.60a}$$

where

$$(\Delta_\mu)_{nm} = \int_{-\pi}^{\pi} \frac{d^4\hat{p}}{(2\pi)^4} 2i \sin\frac{\hat{p}_\mu}{2} e^{i\hat{p}\cdot(n-m)} \tag{4.60b}$$

connects arbitrary sites along the μ-direction. The reason for this is the appearance of the $\hat{p}_\mu/2$ (instead of $\hat{p}_\mu$) in the argument of the sine function. If $\hat{p}_\mu/2$ were replaced by $\hat{p}_\mu$ in (4.60b), then this expression would equal $\hat{\partial}_\mu \delta_{nm} = (\delta_{n+\hat{\mu},m} - \delta_{n-\hat{\mu},m})/2$ and we would be left with an expression for the action involving only nearest neighbour couplings of the fields.

The above discussion suggests that the fields $\hat{Q}(n)$ are related to the fields $\hat{\psi}(n)$ by a non-local transformation. This is indeed the case. Thus the fields $\hat{\psi}(\hat{p})$ and $\hat{Q}(\hat{p})$ appearing in (4.53) and (4.60a) are connected by a unitary transformation which depends on the momentum $\hat{p}$ (see e.g., Kluberg-Stern et al. (1983).

Because of this dependence, the field $\hat{Q}$ at the lattice site N will be given by a linear combination of the "quark" variables $\hat{\psi}$ attached to lattices sites which are arbitrarily far from n.

This concludes our discussion of staggered fermions. As the reader is probably convinced by now, a thorough discussion of this subject, including all lattice symmetries, becomes quite technical. We believe, however, that the material presented in the last two sections will enable the reader to follow the literature on this subject without too much difficulties.

4.7 Ginsparg-Wilson Fermions

Of the two lattice regularizations discussed above, only the Wilson fermions allow one to study models with an arbitrary number of quark flavours. Wilson fermions do however break the chiral symmetry of the continuum fermion action for massless quarks. This makes it difficult to explore the regime of small quark masses in numerical simulations and to study spontaneous chiral symmetry breaking on the lattice. In fact, as we have already pointed out, the Nielsen-Ninomiya theorem tells us that we cannot get around breaking the symmetry (4.8) of the (massless) fermion action, unless we give up at least some important property, like e.g. locality.

But there is another way of breaking chiral symmetry on the lattice in a particular mild and controlled way. It was proposed a long time ago by Ginsparg and Wilson (1982), but has not been seriously considered for 10 years. Ginsparg and Wilson had searched for a lattice Dirac operator by starting from a chirally symmetric action and following a renormalization group blocking procedure, using a chirally non-invariant Kernel. For Ginsparg-Wilson (GW) fermions the fermionic action is of the form

$$S_{ferm} = \sum_{x,y} \bar{\psi}(x)(D(x,y) + m\delta_{xy})\psi(y) \ , \tag{4.61}$$

where the "Dirac Operator" $D(x,y)$ is a 4×4 matrix in Dirac space which breaks the standard chiral symmmetry in a very special way. Since every lattice action must possess the correct naive continuum limit, it follows that for $a \to 0$ $D(x,y)$ becomes the usual continuum Dirac operator. While in the continuum, or in its naively discretized form, this operator anticommutes with γ_5, the GW-Dirac operator satisfies the following GW-relation: *

$$\gamma_5 D(x,y) + D(x,y)\gamma_5 = a \sum_z D(x,z)\gamma_5 D(z,y) \ . \tag{4.62}$$

* Actually this is only the simplest version of the GW relation. A more general

We may write this expression in the more compact form, as is usually the case in the literature,

$$\{\gamma_5, D\} = aD\gamma_5 D \ , \tag{4.63}$$

where D is now a matrix whose rows and colums are labeled by a spin and space-time index. Note that the rhs is of order a. From here one trivially obtains that

$$\{\gamma_5, D^{-1}\} = a\gamma_5 \ , \tag{16.64}$$

which shows that the anticommutator $\{\gamma_5, D^{-1}\}$ breaks chiral symmetry in an ultralocal way. A Dirac operator satisfying the GW relation does however not ensure the absence of species doubling. Any matrix of the form

$$D = \frac{1}{a}(1 - V) \ , \tag{4.65a}$$

where V satisfies

$$\begin{aligned} V^\dagger V &= 1 \\ V^\dagger &= \gamma_5 V \gamma_5 \ , \end{aligned} \tag{4.65b}$$

($\gamma_5^2 = 1$) solves the GW relation. But it was only in 1998 that an explicit expression for D was given (Neuberger, 1998), which is free of doublers and local in a more general sense (Hernandez, 1998). It also satisfies and exact index theorem (Hasenfratz, 1998), a lattice version of the Atiyah-Singer index theorem (Atiyah, 1971). *

The Neuberger solution is obtained by choosing

$$V = A(A^\dagger A)^{-\frac{1}{2}} \tag{4.66a}$$

with

$$A = 1 - aK^{(W)} \ , \tag{4.66b}$$

where $K^{(W)}$ is the Wilson-Dirac operator (4.28b). In the case where the fermions are coupled to gauge fields, $K^{(W)}$ is replaced by an expression depending on the gauge potentials to be discussed in the following chapter.

---

version reads (in matrix notation): $\gamma_5 D + D\gamma_5 = aD\gamma_5 RD$, where R is a non singular hermitean operator which is local in position space and proportional to the unit matrix in Dirac space.

* In the continuum the Dirac operator for massless fermions in a smooth background field carrying non-vanishing topological charge Q (see section (17.6)) possesses left and right-handed zero modes. The difference $n_L - n_R$ in the number of these zero modes equals the topological charge of the backrground field.

Clearly numerical simulations with GW-fermions are much more costly than with Wilson fermions. So why is one so interested in GW-fermions? After all, they also break the standard chiral symmetry (4.8). What makes them interesting in particular, is that the GW-action still possesses an *exact* chiral symmetry which differs from (4.8) by $\mathcal{O}(a)$ lattice artefacts, as was shown by Lüscher (1999a). And this is true for the free theory as well as interacting case. The only thing that matters is the bilinear structure of the action (4.61) in the fermion fields, and that D satisfies the GW-relation. The emphasis above is on the word *exact.* The existence of an exact chiral symmetry should allow one to study not only the regime of small quark masses, but may possibly also resolve a long standing problem of putting chiral gauge theories (like the electroweak theory) on the lattice.

The exact symmetry referred to above corresponds to the transformations

$$\psi \to e^{i\theta\gamma_5(1-\frac{a}{2}D)}\psi\,,$$

$$\bar{\psi} \to \bar{\psi}e^{i\theta\gamma_5(1-\frac{a}{2}D)}\,, \tag{4.67}$$

or infinitessimally

$$\psi \to \psi' = \psi + \delta\psi\,,\quad \bar{\psi} \to \bar{\psi}' = \bar{\psi} + \delta\bar{\psi}\,, \tag{4.68a}$$

where*

$$\delta\psi(x) = i\epsilon\gamma_5\Big[(1 - \frac{a}{2}D)\psi\Big](x)\,,$$

$$\delta\bar{\psi}(x) = i\epsilon\Big[\bar{\psi}(1 - \frac{a}{2}D)\Big](x)\gamma_5\,. \tag{4.68b}$$

One readily finds that the variation of the action is given by

$$\delta S_{ferm} = i\epsilon\sum_x\Big[F(x) + 2m\bar{\psi}(x)\gamma_5\psi(x) + \bar{\Delta}(x)\Big]\,, \tag{4.69a}$$

where $\sum_x = \sum_n a^4$, and

$$F(x) = (\bar{\psi}D)(x)\gamma_5\psi(x) + \bar{\psi}(x)\gamma_5(D\psi)(x) - a(\bar{\psi}D)(x)\gamma_5(D\psi)(x) \tag{4.69b}$$

$$\bar{\Delta}(x) = -\frac{ma}{2}\Big[(\bar{\psi}D)(x)\gamma_5\psi(x) + \bar{\psi}(x)\gamma_5(D\psi)(x)\Big]\,. \tag{4.69c}$$

---

* $\Big[(1 - \frac{a}{2}D)\psi\Big](x)$ stands for $\sum_y(\delta_{xy} - \frac{a}{2}D(x,y))\psi(y)$. Furthermore $\Big[\bar{\psi}(1 - \frac{a}{2}D)\Big](x) = \sum_y \bar{\psi}(y)\Big[\delta_{yx} - \frac{a}{2}D(y,x)\Big]$.

We now note that

$$\sum_x F(x) = \sum_{x,y} \bar{\psi}(x)\Big[\{\gamma_5, D\} - aD\gamma_5 D\Big](x,y)\psi(y) \equiv 0 \,, \tag{4.70}$$

where we made use use of (4.63). Hence for $m = 0$ the action is invariant under the global transformations (4.67), which verifies the observation made by Lüscher (1998).

This is all we will say about GW-fermions at this point. We shall return to them once more in chapter 14, when we discuss the ABJ axial anomaly.

CHAPTER 5

ABELIAN GAUGE FIELDS ON THE LATTICE AND COMPACT QED

5.1 Preliminaries

In 1971 F. Wegner studied a class of Ising models, where the global $Z(2)$ symmetry of the Hamiltonian was promoted to a local one. Although the models did not possess a local order parameter, they did exhibit a phase transition. In constructing the models, Wegner (1971) introduced a number of important concepts which turned out to play also a fundamental role in the lattice formulation of gauge field theories like QED and QCD. In particular he was led to construct a non-local gauge invariant order parameter, whose analog in QCD was later introduced by K.G. Wilson (1974), and provides a criterium for confinement. In QCD this order parameter is known under the name of Wilson loop, although the name "Wegner-Wilson loop" would seem more appropriate. Nevertheless we shall adhere to the general praxis and refer to it simply as *Wilson-loop*.

A common feature of the above mentioned theories is that they possess a local symmetry. In the case of QED or QCD the local symmetry group is a continuous one. The action in these theories is obtained by gauging the global symmetry of the free fermionic action and adding a kinetic term for the gauge fields. In the continuum formulation the construction principle is well known, and we will recapitulate it below for the case of QED. The lattice version of the action can be obtained following the same general type of reasoning, but it will differ in some important details from the naive discretization.

Let us briefly review how one arrives at the gauge-invariant action in continuum QED. The starting point is the action of the free Dirac field:

$$S_F^{(0)} = \int d^4x \bar{\psi}(x)(i\gamma^\mu \partial_\mu - M)\psi(x).$$

This action is invariant under the transformation

$$\begin{aligned} \psi(x) &\to G\psi(x) \\ \bar{\psi}(x) &\to \bar{\psi}(x)G^{-1} \end{aligned}$$

where G is an element of the abelian $U(1)$ group, i.e.

$$G = e^{i\Lambda},$$

with Λ independent of x (i.e. a global transformation). The next step consists in requiring the action to be also invariant under local $U(1)$ transformations with ψ ($\bar{\psi}$) transforming independently at different space-time points. This is accomplished by introducing a four-vector potential $A_\mu(x)$ and replacing the ordinary four-derivative ∂_μ by the covariant derivative D_μ, defined by

$$D_\mu = \partial_\mu + ieA_\mu. \tag{5.1}$$

The resulting new action

$$S_F = \int d^4x \bar{\psi}(i\gamma^\mu D_\mu - M)\psi, \tag{5.2}$$

is then invariant under the following set of local transformations

$$\begin{aligned} \psi(x) &\to G(x)\psi(x), \\ \bar{\psi}(x) &\to \bar{\psi}(x)G^{-1}(x), \end{aligned} \tag{5.3a}$$

$$A_\mu(x) \to G(x)A_\mu(x)G^{-1}(x) - \frac{i}{e}G(x)\partial_\mu G^{-1}(x), \tag{5.3b}$$

where

$$G(x) = e^{i\Lambda(x)} \tag{5.3c}$$

In the present case, (5.3b) is just another form of writing the familiar transformation law, $A_\mu \to A_\mu - \frac{1}{e}\partial_\mu \Lambda$. Since A_μ and G commute, we could have written just as well A_μ instead of $GA_\mu G^{-1}$. In the non-abelian case to be considered later, however, this will be the relevant structure of the gauge transformations. The crucial property which ensures the gauge invariance of (5.2) is that, while A_μ transforms inhomogeneously, the transformation law for the covariant derivative (5.1) is homogeneous:

$$D_\mu \longrightarrow GD_\mu G^{-1} \quad .$$

Having ensured the local gauge invariance of the action by introducing a four-vector field A_μ, we now must add to the expression (5.2) a kinetic term which allows A_μ to propagate. This term must again be invariant under the local transformations (5.3c), and is given by the familiar expression

$$S_G = -\frac{1}{4}\int d^4x F_{\mu\nu}F^{\mu\nu}, \tag{5.4}$$

where $F_{\mu\nu} = \partial_\mu A_\nu - \partial_\nu A_\mu$ is the gauge invariant field strength tensor. The full gauge invariant action describing the dynamics of ψ, $\bar{\psi}$ and A_μ is then given by

$$S_{QED} = -\frac{1}{4}\int d^4x F^{\mu\nu}F_{\mu\nu} + \int d^4x \bar{\psi}(i\gamma^\mu D_\mu - M)\psi. \tag{5.5}$$

The Green functions of the corresponding quantum theory (QED) are (formally) computed from the generating functional

$$Z[J,\eta,\bar{\eta}] = \int DAD\bar{\psi}D\psi e^{iS_{QED}+i\int d^4xJ^\mu A_\mu + i\int d^4x(\bar{\eta}\psi+\bar{\psi}\eta)} \tag{5.6}$$

by differentiating this expression with respect to the sources $J^\mu(x)$, $\eta(x)$ and $\bar{\eta}(x)$, where $\psi, \bar{\psi}$, η and $\bar{\eta}$ are Grassmann-valued fields. The integral (5.6) may be given a meaning within perturbation theory. For a non-perturbative formulation, however, we should define the generating functional on a euclidean space- time lattice. Hence let us do this here for QED and generalize it later to the more complicated case of a non-abelian gauge theory. Our construction procedure will parallel very closely the one described above and is based on the following two requirements: i) The lattice action should be invariant under local $U(1)$ transformations and ii) reduce in the naive continuum limit to the classical continuum action.

Before we carry out this program, let us first obtain the euclidean version of (5.4) and (5.2); to this effect we let x^0 become purely imaginary ($x^0 \to -ix_4$) and replace at the same time A^0 by $+iA_4$; the prescription that A^0 should be replaced by $+iA_4$ is made plausible by considering the case where A_μ is a pure gauge field configuration: $A_\mu = \partial_\mu \Lambda(x)$; thus the replacement $x^0 \to -ix_4$ implies $\partial_0 \to +i\partial_4$. With this formal substitution (5.4) becomes

$$S_G \to \frac{i}{4}\int d^4xF_{\mu\nu}F_{\mu\nu} \quad , \tag{5.7}$$

where a sum over μ and ν ($\mu,\nu = 1,2,3,4$) is understood. Hence $\exp(iS_G)$ goes over into an exponentially damped functional of A_μ, (as it should).

The transition from (5.2) to the imaginary time formulation is also carried out immediately by substituting the euclidean derivative ∂_μ in eq. (4.2) by the corresponding covariant derivative $D_\mu = \partial_\mu + ieA_\mu$. * Hence the action (5.5) goes over into $iS^{(eucl)}$, where

$$S^{(eucl)}_{QED} = S^{(eucl)}_G + S^{(eucl)}_F, \tag{5.8a}$$

with

$$S^{(eucl)}_G = \frac{1}{4}\int d^4xF_{\mu\nu}F_{\mu\nu}, \tag{5.8b}$$

* Notice that the structure of the covariant derivative remains unchanged when making the transition, since both ∂_0 and A_0 follow the rule: $\partial_0 \to i\partial_4$ and $A_0 \to iA_4$.

$$S_F^{(eucl)} = \int d^4x \bar{\psi}(\gamma_\mu D_\mu + M)\psi. \tag{5.8c}$$

Here γ_μ $(\mu = 1, \ldots, 4)$ are the euclidean γ–matrices introduced in chapter 4. Since from here on we shall always work with the euclidean formulation, we shall drop in the following any labeling reminding us of this.

5.2 Lattice Formulation of QED

We start our construction program of lattice QED by considering first the lattice action for a free Dirac field. To parallel as closely as possible the steps in the continuum formulation, we shall work with Wilson fermions, where every lattice site may be occupied by all Dirac components ψ_α. The corresponding action is given by (4.28) which, after making a shift in the summation variable, can be written in the form

$$\begin{aligned} S_F^{(W)} &= (\hat{M} + 4r)\sum_n \bar{\psi}(n)\psi(n) \\ &-\frac{1}{2}\sum_{n,\mu}[\bar{\psi}(n)(r - \gamma_\mu)\psi(n+\hat{\mu}) + \bar{\psi}(n+\hat{\mu})(r+\gamma_\mu)\psi(n)]. \end{aligned} \tag{5.9}$$

Here we have dropped the "hat" on the fermionic variables for simplicity. It will be always evident from our notation, whether we are considering the dimensionless lattice- or the dimensioned continuum formulation.

The action (5.9) is invariant under the global transformations

$$\begin{aligned} \psi(n) &\to G\psi(n), \\ \bar{\psi}(n) &\to \bar{\psi}(n)G^{-1}, \end{aligned}$$

where G is an element of the $U(1)$ group. The next step of our program consists in requiring the theory to be invariant under local $U(1)$ transformations, with the group element G depending on the lattice site. Let us denote it by $G(n)$. Because of the non-diagonal structure of the second term in eq. (5.9) (whose origin is the derivative in the continuum formulation) we are forced to introduce new degrees of freedom. Since the group elements $G(n)$ do not act on the Dirac indices, it is sufficient for the following argument to focus our attention on a typical bilinear term, $\bar{\psi}(n)\psi(n+\hat{\mu})$.

In the continuum formulation it is well known how such bilinear terms should be modified in order to arrive at a gauge-invariant expression. Since according to (5.3a) $\bar{\psi}(x)\psi(y)$ transforms as follows,

$$\bar{\psi}(x)\psi(y) \to \bar{\psi}(x)G^{-1}(x)G(y)\psi(y),$$

we must include a factor depending on the gauge potential which compensates the above gauge variation. This factor, known as the *Schwinger line integral*, is well known, and is given by

$$U(x,y) = e^{ie\int_x^y dz_\mu A_\mu(z)}, \tag{5.10}$$

where the line integral is carried out along a path C connecting x and y and a summation over μ is understood. Notice that $U(x,y)$ is an element of the $U(1)$ group. Under a gauge transformation, $A_\mu \to A_\mu - \frac{1}{e}\partial_\mu\Lambda$, (5.10) transforms as follows

$$U(x,y) \to G(x)U(x,y)G^{-1}(y), \tag{5.11}$$

where $G(x)$ is given by (5.3c). From the above considerations we conclude that the following bilinear expression in the fermion fields ψ and $\bar{\psi}$ is gauge-invariant:

$$\bar{\psi}(x)U(x,y)\psi(y) = \bar{\psi}(x)e^{ie\int_x^y dz_\mu A_\mu(z)}\psi(y). \tag{5.12}$$

Suppose now that $y = x + \epsilon$. Then we conclude from (5.12) that $\bar{\psi}(x)\psi(x+\epsilon)$ and $\bar{\psi}(x+\epsilon)\psi(x)$ must be modified as follows:

$$\bar{\psi}(x)\psi(x+\epsilon) \to \bar{\psi}(x)U(x,x+\epsilon)\psi(x+\epsilon),$$

$$\bar{\psi}(x+\epsilon)\psi(x) \to \bar{\psi}(x+\epsilon)U^\dagger(x,x+\epsilon)\psi(x),$$

where

$$U(x,x+\epsilon) = e^{ie\epsilon\cdot A(x)}$$

and $\epsilon \cdot A = \sum_\mu \epsilon_\mu A_\mu$.

The above considerations suggest that to arrive at a gauge-invariant expression for the fermionic action on the lattice, we should make the following substitutions in (5.9)

$$\bar{\psi}(n)(r-\gamma_\mu)\psi(n+\hat{\mu}) \to \bar{\psi}(n)(r-\gamma_\mu)U_{n,n+\hat{\mu}}\psi(n+\hat{\mu}), \tag{5.13a}$$

$$\bar{\psi}(n+\hat{\mu})(r+\gamma_\mu)\psi(n) \to \bar{\psi}(n+\hat{\mu})(r+\gamma_\mu)U_{n+\hat{\mu},n}\psi(n), \tag{5.13b}$$

where

$$U_{n+\hat{\mu},n} = U^\dagger_{n,n+\hat{\mu}}, \tag{5.14}$$

and $U_{n,n+\hat{\mu}}$ is an element of the $U(1)$ gauge group. It can therefore be written in the form

$$U_{n,n+\hat{\mu}} = e^{i\phi_\mu(n)}, \tag{5.15}$$

where $\phi_\mu(n)$ is restricted to the compact domain $[0, 2\pi]$. The right-hand side of (5.13a) and (5.13b) are now invariant under the following set of local transformations

$$\begin{aligned} \psi(n) &\to G(n)\psi(n), \\ \bar{\psi}(n) &\to \bar{\psi}(n)G^{-1}(n), \\ U_{n,n+\hat{\mu}} &\to G(n)U_{n,n+\hat{\mu}}G^{-1}(n+\hat{\mu}), \\ U_{n+\hat{\mu},n} &\to G(n+\hat{\mu})U_{n+\hat{\mu},n}G^{-1}(n), \end{aligned} \tag{5.16}$$

Notice that in contrast to the matter fields discussed before, the group elements $U_{n,n+\hat{\mu}}$ live on the links connecting two neighbouring lattice sites; hence we will refer to them as link variables and sometimes simply as links. Because of (5.14) they are directed quantities, and we shall use the following graphical representation:

n —→— $n+\hat{\mu}$ $\qquad$ n —←— $n+\hat{\mu}$

$U_{n,n+\hat{\mu}}$ $\qquad$ $U_{n+\hat{\mu},n} = U^\dagger_{n,n+\hat{\mu}}$

Making the substitutions (5.13) in eq. (5.9), we obtain the following gauge-invariant lattice action for Wilson fermions

$$\begin{aligned} S_F^{(W)}[\psi, \bar{\psi}, U] = & (\hat{M} + 4r)\sum_n \bar{\psi}(n)\psi(n) \\ & - \frac{1}{2}\sum_{n,\mu}[\bar{\psi}(n)(r - \gamma_\mu)U_{n,n+\hat{\mu}}\psi(n+\hat{\mu}) \\ & + \bar{\psi}(n+\hat{\mu})(r+\gamma_\mu)U^\dagger_{n,n+\hat{\mu}}\psi(n)]. \end{aligned} \tag{5.17}$$

Let us pause here for a moment and forget the arguments which led us to eq. (5.17). By requiring that $U_{n,n+\hat{\mu}}$ and $U_{n+\hat{\mu},n}$ transform according to (5.16), this expression represents a natural way of implementing $U(1)$-gauge invariance. That the link variables should be elements of the $U(1)$ gauge group, follows from the requirement that their gauge transforms must also be elements of $U(1)$. One then would have to show that in the continuum limit (5.17) can be cast into the form (5.8c) by establishing a relation between the link variables and the vector potential $A_\mu(n)$. The procedure would then be the following. The vector potential $A_\mu(n)$ at the lattice site n is real-valued and carries a Lorentz index. The same is true for $\phi_\mu(n)$, which parametrizes the link variable (5.15). But $\phi_\mu(n)$ takes only values in the interval $[0, 2\pi]$, while the values taken by the vector potential $A_\mu(x)$ in the continuum theory extends over the entire real line. This is no problem, for we must remember that A_μ carries the dimension of inverse length, while ϕ_μ is dimensionless. Let us therefore make the Ansatz $\phi_\mu(n) = caA_\mu(n)$, where a is the lattice spacing, and c is a constant to be determined. For $a \to 0$ the range of A_μ will hence be infinite. With this ansatz it is now a simple matter to check that by scaling $\hat{M}$, ψ and $\bar{\psi}$ appropriately with a (i.e. $\hat{M} \to aM$, $\psi \to a^{3/2}\psi$, $\bar{\psi} \to a^{3/2}\bar{\psi}$) and replacing $U_{n,n+\hat{\mu}}$ for small values of the lattice spacing by

$$U_{n,n+\hat{\mu}} \approx 1 + icaA_\mu(n)$$

expression (5.17) reduces to (5.8c) in the naive continuum limit, if we choose $c = e$. Because of this connection between $U_{n,n+\hat{\mu}}$ and $A_\mu(n)$ we shall use from here on the more suggestive notation

$$U_\mu(n) \equiv U_{n,n+\hat{\mu}} = e^{ieaA_\mu(n)} \quad . \tag{5.18}$$

With this identification* it is now an easy matter to verify that $U_\mu(n)$ transforms as follows under gauge transformations

$$U_\mu(n) \to G(n)U_\mu(n)G^{-1}(n+\hat{\mu}) = e^{ieaA^G_\mu(n)},$$

where $A^G_\mu(n)$ is a discretized version of $A_\mu(x) - \frac{1}{e}\partial_\mu\Lambda(x)$. Hence so far the lattice action (5.17) with $A_\mu(n)$ defined by (5.18) satisfies the basic requirements stated at the beginning of this chapter. To complete our construction of the lattice action for QED, we must obtain the lattice version of (5.8b) which again should be strictly gauge-invariant, and be a functional of the link variables only. Such gauge-invariant functionals are easily constructed by taking the product of link variables around closed loops on the euclidean space-time lattice. Furthermore, because of the local structure of the integrand in (5.8b), it is clear that we should focus our attention on the smallest possible loops. Hence we are led to consider the product of link variables around an elementary plaquette, as shown in Fig. (5-1). Let this plaquette lie in the $\mu - \nu$ plane. We then define

$$U_{\mu\nu}(n) = U_\mu(n)U_\nu(n+\hat{\mu})U^\dagger_\mu(n+\hat{\nu})U^\dagger_\nu(n), \tag{5.19}$$

where we have path-ordered the link variables. Although this path ordering is irrelevant in the abelian case considered here, it will become important when we study QCD. Inserting the expression (5.18) into (5.19), one finds that

$$U_{\mu\nu}(n) = e^{iea^2F_{\mu\nu}(n)}, \tag{5.20}$$

---

* Of course, this identification of $A_\mu(n)$ with the vector potential is only strictly correct in the continuum limit.

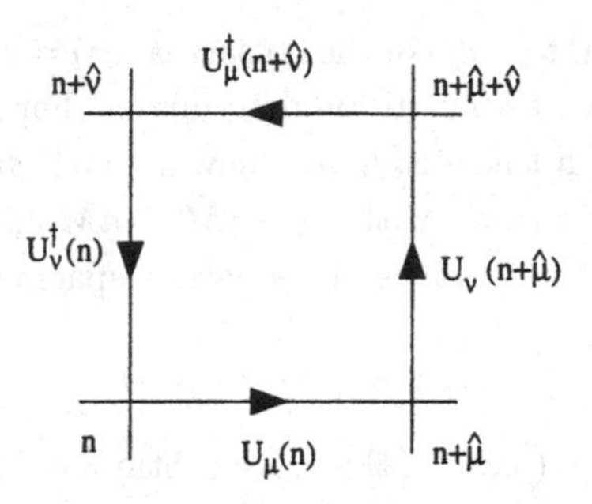

Fig. 5-1 The contribution $U_{\mu\nu}(n)$ of an elementary plaquette with base at n lying in the $\mu\nu$-plane.

where $F_{\mu\nu}(n)$ is a discretized version of the continuum field strength tensor:

$$F_{\mu\nu}(n) = \frac{1}{a}[(A_\nu(n+\hat{\mu}) - A_\nu(n)) - (A_\mu(n+\hat{\nu}) - A_\mu(n))].$$

It then follows immediately from (5.20) that for small lattice spacing

$$\frac{1}{e^2}\sum_n \sum_{\substack{\mu,\nu \\ \mu<\nu}} [1 - \frac{1}{2}(U_{\mu\nu}(n) + U^\dagger_{\mu\nu}(n))] \approx \frac{1}{4}\sum_{\substack{n \\ \mu,\nu}} a^4 F_{\mu\nu}(n) F_{\mu\nu}(n),$$

where the sum appearing on the left-hand side extends over the contributions coming from all distinct plaquettes on the lattice.* Hence from now on we shall write the lattice action for the gauge potential in the compact form

$$S_G[U] = \frac{1}{e^2}\sum_P [1 - \frac{1}{2}(U_P + U_P^\dagger)], \tag{5.21}$$

where U_P (plaquette variable) stands for the product of link variables around the boundary of a plaquette "P" taken (say) in the counterclockwise direction.

At this point we want to mention an interesting property of the lattice formulation. In contrast to the continuum formulation where the coupling constant

* Notice that on the right-hand side of this equation the sum extends over all μ and ν.

e enters linearly in the fermionic piece of the action (cf. eqs. (5.8c) and (5.1)), the coupling now appears with an inverse power in the action for the gauge field! Thus on the lattice, the strong coupling expansion turns out to be the natural one.

This completes the construction of the lattice action for QED. For Wilson fermions it is given by

$$\begin{aligned} S_{QED}[U,\psi,\bar\psi] =& \frac{1}{e^2}\sum_P [1-\frac{1}{2}(U_P+U_P^\dagger)] \\ &+(\hat M+4r)\sum_n \bar\psi(n)\psi(n) - \frac{1}{2}\sum_{n,\mu}[\bar\psi(n)(r-\gamma_\mu)U_\mu(n)\psi(n+\hat\mu) \\ &+\bar\psi(n+\hat\mu)(r+\gamma_\mu)U_\mu^\dagger(n)\psi(n)]. \end{aligned} \tag{5.22}$$

The action (5.22) is to be used in a path-integral formulation, from which any correlation function of the fermionic and link variables can be computed. This path integral will involve an integration over all link variables $U_\mu(n)$, which, as we have emphasized, are elements of a unitary group. Hence the integration is to be performed over the (compact) group manifold which in the present $U(1)$ case is parametrized by a single real angular variable restricted to the range $[0,2\pi]$.* Now we have made a great effort in ensuring the exact gauge invariance of the action. Hence this gauge invariance should not be destroyed in the integration process; i.e. we must define a gauge-invariant integration measure! This is quite trivial in the present case, for under a gauge transformation the link variables are transformed according to (5.16). But since $G(n)$ is an element of the abelian $U(1)$ group, a gauge transformation merely amounts to a site-dependent shift in the phase of $U_\mu(n)$. With the parametrization $U_\mu(n)=e^{i\phi_\mu(n)}$, the gauge-invariant measure to be used in a path integral expression is therefore given by

$$DU \equiv \prod_{n,\mu} d\phi_\mu(n), \tag{5.23}$$

and correlation functions of the link variables and Dirac fields are computed from the following path integral expression

$$\begin{aligned} &\langle\psi_\alpha(n)\cdots\bar\psi_\beta(m)\cdots U_\mu(N)\cdots\rangle = \\ &\qquad = \frac{\int DU D\bar\psi D\psi(\psi_\alpha(n)\cdots\bar\psi_\beta(m)\cdots U_\mu(N)\cdots)e^{-S_{QED}}}{\int DU D\bar\psi D\psi e^{-S_{QED}}}. \end{aligned} \tag{5.24}$$

---

* Since the integration range is compact, one also speaks of compact QED when referring to the lattice formulation.

These correlation functions depend on the parameters $\hat{M}$ and e which enter the expressions for the fermionic and gauge lattice actions (5.17) and (5.21). In the interacting quantum theory defined by (5.24), these parameters can no longer be identified with the physical fermion mass and charge, and must be viewed upon as bare parameters, having no direct physical meaning. To emphasize this point, one usually writes M_0 and e_0 instead of M and e. We have not done so in this chapter, since we have merely constructed the lattice action starting from the free fermion theory. In the following chapter, where we discuss the non-abelian case, we shall, however, make use of this notation.

One other remark must be made. For the euclidean lattice action (5.22) to be a bonafide candidate for defining a quantum theory, it must satisfy the criterion of *reflection positivity.* Reflection positivity is a necessary ingredient for the existence of a non-negative hermitean Hamiltonian, with a positive transfer matrix, and a Hilbert space formulation. This is an important technical detail which we only mention here. The action (5.22) can be shown to satisfy reflection positivity, which within the continuum formulation was first discussed by Osterwalder and Schrader (1973). For details the reader may consult the book by Montvay and Münster (1994), and in particular the references quoted there.

This completes the formulation of lattice QED. As we have seen, the group aspect has played a central role in the above discussion. In a lattice formulation it is not the vector potential which emerges naturally in the process of gauging the free Dirac theory, but rather the group elements $U_\mu(n)$ which live on the links connecting two neighbouring sites. Thus the connection between lattice and continuum variables is much more subtle than in the case of the matter fields. In fact one may easily verify that a naive lattice translation of the minimal substitution rule $\partial_\mu \to D_\mu$ (see eq. (5.1)) will lead to a fermionic action S_F which violates gauge invariance in higher orders of the lattice spacing. For reasons we have already mentioned, however, we have insisted on the strict gauge invariance of the lattice action.

CHAPTER 6

NON-ABELIAN GAUGE FIELDS ON THE LATTICE COMPACT QCD

The lattice gauge theory we discussed in chapter 5 can be easily extended to the case where the abelian group $U(1)$ is replaced by a non-abelian unitary group. Thus suppose that instead of a single free Dirac field we have N such fields ψ^a $(a = 1, \cdots, N)$ of mass M_0. Then the euclidean fermionic action, replacing (5.9), is given by*

$$S_F = (\hat{M}_0 + 4r)\sum_n \sum_{a=1}^{N} \bar{\psi}^a(n)\psi^a(n) - \frac{1}{2}\sum_{n,\mu}\sum_{a=1}^{N}[\bar{\psi}^a(n)(r - \gamma_\mu)\psi^a(n+\hat{\mu}) + \\ + \bar{\psi}^a(n+\hat{\mu})(r+\gamma_\mu)\psi^a(n)]. \tag{6.1}$$

This action is invariant under global unitary transformations in N dimensions, and in particular under the non-abelian subgroup $SU(N)$.** Introducing the following N-component column and row vectors

$$\psi = \begin{pmatrix} \psi^1 \\ \vdots \\ \psi^N \end{pmatrix}, \qquad \bar{\psi} = (\bar{\psi}^1, \cdots, \bar{\psi}^N), \tag{6.2}$$

these transformations read

$$\psi(x) \longrightarrow \underset{\sim}{G}\psi(x) ,$$
$$\bar{\psi}(x) \longrightarrow \bar{\psi}(x)\underset{\sim}{G}^{-1} ,$$

where $\underset{\sim}{G}$ is an element of $SU(N)$. We now want to generalize (5.22) to the non–abelian case. This is straightforward. We only have to replace the Dirac fields ψ and $\bar{\psi}$ by the N–component vectors (6.2) and the link variables $U_\mu(n)$ by the corresponding group elements of $SU(N)$ in the fundamental (N–dimensional) representation. Let us denote the matrix–valued link variables by $\underset{\sim}{U}_\mu(n)$. They can be written in the form

$$\underset{\sim}{U}_\mu(n) = e^{i\underset{\sim}{\phi}_\mu(n)} , \tag{6.3}$$

---

* We omit the "hat" on the dimensionless fields ψ, $\bar{\psi}$ since it is clear that we are discussing the dimensionless formulation.

** This group consists of all unitary $N \times N$ matrices with determinant equal to one.

where $\phi_\mu(n)$ is a hermitean matrix belonging to the Lie algebra of $SU(N)$. Making the above substitutions in (5.17), we obtain the gauged version of (6.1):

$$\begin{aligned} S_F^{(W)} =&(\hat{M}_0 + 4r)\sum_n \bar{\psi}(n)\psi(n) \\ &-\frac{1}{2}\sum_{n,\mu}[\bar{\psi}(n)(r-\gamma_\mu)U_\mu(n)\psi(n+\hat{\mu})+ \\ &+\bar{\psi}(n+\hat{\mu})(r+\gamma_\mu)U_\mu^\dagger(n)\psi(n)]. \end{aligned} \tag{6.4}$$

For the reasons stated at the end of chapter 5, we have now written $\hat{M}_0$ instead of $\hat{M}$. Written out explicitly, a typical term in (6.4) reads

$$\bar{\psi}(n)(r-\gamma_\mu)U_\mu(n)\psi(n+\hat{\mu}) = \sum_{\alpha,\beta,a,b} \bar{\psi}_\alpha^a(n)(r-\gamma_\mu)_{\alpha\beta}(U_\mu(n))_{ab}\psi_\beta^b(n+\hat{\mu}).$$

The action (6.4) is invariant under the following local transformations

$$\begin{aligned} \psi(n) &\to G(n)\psi(n), \\ \bar{\psi}(n) &\to \bar{\psi}(n)G^{-1}(n), \end{aligned} \tag{6.5a}$$

$$\begin{aligned} U_\mu(n) &\to G(n)U_\mu(n)G^{-1}(n+\hat{\mu}), \\ U_\mu^\dagger(n) &\to G(n+\hat{\mu})U_\mu^\dagger(n)G^{-1}(n) \end{aligned} \tag{6.5b}$$

Here $G(n)$ is an element of $SU(N)$ in the fundamental representation. It can therefore be written in the form

$$G(n) = e^{i\Lambda(n)}, \tag{6.5c}$$

where $\Lambda(n)$ is a hermitean matrix belonging to the Lie algebra of $SU(N)$.

It is now a simple matter to construct the other piece of the action analogous to (5.21). Clearly S_G should be gauge-invariant. Since $U_\mu(n)$ transforms according to (6.5b), the simplest gauge-invariant quantity one can build from the group elements $U_\mu(n)$, is the trace of the path ordered product of link variables along the boundary of an elementary plaquette; this path-ordered product is the generalization of (5.19) to the non-abelian case and reads:

$$U_{\mu\nu}(n) = U_\mu(n)U_\nu(n+\hat{\mu})U_\mu^\dagger(n+\hat{\nu})U_\nu^\dagger(n). \tag{6.6}$$

Notice that the trace, and the path-ordering of the links in (6.6) are important now, since the group elements do not commute. In analogy to the abelian case discussed in chapter 5 we now replace (5.21) by

$$S_G = cTr \sum_{\substack{n \\ \mu<\nu}} [1 - \frac{1}{2}(\underset{\sim}{U}_{\mu\nu}(n) + \underset{\sim}{U}^{\dagger}_{\mu\nu}(n))], \tag{6.7}$$

where c is a constant which will be fixed below.

So far we have merely extended the analysis in chapter 5, to the case of a non-abelian group. Undoubtedly the lattice theory constructed in this way describes a quite non-trivial system. But does it have any relevance for physics, and in particular for elementary particle physics? To answer this question we must see whether it has a chance of describing in the continuum limit an interesting field theory. To this effect let us first fix the gauge group that we expect to be relevant for describing the strong interactions of quarks and gluons. It has been known for a long time that quarks (and antiquarks) must come in three "coloured" versions, ($\to \psi^a$, $a = 1,2,3$) and that the observed strongly interacting particles (hadrons) are colour singulets with respect to the group $SU(3)$.* Hence we expect this group to be the one of interest. Now any element $\underset{\sim}{\Theta}$ lying in the Lie algebra of $SU(3)$ can be written in the form

$$\underset{\sim}{\Theta} = \sum_{B=1}^{8} \Theta^B \frac{\lambda^B}{2},$$

where the eight group generators λ^B are usually chosen to be the (3×3) Gell-Mann matrices, satisfying the commutation relations**

$$[\lambda^A, \lambda^B] = 2i \sum_{C=1}^{8} f_{ABC} \lambda^C. \tag{6.8}$$

Here f_{ABC} are the completely antisymmetric structure constants of the group, corresponding to this choice of generators.

---

* These hadrons are built from different "flavoured" quarks (i.e. up, down, strange, charm, top, bottom). Each of these quarks comes in three colours, and they must be combined in such a way that the hadron transforms trivially under $SU(3)$. The reader who is not acquainted with these concepts, and is interested in learning more about it, may consult the book by Close (1979).

** See e.g. the book quoted in the previous footnote.

Let us now study the naive continuum limit of the action (6.4) for the case $N = 3$, proceeding in a way analogous to the abelian case. To this effect we introduce a dimensioned matrix valued lattice field $\underset{\sim}{A}_\mu(n)$ as follows,

$$\phi_\mu(n) = g_0 a \underset{\sim}{A}_\mu(n). \tag{6.9}$$

Here $\phi_\mu(n)$ is defined in (6.3), a is the lattice spacing, and g_0 is a bare coupling constant. Again we have written g_0 instead of g to emphasize that in an interacting theory this coupling constant is one of the *bare* parameters on which the action depends. Since $\underset{\sim}{A}_\mu(n)$ is an element of the Lie algebra of $SU(3)$ it is of the form

$$\underset{\sim}{A}_\mu(n) = \sum_{B=1}^{8} A_\mu^B(n)\frac{\lambda^B}{2}, \tag{6.10}$$

where $A_\mu^B(n)$ are eight real-valued vector fields corresponding to the eight generators of $SU(3)$. Inserting (6.9) into (6.3) and expanding the exponential to leading order in a, one finds, after scaling $\hat{M}_0$, $\underset{\sim}{\psi}$ and $\underset{\sim}{\bar{\psi}}$ appropriately with the lattice spacing, that (6.4) reduces to the following continuum action for $a \to 0$:

$$S_F^{(cont.)} = \int d^4x \underset{\sim}{\bar{\psi}}(x)(\gamma_\mu(\partial_\mu + ig_0\underset{\sim}{A}_\mu) + M_0)\underset{\sim}{\psi}(x).$$

Next we consider the naive continuum limit of eq. (6.7). To this end we define in analogy to (5.20) the matrix-valued lattice field tensor $\underset{\sim}{\mathcal{F}}_{\mu\nu}$ by

$$\underset{\sim}{U}_{\mu\nu}(n) = e^{ig_0 a^2 \underset{\sim}{\mathcal{F}}_{\mu\nu}(n)}. \tag{6.11}$$

Clearly, the relation between $\underset{\sim}{\mathcal{F}}_{\mu\nu}(n)$ and $\underset{\sim}{A}_\mu(n)$ is now much more complicated than in the abelian case. The reason is that the link variables appearing in the product (6.6) are now matrices which do not commute. In order to arrive at the connection between $\underset{\sim}{\mathcal{F}}_{\mu\nu}$ and $\underset{\sim}{A}_\mu$, one needs to make use of the Baker-Campbell-Hausdorff formula

$$e^A e^B = e^{A+B+\frac{1}{2}[A,B]+\cdots}, \tag{6.12}$$

where the "dots" will in general involve an infinite number of terms. But because $\phi_\mu(n)$ is proportional to the lattice spacing (cf. eq. (6.9)), one only needs to compute a few terms in the exponent of (6.12), when this formula is used to calculate the product of link variables. By making use of such relations as

$$\underset{\sim}{\phi}_\mu(n+\nu) \approx \underset{\sim}{\phi}_\mu(n) + a\partial_\nu \underset{\sim}{\phi}_\mu(n) + \cdots \approx g_0 a \underset{\sim}{A}_\mu(n) + g_0 a^2 \partial_\nu \underset{\sim}{A}_\mu(n) + \cdots,$$

one finds that

$$\underset{\sim}{\mathcal{F}}_{\mu\nu} \underset{a\to 0}{\longrightarrow} \underset{\sim}{F}_{\mu\nu} = \partial_\mu \underset{\sim}{A}_\nu - \partial_\nu \underset{\sim}{A}_\mu + ig_0[\underset{\sim}{A}_\mu, \underset{\sim}{A}_\nu]. \tag{6.13}$$

This is the well-known expression for the matrix-valued field tensor in continuum QCD. Since (6.13) is again an element of the Lie algebra, it can be written in the form

$$\underset{\sim}{F}_{\mu\nu} = \sum_{B=1}^{8} F^B_{\mu\nu} \frac{\lambda^B}{2}. \tag{6.14}$$

Making use of (6.8) and of the orthogonality relation of the Gell-Mann matrices,

$$Tr(\lambda^B \lambda^C) = 2\delta_{BC}, \tag{6.15}$$

one arrives at the following connection between the eight components of $F^B_{\mu\nu}$ and A^B_μ, defined in (6.14) and (6.10), respectively:

$$F^B_{\mu\nu} = \partial_\mu A^B_\nu - \partial_\nu A^B_\mu - g_0 f_{BCD} A^C_\mu A^D_\nu. \tag{6.16}$$

Having motivated the introduction of the lattice field strength tensor $\underset{\sim}{\mathcal{F}}_{\mu\nu}$ according to (6.11), we now compute S_G in the naive continuum limit. Approximating (6.11) for small lattice spacing by the first two non-trivial terms in the expansion of the exponential and inserting this expression into (6.7), one finds that

$$S_G \longrightarrow c\frac{g_0^2}{2} S_G^{(cont.)},$$

where

$$S_G^{(cont.)} = \frac{1}{2} Tr \int d^4x \underset{\sim}{F}_{\mu\nu} \underset{\sim}{F}_{\mu\nu}, \tag{6.17}$$

and a sum over μ, ν is understood. This is the well-known gauge field action of QCD. Hence we must choose $c = 2/g_0^2$. The continuum field strength tensor for $SU(N)$ has the same form as given in (6.13) with $\underset{\sim}{A}_\mu$ an element of the Lie algebra of $SU(N)$. Following the same procedure as above one finds that the gauge part of the lattice action is given for all $N > 1$ by

$$S_G^{(SU(N))} = \beta \sum_P [1 - \frac{Tr}{2N}(\underset{\sim}{U}_P + \underset{\sim}{U}_P^\dagger)] \tag{6.18a}$$

where

$$\beta = \frac{2N}{g_0^2} \quad . \tag{6.18b}$$

As in the abelian case, the sum in (6.18a) extends over all distinct plaquettes on the lattice, and we have introduced the notation $\underset{\sim}{U}_P$ for the path-ordered product (6.6) of link variables around the boundary of a plaquette P. Both orientations for this product are taken into account, thus ensuring the hermiticity of the action.

The action (6.18) is invariant under the local transformations (6.5b). Inserting for $\underset{\sim}{U}_\mu(n)$ the expression

$$\underset{\sim}{U}_\mu(n) = e^{ig_0 a \underset{\sim}{A}_\mu(n)} \quad , \tag{6.19}$$

one finds that (6.5b) implies the following transformation law for $\underset{\sim}{A}_\mu$ in the continuum limit:

$$\underset{\sim}{A}_\mu(x) \to \underset{\sim}{G}(x)\underset{\sim}{A}_\mu(x)\underset{\sim}{G}^{-1}(x) - \frac{i}{g_0}\underset{\sim}{G}(x)\partial_\mu \underset{\sim}{G}^{-1}(x). \tag{6.20}$$

This is the non–abelian analog of (5.3b).

For those readers not familiar with continuum QCD we want to make the following remark. Using the relation (6.15), the expression (6.17) may be written in the form

$$S_G^{(cont.)} = \frac{1}{4}\int d^4x F_{\mu\nu}^B F_{\mu\nu}^B, \tag{6.21}$$

where $F_{\mu\nu}^B$ is related to the coloured gauge potentials by (6.16). Hence, in contrast to the abelian case, the pure gauge sector of QCD describes a highly non-trivial interacting theory, which involves tripel and quartic interactions of the fields, A_μ^B. This is the reason why a study of the pure gauge sector of QCD is of great interest. In fact, the self-couplings of the gauge potentials are believed to be responsible for quark confinement. The first non-abelian gauge theory was proposed by Yang and Mills (1954), and was based on $SU(2)$. For this reason one usually refers to (6.18), or (6.21), as the Yang-Mills action.

So far we have constructed the lattice action which possesses the desired naive continuum limit. We must now define the quantum theory by specifying the path integral expression from which correlation functions may be computed. This expression will be of the form (5.24) except that now ψ_α, $\bar{\psi}_\alpha$ and $U_\mu(n)$ will carry additional colour indices corresponding to the three-coloured quarks lying in the fundamental representation of $SU(3)$. What concerns the integration measure DU

on the other hand, it will now depend on the eight real variables parametrizing the group elements of $SU(3)$, and the integration is to be performed over the group manifold. For the same reasons mentioned before in connection with the abelian theory, this integration measure must be gauge-invariant if quantum fluctuations are not to destroy this important principle. Denoting by $\alpha_\ell^A (A = 1, \cdots, 8)$ the group parameters on which the ℓ'th link variable depends, the corresponding integration measure will be of the form*

$$DU = \prod_\ell J(\alpha_\ell)(d\alpha_\ell), \tag{6.22a}$$

where α_ℓ stands for the set $\{\alpha_\ell^A\}$, and

$$(d\alpha_\ell) \equiv \prod_{A=1}^{8} d\alpha_\ell^A. \tag{6.22b}$$

The structure of the Jacobian $J(\alpha_\ell)$ in (6.22a) is determined from the requirement of gauge invariance. For our purpose it will suffice to know some of the standard integrals involving polynomials of the link variables and we shall only list a few of them without proof. A general rule, however, is the following: only those integrals involving products of the link variables will give a non-vanishing contribution, for which the direct product of the corresponding representations contains the identity element. With dU_ℓ defined by

$$dU_\ell \equiv J(\alpha_\ell)(d\alpha_\ell) \quad ,$$

some useful $SU(3)$ group integrals are:

$$\int dU\; U^{ab} = 0, \tag{6.23a}$$

$$\int dU\; U^{ab}\; U^{cd} = 0, \tag{6.23b}$$

$$\int dU\; U^{ab} (U^\dagger)^{cd} = \frac{1}{3}\delta_{ad}\delta_{bc}, \tag{6.23c}$$

$$\int dU\; U^{a_1b_1} U^{a_2b_2} U^{a_3b_3} = \frac{1}{3!}\epsilon_{a_1a_2a_3}\;\epsilon_{b_1b_2b_3}. \tag{6.23d}$$

---

* The integration measure DU is the so called Haar measure. We will discuss this measure in detail in chapter 15, where we shall require its explicit form in order to perform the weak coupling expansion of lattice QCD.

Here U stands generically for any given link variable. The general rules for evaluating arbitrary integrals of the above type have been discussed by Creutz (1978).

An arbitrary correlation function involving the fermionic and link variables can be computed from the following path integral expression

$$\begin{aligned}&\langle \psi^{a_1}_{\alpha_1}(n_1)\cdots\bar{\psi}^{b_1}_{\beta_1}(m_1)\cdots U^{cd}_{\mu_1}(k_1)\cdots\rangle = \\ &=\frac{1}{Z}\int DUD\bar{\psi}D\psi\psi^{a_1}_{\alpha_1}(n_1)\cdots\bar{\psi}^{b_1}_{\beta_1}(m_1)\cdots U^{cd}_{\mu_1}(k_1)e^{-S_{QCD}},\end{aligned} \tag{6.24a}$$

where

$$Z=\int DUD\bar{\psi}D\psi e^{-S_{QCD}}, \tag{6.24b}$$

and where S_{QCD} is given by the sum of the actions (6.18) and (6.4) with $N=3$. For later convenience we summarize the results for QCD below:

$$S_{QCD}=S_G[U]+S^{(W)}_F[U,\psi,\bar{\psi}], \tag{6.25a}$$

$$S_G=\frac{6}{g_0^2}\sum_P[1-\frac{Tr}{6}(\underset{\sim}{U}_P+\underset{\sim}{U}^\dagger_P)], \tag{6.25b}$$

$$\begin{aligned}S^{(W)}_F=&(\hat{M}_0+4r)\sum_n\bar{\underset{\sim}{\psi}}(n)\underset{\sim}{\psi}(n)\\ &-\frac{1}{2}\sum_{n,\mu}[\bar{\underset{\sim}{\psi}}(n)(r-\gamma_\mu)\underset{\sim}{U}_\mu(n)\underset{\sim}{\psi}(n+\hat{\mu})+\\ &+\bar{\underset{\sim}{\psi}}(n+\hat{\mu})(r+\gamma_\mu)\underset{\sim}{U}^\dagger_\mu(n)\underset{\sim}{\psi}(n)].\end{aligned} \tag{6.25c}$$

We have concentrated here on the case of Wilson fermions. The generalization of the free staggered fermion action (4.36) to QCD is obvious. The fields χ and $\bar{\chi}$ become 3–component vectors in colour space and must be coupled to the matrix–valued link variables in a gauge invariant way. Each lattice site can accomodate the three colours. Denoting by $\underset{\sim}{\chi}$ the vector (χ^1,χ^2,χ^3) in colour space, we have that

$$\begin{aligned}S^{(stag.)}_F=&\frac{1}{2}\sum_{n,\mu}\eta_\mu(n)\bar{\underset{\sim}{\chi}}(n)\left(\underset{\sim}{U}_\mu(n)\underset{\sim}{\chi}(n+\hat{\mu})-\underset{\sim}{U}^\dagger_\mu(n-\hat{\mu})\underset{\sim}{\chi}(n-\hat{\mu})\right)\\ &+\hat{M}_0\sum_n\bar{\chi}(n)\chi(n).\end{aligned} \tag{6.26}$$

For each colour, the different Dirac and flavour components of the quark fields are then constructed from the χ–variables at the sixteen lattice sites within a hypercube in the way described in chapter 4.

This completes our construction program for lattice QCD. In the following chapter we introduce an important observable which will play a central role in the study of confinement later on.

CHAPTER 7

THE WILSON LOOP AND THE STATIC QUARK-ANTIQUARK POTENTIAL

One of the crucial tests of QCD is whether it accounts for the fact that isolated quarks have never been seen in nature. It is generally believed that quark confinement is a consequence of the non-abelian nature of the gauge interaction in QCD. In contrast to QED where the field lines connecting a pair of opposite charges are allowed to spread, one expects that the quarks within a hadron * are the sources of chromoelectric flux which is concentrated within narrow tubes (strings) connecting the constituents in the manner shown in fig. (7-1).

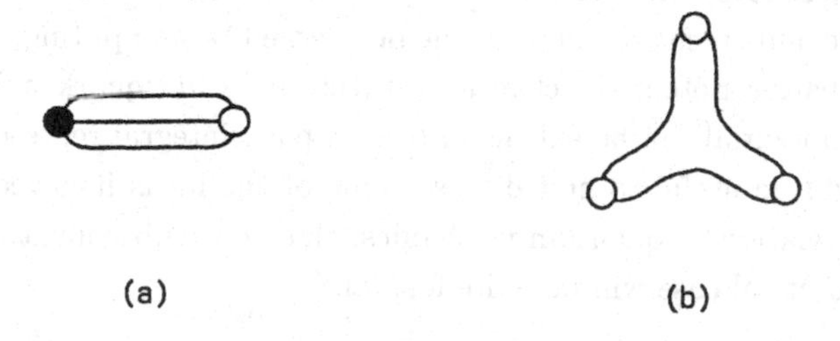

Fig. 7-1 (a) Picture of a meson built from a quark and antiquark which are held together by a string-like colour electric field; (b) corresponding picture of a baryon built from three quarks.

Since the energy is not allowed to spread, the potential of a quark-antiquark ($q\bar{q}$) pair will increase with their separation, as long as vacuum polarization effects do not screen their colour charge. For sufficiently large separations of the quarks, the energy stored in the string will suffice to produce real quark pairs, and the system will lower its energy by going over into a new hadronic state, consisting of colour neutral hadrons. In fig. (7-2) we give a qualitative picture of this hadronization process for the case when the quarks are bound within a meson ($q\bar{q}$-system) or baryon (qqq-system).

---

* We shall often refer to the constituents of hadrons simply as quarks, without distinguishing explicitly between quarks and antiquarks.

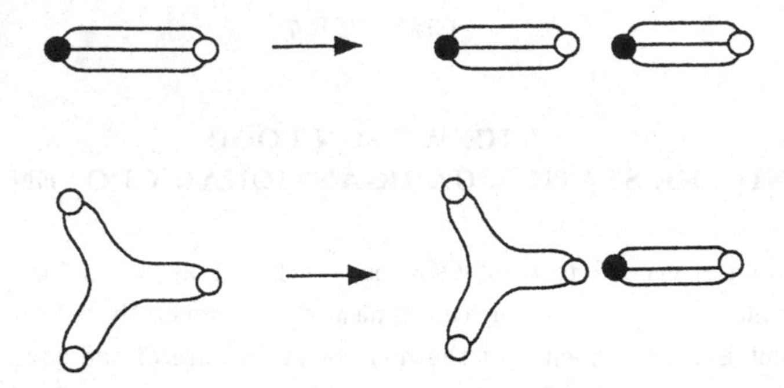

Fig. 7-2 Hadronization of the $q\bar{q}$ and qqq systems as the quarks are pulled apart

Once the colour charges of the quarks and antiquarks have been screened, the remaining Van der Waals type interaction between the colour neutral hadrons becomes the short-range interaction characteristic of the known hadronic world. This picture of confinement can in principle be checked by computing, for example, the non-perturbative potential between a static quark-antiquark pair. We now show how this potential can be extracted from a path integral representation. To this effect, it will be useful to first discuss some of the ideas involved within the context of non-relativistic quantum mechanics, since our subsequent presentation of the field theoretical case will be quite formal.

7.1 A Look at Non-Relativistic Quantum Mechanics

Consider a particle of mass m moving in a potential $V(x)$ in one space dimension. Its propagation is described by the amplitude

$$K(x', t; x, 0) = \langle x' | e^{-iHt} | x \rangle, \tag{7.1}$$

where $H = p^2/2m + V(x)$. Next consider the static limit of (7.1); letting $m \to \infty$, the kinetic term in the Hamiltonian may be dropped and H can be replaced by the potential. Hence (7.1) takes the simple form

$$K(x', t; x, 0) \underset{m\to\infty}{\longrightarrow} \delta(x - x') e^{-iV(x)t}. \tag{7.2}$$

Continuing this expression to imaginary times ($t \to -iT$), we see that the potential $V(x)$ may be determined from the exponential decay of (7.1) as a function of euclidean time T. The δ-function appearing in (7.2) just tells us that an infinitely massive particle does not propagate. In fact the only change in the wave function with time consists in the accumulation of a phase. Thus in the static limit the wave function $\psi(x,t)$ is a solution to the following equation

$$i\partial_t\psi(x,t) = V(x)\psi(x,t),$$

which may immediately be integrated to give

$$\psi(x,t) = e^{-iV(x)t}\psi(x,0). \tag{7.3}$$

The phase $\exp(-iV)$ is just the one appearing in eq. (7.2). To substantiate the formal arguments given above, we illustrate the result (7.2) for the case of the one-dimensional harmonic oscillator whose Hamiltonian is given by $H = p^2/2m + \kappa x^2/2$. The corresponding propagation kernel has the form*

$$K(x',t;x,0) = \left(\frac{m\omega}{2\pi i \sin\,\omega t}\right)^{1/2} e^{\frac{im\omega}{2\sin\omega t}[(x^2+x'^2)\cos\omega t - 2xx']}, \tag{7.4}$$

where $\omega = \sqrt{\kappa/m}$ is the frequency of the oscillator. In order to extract the potential from (7.4), we now take the limit $m \to \infty$, holding κ (i.e. the potential) fixed. This implies that ω must vanish like $1/\sqrt{m}$. It is then a trivial matter to show that

$$K(x',t;\,x,0) \underset{\substack{m\to\infty \\ \kappa\ fixed}}{\longrightarrow} \left[\left(\frac{1}{2\pi i\epsilon}\right)^{1/2} e^{\frac{i(x-x')^2}{2\epsilon}}\right] e^{-\frac{i}{2}(V(x')+V(x))t} \tag{7.5}$$

where $\epsilon = t/m$. In the limit $\epsilon \to 0$ $(m \to \infty)$, the factor appearing within square brackets just becomes $\delta(x - x')$; hence we arrive at the result (7.2).

7.2 The Wilson Loop and the Static $q\bar{q}$-Potential in QED

We now generalize this discussion to the case of a gauge field theory. To keep things as simple as possible, we shall restrict us for the moment to the case of an abelian $U(1)$ gauge theory, and in particular to QED. Furthermore, we shall argue entirely within the framework of the continuum formulation where the physical picture is most transparent. Our presentation is based on the work by Brown and Weisberger (1979), and on the review article by Gromes (1991).

---

* See e.g. Feynman and Hibbs (1965) for a derivation of (7.4) within the path integral framework. We have set $\hbar = 1$.

Consider a heavy quark (Q) and antiquark ($\bar{Q}$), which are introduced into the ground state of a quantum system whose dynamics is described by the action (5.5).* We want to study the energy of this (infinitely) heavy pair when it is coupled to the gauge potential in the usual minimal way. To this effect consider the following gauge invariant state

$$|\phi_{\alpha\beta}(\vec{x},\vec{y})\rangle = \bar{\Psi}_\alpha^{(Q)}(\vec{x},0)U(\vec{x},0;\vec{y},0)\Psi_\beta^{(Q)}(\vec{y},0)|\Omega\rangle, \tag{7.6}$$

where $|\Omega>$ denotes the ground state, and where, for arbitrary time $U(\vec{x},t;\vec{y},t)$ is defined by

$$U(\vec{x},t;\vec{y},t) = e^{ie\int_{\vec{x}}^{\vec{y}} dz^i A_i(\vec{z},t)}, \tag{7.7}$$

with the line integral extending along the straight line path connecting $\vec{x}$ and $\vec{y}$. This phase ensures the gauge invariance of the state (7.6) which describes a quark and antiquark located at $\vec{x}$ and $\vec{y}$ at time $t = 0$. In order to distinguish the heavy quarks, serving as test charges, from the (light) dynamical quarks responsible for the vacuum polarization effects referred to at the beginning of this chapter, we have attached the label "Q" to the corresponding Dirac fields. The state (7.6) is not an eigenstate of the Hamiltonean H. It serves however as a trial state to extract the energy of the lowest eigenstate of H having a non–vanishing projection on $|\phi_{\alpha\beta}\rangle$. This energy will be a function of the separation of the quark and antiquark, and is the quantity that we are interested in. As in the case of our quantum mechanical example we can extract this ground state energy by studying the propagation of the state (7.6). But the procedure will not completely parallel the quantum mechanical case. The difference is that whereas the state $|\vec{x}\rangle$, whose propagation we have studied there, becomes an eigenstate of the Hamiltonian $H = p^2/2m+V(x)$ in the infinite mass limit, this is not true for the trial state (7.6). In the field–theoretic case, where one is dealing with a system having an infinite number of degrees of freedom, there will be many eigenstates of H which have a non–vanishing projection on (7.6); but of all these states we are only interested in that state having the lowest energy. In our quantum mechanical example we would be confronted with a similar situation if we were studying the propagation of a particle of *finite* mass in the potential $V(x)$. In this case the state $|x\rangle$ is no longer an eigenstate of the Hamiltonian and the propagation amplitude will no longer have the simple form (7.2). Instead, we must consider its general spectral

* Although we are studying the $U(1)$ gauge theory, we will refer to the charged particles as quarks and antiquarks.

decomposition

$$K(x',t;x,0) = \sum_n \langle x'|n\rangle\langle n|x\rangle e^{-iE_n t}, \tag{7.8}$$

where E_n are the eigenvalues of $H = p^2/2m + V(x)$, and $|n\rangle$ the corresponding eigenstates. But from the structure of the right hand side of (7.8) we see that we can extract the contribution of the state of lowest energy by studying the propagation amplitude for large euclidean ($t \to -iT$) times:

$$K(x',-iT;x,0) \underset{T\to\infty}{\longrightarrow} \langle x'|0\rangle\langle 0|x\rangle e^{-E_0 T}.$$

This is the well-known Feynman-Kac formula. As an example consider the harmonic oscillator, where $K(x',t;x,0)$ is given by (7.4). Taking the above limit, holding the mass fixed, one readily finds that $E_0 = \frac{1}{2}\omega$. This is of course not the quantity that we were interested in; but the example illustrates the point that in general we must supplement the infinite mass limit with another limit involving the euclidean time. Furthermore, the order of these limits is important! To extract the ground state energy of a quark–antiquark pair, we must first study the propagation of the state (7.6) in the infinite mass limit, and then examine the behaviour of the propagation amplitude for large euclidean times. With this in mind, consider now the following Green function describing the propagation of the state (7.6):

$$\begin{aligned} &G_{\alpha'\beta',\alpha\beta}(\vec{x}\,',\vec{y}\,';\vec{x},\vec{y};t) \\ &= \langle\Omega|T(\bar{\Psi}^{(Q)}_{\beta'}(\vec{y}\,',t)U(\vec{y}\,',t;\vec{x}\,',t)\Psi^{(Q)}_{\alpha'}(x',t)\bar{\Psi}^{(Q)}_{\alpha}(\vec{x},0)U(\vec{x},0;\vec{y},0)\Psi^{(Q)}_{\beta}(\vec{y},0))|\Omega\rangle \end{aligned} \tag{7.9}$$

where "T" is the time-ordering operator. Since in the limit of infinite quark masses*, the positions of the quark and antiquark are frozen, we expect that (7.9) will show the following behaviour

$$G_{\alpha'\beta',\alpha\beta}(\vec{x}\,',\vec{y}\,';\vec{x},\vec{y};-iT) \underset{\substack{1)M_Q\to\infty \\ 2)T\to\infty}}{\longrightarrow} \delta^{(3)}(\vec{x}-\vec{x}\,')\delta^{(3)}(\vec{y}-\vec{y}\,')C_{\alpha'\beta',\alpha\beta}(\vec{x},\vec{y})e^{-E(R)T}, \tag{7.10}$$

where M_Q is the quark (antiquark) mass, $C_{\alpha'\beta',\alpha\beta}(\vec{x},\vec{y})$ is a function describing the overlap of our trial state (7.6) with the ground state of the Hamiltonian in the presence of the static pair, and $E(R)$ is the ground state energy** of the static pair separated by the distance $R = |\vec{x}-\vec{y}|$.

---

* Actually we shall keep the masses finite, in order to control any divergencies that might arise.

** As we shall see in the next chapter, $E(R)$ also includes self-energy effects which need to be subtracted when calculating the interquark potential.

The right-hand side of (7.9) has the (formal) path integral representation

$$G_{\alpha'\beta',\alpha\beta} = \frac{1}{Z}\int DA\, D\psi\, D\bar{\psi}\, D\psi^{(Q)} D\bar{\psi}^{(Q)} (\bar{\psi}^{(Q)}_{\beta'}(\vec{y}\,',t)\ldots\psi^{(Q)}_{\beta}(\vec{y},0))e^{iS}, \quad (7.11)$$

where the expression within brackets stands for the quantity whose expectation value we are calculating, Z is the normalization constant given by the integral (7.11) omitting the above mentioned expression, and S is the action describing the dynamics of the light and heavy quarks, and of the gauge potential:

$$S = S_G[A] + S_F[\psi,\bar{\psi},A] + S_Q[\psi^{(Q)},\bar{\psi}^{(Q)},A].$$

Here S_G and S_F are defined in (5.4) and (5.2) respectively, and*

$$S_Q[\psi^{(Q)},\bar{\psi}^{(Q)},A] = \int d^4x\bar{\psi}^{(Q)}(x)(i\gamma^\mu D_\mu - M_Q)\psi^{(Q)}(x). \quad (7.12)$$

Since this action is quadratic in the fields $\psi^{(Q)}$ and $\bar{\psi}^{(Q)}$, one can immediately perform the integration over these Grassmann variables in (7.11) (see chapter 2):

$$\begin{aligned} G_{\alpha'\beta',\alpha,\beta} = -\frac{1}{Z}\int DA\, D\psi\, D\bar{\psi}\Big[&S_{\beta\beta'}(y,y';A)S_{\alpha'\alpha}(x',x;A) \\ &-S_{\alpha'\beta'}(x',y';A)S_{\beta\alpha}(y,x;A)\Big]\cdot \\ &\cdot U(\vec{x},0;\vec{y},0)U(\vec{y}\,',t;\vec{x}\,',t)\det K^{(Q)}[A]e^{iS_G+iS_F}. \end{aligned} \quad (7.13)$$

Here x, y, x' and y' are the following four–component vectors

$$\begin{aligned} x = (\vec{x},0)\ ,\ y = (\vec{y},0), \\ x' = (\vec{x}\,',t)\ ,\ y' = (\vec{y}\,',t), \end{aligned} \quad (7.14)$$

$S(z,z';A)$ is the Green function describing the propagation of a quark in the external field A_μ,

$$[i\gamma^\mu(\partial_\mu + ieA_\mu(z)) - M_Q]S(z,z';A) = \delta^{(3)}(\vec{z} - \vec{z}\,')\delta(t-t'), \quad (7.15)$$

and $\det K^{(Q)}[A]$ is the determinant of the matrix

$$K^{(Q)}_{\alpha x,\beta y}[A] = [i\gamma^\mu(\partial_\mu + ieA_\mu(x)) - M_Q]_{\alpha\beta}\delta^{(4)}(x-y),$$

---

* Recall that we are still in Minkowski space. Hence γ^μ ($\mu = 0,1,2,3$) are the usual Dirac matrices.

arising from the integration over the heavy quark fields. In perturbation theory the logarithm of this determinant is given by the sum of Feynman graphs consisting of a fermion loop with an arbitrary number of fields A_μ attached to it. This determinant approaches an (infinite) constant for $M_Q \to \infty$ which is however canceled by a corresponding factor contained in Z. Hence in what follows we can set $\det K^{(Q)} = 1$.

For the same reasons as stated above, we can of course also perform the integration over the dynamical quark fields ψ and $\bar{\psi}$ in eq. (7.13). This gives rise to a similar determinant; but its dependence on the gauge potential can no longer be ignored, since these fields have finite mass. It is this determinant which is responsible for the vacuum polarization effects mentioned at the beginning of this chapter.

So far the result (7.13) holds for arbitrary quark mass M_Q. We now want to study this expression for $M_Q \to \infty$. Following Brown and Weisberger (1979), we drop the spatial part of the covariant derivative, but keep its time component. Thus gauge invariance is maintained in this approximation:

$$[i\gamma^0(\partial_0 + ieA_0(z)) - M_Q]S(z, z'; A) = \delta^{(4)}(z - z'). \tag{7.16}$$

Here the derivative acts on z. Equation (7.16) can be easily integrated. Making the Ansatz

$$S(z, z'; A) = e^{ie\int_{z_0}^{z'_0} dt\ A_0(\vec{z},t)} \hat{S}(z - z') \tag{7.17}$$

one finds that $\hat{S}(z - z')$ satisfies a differential equation, which does not involve the gauge potential:

$$(i\gamma^0\partial_0 - M_Q)\hat{S}(z - z') = \delta^{(4)}(z - z'). \tag{7.18}$$

This equation can be readily solved by making a Fourier ansatz for $\hat{S}$, and leads to the following expression for (7.17):

$$\begin{aligned} iS(z, z'; A) = \delta^{(3)}(\vec{z} - \vec{z}\,')e^{ie\int_{z_0}^{z'_0} dt\ A_0(\vec{z},t)} \Big\{ &\Theta(z_0 - z'_0)\left(\frac{1+\gamma_0}{2}\right) e^{-iM_Q(z_0 - z'_0)} \\ &+ \Theta(z'_0 - z_0)\left(\frac{1-\gamma_0}{2}\right) e^{iM_Q(z_0 - z'_0)} \Big\}. \end{aligned} \tag{7.19}$$

This expression shows that the time evolution of the (infinitely) heavy quark fields merely consists in the accumulation of a phase determined by A_0 and the quark mass. This is the statement analogous to (7.3) in our quantum mechanical example. We next insert (7.19) into (7.13). Because of the appearance of the spatial delta function, which merely tells us that an infinitely heavy quark cannot propagate in space, only the first term in eq. (7.13) contributes since $\vec{x} \neq \vec{y}$. Recalling the definitions of x, y, x' and y' given in eq. (7.14), one finds that

$$G_{\alpha'\beta',\alpha\beta} \underset{M_Q\to\infty}{\longrightarrow} \delta^{(3)}(\vec{x}-\vec{x}\,')\delta^{(3)}(\vec{y}-\vec{y}\,')(P_+)_{\alpha'\alpha}(P_-)_{\beta\beta'}e^{-2iM_Qt}\langle e^{ie\oint dz^\mu A_\mu(z)}\rangle, \tag{7.20a}$$

where

$$P_\pm = \frac{1}{2}(1\pm\gamma^0), \tag{7.20b}$$

and where the line integral extends over a closed rectangular path with spatial and temporal extension $R = |\vec{x}-\vec{y}|$ and t, respectively, whose corners are located at the points (7.14). The bracket $\langle\ \rangle$ denotes the ground state expectation value in the absence of the static quark–antiquark source. It is formally given by

$$\langle e^{ie\oint dz^\mu A_\mu(z)}\rangle = \frac{\int DA\, D\psi\, D\bar{\psi}\; e^{ie\oint dz^\mu A_\mu(z)}e^{iS_{QED}}}{\int DA\, D\psi\, D\bar{\psi}\; e^{iS_{QED}}}, \tag{7.21}$$

where S_{QED} is the action defined in eq. (5.5).

Finally let us continue the expression (7.20) to imaginary times, $t\to -iT$. This is accomplished by replacing $-iS_{QED}$ in (7.21) by its euclidean counterpart (5.8), and continuing the exponentials $\exp(-2iM_Qt)$ and $\exp(ie\oint dz^\mu A_\mu(z))$ to imaginary times. By writing out explicitely the contour integral, and recalling that A_0 must be replaced by iA_4 in this continuation process, one finds that

$$\begin{aligned}[G_{\alpha'\beta';\alpha\beta}]_{t\to iT} \underset{M_Q\to\infty}{\longrightarrow}\ & \delta^{(3)}(\vec{x}-\vec{x}\,')\delta^{(3)}(\vec{y}-\vec{y}\,')(P_+)_{\alpha'\alpha}(P_-)_{\beta\beta'} \\ & \cdot e^{-2M_QT}\langle W_C[A]\rangle_{eucl.},\end{aligned} \tag{7.22a}$$

where

$$W_C[A] = e^{ie\oint dz_\mu A_\mu(z)}, \tag{7.22b}$$

and

$$\langle W_C[A]\rangle_{eucl.} = \frac{\int DA\, D\psi\, D\bar{\psi}\; W_C[A]e^{-S^{(eucl.)}_{QED}}}{\int DA\, D\psi\, D\bar{\psi}\; e^{-S^{(eucl.)}_{QED}}}. \tag{7.22c}$$

In (7.22b) the line integral is carried out along a rectangular contour C in euclidean space time, with corners given by $(\vec{x},0), (\vec{y},0), (\vec{y},T)$ and $(\vec{x},T)$. This is the famous Wilson loop.

Finally, to obtain the static quark–antiquark potential we must study the behaviour of (7.22) for large euclidean times T. Comparing this expression with eq. (7.10), we see that the exponential factor $\exp(-2M_Q T)$ just accounts for the fact that the energy of the quark–antiquark system includes the rest mass of the pair. Hence we expect that $\langle W_C[A]\rangle$ behaves as follows for large T

$$W(R,T) \equiv \langle W_C[A]\rangle \underset{T\to\infty}{\longrightarrow} F(R)e^{-E(R)T},$$

where $E(R)$ is the interaction energy of the static quark–antiquark pair separated by a distance R, and $F(R)$ reflects the overlap of our state (7.6) with the ground state of the system in the presence of this pair. Hence we conclude that this energy can be calculated as the following limit:

$$E(R) = -\lim_{T\to\infty} \frac{1}{T} \ln\langle W_C[A]\rangle. \tag{7.23}$$

We want to emphasize the formal simplicity of the above result. To compute the static interquark potential we "merely" need to calculate the expectation value of a gauge invariant quantity built only from the gauge potential. Admittedly the derivation of the result involved a bit of handwaving. Furthermore, we have used a special trial state constructed from the quark fields and the string like operator (7.7), with the line integral taken along a straight line path connecting the quark and antiquark. Especially, in QED, where field lines are allowed to spread all over space, we expect that there are other trial states which have a better overlap with the ground state of the QED–Hamiltonian in the presence of a static source. But also in QCD where the field lines are expected to be squeezed into a tube connecting the two quarks, the use of other trial states can allow one to determine the potential from Wilson loops with a relatively small temporal extension. *

So far we have argued entirely within the continuum formulation where the path integral (7.22c) only has a formal meaning. To define it, we must obtain the lattice version of (7.22b) and (7.22c). This can be easily done. Thus on the lattice the exponential of the line integral in (7.22b) just corresponds to the product of the link variables (5.18) along the rectangular contour C shown in fig. (7-3). Let U_ℓ denote such a link variable. Then we define the Wilson loop operator by**

$$W_C[U] = \prod_{\ell\in C} U_\ell \tag{7.24}$$

* see e.g., Griffith, Michael and Rakow (1983)

** In order not to introduce too many symbols, we use the same symbol W as in the continuum formulation. The argument of W will tell us which formulation we are talking about; notice also that in the abelian case the ordering of the link variables is irrelevant.

Its ground state expectation value

$$W(\hat{R},\hat{T}) \equiv \langle W_C[U] \rangle \tag{7.25a}$$

is given by

$$W(\hat{R},\hat{T}) = \frac{\int DUD\bar{\psi}D\psi W_C[U]e^{-S_{QED}[U,\psi,\bar{\psi}]}}{\int DUD\bar{\psi}D\psi e^{-S_{QED}[U,\psi,\bar{\psi}]}} \tag{7.25b}$$

where in the case of Wilson fermions, S_{QED} is given by (5.22). Notice that $W(\hat{R},\hat{T})$ is a function of the dimensionless ratios $\hat{R} = R/a$ and $\hat{T} = T/a$, with a the lattice spacing.

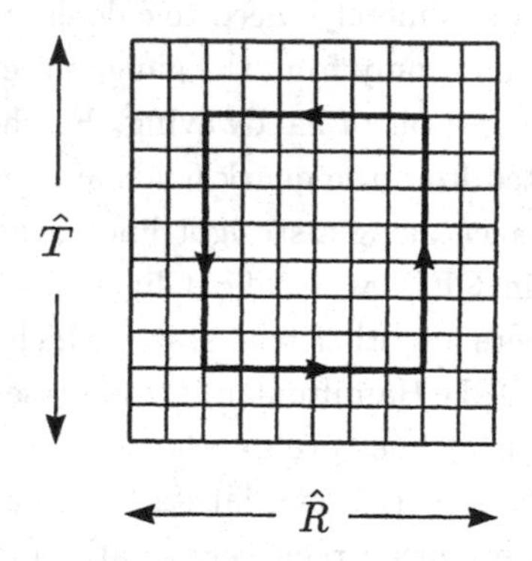

Fig.7-3 Integration contour appearing in eq. (7.22b), relevant for calculating the static interquark potential.

On the basis of the arguments presented in this chapter, we now define the energy of a static $q\bar{q}$-pair measured in lattice units, $\hat{E}(R)$, by an expression analogous to (7.23),

$$\hat{E}(\hat{R}) = -\lim_{\hat{T}\to\infty} \frac{1}{\hat{T}} \ln W(\hat{R},\hat{T}), \tag{7.26}$$

where, as we have pointed out before, $\hat{E}(\hat{R})$ will still contain R-independent self-energy contributions, which have to be substracted when calculating the interquark potential. Relation (7.26) will allow us to compute, at least in principle, the interquark potential using numerical methods.

7.3 The Wilson Loop in QCD

The non–abelian case can be treated in a very similar manner to that discussed in the previous section. The starting point is again a state of the form (7.6), except that now the Dirac fields are replaced by (6.2) with $N = 3$, and we must substitute for $U(\vec{x}, t; \vec{y}, t)$ the operator

$$\underset{\sim}{U}(\vec{x}, t; \vec{y}, t) = P\, e^{ig \int_{\vec{x}}^{\vec{y}} dz^i \underset{\sim}{A}_i(\vec{z}, t)} , \tag{7.27}$$

where $\underset{\sim}{A}_i(\vec{z}, t)$ is the matrix valued field defined in (6.10), and P denotes the path–ordering operation. This path ordering is important to ensure the gauge invariance of the state. Let us see why this is so.

Consider the following generalization of (7.27) to exponentials of line integrals performed along an arbitrary path C connecting two different space–time points x and y:

$$\underset{\sim}{U}(x, y)_C = P\, e^{ig \int_x^y dz^\mu \underset{\sim}{A}_\mu(z)} . \tag{7.28}$$

The path ordering in (7.28) is defined as follows: divide the path C into n infinitesimal segments and let $x_1, x_2, \ldots, x_{n-1}$ denote the intermediate space–time points going from x to y. Furthermore define $dx_\ell = x_\ell - x_{\ell-1}$, with x_0 and x_n identified with x and y, respectively. On each of the infinitesimal segments the exponential in (7.28) can be approximated by the first term in the Taylor expansion. Then (7.28) is given as the limit $\delta x_\ell \to 0$ of the ordered product of these (non–commuting) expressions along the path from x to y:

$$\underset{\sim}{U}(x, y)_C = \lim_{dx_\ell \to 0} [1 + ig\underset{\sim}{A}_\mu(x_0) dx_1^\mu] \ldots [1 + ig\underset{\sim}{A}_\mu(x_{n-1}) dx_n^\mu]. \tag{7.29}$$

Consider now an infinitesimal gauge transformation. According to (6.20), $\underset{\sim}{A}_\mu$ transforms as

$$\underset{\sim}{A}_\mu(x) \to \underset{\sim}{A}_\mu(x) + i[\underset{\sim}{\theta}(x), \underset{\sim}{A}_\mu(x)] - \frac{1}{g}\partial_\mu \underset{\sim}{\theta}(x),$$

where $\underset{\sim}{\theta}(x)$ is an infinitesimal matrix belonging to the Lie–Algebra of $SU(3)$. Up to terms linear in dx_ℓ and $\underset{\sim}{\theta}(x)$ we have that

$$\begin{aligned} 1 + ig\underset{\sim}{A}_\mu(x_{\ell-1}) dx_\ell^\mu \to 1 &+ ig\underset{\sim}{A}_\mu(x_{\ell-1}) dx_\ell^\mu \\ &- g[\underset{\sim}{\theta}(x_{\ell-1}), \underset{\sim}{A}_\mu(x_{\ell-1})] dx_\ell^\mu \\ &- i\left(\partial_\mu \underset{\sim}{\theta}(x_{\ell-1})\right) dx_\ell^\mu . \end{aligned} \tag{7.30}$$

But $\left(\partial_\mu \underset{\sim}{\theta}(x_{\ell-1})\right) dx^\mu_\ell = \underset{\sim}{\theta}(x_\ell) - \underset{\sim}{\theta}(x_{\ell-1})$. Hence up to leading order in θ, (7.30) can also be written in the form

$$1 + igA_\mu(x_{\ell-1})dx^\mu_\ell \to e^{i\theta(x_{\ell-1})}[1 + ig\underset{\sim}{A}_\mu(x_{\ell-1})dx^\mu_\ell]e^{-i\theta(x_\ell)}.$$

We therefore conclude that (7.29) transforms as follows under finite gauge transformations

$$\underset{\sim}{U}(x,y)_C \to \underset{\sim}{G}(x)\underset{\sim}{U}(x,y)\underset{\sim}{G}^{-1}(y), \tag{7.31}$$

where $G(x)$ is an element of the gauge group. The transformation law (7.31) is the analog of (5.11) for the non–abelian case. It guarantees that the state analogous to (7.6) is gauge invariant.

On the lattice, the path ordered exponential (7.28) is just the ordered product of link variables along a path connecting the lattice sites corresponding to x and y. Let us denote these sites by n and m, respectively, and by C_L a path on the lattice connecting n and m. Then the transition from (7.28) to the lattice reads

$$\underset{\sim}{U}(x,y)_C \to \underset{\sim}{U}(n,m)_{C_L} = \prod_{\ell \in C_L} \underset{\sim}{U}_\ell, \qquad \text{(path ordered)}, \tag{7.32}$$

where $\underset{\sim}{U}_\ell$ denotes generically a link variable on C_L. Since under gauge transformations the link variables transform according to (6.5b), it follows that the right–hand side of (7.32) transforms according to

$$\underset{\sim}{U}(n,m)_{C_L} \to \underset{\sim}{G}(n)\underset{\sim}{U}(n,m)_{C_L}\underset{\sim}{G}^{-1}(m).$$

Taking for C_L the Wilson loop, we conclude that

$$W_C[U] = Tr \prod_{\ell \in C_L} \underset{\sim}{U}_\ell \qquad \text{(path ordered)} \tag{7.33}$$

is gauge invariant. This is the analog of (7.24). The corresponding expectation value $\langle W_C[U]\rangle \equiv W(\hat{R},\hat{T})$, is then calculated as before according to a path integral expression analogous to (7.25b):

$$W(\hat{R},\hat{T}) = \frac{\int DUD\bar{\psi}D\psi\, W_C[U]e^{-S_{QCD}[U,\psi,\bar{\psi}]}}{\int DUD\bar{\psi}D\psi e^{-S_{QCD}[U,\psi,\bar{\psi}]}}, \tag{7.34}$$

where for Wilson fermions S_{QCD} is the QCD action given by (6.25), and where DU is the gauge-invariant measure discussed in chapter 6. The interquark potential is then computed according to (7.26).

The right-hand side of (7.34) can be written in a form which involves only an integration over the link variables and hence will be suited for numerical, Monte Carlo, calculations. Indeed, the fermionic contribution to the action, given in (6.25c), is bilinear in the fermion fields and has the form*

$$S_F^{(W)}[U,\psi,\bar{\psi}] = \sum_{n,m} \bar{\psi}_\alpha^a(n) K_{n\alpha a,m\beta b}[U] \psi_\beta^b(m), \tag{7.35}$$

where (a,b) and (α,β) are colour and Dirac-spinor indices, respectively. Hence we can immediately perform the Grassmann integration (see chapter 2) and obtain

$$\langle W_C[U]\rangle = \frac{\int DU\ W_C[U] e^{-S_{eff}[U]}}{\int DU\ e^{-S_{eff}[U]}}, \tag{7.36a}$$

where the effective action, S_{eff}, is given by

$$S_{eff}[U] = S_G[U] - ln\ det K[U]. \tag{7.36b}$$

Here $K[U]$ is the matrix in Dirac, $SU(3)$-colour, and x-space defined in (7.35). In the continuum formulation the matrix elements of K are given by **

$$K_{\alpha x,\beta y}[A] = (\gamma_\mu(\partial_\mu + ig_0 \underset{\sim}{A}_\mu) + M_0)_{\alpha\beta}\ \delta^{(4)}(x-y)\ .$$

The meaning of ln detK[A] is well known in perturbation theory. It is given by the sum of Feynman diagrams consisting of a single fermion loop, with an arbitrary number of external gluon fields attached to it. Hence this term gives rise to the vacuum polarization effects, referred to at the beginning of this chapter. Ignoring these effects amounts to setting $det K = 1$. This is the so-called *quenched approximation.* In this approximation one expects that the static $q\bar{q}$-potential rises with the separation of the quarks. This, as we have pointed out before, is a prerequisite for the hadronization picture mentioned at the beginning of this chapter. Hence calculating the $q\bar{q}$-potential in the quenched approximation is an important first check of confinement.

The above analysis does not yield any information about the spin-dependent forces between quarks since we have merely studied the static limit. To obtain information about the spin-dependent terms in the potential, one must allow the

---

* The same is true for the staggered fermion action (6.26).

** For simplicity we use the same notation for the Wilson loop, fermionic matrix, and effective action as above.

quarks to propagate in space. This means that one has to take into account the spatial part of the covariant derivative in the Dirac equation. This program has been first carried out by Eichten and Feinberg (1981), who treat this term perturbatively. Alternative derivations of the results obtained by these authors have been carried out subsequently by Peskin (1983), Gromes (1984), Barchielli, Montaldi and Prosperi (1988). The reader may consult the recent review article by Gromes (1991) for details on this subject.

Our above discussion has been quite formal. But simplified arguments based on physical intuition often lead to the correct result. So let us verify our conclusions at least within the framework of some simple models. After all, our understanding of quark confinement will depend on our ability of calculating the non-perturbative inter-quark potential, which - at the present state of the art - can only be determined by studying the expectation value of the Wilson loop numerically. Hence let us get some confidence in this procedure by studying some solvable field theories.

CHAPTER 8

THE $Q\bar{Q}$-POTENTIAL IN SOME SIMPLE MODELS

In this chapter we study the potential of a static $q\bar{q}$ pair in two soluble models within the quenched approximation: 1) QED in four space-time dimensions, and 2) compact (lattice) quantum electrodynamics in two dimensions (QED_2). The latter model will also provide us with another opportunity to study the continuum limit and to compare our results with those obtained in a continuum calculation.

8.1 The Potential in Quenched QED

Consider the expectation value of the Wilson loop in QED. In the continuum formulation it is formally given by eq. (7.22c). Performing the fermion integration, we obtain

$$\langle W_C[A] \rangle = \frac{\int DA\, W_C[A] e^{-S_{eff}[A]}}{\int DA\, e^{-S_{eff}[A]}}, \tag{8.1a}$$

where

$$S_{eff}[A] = S_G[A] - \ln\, det\, K[A], \tag{8.1b}$$

$S_G[A]$ is given by (5.8b), and $K[A]$ is a matrix in space-time and Dirac space:

$$K_{x\alpha,y\beta}[A] = [\gamma_\mu(\partial_\mu + ieA_\mu) + M]_{\alpha\beta}\delta^{(4)}(x-y). \tag{8.1c}$$

We now want to calculate the integral (8.1a) in the quenched approximation where vacuum polarization effects, arising from the presence of dynamical fermions, are neglected. This, as we have seen, amounts to setting $det\ K = 1$, and hence replacing S_{eff} by $S_G[A]$. The latter action can be written in the following convenient form:

$$S_G[A] = -\frac{1}{2}\int d^4x A_\mu(x)\Omega_{\mu\nu}A_\nu(x), \tag{8.2a}$$

where

$$\Omega_{\mu\nu} = \delta_{\mu\nu}\Box - \partial_\mu\partial_\nu, \tag{8.2b}$$

and $\Box$ is the four-dimensional Laplacean. Hence in the quenched approximation (q.a.) the expectation value of the Wilson loop is given by

$$\langle W_C[A] \rangle_{q.a.} = \frac{\int DA\, e^{\frac{1}{2}\int d^4x A_\mu \Omega_{\mu\nu} A_\nu + ie\oint dz_\mu A_\mu}}{\int DA\, e^{\frac{1}{2}\int d^4x A_\mu \Omega_{\mu\nu} A_\nu}}. \tag{8.3}$$

Since the integrands in the above expression are exponentials of quadratic forms in the potentials, the integrals can also be carried out in the continuum formulation. Nevertheless, we cannot perform the Gaussian integration in the present form. The reason for this is well known: because of the gauge invariance of the action (8.2) the inverse of the operator $\Omega_{\mu\nu}$ does not exist, since it annihilates all field configurations A_μ which are pure gauge (i.e. of the form $A_\mu = \partial_\mu \Lambda$). This means that the integrands appearing in (8.3) take the same value for all field configurations which only differ from each other by a gauge transformation (notice that the closed line integral is also gauge invariant). These gauge equivalent potentials define an *orbit* for every given field strength $F_{\mu\nu}$, and the integration along any such orbit will give rise to a divergent integral in the numerator *and* denominator of (8.3). The ratio, however, will be finite. To show this, one must have a method for controlling this infinity. An elegant procedure has been given by Faddeev and Popov and amounts to selecting one representative field configuration from each set of gauge-equivalent potentials.* This is done by imposing a gauge condition. Since we are computing the expectation value of a gauge-invariant quantity, the choice of gauge is immaterial. A particularly simple choice is the so-called Feynman gauge and amounts to making the replacement

$$\Omega_{\mu\nu} \longrightarrow \delta_{\mu\nu}\Box \, .$$

The Gaussian integrations in (8.3) may then be performed immediately and one obtains

$$\langle e^{ie \oint dz_\mu A_\mu} \rangle = e^{\frac{e^2}{2} \oint dz_\mu \oint dz'_\nu \delta_{\mu\nu} D(z-z')} , \tag{8.4a}$$

where $D(z - z')$ is the Green's function for the operator $\Box$,

$$\Box D(z - z') = \delta^{(4)}(z - z') \tag{8.4b}$$

i.e.

$$D(x) = -\frac{1}{4\pi^2} \frac{1}{x^2} \quad . \tag{8.4c}$$

Because the integrand in (8.4a) is proportional to $\delta_{\mu\nu}$, the double integral will only receive a contribution when z and z' are located on segments of the integration contour which are parallel to each other. In fig. (8-1) we show various types of

---

* Those readers which are not familiar with the Fadeev-Popov trick may consult the review article by Abers and Lee (1973) or any modern field theory book. In chapter 15 we shall demonstrate this trick for lattice QCD.

diagrams which contribute to the exponential (8.4a). Clearly the leading contribution for large (euclidean) times comes from the diagrams shown in figs. (8-1a,d). Of these, however, the latter one represents the self-energy contribution to the energy of the static $q\bar{q}$ pair. It must therefore be subtracted when computing the inter-"quark" potential. Hence the relevant diagram is that shown in fig. (8-1a). The corresponding integral is easily evaluated and one obtains *

$$\langle e^{ie\oint dz_\mu A_\mu}\rangle_{subtr.} = e^{\frac{e^2}{4\pi R}Tf(R,T)} \underset{T\to\infty}{\longrightarrow} N\, e^{-V(R)T}$$

where

$$f(R,T) = \frac{2}{\pi}[\arctan\frac{T}{R} - \frac{R}{2T}\ln(1+\frac{T^2}{R^2})].$$

Since $f(R,T) \to 1$ for $T \to \infty$, we find that $V(R)$ is just the usual Coulomb potential.

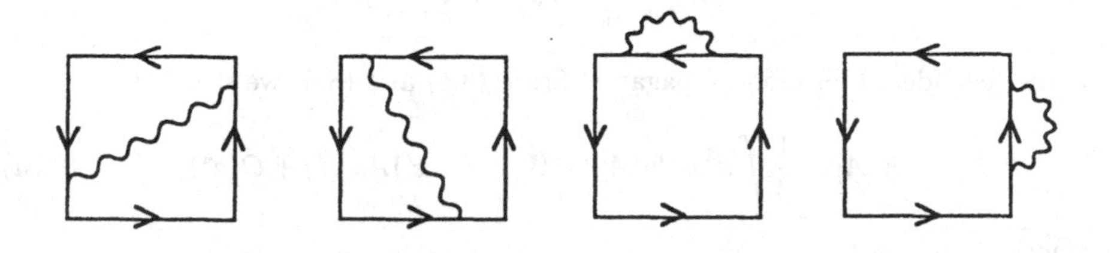

Fig. 8-1 Diagrams contributing to the argument of the exponential in (8.4a)

The potential calculated from (8.3) does not include vacuum polarization effects. When one takes into account dynamical fermions (i.e. fermions of finite mass coupled to the gauge potential), then one must also calculate diagrams containing virtual fermion loops. These loops arise from the contribution of the determinant of the operator $K[A] = \not{\partial} + M + ie\not{A}$ to the effective action (8.1b).** As an example let us calculate the leading order contribution of $ln det K[A]$. To this effect we write $ln det K[A]$ in the form $Tr ln K[A]$, where the trace is taken with respect to the space-time coordinates as well as Dirac spinor indices. Then

$$\begin{aligned} Tr\ ln\ K[A] &= Tr\ ln[(\not{\partial} + M)(1 + \frac{1}{\not{\partial}+M}ie\not{A})] \\ &= c + Tr\ ln\left(1 + \frac{1}{\not{\partial}+M}ie\not{A}\right), \end{aligned} \tag{8.5}$$

* see e.g., Kogut (1979).

** We use the standard notation $\not{b} = \sum_\mu \gamma_\mu b_\mu$.

where $c = Tr\, ln(\not\partial + M)$ is an irrelevant constant, which drops out when calculating the ratio (8.1a). Expanding the logarithm in (8.5) in a formal series, one finds that the leading contribution is of $O(e^2)$:*

$$Tr\ ln\ K[A] = -\frac{1}{2}Tr\left[\frac{1}{\not\partial + M} ie\not{A} \frac{1}{\not\partial + M} ie\not{A}\right] + O(e^4).$$

The right-hand side stands for the following expression

$$\begin{aligned}\frac{1}{2}Tr\left[\frac{1}{\not\partial + M} ie\not{A} \frac{1}{\not\partial + M} ie\not{A}\right] &= \frac{(ie)^2}{2} Tr_D \int d^4x \langle x| \frac{1}{\not\partial + M} \not{A} \frac{1}{\not\partial + M} \not{A} |x\rangle \\ &= \frac{(ie)^2}{2} \int d^4x \int d^4x' Tr_D \left\{ \langle x| \frac{1}{\not\partial + M} |x'\rangle \not{A}(x') \langle x'| \frac{1}{\not\partial + M} |x\rangle \not{A}(x) \right\},\end{aligned} \tag{8.6}$$

where Tr_D denotes the trace in the Dirac indices, and where

$$\langle x| \frac{1}{\not\partial + M} |x'\rangle = S_F(x - x')$$

is the (euclidean) fermion propagator. From (8.5) and (8.6) we see that

$$Tr\ ln\ K[A] = \frac{1}{2} \int d^4x d^4x' A_\mu(x) \Pi_{\mu\nu}(x - x') A_\nu(x') + O(e^4), \tag{8.7a}$$

where

$$\Pi_{\mu\nu}(x - x') = -(ie)^2 Tr_D \{\gamma_\mu (S_F(x - x') \gamma_\nu S_F(x' - x)\} \tag{8.7b}$$

is the vacuum polarization tensor in one loop order. Substituting (8.7a) and (8.2a) into (8.1b) we are led to the following expression for (8.1a) in the Feynman gauge,

$$\langle W_C[A] \rangle = \frac{\int DA e^{ie \oint dz_\mu A_\mu(z)} e^{\frac{1}{2} \int d^4x d^e x' A_\mu(x) \tilde{\Omega}_{\mu\nu}(x - x') A_\nu(x')}}{\int DA e^{\int d^4x d^4x' A_\mu(x) \tilde{\Omega}_{\mu\nu}(x - x') A_\nu(x')}}, \tag{8.8a}$$

where

$$\tilde{\Omega}_{\mu\nu}(z) = \Omega_{\mu\nu}(z) + \Pi_{\mu\nu}(z), \tag{8.8b}$$

$$\Omega_{\mu\nu}(z) = \delta_{\mu\nu} \Box \delta^{(4)}(z). \tag{8.8c}$$

We now perform the Gaussian integrals in (8.8a) and obtain

$$\langle W_c[A] \rangle = e^{\frac{e^2}{2} \oint dz_\mu \oint dz'_\nu \tilde{\Omega}^{-1}_{\mu\nu}(z - z')}. \tag{8.9}$$

---

* Furry's theorem tells us that there are no contributions coming from an odd number of external photon lines.

Here $\tilde{\Omega}^{-1}_{\mu\nu}$ is the inverse of the matrix (8.8b). It is defined by

$$\int d^4z''\tilde{\Omega}_{\mu\lambda}(z-z'')\tilde{\Omega}^{-1}_{\lambda\nu}(z''-z') = \delta_{\mu\nu}\delta^{(4)}(z-z'),$$

and can be easily computed up to $O(e^2)$.

$$\tilde{\Omega}^{-1}_{\mu\nu}(x-y) = \Omega^{-1}_{\mu\nu}(x-y) - \int d^4zd^4z'\Omega^{-1}_{\mu\lambda}(x-z)\Pi_{\lambda\lambda'}(z-z')\Omega^{-1}_{\lambda'\nu}(z'-y) + O(e^4).$$

The two point correlation function of the gauge potential in $O(e^o)$ and $O(e^2)$ is given by $-\Omega^{-1}_{\mu\nu}$ and $-\tilde{\Omega}^{-1}_{\mu\nu}$, respectively. Hence in terms of Feynman diagrams the right hand side of the above expression is given by:

x,μ y,ν $+$ x,μ y,ν $+O(e^4)$.

Thus a typical diagram contributing to (8.9) is that depicted in fig. (8-2).

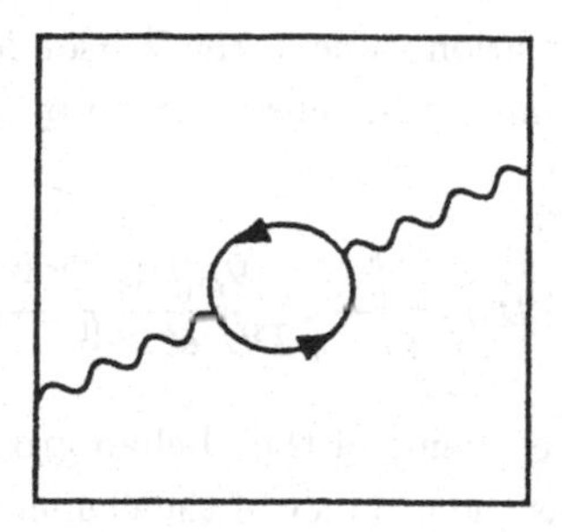

Fig. 8-2 Vacuum polarization graph arising from the fermionic determinant contributing to (8.9).

By carrying out the expansion of (8.5) to higher orders, we arrive at a sum of one-fermion loop contributions with an arbitrary number of external photon lines attached to it.

Unfortunately we are only able to compute the effects arising from dynamical quarks analytically within the framework of perturbation theory. For a non-perturbative treatment we are forced to recur to numerical methods. This, as we shall see later on, turns out to be a quite non-trivial task

8.2 The Potential in Quenched Compact QED_2

We now perform a similar calculation of the potential between two opposite charges but starting from a lattice formulation. For this purpose we consider the case of compact QED in 2 space-time dimensions, which, if we neglect dynamical fermions, may be solved in closed form. The lattice action in the pure gauge sector is given by *

$$S_G = \beta \sum_P [1 - \frac{1}{2}(U_P + U_P^\dagger)], \tag{8.10}$$

where β is some parameter which, in analogy to (5.20), we shall relate to the dimensionless bare coupling $\hat{e}$ by

$$\beta = \frac{1}{\hat{e}^2} \quad , \tag{8.11}$$

and where U_P is given by the product of the link variables

$$U_\mu(n) = e^{i\theta_\mu(n)}$$

taken around an elementary plaquette "P" as discussed in chapter 5.

Next consider the expectation value of the Wilson loop (7.25b), with the contour C having spatial and temporal extension given by $\hat{R}$ and $\hat{T}$. In the quenched approximation it is given by

$$\langle W_C[U] \rangle = \frac{\int DU \; W_C[U] e^{-S_G[U]}}{\int DU \; e^{-S_G[U]}}. \tag{8.12}$$

Since the link variables are elements of the abelian group $U(1)$, it is evident that $W_C[U]$ can also be written as the product of the elementary Wilson loops ($\rightarrow$ plaquette variables) contained within the region R_C, bounded by the square contour C, as shown in fig. (8-3).

$$W_C[U] = \prod_{P \in R_C} U_P. \tag{8.13}$$

Hence**

$$\langle W_C[U] \rangle = \frac{\int DU (\prod_{P \in R_C} U_P) e^{\frac{\beta}{2} \sum_P (U_P + U_P^\dagger)}}{\int DU \; e^{\frac{\beta}{2} \sum_P (U_P + U_P^\dagger)}}. \tag{8.14}$$

---

* Its structure is the same as that discussed in chapter 5 (see eq. (5.21)).

** We have dropped the constant term in the action (8.10) since it cancels in the numerator and denominator of (8.12).

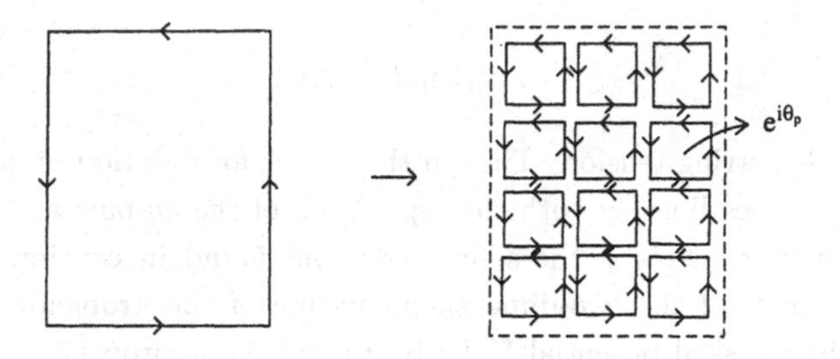

Fig. 8-3 Writing the Wilson loop (8.13) as a product of elementary plaquette contributions. The dashed line stands for the original product of link variables along the contour C.

To carry out this integral it is convenient to choose a gauge where all link variables pointing along the time direction are rotated to the unit element. This can always be achieved by performing an appropriate gauge transformation under which the link variables transform according to (5.16). Hence the contribution of a particular plaquette with origin at $n = (n_1, n_2)$ will be of the form $U_P = exp(i\theta_P)$, where θ_P is given by the difference of the phase-angles associated with two neighbouring links lying on consecutive time slices:

$$\theta_P = \theta_1(n_1, n_2) - \theta_1(n_1, n_2 + 1).$$

Making use of the periodic structure of the integrands in (8.14), one finds that

$$W(\hat{R}, \hat{T}) = \prod_{P \in R_C} \frac{\int_{-\pi}^{\pi} d\theta_P \; e^{i\theta_P} e^{\beta \cos\theta_P}}{\int_{-\pi}^{\pi} d\theta_P \; e^{\beta \cos\theta_P}}, \tag{8.15}$$

where $W(\hat{R}, \hat{T}) = \langle W_C[U] \rangle$. Performing the integral (8.15) one obtains

$$W(\hat{R}, \hat{T}) = \left(\frac{I_1(\beta)}{I_0(\beta)} \right)^{\hat{R}\hat{T}}, \tag{8.16}$$

where $I_n(\beta)$ are the modified Bessel functions of integer order. From (8.16) we read off the $q\bar{q}$ potential in units of the lattice spacing,

$$\hat{V}(\hat{R}) = - \lim_{\hat{T} \to \infty} \frac{1}{\hat{T}} \ln W(\hat{R}, \hat{T}) = \hat{\sigma} \hat{R}, \tag{8.17a}$$

where

$$\hat{\sigma} = \ln\left(\frac{I_0(\beta)}{I_1(\beta)}\right) \tag{8.17b}$$

is the so-called string tension. Thus in the lattice formulation of quenched QED_2 the potential rises linearly with the separation of the $q\bar{q}$-pair and hence confines the charged pair. This is the same behaviour found in continuum QED_2 and is a consequence of the two-dimensional nature of the problem. In fact, let us compute the physical potential $V(R)$ by taking the appropriate continuum limit of the lattice version (8.17). Since continuum QED_2 is a superrenormalizable theory we expect that a simple rescaling of the variables with the lattice spacing a will suffice. This rescaling, however, requires some care. Thus we must clarify first of all which quantities must be kept fixed as we let the lattice spacing go to zero. Since the physical potential has the dimension of inverse length, we must scale $\hat{V}$ with the inverse lattice spacing. Furthermore, $\hat{R}$ is to be replaced by R/a; we therefore consider the expression

$$V(R;\beta,a) \equiv \frac{1}{a^2}\hat{\sigma}(\beta)R. \tag{8.18}$$

From (8.18) we see that if we keep β fixed as $a \to 0$, then V diverges like $1/a^2$! Therefore, this cannot be the correct continuum limit. So let us take a closer look at the meaning of the bare coupling $\hat{e}$ defined in (8.11) by studying the naive continuum limit of (8.10) in the manner described in chapter 5. Making the replacement $U_P \to \exp\,(i\hat{e}a^2F_{\mu\nu})$ for a plaquette "P" lying in the $\mu\nu$-plane (see eq. (5.20)) and expanding the exponential in powers of the lattice spacing squared, one finds that the coupling constant in physical units, e, is related to $\hat{e}$ by

$$e = \frac{1}{a}\hat{e}\,. \tag{8.19}$$

This we could of course have guessed immediately, since the coupling constant in QED_2 carrys the dimension of a mass.

From the above discussion it is evident that (8.11) is a function of the lattice spacing, and that the physical potential should be calculated as the following limit,*

$$V(R,e) = \lim_{a\to 0} V(R,\beta(a),a) \tag{8.20a}$$

---

* We have assumed that (8.19) is a physical coupling constant, which is to be held fixed when performing the continuum limit. The fact that this limit turns out to be finite, justifies a posteriori this assumption, and agrees with what is known from the continuum theory.

where

$$\beta(a) = \frac{1}{e^2 a^2}. \tag{8.20b}$$

Hence β diverges in the continuum limit! This makes it plausible why the lattice formulation (8.12) reproduces the correct continuum limit, as we shall see below. Thus it is evident that for large β the Boltzmann factor appearing in the integrand of (8.12) will ensure that the integral is dominated by those link configurations for which $U_P \approx 1$. Because of (5.20), this implies that the fluctuations in $eF_{\mu\nu}$ are small compared to the inverse lattice spacing squared.

With these remarks let us now compute the continuum limit (8.20a). Inserting in (8.17b) the following asymptotic expansions for $I_1(\beta)$ and $I_0(\beta)$, valid for large β,

$$I_0(\beta) = \frac{e^\beta}{\sqrt{2\pi\beta}}(1 + \frac{1}{8\beta} + \cdots),$$
$$I_1(\beta) = \frac{e^\beta}{\sqrt{2\pi\beta}}(1 - \frac{3}{8\beta} + \cdots),$$

one finds that

$$V(R) = \frac{1}{2}e^2 R.$$

This is the classical energy of a pair of opposite charges separated by a distance R, for electrodynamics in one space dimension.

In the special case considered here, the energy of a $q\bar{q}$-pair is a linear function of their separation for any coupling. In particular, in the strong coupling limit ($\beta \to 0$), the string tension (8.17b) is given by $-ln\,(\beta/2)$. In a four-dimensional gauge theory (without dynamical fermions), the confining nature of the potential obtained in the strong coupling limit is a consequence of the fact that the flux lines connecting the quark and antiquark are squeezed into narrow tubes (strings) along the shortest path joining the $q\bar{q}$-pair (see chapter 11). This string is not allowed to fluctuate for $\hat{g}_0 \to \infty$. Fluctuations may, however, destroy confinement, when one studies the continuum limit. In the above two-dimensional example, the persistence of confinement in the continuum limit ($\hat{e}_0 \to 0$), is not surprising since in one-space dimension there is no way the string can fluctuate. In QCD, however, there is no a priori reason why confinement could not be lost in the continuum limit (which, as we shall see, is also realized at vanishing bare coupling). Should it persist in this limit, it must be a consequence of the non-trivial dynamics.

This completes our demonstration of how the potential of a static $q\bar{q}$ pair may be extracted by studying the expectation value of the Wilson loop for large euclidean times. In the simple examples considered, the calculation could be done exactly. In general, however, we must rely on numerical methods and the starting point will be the lattice version. In this respect the second case treated above exhibited already some interesting features which we shall meet again when studying the continuum limit of QCD. Thus we have seen that taking the continuum limit required β to be a function of the lattice spacing. This functional dependence was very simple in the case considered here, and we could actually determine it from dimensional arguments alone. In the case of QCD, on the other hand, this dependence will not be trivial, and will be determined from the short distance dynamics of QCD.

CHAPTER 9

THE CONTINUUM LIMIT OF LATTICE QCD

9.1 Critical Behaviour of Lattice QCD and the Continuum Limit

In chapter 6 we constructed a lattice gauge theory based on the non-abelian group $SU(3)$ and have given arguments which suggest that in the continuum limit it describes QCD. These arguments were based on the observation that the lattice action (6.25) reduces to the correct expression in the naive continuum limit. But as we have emphasized before, there exist an infinite number of lattice actions which have the same naive continuum limit. We have merely chosen the simplest one, proposed originally by Wilson.* There is, however, no a priori reason why any choice of lattice action satisfying the above mentioned requirement will ensure that the theory processes a continuum limit corresponding to QCD or some other field theory. For this to be the case the lattice theory must exhibit first of all a critical region in parameter space where correlation lengths diverge. To see this, let us consider the case of a pure $SU(3)$ gauge theory, which in the lattice formulation resembles a statistical mechanical system described by the partition function **

$$Z = \int DU \, e^{\frac{1}{g_0^2} Tr \sum_P (U_P + U_P^\dagger)} . \tag{9.1}$$

Suppose that this lattice theory possesses a continuum limit, and that we wanted to extract from it the mass spectrum of the corresponding field theory by studying the appropriate correlation functions for large euclidean times (see chapter 16 for more details). The largest correlation length is then determined by the lowest mass in the problem. If the corresponding physical mass, m, is to be finite, then the mass measured in lattice units, $\hat{m}$, must necessarily vanish in the continuum limit. This in turn implies that the correlation length measured in lattice units, $\hat{\xi}$, must diverge. Hence the continuum field theory can only be realized at a critical point of the statistical mechanical system described by the partition function (9.1).

---

* One can make use of the ambiguity in the action to construct so-called "improved actions", which lead to a suppression of lattice artefacts contributing to observables for finite lattice spacing. This allows one to extract continuum physics already for larger lattice spacings (Symanzik, 1982 and 1983)

** We have dropped the constant term in the action (6.25b) since it is irrelevant when calculating expectation values.

This, of course, is to be expected, since only if the correlation lengths diverge does the system loose its memory of the underlying lattice structure. It follows that if the above system is not critical for any value of the coupling, it cannot possibly describe QCD or any other continuum field theory.

Now studying a system near criticality means tuning the parameters accordingly. In the case considered above, the only parameter is the bare coupling g_0, a dimensionless quantity which is void of any direct physical meaning. The correlation length $\hat{\xi}$ measured in lattice units will depend on this parameter. Hence the continuum limit will be realized for $g_0 \to g_0^\star$, where correlation lengths diverge:

$$\hat{\xi}(g_0) \underset{g_0 \to g_0^\star}{\longrightarrow} \infty. \tag{9.2}$$

We want to emphasize that (9.2) followed from the general requirement that physical quantities should be finite in the limit of zero lattice spacing a. To arrive at the above conclusion we have implicitly introduced a scale from the outside, in terms of which dimensioned observables can be measured.* This scale must clearly be correlated with g_0. The relationship between the two may in principle be determined in the following way. Consider an observable Θ, such as the correlation length or the string tension $\hat{\sigma}$ defined in (8.17a), with mass dimension d_Θ. Let $\hat{\Theta}$ denote the corresponding lattice quantity which may in principle be determined numerically. $\hat{\Theta}$ will depend on the bare parameters of the theory (coupling, masses etc.) which in the simple case considered here is just the dimensionless coupling g_0. The existence of a continuum limit then implies that

$$\Theta(g_0, a) = \left(\frac{1}{a}\right)^{d_\Theta} \hat{\Theta}(g_0) \tag{9.3}$$

approaches a finite limit for $a \to 0$, *if* g_0 is tuned with a in an appropriate way, with $g_0(a)$ approaching the critical coupling $g_0^\star$ defined in eq. (9.2):

$$\Theta(g_0(a), a) \underset{a \to 0}{\longrightarrow} \Theta_{phys.} \quad . \tag{9.4}$$

Hence if the functional dependence of $\hat{\Theta}$ on g_0 is known, we can determine $g_0(a)$ from (9.3) for sufficiently small lattice spacing by fixing the left-hand side at its physical value $\Theta_{phys.}$. This determines g_0 as a function of $a(\Theta_{phys})^{1/d_\Theta}$. In the case of the free scalar field and QED in two dimensions, dimensional arguments

* There exists a priori no such scale in a pure lattice formulation!

alone determined the a-dependence of the bare parameters. In the present case, however, we are faced with a quite nontrivial theory, and the answer is not so simple.

The above discussion did not make use of any particular observable. From (9.3) and (9.4) it may appear, however, that the functional dependence of $g_0(a)$ will depend on the observable considered. For finite lattice spacing this will in fact be true. For sufficiently small a, however, a universal function $g_0(a)$ should exist, which ensures the finiteness of any observable. A corresponding statement is expected to hold if the action depends on several parameters, (e.g. bare coupling constant and quark masses).

We want to emphasize that it is not surprising that the bare parameters will depend in general on the lattice spacing: by making the lattice finer and finer (see fig.(9-1)), the number of lattice sites and links within a given physical volume increases. Hence if physics is to remain the same, the bare parameters must be tuned to a in a way depending in general on the dynamics of the theory.

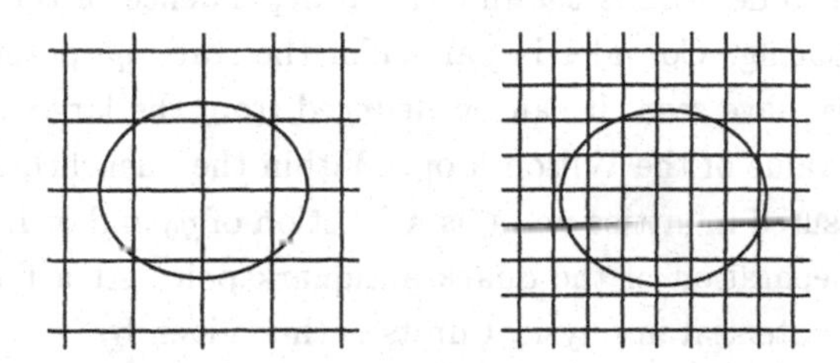

Fig. 9-1 Making the lattice finer by tuning the coupling with the lattice spacing so as to keep physics the same.

Suppose now that the lattice theory describes some field theory in the continuum limit. How do we know whether it is QCD? And how do we know that we are extracting continuum physics in a numerical calculation where we shall always be forced to work on a modest-sized lattice, and hence also at finite lattice spacing? Clearly a first requirement in any numerical calculations should be that the scales which are relevant to the particular problem under investigation are large compared to the lattice spacing, but small compared to the extension of the lattice. Thus on the one hand correlation lengths measured in lattice units should be large; but on the other hand, they are not allowed to exceed the lattice, whose size is limited by the available computer facilities! Hence we must have some clear signal which tells us whether we are extracting continuum QCD, or merely performing an academic exercise.

In the following we shall show that in the case of QCD one can actually determine the functional dependence of g_0 on the lattice spacing for sufficiently small a. We shall restrict our discussion to the case discussed above, where the effects of dynamical fermions are ignored and the action only depends on the bare coupling g_0. Having established the relation between the lattice spacing and g_0, the dependence of any lattice observable on the bare coupling near criticality will be known and can be used as signal for testing the continuum in a numerical calculation performed on a lattice of finite extent.

9.2 Dependence of the Coupling Constant on the Lattice Spacing and the Renormalization Group β-Function

As we have pointed out above, we expect that close to the continuum limit a single function $g_0(a)$ will ensure the finiteness of any observable. Hence we can use any observable to determine the functional dependence of the bare coupling g_0 on the lattice spacing. Consider in particular the static $q\bar{q}$ potential discussed in chapter 7. As we have seen, it can be deduced from the large time behaviour of the expectation value of the Wilson loop. Within the quenched approximation this potential, measured in lattice units, is a function of g_0 and of $\hat{R} = R/a$, where R is the physical separation of the quark-antiquark pair. At a finite, but small lattice spacing the potential in physical units is then given by

$$V(R, g_0, a) = \frac{1}{a}\hat{V}\left(\frac{R}{a}, g_0\right), \tag{9.5}$$

where g_0 must be tuned to a in such a way that for sufficiently small lattice spacing (9.5) becomes independent of a. Hence $V(R, g_0, a)$ must satisfy the so-called renormalization group (RG) equation

$$\left[a\frac{\partial}{\partial a} - \beta(g_0)\frac{\partial}{\partial g_0}\right] V(R, g_0, a) = 0 \tag{9.6a}$$

where

$$\beta(g_0) = -a\frac{\partial g_0}{\partial a} \tag{9.6b}$$

is the Callan-Symanzik β-function (Callan 1970; Symanzik 1970).* Thus if $\beta(g_0)$ would be known, we could integrate (9.6b) to obtain $g_0(a)$. Of course, we cannot calculate $\beta(g_0)$ exactly, but we may determine it in perturbation theory, where (9.6a) must also hold in every order. In the continuum formulation this can e.g. be accomplished by expanding the following expression for the potential

$$V(R, g_0, a) = -\lim_{T\to\infty} \frac{1}{T} \ln\langle P e^{ig_0 \oint dz_\mu \underset{\sim}{A}_\mu(z)} \rangle \tag{9.7}$$

in powers of the coupling constant g_0, and inserting the expression into the RG equation (9.6).

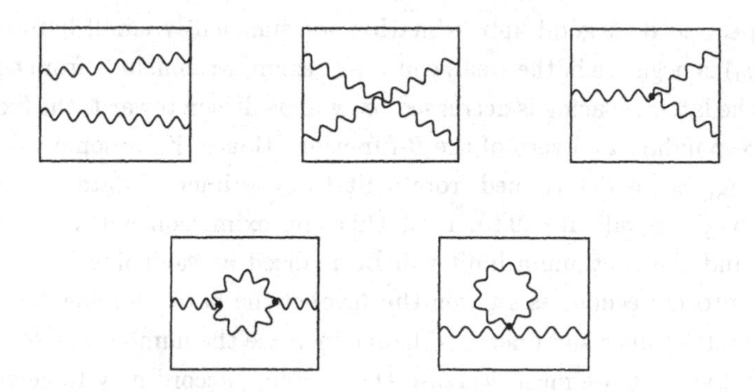

Fig. 9-2 Classes of diagrams contributing to the potential in order g^4. The lines connect to arbitrary points on the Wilson contour

Because of the non-abelian nature of the gauge potential and the path ordering prescription, the calculation is much more involved than in the abelian case. In fig.(9-2) we show the diagrams contributing to (9.7) in order g_0^4. On the lattice the $SU(N)$ potential has been computed by Kovacs (1982), and by Heller and Karsch (1985) up to $\mathcal{O}(g_0^4)$. Because of the complicated structure of the Feynman rules (see chapter 15) these computations are quite involved. Up to the above order the potential is given by **

$$V(R) \approx -\frac{g_0^2(a)}{4\pi R} C_2(F)\Big[1 + g_0^2(a)\frac{11N}{24\pi^2}\ln(7.501\frac{R}{a}) + \frac{1}{4}g_0^2(a)C_2(F)\Big] \tag{9.8}$$

* Suppose we determine $g_0(a)$ by holding a hadron mass M fixed at its physical value; then $M = \frac{1}{a}\hat{M}(g_0(a))$. Hence g_0 is a function of Ma, and $adg_0/da = f(Ma) = -\beta(g_0(a))$, where $\beta(g_0)$ is the lattice version of the Callan-Symanzik β-function.

** This expression differs from an earlier approximate calculation (Susskind 1976) where $C_2(F)$ is the quadratic Casimir operator in the fundamental representation. For $SU(N)$ it is given by $C_2(F) = \frac{N^2-1}{2N}$. Next, we demonstrate how one may use the perturbative expression (9.8) to determine the non-perturbative relation between g_0 and the lattice spacing a, which ensures that the full potential V becomes independent of the lattice spacing for sufficiently small a. To this effect we first determine the β-function to lowest order in g_0 by inserting (9.8) into the RG equation (9.6). One readily finds that for SU(N)

$$\beta(g_0) \approx -\frac{11N}{48\pi^2} g_0^3. \tag{9.9}$$

This we expect to be a good approximation for sufficiently small bare coupling. Because $\beta(g_0)$ is negative in the small coupling region, we conclude from eq. (9.6b) that, when the lattice spacing is decreased, g_0 will be driven towards the fixed point $g_0^\star = 0$, corresponding to a zero of the β-function. Hence, if for some value of the lattice spacing, g_0 (as determined from a fit to experimental data) turns out to be small enough to validate (9.9), then this approximation will improve as we decrease a, and the continuum limit will be realized at vanishing bare coupling! This is asymptotic freedom as seen on the level of the bare coupling constant: as we make the lattice finer and finer, and hence increase the number of sites within a given physical volume, we must decrease the coupling accordingly to keep physics the same.

Integrating eq. (9.6b) one now obtains a relation between g_0 and a:*

$$a = \frac{1}{\Lambda_L} e^{-\frac{1}{2\beta_0 g_0^2}}, \tag{9.10a}$$

where

$$\beta_0 = 11N/48\pi^2, \tag{9.10b}$$

and Λ_L is an integration constant with the dimension of a mass.

The above derivation of the β-function in leading order was based on the perturbative expression for the potential (9.8) and the renormalization group

---

* We remind the reader that we have only considered the leading term in the β-function. This term as well as that of order g_0^5 determines the behaviour of the theory near the *RG*-fixed point. Their structure is independent of the observable that one uses to compute it; in the usual continuum language: The first two coefficients of the perturbative β-function are independent of the renormalization scheme.

equation. An alternative procedure is to relate the bare coupling constant to the renormalized coupling constant in perturbation theory. By holding the latter fixed, while varying the lattice cutoff, one then obtains a perturbative expression for $a\partial g_0/\partial a$, and hence the β-function. We feel, however, that the above approach (Kogut, 1983) is more transparent, since the potential has a direct physical meaning.

Having obtained the β-function, let us now use the RG-equation (9.6) to obtain an improved expression for the potential. This will not only be very instructive to the reader, but will also serve to illustrate the basic ideas discussed in the previous section.

Consider the potential as given by the right-hand side of (9.5). Inserting this expression* into (9.6), one readily arrives at the following alternative RG-equation in which the derivative $\partial/\partial a$ has been traded in favour of the physical separation of the $q\bar{q}$-pair:

$$\left[R\frac{\partial}{\partial R} + \beta(g_0)\frac{\partial}{\partial g_0}\right] V(R, g_0, a) = -V(R, g_0, a).$$

If we define the dimensionless quantity

$$\tilde{V}(R, g_0, a) = RV(R, g_0, a), \tag{9.11}$$

then $\tilde{V}$ satisfies the following differential equation:

$$\left[R\frac{\partial}{\partial R} + \beta(g_0)\frac{\partial}{\partial g_0}\right] \tilde{V}(R, g_0, a) = 0. \tag{9.12}$$

This is an interesting equation, for it tells us how an infinitesimal change in R can be compensated by a corresponding change in the bare coupling constant, keeping the lattice spacing fixed. In other words, any change in R can be absorbed into a (R-dependent) redefinition of the coupling strength. Let us be more explicit. Suppose we know $V(R, g_0, a)$ for some given separation R_0 of the quark-antiquark pair. Question: can we determine the potential for separations $R = \lambda R_0$?. Consider $\tilde{V}(\lambda R_0, g_0, a)$, where $\tilde{V}$ has been defined in (9.11). Then (9.12) leads to the following equation involving dimensionless variables only:

$$\left[\lambda\frac{\partial}{\partial \lambda} + \beta(g_0)\frac{\partial}{\partial g_0}\right] \tilde{V}(\lambda R_0, g_0, a) = 0. \tag{9.13}$$

---

* Our presentation parallels closely that of Kogut (1983).

One now easily verifies that the solution to (9.13) is given by

$$\tilde{V}(\lambda R_0, g_0, a) = \tilde{V}(R_0, \bar{g}_0(\lambda), a), \tag{9.14}$$

where the "running" coupling constant $\bar{g}_0(\lambda)$ satisfies an equation analogous to (9.6b):

$$\lambda \frac{\partial \bar{g}_0}{\partial \lambda} = -\beta(\bar{g}_0(\lambda)), \tag{9.15}$$

with

$$\bar{g}_0(1) = g_0.$$

Inserting for $\beta(g_0)$ the expression (9.9), one finds upon integrating (9.15) that

$$\lambda = e^{-\frac{1}{2\beta_0}\left[\frac{1}{\bar{g}_0^2(\lambda)} - \frac{1}{g_0^2}\right]}, \tag{9.16}$$

where β_0 is given by (9.10b). Solving (9.16) for $g_0^2(\lambda)$, we obtain

$$\bar{g}_0^2(\lambda) = \frac{g_0^2}{1 - 2\beta_0 g_0^2 \ln \lambda}. \tag{9.17}$$

Hence what concerns the dimensionless quantity (9.14), scaling R_0 with a factor λ is equivalent to replacing the bare coupling constant g_0 by (9.17). The corresponding statement for the interquark potential now follows immediately from (9.11):

$$\begin{aligned} V(\lambda R_0, g_0, a) &= \frac{1}{\lambda R_0} \tilde{V}(\lambda R_0, g_0, a) \\ &= \frac{1}{\lambda} V(R_0, \bar{g}_0(\lambda), a). \end{aligned} \tag{9.18}$$

Let us now use this relation to obtain a renormalization-group improved expression for the potential, replacing the perturbative expression (9.8). This expression suggests that we should normalize the potential as follows:

$$V(a, g_0, a) = C \frac{g_0^2}{4\pi a}.$$

It then follows from (9.18) that $V(\lambda a, g_0, a) = C\bar{g}_0^2(\lambda)/(4\pi\lambda a)$. By choosing $\lambda = R/a$, we therefore find that

$$V(R, g_0, a) = C \frac{\bar{g}_0^2(R/a)}{4\pi R},$$

which, upon substituting for $\bar{g}_0^2(R/a)$ the expression (9.17), becomes

$$V(R, g_0, a) = \frac{C}{4\pi R}\left[\frac{g_0^2}{1 - 2\beta_0 g_0^2 \ln\left(\frac{R}{a}\right)}\right], \tag{9.19a}$$

or

$$V(R, g_0, a) = -\frac{C}{4\pi R}\frac{1}{2\beta_0 \ln\left(\frac{R}{a}e^{-\frac{1}{2\beta_0 g_0^2}}\right)} \tag{9.19b}$$

Expanding the denominator in (9.19a) to leading order in g_0^2, we recover eq. (9.8). Furthermore, requiring $V(R, g_0, a)$ to be independent of a, we arrive at the non-perturbative relation (9.10), as follows immediately from eq. (9.19b). Hence Λ_L is related to the strength of the potential V by

$$V(R) = -\frac{C}{4\pi R}\frac{1}{2\beta_0 \ln(R\Lambda_L)}$$

In contrast to the lattice spacing, Λ_L is a physical scale in terms of which dimensioned quantities can be measured. Thus by construction the quantity

$$\Lambda_L = \frac{1}{a}\, e^{-\frac{1}{2\beta_0 g_0^2}} \tag{9.20}$$

satisfies an equation analogous to (9.6) and hence is a renormalization group invariant quantity. Solving (9.20) for g_0 we see that the bare coupling vanishes like $1/\ln(a\Lambda_L)$ as $a \to 0$:

$$g_0^2(a) = -\frac{1}{2\beta_0 \ln(a\Lambda_L)}.$$

For the sake of completeness we also give here the relation between g_0 and the lattice spacing, derived from the first two (universal) coefficients in the power series expansion of the β-function for N_F flavours of massless quarks: *

$$\beta(g_0) = -\beta_0\, g_0^3 - \beta_1 g_0^5, \tag{9.21a}$$

$$\beta_0 = \frac{1}{16\pi^2}(11 - \frac{2}{3}N_F),\ \beta_1 = \frac{1}{(16\pi^2)^2}(102 - \frac{38}{3}N_F) \tag{9.21b}$$

$$a = \frac{1}{\Lambda_L}R(g_0), \tag{9.21c}$$

* In one loop order the β -function was computed by Gross and Wilczek (1973) and by Politzer (1973); the computation to second order has been carried out by Caswell (1974), Jones (1974), and Belavin and Migdal (1974)

$$R(g_0) = (\beta_0 g_0^2)^{-\beta_1/2\beta_0^2}\, e^{-\frac{1}{2\beta_0 g_0^2}}. \tag{9.21d}$$

Let us pause here for a moment. Our renormalization group arguments have shown that the potential is of the form

$$V(R) = -C\frac{\alpha(R)}{R}$$

where $\alpha(R)$ increases with increasing separation of the quark and antiquark:

$$\alpha(R) = \frac{g_0^2}{1 - 2\beta_0 g_0^2 \ln\left(\frac{R}{a}\right)} \tag{9.22}$$

Clearly this result can only be meaningful if R is larger than the lattice spacing, but much smaller than the inverse lattice cutoff. For $R = \Lambda_L^{-1}$ the effective coupling strength diverges! This behaviour is quite different from that encountered in QED. Thus when a charge is inserted into the vacuum of QED it will polarize the medium in such a way that the effective charge measured at a distance R is less than the original charge. In QCD, on the other hand, the opposite phenomenon takes place. The non-abelian couplings of the gauge potentials lead to antiscreening. This suggests that for large separations of the quark and antiquark the interaction may become strong. Unfortunately, we have no way at present to calculate the $q\bar{q}$-potential for large separations analytically, and we must take recourse to numerical methods.

But how do we know whether we are extracting continuum physics when performing calculations on finite (rather small!) lattices?. The answer to this question is found in the relation (9.21c,d). This relation tells us how the bare coupling constant controls the lattice spacing. Indeed, inserting (9.21c) into (9.3), the requirement (9.4) implies that for $g_0 \approx g_0^\star = 0$, $\hat{\Theta}(g_0)$ must behave as follows:

$$\hat{\Theta}(g_0) \underset{g_0 \to 0}{\approx} \hat{C}_\Theta (R(g_0))^{d_\Theta} \tag{9.23}$$

where $\hat{C}_\Theta$ is a dimensionless constant. Quantities behaving like (9.23) are said to show "asymptotic scaling". By studying the ratio $\hat{\Theta}(g_0)/(R(g_0))^{d_\Theta}$ as a function of g_0 in the scaling region one then determines the constant $\hat{C}_\Theta$.

In an actual numerical calculation on a lattice of finite size there will exist in general only a narrow region in coupling constant space where $\hat{\Theta}(g_0)$ scales according to (9.23). This region is called the "scaling window". Thus since the lattice spacing is controlled by the bare coupling according to (9.21c,d), physics will no longer fit on the lattice if g_0 (and hence a) becomes too small ($\to$ finite size effects). On the other hand, by increasing the bare coupling, the lattice may become too coarse to account for fluctuations taking place on a small scale, and we are leaving the continuum region. This is the reason for the narrow window.

Returning now to (9.23), we find upon inserting this expression together with (9.21c) into (9.3), that the observable defined in (9.4) can be expressed in terms of the lattice scale Λ_L by

$$\Theta_{phys.} = \hat{C}_\Theta (\Lambda_L)^{d_\Theta} \quad . \tag{9.24}$$

This shows that physical quantities can be calculated in units of the undetermined mass scale Λ_L. Hence a lattice calculation can only determine dimensionless ratios of physical quantities (e.g. ratios of particle masses).

A particularly interesting example is the string tension σ which is the coefficient of the linearly rising part of the interquark potential. Measured in lattice units it is only a function of the bare coupling: $\hat{\sigma}(g_0)$. In physical units, however, it has the dimension of (mass)^2, so that the physical string tension is given by

$$\sigma = \lim_{a\to 0} \frac{1}{a^2} \hat{\sigma}(g_0(a)) \quad .$$

From the above discussion we hence conclude that for $g_0 \to 0$, $\hat{\sigma}(g_0)$ must depend as follows on g_0

$$\hat{\sigma}(g_0) \approx \hat{C}_\sigma (R(g_0))^2, \tag{9.25}$$

which in view of (9.21d) tells us that σ is a non-perturbative observable.

Finally we remark that the appearance of Λ_L in a theory which a priori is free of any scale (like the case considered here) is well known from perturbative continuum QCD, where the necessity of renormalizing the theory also requires the introduction of a scale Λ_{QCD}. The numerical values of Λ_L and Λ_{QCD} are not the same and indeed differ substantially (see chapter 15).

In the following chapter we will make use of scaling arguments of the type discussed above, to derive a relation between the potential of a quark-antiquark pair and the energy stored in the colour electric and magnetic field.

CHAPTER 10

LATTICE SUM RULES

In chapter 7 we have shown that the static quark-antiquark potential can be determined from the exponential decay of the expectation value of the Wilson loop for large euclidean times. In the pure Yang-Mills theory we expect that this potential rises linearly for large quark-antiquark separations, leading to quark confinement. As we have already pointed out, this linear rise is believed to be due to the formation of a narrow flux tube connecting the quark-antiquark pair, in which the colour field energy is concentrated. The energy stored in the colour fields should match, after subtracting the self energy contributions of the quark and antiquark, the interquark potential, as determined from the Wilson loop. In order to be able to study the distribution of the field energy surrounding the quark-antiquark pair we need a non-perturbative expression for the field energy which is suited for Monte Carlo simulations. To this effect we shall derive an energy sum rule which relates the potential to the expectation value of an operator which can be identified with the field energy of a quark-antiquark pair. The same line of reasoning leads to a similar sum rule relating the mass of a glueball * to the energy stored in the chromoelectric and magnetic fields. Such sum rules have been first obtained by Michael (1987), whose derivation, however, turned out to be incomplete [Rothe (1995a,b); Michael (1996)]. Before deriving these sum rules it is instructive to illustrate the basic idea that goes into their derivation in a simple quantum mechanical example. Although in principle we can chose any potential for purposes of illustration, we prefer to be specific, and consider the harmonic oscillator, where the sum rule can be checked exactly.

10.1 Energy Sum Rule for the Harmonic Oscillator

Consider the imaginary time Green function, $< q'|e^{-H\tau}|q >$, whose path integral representation has been discussed in chapter 2. By expanding $|q >$ in a complete set of energy eigenstates one immediately concludes that this matrix element is dominated for large τ by the contribution of the ground state, i.e.,

$$< q'|e^{-H\tau}|q > \rightarrow \psi_0(q')\psi_0^*(q)e^{-E_0\tau} \ , \tag{10.1}$$

* Glueballs constitute the particle spectrum of the pure SU(3) Yang-Mills theory.

where ψ_0 is the wave function with energy E_0. By setting $q' = q$ and integrating over q we therefore have that

$$\int dq < q|e^{-H\tau}|q > \underset{\tau\to\infty}{\longrightarrow} e^{-E_0\tau} \ . \tag{10.2}$$

Hence by studying the behaviour of the lhs for large euclidean times we can in principle extract the ground state energy. What we are interested in, however, is an expression which relates the ground state energy to an ensemble average of the kinetic and potential energy. In the following we now illustrate the basic ideas for accomplishing this program for the case of the harmonic oscillator whose Hamiltonian is given by

$$H = \frac{p^2}{2m} + \frac{1}{2}\kappa q^2 \ .$$

The first step in deriving an energy sum rule consists in expressing the imaginary time Green function in (10.2) as a configuration space path integral. Proceeding as in chapter two, the lhs of (10.2) is given for small time step ϵ by

$$\int dq < q|e^{-H\tau}|q > \approx (\frac{m}{2\pi\epsilon})^{\frac{N}{2}} \int \prod_{n=0}^{N-1} dq_n e^{-S[q;m,\kappa,N,\epsilon]}|_{q_N=q_0} \ , \tag{10.3a}$$

where

$$S[q;m,\kappa,N,\epsilon] = \sum_{n=0}^{N-1} \epsilon[\frac{1}{2}m(\frac{q_{n+1}-q_n}{\epsilon})^2 + \frac{1}{2}\kappa q_n^2] \ , \tag{10.3b}$$

and we have set $q = q_0$. The euclidean time τ is given by $N\epsilon$. Notice that we did not take the continuum limit ($N \to \infty$, $\epsilon \to 0$, with $N\epsilon = \tau$ fixed) on the rhs. For sufficiently small ϵ, and hence large $N = \frac{\tau}{\epsilon}$ the rhs will be a good approximation to the Green function. How small ϵ has to be chosen to approximate continuum physics will depend on the values of the dimensioned parameters m and κ which determine the relevant scales in the problem. Now for a fixed non-vanishing value of ϵ, the limit $\tau \to \infty$ in (10.2) is realized for $N \to \infty$. We are therefore led to the statement that

$$(\frac{m}{2\pi\epsilon})^{\frac{N}{2}} \int \prod_{n=0}^{N-1} dq_n e^{-S[q;m,\kappa,N,\epsilon]}|_{q_N=q_0} \underset{N\to\infty}{\longrightarrow} e^{-\hat{E}_0 N} \ , \tag{10.4}$$

where $\hat{E}_0 = \epsilon E_0$ is the energy measured in units of the lattice spacing ϵ. By scaling all the dimensioned variables with ϵ according to their canonical dimension, we arrive at the following dimensionless lattice version of (10.4)

$$\hat{G}(\hat{m},\hat{\kappa},N) \underset{N\to\infty}{\longrightarrow} e^{-\hat{E}_0(\hat{m},\hat{\kappa})N} \ , \tag{10.5a}$$

where

$$\hat{G}(\hat{m},\hat{\kappa},N) = (\frac{\hat{m}}{2\pi})^{\frac{N}{2}} \int \prod_{n=0}^{N-1} d\hat{q}_n e^{-\hat{S}[\hat{q};\hat{m},\hat{\kappa},N]}|_{\hat{q}_N=\hat{q}_0} , \tag{10.5b}$$

and

$$\hat{S}[\hat{q};\hat{m},\hat{\kappa},N] = \sum_{n=0}^{N-1}[\frac{1}{2}\hat{m}(\hat{q}_{n+1}-\hat{q}_n)^2 + \frac{1}{2}\hat{\kappa}\hat{q}_n^2)] . \tag{10.5c}$$

Here $\hat{m} = m\epsilon$ and $\hat{\kappa} = \epsilon^3\kappa$ are dimensionless parameters. In the continuum limit $\hat{m} \to 0$, $\hat{\kappa} \to 0$ with $\frac{\hat{m}^3}{\hat{\kappa}} = \frac{m^3}{\kappa}$ fixed. From (10.5) it follows that the energy measured in lattice units can be computed as the following limit

$$\hat{E}_0(\hat{m},\hat{\kappa}) = -\lim_{N\to\infty}\frac{1}{N}\ln\hat{G}(\hat{m},\hat{\kappa},N) . \tag{10.6}$$

Let us now repeat this excercise using another (fine) discretization $\epsilon' = \frac{1}{\xi}\epsilon$. Then the path integral representation of the lhs of (10.2) will differ from that in (10.3a) in that ϵ and N are replaced by $\epsilon' = \frac{1}{\xi}\epsilon$, and $N' = \xi N$ (recall that $N'\epsilon' = N\epsilon = \tau$). After scaling all variables with the *original* lattice spacing ϵ according to their canonical dimensions (so that the values of the parameters $\hat{m}$ and $\hat{\kappa}$ are the same as before), the new path integral expression for the lhs of (10.2) reads

$$\hat{G}(\hat{m}(\xi),\hat{\kappa}(\xi),N') = [\frac{\hat{m}(\xi)}{2\pi}]^{\frac{N'}{2}} \int \prod_{n=0}^{N'-1} d\hat{q}_n e^{-\hat{S}[\hat{q},\hat{m}(\xi),\hat{\kappa}(\xi),N']}|_{\hat{q}_N=\hat{q}_0} , \tag{10.7a}$$

where

$$\hat{S}[\hat{q},\hat{m}(\xi),\hat{\kappa}(\xi),N'] = \sum_{n=0}^{N'-1}[\frac{1}{2}\hat{m}(\xi)(\hat{q}_{n+1}-\hat{q}_n)^2 + \frac{1}{2}\hat{\kappa}(\xi)\hat{q}_n^2] , \tag{10.7b}$$

and

$$\hat{m}(\xi) = \xi\hat{m} \quad ; \quad \hat{\kappa}(\xi) = \frac{1}{\xi}\hat{\kappa} . \tag{10.7c}$$

Hence the parameters $\hat{m}$ and $\hat{\kappa}$ have now acquired a ξ-dependence, which is very simple in the present case. Certainly for $\xi > 1$ the above path integral expression for the Green function is at least as good as the previous one, since we have made the lattice even finer. For $N' \to \infty$ (10.7a) will now behave like

$$\hat{G}(\hat{m}(\xi),\hat{\kappa}(\xi),N') \underset{N'\to\infty}{\longrightarrow} e^{-\hat{E}_0(\hat{m}(\xi),\hat{\kappa}(\xi))N'} . \tag{10.8}$$

as follows from (10.5a). But for sufficiently small ϵ we must have that $\hat{G}(\hat{m},\hat{\kappa},N) = \hat{G}(\hat{m}(\xi),\hat{\kappa}(\xi),\xi N)$, where $\hat{m} = \hat{m}(1)$, $\hat{\kappa} = \hat{\kappa}(1)$. We hence conclude that

$$\xi\hat{E}_0(\hat{m}(\xi),\hat{\kappa}(\xi)) = \hat{E}_0(\hat{m},\hat{\kappa}) , \tag{10.9}$$

i.e., the lhs must be independent of ξ. By taking the derivative of the lhs with respect to ξ we therefore arrive at the following alternative equation for the ground state energy

$$\hat{E}_0(\hat{m},\hat{\kappa}) = -[\frac{\partial \hat{E}_0(\hat{m}(\xi),\hat{\kappa}(\xi))}{\partial \xi}]_{\xi=1} \ . \tag{10.10}$$

But according to (10.8)

$$\hat{E}_0(\hat{m}(\xi),\hat{\kappa}(\xi)) = -\lim_{N'\to\infty}\frac{1}{N'}\ln \hat{G}(\hat{m}(\xi),\hat{\kappa}(\xi),N') \ . \tag{10.11}$$

Taking the derivative of (10.7a) with respect to ξ, making use of (10.7c), one therefore finds that (10.10) translates into

$$\hat{E}_0(\hat{m},\hat{\kappa}) = \frac{1}{2} + \lim_{N'\to\infty}\frac{1}{N'}\sum_{n=0}^{N'-1} < -\frac{1}{2}\hat{m}(\hat{q}_{n+1}-\hat{q}_n)^2 + \frac{1}{2}\hat{\kappa}\hat{q}_n^2 > \ , \tag{10.12a}$$

where generically

$$< O(\hat{q}) > = \frac{\int D\hat{q} O(\hat{q}) e^{-\hat{S}}}{\int D\hat{q} e^{-\hat{S}}} \tag{10.12b}$$

with $\hat{S}$ the action (10.5c). The expression (10.12a) can be further simplified. Because of the periodic boundary condition $\hat{q}_0 = \hat{q}_N$, we have compactified the imaginary time direction. The value of the expectation value in (10.12a) will be independent of the time slice labeled by n. Hence (10.12a) can be simplified to read

$$\hat{E}_0(\hat{m},\hat{\kappa}) = \frac{1}{2} + < -\frac{1}{2}\hat{m}\dot{\hat{q}}_\ell^2 + \frac{1}{2}\hat{\kappa}\hat{q}_\ell^2 > \ , \tag{10.13a}$$

where

$$\dot{\hat{q}}_\ell = \hat{q}_{\ell+1} - \hat{q}_\ell \ , \tag{10.13b}$$

and ℓ denotes some arbitrarily chosen temporal lattice site. For the ground state energy measured in physical units, $E_0 = \frac{1}{\epsilon}\hat{E}_0$, one then obtains

$$E_0(m,\kappa) = \frac{1}{2\epsilon} + < -\frac{1}{2}m\dot{q}^2(\tau) + \frac{1}{2}\kappa q^2(\tau) > \ , \tag{10.14}$$

where all variables are now dimensionful, and where we have used the more suggestive notation $q(\tau)$ instead of q_ℓ.

Equation (10.14) relates the ground state energy to an ensemble average of the kinetic and potential energy, calculated with the Boltzmann distribution $e^{-\hat{S}}$. In the euclidean version, however, the kinetic term is seen to yield a negative contribution! Although this is naively understood by recalling that the transition from real to imaginary time will map $\frac{1}{2}(\frac{dq}{dt})^2$ into -$\frac{1}{2}(\frac{dq}{d\tau})^2$, the above result still looks surprising at first sight. After all, the contribution of the kinetic energy should be positive. A detailed calculation shows that $< -\frac{1}{2}(\frac{dq}{d\tau})^2 >$ is itself divergent in the limit $\epsilon \to 0$. The divergence is cancelled by the first term in (10.14) and leaves a positive contribution, which has precisely the expected form.* The fact that the kinetic contribution diverges does not come as a surprise. Looking at the integrand on the rhs of (10.3a) we see that the width of the distribution in $q_{\ell+1} - q_\ell$ is only of $\mathcal{O}(\sqrt{\epsilon})$, so that $< \dot{q}_\ell^2 >=< (q_{\ell+1} - q_\ell)^2/\epsilon^2 >$ is expected to be of $\mathcal{O}(\frac{1}{\epsilon})$.

Since the appearance of the minus sign in front of the kinetic term is so characteristic of the euclidean formulation, and, in fact, will pop up again when we discuss the $SU(N)$ gauge theory in the following section, it is instructive to derive the above result for the kinetic contribution in an alternative way.

The energy E_0 is given by the ground state expectation value of the Hamiltonian, i.e,

$$E_0 =< 0|\frac{1}{2m}P^2 + V(Q)|0 > ,$$

where P and Q are the momentum and coordinate operators, and $|0>$ is the ground state (or "physical vacuum" in the language of field theory). We are interested, in particular, in expressing $< 0|P^2|0 >$ as a euclidean path integral. To this effect consider the (imaginary time) Heisenberg operator

$$Q(\tau) = e^{H\tau} Q e^{-H\tau} ,$$

and the corresponding conjugate momentum

$$P(\tau) = e^{H\tau} P e^{-H\tau} .$$

They satisfy the canonical commutation relation

$$[Q(\tau), P(\tau)] = i .$$

The equation of motion for the operator $Q(\tau)$ reads

$$\dot{Q}(\tau) = e^{H\tau}[H, Q]e^{-H\tau} = -\frac{i}{m}P(\tau) .$$

---

* The cancellation will be demonstrated in section 4 of chapter 18 for the more general case of a harmonic oscillator in contact with a heat bath, where a similar energy sum rule holds.

Since $|0>$ is an eigenstate of the Hamiltonian it follows that $<0|P^2|0>=<0|P^2(\tau)|0>$. Consider now the ground state expectation value of the euclidean time ordered product of $Q(\tau)Q(\tau')$. This time ordered product has a path integral representation, given by (2.26a), i.e.,

$$<0|\theta(\tau-\tau')Q(\tau)Q(\tau')+\theta(\tau'-\tau)Q(\tau')Q(\tau)|0>=\frac{1}{Z}\int Dq\, q(\tau)q(\tau')e^{-\int d\tau[\frac{1}{2}m\dot{q}^2+V(q)]}$$

where

$$Z=\int Dq\; e^{-\int d\tau[\frac{1}{2}m\dot{q}^2(\tau)+V(q(\tau))]}$$

Taking the derivative * with respect to τ and τ' on both sides of this equation, and making use of the fact that the operators $Q(\tau)$ commute for equal times, one finds, after setting $\dot{Q}(\tau)=-\frac{i}{m}P(\tau)$ that

$$\begin{aligned}-\frac{i}{m}\delta(\tau-\tau')[Q(\tau),P(\tau)]-\frac{1}{m^2}&<0|T(P(\tau)P(\tau'))|0>\\ &=\frac{1}{Z}\int Dq\dot{q}(\tau)\dot{q}(\tau')e^{-\int d\tau[\frac{1}{2}m\dot{q}^2(\tau)+V(q(\tau))]}\,.\end{aligned}$$

Since the operators appearing in the time ordered product commute, we can omit the time ordering operation. Now on a discretized time lattice, the continuum δ-function is replaced by (see eq. (2.65))

$$\delta(\tau-\tau')\to\frac{1}{\epsilon}\delta_{nn'}=\int_{-\frac{\pi}{\epsilon}}^{\frac{\pi}{\epsilon}}\frac{dp_4}{2\pi}e^{ip_4(n-n')\epsilon}\,.$$

Hence setting $\tau=\tau'$ (i.e. $n=n'$) implies $\delta(0)\to\frac{1}{\epsilon}$. After replacing the equal time commutator of $Q(\tau)$ and $P(\tau)$ by "i", we find that

$$\frac{1}{2m}<0|P^2|0>=\frac{1}{2\epsilon}-\frac{1}{Z}\int Dq\frac{1}{2}m\dot{q}(\tau)^2e^{-\int d\tau[\frac{1}{2}m\dot{q}^2(\tau)+V(q(\tau))]}\,.$$

Hence the kinetic contribution to E_0 has precisely the form given by the first two terms in (10.14)

---

* Of course when writing $\dot{q}(\tau)$ within a path integral, we always mean its discretized version.

10.2 The SU(N) Gauge Action on an Anisotropic Lattice

In the case of our quantum mechanical example, the dependence of the dimensionless parameters $\hat{m}$ and $\hat{\kappa}$ on the asymmetry parameter $\xi = \frac{\epsilon}{\epsilon'}$ was very simple. In lattice gauge field theories the analogous parameter is given by $\xi = \frac{a}{a_\tau}$, where a and a_τ are the spatial and temporal lattice spacings. On an isotropic lattice $a_\tau = a$. The spatial lattice spacing now plays the role of the reference "lattice spacing" ϵ in our quantum mechanical example. The parameterization of the $SU(N)$ gauge action on an anisotropic lattice turns out to be more subtle than in the quantum mechanical case, and does not follow from naive considerations alone.

Consider the continuum action (6.17) for the pure SU(N) gauge theory. In the temporal gauge, $A_4^B = 0$, where $F_{4i}^B = \partial_4 A_i^B$ it has the form

$$S_G[A] = \int d\tau \int d^3x \frac{1}{2} \sum_{B,i} (\dot{A}_i^B(\vec{x},\tau))^2 + S_I[A] \ ,$$

where

$$S_I[A] = \frac{1}{4} \int d\tau \int d^3x \sum_{i,j,B} F_{ij}^B(\vec{x},\tau) F_{ij}^B(\vec{x},\tau) \ ,$$

and "B" are the colour indices. In this gauge the action thus has a similar form as that of a quantum mechanical system, except that we are dealing with a system with an infinite number of degrees of freedom. A naive discretization of the degrees of freedom, labeled by $\vec{x}$, and of the euclidean time integral, yields

$$S_G[A] = \sum_n a_\tau a^3 \{ \frac{1}{2a_\tau^2} \sum_{i,B} [A_i^B(\vec{n}, n_4 + 1) - A_i^B(\vec{n}, n_4)]^2 + \frac{1}{4} \sum_{i,j,B} F_{ij}^B(n) F_{ij}^B(n) \} \ ,$$

where $A_i^B(n)$ and $F_{ij}^B(n)$ are the potentials and colour-magnetic fields at the lattice site $n = (\vec{n}, n_4)$. The discretized colour-magnetic contribution $\sum_{i,j,B} F_{ij}^B(n) F_{ij}^B(n)$ does not involve an explicit a_τ-dependence. Having discretized the degrees of freedom, we have thereby introduced an additional scale, given by the spatial lattice spacing. By scaling the potentials with this lattice spacing according to their canonical dimension, i.e., introducing the dimensionless potentials $\hat{A}_i^B = aA_i^B$, the above action can be written in the form

$$S_G = \sum_n \{ \frac{\xi}{2} \hat{F}_{i4}^B(n) \hat{F}_{i4}^B(n) + \frac{1}{4\xi} \hat{F}_{ij}^B(n) \hat{F}_{ij}^B(n) \} \ ,$$

where $\hat{F}_{4i}^B = aa_\tau F_{4i}^B$, $\hat{F}_{ij}^B = a^2 F_{ij}^B$, and where a sum over repeated indices is now understood. Written in terms of the field strengths, the expression appears to be gauge invariant. For finite lattice spacing this is however not the case. To ensure gauge invariance for arbitrary lattice spacings, we must introduce the potentials in the form of link variables, as discussed in chapter 6. The matrix valued fields strength tensor $\hat{F}_{\mu\nu}$ is then related to the plaquette variables by $U_{\mu\nu}(n) \approx \exp[ig_0\hat{F}_{\mu\nu}(n)]$. One then readily verifies that the following $SU(N)$ action possesses the correct *naive* continuum limit

$$S_G[U] = \frac{2N}{g_0^2}\frac{1}{\xi}\mathcal{P}_s + \frac{2N}{g_0^2}\xi\mathcal{P}_\tau \,, \tag{10.15a}$$

where

$$\mathcal{P}_s = \sum_{P_s}[1 - \frac{1}{2N}Tr(U_{P_s} + U_{P_s}^\dagger)] \,, \tag{10.15b}$$

$$\mathcal{P}_\tau = \sum_{P_\tau}[1 - \frac{1}{2N}Tr(U_{P_\tau} + U_{P_\tau}^\dagger)] \,, \tag{10.15c}$$

Here U_{P_s} and U_{P_τ} are the spatial and temporal plaquette variables which are related to the colour electric and magnetic fields. Notice that the ξ-dependence of the kinetic (electric) and potential (magnetic) term is the same as in the case of the harmonic oscillator. This form for the action had been proposed by Engels et. al., and by Kuti et. al. [Engels (1981b), Kuti (1981)]. The naive ξ-dependence of the above action is however too simple. The reason is that this action describes a complicated interacting system. As a consequence of quantum fluctuations the couplings associated with the temporal and spatial plaquettes must be separately tuned with the spatial lattice spacing and the asymmetry parameter ξ to ensure that physical observables become independent of a and ξ close to the continuum limit. Instead of the spatial lattice spacing one can also choose the bare coupling constant g_0 defined on an isotropic lattice. We therefore consider the following more general form for the action [Hasenfratz and Hasenfratz (1981b); Karsch (1982)],

$$S_G[U] = \hat{\beta}_s\mathcal{P}_s + \hat{\beta}_\tau\mathcal{P}_\tau \,, \tag{10.16a}$$

where

$$\hat{\beta}_s = \frac{2N}{g_s^2(g_0,\xi)}\frac{1}{\xi} \,, \qquad \hat{\beta}_\tau = \frac{2N}{g_\tau^2(g_0,\xi)}\xi \,, \tag{10.16b}$$

with $g_s(g_0,\xi)$ and $g_\tau(g_0,\xi)$ satisfying the condition

$$g_s(g_0,1) = g_\tau(g_0,1) = g_0 \,. \tag{10.16c}$$

Hence*

$$\hat{\beta}_\sigma \xrightarrow[\xi\to 1]{} \hat{\beta} = 2N/g_0^2\,.$$

Let us pause here for a moment and return to the case of the harmonic oscillator. There the ξ-dependence of the action was absorbed into a mass and coupling parameter. This ξ-dependence was given by naive arguments alone: $\hat{m}(\xi) = \xi\hat{m}$ and $\hat{\kappa}(\xi) = \frac{1}{\xi}\hat{\kappa}$. Hence $\hat{m}(\xi)$ and $\hat{\kappa}(\xi)$ are functions of ξ and the dimensionless mass and κ parameters $\hat{m} = \hat{m}(1)$ and $\hat{\kappa} = \hat{\kappa}(1)$. When taking the continuum limit, these dimensionless parameters must be tuned with the lattice spacing ϵ according to $\hat{m} = m\epsilon$ and $\hat{\kappa} = \epsilon^3\kappa$, where m and κ are physical parameters which are held fixed when taking the continuum limit. The naive ξ dependence of the parameters in the quantum mechanical case correspond to the explicit ξ-dependence of the couplings (10.16b), while the ξ-dependence of $g_s(g_0,\xi)$ and $g_\tau(g_0,\xi)$ is a consequence of quantum fluctuations. As has been shown by Karsch (1982), $\hat{\beta}_s$ and $\hat{\beta}_\tau$ can be related in the weak coupling limit ($g_0 \to 0$, or $\hat{\beta} \to \infty$) to the coupling $\hat{\beta}$ on an isotropic lattice as follows

$$\begin{aligned}\frac{1}{\xi}\hat{\beta}_\tau(\hat{\beta},\xi) &= \hat{\beta} + 2Nc_\tau(\xi) + \mathcal{O}(\hat{\beta}^{-1}),\\ \xi\hat{\beta}_s(\hat{\beta},\xi) &= \hat{\beta} + 2Nc_s(\xi) + \mathcal{O}(\hat{\beta}^{-1}),\end{aligned} \tag{10.17}$$

where $c_\sigma(1) = 0$, $\sigma = s,\tau$. The ξ-dependence of the functions $c_\sigma(\xi)$ have been studied in detail by this author. When taking the continuum limit at a fixed value of ξ, the coupling constant g_0 in (10.16b), defined on a isotropic lattice, must be tuned with the spatial lattice spacing as dictated by the renormalization group. As we shall see in the following sections, the ξ-dependence of the couplings g_σ in (10.16b) will lead to an energy sum rule for the $q\bar{q}$-potential, and the glueball mass, which differs in an important way from that which one would expect naively.

10.3 Sum Rules for the Static $q\bar{q}$-Potential

Having parametrized the $SU(N)$-Yang-Mills action on an anisotropic lattice, we can now proceed along similar lines as in section 2, to derive an energy sum rule for the static quark-antiquark potential, which relates this quantity to the energy stored in the chromoelectric and chromomagnetic fields of a quark-antiquark pair. To guide the readers attention, let us briefly outline the general strategy we shall

---

* We have denoted β with a "hat", in order not to confuse it with the renormalization group β-function (9.6b) which will be relevant further below

follow. We first derive a sum rule for the static $q\bar{q}$-potential in the pure SU(N) gauge theory, which relates the potential to the *action* stored in the chromoelectric and magnetic fields. For this we will only need to know the expression of the action on an isotropic lattice, i.e., a lattice with equal spacings in the space and time directions. This action sum rule will play an important role in our subsequent discussion of the energy sum rule. For the derivation of the latter sum rule we shall need the expression for the action on an anisotropic lattice, obtained in the previous section. As in the case of the quantum mechanical example, the energy sum rule then follows by requiring that in the continuum limit the potential calculated from a Wilson loop on an anisotropic lattice should be independent of the anisotropy parameter $\xi = \frac{a}{a_\tau}$. The very same requirement, when applied to the special case of a confining, linearly rising, potential leads to a "coupling constant sum rule" [Karsch (1982)], which will allow us to confirm that the energy stored in the chromoelectric and magnetic field matches the interquark potential. In particular, it will allow us to identify the contribution to the field energy arising from the trace anomaly, which will be shown to account for 1/2 of the field energy stored in the flux tube.

Let us briefly state what is meant by the trace anomaly. The colour-electric and magnetic field energy density is given by T_{00}, where $T_{\mu\nu}$ is the energy momentum tensor. On the classical level this tensor is traceless and symmetric. On the quantum level, however, it is known from perturbation theory that quantum fluctuations give rise to the so called trace anomaly: the energy momentum tensor $T_{\mu\nu}$ is no longer traceless. Now $T_{\mu\nu}$ can be trivially decomposed into a traceless and trace part,

$$T_{\mu\nu} = (T_{\mu\nu} - \frac{1}{4}g_{\mu\nu}T) + \frac{1}{4}g_{\mu\nu}T \ , \tag{10.18}$$

where T is the trace, $T = \sum_\mu T^\mu_\mu$. One therefore expects that in addition to the naive field energy density (which is given by an expression analogous to that in electrodynamics) the potential receives a contribution arising from the non-vanishing trace of the energy momentum tensor. This trace is given in Minkowski space by [Collins (1977)]

$$T(x) = \frac{2\beta(g)}{g}\mathcal{L}(x) \ , \tag{10.19a}$$

where

$$\mathcal{L}(x) = \frac{1}{4}F^{A\mu\nu}(x)F^A_{\mu\nu}(x) \ . \tag{10.19b}$$

A summation over repeated indices is always understood. $\beta(g) = \mu\partial g/\partial\mu$ is the β-function of the continuum SU(N)-gauge theory with μ the renormalization scale.

Hence from (10.18), we expect that the interquark potential receives an anomalous contribution having the form

$$V_{anom}(R) = \frac{1}{4} < \int d^3x \; T(x) >_{q\bar{q}} = \frac{\beta(g)}{2g} < \int d^3x \; \mathcal{L}(x) >_{q\bar{q}} \; . \qquad (10.20)$$

In the euclidean formulation, the action density will be replaced by its euclidean counterpart $\frac{1}{4}F^A{}_{\mu\nu}(x)F^A_{\mu\nu}(x)$. Since the lattice provides a non-perturbative regularization of the partition function, it is the appropriate framework for deriving a non-perturbative expression for the contribution of the normal and anomalous part of the field energy to the interquark potential. Originally it was C. Michael (1987) who derived an action and energy sum rule for the $q\bar{q}$-potential in an SU(N) gauge theory. The derivation was however incomplete. In the following we shall derive the corrected version of these sum rules. As we shall see, the derivation is straightforward and leads in a very natural way to a decomposition of the field energy into a normal and anomalous part [Rothe (1995a,b)].

i) *Action Sum Rule*

Consider the ground state energy of a quark-antiquark pair separated by a distance $\hat{R}$. As always, quantities denoted with a "hat" are understood to be measured in units of the lattice spacing. The energy $\hat{E}_0(\hat{R})$ can be calculated from the expectation value of the Wilson loop according to

$$\hat{E}_0(\hat{R}) = - \lim_{\hat{T}\to\infty} \frac{1}{\hat{T}} \ln < W(\hat{R},\hat{T}) > \; . \qquad (10.21a)$$

where

$$< W(\hat{R},\hat{T}) > = \frac{\int DU W(\hat{R},\hat{T}) e^{-S_G}}{\int DU e^{-S_G}} \; . \qquad (10.21b)$$

Recall that $W(\hat{R},\hat{T})$ is given by the trace of the path ordered product of link variables around a rectangular loop with $\hat{R}$ and $\hat{T}$ lattice spacings in the spatial and euclidean time directions (Our notation here deviates from that in chapter 7). On an isotropic lattice $< W(\hat{R},\hat{T}) >$ is calculated with the action (10.15) with $\xi = 1$. This action has the form

$$S_G = \hat{\beta}(\mathcal{P}_\tau + \mathcal{P}_s) \qquad (10.22)$$

where $\hat{\beta} = \frac{2N}{g_0^2}$ for $SU(N)$. The energy $\hat{E}_0(\hat{R})$, defined by (10.21a), is a function of $\hat{R}$ and $\hat{\beta}$ and includes the self-energy contributions of the quark and antiquark.

Since these contributions do not depend on $\hat{R}$, they can be eliminated by considering the difference $\hat{E}_0(\hat{R},\hat{\beta}) - \hat{E}_0(\hat{R}_0,\hat{\beta})$, where $\hat{R}_0$ is some reference $q\bar{q}$-separation. We then define the $q\bar{q}$ potential by

$$\hat{V}(\hat{R},\hat{\beta}) = -\lim_{\hat{T}\to\infty} \frac{1}{\hat{T}}[\ln < W(\hat{R},\hat{T}) >]_{subtr} \,. \tag{10.23}$$

From here on we will always assume that such a subtraction has been carried out, and shall drop the subscript "subtr" for simplicity. Following Michael (1987) we now take the derivative of (10.23) with respect to $\hat{\beta}$ and obtain

$$\frac{\partial\hat{V}(\hat{R},\hat{\beta})}{\partial\hat{\beta}} = \lim_{\hat{T}\to\infty} \frac{1}{\hat{T}} < \mathcal{P}_\tau + \mathcal{P}_s >_{q\bar{q}-0} \,, \tag{10.24a}$$

where $< O >_{q\bar{q}-0}$ is defined generically by

$$< O >_{q\bar{q}-0} = \frac{< W(\hat{R},\hat{T})O >}{< W(\hat{R},\hat{T}) >} - < O > . \tag{10.24b}$$

The rhs of this expression is the expectation value of the operator O in the $q\bar{q}$-state measured relative to the vacuum. In the limit $\hat{T}\to\infty$, the rhs of (10.24a) can be approximated by*

$$< \mathcal{P}_\sigma >_{q\bar{q}-0} \underset{\hat{T}\to\infty}{\approx} \hat{T} < \mathcal{P}'_\sigma >_{q\bar{q}-0}, \tag{10.25a}$$

where

$$\mathcal{P}'_\sigma = \sum_P [1 - \frac{1}{2N} Tr(U_{P_\sigma} + U^\dagger_{P_\sigma})]_{n_\tau\ fixed}, \quad ;(\sigma = \tau, s) \tag{10.25b}$$

is the contribution to $\mathcal{P}_\sigma$ arising from plaquettes located on a *fixed* time slice, and with the Wilson loop extending from $n_4 = -\frac{\hat{T}}{2}$ to $n_4 = \frac{\hat{T}}{2}$, with $\hat{T}\to\infty$. This time slice is conveniently chosen to be the $n_4 = 0$ plane. Hence,

$$\begin{aligned} \hat{\beta}\frac{\partial\hat{V}(\hat{R},\hat{\beta})}{\partial\hat{\beta}} &= \hat{\beta} < \mathcal{P}'_\tau + \mathcal{P}'_s >_{q\bar{q}-0} \\ &\underset{naive}{\longrightarrow} a\sum_{\vec{x}} a^3 \frac{1}{2} < E^2(\vec{x}) + B^2(\vec{x}) >_{q\bar{q}-0} \,, \end{aligned} \tag{10.26}$$

where in the last step we have taken the naive continuum limit. E^2 and B^2 are the square of the euclidean (!) colour electric and magnetic fields E_i^A, and B_i^A

---

* See the argument given by Michael (1987). Note that $< \mathcal{P}' >_{q\bar{q}-0}$ is calculated with a Wilson loop of very large extension in euclidean time.

summed over $i = 1, 2, 3$, and colours. The expectation value $\frac{1}{2} < E^2(\vec{x}) + B^2(\vec{x}) >$ is not to be confused with the Minkowski field energy. In fact, $\hat{\beta}(\mathcal{P}'_\tau + \mathcal{P}'_s)$ is the contribution to the *action* coming from a fixed time slice.

We next make use of the renormalization group to cast the lhs of (10.26) in a form involving the potential and its derivative with respect to $\hat{R}$. In the limit of vanishing lattice spacing "a" we have that

$$\frac{1}{a}\hat{V}\left(\frac{R}{a}, \hat{\beta}(a)\right) \underset{a\to 0}{\longrightarrow} V(R), \tag{10.27}$$

where $V(R)$ is the interquark potential in physical units. The dependence of $\hat{\beta}(a) = 2N/g_0^2(a)$ on the lattice spacing is given, close to the continuum limit, by the renormalization group relation (9.21c,d). The invariance of the lhs of (10.27) with regard to changes in the lattice spacing leads to

$$\frac{\partial\hat{\beta}}{\partial lna}\frac{\partial\hat{V}(\hat{R},\hat{\beta})}{\partial\hat{\beta}} = \hat{V}(\hat{R},\hat{\beta}) + \hat{R}\frac{\partial\hat{V}(\hat{R},\hat{\beta})}{\partial\hat{R}} \ ,$$

where it is understood that this relation holds close to the continuum limit. Making use of this expression, equation (10.26) takes the following form*

$$\hat{V}(\hat{R},\hat{\beta}) + \hat{R}\frac{\partial\hat{V}(\hat{R},\hat{\beta})}{\partial\hat{R}} = \frac{\partial\hat{\beta}}{\partial lna} < \mathcal{P}'_\tau + \mathcal{P}'_s >_{q\bar{q}-0} \ . \tag{10.28}$$

In the case of a confining potential, $\hat{V}(\hat{R},\hat{\beta}) = \hat{\sigma}(\hat{\beta})\hat{R}$, this equation reduces to

$$2\hat{\sigma}(\hat{\beta})\hat{R} = \frac{\partial\hat{\beta}}{\partial lna} < \mathcal{P}'_\tau + \mathcal{P}'_s >_{q\bar{q}-0} \ , \tag{10.29}$$

while for a pure Coulomb-type potential the lhs of (10.28) vanishes. Thus in an $SU(N)$ gauge theory, where $\frac{\partial\hat{\beta}}{\partial lna} \neq 0$, the potential cannot be of the pure Coulomb type.

ii) *Energy Sum Rule*

Let us now turn to our second objective, and derive a sum rule, relating the interquark potential to the field energy. This sum rule is obtained by requiring that a lattice regularization involving different lattice spacings in the temporal and spatial directions should lead to the same potential as that computed from

* As was first noted by H.G. Dosch the action sum rule of Michael (1987) contained an error (private communication).

an isotropic lattice. This is analogous to the requirement which in the case of our quantum mechanical example led us to an energy sum rule for the ground state energy. Here it is the ground state energy of a static quark-antiquark pair interacting via Young-Mills fields.

Consider the expectation value of a Wilson loop on an isotropic lattice. The expectation value is computed with the action (10.22). Next consider a Wilson loop on an anisotropic lattice with the *same* physical extension in the spatial and temporal directions. The expectation value must now be calculated with the ξ-dependent action (10.16a,b), which we shall denote by $S(\xi)$. Hence the number of lattice sites in the euclidean time direction is now $\xi\hat{T}$. Since both Wilson loops have the same physical extension, their expectation values must be the same, if the lattice is fine enough to approximate continuum physics. This leads to the requirement that

$$< W(\hat{R},\hat{T}) >_{S(1)} = < W(\hat{R},\xi\hat{T}) >_{S(\xi)} \ .$$

From (10.23) it then follows that*

$$\hat{V}(\hat{R},\hat{\beta}) = \xi\tilde{V}(\hat{R},\beta_s(\xi),\beta_\tau(\xi)) \ , \tag{10.30a}$$

where

$$\tilde{V}(\hat{R},\hat{\beta}_s(\xi),\hat{\beta}_\tau(\xi)) = -\lim_{\hat{T}'\to\infty}\frac{1}{\hat{T}'}\ln < W(\hat{R},\hat{T}') >_{S(\xi)} \ . \tag{10.30b}$$

and

$$< W(\hat{R},\hat{T}) >_{S(\xi)} = \frac{\int DU\, W(\hat{R},\hat{T})e^{-[\hat{\beta}_s(\xi)\mathcal{P}_s+\hat{\beta}_\tau(\xi)\mathcal{P}_\tau]}}{\int DU\, e^{-[\hat{\beta}_s(\xi)\mathcal{P}_s+\hat{\beta}_\tau(\xi)\mathcal{P}_\tau]}} \ . \tag{10.30c}$$

Clearly

$$\tilde{V}(\hat{R},\beta_s(\xi),\beta_\tau(\xi))|_{\xi=1} = \hat{V}(\hat{R},\hat{\beta}) \ .$$

Equation (10.30a) implies that

$$\frac{d}{d\xi}\left[\xi\tilde{V}\left(\hat{R},\beta_s(\xi),\beta_\tau(\xi)\right)\right] = 0 \ . \tag{10.31}$$

This equation is the basis for deriving the desired energy sum rule. From (10.30b,c) one finds that

$$\frac{\partial\tilde{V}}{\partial\beta_\sigma} = < \mathcal{P}'_\sigma >_{q\bar{q}-0}; \quad \sigma = \tau, s \ .$$

* We suppress the dependence of β_σ on g_0, or alternatively on a, on which the coupling g_0 depends, since it is held fixed in the analysis.

Upon carrying out the differentiation (10.31), and then returning to the isotropic lattice $\xi = 1$, we are led to the relation

$$\hat{V}(\hat{R},\hat{\beta}) = -\left[\left(\frac{\partial\hat{\beta}_\tau}{\partial\xi}\right) < \mathcal{P}'_\tau >_{q\bar{q}-0} + \left(\frac{\partial\hat{\beta}_s}{\partial\xi}\right) < \mathcal{P}'_s >_{q\bar{q}-0}\right]_{\xi=1} .$$

Here $\hat{V}(\hat{R},\hat{\beta})$ is the potential in lattice units computed on an isotropic lattice. This expression can be written in the form

$$\hat{V}(\hat{R},\hat{\beta}) = \eta_- < -\mathcal{P}'_\tau + \mathcal{P}'_s >_{q\bar{q}-0} - \eta_+ < \mathcal{P}'_\tau + \mathcal{P}'_s >_{q\bar{q}-0} , \qquad (10.32a)$$

where

$$\eta_\pm = \frac{1}{2}\left[\left(\frac{\partial\hat{\beta}_\tau}{\partial\xi}\right)_{\xi=1} \pm \left(\frac{\partial\hat{\beta}_s}{\partial\xi}\right)_{\xi=1}\right] . \qquad (10.32b)$$

Consider the first term appearing on the rhs of (10.32a). In the continuum limit $g_0 \to 0$ (or $\hat{\beta} \to \infty$). From the weak coupling relations (10.17) one finds that

$$\eta_- \underset{\hat{\beta}\to\infty}{\longrightarrow} \hat{\beta} . \qquad (10.33)$$

Hence

$$\begin{aligned} \eta_- < -\mathcal{P}'_\tau + \mathcal{P}'_s >_{q\bar{q}-0} &\underset{\hat{\beta}\to\infty}{\longrightarrow} \hat{\beta} < -\mathcal{P}'_\tau + \mathcal{P}'_s >_{q\bar{q}-0} \\ &\underset{naive}{\longrightarrow} a \sum_{\vec{x}} a^3 \frac{1}{2} < -E^2(\vec{x}) + B^2(\vec{x}) >_{q\bar{q}-0} , \end{aligned} \qquad (10.34)$$

where we have taken the naive continuum limit in the last step. This suggests that the first term appearing on the rhs of (10.32a) is the (euclidean) lattice version of the usual Minkowsky field energy of a quark-antiquark pair (measured relative to the vacuum), coming from the traceless part of the energy momentum tensor. Notice that just as in our quantum mechanical example, the contribution arising from the "kinetic term", involving the time like plaquettes (which is associated with the electric field energy density), carries a negative sign in the euclidean formulation.

We now show that for the case of a linearly rising (confining) potential one can actually determine what fraction of the potential is made from "ordinary" field energy. By making use of the action sum rule (10.28), we can cast the energy sum rule (10.32) in the form

$$\hat{V}(\hat{R},\hat{\beta}) + \eta_+ \frac{\partial ln a}{\partial\hat{\beta}}\left[\hat{V}(\hat{R},\hat{\beta}) + \hat{R}\frac{\partial\hat{V}(\hat{R},\hat{\beta})}{\partial\hat{R}}\right] = \eta_- < -\mathcal{P}'_\tau + \mathcal{P}'_s >_{q\bar{q}-0} . \qquad (10.35)$$

The combination η_+ defined in (10.32b) has been determined non-perturbatively by Karsch (1982), by requiring that the string tension determined either from a space-like or time like Wilson loop on an anisotropic lattice should yield the same result. Karsch finds that

$$\eta_+ = -\frac{1}{4}\frac{\partial\hat{\beta}}{\partial \ln a} = -\frac{\beta_L(g_0)}{2g_0}\hat{\beta}\,, \tag{10.36a}$$

where

$$\beta_L(g_0) = -a\frac{\partial g_0}{\partial a} \tag{10.36b}$$

is the β- function discussed in chapter 9. Hence $\eta_+(\partial \ln a/\partial\hat{\beta}) = -1/4$, so that (10.32a) becomes

$$\hat{V}(\hat{R},\hat{\beta}) - \frac{1}{4}\left[\hat{V}(\hat{R},\hat{\beta}) + \hat{R}\frac{\partial\hat{V}(\hat{R},\hat{\beta})}{\partial\hat{R}}\right] = \eta_- < -\mathcal{P}'_\tau + \mathcal{P}'_s >_{q\bar{q}-0} \tag{10.37}$$

For the case of a linearly rising (confining) potential this equation reduces to

$$\frac{1}{2}\hat{V}_{conf}(\hat{R},\hat{\beta}) = \eta_- < -\mathcal{P}'_\tau + \mathcal{P}'_s >_{q\bar{q}-0}\,, \tag{10.38}$$

which, according to (10.34) suggests that the "normal" field energy accounts for only one half of the interquark potential. The other half must therefore be provided by the second term in (10.32a). If the first term in (10.32a) is the lattice version of the semiclassical field energy, then the second term should be the contribution coming from the trace anomaly. Thus making use of (10.36a) we have that

$$-\eta_+ < \mathcal{P}'_\tau + \mathcal{P}'_s >_{q\bar{q}-0} = \frac{1}{4}\left(\frac{2\beta_L}{g_0} < \hat{L} >_{q\bar{q}-0}\right)\,, \tag{10.39a}$$

where

$$\hat{L} = \hat{\beta}(\mathcal{P}'_\tau + \mathcal{P}'_s) \tag{10.39b}$$

is the (dimensionless) lattice version of the euclidean continuum Lagrangian density integrated over all space at a fixed time. The energy sum rule (10.32a) therefore becomes

$$\hat{V}(\hat{R},\hat{\beta}) = \eta_- < -\mathcal{P}'_\tau + \mathcal{P}'_s >_{q\bar{q}-0} + \frac{1}{4}\left(\frac{2\beta_L}{g_0} < \hat{L} >_{q\bar{q}-0}\right)$$

The quantity $(2\beta_L/g_0)\hat{L}$ in (10.40a) has the form of the anomalous contribution to the Hamiltonian following from the trace anomaly as computed in lattice perturbation theory by Caracciolo, Menotti and Pelisetto (Caracciolo, 1992). For weak coupling the anomalous contribution to the potential (10.40), in physical units, takes the form

$$V_{anom}(R,\hat{\beta}(a),a) = \frac{\beta_L(g_0)}{2g_0}\int d^3x\frac{1}{2} < [E^2(x)+B^2(x)] >_{q\bar{q}-0} \ . \qquad (10.41)$$

The rhs is a finite, renormalization group invariant expression. It can be expressed in terms of a renormalized coupling constant g, and renormalized squared colour electric and magnetic fields. The form remains the same, except that $\beta_L(g_0)/g_0$ is replaced by $\beta(g)/g$, where $\beta(g) = \mu\partial g/\partial\mu$ is the continuum beta-function, with μ the renormalization scale.* The right hand side of (10.41) then just becomes 1/4 of the space integral of the trace anomaly (10.19) expressed in terms of the euclidean fields.

Finally let us return to the action sum rule in the form (10.28). From (10.36a) and (10.39b) we see that the rhs of (10.28) is directly related to the trace anomaly. Thus we can write the action sum rule in the form

$$\hat{V}(\hat{R},\hat{\beta}) + \hat{R}\frac{\partial\hat{V}(\hat{R},\hat{\beta})}{\partial\hat{R}} = \frac{2\beta_L}{g_0} < \hat{L} >_{q\bar{q}-0} \ . \qquad (10.42)$$

The rhs of this equation is just the trace of the energy momentum tensor summed over the spatial lattice sites at fixed (euclidean) time. Hence for a confining potential of the form $\hat{V}_{conf} = \hat{\sigma}\hat{R}$ the second term appearing on the rhs of (10.40) yields, as expected, just 1/2 of the potential.

10.4 Determination of the Electric, Magnetic and Anomalous Contribution to the $q\bar{q}$ Potential

The lattice energy sum rule has been checked in lattice perturbation theory by Feuerbacher (2003a,b) up to $\mathcal{O}(g_0^4)$. The computations are very involved. Some technical details are given in Appendix A.

The reader may wonder: why check the sum rule? Is it not exact? While the action sum rule (10.24a) is an identity following from the definition of the $q\bar{q}$-potential via the Wilson loop, the derivation of the energy sum rule (10.40) relies on a number of input informations which, although quite plausible, are not self evident: i) The structure of the action on an anisotropic lattice is taken to be given by (10.16). This is the standard form of the action considered in the literature;

* See e.g. Dosch et al. (1995)

ii) Close to the continuum limit the potential should become independent of the anisotropy of the lattice, ξ, and satisfy (10.31). This must of course be so, if it is to be an observable. It is however not self evident that this is indeed the case; iii) A perturbative check of the sum rule requires a perturbative expression for $\eta_\pm$. This expression has been given by Karsch (Karsch, 1985). Hence checking the sum rule in perturbation theory would not only confirm the presence of a contribution to the potential arising from the trace anomaly of the energy momentum tensor, but also confirm indirectly the perturbative relations obtained from other considerations in the literature. Having checked the sum rule (at least in perturbation theory), one can extract the electric, magnetic and anomalous contributions to the potential. To this effect let us first write the energy sum rule (10.40) in the form

$$\hat{V}(\hat{R},\hat{\beta}) = \lim_{\hat{T}\to\infty} \frac{1}{\hat{T}}\left[\eta_- < -\mathcal{P}_\tau + \mathcal{P}_s >_{q\bar{q}-0} + \frac{\beta_L(g_0)}{2g_0}\hat{\beta} < \mathcal{P}_\tau + \mathcal{P}_s >_{q\bar{q}-0}\right] , \tag{10.43a}$$

where we made use of the definition (10.39b), and, in accordance with (10.25a), we have made the replacement

$$< \mathcal{P}'_\sigma > = \lim_{T\to\infty} \frac{1}{T} < \mathcal{P}_\sigma > . \tag{10.43b}$$

By combining this expression with the action sum rule (10.24a) one finds for $SU(N)$

$$\begin{aligned}
V_{elec} &\equiv \lim_{\hat{T}\to\infty} \frac{1}{T}\eta_- < -\mathcal{P}_\tau >_{q\bar{q}-0} = \frac{1}{2}V + \frac{1}{4}g_0\beta_L(g_0)\frac{\partial \hat{V}}{\partial g_0^2} + \eta_- \frac{g_0^4}{4N}\frac{\partial \hat{V}}{\partial g_0^2} , \\
V_{magn} &\equiv \lim_{\hat{T}\to\infty} \frac{1}{T}\eta_- < \mathcal{P}_s >_{q\bar{q}-0} = \frac{1}{2}V + \frac{1}{4}g_0\beta_L(g_0)\frac{\partial \hat{V}}{\partial g_0^2} - \eta_- \frac{g_0^4}{4N}\frac{\partial \hat{V}}{\partial g_0^2} , \\
V_{anom} &\equiv -\frac{1}{2}g_0\beta_L(g_0)\frac{\partial \hat{V}}{\partial g_0^2} ,
\end{aligned} \tag{10.44}$$

where, according to (10.17), and (10.32b), and making use of $c_\tau(1) = c_s(1) = 0$ (see sec. 10.2),

$$\eta_- = \frac{2N}{g_0^2} + N(c'_\tau - c'_s) + \mathcal{O}(g_0^2) . \tag{10.45}$$

The derivatives $c'_\sigma \equiv [dc_\sigma(\xi)/d\xi]_{\xi=1}$ have been determined by Karsch (1985).

Fig. 10-1, taken from Feuerbacher (2003) shows the electric, magnetic and anomalous contributions to the $SU(3)$ potential computed from (10.44) and the expression (9.8). Notice that while the leading electric contribution to the potential

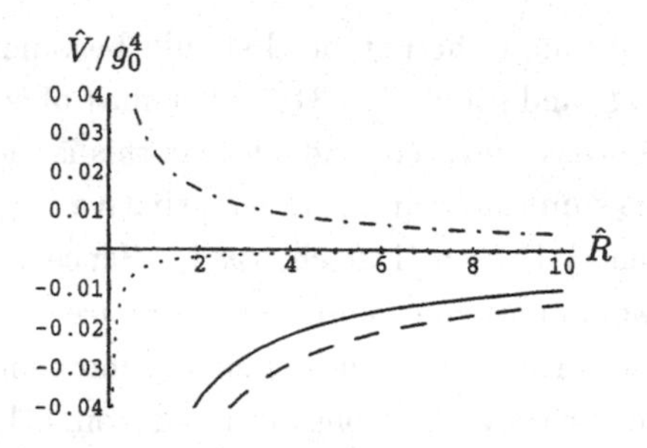

Fig. 10-1 The next to leading order contributions, divided by g_0^2, of the electric field energy (dashed), magnetic field energy (dot-dashed) and trace anomaly (dotted) to the $SU(3)$ $q\bar{q}$ potential (solid line). The figure is taken from Feuerbacher (2003).

is of $\mathcal{O}(g_0^2)$, the magnetic contribution is of $\mathcal{O}(g_0^4)$, and same is true for the anomalous part. Note also that in the perturbative regime the trace anomaly contributes significantly only for small $\hat{R}$, and that the electric and magnetic contributions are of opposite sign, while in lattice simulations carried out in the *non-perturbative* region for large quark-antiquark separations they are found to be of the same sign. The solid curve in fig. (10-1) is the full potential in $\mathcal{O}(g_0^4)$. This curve agrees with that obtained from a perturbative calculation of the rhs of the energy sum rule (10.43).

10.4 Sum Rules for the Glueball Mass

The interquark potential is not the only observable for which one can derive an action and energy sum rule. In fact, the way the sum rules (10.40) and (10.42) were derived, it is evident that similar expressions will hold for any observable which can be determined from the exponential decay in euclidean time of the expectation value of some operator. In the pure SU(N) gauge theory, such observables are the masses of glueballs states, which are eigenstates of the SU(N) Hamiltonian. In numerical simulations particle masses are determined from the exponential decay in euclidean time of correlators of operators which excite the state of interest. In principle the specific form of the operator is not important, as long as it excites states having a non-vanishing projection onto the state of interest. In practice, however, a judicious choice needs to be made in order to enhance the signal in Monte Carlo calculations. The anomalous contribution to the Hamiltonian, arising from the trace anomaly, should also manifest itself in the energy and action sum rules for the glueball mass. To determine the mass of a glueball with a given set of quantum numbers one can construct such correlators from combinations of space-like Wilson loops located at times $-\frac{\hat{T}}{2}$ and $\frac{\hat{T}}{2}$, averaged over all spatial lattice sites to project out a zero-momentum state [Michael (1987)]. Let us denote the corresponding operators by $G(-\hat{T})$ and $G(\hat{T})$. The lowest glueball mass with non-vacuum quantum numbers is then given by

$$\hat{M} = -\lim_{\hat{T}\to\infty} \frac{1}{\hat{T}} \ln < G(\hat{T})G(-\hat{T}) > . \tag{10.46}$$

which is the analog of (10.21a). Proceeding in exactly the same way as for the case of the $q\bar{q}$ potential one then arrives at the following action and energy sum rules for the glueball mass [Rothe (1995b)],

$$\hat{M} = \frac{2\beta_L(g_0)}{g_0} < \hat{L} >_{1-0} , \tag{10.47a}$$

$$\hat{M} = \eta_- < -\mathcal{P}'_\tau + \mathcal{P}'_s >_{1-0} + \frac{\beta_L(g_0)}{2g_0} < \hat{L} >_{1-0} , \tag{10.47b}$$

where $\hat{L}$ has been defined in (10.39b), and where the bracket $< \mathcal{O} >_{1-0}$ stands generically for the following correlator

$$< \mathcal{O} >_{1-0} = \lim_{\hat{T}\to\infty} \frac{< G(\hat{T})\mathcal{O}\mathcal{G}(-\hat{T}) >}{< \mathcal{G}(\hat{T})\mathcal{G}(-\hat{T}) >} - < \mathcal{O} > .$$

Notice that the energy sum rule (10.44b) has exactly the same form as for the $q\bar{q}$-potential given by (10.40). The form of the action sum rule (10.44a) however differs from (10.42), since in the case of the glueball mass, there is no analog of the derivative term. This manifests itself in that the contribution of the anomalous field energy to the glueball mass accounts for only 1/4 of the glueball mass, as follows immediately from (10.44a). This result is consistent with that obtained by Michael (1996) for the contribution of the "normal field energy" to the glueball mass.

As we have seen in this chapter, the lattice formulation of the pure SU(N) gauge theory on an anisotropic lattice has allowed us to derive in a straight forward way a non-perturbative expression for the energy stored in the chromoelectric and magnetic fields of a static $q\bar{q}$-pair and glueball, in which the contributions arising from the traceless and trace part of the energy momentum tensor are clearly exhibited. In principle these sum rules provide us with an alternative way to determine the string tension or glueball mass. But what is more important, these sum rules allow us to obtain detailed information about the distribution of the field energy in a flux tube connecting a static quark and antiquark. In most Monte Carlo simulations only the action density has been measured with good precision, since it is more accessible to Monte Carlo simulations (For a recent computation see Bali (1995)). The energy density, on the other hand, including the anomalous contribution, has so far not been studied in such detail in the literature.

CHAPTER 11

THE STRONG COUPLING EXPANSION

In chapters 7 and 8 we have shown how the static $q\bar{q}$-potential $V(R)$ can be determined by studying the expectation value of the Wilson loop for large euclidean times. In QCD one believes that this potential confines quarks; more precisely, one expects that for large separations of the quark-antiquark pair, $V(R)$ rises linearly with R up to distances where vacuum polarization effects, due to the presence of dynamical fermions, screen the interaction. As we have seen, such a behaviour of the potential cannot be generated within perturbation theory. For this reason, we have no way (at present) to calculate it analytically, and hence are forced to determine it numerically. On the other hand, analytic statements can be made in the strong coupling region. Indeed, in the absence of dynamical fermions the structure of the action (5.21) for QED, and (6.25b) for QCD suggests a natural expansion in powers of the inverse coupling. This is the analog of the high temperature expansion in statistical mechanics. In the following section we shall concentrate on the leading strong coupling approximation to the static $q\bar{q}$ potential, ignoring vacuum polarization effects. As was first shown by Wilson (1974), this potential confines quarks. In fact it was this observation which has stimulated the great interest in lattice gauge theories.

11.1 The $q\bar{q}$-Potential to Leading Order in Strong Coupling

Consider the $SU(N)$ lattice gauge theory in the pure gauge sector *. The corresponding lattice action is given by

$$S = -\beta \sum_P S_P \; + \; const.,$$

where

$$\beta = 2N/g_0^2,$$

and **

$$S_P = \frac{1}{2N} Tr(\underset{\sim}{U}_P + \underset{\sim}{U}_P^\dagger)$$

---

* We give the formulation for $SU(N)$, since we are interested later also in the case of $SU(2)$.

** S_P has been normalized in such a way that $S_P \to 1$ if $U_P \to 1$, i.e. close to the continuum limit. The links $U_\mu(n)$ and plaquette variables U_P lie in the N-dimensional fundamental representation of $SU(N)$.

is the contribution to the action associated with a plaquette P (see chapter 6). The corresponding partition function reads

$$Z = \int DU\ e^{\beta \sum_P S_P} \quad , \tag{11.1}$$

and the expectation value of the Wilson loop with spatial and temporal extention $\hat{R}$ and $\hat{T}$, respectively, is given by

$$\langle W_C[U] \rangle = \frac{\int DU\ W_C[U] \prod_P\ e^{\beta S_P}}{\int DU\ \prod_P\ e^{\beta S_P}} \quad , \tag{11.2}$$

where $W_C[U]$ is defined in (7.33). We next expand the exponential in (11.2) in powers of the coupling β:

$$e^{\beta \sum_P S_P} = \prod_P \left[\sum_n \frac{\beta^n}{n!} (S_P)^n \right] \quad . \tag{11.3}$$

Since each plaquette in the expansion costs a factor β, the leading contribution, for $\beta \to 0$ to the numerator in (11.2) is obtained by paving the inside of the Wilson loop with the smallest number of elementary plaquettes yielding a non-vanishing value for the integral. Consider the case of $SU(3)$. As is evident from the integration rules (6.23), the relevant configuration is the one shown in fig. (11-1). Hence the leading term in the strong coupling expansion of the numerator is proportional to $\beta^{\hat{A}}$, where $\hat{A}$ is the minimal area bounded by the rectangular contour C: $\hat{A} = \hat{R}\hat{T}$. On the other hand, the leading contribution to the denominator is obtained by making the replacement $\exp(\beta \sum S_P) \to 1$. Hence for small β, (11.2) will be proportional to $\left(\frac{\beta}{6}\right)^{\hat{R}\hat{T}}$.

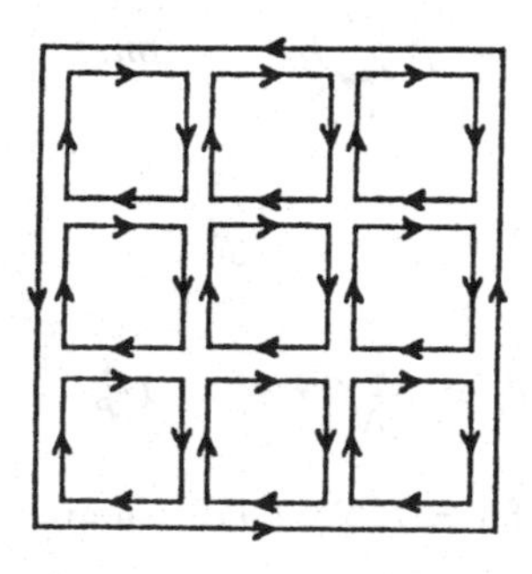

Fig. 11-1 Leading contribution to $\langle W_C \rangle$ in the strong coupling approximation.

The factor multiplying this expression may also be readily calculated by noting that according to (6.23c) the colour indices of the link variables are identified at each of the lattice sites, and that the integration over each pair of oppositely oriented links yields a factor of 1/3. Now the number of link integrations for a given set of colour indices is $2\hat{R}\hat{T}+\hat{R}+\hat{T}$. Hence there is a factor $(1/3)^{2\hat{R}\hat{T}+\hat{R}+\hat{T}}$ coming from the integrations. But there are three possible colours associated with the $(\hat{R}+1)(\hat{T}+1)$ lattice sites. Consequently the factor multiplying $(\beta/6)^{\hat{A}}$ is given by $(1/3)^{\hat{A}-1}$. We hence conclude that in leading order of strong coupling

$$\langle W_c[U]\rangle \approx 3\left(\frac{\beta}{18}\right)^{\hat{R}\hat{T}} .$$

The $q\bar{q}$-potential in the strong coupling limit is therefore given by

$$\hat{V}(\hat{R}) = -\lim_{\hat{T}\to\infty}\frac{1}{\hat{T}}\ln\langle W_C[U]\rangle = \hat{\sigma}(g_0)\hat{R}, \tag{11.4a}$$

where

$$\hat{\sigma} = -\ln\left(\frac{\beta}{18}\right) \tag{11.4b}$$

is the string tension measured in lattice units. Thus in the leading strong coupling approximation, QCD confines quarks: the expectation value of the Wilson loop exhibits an area law behaviour, $\langle W_C\rangle \to \exp(-\hat{\sigma}\hat{R}\hat{T})$. If this confining property persists into the small coupling regime, where continuum physics is (hopefully) observed, then $\hat{\sigma}(g_0)$ must depend on g_0 according to (9.25). In the one-loop approximation to the β-function this dependence on g_0 is given by

$$\hat{\sigma}(g_0) \underset{g_0\to 0}{\approx} C_\sigma\, e^{-\frac{1}{\beta_0 g_0^2}} . \tag{11.5}$$

Its explicitly known form provides us with a signal which will tell us whether we are extracting continuum physics from a numerical calculation.

We want to point out that the same calculation performed for compact QED would also have given a potential of the form (11.4a). But in QED we know that this potential is given by the Coulomb law! Hence compact QED must exhibit at least two phases. In fact it has been shown by Guth (1980), using an action of the Villain form, that the lattice $U(1)$ gauge theory possesses a weak coupling Coulomb phase. That the strong coupling and weak coupling regions are separated by a phase transition, has also been verified in numerical simulations. (see e.g., Lautrup and Nauenberg, 1980). It is therefore important to check that in the case of QCD there exists no such phase transition to a weak coupling regime.

11.2 Beyond the Leading Approximation

Higher order corrections to the potential (11.4) can in principle be computed by expanding the exponentials appearing in (11.2) in powers of the coupling β and performing the corresponding (Haar) integrals over the link variables. As is evident from the expansion (11.3), the contribution of order β^n will involve (before integration) all possible sets consisting of n plaquettes, including also multiples of the same plaquette. To each oriented plaquette P in a given diagram we associate a factor $\left(\frac{\beta}{6}\right)\ Tr\ \underset{\sim}{U}_P$, or $\left(\frac{\beta}{6}\right)\ Tr\ \underset{\sim}{U}_P^\dagger$, depending on its orientation, and include a factor $1/n!$, where n is the multiplicity of P. However, of all the possible sets contributing to the sum (11.3), only a certain subset of plaquette configurations will survive the integrations in (11.2) because of the integration rules (6.23).

Thus we are faced with two problems in computing higher order corrections to the potential: i) enumerating all the diagrams contributing in a given order, and ii) calculating the Haar integrals. As the reader can imagine, already the bookkeeping problem associated with the first step will become non-trivial as we go to higher and higher orders.

To keep the discussion as simple as possible, let us exemplify the method for the two dimensional abelian model discussed in chapter 8. As we shall see it will serve to illustrate the basic point we wish to make. Consider QED_2 in the quenched approximation. The partition function (11.1) has the form

$$Z = \int DU\ e^{\frac{\beta}{2}\sum_P (U_P + U_P^\dagger)}, \tag{11.6a}$$

where

$$U_P = e^{i\theta_P} \tag{11.6b}$$

is an element of the $U(1)$ group associated with the plaquette P. The expectation value of the Wilson loop is then given by (8.15). Expanding the exponentials appearing in the integrands of this expression in powers of the coupling β, one finds that in the next to leading order the only plaquette configurations which contribute to the numerator and denominator are those shown in Fig. (11-2), since diagrams containing any unpaired link will not contribute after performing the integrals over the link variables. *

* Notice that link variables of opposite orientations must appear paired.

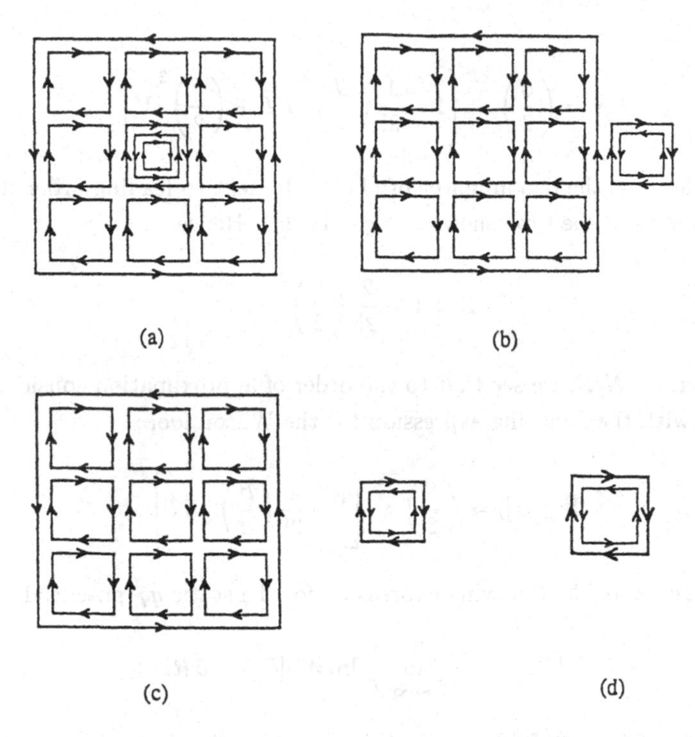

Fig. 11-2 Next to leading order contributions to $\langle W_C \rangle$ in the strong coupling expansion. Diagrams (a-c) contribute to the numerator (11.2), while diagram (d) contributes to the denominator.

Hence we have three types of diagrams contributing to the numerator in this non-leading order: i) diagrams where the inside of the Wilson loop contains a triple of the same plaquette; the number of such diagrams is given by $3\hat{R}\hat{T}$;** ii) diagrams where a multiple plaquette is attached to the contour C, as shown in Fig. (11-2b). The number of these diagrams is $N_c \approx 4(\hat{R}+\hat{T})$; and finally iii) disconnected diagrams of the type shown in Fig. (11-2c). The number of such diagrams is approximately given by $2(\hat{V} - \hat{R}\hat{T} - N_c/2)$, where $\hat{V}$ is the "volume" of the lattice. Since in the $U(1)$ case the group integrals are trivial, we obtain the following expression for the numerator, including the leading contribution,

** In counting the number of each type of diagram we must take into account the various possible relative orientations of the multiple plaquettes. Thus the factor 3 arises from the expansion of $(U + U^\dagger)^3 = 3U^\dagger UU + \cdots$.

$(\beta/2)^{\hat{R}\hat{T}}$:

$$N \approx \left(\frac{\beta}{2}\right)^{\hat{R}\hat{T}} \left[1 - \frac{1}{2!}\left(\frac{\beta}{2}\right)^2 \hat{R}\hat{T} + \left(\frac{\beta}{2}\right)^2 \hat{V}\right].$$

Consider now the denominator of (11.2). In next to leading order it is given by the diagrams of the type shown in fig. (11-2d). Hence

$$Z \approx 1 + \frac{2}{2!}\left(\frac{\beta}{2}\right)^2 \hat{V}.$$

Taking the ratio N/Z, we see that to the order of approximation considered here, we are left with the following expression for the Wilson loop:

$$\langle W_C[U]\rangle \approx \left(\frac{\beta}{2}\right)^{\hat{R}\hat{T}} \left[1 - \frac{1}{2!}\left(\frac{\beta}{2}\right)^2 \hat{R}\hat{T}\right].$$

Hence we are led to the following expression for the static $q\bar{q}$–potential,

$$V(\hat{R}) = -\lim_{\hat{T}\to\infty} \frac{1}{\hat{T}} \ln\langle W_C[U]\rangle = \hat{\sigma}\hat{R},$$

where the string tension $\hat{\sigma}$ is given in this approximation by

$$\hat{\sigma} = -\ln\frac{\beta}{2} + \frac{\beta^2}{8}. \tag{11.7}$$

Notice that this potential is determined solely by connected plaquette configurations attached to the minimal surface enclosed by the contour C.

The $U(1)$ case discussed above is of course the simplest example we could choose. Nevertheless, the number of connected and disconnected diagrams contributing to (11.2) will rapidly increase with the order. In particular, the occurrence of multiples of a given plaquette will complicate the bookkeeping and the computation of the Haar integrals in the non-abelian gauge theory. This latter difficulty may, however, be avoided by making use of the so-called character expansion of the exponential of the action. We again demonstrate the procedure for the case of quenched QED_2, where the character expansion is just the well-known Fourier-Bessel expansion of $\exp(\beta\cos\theta_P)$:

$$e^{\frac{\beta}{2}(U_P+U_P^\dagger)} = \sum_{\nu=-\infty}^{\infty} I_\nu(\beta)e^{i\nu\theta_P}. \tag{11.8}$$

$I_\nu(\beta)$ is the modified Bessel function and $exp(i\nu\theta_P)$ is the character of the plaquette variable (11.6b) in the ν'th irreducible representation of the compact $U(1)$ group. Hence instead of having to deal with contributions to (11.3) arising from multiples of the same plaquettes, every plaquette will now occur only once in the expansion (11.8), but in all possible irreducible representations of the compact $U(1)$ group. But because of the orthogonality relations satisfied by $exp(i\nu\theta)$, only a few terms in the sum (11.8) will contribute to (11.2). As an example let us derive the result (11.7) using the expansion (11.8). Making use of the fact that in the present case the Wilson loop can be written in the form (8.15), one finds upon inserting (11.8) in (8.15), that the numerator N is given by*

$$N = \prod_{P \in R_c} \int_{-\pi}^{\pi} \frac{d\theta_P}{2\pi} \left\{ e^{i\theta_P} \sum_\nu I_\nu(\beta) e^{i\nu\theta_P} \right\} \prod_{P \notin R_c} \int_{-\pi}^{\pi} \frac{d\theta_P}{2\pi} \sum_\nu I_\nu(\beta) e^{i\nu\theta_P}$$

where R_c denotes the region enclosed by the loop C. The integrals can be performed immediately and one obtains

$$N = [I_1(\beta)]^{\hat{R}\hat{T}} [I_0(\beta)]^{\hat{V} - \hat{R}\hat{T}}.$$

The corresponding expression for the denominator in (8.15) reads:

$$\begin{aligned} Z = & \prod_{P \in \hat{V}} \int_{-\pi}^{\pi} \frac{d\theta_P}{2\pi} \sum_\nu I_\nu(\beta) e^{i\nu\theta_P} \\ = & [I_0(\beta)]^{\hat{V}}. \end{aligned}$$

Hence the volume-dependent factor cancels in the ratio N/Z and we obtain the exact result

$$\langle W_C[U] \rangle = \left(\frac{I_1(\beta)}{I_0(\beta)} \right)^{\hat{R}\hat{T}} \tag{11.9}$$

which coincides with (8.17b). Now for small β we have that

$$\begin{aligned} I_0(\beta) \approx & 1 + \left(\frac{\beta}{2}\right)^2 \\ I_1(\beta) \approx & \left(\frac{\beta}{2}\right) \left[1 + \frac{1}{2!} \left(\frac{\beta}{2}\right)^2 \right] \end{aligned} .$$

Inserting these expressions into (11.9), we come back to our result (11.7).

---

* We have normalized the integration measure so that $\int dU = 1$.

The fact that we were able to solve the above problem to any order in β is connected with the 2-dimensional abelian nature of the model. In four-dimensional gauge theories more sophisticated methods are required for carrying out the strong coupling expansion. The interested reader may consult the extensive work carried out by Münster (1981) who has calculated the string tension for the case of an $SU(2)$ gauge theory up to twelvth order in β.

11.3 The Lattice Hamiltonian in the Strong Coupling Limit and the String Picture of Confinement

We have seen in the previous sections that for strong coupling the pure $SU(3)$ gauge theory confines quarks. The usual picture of confinement is that the chromoelectric flux linking a quark and antiquark is squeezed within a narrow tube (string) carrying constant energy density. As a consequence, the energy of the system increases linearly with the separation of the quarks. The purpose of this section is to verify the string picture of confinement in the strong coupling approximation within the framework of the Hamiltonian formulation as discussed originally by Kogut and Susskind (1975).

In the continuum formulation, the Hamiltonian is the generator of infinitesimal time translations and is obtained by a Legendre transformation from the Lagrangean of the theory. Within the framework of the lattice formulation the natural way of introducing the Hamiltonian is via the transfer matrix. In chapter 2 we defined the transfer matrix for the case of a quantum mechanical system with generalized coordinates $q_\alpha (\alpha = 1, ..., n)$ by

$$T_{q'q} = \langle q'|e^{-\epsilon H}|q\rangle. \tag{11.10}$$

Here H is the Hamiltonian of the system, which, in the particular example considered, had the form (2.8), and ϵ is an infinitesimal time–step. The corresponding transfer matrix was given by (2.37b). Expressed in terms of the transfer matrix, the partition function (2.32b) took the form

$$Z = \int \prod_{\alpha,\ell'} dq_\alpha^{(\ell')} \prod_\ell T_{q^{(\ell+1)}q^{(\ell)}} \tag{11.11}$$

where $q^{(k)} = \{q_\alpha^{(k)}\}$ could be interpreted as the coordinates of the system at "time" τ_k on the euclidean time lattice. We then discussed a method which allowed us to construct the Hamiltonian from the knowledge of the transfer matrix. In this section, we want to apply this method to the case of a lattice gauge theory. To this effect we will have to choose a gauge. The reason for this is the following. In the lattice formulation of a gauge theory, the action is a function of the link variables which live on all the links of the euclidean space-time lattice. Therefore the action not only depends on the group parameters associated with the links lying in fixed time slices, but also on those parametrizing the link variables that live on links connecting sites at different times. We can, however, eliminate the latter (bothersome) variables by choosing a gauge where all time-like oriented links* are replaced by the unit matrix. This can always be achieved by an appropriate gauge transformation, and corresponds in the continuum formulation to choosing the temporal gauge, $A_4(x) = 0$. This we are allowed to do, since we are computing a gauge-invariant observable, namely the Hamiltonian.

Of course, setting $A_4(x) = 0$ does not fix the gauge completely. We are still free to perform time-independent gauge transformations. This allows us to fix a subset of the remaining link variables oriented along the space directions. The variables that can be fixed by a time-independent gauge transformation are restricted by the requirement that there exist no closed paths on the lattice along which all the degrees of freedom are "frozen" (see Fig. (11-3)). This is obvious since the trace of the path-ordered product of link variables along closed paths is gauge-invariant.

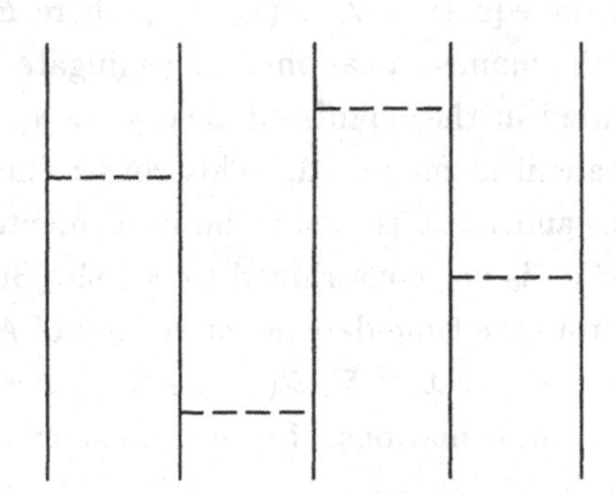

Fig. 11-3 The temporal gauge corresponds to fixing the link variables on the solid lines to unity. The dashed links can be fixed by time independent gauge transformations.

That the temporal gauge is the appropriate one for constructing the Hamil-

* Sometimes we use the word "link" instead of "link variable" if it is clear from the context what is meant.

tonian via the transfer matrix is also strongly suggested by looking at the continuum action of the pure $SU(3)$ gauge theory. This action is given by (6.21), where $F^B_{\mu\nu}(B = 1,...8)$ has been defined in (6.16). Setting $A_4 = 0$, the action takes the following simple form

$$S_G[A] = \int d\tau \int d^3x \left[\frac{1}{2} \sum_{i,B} (\dot{A}^B_i(\vec{x},\tau))^2 + \frac{1}{4} \sum_{i,j,B} F^B_{ij}(\vec{x},\tau) F^B_{ij}(\vec{x},\tau) \right],$$

where we have set $x_4 = \tau$ and where $\dot{A}^B_i$ denotes the "time"–derivative of A^B_i. Notice the striking similarity between the above expression and (2.11).

The lattice version of the action in the temporal gauge is of course more complicated (see previous chapter). But the above observation suggests that also there the action will acquire a structure which will allow us to write the partition function in the form (11.11).

There is another important point that must be mentioned. By eliminating the time component of the gauge potential in the action, we are loosing one of "Maxwell's" equations, namely Gauss's law. Consider for simplicity continuum electrodynamics in the absence of sources. The following discussion can be readily extended to the case of a non–abelian theory and to euclidean space–time. Let us not fix the gauge. By varying the action with respect to the time component of the potential, one arrives at the equation $\vec{\nabla} \cdot \vec{E}(x) = 0$, where $\vec{E}$ is the electric field. The components E_i are the momenta canonically conjugate to A_i. Let us denote the corresponding operators in the quantized theory by π_i. Then $\vec{\nabla} \cdot \vec{\pi}(x) = 0$ is a constraint for the canonical momenta. This constraint is lost when we set $A_4(x) = 0$ in the action, and must therefore be implemented on the states. In the temporal gauge $\vec{\nabla} \cdot \vec{\pi}(x)$ is not constrained to vanish. But its time derivative still vanishes, as follows from the time-dependent version of Ampère's law. Now it can be easily shown that the operators $\vec{\nabla} \cdot \vec{\pi}(x)$ are the generators of infinitesimal time-independent gauge transformations. Indeed, consider the following unitary operator

$$T[\Lambda] = e^{-i\int d^3x \Lambda(\vec{x}) \vec{\nabla}\cdot\vec{\pi}(x)}.$$

Because $\vec{\nabla} \cdot \vec{\pi}$ is independent of the time, one finds, upon using the canonical commutation relations of the momenta and gauge potentials, that

$$T[\Lambda] A_i(x) T[\Lambda]^{-1} = A_i(x) + \partial_i \Lambda(\vec{x}).$$

Hence imposing Gauss's law on the states is equivalent to restricting the physical states to be invariant under the residual group of time-independent gauge transformations. This observation will play an important role when we discuss the energy spectrum of the Hamiltonian. In the presence of sources $\vec{\nabla} \cdot \vec{\pi}(x)$ is replaced by $\vec{\nabla} \cdot \vec{\pi}(x) - \rho(x)$, where ρ is the charge density. This latter quantity, constructed from the fermion fields, generates the corresponding transformations on the fermionic variables.

After these remarks, we now turn to the construction of the lattice Hamiltonian in a pure gauge theory. To keep the discussion as simple as possible, we will consider the abelian $U(1)$ theory discussed in chapter 5. This theory will already exhibit an important feature encountered in the non-abelian case. For the derivation of the lattice Hamiltonian in the $SU(3)$ gauge theory, using the transfer matrix approach, we refer the reader to the paper by Creutz (1977).

Our procedure for constructing the lattice Hamiltonian parallels the quantum mechanical case discussed in chapter 2. The first step consists in identifying the transfer matrix, and writing the partition function of the $U(1)$ gauge theory in a form analogous to (11.11). This is done by working in the temporal gauge. We then obtain the Hamiltonian by studying the transfer matrix for infinitesimal *temporal* lattice spacing. Since the spatial lattice spacing is to be kept fixed while taking the temporal lattice spacing to zero, we must first rewrite the action (5.21) on an asymmetric lattice. Following Creutz (1977) we take the U(1) action to be of the form

$$S_G[U] = \frac{1}{2g^2\rho}\sum_{P_\tau}[2 - (U_{P_\tau} + U^\dagger_{P_\tau})] + \frac{\rho}{2g^2}\sum_{P_s}[2 - (U_{P_s} + U^\dagger_{P_s})], \qquad (11.12)$$

where $\rho = a_T/a$, and where P_s and P_τ denote space-like and time-like oriented plaquettes, respectively. This is the analog of (10.15) with $\rho = 1/\xi$. This action possesses the correct naive continuum limit. We next choose the temporal gauge; i.e., we set all the link variables associated with time-like links equal to one. Then a plaquette variable with base $(\mathbf{n}, n_4)$, lying in the $i4$-plane, is given by

$$U_{i4}(n) = \square = U_i(\mathbf{n}, n_4)U_i^\dagger(\mathbf{n}, n_4 + 1) = e^{i(\theta_i(\mathbf{n},n_4) - \theta_i(\mathbf{n},n_4+1))},$$

where the dashed lines stand for link variables that have been set equal to one in the temporal gauge. The coordinate degrees of freedom of the system are labeled by the spatial lattice site $\mathbf{n}$ and the spatial direction of the link variables. Let us collect these labels into a single index $\alpha = (\mathbf{n}, i)$ and set $n_4 = \ell$. Furthermore, to parallel our quantum mechanical example, we relabel the group parameters, parametrizing the link variables, as follows:

$$\theta_\alpha^{(\ell)} \equiv \theta_i(\mathbf{n}, \ell).$$

Then the partition function associated with the action (11.12) can be written in a form analogous to (11.11):

$$Z = \int DU e^{-S_G[U]} = \int \prod_{\alpha, \ell'} d\theta_\alpha^{(\ell')} \prod_\ell T_{\theta^{(\ell+1)}\theta^{(\ell)}}, \tag{11.13a}$$

where

$$T_{\theta^{(\ell+1)}\theta^{(\ell)}} = e^{-a_\tau V[\theta^{(\ell)}]} \prod_\alpha e^{-\frac{1}{g^2\rho}[1-\cos(\theta_\alpha^{(\ell+1)} - \theta_\alpha^{(\ell)})]}, \tag{11.13b}$$

with the potential V defined by

$$V[\theta^{(\ell)}] = \frac{1}{g^2 a} \sum_{P_s(\ell)} [1 - \frac{1}{2}(U_{P_s(\ell)} + U^\dagger_{P_s(\ell)})]. \tag{11.13c}$$

The sum extends over all plaquettes located on the ℓ'th time slice. The potential (11.13c) is therefore a function of the group parameters labeling the link variables located on this time slice. We next introduce a set of commuting operators $\{\Theta_\alpha\}$ and simultaneous eigenstates of these by

$$\Theta_\alpha |\theta\rangle = \theta_\alpha |\theta\rangle,$$
$$\langle\theta'|\theta\rangle = \prod_\alpha \delta(\theta'_\alpha - \theta_\alpha).$$

The states $|\theta\rangle$ are the analogue of $|q\rangle$ in the quantum mechanical example discussed in chapter 2. Let P_α be the momentum canonically conjugate to Θ_α:

$$[\Theta_\alpha, P_\beta] = i\delta_{\alpha\beta}.$$

Then $exp(-i\xi \cdot P)$, with $\xi \cdot P = \sum_\alpha \xi_\alpha P_\alpha$, generates translations by ξ:

$$e^{-i\xi\cdot P}|\theta\rangle = |\theta + \xi\rangle.$$

From here, and the orthogonality of the states $|\theta\rangle$ we conclude that

$$\langle\theta^{(\ell+1)}|e^{-i\xi\cdot P}|\theta^{(\ell)}\rangle = \prod_\alpha \delta(\theta_\alpha^{(\ell+1)} - \theta_\alpha^{(\ell)} - \xi_\alpha).$$

Hence (11.13b) can also be written as follows

$$T_{\theta^{(\ell+1)}\theta^{(\ell)}} = \langle \theta^{(\ell+1)} | T | \theta^{(\ell)} \rangle, \tag{11.14a}$$

where

$$T = \prod_\alpha \int d\xi_\alpha e^{-i\xi_\alpha P_\alpha - \frac{1}{g^2\rho}(1-\cos\xi_\alpha)} e^{-a_\tau V[\Theta]}. \tag{11.14b}$$

To obtain the expression for the lattice Hamiltonian, we must now take the temporal lattice spacing to zero, while keeping the spatial lattice spacing fixed. This means that $\rho = a_\tau / a$ approaches zero. But for small ρ the integrand in (11.14b) is dominated by those values of ξ_α for which $\cos\xi_\alpha \approx 1$. We are then allowed to replace $1 - \cos\xi_\alpha$ by $\xi_\alpha^2/2$. Performing the remaining Gaussian integral, we therefore find that, apart from an irrelevant constant,

$$T = e^{-a_\tau(\frac{g^2}{2a}\sum_\alpha P_\alpha^2 + V[\Theta])}.$$

The quantity appearing within brackets is the lattice Hamiltonian we were looking for:

$$H = \frac{g^2}{2a} \sum_\alpha P_\alpha^2 + V[\Theta]. \tag{11.15}$$

In the coordinate representation the canonical momenta P_α are given by

$$P_\alpha = -i\frac{\partial}{\partial\theta_\alpha}.$$

Recall that α stands collectively for the set (n, i), where n is the spatial location of the lattice site and i is the direction of the link with base at n.* To make this explicit, we shall write $\theta_i(n)$ instead of θ_α. Then the Hamiltonian (11.15) becomes

$$H = -\frac{g^2}{2a} \sum_{n,i} \frac{\partial^2}{\partial\theta_i(n)^2} + V[\theta], \tag{11.16a}$$

where $V[\theta]$ has the following structure:

$$\begin{aligned} V[\theta] &= -\frac{1}{2g^2 a}\sum_{P_s}(\circlearrowleft + \circlearrowright) + \text{ const.} \\ &= -\frac{1}{2g^2 a}\sum (UUUU + \ h.c.) + \text{ const.} \end{aligned} \tag{11.16b}$$

* From now on it will be understood that n and x denote spatial positions.

Here $UUUU$ denote the product of link variables along the boundary of a plaquette on the spatial lattice. These link variables are parametrized by the angular variables $\theta_i(n)$. Thus, the contribution of a plaquette lying in the ij-plane with base at n is given by

$$\square_{ij} = e^{i\theta_i(n)} e^{i\theta_j(n+\hat{e}_i)} e^{-i\theta_i(n+\hat{e}_j)} e^{-i\theta_j(n)} .$$

Before we proceed with the discussion of (11.16), let us establish the connection with the Hamiltonian in the continuum formulation. To this effect we introduce the gauge potentials in the by now familiar way:

$$\theta_i(n) \to agA_i(x), \quad x = na.$$

The Hamiltonian (11.16) then takes the following form in the continuum limit

$$H \underset{a\to 0}{\longrightarrow} -\frac{1}{2}\int d^3x \sum_i \left(\frac{\delta}{\delta A_i(x)}\right)^2 + \tilde{V}[A], \tag{11.17a}$$

where

$$\tilde{V}[A] = \frac{1}{4}\sum_{i,j}\int d^3x F_{ij}(x)F_{ij}(x) \tag{11.17b}$$

is the contribution arising from the spatial plaquettes in (11.16b) and $\delta/\delta A_i(x)$ denotes the functional derivative with respect to $A_i(x)$. Its action on $A_j(y)$ is defined by

$$\frac{\delta}{\delta A_i(x)} A_j(y) = \delta_{ij}\delta^{(3)}(x-y).$$

The right-hand side of (11.17a) is nothing but the familiar expression for the temporal gauge-Hamiltonian in the so-called field representation, where the potentials $A_i(x)$ are ordinary functions of the spatial coordinates and where the components E_i of the electric field (which are the momenta canonically conjugate to A_i) are represented by the functional derivative $-i\delta/\delta A_i$. This Hamiltonian acts on the space of "wave functions", which are functionals of the spatial components of the vector potential. We therefore see that in the case of the $U(1)$ gauge theory the kinetic part of the lattice Hamiltonian has a structure which is similar to that of the continuum formulation. There is, however, an important difference. While in the continuum formulation the potentials can take arbitrary values, the group parameters $\theta_i(n)$ take values in the interval $[0, 2\pi]$. The wave functions are single valued functions of these variables. As a consequence the eigenvalue spectrum of the kinetic term in (11.16a) is discrete. As seen from (11.16a,b), it is this term which dominates the Hamiltonian in the strong coupling limit:

$$H \to H_0 = -\frac{g^2}{2a}\sum_{n,i}\frac{\partial^2}{\partial\theta_i(n)^2}\,. \tag{11.18}$$

This shows that in this limit the relevant contribution to the Hamiltonian comes from the electric field.

Let us now study the eigenvalue spectrum of H_0. The eigenfunctions of H_0 are given by

$$\psi_{\{N_i(n)\}}[\theta] = \prod_{i,n}[U_i(n)]^{N_i(n)}, \tag{11.19}$$

where

$$U_i(n) = e^{i\theta_i(n)}$$

is a link variable with base at n, pointing in the i'th spatial direction, and where $N_i(n)$ is the excitation number of the link connecting the sites n and $n+\hat{e}_i$. These wave functions are normalized as follows:

$$\int\prod_{k,m}\frac{d\theta_k(m)}{2\pi}\psi^*_{\{N_i(n)\}}[\theta]\psi_{\{N_j(m)\}}[\theta] = \delta_{\{N_i(n)\},\{N_j(m)\}}$$

The energy associated with the state (11.19) is given by

$$E_{\{N_i(n)\}} = \frac{g^2}{2a}\sum_{i,n}(N_i(n))^2. \tag{11.20}$$

It must, however, be remembered that not all such states are physical. Thus we have emphasized before that only gauge-invariant states belong to the physical Hilbert space. Hence only wave functions built from products of link variables along one or several closed loops have a physical meaning.

Let us consider a few states. The lowest energy state is the one where no link is excited. The wave function of the next higher energy state is given by an elementary plaquette variable located anywhere on the lattice. The energy associated with this excitation is given according to (11.20) by $4(g^2/2a)$, since there are four links on the boundary of an elementary plaquette. By exciting larger loops or several loops, we obtain states of increasing energy. In Fig. (11-4) we show various types of excitations on a two-dimensional spatial lattice.

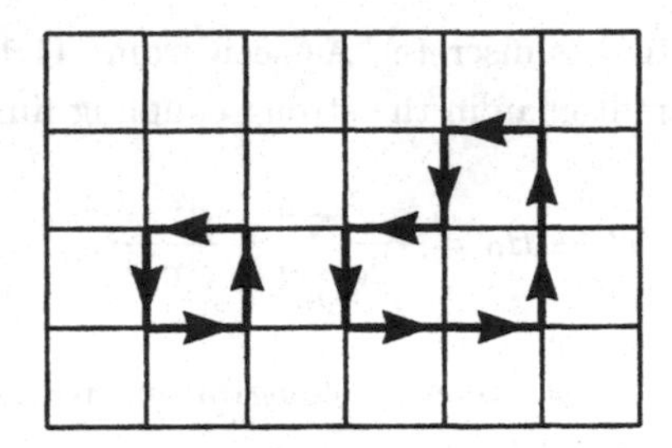

Fig. 11-4 Two types of excitations on a two dimensional spatial lattice.

The above strong coupling picture gets modified when fermions are coupled to the link variables. Suppose that we introduce into the pure gauge medium a very heavy pair of opposite charges at the lattice sites n and m, respectively. Then we can build a gauge-invariant state by connecting the two charges by a string built from the product of link variables. The lowest energy state of the system is obtained by exciting the links along the shortest path connecting the two charges. Each excited link contributes an energy $g^2/2a$. Consider, in particular, two charges located on a straight line path on the lattice, as shown in Fig. (11-5a). Their energy is given by

$$E_0 = \frac{g^2}{2a}\hat{L}, \tag{11.21}$$

where $\hat{L}$ is the separation of the pair measured in lattice units. Thus in the strong coupling limit we have confinement already in the $U(1)$ theory!

The above result was obtained in the strong coupling limit and in the absence of dynamical (finite mass) fermions. The effect of the potential in (11.16) for finite (but large) coupling can be calculated in perturbation theory. Clearly the states discussed above are no longer eigenstates of the Hamiltonian when the potential is turned on. In this case the string connecting the two charges will be allowed to fluctuate. To see this consider the action of the potential on the state depicted in Fig. (11-5a). In particular consider the effect arising from those plaquettes in the potential (11.16b) having a link in common with the string connecting the two charges. In Fig. (11-5b,c) we show the possibilities corresponding to overlapping flux lines. The wave functions associated with these states are those constructed from the link variables shown in Fig. (11-5d,e), where the darkened line corresponds to a doubly excited link. For finite coupling the lowest energy state of the charged pair will include these (and many other) excitations. Thus

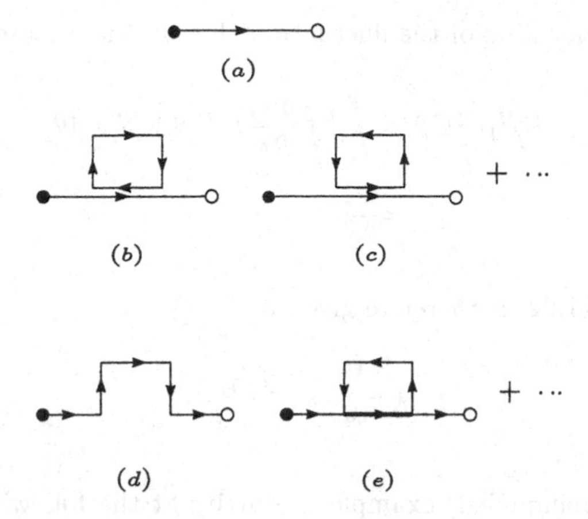

Fig. 11-5 (a) Eigenstate of the Hamiltonian in the strong coupling limit; (b,c) Configurations created by the action of the potential on the state depicted in (a); (d,e) Corresponding link configurations from which the wave functionals (11.19) are constructed. The darkened line in (e) denotes a doubly excited link. The dots stand for the other three possible ways of attaching the plaquettes on the three dimensional spatial lattice.

the original rigid string begins to fluctuate when the potential is turned on.

Let us compute the change in energy ΔE arising from fluctuations of the type depicted in Fig. (11-5d,e). They can be computed by standard perturbation theory: Let $|E_0\rangle$ denote the ground state of the $q\bar{q}$-pair in the strong coupling limit, depicted in Fig. (11-5a), and $|E_f\rangle$ the eigenstate of H_0 corresponding to link configurations of the type shown in Fig. (11-5d,e). Then

$$\Delta E = \sum_{f \neq 0} \frac{|\langle E_0|V|E_f\rangle|^2}{E_0 - E_f}, \tag{11.22}$$

where the sum extends over all states $|E_f\rangle$ with the fluctuation taking place anywhere along the line connecting the two charges . The energy E_0 is given by (11.21). Furthermore, from (11.20) we obtain for the states of type d and e,

$$E_f^{(d)} = \frac{g^2}{2a}(\hat{L} + 2),$$
$$E_f^{(e)} = \frac{g^2}{2a}(\hat{L} + 6),$$

irrespective of the location of the fluctuation. For all these states

$$\langle E_f^{(i)}|V|E_0\rangle = \int \prod_\alpha \frac{d\theta_\alpha}{2\pi} \psi^{(i)}[\theta] V[\theta] \psi_0[\theta]$$
$$= \frac{1}{2g^2 a}$$

The energy shift (11.22) is therefore given by

$$\Delta E_{\bar{q}q} = -\frac{4}{3} \frac{\hat{L}}{a g^6} .$$

Hence in this (oversimplified) example, we arrive at the following expression for the string tension measured in lattice units:

$$\hat{\sigma} = \frac{g^2}{2} \left(1 - \frac{8}{3g^8}\right) .$$

For the case of the pure $SU(3)$ gauge theory, Kogut, Pearson, and Shigemitsu (1981) have computed the string tension up to $O(g^{-24})$. These computations (which are quite non-trivial) suggest that the strong coupling expansion, when carried out to sufficiently high orders, yields a string tension in the intermediate coupling region which can be fitted at the low coupling end with the square of the function $R(g)$ defined in (9.21d). In the strong coupling regime the flux tube connecting the two charges will have a finite width of the order of the lattice spacing. But as the coupling is decreased, not only the width of the flux tube will increase, but also its shape will eventually undergo strong changes. Furthermore, the number of string configurations of a given length $\hat{L}$ will increase dramatically with $\hat{L}$.

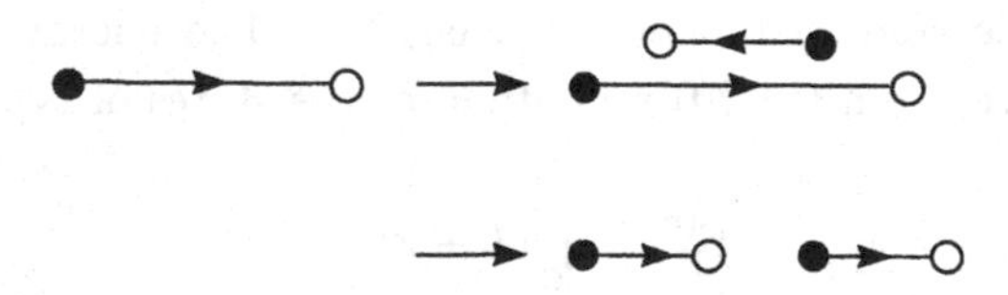

Fig. 11-6 Breaking of the string due to pair production.

In the presence of dynamical fermions this picture is further modified. Thus for sufficiently large separations of the two charges, the system can lower its energy by creating an oppositely charged pair connected by a string as shown in Fig. (11-6). This corresponds in the strong coupling approximation to the hadronization process mentioned at the beginning of chapter 7. This concludes our discussion of the string picture of confinement in the strong coupling Hamiltonian formulation. We have seen that for strong coupling, the $U(1)$ theory leads to a non-vanishing string tension. This result is in agreement with that obtained by studying the Wilson loop. In the continuum limit confinement should of course be lost. On the other hand, the $SU(3)$ gauge theory should still exhibit confinement in this limit. We have concentrated our attention on the $U(1)$ gauge theory since it allowed us to demonstrate in a simple way how a string picture of confinement emerges in the strong coupling limit, if the theory is compactified. The non-abelian case is of course more complicated but the basic ideas are the same. For a discussion of the $SU(3)$ gauge theory, the reader may confer the paper by Creutz (1977).

Finally we want to mention that the construction of the lattice Hamiltonian via the transfer matrix is not the only method. For an alternative procedure based on canonical methods the reader may consult the review article by Kogut (1983) or the lectures of this author given at the International School of Physics, Enrico Fermi (Kogut, 1984).

CHAPTER 12

THE HOPPING PARAMETER EXPANSION

The inclusion of fermions in lattice calculations is a very non–trivial problem. In the pure gauge theory correlation functions can be calculated by Monte Carlo methods (see chapter 16). Such methods cannot be applied directly to path integrals involving Grassmann variables. To overcome this difficulty one must first integrate out the fermions. The resulting path integral expression then only involves bosonic variables and one can evaluate them, in principle, using statistical methods. But the Boltzmann distribution is now determined by an effective action which is a non–local function of the link variables. Not only that. When calculating fermionic correlation functions, the ensemble average must be performed over expressions which are themselves non–local functions of these variables. For this reason most of the numerical calculations have been performed over many years in the pure gauge sector, or in the so–called quenched approximation, where the effects of pair production processes are neglected. Computations in full QCD were restricted to very small lattices. With the advent of the supercomputers the situation has improved substantially, and numerical calculations with dynamical fermions performed on larger lattices have become feasible. But the computer times required are still astronomical.

A brute–force numerical calculation does not provide us with much insight into the detailed dynamics. One would therefore like to have some analytic way of estimating the effects of dynamical fermions on physical observables. The hopping parameter expansion (HPE) allows one at least to study these effects for large bare lattice–masses of the quarks (it is therefore only useful far away from the continuum limit). Furthermore, when combined with the strong coupling expansion, it also provides us with a physical picture of how hadrons propagate on a lattice, and how pair production processes influence the observables.

The purpose of this chapter is two–fold. We first want to show how the calculation of arbitrary correlation functions of the fermion fields and the link variables can be reduced to a pure bosonic problem, which can in principle be handled by numerical methods. This will be the subject of the following section. In section 2 we then discuss the hopping parameter expansion of the two-point fermion correlation function, and show how hadrons propagate on the lattice in a combined HPE and strong coupling expansion. Section 3 is devoted to the hopping parameter expansion of the effective action. This allows one to include the effects of pair production processes. Finally, in section 4 we demonstrate in some examples how the HPE expansion respects the Pauli exclusion principle, which forbids that two identical quarks (or antiquarks) can occupy the same lattice site.

12.1 Path Integral Representation of Correlation Functions in Terms of Bosonic Variables

In the following we shall consider Wilson fermions of a single flavour. Let A, B, $C\dots$ be collective indices for the colour and Dirac degrees of freedom labeling the quark fields. Thus ψ_A, ψ_B will stand for ψ^a_α and $\bar\psi^b_\beta$, respectively. The path integral representation of a general correlation function of the fermionic and link variables is then given by

$$\begin{aligned}&\langle\psi_{A_1}(n_1)\cdots\psi_{A_N}(n_N)\bar\psi_{B_1}(m_1)\cdots\bar\psi_{B_N}(m_N)U^{c_1d_1}_{\mu_1}(k_1)\cdots U^{c_ld_l}_{\mu_l}(k_l)\rangle\\&=\frac{\int DUD(\bar\psi\psi)\psi_{A_1}(n_1)\cdots\bar\psi_{B_1}(m_1)\cdots U^{c_1d_1}_{\mu_1}(k_1)\cdots e^{-S_{QCD}[U,\psi,\bar\psi]}}{\int DUD(\bar\psi\psi)e^{-S_{QCD}[U,\psi,\bar\psi]}}\end{aligned}\tag{12.1a}$$

where *

$$D(\bar\psi\psi)=\prod_{A,n}d\bar\psi_A(n)d\psi_A(n),\tag{12.1b}$$

and where $U^{ab}_\mu(n)$ denotes a matrix element of the link variables $\underset{\sim}{U}_\mu(n)$. Notice that (12.1a) contains the same number of ψ and $\bar\psi$ fields since all other correlation functions vanish because of the Grassmann integration rules discussed in chapter two.

Consider the fermionic contribution to S_{QCD}. For Wilson fermions it is given by (6.25c). Let us write this contribution in the form

$$S^{(W)}_F=\frac{1}{2\kappa}\sum_{n,m}\bar\psi(n)\underset{\sim}{K}_{nm}[U]\psi(m),\tag{12.2a}$$

where

$$\kappa=\frac{1}{8r+2\hat M_0}\tag{12.2b}$$

* For later convenience we have chosen to write the measure in this form, rather than $D\bar\psi D\psi$. This has of course no influence on the ratio (12.1a), but eliminates some unpleasant minus signs which would otherwise arise in intermediate steps of the calculation.

is the so-called hopping parameter (for reasons which will become clear below), and $\underset{\sim}{K}_{nm}$ are matrices in Dirac and colour space:

$$\underset{\sim}{K}_{nm}[U] = \delta_{nm}\mathbb{1} - \kappa \sum_{\mu>0} [(r-\gamma_\mu)\underset{\sim}{U}_\mu(n)\delta_{n+\hat{\mu},m} + (r+\gamma_\mu)\underset{\sim}{U}_\mu^\dagger(n-\hat{\mu})\delta_{n-\hat{\mu},m}]. \quad (12.2c)$$

Thus $\underset{\sim}{K}_{nm}$ is of the form

$$\underset{\sim}{K}_{nm} = \delta_{nm}\mathbb{1} - \kappa \underset{\sim}{M}_{nm}[U], \quad (12.3a)$$

where the only non-vanishing matrices $\underset{\sim}{M}_{nm}$ are those connecting neighbouring lattice sites:

$$\begin{aligned} \underset{\sim}{M}_{n\ n+\hat{\mu}}[U] &= (r-\gamma_\mu)\underset{\sim}{U}_\mu(n), \\ \underset{\sim}{M}_{n\ n-\hat{\mu}}[U] &= (r+\gamma_\mu)\underset{\sim}{U}_\mu^\dagger(n-\hat{\mu}). \end{aligned} \quad (12.3b)$$

The corresponding expressions for the matrix elements read:

$$\begin{aligned} (\underset{\sim}{M}_{n,n+\hat{\mu}})_{\alpha a,\beta b} &= (r-\gamma_\mu)_{\alpha\beta}(\underset{\sim}{U}_\mu(n))_{ab}, \\ (\underset{\sim}{M}_{n,n-\hat{\mu}})_{\alpha a,\beta b} &= (r+\gamma_\mu)_{\alpha\beta}(\underset{\sim}{U}_\mu^\dagger(n-\hat{\mu}))_{ab}. \end{aligned} \quad (12.3c)$$

In the literature it is customary to eliminate the factor $1/2\kappa$ appearing in (12.2a)by scaling the fermion fields with $\sqrt{2\kappa}$: $\psi \to \sqrt{2\kappa}\psi, \bar{\psi} \to \sqrt{2\kappa}\bar{\psi}$. The corresponding Jacobian in the fermion integration measure drops out in the ratio (12.1a). Hence by eliminating the factor $1/2\kappa$ in (12.2a),the path integral expression (12.1) yields the original correlation function multiplied by $(1/2\kappa)^N$. With this in mind, the action we shall be working with is given by

$$S_{QCD} = S_G[U] + S_F^{(W)}[U, \psi, \bar{\psi}], \quad (12.4a)$$

where S_G has been defined in (6.25b), and where

$$S_F^{(W)} = \sum_{A,B} \bar{\psi}_A(n) K_{nA,mB}[U] \psi_B(m), \quad (12.4b)$$

with

$$K_{nA,mB}[U] = \delta_{nm}\delta_{AB} - \kappa(\underset{\sim}{M}_{nm})_{AB}. \quad (12.4c)$$

Here we have made use of the collective notation for the Dirac and colour indices introduced above. Thus $\delta_{AB} \equiv \delta_{\alpha\beta}\delta_{ab}$, and $(\underset{\sim}{M}_{nm})_{AB} \equiv (\underset{\sim}{M}_{nm})_{\alpha a,\beta b}$. Using the fact that (see chapter 2),

$$\det K[U] = \int D(\bar{\psi}\psi) e^{-S_F^{(W)}[U,\psi,\bar{\psi}]}, \quad (12.5)$$

one readily verifies that (12.1a) can also be written in the form

$$\langle\psi_{A_1}(n_1)\cdots\bar{\psi}_{B_1}(m_1)\cdots U^{c_1d_1}_{\mu_1}(k_1)\cdots\rangle = \\ = \frac{\int DU\langle\psi_{A_1}(n_1)\cdots\bar{\psi}_{B_1}(m_1)\cdots\rangle_{S_F} U^{c_1d_1}_{\mu_1}(k_1)\cdots \det K[U]e^{-S_G[U]}}{\int DU \det K[U]e^{-S_G[U]}}, \tag{12.6a}$$

where $\langle\psi_{A_1}(n_1)\cdots\bar{\psi}_{B_1}(m_1)\cdots\rangle_{S_F}$ is the purely fermionic correlation function in the "external" field defined by the link variable configuration $\{U_\mu(n)\}$; i.e.

$$\langle\psi_{A_1}(n_1)\cdots\bar{\psi}_{B_1}(m_1)\cdots\rangle_{S_F} = \frac{\int D(\bar{\psi}\psi)\psi_{A_1}(n_1)\cdots\bar{\psi}_{B_1}(m_1)\cdots e^{-S^{(W)}_F[U,\psi,\bar{\psi}]}}{\int D(\bar{\psi}\psi)e^{-S^{(W)}_F[U,\psi,\bar{\psi}]}} \tag{12.6b}$$

In the so-called quenched approximation (q.a.), where $\det K[U]$ is replaced by a constant, the correlation function (12.6a) is just the ensemble average of the external field correlation function (12.6b) multiplied by the link variables appearing on the left hand side of eq.(12.6a), and averaged with the Boltzmann distribution of the pure gauge theory, $exp(-S_G[U])$.

In chapter 2 we have shown that path integrals of the type (12.6b) can be expressed in terms of products of the two-point correlation functions (2.57) according to (2.59). A simple translation of these expressions to the path integral (12.6b) leads to the following result:

$$\langle\psi_{A_1}(n_1)\cdots\psi_{A_\ell}(n_\ell)\bar{\psi}_{B_1}(m_1)\cdots\psi_{B_\ell}(m_\ell)\rangle_{S_F} \\ = \sum_{contr.}\psi_{A_1}(n_1)\cdots\psi_{A_\ell}(n_\ell)\bar{\psi}_{B_1}(m_1)\cdots\bar{\psi}_{B_\ell}(m_\ell). \tag{12.7}$$

The right-hand side is meant to be the sum over all possible complete contractions of $\psi-\bar{\psi}$ pairs,* where a contraction is defined generically by

$$\underbrace{\psi_A(n)\bar{\psi}_B(m)} = K^{-1}_{nA,mB}[U]. \tag{12.8}$$

Hence in the quenched approximation we "only" need to compute the ensemble average of products involving the two-point function (12.8) and the link variables with a Boltzmann distribution corresponding to that of the pure gauge theory:

$$\langle\psi_{A_1}(n_1)\cdots\bar{\psi}_{B_1}(m_1)U^{c_1d_1}_{\mu_1}(k_1)\cdots\rangle_{q.a.} \\ =\langle\sum_{contr.}(\psi_{A_1}(n_1)\cdots\bar{\psi}_{B_1}(m_1)\cdots)U^{c_1d_1}_{\mu_1}(k_1)\cdots\rangle_{S_G}, \tag{12.9}$$

---

* For the precise meaning of the right-hand side of (12-7) see the discussion in chapter 2 after eq. (2.59).

where $\langle\ \rangle_{S_G}$ denotes the ensemble average taken with the Boltzmann factor of the pure gauge theory. But neither the integral (12.6a) with $\det K = 1$, nor $K^{-1}[U]$ alone, can be calculated in closed form, and we are forced in general to compute these quantities numerically. We may, however, calculate (12.8) within the so-called hopping parameter expansion. This we will do in the following section.

12.2 Hopping Parameter Expansion of the Fermion Propagator in an External Field

Consider the matrix K defined in (12.4c).

$$K[U] = 1 - \kappa M[U] \tag{12.10}$$

Its inverse is the fermion propagator for a given link-variable configuration. For small κ (i.e., large bare quark mass $\hat{M}_0$) * we can expand K^{-1} in powers of the hopping parameter as follows:

$$K^{-1} = (1 - \kappa M)^{-1} = \sum_{\ell=0}^{\infty} \kappa^\ell M^\ell.$$

The corresponding expression for the matrix elements of K^{-1} reads

$$K^{-1}_{nA,mB}[U] = \delta_{nm}\delta_{AB} + \kappa(\underset{\sim}{M}_{nm})_{AB} + \sum_{\ell=2}^{\infty} \kappa^\ell \sum_{\{n_i\}} (\underset{\sim}{M}_{nn_1}\underset{\sim}{M}_{n_1n_2}\cdots\underset{\sim}{M}_{n_{\ell-1}m})_{AB}, \tag{12.11}$$

where the non-vanishing matrices $\underset{\sim}{M}_{nm}$ have been given in (12.3b). It follows from (12.11) and the fact that $\underset{\sim}{M}_{k\ell}$ connects only nearest neighbours on the lattice, that the contributions to $K^{-1}_{nA,mB}[U]$ in order κ^ℓ can be computed according to the following rules:

i) Consider all possible paths of length ℓ on the lattice starting at the lattice site n occupied by ψ_A (open circles) and terminating at the site m occupied by $\bar{\psi}_B$ (black blobs). As shown in fig. (12-1), these paths may intersect each other, and turn back on themselves, creating for example appendix-like structures.

ii)Associate with each link with base at k, and pointing in the $\pm\mu$ direction, the matrices

$$\underset{\sim}{M}_{k,k+\hat{\mu}} = (r - \gamma_\mu)\underset{\sim}{U}_\mu(k)$$

---

* We are thus necessarily away from the continuum limit, which is realized for $\hat{M}_0 \to 0$.

$$\underset{\sim}{M}_{k,k-\hat{\mu}} = (r+\gamma_\mu)\underset{\sim}{U}^\dagger_\mu(k-\hat{\mu})$$

iii) Take the ordered product of all these matrices along a path following the arrow pointing from n to m, and take the AB-matrix element of this expression

iv) Sum over all possible paths leading from n to m.

The number of diagrams reduces drastically if we choose the Wilson parameter to be $r = 1$, for in this case diagrams with appendices, such as shown in fig. (12-1c), will not contribute. This is a consequence of the fact that $(1-\gamma_\mu)(1+\gamma_\mu) = 0$. In the following we shall choose $r = 1$.

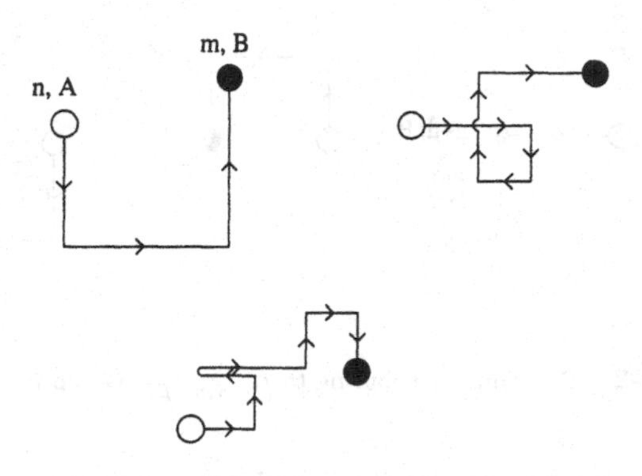

Fig. 12-1 Diagrams contributing to $K^{-1}_{nA,mB}[U]$ in the hopping parameter expansion.

As an example let us calculate in order κ^4 the correlation function describing the propagation of a colour-singlet scalar particle, consisting of a local $q\bar{q}$-pair, between two neighbouring lattice sites n, and $n+\hat{\mu}$. Let us denote the corresponding composite field by $\phi(n)$:

$$\phi(n) \equiv \sum_{a,\alpha} \bar{\psi}^a_\alpha(n)\psi^a_\alpha(n).$$

The correlation function of ϕ in the quenched approximation is given by:

$$\begin{aligned}\langle\phi(n+\hat{\mu})\phi(n)\rangle_{q.a} &= \sum_{\substack{a,\alpha\\ b,\beta}} \langle \bar{\psi}^b_\beta(n+\hat{\mu})\psi^b_\beta(n+\hat{\mu})\bar{\psi}^a_\alpha(n)\psi^a_\alpha(n)\rangle_{S_G} \\ &+ \sum_{\substack{a,\alpha\\ b,\beta}} \langle \bar{\psi}^b_\beta(n+\hat{\mu})\psi^b_\beta(n+\hat{\mu})\bar{\psi}^a_\alpha(n)\psi^a_\alpha(n)\rangle_{S_G} \quad .\end{aligned}$$

Consider the first term on the right-hand side which is the contribution describing the propagation of a quark (and antiquark) between two neighbouring lattice sites:

$$\langle\langle\phi(m)\phi(n)\rangle\rangle_{q.a.} = -\sum_{A,B}\langle K^{-1}_{nA,mB}[U]K^{-1}_{mB,nA}[U]\rangle_{S_G}. \tag{12.12}$$

Here $m = n + \hat{\mu}$, and A and B stand for the set (a, α) and (b, β), respectively. To compute the contribution in order κ^4 of (12.12), we must expand K^{-1} up to order κ^3. In fig. (12-2) we show, for the case of a two-dimensional lattice, the various paths C_1, C_2, C_3 of interest.

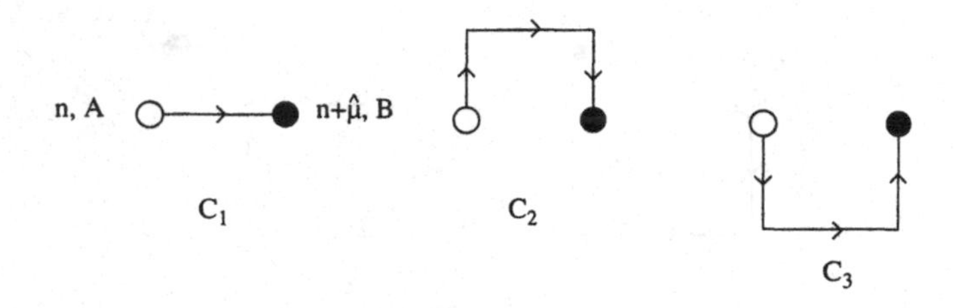

Fig. 12-2 Diagrams contributing to $K^{-1}_{nA,n+\hat{\mu}B}[U]$ up to order κ^3.

The corresponding expressions obtained by applying the general rules stated above with $r = 1$, then read

$$K^{-1}_{nA;n+\hat{\mu}B} = \sum_{\ell=1}^{3}(\underset{\sim}{\Gamma}_{C_\ell})_{\alpha\beta}(\underset{\sim}{U}_{C_\ell})^{ab}$$

where

$$\begin{aligned}\underset{\sim}{\Gamma}_{C_1} &= (1-\gamma_\mu)\\ \underset{\sim}{U}_{C_1} &= \underset{\sim}{U}_\mu(n)\end{aligned}$$

and

$$\begin{aligned}\underset{\sim}{\Gamma}_{C_2} &= (1-\gamma_\nu)(1-\gamma_\mu)(1+\gamma_\nu),\\ \underset{\sim}{U}_{C_2} &= \underset{\sim}{U}_\nu(n)\underset{\sim}{U}_\mu(n+\hat{\nu})\underset{\sim}{U}^\dagger_\nu(n+\hat{\mu}),\\ \underset{\sim}{\Gamma}_{C_3} &= (1+\gamma_\nu)(1-\gamma_\mu)(1-\gamma_\nu),\\ \underset{\sim}{U}_{C_3} &= \underset{\sim}{U}^\dagger_\nu(n-\hat{\nu})U_\mu(n-\hat{\nu})U_\nu(n+\hat{\mu}-\hat{\nu}) \quad .\end{aligned}$$

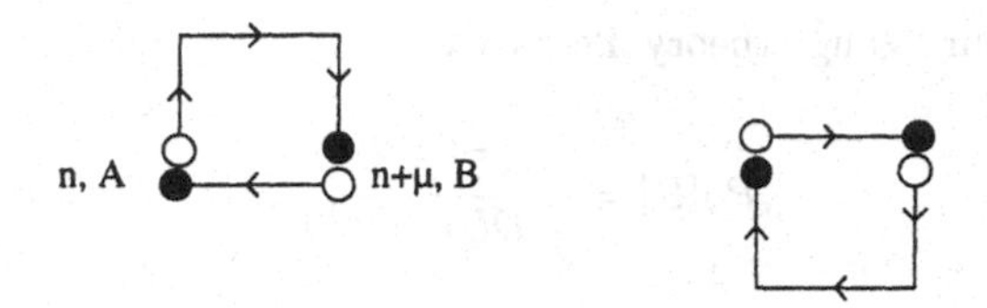

Fig. 12-3 Diagrams contributing in order κ^4 to the external field correlation function (12.12).

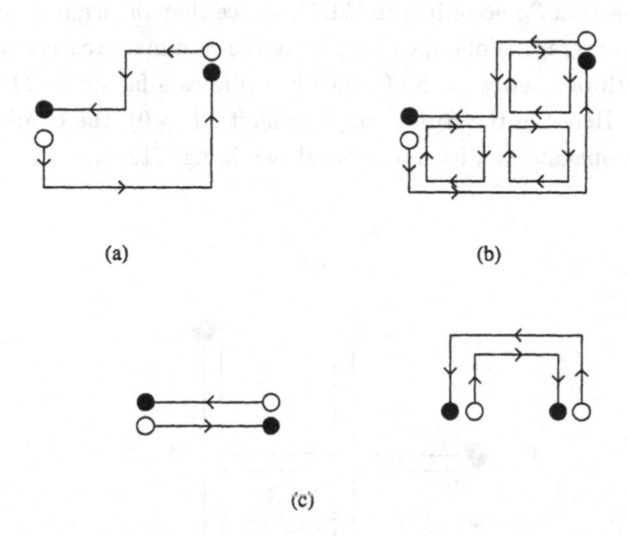

Fig. 12-4 (a) Diagrams contributing to an external field 4-point correlation function; (b) contribution to the correlation function in the leading strong coupling approximation; (c) type of diagrams contributing to (12.12) in the limit $\beta \to 0$.

Hence the relevant diagrams associated with the correlation function (12.12) in order κ^4 are those shown in fig. (12-3). Their contribution is obtained by taking the product of the individual propagators K^{-1} making up the diagram.

So far we have considered quark-correlation functions evaluated for a given configuration of the link variables. In the quenched approximation ($\det K[U] \to$ *const.*), these correlation functions must still be weighted with a Boltzmann distribution of the pure gauge theory, i.e. with

$$P_G[U] = \frac{e^{-S_G[U]}}{\int DUe^{-S_G[U]}} \, . \tag{12.13}$$

It is instructive to calculate the ensemble average of some simple correlation functions in leading order of the strong coupling approximation. Consider for example the contribution of $O(\kappa^8)$ to the correlation function (12.12) depicted in fig. (12-4a). This contribution must be averaged with the Boltzmann distribution (12-13). Expanding the action S_G according to (11.3), we see that the leading contribution to the ensemble average is obtained by paving the minimal area enclosed by the quark paths with plaquettes, each of which contributes a factor β. This is shown in fig. (12-4b). Hence in the strong coupling limit ($\beta \to 0$), the quark-antiquark pair can only propagate as a local unit, as shown in fig. (12-4c).

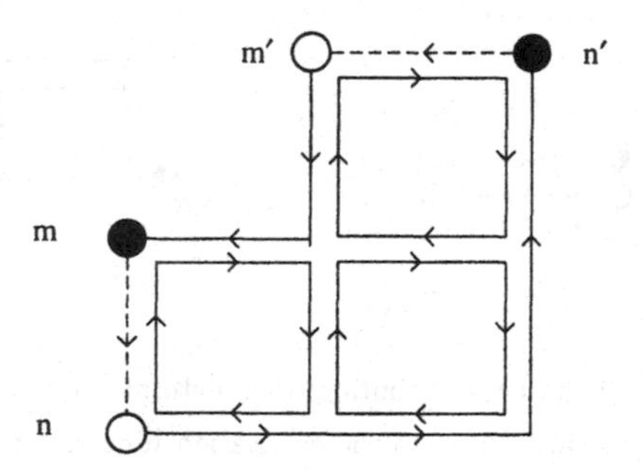

Fig. 12-5 Diagram contributing to the gauge invariant correlation function $< \bar{\psi}(m)U...U\psi(n)\bar{\psi}(n')U...U\psi(m') >$ in order κ^6, for the case where n and m (n' and m') are neighbouring lattice sites.

On the other hand consider the propagation of an extended meson. Then we must study correlation functions of a gauge-invariant composite field having the following schematic structure

$$\phi_{q\bar{q}} = \bar{\psi}(m)\underset{\sim}{U} \cdots \underset{\sim}{U}\psi(n),$$

where $\underset{\sim}{U} \cdots \underset{\sim}{U}$ denotes the matrix product of link variables along a path connecting the lattice sites n and m. In fig. (12-5) we show a diagram contributing to the "meson" propagator

$$\Gamma = \langle \bar{\psi}(m)\underset{\sim}{U} \cdots \underset{\sim}{U}\psi(n)\bar{\psi}(n')\underset{\sim}{U} \cdots \underset{\sim}{U}\psi(m')\rangle$$

in $O(\kappa^6)$ of the HPE, and $O(\beta^3)$ of the strong coupling expansion, for the case where m and n (m' and n') are neighbouring lattice sites. The dashed lines stand for the product of link variables appearing in the above correlation function. Notice that the "strings" connecting the quarks and antiquarks (dashed lines) are essential, since the group integral over a single link variable vanishes. For the same reason a single quark cannot propagate in any order of the HPE. This can be viewed as another statement of confinement.

12.3 Hopping Parameter Expansion of the Effective Action

As we have just seen, we can in principle compute a general correlation function in the quenched approximation by calculating the ensemble average of products of fermionic two-point correlation functions and the link variables, with a Boltzmann distribution given by (12.13). In full QCD, however, we must also include the fermionic determinant, $\det K[U]$, in (12.6a). This is equivalent to taking the ensemble average with the following probability distribution

$$P[U] = \frac{e^{-S_{eff}[U]}}{\int DUe^{-S_{eff}[U]}}, \tag{12.14a}$$

where the effective action, S_{eff}, is defined by

$$S_{eff}[U] = S_G[U] - \ln\det K[U]. \tag{12.14b}$$

Correspondingly, (12.9) is now replaced by

$$\langle \psi_{A_1}(n_1)..\bar{\psi}_{B_1}(m_1)..U_{\mu_1}^{c_1 d_1}(k_1)..\rangle = \langle \sum_{contr.} (\psi_{A_1}(n_1)..\bar{\psi}_{B_1}(m_1)..)U_{\mu_1}^{c_1 d_1}(k_1)..\rangle_{S_{eff}}.$$

We remark that it is not obvious that (12.14a) can be interpreted as a probability distribution. For this to be the case one must show that $\ln\det\ K[U]$ is real and bounded from above. That $\det K[U]$ is real can be shown rather easily. We first notice that $K[U]$ is not a hermitean matrix. From the definition (12.2c) and the hermiticity of the γ-matrices, it follows that

$$\begin{aligned} K^*_{mB,nA} = \delta_{nm}\delta_{\alpha\beta}\delta_{ab} - \kappa\sum_\mu \Big[&(r+\gamma_\mu)_{\alpha\beta}(\underset{\sim}{U}_\mu(n))_{ab}\delta_{n+\hat{\mu},m} \\ &+ (r-\gamma_\mu)_{\alpha\beta}(\underset{\sim}{U}^\dagger_\mu(n-\hat{\mu}))_{ab}\delta_{n-\hat{\mu},m}\Big], \end{aligned}$$

i.e.,

$$K^*_{mB,nA} \neq K_{nA,mB}.$$

In fact the hermitian adjoint of the matrix K is obtained from K by replacing γ_μ by $-\gamma_\mu$. This observation can be stated in the form

$$K^\dagger = \gamma_5 K \gamma_5 \ ,$$

where γ_5 is a matrix in Dirac space which anticommutes with all γ_μ's, and whose square is the unit matrix. Since $\det K^\dagger = \det\ \gamma_5 K \gamma_5 = \det K$, it follows that $\det K$ is real. Furthermore it has been shown by Seiler (1982) that for $\kappa < 1/8$ (which is realized for the usual choice of Wilson parameter $r = 1$) $0 < \det K[U] < 1$. Hence we can interpret the expression (12.14a) as a probability distribution.

In contrast to $S_G[U]$, the effective action (12.14b) is a non-local function of the link variables, and its numerical calculation demands an enormous amount of computer time. Before the advent of supercomputers, one has therefore mainly concentrated on calculating fermionic correlation functions in the quenched approximation. This may be a reasonable approximation to estimate such quantities as hadron masses. On the other hand, there are problems where vacuum polarization effects play a crucial role. Thus for example the screening of the quark-antiquark potential at large distances is due to processes involving the creation of quark-antiquark pairs. These effects arise from the determinant of the fermionic matrix $K[U]$.

We now derive graphical rules for computing $\ln\ \det K[U]$ in the hopping parameter expansion. To this effect we first rewrite $\ln\ \det K$ as follows

$$\ln \det K[U] = Tr \ln K[U],$$

where "Tr" denotes the trace in the internal space as well as in space-time. Substituting for $K[U]$ the expression (12.10) and expanding the logarithm in powers of κM, we have that*

$$Tr \ln K[U] = -\sum_{\ell=1}^{\infty} \frac{\kappa^\ell}{\ell} Tr(M^\ell). \tag{12.15}$$

* Actually, there is no contribution to the trace coming from $\ell = 1$, since $\underset{\sim}{M}_{nm}$ connects different lattice sites.

To get a diagramatic representation of the sum in (12.15), we write out the space-time trace explicitly:

$$Tr \ln K[U] = -\sum_{\ell=2}^{\infty} \frac{\kappa^\ell}{\ell} \sum_{\{n_i\}} tr\, \underset{\sim}{M}_{n_1 n_2} \underset{\sim}{M}_{n_2 n_3} \cdots \underset{\sim}{M}_{n_\ell n_1}, \tag{12.16}$$

where "*tr*" now stands for the trace in the internal space. The non-vanishing matrices $\underset{\sim}{M}_{nm}$ have been defined in (12.3b). From the structure (12.16) it follows immediately that the contributions of order κ^ℓ can be associated with closed paths of length ℓ on the lattice with an arbitrary sense of circulation. These contributions are calculated as follows. Consider a given closed geometrical contour C_{ℓ_0} on the lattice, with ℓ_0 the perimeter of the contour measured in lattice units. Because of the space-time trace in (12.15) a path winding around the contour one or more times can start at any one of the ℓ_0 lattice sites on C_{ℓ_0}. Hence each path is weighted with a factor ℓ_0. A path tracing out the contour C_{ℓ_0} n times will however contribute in order $\kappa^{n\ell_0}$. Hence the contribution to (12.16) associated with all possible paths on the closed contour C_{ℓ_0} is given by

$$-\sum_{n=1}^{\infty} \frac{\kappa^{n\ell_0}}{n} tr \underset{\sim}{M}^n_{C_{\ell_0}} ,$$

where $\underset{\sim}{M}_{C_{\ell_0}}$ is the path ordered product of the matrices (12.3b) along the geometrical contour C_{ℓ_0}. The rhs of (12.16) is now obtained by summing the above expression over all possible geometrical contours of arbitrary shape, and arbitrary perimeter:

$$Tr \ln K[U] = \ln \det K[u] = -\sum_{\ell_0} \sum_{\{C_{\ell_0}\}} \sum_{n=1}^{\infty} \frac{(\kappa^{\ell_0})^n}{n} tr \underset{\sim}{M}^n_{C_{\ell_0}} . \tag{12.17}$$

From here we can further obtain an elegant expression for the determinant itself [Stamatescu (1982, 1992)]. Thus (12.17) can also be written in the form

$$\ln \det K[U] = \sum_{\ell_0} \sum_{\{C_{\ell_0}\}} tr \ln(1 - \kappa^{\ell_0} \underset{\sim}{M}_{C_{\ell_0}}) .$$

It therefore follows that

$$\det K[U] = \prod_{\ell_0} \prod_{\{C_{\ell_0}\}} \det(1 - \kappa^{\ell_0} \underset{\sim}{M}_{C_{\ell_0}}) . \tag{12.18}$$

Note that the determinant appearing in this expression is that of a finite matrix in colour and Dirac space.

In fig. (12-6) we show some diagrams contributing to (12.17). As in the case of the external field Green's function discussed in section 2, they also include paths passing through a given lattice site an arbitrary number of times, as well as paths having appendix-like structures. As before these appendices do not contribute if the Wilson parameter is chosen to be $r = 1$.*

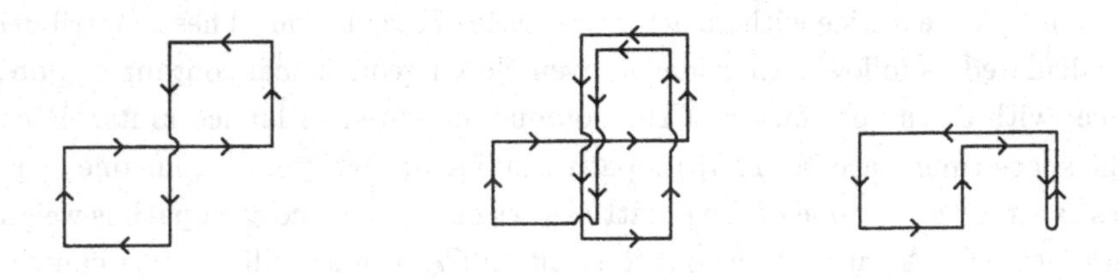

Fig. 12-6 Diagrams contributing to ln detK in the hopping parameter expansion.

The above hopping parameter expansion for the effective action (12.14b) can be combined with the corresponding hopping expansion for the fermionic correlation function discussed in section 2. In those regions of parameter space where $\ln \det K$ and K^{-1} can be approximated by a few terms in the expansion (i.e. small hopping parameter, or large $\hat{M}_0$), the computational effort is thereby drastically reduced. In practice, however, where one is interested in working in the scaling region with small quark masses, non-perturbative methods for computing K^{-1} and $\det K$ are required.

The hopping parameter expansion can be used to obtain a qualitative picture of the screening of the static quark-antiquark potential due to "pair-production" processes (see e.g. Joos and Montvay, 1983). The following picture is very crude. Within the framework of the strong coupling expansion, the Wilson loop will be paved not only with the plaquettes arising from the action $S_G[U]$, but also with closed loops built from link variables, arising from the fermionic determinant. In the HPE, these latter contributions come with a power of κ, determined by the

* In the continuum limit physics should be independent of the choice for r, as long as $r \neq 0$.

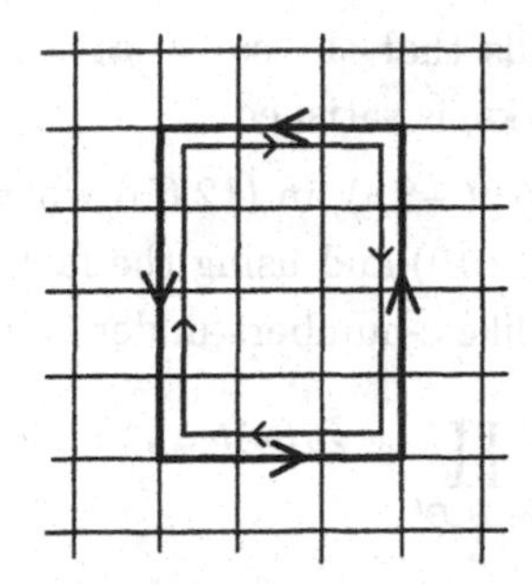

Fig. 12-7 Contribution to $\langle W_C \rangle$ coming from the fermionic determinant (inside loop) which screens the $q\bar{q}$ -potential in the strong coupling limit and in lowest order of the hopping parameter expansion.

length of the loop. These different loop contributions will compete with each other in determining the interquark potential. In particular, in the strong coupling limit, $\beta \to 0$, the Wilson loop will be paved with loops arising only from the fermionic determinant. In fig. (12-7) we show a diagram which corresponds to a dynamical quark and antiquark being created at the positions of the (infinitely) heavy antiquark and quark, respectively. The contribution of this diagram is of the form

$$\langle W_C[U] \rangle \approx \kappa^{2(\hat{R}+\hat{T})} = e^{(\hat{R}+\hat{T})\ln \kappa^2},$$

which shows that the dynamical $q\bar{q}$-pair leads to screening of the heavy quark-antiquark potential, which in this crude approximation is just a constant: $V_{q\bar{q}} = -\ln \kappa^2$.

12.4 The HPE and the Pauli Exclusion Principle

In the previous sections we obtained graphical rules for calculating the two-point fermion correlation function and the effective action in any order of the hopping parameter expansion. In particular the rules for calculating the quark propagators in a background field, corresponding to a given link–variable configuration, where based on eq. (12.7), derived in chapter 2 using the concept of a generating functional. We now want to rederive the results obtained above by applying the Grassmann integration rules (discussed in chapter 2) directly to the path integral expressions for the external field correlation function (12.6b) and the determinant of $K[U]$. In this way we shall demonstrate explicitly that the Pauli exclusion principle, which forbids that the same lattice site can be occupied by two identical quarks or antiquarks, is satisfied.

Consider the exponential, $\exp(-S_F)$, in (12.6b) where S_F is given by (12.4b). Inserting for K the expression (12.10) and using the fact that bilinear expressions in Grassmann variables behave like *c*-numbers under commutation, we have that

$$e^{-S_F} = \prod_{n,A} e^{-\bar{\psi}_A(n)\psi_A(n)} \prod_{\substack{m,B,B' \\ \hat{\mu}}} e^{\kappa\bar{\psi}_B(m)M_{mB,m+\hat{\mu}B'}\psi_{B'}(m+\hat{\mu})} \quad . \tag{12.19}$$

where $M_{mB,nB'} \equiv (M_{mn})_{BB'}$. Here, and in the following, the unit vector $\hat{\mu}$ can point in the positive and negative directions. Now, since products of a given Grassmann variable vanish, only the first two terms in the expansion of the exponentials (12.19) will contribute; hence

$$e^{-S_F} = \prod_{n,A}[1 + m_A(n)] \prod_{\substack{m,B,B' \\ \mu}} [1 + \kappa d_{BB'}(m, \hat{\mu})], \tag{12.20a}$$

where*

$$m_A(n) = \psi_A(n)\bar{\psi}_A(n), \tag{12.20b}$$

$$d_{BB'}(m, \hat{\mu}) = \bar{\psi}_B(m)M_{mB,m+\hat{\mu}B'}[U]\psi_{B'}(m + \hat{\mu}). \tag{12.20c}$$

For convenience we shall refer to (12.20b) and (12.20c) as "monomers" and "dimers", respectively. These names have also been used elsewhere in the literature, although in a somewhat different way, and we apologize for borrowing these suggestive names for our purpose.**

According to (12.20), the basic elements appearing in the integrand of the numerator and denominator of (12.6b) in any order of the hopping parameter expansion are:

i) The quark (antiquark) fields $\psi_{A_1}(n_1)\cdots(\bar{\psi}_{B_1}(m_1)\cdots)$ of the correlation function to be calculated. We denote these graphically by an open circle ($\psi \to o$) and an extended dot ($\bar{\psi} \to \bullet$) and shall refer to them as the *external* fields.

ii) Monomers, consisting of a quark and an antiquark located at the same lattice site.

---

* Notice that we have interchanged the order of the Grassmann variables in (12.20b). This gives rise to a minus sign.

** See e.g., Gruber and Kunz (1971); Rossi and Wolff (1984); Burkitt, Mütter and Überholz (1987).

13.2 Weak Coupling Expansion of Correlation Functions in the ϕ^3–Theory

Consider the following euclidean continuum action for a real scalar field:

$$S[\phi] = \frac{1}{2}\int d^4x\phi(x)(-\Box + M^2)\phi(x) + \frac{g_0}{3!}\int d^4x(\phi(x))^3.$$

Here $\Box$ is the four-dimensional Laplacean, M is the bare mass, and g_0 is the bare coupling constant carrying the dimension of a mass. As usual, the combinatorial factor 3! has been introduced to simplify the Feynman rules. Introducing a space time lattice, and scaling $\phi, \Box, M$ and g_0 with the lattice spacing a according to their canonical dimension*, the lattice action takes the form

$$S[\hat{\phi}] = \sum_{n,m}\hat{\phi}_n K_{nm}\hat{\phi}_m + \frac{\hat{g}_o}{3!}\sum_n \hat{\phi}_n^3 \ ,$$

where K is the matrix defined in (3.10b), and where n and m are 4-component vectors labeling the lattice sites. Correlation functions of the fields $\hat{\phi}_n$ can be computed from the generating functional

$$Z[\hat{J}] = \int D\hat{\phi}e^{-S[\hat{\phi}]+\sum \hat{J}_n\hat{\phi}_n}$$

by differentiating this expression with respect to the currents $\hat{J}_n$:**

$$<\hat{\phi}_{n_1}\hat{\phi}_{n_2}\dots\hat{\phi}_{n_l}> = \frac{1}{Z[0]}\left\{\frac{\partial^l Z[\hat{J}]}{\partial\hat{J}_{n_1}\dots\partial\hat{J}_{n_l}}\right\}_{J=0} . \qquad (13.1)$$

As has been shown in section 2 of the second chapter, $Z[\hat{J}]$ can be computed in perturbation theory as follows:

$$\begin{aligned} Z[\hat{J}] &= e^{-S_{int}[\frac{\partial}{\partial\hat{J}}]}Z_0[\hat{J}] \\ &= \sum_{N=0}^{\infty}\frac{1}{N!}\left(-S_{int}[\frac{\partial}{\partial\hat{J}}]\right)^N Z_0[\hat{J}], \end{aligned} \qquad (13.2a)$$

---

* i.e., we define the dimensionless quantities $\hat{\phi} = a\phi$, $\hat{M} = aM$, $\hat{\Box} = a^2\Box$, and $\hat{g}_0 = ag_0$.

** Notice that the subscript on n_i labels different four-component vectors, and is not to be confused with the i'th coordinate of a lattice site.

where

$$S_{int}[\hat{\phi}] = \frac{\hat{g}_0}{3!}\sum_n \hat{\phi}_n^3, \tag{13.2b}$$

and $S_{int}[\partial/\partial\hat{J}]$ is obtained from (13.2b) by making the replacement $\hat{\phi}_n \to \partial/\partial\hat{J}_n$. $Z_0[\hat{J}]$ is the generating functional of the free theory, given by eq. (3.12), i.e.,

$$Z_0[\hat{J}] = \frac{1}{\sqrt{detK}} e^{\frac{1}{2}\sum_{n,m} \hat{J}_n K_{nm}^{-1} \hat{J}_m}. \tag{13.2c}$$

From (13.1) and (13.2) one derives the Feynman rules in the standard way. In every given order of perturbation theory the contribution to the correlation function (13.1) can be represented by a set of Feynman diagrams built from the interaction vertices with coupling $-\hat{g}_0$ and propagators $\hat{\Delta}_{nm} = K_{nm}^{-1}$, represented graphically as follows:

$$\underset{n \qquad\qquad m}{\bullet\!\!-\!\!-\!\!-\!\!-\!\!\bullet} \longrightarrow \hat{\Delta}_{nm} = K_{nm}^{-1}$$

$$-\!\!\!<\ \longrightarrow -\hat{g}_0$$

There are two types of lines (propagators) associated with a general Feynman diagram: a) lines that connect one of the external lattice sites appearing in the correlation function with an interaction vertex, and b) lines connecting two vertices. We shall refer to the former as *external* lines, and to the latter as *internal* lines.

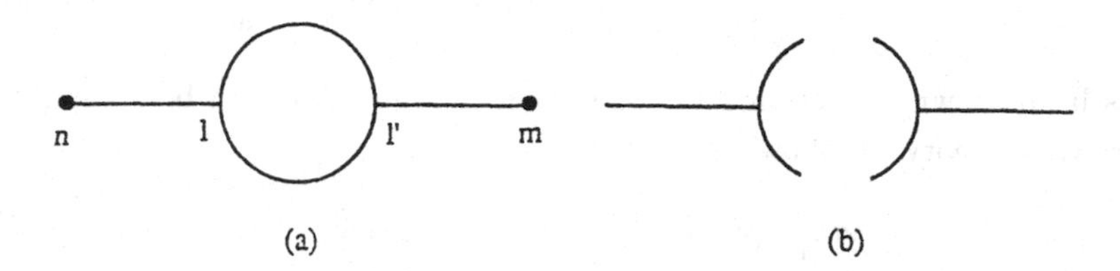

Fig 13-1 (a) Diagram contributing to $\langle\phi_n\phi_m\rangle$ in $O(g_0^2)$; (b) Lines emanating from the two vertices which must be contracted to form the diagram in (a).

As an example consider the contribution of order g_0^2 to the two–point correlation function $<\phi_n\phi_m>$ shown in fig. (13-1a). It is given by*

$$<\hat{\phi}_n\hat{\phi}_m>_{(a)}=\frac{1}{2}\hat{g}_0^2\sum_{l,l'}\hat{\Delta}_{nl}\hat{\Pi}_{ll'}\hat{\Delta}_{l'm}, \tag{13.3a}$$

where

$$\hat{\Pi}_{ll'}=(\hat{\Delta}_{ll'})^2. \tag{13.3b}$$

The "symmetry factor" of 1/2 multiplying (13.3a) arises as follows. Let us count the number of ways that the six lines emanating from the two vertices shown in fig. (13-1b) can be connected to make up the diagram depicted in fig. (13-1a). There are six ways of choosing the endpoint n. This leaves us with three possibilities for choosing m. The number of possibilities to connect the four remaining lines is two. Hence there are $6\cdot 3\cdot 2=(3!)^2$ ways of building up the diagram. Now each vertex yields a factor $1/3!$. This leaves us with a factor of $1/2!$ arising from the second order term in (13.2a). Let us now write the right–hand side of (13.3a) in momentum space.

As we have learned in chapter 3, the free propagator $\hat{\Delta}_{nm}$ has the following Fourier representation (cf. eq. (3.17))

$$\hat{\Delta}_{nm}=\int_{-\pi}^{\pi}\frac{d^4\hat{k}}{(2\pi)^4}\frac{e^{i\hat{k}\cdot(n-m)}}{\widehat{k^2}+\hat{M}^2}, \tag{13.4a}$$

where

$$\widehat{k^2}-\sum_{\mu=1}^{4}\widehat{k_\mu^2}, \tag{13.4b}$$

and

$$\widehat{k_\mu}=2\sin\frac{\hat{k}_\mu}{2}. \tag{13.4c}$$

Notice that while $\hat{k}_\mu$ denotes the momentum measured in lattice units, $\widehat{k^\mu}$ (with the extended "hat") denotes the dimensionless periodic function (13.4c). Because of the appearance of the square of $\widehat{k_\mu}$ in the integral (13.4a), the integrand is a periodic function of $\hat{k}_\mu$ with periodicity 2π. Inserting the expression (13.4) into (13.3) and performing the sum over l and l' using the representation (2.64) for the periodic δ-function (with $a=1$) one finds that

$$<\hat{\phi}_n\hat{\phi}_m>=\int_{BZ}\frac{d^4\hat{k}}{(2\pi)^4}\frac{d^4\hat{k}'}{(2\pi)^4}\hat{G}(\hat{k},\hat{k}';\hat{M})e^{i\hat{k}\cdot n-i\hat{k}'\cdot m}, \tag{13.5a}$$

---

* As always, we denote dimensionless (lattice) variables and functions with a small hat.

where

$$\hat{G}(\hat{k};\hat{k}';\hat{M}) = \frac{1}{\widehat{k^2}+\hat{M}^2}\hat{\Pi}(\hat{k},\hat{k}';\hat{M})\frac{1}{\widehat{k'^2}+\hat{M}^2}, \tag{13.5b}$$

and

$$\hat{\Pi}(\hat{k},\hat{k}';\hat{M}) = \frac{\hat{g}_0^2}{2}\int_{BZ}\prod_{i=1}^{2}\frac{d^4\hat{l}_i}{(2\pi)^4}[(2\pi)^4]^2\delta_P^{(4)}(\hat{k}-\hat{l}_1-\hat{l}_2)\delta_P^{(4)}(\hat{k}'-\hat{l}_1-\hat{l}_2)\prod_{i=1}^{2}\frac{1}{\left(\widehat{l_i^2}+\hat{M}^2\right)}. \tag{13.5c}$$

Here $\hat{l}_i$ denote the "line" momenta carried by the internal lines of the momentum space diagram corresponding to fig. (13.1a). Notice that the general structure of the expressions (13.5) is the same as that of the continuum formulation if $\widehat{k_\mu}$ is replaced by $\hat{k}_\mu$, except that all the variables are expressed in lattice units. But because of the appearance of the periodic δ-functions, momenta are conserved at the vertices modulo $2n\pi$ (n an integer). This is important, since, e.g., for certain values of $\hat{k}$ and $\hat{l}_1$ in the (first) BZ, the argument of $\delta_P^{(4)}(\hat{k}-\hat{l}_1-\hat{l}_2)$ will vanish only for $\hat{l}_2$ in the next BZ. Consider, for example, the integration in (13.5c) over $\hat{l}_2$. Then $\hat{l}_2$ is fixed to be $\hat{l}_2 = \hat{k}-\hat{l}_1+2N\pi$, where the integer N is determined by the requirement that $\hat{l}_2$ lies within the integration interval. But since the integrand is itself a periodic function of the momenta, integrating over $\hat{l}_2$ is equivalent to setting $\hat{l}_2 = \hat{k}-\hat{l}_1$. In other words, we can implement momentum conservation in the way familiar from the continuum formulation. We are therefore left with the following expression for (13.5c):

$$\hat{\Pi}(\hat{k},\hat{k}';\hat{M}) = (2\pi)^4\delta_P^{(4)}(\hat{k}-\hat{k}')\hat{\Pi}(\hat{k},\hat{M}),$$

where

$$\hat{\Pi}(\hat{k},\hat{M}) = \frac{\hat{g}_0^2}{2}\int_{BZ}\frac{d^4\hat{q}}{(2\pi)^4}\frac{1}{[\widehat{q^2}+\hat{M}^2][\widehat{(k-q)}^2+\hat{M}^2]},$$

and

$$\widehat{(k-q)}_\mu = 2\sin[\frac{1}{2}(\hat{k}-\hat{q})_\mu].$$

Let us next compute the contribution of the diagram in fig. (13-1a) to the physical correlation function $\langle\phi(x)\phi(y)\rangle$ by scaling all lattice variables appropriately with a. Proceeding as in chapter 3, we introduce the dimensioned variables $\phi = \hat{\phi}/a$, $x = na$, $y = ma$, $M = \hat{M}/a$, $k = \hat{k}/a$, $k' = \hat{k}'/a$, $q = \hat{q}/a$, as well as the dimensioned coupling constant $g_0 = \hat{g}_0/a$, and study the behaviour of the integral as the lattice spacing is decreased, keeping x, y, M and g_0 fixed. One finds that formally*

$$\langle\phi(x)\phi(y)\rangle = \lim_{a\to 0}\int_{-\pi/a}^{\pi/a}\frac{d^4k}{(2\pi)^4}\frac{d^4k'}{(2\pi)^4}G(k,k';M,a)e^{ik\cdot x}e^{-ik'\cdot y}, \tag{13.6a}$$

where

$$G(k,k';M,a) = (2\pi)^4\delta_P^{(4)}(k-k')\Big[\frac{1}{\tilde{k}^2+M^2}\Pi(k;M,a)\frac{1}{\tilde{k}^2+M^2}\Big], \tag{13.6b}$$

and

$$\Pi(k;M,a) = \frac{g_0^2}{2}\int_{-\pi/a}^{\pi/a}\frac{d^4q}{(2\pi)^4}\frac{1}{[\tilde{q}^2+M^2][\widetilde{(k-q)}^2+M^2]}. \tag{13.6c}$$

Here the dimensioned variables denoted with a "tilde" are defined generically by

$$\tilde{p}_\mu = \frac{2}{a}\sin\frac{p_\mu a}{2} \tag{13.7a}$$

$$\tilde{p}^2 = \sum_{\mu=1}^{4}\tilde{p}_\mu^2. \tag{13.7b}$$

The graphical representation of the quantity appearing within square brackets in (13.6b) is given in fig. (13-2).

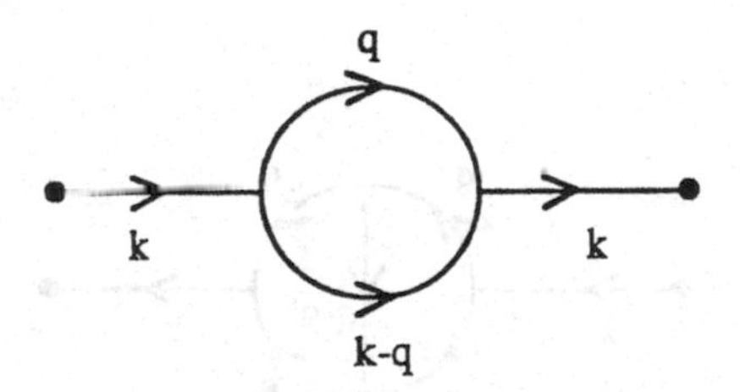

Fig. 13-2 Contribution of $O(g_0^2)$ to the propagator in momentum space

Let us summarize the important properties of the Fourier transform of the correlation function (13.6a):

i) The general structure of $G(k,k';M,a)$ is the same as in the continuum formulation, except that propagators are replaced by their lattice analogues,

$$\Delta(p) = \frac{1}{\tilde{p}^2+M^2}, \tag{13.8}$$

---

* Actually this limit does not exist and one must invoke renormalization before taking it. We shall come back to this point later; here we are only interested in discussing some formal aspects.

and that the momentum integrations are carried out over the Brillouin zone: $[-\frac{\pi}{a}, \frac{\pi}{a}]^4$. Hence apart from these modifications the Feynman rules are the same as those of the continuum formulation.

ii) In the limit $a \to 0$ the lattice propagators reduce to those of the continuum theory.

iii) $G(k, k'; M, a)$ is a periodic function in each of the components of the momenta, with periodicity $2\pi/a$.

iv) The integrand of the lattice Feynman integral (13.6c) is a periodic function of the loop momentum q, with periodicity $2\pi/a$. Furthermore, it possesses a finite continuum limit.

v) If the integrand of the lattice Feynman integral (13.6c) is replaced by its naive continuum limit, then the resulting integral is given by the continuum Feynman rules with a momentum cutoff π/a.

Although we have only discussed a particular example, these properties hold for any Feynman diagram if we choose an appropriate set of loop integration variables. A natural set of integration variables is obtained by identifying these with a subset of the line momenta.* This is the choice we shall make in this and the following two chapters. In fig. (13-3) we show such a natural choice of integration variable for a diagram contributing to the two–point function in $0(g_0^4)$.

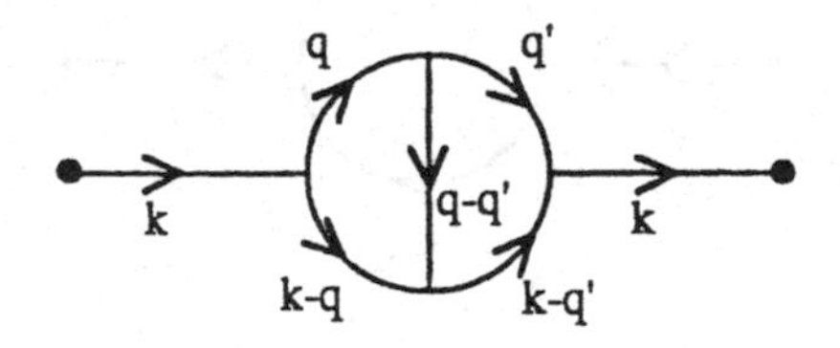

Fig. 13-3 A natural choice for the loop integration variables

Given a lattice Feynman integral having the above properties, we now want to learn something about its continuum limit. In general this limit cannot be be calculated by first evaluating the integrals for finite lattice spacing and then taking the limit $a \to 0$. But since the integrands are finite in this limit, one may expect that under certain conditions, one can replace the integrands by their naive continuum limit, and hence get rid of the periodic structure of the integrands

* See Reisz (1988c) for a discussion of more general integration variables.

which complicate enormously the computations. The power counting theorem of Reisz (1988c), which we discuss in the following section, will tell us when this can be done. It also plays a central role for formulating a renormalization program.*

13.3 The Power Counting Theorem of Reisz

Consider a general lattice Feynman integral in the scalar ϕ^3 theory. We assume that we have scaled all variables with the lattice spacing a in an appropriate way so that this integral is the lattice-regulated version of a continuum Feynman integral**. Furthermore we assume that

a) all trivial integrations asociated with the conservation of energy and momenta at the vertices have been performed, and

b) that the loop integration variables $q_i (i = 1, ..., L)$ have been chosen in such a way that the integrand is a periodic function in each component of q_i, with periodicity $2\pi/a$. The domain of integration is the first BZ.

Let k and l denote collectively the set of momenta associated with the external and internal lines of the diagram, respectively, and q the collection of independent loop integration variables. A general Feynman integral in the ϕ^3-theory then has the following structure,

$$F(k; M, a) - \int_{BZ} \prod_{i=1}^{L} \frac{d^4 q_i}{(2\pi)^4} \frac{1}{D(k, q; M, a)},$$

where the integrand is given by a product of the propagators (13.8) associated with the internal lines of the diagram:

$$D(k, q; M, a) = \prod_{i=1}^{I} (\tilde{l}_i^2(k, q) + M^2).$$

Here I is the number of internal lines, and the μ'th component of $\tilde{l}_i$ is defined by an expression analogous to (13.7a). With a natural choice of loop integration

* A summary can be found in the lectures by Lüscher at Les Houches (1988).

** In lattice gauge theories we shall also have to deal with Feynman integrals which have no continuum analogue.

variables l_i will then be of the form*

$$l_i(k,q) = \sum_{j=1}^{L} c_{ij} q_j + Q_j(k),$$

where c_{ij} are either ± 1 or 0. We will be interested, however, in integrals of a more general structure. The reason is the following: take for example the integral (13.6c). It actually diverges logarithmically for $a \to 0$. But by subtracting from it its contribution at $k = 0$, we arrive at an expression which possesses a finite continuum limit. We therefore decompose (13.6c) as follows:

$$\Pi(k; M, a) = \Pi(0, M, a) + \tilde{\Pi}(k, M, a) \tag{13.9a}$$

where

$$\tilde{\Pi}(k, M, a) = \frac{g_0^2}{2} \int_{BZ} \frac{d^4 q}{(2\pi)^4} \frac{1}{(\tilde{q}^2 + M^2)} \left[\frac{1}{(\widetilde{q-k})^2 + M^2} - \frac{1}{\tilde{q}^2 + M^2} \right], \tag{13.9b}$$

The first term appearing on the right-hand side of (13.9a) diverges in the limit $a \to 0$. This divergent constant can be absorbed into the bare mass parameter M in the way familiar from continuum perturbation theory. The remaining integral will be shown to possess a finite continuum limit. This integral has the following form

$$\tilde{\Pi}(k; M, a) = \frac{g_0^2}{2} \int_{BZ} \frac{d^4 q}{(2\pi)^4} \frac{N(k, q; M, a)}{D(k, q; M, a)}, \tag{13.10a}$$

where**

$$N(k, q; M, a) = \tilde{q}^2 - (\widetilde{q-k})^2, \tag{13.10b}$$

and

$$D(k, q; M, a) = (\tilde{q}^2 + M^2)^2 [(\widetilde{q-k})^2 + M^2]. \tag{13.10c}$$

But this is not the only motivation for studying integrals of a more general structure. In lattice QED or QCD the integrals associated with Feynman diagrams will be already of the type

$$I_F(k; M, a) = \int_{BZ} \prod_{i=1}^{L} \frac{d^4 q_i}{(2\pi)^4} \frac{N(k, q; M, a)}{D(k, q; M, a)} \tag{13.11}$$

---

* We want to emphasize that it is important that the loop momenta are chosen in such a way that the coefficients c_{ij} are integers. Only then does the power counting theorem of Reisz apply for Feynman integrals involving an arbitrary number of loop integrations. Such a choice is always possible

** In the present case the numerator function N does not depend on M.

before invoking any renormalization procedure.* We therefore are interested in an answer to the following questions: a) When does the integral (13.11) possess a finite continuum limit, and b) if so, what can we say about this limit?

These questions have been answered by Reisz (1988c), who proved a power counting theorem for lattice theories, analogous to that familiar from continuum perturbation theory. This theorem applies to integrals of the type (13.11) with N and D satisfying the following requirements:

i) There exists an integer κ such that **

$$N(k,q;M,a) = a^{-\kappa}\hat{N}(ka,qa;Ma) \tag{13.12a}$$

where $\hat{N}$ is a smooth function of the variables ka, qa, Ma. Furthermore, $\hat{N}$ is periodic in each component of the dimensionless loop momenta $\hat{q} \equiv qa$ with periodicity 2π, and a polynomial in Ma.

ii) The continuum limit of $N(k,q;M,a)$ exists. We shall denote it by $P(k,q;M)$:

$$\lim_{a\to 0} N(k,q;M,a) = P(k,q;M) \tag{13.12b}$$

iii) The denominator $D(k,q;M,a)$ is of the form

$$D(k,q;M,a) = \prod_{i=1}^{I} D_i(l_i(k,q);M_i,a). \tag{13.12c}$$

Furthermore, there exists a smooth function $\hat{F}_i(\hat{l}_i;\hat{M}_i)$, which is periodic in $\hat{l}_i$ with periodicity 2π and a polynomial in $\hat{M}_i$, such that

$$D_i(l_i;M_i,a) = \frac{1}{a^2}\hat{F}_i(l_i a;M_i a). \tag{13.12d}$$

iv) The continuum limit of $D_i(l_i;M_i,a)$ exists and is given by

$$\lim_{a\to 0} D_i(l_i;M_i,a) = l_i^2 + M_i^2. \tag{13.12e}$$

v) There exist positive constants a_0 and K such that

$$|D_i(l_i;M_i,a)| \geq K(\tilde{l}_i^2 + M_i^2) \tag{13.12f}$$

* If I_F depends on several masses, then M stands collectively for all of them.

** Recall that k and q denote collective variables. To be general we also include the case where several masses $\{M_i\} \equiv M$ are involved.

for every $a < a_0$, and l_i in the BZ.

Notice that this condition is automatically satisfied for scalar particles. But for naive fermions it is violated for momenta ℓ_i at the edges of the BZ. On the other hand, for Wilson fermions, the denominator function appearing under the integral in (4.29a) is given by

$$D(l; M, a) = \sum_{\mu} \frac{1}{a^2} \sin^2(l_\mu a) + M(l)^2,$$

where $M(l)$ has been defined in (4.29b). Expressed in terms of the momenta $\tilde{l}_\mu$, $D(l; M, a)$ takes the following form for $r = 1$:

$$D(l; M, a) = (\tilde{l}^2 + M^2) + Ma\tilde{l}^2 + \frac{a^2}{4} \sum_{\mu \neq \nu} \tilde{l}_\mu^2 \tilde{l}_\nu^2 .$$

Hence $D(l, M, a) \geq (\tilde{l}^2 + M^2)$ and (13.12f) is satisfied. Now if the integrand in (13.11) has the properties (13.12) and if it satisfies the power counting theorem of Reisz (which we discuss below), then a) the integral (13.11) possesses a finite continuum limit, and b) this limit coincides with the expression obtained by replacing the integrand by its continuum limit, and sending the cutoff π/a to infinity; i.e.

$$\lim_{a \to 0} I_F(k; M, a) = \int_{-\infty}^{\infty} \prod_{i=1}^{L} \frac{dq_i}{(2\pi)^4} \frac{P(k, q; M)}{\prod_{i=1}^{I} (l_i^2 + M_i^2)}, \tag{13.13}$$

where $P(k, q; M)$ has been defined in (13.12b). This is a very nice result, for in this case we get rid of the periodic structure of the integrand in (13.11) and are left with an ordinary continuum Feynman integral.

We want to emphasize that the conditions i)—iv) are rather weak. In fact we know of no example when they are not satisfied. On the other hand, condition v) imposes a non-trivial constraint on the structure of the denominator for it implies that for momenta l_i lying at the edges of the BZ, $D_i(l_i, M_i, a)$ must diverge like $1/a^2$ in the continuum limit. Hence naive fermions are excluded, while Wilson fermions satisfy condition (13.12f). The following example illustrates the role played by this condition.

Consider the one-dimensional integral

$$f_N(a) = \int_{-\pi/a}^{\pi/a} dk \frac{1}{\frac{N^2}{a^2} \sin^2 \frac{ka}{N} + M^2} \tag{13.14}$$

where $N = 1$ or 2. For $N = 2$ the integrand is the one-dimensional analogue of the scalar propagator (13.8). On the other hand for $N = 1$ the denominator in (13.14) is the analogue of that encountered for naive fermions. Consider first the integral which is obtained from (13.14) by replacing the integrand by its continuum limit and sending the cutoff to infinity. Independent of the choice of N, the result is π/M. Next let us calculate the integral exactly in the limit $a \to 0$. To this end we make use of the following integral representation for the integrand

$$\frac{1}{x^2 + M^2} = \int_0^\infty d\rho e^{-(x^2+M^2)\rho},$$

where $x^2 = \frac{N^2}{a^2}\sin^2\frac{ka}{N} = \frac{N^2}{2a^2}(1-\cos\frac{2ka}{N})$. Substituting this expression into (13.14), one finds that for $N = 1$ or 2, the integral can be written in the form

$$f_N(a) = \frac{4\pi a}{N^2}\int_0^\infty dy e^{-(1+\frac{2a^2M^2}{N^2})y} I_0(y), \tag{13.15}$$

where $I_0(y)$ is the modified Bessel function. For large arguments, $I_0(y)$ behaves as follows:

$$I_0(y) \underset{y\to\infty}{\longrightarrow} \frac{1}{\sqrt{2\pi y}} e^y. \tag{13.16}$$

Hence for any finite a the integral (13.15) exist. For $a \to 0$, however, the integral diverges, as it must, if (13.15) is to possess a finite limit. We can therefore calculate this limit by substituting the right–hand side of (13.16) into (13.15). The resulting integral can be immediately performed, and one obtains

$$f_N(a) \underset{a\to 0}{\longrightarrow} \frac{2}{N}\left(\frac{\pi}{M}\right).$$

Hence the result coincides with the one obtained in the naive approach for $N = 2$, while it is twice as large for $N = 1$. This does not come as a surprise, since for $N = 1$, and k within the BZ, $\sin ka$ not only vanishes for $k = 0$, but also at the corner of the Brillouin zone. This is the analog of the familiar doubling problem!

The above example shows that even if a lattice integral, and its naive approximation, both possess a finite limit for $a \to 0$, their continuum limits will not necessarily coincide. In fact for $N = 1$ the denominator appearing in the integrand of (13.14) violates condition (13.12f).

After these preliminaries, we are now in the position to discuss the power counting theorem of Reisz, which applies to integrals having the form (13.11), with the integrand satisfying conditions (13.12a-f). In order to establish the existence, or non–existence, of the continuum limit, we need a definition for the lattice degree of divergence (LDD) analogous to that introduced in continuum perturbation theory to study the convergence of Feynman integrals. But in contrast to the continuum formulation, the LDD not only refers to the behaviour of the integrand for large loop momenta, but it characterizes its behaviour under a simultaneous scale transformation of the loop momenta and lattice spacing. This is connected with the periodic structure of the integrand. Let us first introduce the lattice degree of divergence for an integral involving a single loop momentum, and generalize the concept afterwards.

Consider a function $W(k,q;M,a)$ depending on a single loop momentum q. Then according to Reisz (1988), the LDD is given by the exponent α defined by

$$W(k,\lambda q;M,a/\lambda) \underset{\lambda\to\infty}{\approx} W_0(k,q;M,a)\lambda^{\alpha} + O(\lambda^{\alpha-1}). \tag{13.17}$$

To make it explicit that the LDD is obtained by scaling q and a in such a way that the product $\hat{q} \equiv qa$ is held fixed, we shall use the following notation:

$$\mathrm{degr}_{\hat{q}} W = \alpha.$$

Applying this definition to the numerator and denominator functions, N and D, of a one loop integral of the type (13.11) with $L = 1$, one finds that the LDD of the integrand is given by $\mathrm{degr}_{\hat{q}} N - \mathrm{degr}_{\hat{q}} D$. The LDD of the corresponding integral (which includes the behaviour of the integration measure under the scale transformation $q \to \lambda q$) is then defined by $\mathrm{degr} I_F = 4 + \mathrm{degr}_{\hat{q}} N - \mathrm{degr}_{\hat{q}} D$. The power counting theorem of Reisz then states that if $\mathrm{degr} I_F < 0$, then the continuum limit of the lattice integral exists and coincides with its naive continuum limit.

As an example consider the integral (13.10). A simple calculation shows that $\mathrm{degr}_{\hat{q}} N = 1$ and $\mathrm{degr}_{\hat{q}} D = 6$, so that $\mathrm{degr}_{\hat{q}} \tilde{\Pi} = -1$. Hence the integral converges for $a \to 0$ to

$$\tilde{\Pi}(k;M,a) \underset{a\to 0}{\longrightarrow} \frac{g_0^2}{2}\int_{-\infty}^{\infty} \frac{d^4q}{(2\pi)^4} \frac{2q\cdot k - k^2}{(q^2+M^2)^2[(q-k)^2+M^2]}.$$

Consider next Feynman integrals involving L loop momenta $q_1,\ldots,q_L$, where $L > 1$. The *overall* LDD is obtained by scaling all these momenta with the same factor λ, while reducing at the same time the lattice spacing by a factor $1/\lambda$. But even if this LDD is negative, the Feynman integral need not converge. The reason is that divergencies may arise from integration regions, where only a subset of the line momenta becomes large, while the others are kept finite. Consider for example the diagram depicted in fig. (13-3) for a fixed external momentum k. By keeping q, or q', or q-q' fixed, large momenta are only allowed to flow in the loops of figs. (13-4a,b,c) denoted by a solid line. In the remaining diagram, none of these momenta are kept fixed.

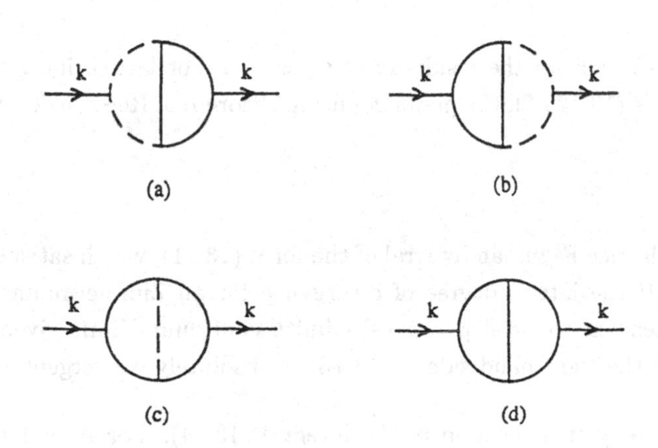

Fig. 13-4 Diagrams displaying the four Zimmermann subspaces corresponding to holding (a) q, (b) q', (c) q-q' fixed or (d) none of these in fig. 13-3.. The solid lines can carry arbitrarily large loop momenta

This defines four Zimmermann subspaces. To each of these subspaces we can associate an LDD by studying the behaviour of the integral associated with the diagrams obtained by omitting the propagators denoted by the dashed lines in fig. (13-4), and fixing the momenta carried by these lines. Consider for example the diagram depicted in fig. (13-4c). Let u be the momentum in the dashed line which is held fixed. Then the relevant integral corresponding to this Zimmermann subspace H is given by

$$I_{(H)} = \int_{BZ} \frac{d^4q}{(2\pi)^4} \frac{1}{(\tilde{q}^2 + M^2)[\widetilde{(q-u)}^2 + M^2][\widetilde{(q-k)}^2 + M^2][\widetilde{(q-k-u)}^2 + M^2]} .$$

The LDD corresponding to the subspace H is obtained by scaling the loop variable q and lattice spacing a with λ and $1/\lambda$, respectively. This LDD is given by $\mathrm{degr}_H I_{(N)} = -4$.

The generalization of these ideas to Feynman integrals involving any number of loop integration variables is straight-forward: Given a Zimmermann subspace, we write the line momenta as a function of those momenta which are to be held fixed and a set of loop momenta which are scaled with a common factor λ. At the same time the lattice spacing is multiplied by λ^{-1}. One then determines the LDD of the integrand according to (13.17) where q now stands collectively for all momenta that are integrated over the BZ. The LDD of the integral is obtained by including the behaviour of the integration measure under the scale transformation $q \to \lambda q$.

Let us summarize the results of this section. For lattice integrals satisfying the conditions (13.12a-f), the power counting theorem of Reisz makes the following assertion:

Theorem

Let I_F be a lattice Feynman integral of the form (13.11), which satisfies conditions (13.12a-f). If the lattice degree of divergence for all Zimmermann subspaces is negative, then the integral possesses a finite continuum limit given by (13.13). Furthermore the right-hand side of (13.13) is absolutely convergent.

Let us apply this theorem to the integral (13.14). For $N = 1$ this theorem does not apply since the integrand violates (13.12f). For $N = 2$ this condition is fulfilled. Furthermore the degree of divergence of the denominator is 2. Hence the LDD of $f_2(a)$ is -1, and the continuum limit is given by

$$\lim_{a\to 0} f_2(a) = \int_{-\infty}^{\infty} dk \frac{1}{k^2 + M^2}.$$

This agrees with our earlier observation.

For a proof of the above theorem we refer the reader to the work of Reisz (1988c). As we have already mentioned, this theorem plays a central role in developing a renormalization program for lattice field theories. But it is clearly also very useful for studying such problems as we have mentioned in the introduction, since it allows us to replace all lattice integrals which satisfy the conditions of the theorem by ordinary Feynman integrals whose symmetries are manifest.

This concludes our discussion of some of the basic lattice concepts which are relevant to the perturbative study of any lattice field theory. In the following two chapters we discuss lattice gauge theories which will be burdened by a number of additional problems.

CHAPTER 14

WEAK COUPLING EXPANSION (II) LATTICE QED

The scalar ϕ^3–theory we discussed in the previous chapter was a good laboratory for introducing a number of important concepts in weak coupling perturbation theory which are relevant to all lattice field theories of interest to elementary particle physics. We now extend our discussion to the case of lattice gauge theories, which present some problems of their own. Since the perturbative treatment of lattice QCD involves a number of technicalities arising from its non–abelian structure, we will begin with a discussion of lattice QED, where the Feynman rules can be easily derived.

As we shall see, the lattice regularization of a gauge field theory gives rise to an infinite number of so–called "irrelevant" interaction vertices which vanish in the limit of zero lattice spacing. Nevertheless, some of these vertices can contribute to correlation functions in the continuum limit through divergent loop corrections. For QED in a linear covariant gauge, these vertices originate only from the lattice regulated action. The purpose of this chapter is to demonstrate i) how the structure of the interaction vertex in the continuum formulation is modified by the lattice regularization, and ii) to elucidate the role played by irrelevant vertices in cancelling ultra–violet divergencies in lattice Feynman integrals, which cannot be removed by renormalization.

14.1 The Gauge Fixed Lattice Action

In lattice QED, the link variables are elements of the abelian U(1) group. Their parametrization in terms of a single angular variable is given by $U_\mu(n) = \exp(i\phi_\mu(n))$. Correlation functions of the link variables and the fermion fields are computed according to (5.24), where, because of the abelian nature of the link variables the integration measure DU has the simple form (5.23).

Since we are dealing with a gauge theory, we will eventually be interested in studying the ground state expectation value of gauge invariant functionals of the dimensionless fields ϕ, $\hat{\psi}$, and $\hat{\bar{\psi}}$. We denote these functionals by $\Gamma[\phi; \hat{\psi}, \hat{\bar{\psi}}]$. The ground state expectation value of Γ is given by

$$< \Gamma >= \frac{1}{Z} \int D\phi D\hat{\bar{\psi}} D\hat{\psi} \Gamma[\phi, \hat{\psi}, \hat{\bar{\psi}}] e^{-S_{QED}[\phi, \hat{\psi}, \hat{\bar{\psi}}]}, \qquad (14.1a)$$

where

$$Z = \int D\phi D\bar{\hat{\psi}} D\hat{\psi} e^{-S_{QED}[\phi,\hat{\psi},\bar{\hat{\psi}}]} \tag{14.1b}$$

is the "partition" function for lattice QED, and $S_{QED}[\phi, \hat{\psi}, \bar{\hat{\psi}}]$ is the gauge invariant action expressed in terms of the dimensionless fields $\phi, \hat{\psi}$, and $\bar{\hat{\psi}}$. For Wilson fermions it is given by (5.22),* with the link and plaquette variables expressed in terms of $\{\phi_\mu(n)\}$. One readily verifies that

$$S_{QED}[\phi, \hat{\psi}, \bar{\hat{\psi}}] = S_G[\phi] + S_F^{(W)}[\phi, \hat{\psi}, \bar{\hat{\psi}}], \tag{14.2a}$$

where **

$$S_G[\phi] = \frac{1}{2e_0^2} \sum_{n,\mu,\nu} [1 - \cos\phi_{\mu\nu}(n)], \tag{14.2b}$$

$$\phi_{\mu\nu}(n) = \hat{\partial}_\mu^R \phi_\nu(n) - \hat{\partial}_\nu^R \phi_\mu(n), \tag{14.2c}$$

and

$$\begin{aligned} S_F^{(W)}[\phi, \hat{\psi}, \bar{\hat{\psi}}] =& (\hat{M}_0 + 4r) \sum_n \bar{\hat{\psi}}(n)\hat{\psi}(n) \\ & - \frac{1}{2} \sum_{n,\mu} [\bar{\hat{\psi}}(n)(r - \gamma_\mu) e^{i\phi_\mu(n)} \hat{\psi}(n + \hat{\mu}) \\ & + \bar{\hat{\psi}}(n + \hat{\mu})(r + \gamma_\mu) e^{-i\phi_\mu(n)} \hat{\psi}(n)]. \end{aligned} \tag{14.2d}$$

The action of the right lattice derivative, $\hat{\partial}_\mu^R$, appearing in eq. (14.2c), is defined by an expression analogous to (4.43b).

Because the coupling constant e_0 occurs with an inverse power in (14.2b), one *naively* expects that the integral (14.1) is dominated for small coupling by those configurations ϕ lying in the immediate neighbourhood of the classical minimum of S_G. *** This minimum is realized for link configurations which are pure gauge,

---

* In chapter 5 we had supressed the "hat" on the Dirac fields, since we were only interested in the dimensionless formulation

** Here, and in the following, we shall use the same symbols for S_{QED}, S_G and $S_F^{(W)}$, irrespective of whether they are considered to be functions of the link variables $U_\mu(n)$, or the angular variables $\phi_\mu(n)$ parametrizing these. The factor 1/2 multiplying the sum in (14.2b) takes account of the fact that we are summing over all values of μ and ν. Notice that there is no contribution coming from $\mu = \nu$.

*** We want to emphasize that this argument is only formal. Large quantum fluctuations in the fields could turn out to play an important role, invalidating perturbation theory.

and therefore is degenerate. This degeneracy must be removed before performing the weak coupling expansion, since otherwise one cannot define the free propagator of the photon. In continuum perturbation theory this is a well–known fact. That it is also true in the lattice formulation can be seen immediately by expanding (14.2b) in powers of $\phi_{\mu\nu}$, and looking at the quadratic contribution:

$$S_G[\phi] = \frac{1}{4e_0^2} \sum_{n,\mu,\nu} \phi_{\mu\nu}(n)\phi_{\mu\nu}(n) + \cdots .$$

Except for the factor $1/e_0^2$, this contribution has a structure analogous to that encountered in the continuum formulation. Hence the free propagator of the ϕ_μ–field cannot be defined, since $\phi_{\mu\nu}$ vanishes for all configurations ϕ which are pure gauge: $\phi_\mu(n) = \hat{\partial}_\mu^R \Lambda(n)$. The solution to this problem is well–known: we have to introduce a gauge condition in (14.1) which selects from each gauge orbit a single representative. Along each such orbit the integrands appearing in the numerator and denominator of (14.1) are constants. This must be done in such a way that gauge invariant correlation functions are not affected by the gauge fixing procedure for *any* finite lattice spacing. An elegant way of introducing a gauge condition was proposed by Faddeev and Popov, and is referred to in the literature as the Faddeev–Popov trick.* Since we shall demonstrate this trick for a generalized Lorentz gauge later on when we discuss the non–abelian theory (where the computations are non–trivial), we shall only state here the result, which in the abelian case is very simple. Consider the following generalized Lorentz gauge**

$$\mathcal{F}_n[\phi;\chi] = \hat{\partial}_\mu^L \phi_\mu(n) - \chi(n) = 0, \tag{14.3}$$

where χ is some given arbitrary field, $\hat{\partial}_\mu^L$ is the left lattice derivative defined by an expression analogous to (4.26b). The reason for having introduced the left derivative will become clear later on. Applying the Faddeev–Popov trick, one finds that the above gauge condition can be implemented by merely introducing a set of δ–functions in the integrands of (14.1a,b) which ensure that only those field configurations ϕ_μ contribute to the integrals which satisfy (14.3). Hence (14.1a) can also be written in the form

$$<\Gamma>= \frac{\int D\phi D\hat{\bar{\psi}} D\hat{\psi} \prod_n \delta(\mathcal{F}_n[\phi;\chi])\Gamma[\phi,\hat{\psi},\hat{\bar{\psi}}]e^{-S_{QED}[\phi,\hat{\psi},\hat{\bar{\psi}}]}}{\int D\phi D\hat{\bar{\psi}} D\hat{\psi} \prod_n \delta(\mathcal{F}_n[\phi;\chi])e^{-S_{QED}[\phi,\hat{\psi},\hat{\bar{\psi}}]}}. \tag{14.4}$$

---

* If the reader is not familiar with this trick in continuum field theory, he may consult the review article by Abers and Lee (1973), or any modern field theory book.

** It will be understood from now on that repeated Lorentz indices are summed.

Next we perform one further standard trick to get rid of the δ–functions. Since we are calculating a gauge invariant correlation function, the choice of χ in (14.3) is immaterial. We can therefore average the numerator and denominator in (14.4) over χ with a Gaussian weight factor $\exp\left(-\frac{1}{2\alpha}\sum_n(\chi(n))^2\right)$. The resulting expression then takes the form

$$<\Gamma>=\frac{\int D\phi D\bar{\hat{\psi}} D\hat{\psi}\Gamma[\phi,\hat{\psi},\bar{\hat{\psi}}]e^{-S^{(tot)}_{QED}[\phi,\hat{\psi},\bar{\hat{\psi}}]}}{\int D\phi D\bar{\hat{\psi}} D\hat{\psi} e^{-S^{(tot)}_{QED}[\phi,\hat{\psi},\bar{\hat{\psi}}]}}, \tag{14.5a}$$

where the "total" action $S^{(tot)}_{QED}$ is given by

$$S^{(tot)}_{QED}[\phi,\hat{\psi},\bar{\hat{\psi}}] = S_G[\phi] + S^{(W)}_F[\phi,\hat{\psi},\bar{\hat{\psi}}] + S_{GF}[\phi], \tag{14.5b}$$

$$S_{GF}[\phi] = \frac{1}{2\alpha}\sum_n \left(\hat{\partial}^L_\mu \phi_\mu(n)\right)^2. \tag{14.5c}$$

The subscript "GF" stands for "gauge fixing".

So far the coupling constant occurs in that piece of the action depending only on the link variables. Furthermore, it appears with an inverse power, which is peculiar to the lattice formulation. In the continuum formulation, on the other hand, this coupling constant enters linearly in the fermionic part of the action and not at all in the kinetic term for the gauge field. To establish the connection between the lattice and continuum action we must introduce a lattice scale a, and a set of dimensioned gauge potentials and fermion fields. This is done in a way analogous to that described in chapters 4 and 5. But since we want our discussion to parallel as much as possible the continuum case, we shall use a slightly modified notation. Let $x_\mu = n_\mu a$ be the coordinates of the lattice sites and $\psi(x), \bar{\psi}(x)$ and $A_\mu(x)$ the dimensioned fermion fields and gauge potentials evaluated at these sites. The action (14.2a) can be written as a functional of these fields by making the following substitutions: $\hat{\psi}(n) \to a^{3/2}\psi(x)$, $\bar{\hat{\psi}}(n) \to a^{3/2}\bar{\psi}(x)$, $\phi_\mu(n) \to e_0 a A_\mu(x)$. Furthermore, we define a dimensioned bare mass parameter M_0 by $\hat{M}_0 = aM_0$ and introduce the following short–hand notation

$$\sum_x \equiv \sum_n a^4.$$

In the continuum limit $\sum_x$ goes over into $\int d^4x$. With these replacements (14.5b) becomes*

$$S^{(tot)}_{QED}[A,\psi,\bar{\psi}] = S_G[A] + S^{(W)}_F[A,\psi,\bar{\psi}] + S_{GF}[A], \tag{14.6a}$$

---

* In order not to introduce new symbols, we keep the old notation for the various contributions to the action.

where

$$S_G[A] = \frac{1}{2e_0^2 a^4} \sum_{x,\mu,\nu} [1 - \cos(e_0 a^2 F_{\mu\nu}(x))], \tag{14.6b}$$

$$\begin{aligned} S_F^{(W)} =& (M_0 + \frac{4r}{a}) \sum_x \bar{\psi}(x)\psi(x) \\ &- \frac{1}{2a} \sum_{x,\mu} \Big[\bar{\psi}(x)(r - \gamma_\mu) e^{ie_0 a A_\mu(x)} \psi(x + a\hat{\mu}) \\ &+ \bar{\psi}(x + a\hat{\mu})(r + \gamma_\mu) e^{-ie_0 a A_\mu(x)} \psi(x) \Big], \end{aligned} \tag{14.6c}$$

$$S_{GF}[A] = \frac{1}{2\alpha_0} \sum_x (\partial_\mu^L A_\mu(x))^2. \tag{14.6d}$$

Here we have introduced the arbitrary parameter α_0 by setting $\alpha = e_0^2 \alpha_0$. The lattice field strength tensor appearing in (14.6b) is given by

$$F_{\mu\nu}(x) = \partial_\mu^R A_\nu(x) - \partial_\nu^R A_\mu(x), \tag{14.7}$$

and the action of the dimensioned "right" and "left" derivatives in (14.7) and (14.6d) are defined by

$$\partial_\mu^R f(x) = \frac{1}{a}(f(x + a\hat{\mu}) - f(x)), \tag{14.8a}$$

$$\partial_\mu^L f(x) = \frac{1}{a}(f(x) - f(x - a\hat{\mu})). \tag{14.8b}$$

The next step consists in expanding the action in powers of the bare coupling. This gives rise to an infinite number of interaction terms contributing to S_G and $S_F^{(W)}$. Of these only those terms survive in the naive continuum limit which are characteristic of the continuum formulation. Nevertheless, as we have pointed out, we cannot simply ignore the irrelevant contributions when performing the weak coupling expansion in a lattice regulated theory. In the following we include only those "irrelevant" interaction vertices which vanish linearly with a in the naive continuum limit.

Consider first the contribution to the action depending only on the link variables, i.e. S_G. Expanding (14.6b) in powers of the lattice spacing one finds that

$$S_G = \frac{1}{4} \sum_x F_{\mu\nu}(x) F_{\mu\nu}(x) + O(a^4). \tag{14.9}$$

By making use of the relation*

$$\sum_x (\partial_\mu^R f(x))g(x) = -\sum_x f(x)\partial_\mu^L g(x)\,, \tag{14.10}$$

we can rewrite (14.6b) up to terms vanishing like a^4 as follows:

$$S_G \approx -\frac{1}{2}\sum_x A_\mu(x)(\delta_{\mu\nu}\Box - \partial_\mu^R\partial_\nu^L)A_\nu(x). \tag{14.11a}$$

Here

$$\Box = \sum_\mu \partial_\mu^R\partial_\mu^L \tag{14.11b}$$

is the hermitean lattice Laplacean in 4 space-time dimensions:

$$\Box f(x) = \frac{1}{a^2}\sum_\mu [f(x+a\hat{\mu}) + f(x-a\hat{\mu}) - 2f(x)].$$

Next, consider that piece of the action arising from gauge fixing δ-function, i.e. (14.6d). Using again relation (14.10) we have that

$$S_{GF} = -\frac{1}{2\alpha_0}\sum_x A_\mu(x)\partial_\mu^R\partial_\nu^L A_\nu(x). \tag{14.12}$$

Notice that because (14.6d) involves the left lattice derivative, the tensor structure of (14.12) is the same as that appearing in (14.11a). Hence by combining (14.11a) and (14.12), we arrive at the following contribution to $S_G^{(tot)}$ quadratic in the gauge potentials

$$S_G^{(0)}[A] = \frac{1}{2}\sum_{x,y} A_\mu(x)\Omega_{\mu\nu}(x,y)A_\nu(y), \tag{14.13a}$$

where

$$\Omega_{\mu\nu}(x,y) = \left(-\delta_{\mu\nu}\Box + (1-\frac{1}{\alpha_0})\partial_\mu^R\partial_\nu^L\right)\delta_P^{(4)}(x-y). \tag{14.13b}$$

Here $\delta_P^{(4)}(z)$ is the periodic δ-function

$$\delta_P^{(4)}(z) = \int_{BZ}\frac{d^4k}{(2\pi)^4}e^{ik.z}, \quad z = na, \tag{14.14}$$

* This relation follows immediately by introducing (14.8a) into the left-hand side of (14.10), and making a shift in the summation variable. We assume here that we are dealing with an infinite lattice, or with a finite lattice, with $f(x)$ and $g(x)$ satisfying periodic boundary conditions.

where from now on BZ stands for the dimensioned Brillouin zone $[-\pi/a, \pi/a]$.

Finally, consider the fermionic contribution (14.6c). Expanding the link variables $U_\mu = \exp(ie_0 a A_\mu)$ up to terms quadratic in the coupling, one finds that

$$S_F^{(W)} = S_F^{(0)} + S_F^{(1)} + S_F^{(2)} + O(a^2), \tag{14.15a}$$

where

$$\begin{aligned} S_F^{(0)} =& (M + \frac{4r}{a}) \sum_x \bar{\psi}(x)\psi(x) \\ & -\frac{1}{2a} \sum_{x,\mu} [\bar{\psi}(x)(r - \gamma_\mu)\psi(x + a\hat{\mu}) + \bar{\psi}(x + a\hat{\mu})(r + \gamma_\mu)\psi(x)] \end{aligned} \tag{14.15b}$$

is the free fermion action expressed in terms of the dimensioned variables, and

$$\begin{aligned} S_F^{(1)} =& -\frac{ie_0}{2} \sum_{x,\mu} [\bar{\psi}(x)(r - \gamma_\mu)A_\mu(x)\psi(x + a\hat{\mu}) \\ & - \bar{\psi}(x + a\hat{\mu})(r + \gamma_\mu)A_\mu(x)\psi(x)], \end{aligned} \tag{14.15c}$$

$$\begin{aligned} S_F^{(2)} =& \frac{e_0^2}{4} a \sum_{x,\mu} [\bar{\psi}(x)(r - \gamma_\mu)A_\mu^2(x)\psi(x + a\hat{\mu}) \\ & + \bar{\psi}(x + a\hat{\mu})(r + \gamma_\mu)A_\mu^2(x)\psi(x)]. \end{aligned} \tag{14.15.d}$$

Collecting our results, we therefore find that the total action (14.6a) is given by

$$S_{QED}^{(tot)}[A, \psi, \bar{\psi}] = S_G^{(0)}[A] + S_F^{(0)}[\psi, \bar{\psi}] + S_{int}[A, \psi, \bar{\psi}] + O(a^2), \tag{14.16a}$$

where

$$S_{int}[A, \psi, \bar{\psi}] = \sum_{\ell=1}^{2} S_F^{(\ell)}[A, \psi, \bar{\psi}] \tag{14.16b}$$

is the contribution arising from the fermion–gauge–field interaction. From (14.15) we see that S_{int} describes not only the interaction of a single photon with the Dirac field, but also includes a contribution involving the coupling of two photons to the fermions. Whereas the former contribution reduces in the naive continuum limit to the familiar interaction term,

$$S^{(1)} \to ie_0 \int d^4x \bar{\psi}(x)\gamma_\mu A_\mu(x)\psi(x),$$

the latter contribution, $S_F^{(2)}$, has no analog in the continuum and in fact vanishes for $a \to 0$. Nevertheless, as we shall see, it plays an important role in canceling divergent contributions to the vacuum polarization, which cannot be eliminated by renormalizing the fields and bare parameters. This is not surprising, since these vertices are a consequence of the lattice regularization which provides us with a gauge invariant cutoff.*

14.2 Lattice Feynman Rules

Assuming that $S_{QED}^{(tot)}$ can be approximated by (14.16), and that the integration range of A_μ in the path–integral expression for the correlation functions can be extended to infinity,** the perturbative expansion of any correlation function involving a product of the fermion fields ψ and $\bar{\psi}$ and gauge potentials A_μ is obtained in the way familiar from continuum perturbation theory. In any given order of the coupling, the contributions to the correlation function can be represented in terms of Feynman diagrams built from the free propagators of the gauge potential and the fermion field, and from the interaction vertices. Their momentum space representations can be easily deduced by writing the action in momentum space. Consider first the contribution (14.13). Introducing the following Fourier decomposition of the fields

$$\begin{aligned} A_\mu(x) &= \int_{BZ} \frac{d^4k}{(2\pi)^4} \tilde{A}_\mu(k) e^{ik\cdot x}, \\ \psi_\alpha(x) &= \int_{BZ} \frac{d^4p}{(2\pi)^4} \tilde{\psi}_\alpha(p) e^{ip\cdot x}, \qquad (x = na) \\ \bar{\psi}_\alpha(x) &= \int_{BZ} \frac{d^4p}{(2\pi)^4} \tilde{\bar{\psi}}(p) e^{-ip\cdot x}, \end{aligned} \tag{14.17}$$

and making use of the relation (2.64), one readily finds that it can be written in the form

$$S_G^{(0)} = \frac{1}{2} \int_{BZ} \frac{d^4k}{(2\pi)^4} \frac{d^4k'}{(2\pi)^4} \tilde{A}_\mu(k') \left[e^{-ik'_\mu \frac{a}{2}} \Omega_{\mu\nu}(k', k) e^{-ik_\nu \frac{a}{2}} \right] \tilde{A}_\nu(k), \tag{14.18a}$$

where

$$\Omega_{\mu\nu}(k', k) = (2\pi)^4 \delta_P^{(4)}(k + k') \Omega_{\mu\nu}(k), \tag{14.18b}$$

and

$$\Omega_{\mu\nu}(k) = \left(\delta_{\mu\nu} \hat{k}^2 - \left(1 - \frac{1}{\alpha_0}\right) \hat{k}_\mu \hat{k}_\nu \right). \tag{14.18c}$$

---

* In continuum perturbation theory it is well–known that Feynman integrals must be regularized in a gauge invariant way.

** We know of no rigorous proof that this is a legitimate procedure.

Here $\tilde{k}_\mu$ is defined by an expression analogous to (13.7a). Notice that because of the appearance of the phase factors in the integrand of (14.18a), the quantity within square brackets is a periodic function in each component of the momenta with periodicity $2\pi/a$.

Next, consider the contribution $S_F^{(0)}$ defined in (14.15b). Its decomposition in momentum space is given by

$$S_F^{(0)} = \int_{BZ} \frac{d^4p}{(2\pi)^4} \frac{d^4p'}{(2\pi)^4} \bar{\tilde{\psi}}_\alpha(p') K_{\alpha\beta}(p', p) \tilde{\psi}_\beta(p), \tag{14.19a}$$

where

$$K_{\alpha\beta}(p', p) = (2\pi)^4 \delta_P^{(4)}(p - p') K_{\alpha\beta}(p), \tag{14.19b}$$

and

$$K_{\alpha\beta}(p) = \frac{i}{a} \sum_\mu (\gamma_\mu)_{\alpha\beta} \sin p_\mu a + M(p) \delta_{\alpha\beta}. \tag{14.19c}$$

The momentum dependent mass $M(p)$ has been defined in (4.29b).

Finally, one easily verifies that (14.15c) and (14.15d) have the following momentum space decomposition:

$$S_F^{(1)} = -\int_{BZ} \frac{d^4k}{(2\pi)^4} \frac{d^4p}{(2\pi)^4} \frac{d^4p'}{(2\pi)^4} \bar{\tilde{\psi}}_\alpha(p') \tilde{A}_\mu(k) \tilde{\psi}_\beta(p) \left[e^{i(p-p')_\mu \frac{a}{2}} \cdot \Gamma^{(1)}_{\mu;\alpha\beta}(p', p, k) \right], \tag{14.20a}$$

$$\Gamma^{(1)}_{\mu;\alpha,\beta}(p', p, k) = (2\pi)^4 \delta_P(p - p' + k) V^{(1)}_{\mu;\alpha\beta}(p + p'), \tag{14.20b}$$

$$V^{(1)}_{\mu;\alpha\beta}(q) = -ie_0 \left[(\gamma_\mu)_{\alpha\beta} \cos \frac{q_\mu a}{2} - ir\delta_{\alpha\beta} \sin \frac{q_\mu a}{2} \right]; \tag{14.20c}$$

$$S_F^{(2)} = -\frac{1}{2!} \int_{BZ} \frac{d^4k}{(2\pi)^4} \frac{d^4k'}{(2\pi)^4} \frac{d^4p}{(2\pi)^4} \frac{d^4p'}{(2\pi)^4} \bar{\tilde{\psi}}_\alpha(p') A_\mu(k) A_\nu(k') \tilde{\psi}_\beta(p)$$

$$\cdot \left[e^{i(p-p')_\mu \frac{a}{2}} \Gamma^{(2)}_{\mu\nu;\alpha\beta}(p', p, k', k) \right], \tag{14.21a}$$

$$\Gamma^{(2)}_{\mu\nu;\alpha\beta}(p, p', k, k') = (2\pi)^4 \delta_P^{(4)}(p - p' + k + k') V^{(2)}_{\mu\nu;\alpha\beta}(p + p') \tag{14.21b}$$

$$V^{(2)}_{\mu\nu;\alpha\beta}(q) = -e_0^2 a \delta_{\mu\nu} \left[r\delta_{\alpha\beta} \cos \frac{q_\mu a}{2} - i(\gamma_\mu)_{\alpha\beta} \sin \frac{q_\mu a}{2} \right]. \tag{14.21c}$$

Notice again, that because of the appearance of the phases, the quantities appearing within square brackets in (14.20a) and (14.21a) are periodic functions in all components of the momenta with periodicity $2\pi/a$. Hence momentum conservation can be implemented in the usual way. Accordingly, we can replace the phase factors in (14.20a) and (14.21a) by $\exp(-ik_\mu a/2)$ and $\exp(-i(k+k')_\mu a/2)$, respectively. These phases can now be absorbed in the Fourier transforms of the gauge potentials, which amounts to redefining the potentials at the midpoints of the links connecting two neighbouring lattice sites. Although we could have avoided these phases right from the start by Fourier decomposing $A_\mu(x)$ as follows

$$A_\mu(x) = \int_{BZ} \frac{d^4k}{(2\pi)^4} \tilde{A}_\mu(k) e^{ik\cdot(x+a\hat{\mu}/2)}, \tag{14.22}$$

we have nevertheless preferred to carry them along in order to exhibit the $2\pi/a$-periodic structure of the above mentioned expressions. But when computing the contribution of a particular Feynman diagram one finds that the phases associated with the interaction vertices, and the photon propagator (deduced from (14.18a)) cancel at each interaction vertex. The only phases that remain are those associated with the gauge potentials appearing in the correlation function. Hence by Fourier decomposing the correlation functions as follows,

$$\begin{aligned}&<\psi_{\alpha_1}(x_1)\dots\psi_{\alpha_n}(x_n)\bar{\psi}_{\beta_1}(y_1)\dots\bar{\psi}_{\beta_n}(y_n)A_{\mu_1}(z_1)\dots A_{\mu_\ell}(z_\ell)> \\ &= \int \prod_{i=1}^n \frac{d^4p_i}{(2\pi)^4} \prod_{i=1}^n \frac{d^4p'_i}{(2\pi)^4} \prod_{j=1}^\ell \frac{d^4k_j}{(2\pi)^4} \tilde{\Gamma}_{\alpha_1\dots\alpha_n\beta_1\dots\beta_n\mu_1\dots\mu_\ell}(\{p_i\},\{p'_i\},\{k_j\}) \\ &\qquad \times e^{i\sum_{i=1}^n (p_i\cdot x_i - p'_i\cdot y_i)} e^{i\sum_{j=1}^\ell k_j\cdot(z_j+a\hat{\mu}_j/2)},\end{aligned} \tag{14.23}$$

we can calculate the contribution of a Feynman diagram to the correlation function in momentum space, $\tilde{\Gamma}_{\alpha_1\dots\mu_\ell}(p_1,\dots k_\ell)$, using the propagators and vertices deduced from (14.18) to (14.21) ignoring the phases factors. The propagators of the gauge potential and of the fermion field are given by the inverse of the matrices (14.18c) and (14.19c), respectively, while the vertices are given by (14.20c) and (14.21c). In the continuum limit $V^{(2)}_{\mu\nu;\alpha\beta}(q)$ vanishes, and $V_{\mu\nu;\alpha\beta}(q)$ reduces to the vertex function of the continuum theory, i.e., $-ie_0(\gamma_\mu)_{\alpha\beta}$.

Except for the fact that on the lattice we must also take into account "irrelevant" interaction vertices, the rules for computing the contribution of a particular Feynman diagram are the same as in the continuum formulation. For finite lattice spacing the corresponding Feynman integrals are however much more complicated than those encountered in continuum perturbation theory, where the integrals are regularized a posteriori, and do not follow from a space–time regulated generating functional.

We now summarize the rules for calculating the contribution of a Feynman diagram to the correlation function $\tilde{\Gamma}_{\alpha_1 \ldots \mu_\ell}(p_1, \ldots, k_\ell)$ defined in (14.23).

i) To an internal fermion or photon line associate the propagators

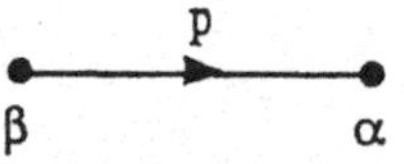

p; β α

$$S_F(p)_{\alpha\beta} = K^{-1}_{\alpha\beta}(p) = \left[\frac{-i\sum_\mu \frac{1}{a}\gamma_\mu \sin p_\mu a + M(p)}{\sum_\mu \frac{1}{a^2}\sin^2 p_\mu a + M^2(p)}\right]_{\alpha\beta}$$

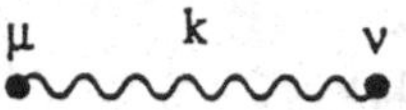

μ k ν

$$D_{\mu\nu}(k) = \Omega^{-1}_{\mu\nu}(k) = \frac{1}{\tilde{k}^2}\left(\delta_{\mu\nu} - (1-\alpha_0)\frac{\tilde{k}_\mu \tilde{k}_\nu}{\tilde{k}^2}\right)$$

ii) For the vertices insert the following expressions:

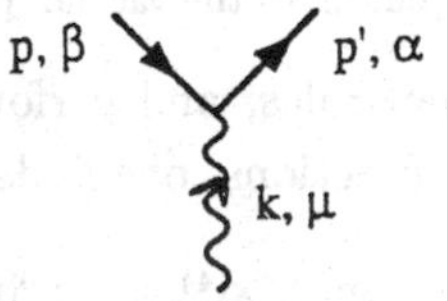

p, β; p′, α; k, μ

$$-ie_0(2\pi)^4\delta_P^{(4)}(p-p'+k)\left[(\gamma_\mu)_{\alpha\beta}\cos\left(\frac{(p+p')_\mu a}{2}\right) - ir\delta_{\alpha\beta}\sin\left(\frac{(p+p')_\mu a}{2}\right)\right]$$

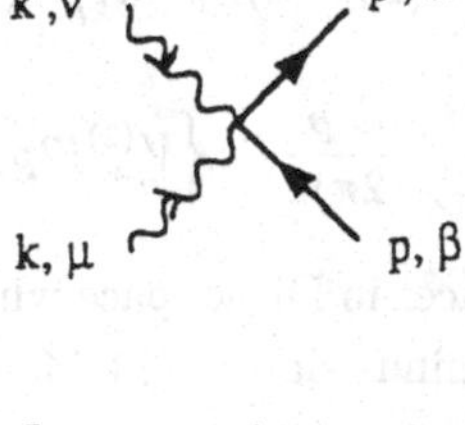

k′,ν; p′, α; k, μ; p, β

$$-e_0^2(2\pi)^4\delta_P^{(4)}(p-p'+k+k')a\delta_{\mu\nu}\left[r\delta_{\alpha\beta}\cos\left(\frac{(p+p')_\mu a}{2}\right) - i(\gamma_\mu)_{\alpha\beta}\sin\left(\frac{(p+p')_\mu a}{2}\right)\right]$$

iii) For every closed fermion loop include a minus sign.

iv) Contract all Dirac indices following the fermion lines, and all Lorentz indices following the photon lines.

v) Integrate the internal momenta over the Brillouin zone with integration measures having the generic form $d^4k/(2\pi)^4$.

As an example consider the vacuum polarization tensor $\pi_{\mu\nu}$ in order e_0^2. In the lattice regularized theory there are two diagrams that contribute. They are depicted in fig. (14-1).

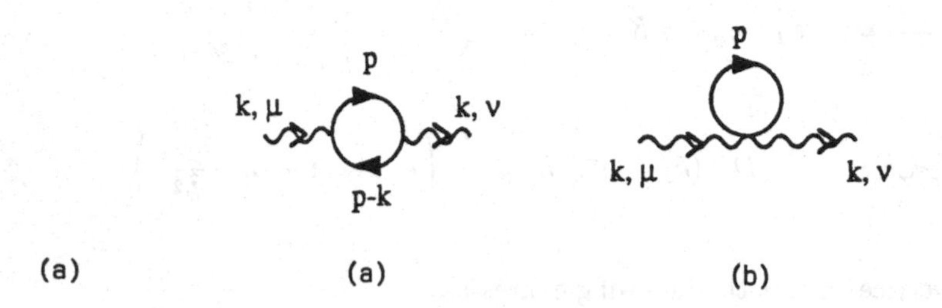

Fig. 14-1 Diagrams contributing to the vacuum polarization tensor $\Pi_{\mu\nu}(k)$

Applying the above Feynman rules, and performing the trivial integrations associated with the (periodic) δ–functions, one finds that

$$\Pi_{\mu\nu}(k,k') = (2\pi)^4\delta_P^{(4)}(k+k')\Pi_{\mu\nu}(k), \tag{14.24a}$$

where

$$\Pi_{\mu\nu}(k) = \Pi_{\mu\nu}^{(a)}(k) + \Pi_{\mu\nu}^{(b)}, \tag{14.24b}$$

$$\Pi_{\mu\nu}^{(a)}(k) = -\int_{BZ}\frac{d^4p}{(2\pi)^4}Tr\left[V_\nu^{(1)}(2p-k)S_F(p)V_\mu^{(1)}(2p-k)S_F(p-k)\right], \tag{14.24c}$$

$$\Pi_{\mu\nu}^{(b)} = -\int_{BZ}\frac{d^4p}{(2\pi)^4}Tr\left\{V_{\mu\nu}^{(2)}(2p)S_F(p)\right\}. \tag{14.24d}$$

Here $V_\mu^{(1)}$ and $V_{\mu\nu}^{(2)}$ are the matrices in Dirac space whose matrix elements are given by (14.20c) and (14.21c). The minus sign in (14.24c,d) takes into account rule iii) given above. Applying the power counting theorem of Reisz discussed in chapter 13, we conclude that the lattice degree of divergence of both integrals is 2. Hence they diverge like $1/a^2$ in the continuum limit. If this divergence would persist after combining the two integrals, then the theory would not be renormalizable, since there is no mass–counterterm available to cancel this divergence. In continuum QED, where only the diagram shown in fig. (14-1a) contributes, the corresponding Feynman integral is also superficially quadratically divergent. But because of the

Ward identities, it actually only diverges logarithmically. This divergence can be eliminated by wave function renormalization. On the other hand, the Ward identities, following from gauge invariance on the lattice, are only satisfied after including the contribution coming from the diagram depicted in fig. (14-1b); hence only then do we expect a cancellation of unwanted divergencies. For vanishing bare fermion mass and Wilson parameter r, this cancellation can be easily demonstrated (Kawai, Nakayama, and Seo, 1981). We first isolate that part of $\Pi^{(a)}_{\mu\nu}$ and $\Pi^{(b)}_{\mu\nu}$ diverging like $1/a^2$ for $a \to 0$ by decomposing these quantities as follows:

$$\Pi^{(i)}_{\mu\nu}(k) = \Pi^{(i)}_{\mu\nu}(0) + \left[\Pi^{(i)}_{\mu\nu}(k) - \Pi^{(i)}_{\mu\nu}(0)\right],$$

where $i = a$ or b. The quadratically divergent part is contained entirely in $\Pi^{(i)}_{\mu\nu}(0)$. Performing the trace in (14.24c) one finds that

$$\Pi^{(a)}_{\mu\nu}(0) = -\frac{e_0^2}{a^2}\int_{-\pi}^{\pi}\frac{d^4\hat{p}}{(2\pi)^4}\Gamma_{\mu\lambda\nu\rho}\frac{\sin\hat{p}_\lambda \sin\hat{p}_\rho \cos\hat{p}_\mu \cos\hat{p}_\nu}{\left(\sum_\sigma \sin^2\hat{p}_\sigma\right)^2}, \tag{14.25a}$$

where

$$\begin{aligned}\Gamma_{\mu\lambda\nu\rho} &= Tr(\gamma_\mu\gamma_\lambda\gamma_\nu\gamma_\rho)\\ &= 4(\delta_{\mu\lambda}\delta_{\nu\rho} - \delta_{\mu\nu}\delta_{\lambda\rho} + \delta_{\mu\rho}\delta_{\nu\lambda}).\end{aligned} \tag{14.25b}$$

By carrying out the summation over the Lorentz indices one easily verifies that (14.25a) is proportional to $\delta_{\mu\nu}$ and can be written in the form

$$\begin{aligned}\Pi^{(a)}_{\mu\nu}(0) &= \frac{4e_0^2}{a^2}\delta_{\mu\nu}\int_{-\pi}^{\pi}\frac{d^4\hat{p}}{(2\pi)^4}\left[\frac{\cos^2\hat{p}_\mu}{\sum_\lambda \sin^2 p_\lambda} + \frac{1}{2}\sin(2\hat{p}_\mu)\frac{\partial}{\partial\hat{p}_\mu}\frac{1}{\sum_\lambda \sin^2 p_\lambda}\right]\\ &= \frac{4e_0^2}{a^2}\delta_{\mu\nu}\int_{-\pi}^{\pi}\frac{d^4\hat{p}}{(2\pi)^4}\frac{\sin^2\hat{p}_\mu}{\sum_\lambda \sin^2 p_\lambda}.\end{aligned} \tag{14.26}$$

On the other hand, for $r = M_0 = 0$, (14.24d) becomes

$$\Pi^{(b)}_{\mu\nu} = \frac{-e_0^2}{a^2}\int_{-\pi}^{\pi}\frac{d^4\hat{p}}{(2\pi)^4}Tr\left(\gamma_\mu\frac{\sum_\lambda \gamma_\lambda \sin\hat{p}_\lambda}{\sum_\rho \sin^2\hat{p}_\rho}\right)\sin\hat{p}_\mu,$$

which, upon making use of the relation $tr(\gamma_\mu\gamma_\nu) = 4\delta_{\mu\nu}$, reduces to the negative of (14.26). This simple example demonstrates the important role played by irrelevant vertices in canceling divergences that cannot be removed by renormalization.

Let us summarize the lesson we have learned. The lattice provides us with a gauge invariant regularization scheme. Although this gauge invariance is broken by the gauge condition, the Faddeev-Popov procedure will leave gauge invariant correlation functions unchanged, if we include the contributions of all irrelevant vertices. This is true for finite lattice spacing. But when studying the continuum limit of correlation functions in a given order of weak coupling perturbation theory, only a subset of these vertices need to be taken into account. These vertices will ensure the renormalizability of the theory and the restauration of the continuum space-time symmetries in the limit of vanishing lattice spacing.

14.3 Renormalization of the Axial Vector Current in One-Loop Order

When computing decays like $\pi^- \to e + \bar{\nu}_e$, one needs to calculate the matrix element $< 0|j_{5\mu}(0)|\pi^-(\vec{p}) >= p_\mu f_\pi$, where f_π is the pion decay constant, and $j_{5\mu}(x)$ is the renormalized axial vector current. We must therefore know how the lattice regularized bare current is renormalized. Here Ward identities will serve as guidelines for determining the renormalization constants. It is instructive to compute these constants in one-loop order perturbation theory.

Let us first exemplify the main idea for QED in the continuum. In the tree graph approximation the axial vector current is $j_{5\mu}|_{tree} = \bar{\psi}(x)\gamma_\mu\gamma_5\psi(x)$. The one loop correction to this current is given by the diagrams depicted in fig. (14-2). In the limit of vanishing fermion mass the QED action is invariant under global γ_5-transformations,

$$\begin{aligned} \psi(x) &\to e^{i\epsilon\gamma_5}\psi(x) \ , \\ \bar{\psi}(x) &\to \bar{\psi}(x)e^{i\epsilon\gamma_5} \ . \end{aligned} \tag{14.27}$$

By performing an infinitessimal *local* γ_5-transformation of the fermion fields in the generating functional of Green functions (i.e., ϵ becomes x-dependent), and making use of the invariance of the measure, one easily derives a Ward identitiy for an n-point vertex function with an the insertion of the divergence of the axial vector current. In the continuum formulation of the path integral the Ward identity is only formally defined. In a lattice regularization of the path integral, this Ward identity is an exact statement, but needs, in general, to be renormalized as the cutoff (lattice spacing) is removed. It is instructive to first derive this *naive* identity in continuum QED in one-loop order directly for the diagram shown in fig. (14-2a). In momentum space its divergence is given by

$$iq_\mu\Lambda_{5\mu}(p,p') = i(-ie)^2 q_\mu \sum_\lambda \int \frac{d^4\ell}{(2\pi)^4}\left[\gamma_\lambda S_0(p'+\ell)\gamma_\mu\gamma_5 S_0(p+\ell)\gamma_\lambda\right]\frac{1}{\ell^2} \ , \tag{14.28}$$

where $q = p - p'$, and S_0 is the free fermion propagator $S_0 = (i\gamma_\mu q_\mu + m_0)^{-1}$; m_0 is the bare fermion mass. Making use of the trivial identity

$$i(p-p')_\mu\gamma_\mu\gamma_5 = 2m_0\gamma_5 - S_0^{-1}(p')\gamma_5 - \gamma_5 S_0^{-1}(p) \ , \tag{14.29}$$

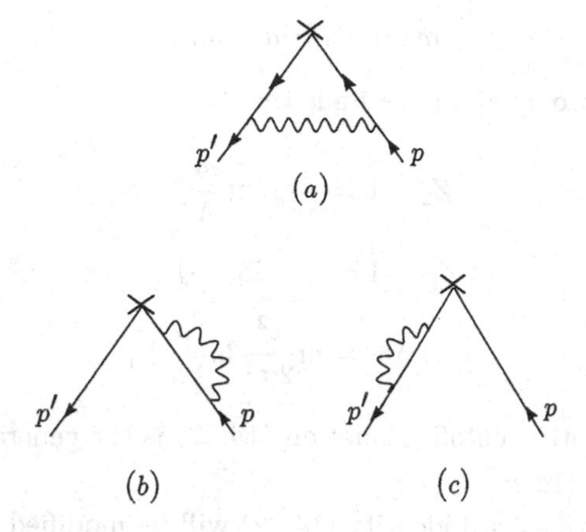

Fig. 14-2 Diagrams contributing in $\mathcal{O}(e^2)$ to the axial vector (pseudoscalar) current in the continuum formulation. The cross stands for the insertion of $\gamma_\mu\gamma_5$ (γ_5).

which is valid for arbitrary momentum ℓ, one immediately verifies that

$$iq_\mu\Lambda_{5\mu}(p,p') = 2m_0\Lambda_5(p,p') + \gamma_5\Sigma(p) + \Sigma(p')\gamma_5 \ . \tag{14.30}$$

Here $\Sigma(p)$ is the self energy, and Λ_5 is given by

$$\Lambda_5(p,p') = i(-ie)^2\sum_\lambda\int\frac{d^4\ell}{(2\pi)^4}\left[\gamma_\lambda S_0(p'+\ell)\gamma_5 S_0(p+\ell)\gamma_\lambda\right]\frac{1}{\ell^2} \ . \tag{14.31}$$

The self energy is divergent and requires mass as well as wave function renormalization. Mass and wave function renormalization as is dicated by QED in fact suffices to render all the terms in (14.30) finite. This can be easily checked in the Pauli-Villars, or dimensional regularization. The renormalized Ward identity for the corresponding vertex function takes the form

$$iq_\mu\Gamma_{5\mu}(p,p')_R = 2m\Gamma_5(p,p')_R - S_F^{-1}(p')_R\gamma_5 - \gamma_5 S_F^{-1}(p)_R \ , \tag{14.32}$$

where $S_F(q)_R$ is the renormalized fermion propagator to one loop order, i.e., $(i\gamma\cdot p + m - \Sigma_R(p))^{-1}$, and

$$\Gamma_{5\mu}(p,p')_R = Z_2\Gamma_{5\mu}(p,p') \ ,$$

$$\Gamma_5(p,p')_R = Z_P\Gamma_5(p,p') \ ,$$

$$m_0 = Z_2^{-1}(m + \delta m) \ . \tag{14.33}$$

In the one-loop approximation one finds that*

$$\begin{aligned} Z_2 - 1 &= \frac{e^2}{8\pi^2} \ln(\frac{m}{\Lambda}) \ , \\ Z_P - 1 &= \frac{e^2}{2\pi^2} \ln(\frac{m}{\Lambda}) \ , \\ \delta m &= m \frac{e^2}{2\pi^2} \ln(\frac{m}{\Lambda}) \ , \end{aligned} \tag{14.34}$$

where Λ is a momentum cutoff. Equation (14.32) is the generalization of the tree level Ward identity (14.29).

On the lattice the Ward identity (14.32) will be modified by terms which are *naively* irrelevant, i.e., which vanish in the naive continuum limit. We shall refer to them in the following simply as "irrelevant". In particular it will involve a contribution arising from a chiral symmetry breaking term in the action, which ensures the absence of fermion doubling. For concreteness sake we will consider the case of Wilson fermions. As we shall see, the irrelevant contribution referred to above leads to an additional *finite* renormalization of the axial vector current (apart from the QED wave function renormalization) in the continuuum limit. This problem has been first discussed by Karsten and Smit (1981). ** Here we dicuss it in some detail.

To derive the lattice regularized Ward identity analogous to (14.32) we consider the following lattice integral,

$$Z_{\mathcal{O}} = \int DUD\psi D\bar{\psi} \ \mathcal{O}(\psi, \bar{\psi}) e^{-S} \ , \tag{14.35a}$$

where

$$\mathcal{O} = \psi_\alpha(y)\bar{\psi}_\beta(z) \ , \tag{14.35b}$$

and S is the lattice action for QED with Wilson fermions, (5.22). The fermionic contribution can be written in the form

$$S_{ferm} = \sum_x \bar{\psi}(x)\{\frac{1}{2}\gamma_\mu(D_\mu^R[U] + D_\mu^L[U]) + m_0\} + \Delta S \ , \tag{14.36a}$$

---

* We are neglecting finite contributions. It therefore corresponds to a minimal substraction scheme.

** The renormalization of the axial vector current for different lattice realizations of the current has been considered by Meyer (1983).

where

$$\Delta S = \frac{r}{2} a \sum_x \bar{\psi}(x) D^L_\mu[U] D^R_\mu[U] \psi(x) \tag{14.36b}$$

is the chiral symmetry breaking Wilson term which vanishes for $a \to 0$, and

$$D^R_\mu[U]\psi(x) = \frac{1}{a}[U_\mu(x)\psi(x + a\hat{\mu}) - \psi(x)] ,$$

$$D^L_\mu[U]\psi(x) = \frac{1}{a}[\psi(x) - U^\dagger_\mu(x - a\hat{\mu})\psi(x - a\hat{\mu})] , \tag{14.37}$$

are the covariant right and left lattice derivatives. r is the Wilson Parameter. Making use of the invariance of the partition function (14.35a) under the following infinitessimal local axial transformation of the fermion fields (i.e. a change of variables),*

$$\delta\psi(x) = i\epsilon(x)\gamma_5\psi(x) ,$$

$$\delta\bar{\psi}(x) = i\epsilon(x)\bar{\psi}(x)\gamma_5 , \tag{14.38}$$

where $\gamma_5 = i\gamma_1\gamma_2\gamma_3\gamma_4$, and $\{\gamma_\mu, \gamma_5\} = 0$ for all μ, one is led to the identity

$$< \mathcal{O}\delta S - \delta\mathcal{O} >= 0 . \tag{14.39}$$

One then readily verifies that for Wilson fermions

$$\delta S = i \sum_x \epsilon(x)[-\partial^L_\mu j^5_\mu(x) + 2m_0 j_5(x) + \Delta(x)] , \tag{14.40}$$

where $\sum_x = \sum_n a^4$, $x_\mu = n_\mu a$ ($n_\mu \in Z$), ∂^L_μ denotes the left lattice derivative,

$$j_{5\mu}(x) = \frac{1}{2}[\bar{\psi}(x)\gamma_\mu\gamma_5 U_\mu(x)\psi(x + a\hat{\mu}) + \bar{\psi}(x + a\hat{\mu})\gamma_\mu\gamma_5 U^\dagger_\mu(x)\psi(x)] \tag{14.41}$$

is the gauge invariant axial vector current, and

$$j_5(x) = \bar{\psi}(x)\gamma_5\psi(x) \tag{14.42}$$

is the pseudoscalar current. Furthermore

$$\begin{aligned}\Delta(x) = &- \frac{r}{2a}\sum_\mu \bar{\psi}(x)\gamma_5[U_\mu(x)\psi(x + a\hat{\mu}) + U^\dagger_\mu(x - a\hat{\mu})\psi(x - a\hat{\mu}) - 2\psi(x)] \\ &- \frac{r}{2a}\sum_\mu [\bar{\psi}(x - a\hat{\mu})U_\mu(x - a\hat{\mu}) + \bar{\psi}(x + a\hat{\mu})U^\dagger_\mu(x) - 2\bar{\psi}(x)]\gamma_5\psi(x) .\end{aligned} \tag{14.43}$$

* The integration measure is invariant under this transformation

is the operator originating from the chiral symmetry breaking Wilson term (14.36b) in the action. Finally, the variation $\delta\mathcal{O}$ in (14.39), with $\mathcal{O}$ given by (14.35b), is given by

$$\delta\mathcal{O} = i\sum_x \epsilon(x)\sum_\delta [\delta_{xy}(\gamma_5)_{\alpha\delta}\psi_\delta(y)\bar{\psi}_\beta(z) + \delta_{xz}\psi_\alpha(y)\bar{\psi}_\delta(z)(\gamma_5)_{\delta\beta}] \,.$$

Since $\epsilon(x)$ in (14.40) is an arbitrary infinitessimal function, one is led to the Ward identity

$$\begin{aligned} < \partial^L_\mu j_{5\mu}(x)\psi_\alpha(y)\bar{\psi}_\beta(z) > &= 2m_0 < j_5(x)\psi_\alpha(y)\bar{\psi}_\beta(z) > + < \Delta(x)\psi_\alpha(y)\bar{\psi}_\beta(z) > \\ &- \sum_\delta \delta_{xy}(\gamma_5)_{\alpha\delta} < \psi_\delta(y)\bar{\psi}_\beta(z) > \\ &- \sum_\delta \delta_{xz} < \psi_\alpha(y)\bar{\psi}_\delta(z) > (\gamma_5)_{\delta\beta} \,. \end{aligned} \tag{14.44}$$

Notice that Δ is an "irrelevant" operator which vanishes in the classical continuum limit. It could however (and in fact does) play a relevant role on quantum level. In the following we will consider (14.44) up to $\mathcal{O}(e^2)$. To this order Δ is given by

$$\begin{aligned} \Delta(x) \approx -\frac{r}{2a}\sum_\mu \Big\{&\bar{\psi}(x)\gamma_5\Big[1 + ieaA_\mu(x+a\hat{\mu}/2) - \frac{e^2a^2}{2!}(A_\mu(x+a\hat{\mu}/2))^2\Big]\psi(x+a\hat{\mu}) \\ &+ \bar{\psi}(x)\gamma_5\Big[1 - ieaA_\mu(x-a\hat{\mu}/2) - \frac{e^2a^2}{2!}(A_\mu(x-a\hat{\mu}/2))^2\Big]\psi(x-a\hat{\mu}) \\ &+ h.c.\Big\} + \frac{2r}{a}\sum_\mu \bar{\psi}(x)\gamma_5\psi(x) \,. \end{aligned} \tag{14.45}$$

Hence up to $\mathcal{O}(e^2)$,

$$\Delta(x) = -a\frac{r}{2}\sum_\mu \bar{\psi}(x)\gamma_5\Box\psi(x) - a\frac{r}{2}\sum_\mu \Box\bar{\psi}(x)\gamma_5\psi(x) + \chi(x) \,, \tag{14.46}$$

where $\chi(x)$ is an "irrelevant" operator involving one and two photon fields. In momentum space the sum of the first two terms on the rhs of (14.46) become proportional to $M_r(p;a) + M_r(p';a)$, where

$$M_r(q;a) = \frac{2r}{a}\sum_\mu \sin^2\frac{q_\mu a}{2} \tag{14.47}$$

is the r-dependent part of the Wilson mass (4.29b).

With the decomposition (14.46), the Ward identity (14.44) implies the following relation between the axial vector and pseudoscalar vertex functions in momentum space

$$i\tilde{q}_\mu \Gamma_{5\mu}(p,p';m_0,a) = 2[M(p;m_0,a) + M(p';m_0,a)]\Gamma_5(p,p';m_0,a) + \Gamma_5^{(\chi)}(p,p';m_0,a) \\ - \gamma_5 S_F^{-1}(p;m_0,a) - S_F^{-1}(p';m_0,a)\gamma_5 \,, \tag{14.48}$$

where $\Gamma^{(\chi)}$ is the two-point fermion vertex function with an insertion of the χ-operator. $M(q;m_0,a)$ is the "Wilson" mass

$$M(q;m_0,a) = m_0 + M_r(q,a) \,, \tag{14.49}$$

and S_F^{-1} is the inverse of the lattice regularized fermion propagator

$$S_F^{-1}(p;m_0,a) = i\gamma_\mu \tilde{p}_\mu + M(p;m_0,a) - \Sigma(p;m_0,a) \,, \tag{14.50a}$$

where

$$\tilde{p}_\mu = \frac{1}{a}\sin p_\mu a \,. \tag{14.50b}$$

In one-loop order, $\Gamma_{5\mu}$ and Γ_5 are given by the sum of Feynman diagrams of the form $a-f$ and a,e,f, respectively, shown in fig. (14-3). The diagrams contributing to the self energy $\Sigma(p)$ are those labeled by g and h. The Ward Identity (14.48) is nothing but the generalization of the following *tree-level* lattice identity analogous to (14.29):

$$\sin\left[\left(\frac{p-p'}{2}\right)_\mu a\right]\Gamma_{5\mu}(p,p';m_0,a)_{tree} = 2[M(p;m_0,a) + M(p';m_0,a)]\gamma_5 \\ - \gamma_5 S_0^{-1}(p;m_0,a) - S_0^{-1}(p';m_0,a)\gamma_5 \,, \tag{14.51a}$$

where

$$\Gamma_{5\mu}|_{tree} = \cos\left(\frac{p+p'}{2}\right)_\mu a\gamma_\mu\gamma_5 \,. \tag{14.51b}$$

In $\mathcal{O}(e^2)$ we have the following relation between the one particle irreducible vertex functions

$$i\tilde{q}_\mu \Lambda_{5\mu}(p,p';m_0,a) = 2[M(p;m_0,a) + M(p';m_0,a)]\Lambda_5(p,p';m_0,a) \\ + \Lambda_5^{(\chi)}(p,p';m_0,a) + \gamma_5\Sigma(p;m_0,a) + \Sigma(p';m_0,a)\gamma_5 \,, \tag{14.52}$$

which is the lattice version of (14.30). Apart from the appearance of the Wilson mass, it involves a naively "irrelevant" contribution $\Lambda_5^{(\chi)}$. In the following we will

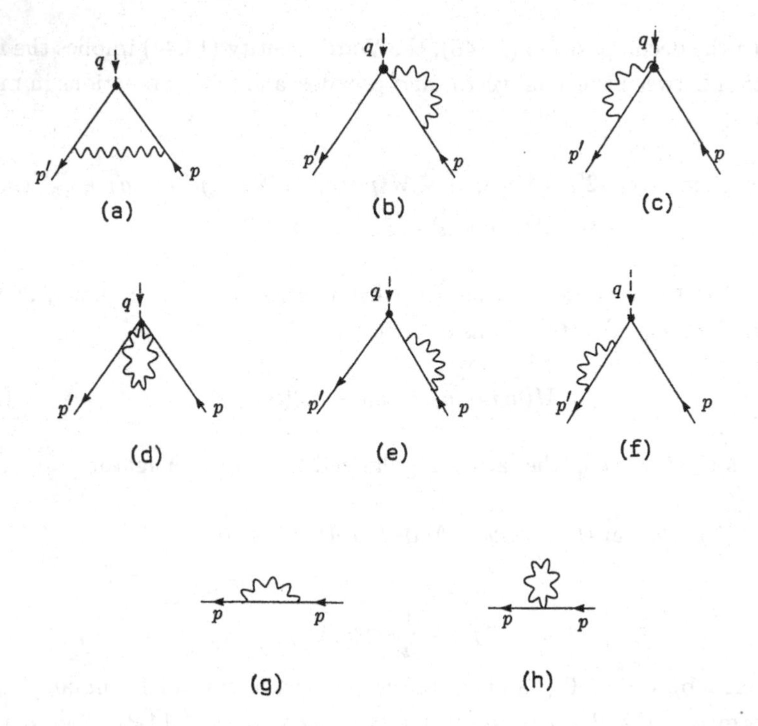

Fig. 14-3 Lattice diagrams contributing in $\mathcal{O}(e^2)$ to the Ward identity (14.48). Diagrams b, c, d and h have no counterpart in the continuum. Diagrams $a - f$ contribute to $\Gamma_{5\mu}$. In this case the big dots stand for the insertion of the axial vector vertices defined via the expansion of (14.41) in powers of the gauge field. Diagrams contributing to $\Gamma_5^{(\chi)}$ are shown in figs. b, c and d, where the big "dot" now stands for the insertion of the vertices defined via (14.45) and (14.46). Only diagrams a, e, f contribute in one loop order to $\Gamma_5(p, p'; a)$ because of the ultralocality of the pseudoscalar current (14.42).

study in detail the renormalized version of the bare lattice Ward identitiy (14.48) to one-loop order. Our emphasis will be placed on the role played by the above mentioned irrelevant contribution. The determination of the renormalization constants involve some tedious, but straight forward calculations. They are extracted by Taylor expanding the expressions for the vertex functions around vanishing momenta. Only first order polynomials in the momenta need to be considered. The coefficients in the expansion are in general complicated integral expressions, whose behaviour for $a \to 0$ can however be determined rather easily. We leave these computations as a lengthy exercise for the reader. In Appendix B we have summarized the relevant vertices in momentum space required for carrying out the computations.

Consider the lattice regularized Ward identity (14.48). Let $Z_2^{1/2}$ be the QED wave function renormalization constant for the fermion fields. Multiplying (14.48) by Z_2 we have that

$$\begin{aligned} i\tilde{q}_\mu Z_2 \Gamma_{5\mu}(p,p';m_0,a) - 2m_0 Z_2 \Gamma_5(p,p';m_0,a) - Z_2\Gamma_5^{(\Delta)}(p,p';m_0,a) \\ = -\gamma_5 Z_2 S_F^{-1}(p;m_0,a) - Z_2 S_F^{-1}(p';m_0,a)\gamma_5 \ , \end{aligned} \tag{14.53}$$

where

$$\begin{aligned} Z_2\Gamma^{(\Delta)}(p,p';m_0,a) = 2Z_2[M_r(p;m_0,a) + M_r(p';m_0,a)]\Gamma_5(p,p';m_0,a) \\ + Z_2\Gamma_5^{(x)}(p,p';m_0,a) \ . \end{aligned} \tag{14.54}$$

Diagrams contributing to $\Gamma_5^{(x)}$ are labeld by b, c, d in fig. (14-3). For the rhs of (14.53) to be finite we must also perform a mass renormalization:

$$m_0 = Z_2^{-1}(m + \delta m) \ . \tag{14.55}$$

Here m is a renormalized mass. With the definitions (14.47), (14.49), (14.50a) and (14.55), we then have to one loop order that

$$\lim_{a\to 0} Z_2 S_F^{-1}(p;m_0,a) = i\gamma_\mu p_\mu + m - \Sigma_R(p,m) \ , \tag{14.56}$$

where

$$\Sigma_R(p;m) = \lim_{a\to 0}[\Sigma(p;m,a) - \delta m - i(Z_2 - 1)\gamma_\mu \tilde{p}_\mu] \tag{14.57}$$

is the renormalized self energy. Z_2 and δm are chosen such as to the render this expression finite.

The in the limit $a \to 0$ divergent parts of δm and Z_2 can be computed by Taylor expanding $\Sigma(p;m,a)$ up to first order in the momentum. From the diagrams g and h in fig. (14-3) one finds after some lenghty, but straight forward algebra that the divergent parts are given by

$$\delta m = \frac{C(0)}{a} + \frac{e^2}{2\pi^2} m \ln(ma) \ , \tag{14.58a}$$

$$Z_2 - 1 = \frac{e^2}{8\pi^2} \ln(ma) \ , \tag{14.58b}$$

where

$$C(ma) = re^2 \int_{-\pi}^{\pi} \frac{d^4\hat{\ell}}{(2\pi)^4} \frac{\eta(\hat{\ell})}{\tilde{\ell}^2 [\tilde{\ell}^2 + \hat{M}^2(\hat{\ell}, ma)]} , \qquad (14.58c)$$

with

$$\eta(\hat{\ell}) = \tilde{\ell}^2 - 2\sum_\rho \sin^2\frac{\hat{\ell}_\rho}{2} \sum_\sigma [\cos^2(\frac{\hat{\ell}_\sigma}{2}) - r^2 \sin^2(\frac{\hat{\ell}_\sigma}{2})] - 4[\tilde{\ell}^2 + \hat{M}^2(\hat{\ell}, ma)] \qquad (14.58d)$$

and

$$\tilde{\tilde{\ell}}_\sigma = 2\sin\frac{\hat{\ell}_\sigma}{2} ,$$

$$\tilde{\ell}_\sigma = \sin\hat{\ell}_\sigma . \qquad (14.58e)$$

$\hat{M}(\hat{\ell}, ma)$ is the Wilson mass (14.49), with $m_0 \to m$, measured in lattice units (Quantities with a "hat" are measured in lattice units), and $C(0)$ is a finite constant. Hence to one loop order and $a \to 0$ (14.55) is given by

$$m_0 = m + \frac{C(0)}{a} + \frac{3e^2}{8\pi^2} \ln(ma) . \qquad (14.59)$$

In contrast to the continuum formulation, m_0 involves a *linearly* divergent contribution proportional to the Wilson parameter. As we shall see this linear divergence will be elliminated by a corresponding divergent contribution of the naively irrelevant term $\Gamma_5^{(\chi)}$ in the Ward identity. The logarithmic divergent expressions $Z_2 - 1$ and δm are the lattice analog of the expressions in the continuum (14.34).

Consider now the lhs of (14.53). Since with the above choice of Z_2 and δm the rhs is finite for $a \to 0$, so is the lhs. Consider first the contribution $i\tilde{q}_\mu Z_2 \Gamma_{5\mu}$. Since the highest "lattice degree of divergence" (LDD; see sec. 13.3) of the one loop diagrams contributing to $\Gamma_{5\mu}$ is zero, this expression is at most logarithmically divergent. In one loop order we have that

$$i\tilde{q}_\mu Z_2 \Gamma_{5\mu}(p, p'; m_0, a) = i\tilde{q}_\mu \gamma_\mu \gamma_5 + i\tilde{q}_\mu [\Lambda_{5\mu}(p, p'; m, a) + (Z_2 - 1)\gamma_\mu \gamma_5] . \quad (14.60)$$

$\Lambda_{5\mu}$ denotes the contribution of the one-loop diagrams $a - d$ shown in fig. (14-3). Of all these diagrams only the triangle diagrams turns out to be (logarithmically) divergent. We can extract the divergent part by studying the small a behaviour of the corresponding lattice Feynman integral at zero external momenta. After some lengthy, but straighforward algebra one finds that the quantity appearing within square brackets in (14.60) is finite for $a \to 0$. Consequently the remaining (pseudoscalar) terms on the lhs of (14.53) are also finite, since the Ward Identity holds for arbitrary a. Consider first the contribution of $2[M_r(p,a)+M_r(p',a)]\Gamma_5$ in (14.54). Since the one-loop contribution to Γ_5 has LDD = 0, and hence diverges at most logarithmically, while M_r vanishes linearly with the lattice spacing, this term will make no contribution in the continuuum limit. Next consider the contribution of $Z_2\Gamma_5^{(\chi)}$ in (14.54). To $\mathcal{O}(e^2)$ $Z_2\Gamma_5^{(\chi)}$ is just $\Gamma_5^{(\chi)}$.* The corresponding Feynman diagrams b, c, d in fig. (14-3) have LDD = 1. We therefore decompose this term follows:

$$\Gamma_5^{(\chi)} = T_0\Gamma_5^{(\chi)} + (T_1 - T_0)\Gamma_5^{(\chi)} + (1 - T_1)\Gamma_5^{(\chi)} \,, \tag{14.61}$$

where T_n denotes the Taylor expansion in the momenta around $p = p' = 0$ up to n'th order. The last term on the rhs has negative lattice degree of divergence and hence possesses a finite continuum limit. This limit may be calculated by making use of the Reisz theorem (see sec. 13.3). Since $\Gamma_5^{(\chi)}$ is a naively irrelevant contribution, this term vanishes in the continuum limit. For $T_0\Gamma_5^{(\Delta)}$ and $(T_1 - T_0)\Gamma_5^{(\Delta)}$ one then finds after some lengthy algebra that

$$\begin{aligned} T_0\Gamma_5^{(\Delta)} &= 2mz_P^{(\Delta)}\gamma_5 \,, \\ (T_1 - T_0)\Gamma_5^{(\Delta)} &= iz_A^{(\Delta)} q_\mu\gamma_\mu\gamma_5 \,, \end{aligned} \tag{14.62a}$$

where

$$z_P^{(\Delta)} = -\frac{C(ma)}{ma} - 2r^2e^2\int_{-\pi}^{\pi}\frac{d^4\hat{\ell}}{(2\pi)^4}\frac{\sum_\sigma \sin^2\frac{\hat{\ell}_\sigma}{2}}{\tilde{\ell}^2[\tilde{\ell}^2 + \hat{M}^2]} \tag{14.62b}$$

and $z_A^{(\Delta)}$ is finite. After a rather lengthy calculation its expression is found to be

$$z_A^{(\Delta)}(\hat{m}) = e^2\int_{-\pi}^{\pi}\frac{d^4\hat{\ell}}{(2\pi)^4}\frac{f_\sigma(\hat{\ell})}{\tilde{\ell}^2[\tilde{\ell}^2 + \hat{M}^2(\hat{\ell},\hat{m})]} + e^2\int_{-\pi}^{\pi}\frac{d^4\hat{\ell}}{(2\pi)^4}\frac{g_\sigma(\hat{\ell})}{\tilde{\ell}^2[\tilde{\ell}^2 + \hat{M}^2(\hat{\ell},\hat{m})]^2} \,, \tag{14.62c}$$

where $\hat{m} = ma$,

$$f_\sigma(\hat{\ell}) = -2r\hat{M}_r(\hat{\ell}) + 2r^2\cos\hat{\ell}_\sigma\sum_\lambda \sin^2\frac{\hat{\ell}_\sigma}{2} + 2r\hat{M}(\hat{\ell},\hat{m})\sin^2\frac{\hat{\ell}_\sigma}{2} \,, \tag{14.62d}$$

$$\begin{aligned} g_\sigma(\hat{\ell}) = 2\hat{M}_r(\hat{\ell})\Big[\eta_\sigma(\hat{\ell}) + r^2\sum_\lambda \sin^2\frac{\hat{\ell}_\lambda}{2}\Big](r\sin^2\hat{\ell}_\sigma - \hat{M}(\hat{\ell},\hat{m})\cos\hat{\ell}_\sigma) \\ + r\cos\hat{\ell}_\sigma(\sin^2\hat{\ell}_\sigma - \sum_\lambda \sin^2\hat{\ell}_\lambda)\Big] \,, \end{aligned} \tag{14.62e}$$

---

* Recall that χ is itself an operator of $\mathcal{O}(e)$. It arises from the expansion of the fermionic contribution to the lattice action in powers of the gauge potentials, and is a lattice artefact.

with

$$\hat{M}(\hat{\ell},\hat{m}) = \hat{m} + 2r\sum_\lambda \sin^2\frac{\hat{\ell}_\lambda}{2}\,, \tag{14.62f}$$

and

$$\eta_\sigma(\hat{\ell}) = 2\cos^2\frac{\hat{\ell}_\sigma}{2} - \sum_\lambda \cos^2\frac{\hat{\ell}_\lambda}{2}\,. \tag{14.62g}$$

As always, quantities with a "hat" are measured in lattice units. The continuum limit corresponds to vanishing $\hat{m} = ma$. The above integrals are finite in this limit.

Summarizing we have that $Z_2\Gamma_5^{(\Delta)}$ in (14.53) is given by

$$Z_2\Gamma_5^{(\Delta)}(p,p') = 2mz_P^{(\chi)}\gamma_5 + iz_A^{(\chi)}q_\mu\gamma_\mu\gamma_5 + \cdots\,, \tag{14.63}$$

where the "dots" stand for terms vanishing in the continuum limit. Note that this expression is completely local and has the form of possible counterterms for the pseudoscalar and axial vector current. This will in fact be just the case.

Finally consider the contribution $2m_0Z_2\Gamma_5(p,p';m_0,a)$ in (14.53), where m_0 is defined in terms of the renormalized mass m by (14.55). In the one-loop approximation it is given by

$$2m_0Z_2\Gamma_5(p,p';m_0,a) = 2m\gamma_5 + 2m[\Lambda_5(p,p';m,a) + \frac{\delta m}{m}\gamma_5]\,, \tag{14.64}$$

where the diagram contributing to Λ_5 is shown in fig. (14-3a). Since $\Lambda_5(p,p';m,a)$ has LDD $= 0$, it diverges at most logarithmically for $a \to 0$. But δm has a linearly divergent part, so that (14.64) diverges even linearly for $a \to 0$. Now comes an important role played by the Δ-term in (14.53). By combining (14.64) with (14.63), and making use of (14.58a) and (14.58b), we have that for $a \to 0$

$$2m_0Z_2\Gamma_5(p,p';m_0,a) + Z_2\Gamma^{(\Delta)}(p,p') \approx 2m[\Gamma_5(p,p';m,a) + \tilde{z}_P\gamma_5] + iz_A^{(\chi)}q_\mu\gamma_\mu\gamma_5\,, \tag{14.65a}$$

where

$$\tilde{z}_P = z_P^{(\chi)} + \frac{\delta m}{m} = \frac{e^2}{2\pi^2}\ln(ma)\,. \tag{14.65b}$$

This is precisely the lattice analog of the renormalization constant $z_P \equiv Z_P - 1$ in (14.34). Hence, after wave function and mass renormalization, the Ward identity (14.53) reads as follows in the continuum limit

$$iq_\mu\Gamma_{5\mu}(p,p')_R = 2m\Gamma_5(p,p')_R + i\tilde{z}_A^{(\chi)}q_\mu\gamma_\mu\gamma_5 - \gamma_5S_F^{-1}(p) - S_F^{-1}(p')\gamma_5\,, \tag{14.66}$$

where $\bar{z}_A^{(x)} = z_A^{(x)}(0)$ is a finite constant determined from (14.62c), and S_F is the renormalized propagator. The extra term $i\bar{z}_A^{(x)} q_\mu \gamma_\mu \gamma_5$ can be removed by a finite renormalization of the axial vector current, so that the final form of the Ward identity becomes

$$iq_\mu \bar{\Gamma}_{5\mu}(p,p')_R = 2m\Gamma_5(p,p')_R - \gamma_5 S_F^{-1}(p) - S_F^{-1}(p')\gamma_5 \ , \tag{14.67a}$$

where

$$\bar{\Gamma}_{5\mu}(p,p')_R = \lim_{a\to 0} {Z_A^{(x)}}^{-1} Z_2(ma)\Gamma_{5\mu}(p,p';ma,a) \ , \tag{14.67b}$$

$$\Gamma_5(p,p')_R = \lim_{a\to 0} Z_P(ma)\Gamma_5(p,p';ma,a) \ , \tag{14.67c}$$

and $Z_A^{(x)} = 1 + \bar{z}_A^{(x)}$, $Z_P = 1 + z_P$. Z_2 and Z_P are logarithmically divergent renormalization constants having a form completely analogous to (14.34) of the continuum formulation. Note that the linear divergent contribution $T_0\Gamma_5^{(x)}$ in (14.61) to the Ward identity (14.53) played a crucial role in rendering the expression finite. This term cancelled the linear divergent contribution arising from δm, which is a consequence of the lattice regularization, and has no counterpart in the continuum. What is new on the lattice, is that the axial vector current requires an additional finite renormalization, in order that the Ward identity retains its naive structure after renormalization. Thus QED renormalization alone does not suffice.

We have considered above the Ward identity relevant for studying the renormalization of the axial vector current. Ward identities involving the insertion of the divergence of the axial vector current in more general Green functions can also be readily be obtained. They can be best summarized by making use of functional methods. Thus consider the generating functional of Green functions,

$$Z[\eta,\bar{\eta},J] = \int DUD\psi D\bar{\psi}\ e^{-S[U,\psi,\bar{\psi}]+S_{source}} \ , \tag{14.68a}$$

where

$$S_{source} = \sum_x [\bar{\eta}(x)\psi(x) + \bar{\psi}(x)\eta(x) + J_\mu(x)U_\mu(x)] \ . \tag{14.68b}$$

Making the infinitessimal change of variables $\psi(x) \to \psi(x) + \delta\psi(x)$, $\bar{\psi}(x) \to \bar{\psi}(x) + \delta\bar{\psi}(x)$, with $\delta\psi$ and and $\delta\bar{\psi}$ given by (14.38), under which the measure is invariant, one imediately concludes that

$$< \delta S - \delta S_{source} >_{\eta,\bar{\eta},J} = 0 \ , \tag{14.69a}$$

where, as before, δS is given by (14.40), and

$$\delta S_{source} = i \sum_x \epsilon(x)[\bar{\eta}(x)\gamma_5\psi(x) + \bar{\psi}(x)\gamma_5\eta(x)] \ . \qquad (14.69b)$$

Expression (14.69a) can also be written as follows

$$\int DUD\psi D\bar{\psi}\Big[- \partial^L_\mu j_{5\mu}(x) + 2m_0 j_5(x) + \Delta(x) \\ - \bar{\eta}_\alpha(x)(\gamma_5)_{\alpha\beta}\frac{\partial}{\partial\bar{\eta}_\beta(x)} - \eta_\beta(x)(\gamma_5)_{\alpha\beta}\frac{\partial}{\partial\eta_\alpha(x)}\Big] e^{-S+S_{source}} = 0 \ ,$$

or compactly

$$< -\partial^L_\mu j_{5\mu}(x) + 2m_0 j_5(x) + \Delta(x) >_{\eta,\bar{\eta},J} = \bar{\eta}(x)\gamma_5\frac{\partial Z}{\partial\bar{\eta}(x)} - \frac{\partial Z}{\partial\eta(x)}\gamma_5\eta(x) \ . \quad (14.70)$$

By differentiating this expression with respect to the sources we generate Ward identities involving the insertion of the divergence of the axial vector current in arbitrary Green functions. The Ward identity we have discussed above follows by differentiating (14.70) with respect to η and $\bar{\eta}$ and setting thereafter $\eta = \bar{\eta} = J_\mu = 0$. A similar analysis as the one discussed above shows that Ward identities involving the insertion of the divergence of the axial vector currrent in higher n-point functions involving $n_F \geq 2$ external fermion lines and $n_A \geq 1$ gauge fields are finite and non-anomalous upon QED renormalization.

14.4 The ABJ Anomaly

In continuum QED or QCD it is well known that Ward identities following from gauge invariance play a very important role in securing the renormalizabilty of these theories. In general, Ward identities relating different unrenormalized Green functions are derived by considering local transformations of the fields in the generating functional. If the "naive" form of these identities retain their structure after renormalization, then we say that the Ward identities are non-anomalous. If their structure is not preserved (on account of quantum fluctuations) then one speaks of anomalous Ward identities. Such a breakdown on quantum level can be desastrous, for it may not allow the quantization of the theory. Thus e.g. in the case of the electroweak theory it is important that the chiral symmetry of the classical action, in the massless quark limit, remains unbroken on quantum level, since the gauge fields are coupled to chiral currents which are the sources for the fields. An example of a harmeless anomaly is the ABJ-anomaly (Adler

1969, Bell and Jackiw 1969), which plays an important role in the description of the electromagnetic decay $\pi^0 \to 2\gamma$. The anomaly is harmless because it manifests itself in the divergence of a current (the axial vector current) which is not the source for the gauge fields in QED or QCD. How does this anomaly arise within the framework of a lattice regularized gauge theory?

Let me first remind the reader of how the axial anomaly arises in continuum QED. Consider first the partition function for *continuum* QED in an external gauge field,

$$Z[A] = \int D\psi D\bar{\psi}\; e^{-S_{ferm}[A,\psi,\bar{\psi}]} \,, \tag{14.71}$$

where S_{ferm} is the fermionic contribution to the action (5.8c). In the limit of vanishing fermion mass this action is invariant under the global transformations $\psi(x) \to exp(i\alpha\gamma_5)\psi(x)$ and $\bar{\psi}(x) \to \bar{\psi}(x)exp(i\alpha\gamma_5)$, where α is an x-independent parameter. Next consider an infinitessimal local transformation with $\alpha \to \epsilon(x)$. The fermion measure in (14.71) is, *at least formally*, invariant under this transformation.* Let δS_{ferm} be the corresponding change of the action. Since the above local transformations of the fermion fields just correspond to a change of variables, we (naively) conclude that

$$< \delta S_{ferm} >_A \equiv \frac{1}{Z}\int D\psi D\bar{\psi}\; \delta S_{ferm} e^{-S_{ferm}} = 0 \,. \tag{14.72}$$

$< \delta S_{ferm} >_A$ is the expectation value of δS_{ferm} in a background gauge field. One readily finds (after a partial integration) that

$$\delta S_{ferm} = \int d^4x\, \epsilon(x)\left[2m\bar{\psi}(x)\gamma_5\psi(x) - \partial_\mu(\bar{\psi}(x)\gamma_\mu\gamma_5\psi(x))\right] .$$

Since $\epsilon(x)$ is an arbitrary function, it follows from (14.72) that

$$< \partial_\mu j_{5\mu}(x) >_A = 2m < j_5(x) >_A \,,\; (naive) \,, \tag{14.73}$$

where $j_{5\mu} = \bar{\psi}\gamma_\mu\gamma_5\psi$. But actually this equation is violated because of quantum fluctuations, as is demonstrated in any book on quantum field theory. The violation is induced by the triangle graphs shown in fig. (14-4), each of which is

---

* Actually, the integration measure needs to be regularized. As has been shown by Fujikawa (Fujikawa, 1979) the ABJ anomaly can be viewed as a consequence of having to regularize this measure.

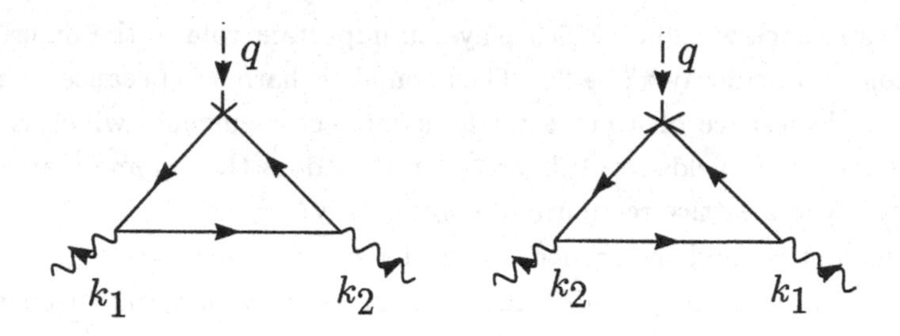

Fig. 14-4 Triangle diagrams which give rise to the ABJ anomaly in the continuum.

linearly divergent and therefore must be regularized. One then finds that (14.73) is modified by an additional term as follows

$$< \partial_\mu j_{5\mu}(x) >_A = 2m < j_5(x) >_A + \frac{e^2}{16\pi^2} F_{\mu\nu} \tilde{F}^{\mu\nu} \, , \tag{14.74}$$

where $\tilde{F}_{\mu\nu}$ is the dual field strength tensor, $\tilde{F}_{\mu\nu} = \frac{1}{2}\epsilon_{\mu\nu\lambda\rho}F_{\lambda\rho}$.

On the lattice the regularization is introduced already on the level of the partition function. Hence any considerations of the above type will leave us with equations which are exact. One then may be left with an anomaly when the cut-off is removed, i.e. upon taking the continuum limit.

In the continuum formulation of QCD (or QED) it is well known that different gauge invariant regularization schemes all yield the same expression for the axial anomaly. Any candidate for a lattice discretization of QED or QCD should also reproduce the correct axial anomaly in the continuum limit. As we have seen in chapter 2, a naive discretization of the fermionic action which is local, hermitean, chirally symmetric for vanishing fermion mass, and having the correct continuum limit, will necessarily lead to the fermion doubling problem. That this must be so is a consequence of the Nielsen-Ninomyia theorem (1981). To avoid the problem of species doubling, the chiral symmetry must be broken explicitely, if one refrains from abandoning at least some of the other properties. As we have seen in chapter 5, a simple way to accomplish this is to add to the naively discretized fermion action a "Wilson term", leading to (5.17), which ensures that in the limit of vanishing lattice spacing the unwanted fermion modes acquire an infinite mass and hence decouple. For Wilson fermions the axial anomaly has been first studied by Karsten and Smit (Karsten, 1981). These authors showed that the origin of the anomaly was an irrelevant term in the lattice Ward identity. The anomaly was also studied by Rothe and Sadooghi (Rothe, 1998), using the small-a-expansion scheme of Wetzel (1985). In the former reference it was shown that, in the limit of vanishing lattice spacing a, this naively irrelevant contribution is indeed given by the $D \to 4$ limit of the dimensionally regulated continuum triangle graph. These computations are very involved and we do not present them here. Subsequently, the anomaly was studied within a more general framework by Reisz and Rothe (Reisz 1999), where the action is not assumed to have the form proposed by Wilson.

As we have pointed out in section (4.7), there is an even milder way of breaking chiral symmetry on the lattice, proposed a long time ago by Ginsparg and Wilson (1982). For Ginsparg-Wilson (GW) fermions the fermionic action is of the form

$$S_{ferm} = \sum_{x,y} \bar{\psi}(x)(D(x,y) + m\delta_{x,y})\psi(y) , \qquad (14.75)$$

where the "Dirac Operator" $D(x,y)$ is a 4×4 matrix in Dirac space which breaks chiral symmmetry in a very special way. While the Dirac operator in the continuum, or its naively discretized version, anticommutes with γ_5, the GW-Dirac operator satisfies the following GW-relation:

$$\{\gamma_5, D\} = aD\gamma_5 D , \qquad (14.76)$$

where D is a matrix whose rows and columns are labeled by the a spin and space-time index. It is a function of the link variables and can be expanded in terms of the gauge potentials,

$$D(x,y) = \sum_{n,\mu_i,a_i,x_i} \frac{1}{n!} D^{(n)}_{\mu_1 \cdots \mu_n}(x,y|x_1 \cdots x_n) A_{\mu_1}(x_1) \cdots A_{\mu_n}(x_n) , \qquad (14.77)$$

where x denotes a lattice site. For Wilson fermions $D(x,y)$ is a strictly local operator connecting only neighbouring lattice sites, and with the gauge potentials living on the corresponding links. On the other hand, the Dirac operator $D(x,y)$ for GW-fermions is non local in the sense that it connects arbitrary lattice sites and does *not* involve only the gauge potentials at sites close to x. For the Neuberger solution to (14.76), given by (4.65a) and (4.66), the non-locality arises from the inverse of $(A^\dagger A)^{1/2}$. Nevertheless, as has been shown by Hernandez et al. (Hernandez, 1999), Neuberger's Dirac operator is still local in the more general sense, that the Dirac operator decays exponentially at large distances, with a decay rate proportional to $1/a$. In the naive continuum limit we have of course that $D(x,y) \to \gamma_\mu D_\mu[A]$, where $D_\mu[A]$ is the covariant derivative in the continuum.

In the following we shall discuss two alternative points of view of how the ABJ anomaly is generated in the case of GW fermions. In the first approach we will make use of the observation made by Lüscher (1998), which was proven in sec. 4.7, that the GW-action possesses an exact global axial-type symmetry for vanishing fermion mass. As we have seen in sec. 4.7, this symmetry differes from the standard one (4.8) by lattice artefacts. Associated with this symmetry is an axial vector current which is conserved on classical level. Following a similar procedure as in the continuum, we then derive a Ward identity and identify the anomalous contribution (ABJ anomaly). Within this approach the anomaly will arise from the non-invariance of the fermionic integration measure under the non-standard local axial transformations. In the second approach we then consider the Ward identity derived for standard local axial transformations under which the measure is invariant, but the action is not. The ABJ anomaly now originates from the explicit chiral symmetry breaking in the GW-action, and hence parallells the approach to the anomaly in the case of Wilson fermions.

Approach 1

Consider the action (14.75), where D satisfies the GW-relation (14.76). The GW-Dirac operator is a function of the link variables. This action is invariant for $m = 0$ under the global infinitessimal transformations (4.68). Consider the case where ϵ in (4.68b) is a function of the space-time coordinates, i.e., a local transformation.:

$$\psi \to \psi' = \psi + \delta\psi \ , \quad \bar{\psi} \to \bar{\psi}' = \bar{\psi} + \delta\bar{\psi} \ , \tag{14.78a}$$

$$\delta\psi(x) = i\epsilon(x)\gamma_5\Big[(1 - \frac{a}{2}D)\psi\Big](x) \ , \tag{14.78b}$$

$$\delta\bar{\psi}(x) = i\epsilon(x)\Big[\bar{\psi}(1 - \frac{a}{2}D)\Big](x)\gamma_5 \ . \tag{14.78c}$$

One then readily verifies that the variation of the action is given by

$$\delta S_{ferm} = i\sum_x \epsilon(x)\Big[F(x) + 2m\bar{\psi}(x)\gamma_5\psi(x) + \bar{\Delta}(x)\Big] \ , \tag{14.79}$$

where $\sum_x = \sum_n a^4$. $F(x)$ and $\Delta(x)$ are defined in (4.69b) and (4.69c), with $F(x)$ satisfying (4.70). Now according to the Poincare Lemma on the lattice (Lüscher (II), 1999), (14.70) implies that there exists an axial vector current $j_{5\mu}(x)$ such that

$$F(x) = -\partial^L_\mu j_{5\mu}(x) \ , \tag{14.80}$$

where ∂^L_μ is the left lattice derivative (the proof of this lemma is quite involved). Since the GW-Dirac operator has the correct continuum limit, it follows that for

$a \to 0$, $j_{5\mu}(x) \to \bar{\psi}(x)\gamma_\mu\gamma_5\psi(x)$. With (14.80) we conclude that (14.79) is given by

$$\delta S_{ferm} = i \sum_x \epsilon(x)[-\partial_\mu^L j_{5\mu}(x) + 2m\bar{\psi}(x)\gamma_5\psi(x) + \bar{\Delta}(x)] \,. \tag{14.81}$$

In contrast to the case of Wilson fermions, where $j_{5\mu}(x)$ is given by (14.41), we do not know the explicit expression for the axial vector current. The last term on the rhs is *not* responsible for the anomaly. In fact its external field expectation value vanishes in the continuum limit. The anomaly arises from the non-invariance of the fermionic measure under the variations (14.78). This leads to a Jacobian. Thus from (14.78) we have that

$$\frac{\delta\psi'(x)}{\delta\psi(y)} \equiv B(x,y) = \Big[1 + i\epsilon(x)\gamma_5(1 - \frac{a}{2}D)\Big](x,y) \,.$$

The determinant of this matrix is given by

$$det\ B = e^{Tr \ln B} \approx 1 + Tr \ln B \,.$$

Here the trace is carried out in Dirac, as well as in coordinate space. Hence the Jacobian of the transformation is

$$J[\frac{\delta\psi'}{\delta\psi}] = det\ B = 1 - i\frac{a}{2}\sum_x \epsilon(x) tr_D(\gamma_5 D)(x,x) \,,$$

where tr_D denotes the trace in Dirac space. A corresponding expression holds for $J[\delta\bar{\psi}'/\delta\bar{\psi}]$. Consider now the partition function in a background link variable configuration,

$$Z[U] = \int D\psi D\bar{\psi}\ e^{-S_{ferm}[\psi,\bar{\psi},U]} \,. \tag{14.82}$$

Making the infinitessimal change of variables (14.78) leaves this expression invariant. Hence we conclude that (taking into account the Jacobian of the transformation) that

$$< \delta S_{ferm} >_U - ia \sum_x \epsilon(x) tr_D(\gamma_5 D(x,x)) = 0 \,. \tag{14.83}$$

To arrive at the final form of the Ward identity we rewrite the contribution $< \bar{\Delta}(x) >_U$ to $< \delta S_{ferm} >$ in (14.81) as follows: from (4.69c), $< \bar{\Delta}(x) >_U$ is given in matrix notation by *

$$< \bar{\Delta}(x) >_U = \frac{am}{2} tr_D \Big[(D+m)^{-1} D\gamma_5 + \gamma_5 D(D+m)^{-1}\Big](x,x) \,,$$

* We have dropped the explicit dependence of the Dirac operator on the link variables, for simplicity.

where we have made use of

$$< \psi_\alpha(x)\bar{\psi}_\beta(y) >_U = (D+m)^{-1}_{\alpha\beta}(x,y) \tag{14.84}$$

Making use of $(D+m)^{-1}D = 1 - m(D+m)^{-1}$ and $tr_D\gamma_5 = 0$, we can further write

$$< \bar{\Delta}(x) >_U = -m^2 a tr_D(\gamma_5(D+m)^{-1})(x,x) = m^2 a < \bar{\psi}(x)\psi(x) > \ .$$

Hence the Ward identity (14.83) takes the form

$$< \partial^L_\mu j_{5\mu}(x) >_U = 2m < j_5(x) >_U - a tr_D(\gamma_5 D(x,x)) \ , \tag{14.85a}$$

where

$$j_5(x) = (1 + \frac{ma}{2})\bar{\psi}(x)\gamma_5\psi(x) \ . \tag{14.85b}$$

The last term on the rhs of (14.85a) is an anomalous contribution and yields in the continuum limit the ABJ anomaly (Hasenfratz, 1998: Lüscher (I), 1998).

We now proceed to derive the above Ward identity from a more conventional point of view which parallels the case of Wilson fermions.

Approach 2

Consider once again the action (14.75) and the partition function (14.82). Let us carry out a change of variables (14.78a) induced by the *standard* local axial transformations, where $\delta\psi$ and $\delta\bar{\psi}$ are given by (14.38). Under this transformation the integration measure is invariant since $tr_D\gamma_5 = 0$. The change in the action is easily computed and now given by

$$\delta S_{ferm} = i\sum_x \epsilon(x)\left[\bar{\psi}(x)\gamma_5(D\psi)(x) + (\bar{\psi}D)(x)\gamma_5\psi(x)\right] + 2im\sum_x \epsilon(x)\bar{\psi}(x)\gamma_5\psi(x) \ . \tag{14.86}$$

We now make use of the definition (4.69b) and of (14.80) to write this expression in the form

$$\delta S_{ferm} = i\sum_x \epsilon(x)[-\partial^L_\mu j_{5\mu}(x) + 2m\bar{\psi}(x)\gamma_5\psi(x) + a(\bar{\psi}D)(x)\gamma_5(D\psi)(x)] \tag{14.87}$$

Since the fermionic measure in (14.82) is invariant under the above local axial transformations, the Ward identity now reads

$$< \delta S_{ferm} >_U = 0 \ . \tag{14.88}$$

Making use of (14.84), the expectation value of the last term in (14.87) can be written in the form

$$a < (\bar{\psi}D)(x)\gamma_5(D\psi)(x) >= -atr_D(\gamma_5 D(x.x)) - m^2 atr_D\left[\gamma_5(D+m)^{-1}(x,x)\right], \tag{14.89}$$

where the last term is just $m^2 a < \bar{\psi}(x)\gamma_5\psi(x) >_U$. Thus one arrives once again at the anomalous Ward Identity (14.85).

We shall not discuss the anomalous contribution any further, since it has been shown by Reisz and Rothe (Reisz 1999) that any lattice action satisfying some very general conditions (which also hold for GW-fermions) will necessarily reproduce the correct anomaly in the continuum limit. For the more complicated case of QCD we will present a proof in section 6 of the following chapter.

CHAPTER 15

WEAK COUPLING EXPANSION (III) LATTICE QCD

Weak coupling perturbation theory in lattice QCD is much more involved than in the U(1) gauge theory discussed in the previous section. The reasons for this are the following: a) The lattice action is a complicated functional of the coloured gauge potential; b) the gauge invariant integration measure associated with the link variables depends non–trivially on the gauge fields, and c) the generalized covariant gauge, analogous to that discussed in the abelian case, can no longer be implemented in a trivial way. This latter feature is of course also true in continuum QCD. But whereas there the Faddeev–Popov determinant, which emerges when the gauge is fixed, can be represented in terms of an effective ghost–gauge field interaction which is linear in the gauge potential, this is no longer true in lattice QCD.

The complexity of the expressions in perturbative lattice QCD, is a consequence of the gauge invariant lattice regularization, which, as in the U(1) case, leads to an infinite number of interaction vertices. But because of the non–abelian structure of the theory, most of these vertices have a very complicated structure. Although in the naive continuum limit only those vertices survive which are characteristic of the continuum formulation, irrelevant contributions to the action do play an important role when studying the continuum limit of Feynman integrals. Hence one must exercise great care in including all lattice artefacts in the action, which contribute to the correlation functions in this limit. In this connection let us recall that the whole point of the lattice formulation was that it provides a regularization scheme, where gauge invariance is ensured for *any* finite lattice spacing. Only by including all lattice artefacts we will therefore be ensured that for any finite lattice spacing gauge invariant correlation functions will be independent of the choice of gauge, and that the gauge fixed theory will possess a BRS–type symmetry, reflecting the original gauge invariance of the theory before fixing the gauge. This is important since this symmetry leads to Ward–identities which play an important role in developing a renormalization program. One therefore should abstain from making any approximation when expressing the link–integration measure and Faddeev–Popov determinant in terms of the gauge potentials before removing the lattice cutoff.

In the following we shall set up the generating functional in a form which is suited for performing the weak coupling expansion in lattice QCD. We then derive lattice expressions for the propagators and vertices relevant for low order perturbative calculations. We will however not discuss lattice Ward identities, nor renormalization in this book. A detailed discussion of these topics can be found in the work by Reisz (1988a,b).

15.1 The Link Integration Measure

In lattice QCD, the link variables are elements of SU(3) in the fundamental representation.* They hence can be written in the form

$$U_\mu(n) = e^{i\phi_\mu(n)} \tag{15.1a}$$

where $\phi_\mu(n)$ is an element of the Lie–algebra of SU(3):

$$\phi_\mu(n) = \sum_{A=1}^{8} \phi_\mu^A(n) T^A. \tag{15.1b}$$

Here T^A $(A = 1, \ldots, 8)$ are the generators of the group in the fundamental representation. We chose them to be given by

$$T^A - \frac{\lambda^A}{2},$$

where λ^A are the Gell-Mann matrices introduced in chapter 6. From (6.8), and (6.15) we have that

$$[T^A, T^B] = i \sum_C f_{ABC} T^C, \tag{15.2a}$$

$$Tr(T^A T^B) = \frac{1}{2}\delta_{AB}. \tag{15.2b}$$

Under a gauge transformation, the link variables transform according to

$$U_\mu(n) \to g(n) U_\mu(n) g^{-1}(n + \hat{\mu}),$$

---

* In contrast to the notation used in chapter 6, we shall not underline matrices in colour space with a "twidle". Except for the generators, all quantities carrying a colour index will be c–numbers. Quantities without a colour index are matrix–valued.

where $g(n)$ and $g(n+\hat{\mu})$ are elements of $SU(3)$. Correlation functions involving the product of link variables and coloured quark fields are computed according to (6.24), where DU is the gauge invariant measure associated with the link variables. Our present objective is to express this integration measure in terms of the group parameters $\phi_\mu^A(n)$ defined in (15.1b). To this effect we first construct the invariant measure associated with a single link variable, following Kawai et al. (1981).

Let U be an element of $SU(3)$ with Lie algebra L. Then U can be written in the form $U = exp(i\phi)$, with $\phi \in L$. Consider the following bilinear differential form

$$d^2 s = Tr(dU^\dagger dU), \tag{15.3}$$

where $dU = U(\phi + d\phi) - U(\phi)$. It is invariant under left or right multiplication of U with a group element of $SU(3)$. Expressed in terms of the coordinates $\{\phi^A\}$, parametrizing ϕ, (15.3) will have the form

$$d^2 s = \sum_{A,B} g_{AB}(\phi) d\phi^A d\phi^B. \tag{15.4}$$

This defines a metric, $g_{AB}(\phi)$ on the group manifold. The gauge invariant integration measure (Haar measure) is then given by

$$d\mu(\phi) = \sqrt{det\ g(\phi)} \prod_A d\phi^A, \tag{15.5}$$

where $g(\phi)$ is the matrix constructed from the elements $g_{AB}(\phi)$. To calculate $g(\phi)$ let us rewrite (15.3) as follows (Kawai et. al. (1981)):

$$d^2 s = Tr\{(U^{-1}dU)^\dagger (U^{-1}dU)\}. \tag{15.6}$$

As is shown in appendix A, $U^{-1}dU$ is an element of the Lie–algebra of SU(3) and has the form

$$U^{-1}dU = i \sum_{A,B} T^A M_{AB}(\phi) d\phi^B, \tag{15.7a}$$

where

$$M_{AB}(\phi) = \left(\frac{1 - e^{-i\Phi}}{i\Phi}\right)_{AB}, \tag{15.7b}$$

$$\Phi = \sum_{A=1}^{8} \phi^A t^A, \tag{15.7c}$$

and t^A are the generators of $SU(3)$ in the adjoint representation*. Their matrix elements are given by

$$t^A_{BC} = -if_{ABC}, \tag{15.8}$$

where f_{ABC} are the structure constants of the group appearing in (15.2a). The generators t^A satisfy the following orthogonality relation:

$$Tr(t^A t^B) = 3\delta_{AB}. \tag{15.9}$$

Inserting the expression (15.7a) into (15.6), and making use of (15.2b), one finds that the metric $g_{AB}(\phi)$, defined in (15.4) is given by

$$\begin{aligned} g_{AB}(\phi) &= \frac{1}{2}(M^\dagger(\phi)M(\phi))_{AB} \\ &= \left(\frac{1-\cos\Phi}{\Phi^2}\right)_{AD} . \end{aligned} \tag{15.10}$$

Hence $g(\phi)$ has the following power series expansion in terms of $\{\phi^A\}$:

$$g(\phi) = \frac{1}{2} + \sum_{\ell=1}^{\infty} \frac{(-1)^\ell}{(2\ell+2)!}(\Phi)^{2\ell}, \tag{15.11}$$

where Φ has been defined in (15.7c). This is the expression we were looking for. Notice that by construction, $g(\phi)$ is a non-negative hermitian matrix. Hence its determinant is real and non-negative.

The invariant integration measure associated with a single link variable $U_\mu(n)$ is obtained from (15.5) and (15.10) by replacing ϕ^A by $\phi^A_\mu(n)$. Taking the product of these measures we arrive at the desired expression for DU:

$$DU = \left\{\prod_{n,\mu} \sqrt{det\left[\frac{1}{2}M^\dagger(\phi_\mu(n))M(\phi_\mu(n))\right]}\right\} D\phi, \tag{15.12a}$$

where

$$M(\phi_\mu(n)) = \frac{1-e^{-i\Phi_\mu(n)}}{i\Phi_\mu(n)}, \tag{15.12b}$$

$$\Phi_\mu(n) = \sum_A t^A \phi^A_\mu(n), \tag{15.12c}$$

* Capital letters always run from one to eight.

and

$$D\phi = \prod_{n,A,\mu} d\phi_\mu^A(n). \tag{15.12d}$$

We next rewrite (15.12) in a way which is convenient for later computation. Since the determinant of $g(\phi)$ can also be written in the form $\exp(Tr \ln g(\phi))$, we obtain the following alternative expression for DU:

$$DU = e^{-S_{meas}[\phi]} D\phi, \tag{15.13a}$$

where, apart from an irrelevant additive constant, $S_{meas}[\phi]$ is given by

$$S_{meas.}[\phi] = -\frac{1}{2}\sum_{n,\mu} Tr\ ln \left[\frac{2(1-\cos\Phi_\mu(n))}{\Phi_\mu^2(n)}\right]. \tag{15.13b}$$

The quantity appearing within square brackets is a polynomial in the matrix Φ_μ. $S_{meas.}[\phi]$ can also be written in the form

$$S_{meas.}[\phi] = -\frac{1}{2}\sum_{n,\mu} Tr\ ln[1 + N(\phi_\mu(n))], \tag{15.14a}$$

where

$$N(\phi_\mu(n)) = 2\sum_{\ell=1}^{\infty} \frac{(-1)^\ell}{(2\ell+2)!}(\Phi_\mu)^{2\ell}, \tag{15.14b}$$

with Φ_μ defined in (15.12c). Notice the difference in the structure of the integration measure (15.13) and its abelian $U(1)$ counterpart, where DU is given by (5.23). Indeed in the abelian case $U^{-1}dU = id\phi$, where ϕ is the single real variable parametrizing U. Hence according to (15.6), the right-hand side is just given by $(d\phi)^2$.

Consider now the ground state expectation value of a gauge invariant functional of the dimensionless fields $\phi_\mu^A(n), \hat{\psi}_\alpha^a(n), \hat{\bar{\psi}}_\alpha^a(n)$.* We denote this functional by $\Gamma[\phi, \hat{\psi}, \hat{\bar{\psi}}]$. For the action we will take the standard Wilson form given in eqs. (6.25), except that now we shall write $S_{QCD}[\phi, \hat{\psi}, \hat{\bar{\psi}}]$ instead of $S_{QCD}[U, \hat{\psi}, \hat{\bar{\psi}}]$ to emphasize that S_{QCD} should be expressed in terms of the fields ϕ_μ^A. Writing the

* Recall that while the capital letters $A, B, \ldots$ run from one to eight, small Latin letters run from one to three, since the quark fields transform under the fundamental representation of $SU(3)$.

link integration measure in the form (15.13a), the ground state expectation value of $\Gamma[\phi, \hat{\psi}, \bar{\hat{\psi}}]$ is given by

$$\langle\Gamma[\phi, \hat{\psi}, \bar{\hat{\psi}}]\rangle = \frac{\int D\phi D\bar{\hat{\psi}} D\hat{\psi} \Gamma[\phi, \hat{\psi}, \bar{\hat{\psi}}] e^{-S_{QCD}[\phi,\hat{\psi},\bar{\hat{\psi}}]-S_{meas}[\phi]}}{\int D\phi D\bar{\hat{\psi}} D\hat{\psi} e^{-S_{QCD}[\phi,\hat{\psi},\bar{\hat{\psi}}]-S_{meas}[\phi]}}. \tag{15.15}$$

This expression is not yet suited for carrying out the weak coupling expansion. As in the $U(1)$ case considered in the previous chapter, we must still fix the gauge.*

15.2 Gauge Fixing and the Faddeev–Popov Determinant

A popular choice for a local gauge condition which is linear in the fields ϕ_μ^A, and respects the discrete lattice symmetries, is the following generalization of (14.3),**

$$\mathcal{F}_n^A[\phi; \chi] = \hat{\partial}_\mu^L \phi_\mu^A(n) - \chi^A(n) = 0, \tag{15.16}$$

where $\chi^A(n)$, A=1,...8, are some arbitrary given fields. We want to introduce this gauge condition into the functional integral (15.15). This must be done in such a way that expectation values of gauge-invariant observables are not affected by the gauge-fixing procedure. Following the prescription given by Faddeev and Popov for the continuum formulation, we consider the integral

$$\Delta_{FP}^{-1}[\phi; \chi] = \int Dg \prod_{n,A} \delta(\mathcal{F}_n^A[{}^g\phi, \chi]), \tag{15.17}$$

where ${}^g\phi$ denotes collectively the gauge transform of the group parameters $\{\phi_\mu^A(n)\}$ which parametrize the link variables $U_\mu(n)$. The above integral is carried out over the gauge group manifold, with the integration measure Dg being given by the product of the invariant Haar measures on SU(3) at every lattice site,

$$Dg = \prod_n d\mu(g_n).$$

Because by definition of the Haar measure $d\mu(gg') = d\mu(g)$, it follows that $\Delta_{FP}[\phi; \chi]$ is gauge invariant:

$$\Delta_{FP}[{}^g\phi; \chi] = \Delta_{FP}[\phi; \chi].$$

---

* Within the framework of continuum perturbation theory, this is an obvious requirement, since otherwise the gluon propagator cannot be defined. In a lattice formulation $S_{meas}[\phi]$ includes a term quadratic in the fields ϕ_μ^A which is not gauge invariant. But since it depends on the coupling, it must be treated as part of the interaction.

** It will be always understood that repeated Lorentz indices are summed.

We next introduce the identity

$$1 = \Delta_{FP}[\phi;\chi] \int Dg \prod_{n,A} \delta(\mathcal{F}_n^A[{}^g\phi;\chi])$$

into the integrands of (15.15). Using the fact that Γ, S_{QCD}, Δ_{FP}, as well as the integration measures $DU = \exp(-S_{meas.})D\phi$ and $D\bar{\hat{\psi}}D\hat{\psi}$ are gauge invariant, we can replace the fields ϕ, $\hat{\psi}$, and $\bar{\hat{\psi}}$ in these expressions by their gauge transforms ${}^g\phi$, ${}^g\hat{\psi}$ and ${}^g\bar{\hat{\psi}}$, respectively. A simple redefinition of the integration variables then leads to the following alternative expression for (15.15):

$$\begin{aligned} < \Gamma[\phi, \hat{\psi}, \bar{\hat{\psi}}] >= \frac{1}{Z} \int D\phi D\bar{\hat{\psi}} D\hat{\psi} \Delta_{FP}[\phi;\chi] \prod_{n,A} \delta(\mathcal{F}_n^A[\phi;\chi]) \\ \cdot \Gamma[\phi, \hat{\psi}, \bar{\hat{\psi}}] e^{-S_{QCD}[\phi,\hat{\psi},\bar{\hat{\psi}}] - S_{meas.}[\phi]}, \end{aligned} \tag{15.18a}$$

where the normalization constant Z is given by

$$Z = \int D\phi D\bar{\hat{\psi}} D\hat{\psi} \Delta_{FP}[\phi;\chi] \prod_{n,A} \delta(\mathcal{F}_n^A[\phi,\chi]) e^{-S_{QCD}[\phi,\hat{\psi},\bar{\hat{\psi}}] - S_{meas.}[\phi]}. \tag{15.18b}$$

We must now compute $\Delta_{FP}[\phi;\chi]$. But because of the gauge fixing δ–function appearing in (15.18), we only need to know $\Delta_{FP}[\phi;\chi]$ for field configurations ϕ satisfying the gauge condition (15.16). Hence it suffices to calculate the integrand of (15.17) for ${}^g\phi$ in the infinitesimal neighbourhood of $g = 1$ with ϕ restricted by (15.16). Accordingly, we must compute the change in the fields $\phi_\mu^A(n)$ induced by an infinitesimal gauge transformation. In the continuum formulation the corresponding change in the potentials A_μ^B is a linear functional of the gauge fields. In a lattice regulated theory this is no longer true. Hence also $\Delta_{FP}[\phi;\chi]$ will acquire a non–trivial structure.

The response of $\phi_\mu^A(n)$ to an infinitesimal gauge transformation has been calculated in appendix B. Let $\delta_{(\epsilon)}\phi_\mu^A(n)$ denote the change in $\phi_\mu^A(n)$ induced by the transformation, i.e.

$$U_\mu(n) \to e^{i\epsilon(n)} U_\mu(n) e^{-i\epsilon(n+\hat{\mu})} = e^{i(\phi_\mu(n) + \delta_{(\epsilon)}\phi_\mu(n))}$$

where $\epsilon(m)$ are elements of the Lie algebra of SU(3) in the fundamental representation:

$$\epsilon(m) = \sum_A T^A \epsilon^A(m).$$

Then it has been shown in appendix B that

$$\delta_{(\epsilon)}\phi^A_\mu(n) = -\sum_B \hat{D}_\mu[\phi]_{AB}\epsilon^B(n) , \tag{15.19a}$$

where

$$\hat{D}_\mu[\phi] = M^{-1}(\phi_\mu(n))\,\hat{\partial}^R_\mu + i\Phi_\mu(n). \tag{15.19b}$$

$M^{-1}(\phi_\mu(n))$ is the inverse of the matrix (15.12b), and $\Phi_\mu(n)$ has been defined in (15.12c). The first few terms in the expansion of M^{-1} in powers of $\phi_\mu(n)$ are given by

$$M^{-1}(\phi_\mu(n)) = 1 + \frac{i}{2}\Phi_\mu(n) - \frac{1}{12}(\Phi_\mu(n))^2 + \dots . \tag{15.19c}$$

The non–linear response of $\phi^A_\mu(n)$ to an infinitesimal gauge transformation is due to lattice artefacts. Indeed, making the replacement

$$\phi^B_\mu(n) \to g_0 a A^B_\mu(x)$$

in (15.19), and using the more suggestive (continuum) notation $\epsilon^B(x)$ instead of $\epsilon^B(n)$, one finds that

$$g_0\delta_{(\epsilon)}A^B_\mu(x) \underset{a\to 0}{\longrightarrow} -\sum_C D_\mu[A]_{BC}\epsilon^C(x) ,$$

where

$$D_\mu[A] = \partial_\mu + ig_0\sum_B t^B A^B_\mu(x)$$

is the matrix valued covariant derivative of the continuum formulation.

Let us now calculate the function $\mathcal{F}^A_n[{}^g\phi,\chi]$ in (15.17) for g in the neighborhood of the identity, and for fields $\phi^A_\mu(n)$ satisfying the gauge condition (15.16). Making use of (15.19) one finds that

$$\mathcal{F}^A_n[{}^g\phi;\chi] \underset{g\approx 1}{\approx} -\sum_{m,B} L_{nA,mB}[\phi]\epsilon^B(m) , \tag{15.20a}$$

where

$$L_{nA,mB}[\phi] = \hat{\partial}^L_\mu \hat{D}_\mu[\phi]_{AB}\delta_{nm} \tag{15.20b}$$

is the analog of the matrix $\partial_\mu D_\mu[A]_{AB}\delta^{(4)}(x-y)$ in the continuum formulation. Note that according to (15.19b), $\hat{D}_\mu[\phi]$ is a local function of the matrix valued field $\Phi_\mu(n)$, and that all lattice derivatives act on the lattice site "n". We next introduce the expression (15.20a) into (15.17). Since for fields $\phi_\mu^A(n)$ which satisfy the gauge condition (15.16) the integral only receives a contribution for group elements g in the immediate neighbourhood of the identity, we may replace the group integration measure by

$$Dg \to \prod_{n,A} d\epsilon^A(n).$$

The integral (15.17) can now be immediately performed and one finds that for field configurations satisfying the gauge condition, Δ_{FP} is independent of χ and given by

$$\Delta_{FP}[\phi] = \det(-L[\phi])$$

where $L[\phi]$ is the matrix defined in (15.20b). Δ_{FP} is referred to in the literature as the Faddeev–Popov determinant.

In principle we could incorporate the effect of $\Delta_{FP}[\phi]$ into an effective action by setting $\Delta_{FP}[\phi] = \exp(Tr\ln(-L[\phi]))$. We had adopted such a procedure in connection with the link integration measure. But whereas $S_{meas.}[\phi]$ is given by a sum over local products of the fields $\phi_\mu^A(n)$, this is not the case for $Tr\ln(-L[\phi])$.* To derive the Feynman rules, however, we want to start from an action where the fields are coupled locally. Hence we cannot incorporate the effect of the Faddeev–Popov determinant into an effective action in the above mentioned way. Using a standard trick however, we can circumvent this difficulty. Thus making use of the formula (2.47) we can write $\det L[\phi]$ in the form

$$\det L[\phi] = \int \prod_{A,n} d\bar{\hat{c}}^A(n) d\hat{c}^A(n) e^{-S_{FP}[\phi,\hat{c},\bar{\hat{c}}]}, \qquad (15.21a)$$

where

$$S_{FP}[\phi,\hat{c},\bar{\hat{c}}] = -\sum_{A,B,n} \bar{\hat{c}}^A(n)\hat{\partial}_\mu^L \hat{D}_\mu[\phi]_{AB}\hat{c}^B(n). \qquad (15.21b)$$

The Grassmann valued fields $\hat{c}^A(n)$ and $\bar{\hat{c}}^A(n)$ $(A = 1,\ldots,8)$ carry a colour but no Dirac–index. They transform according to the adjoint representation of SU(3), and are the lattice analogs of the famous Faddeev–Popov ghost fields.

Finally, we must get rid of the gauge fixing δ–functions in (15.18a,b). We do this in the way described in chapter 14. Since the numerator and denominator in

* This non–local property is of course not peculiar to the lattice formulation.

(15.18a) do not depend on the choice of the fields $\chi^A(n)$, we can average these expressions over $\chi^A(n)$ with a Gaussian weight factor $\exp\left[-\frac{1}{2\alpha}\sum_{n,A}\chi^A(n)\chi^A(n)\right]$. Collecting the results obtained so far, we therefore find that (15.18) can be written in the form*

$$< \Gamma[\phi, \hat{\psi}, \bar{\hat{\psi}}] >= \frac{\int D\phi D\bar{\hat{\psi}} D\hat{\psi} D\bar{\hat{c}} D\hat{c} \Gamma[\phi, \hat{\psi}, \bar{\hat{\psi}}] e^{-S^{(tot)}_{QCD}[\phi,\hat{\psi},\bar{\hat{\psi}},\hat{c},\bar{\hat{c}}]}}{\int D\phi D\bar{\hat{\psi}} D\hat{\psi} D\bar{\hat{c}} D\hat{c} e^{-S^{(tot)}_{QCD}[\phi,\hat{\psi},\bar{\hat{\psi}},\hat{c},\bar{\hat{c}}]}}, \tag{15.22a}$$

where the "total" action, $S^{(tot)}_{QCD}$, is given by

$$S^{(tot)}_{QCD} = S_G[\phi] + S^{(W)}_F[\phi, \hat{\psi}, \bar{\hat{\psi}}] + S_{GF}[\phi] + S_{meas.}[\phi] + S_{FP}[\phi, \hat{c}, \bar{\hat{c}}]. \tag{15.22b}$$

Here $S_G[\phi]$ and $S^{(W)}_F[\phi, \hat{\psi}, \bar{\hat{\psi}}]$ are given by (6.25b) and (6.24c) with the link variables $U_\mu(n)$ replaced by $\exp(i\phi_\mu(n))$, and $S_{GF}[\phi]$ is the non–abelian analog of (14.5c):

$$S_{GF}[\phi] = \frac{1}{2\alpha}\sum_{n,A}\left(\hat{\partial}^L_\mu \phi^A_\mu(n)\right)^2. \tag{15.22c}$$

The expression (15.22a) provides the appropriate starting point for performing a weak coupling expansion analogous to that described for the abelian case considered in the previous chapter. We have gone quite a way to arrive at this expression. And we still have to do some work to derive the Feynman rules from the generating functional

$$Z[\hat{J}, \hat{\eta}, \bar{\hat{\eta}}, \hat{\xi}, \bar{\hat{\xi}}] = \int D\phi D\bar{\hat{\psi}} D\hat{\psi} D\bar{\hat{c}} D\hat{c} e^{-S^{(tot)}_{QCD}[\phi,\hat{\psi},\bar{\hat{\psi}},\hat{c},\bar{\hat{c}}]} \cdot e^{\sum_n\left[\hat{J}^A_\mu(n)\phi^A_\mu(n)+\bar{\hat{\eta}}^a_\alpha(n)\hat{\psi}^a_\alpha(n)+\bar{\hat{\psi}}^a_\alpha(n)\hat{\eta}^a_\alpha(n)+\bar{\hat{\xi}}^A(n)\hat{c}^A(n)+\bar{\hat{c}}^A(n)\hat{\xi}^A(n)\right]}.$$

The reason is that the contribution of $S_G[\phi]$ to S_{QCD}, is a complicated functional of the fields ϕ^A_μ and their derivatives. This is a consequence of the non–abelian character of the link variables. In the continuum formulation this piece of the action gives rise to triple and quartic interactions of the gluon fields. In the lattice formulation, on the other hand, not only do these vertices get modified by lattice artefacts, but there are also an infinite number of additional interaction vertices which contribute to the correlation functions for *finite* lattice spacing. Only a finite number of these vertices, however, contribute in the continuum limit in a given order of perturbation theory.

---

* In chapter 2 we had omitted the "hat" on $\hat{\psi}$ and $\bar{\hat{\psi}}$ for convenience.

15.3 The Gauge Field Action

Consider the action (6.25b). It can be written in the form

$$S_G[\phi] = \frac{1}{g_0^2} Tr \sum_{\substack{n,\mu,\nu \\ \mu\neq\nu}} (\mathbb{1} - U_{\mu\nu}(n)), \tag{15.23}$$

where "$\mathbb{1}$" denotes the 3×3 unit matrix, and $U_{\mu\nu}(n)$ is given by the ordered product of link matrices around an elementary plaquette lying in the $\mu\nu$–plane:

$$U_{\mu\nu}(n) = e^{i\phi_\mu(n)} e^{i\phi_\nu(n+\hat{\mu})} e^{-i\phi_\mu(n+\hat{\nu})} e^{-i\phi_\nu(n)}. \tag{15.24}$$

Clearly,

$$U_{\mu\nu}^\dagger(n) = U_{\nu\mu}(n). \tag{15.25}$$

Since $U_{\mu\nu}(n)$ is an element of SU(3) in the fundamental representation, it can be written in the form

$$U_{\mu\nu}(n) = e^{i\phi_{\mu\nu}(n)}, \tag{15.26}$$

where $\phi_{\mu\nu}(n)$ is an element of the Lie algebra of SU(3) in the above mentioned representation:

$$\phi_{\mu\nu}(n) = \sum_A \phi_{\mu\nu}^A(n) T^A. \tag{15.27}$$

From (15.25) it follows that

$$\phi_{\mu\nu}^A(n) = -\phi_{\nu\mu}^A(n).$$

Consider the trace of $(\mathbb{1} - U_{\mu\nu})$. Expanding (15.26) in powers of $\phi_{\mu\nu}$, and recalling that $TrT^A = 0$, we have that

$$Tr(\mathbb{1} - U_{\mu\nu}) = Tr\left\{\frac{1}{2}(\phi_{\mu\nu})^2 + \frac{i}{3!}(\phi_{\mu\nu})^3 - \frac{1}{4!}(\phi_{\mu\nu})^4 + \ldots\right\}.$$

Because $\phi_{\mu\nu} = -\phi_{\nu\mu}$, it follows that the cubic term will not contribute to the action. Hence

$$S_G[\phi] = \frac{1}{g_0^2} \sum_{n,\mu,\nu} Tr\left\{\frac{1}{2}(\phi_{\mu\nu}(n))^2 - \frac{1}{4!}(\phi_{\mu\nu}(n))^4 + \ldots\right\}. \tag{15.28}$$

The trace can be easily evaluated by making use of the following relations*

$$\begin{aligned}
Tr(T^A T^B) &= \frac{1}{2}\delta_{AB}, \\
Tr(T^A T^B T^C) &= \frac{1}{4}(d_{ABC} + i f_{ABC}), \\
Tr(T^A T^B T^C T^D) &= \frac{1}{12}\delta_{AB}\delta_{CD} - \frac{1}{8} f_{ABE} f_{CDE} + \frac{1}{8} d_{ABE} d_{CDE} \\
&\quad + \frac{i}{8}(f_{ABE} d_{CDE} + f_{CDE} d_{ABE}),
\end{aligned}$$

which can be easily derived using the commutation relations (15.2a) and the following expression for the anticommutator of T^A and T^B:

$$\{T^A, T^B\} = \frac{1}{3}\delta_{AB} I + d_{ABC} T^C.$$

Written in terms of the fields $\phi^1_{\mu\nu}, \ldots, \phi^8_{\mu\nu}$, (15.28) becomes

$$\begin{aligned}
S_G[\phi] = \frac{1}{g_0^2} \sum_{n,\mu,\nu} \Big[\frac{1}{4} \sum_A (\phi^A_{\mu\nu})^2 - \frac{1}{288} \sum_{A,B} (\phi^A_{\mu\nu})^2 (\phi^B_{\mu\nu})^2 \\
- \frac{1}{192} \sum_{A,B,\ldots,E} d_{ABE} d_{CDE} \phi^A_{\mu\nu} \phi^B_{\mu\nu}\, \phi^C_{\mu\nu} \phi^D_{\mu\nu} + \ldots \Big],
\end{aligned} \tag{15.29}$$

where we have suppressed the dependence of the fields on the lattice site. Next we express the right–hand side of this expression in terms of the fields $\{\phi^A_\mu(n)\}$ which, apart from a factor g_0, are the lattice analogues of the coloured gauge potentials. To this effect we first derive a relation between $\phi_{\mu\nu}(n)$ and the matrix valued fields $\phi_\mu(n), \phi_\nu(n+\hat\mu), \phi_\mu(n+\hat\nu)$ and $\phi_\nu(n)$ appearing in the product (15.24). We do this by making repeated use of the Campbell–Baker–Hausdorff (CBH) formula, which states the following: Let G be a Lie group with Lie algebra Ł, and let B_1 and B_2 be elements of Ł. Consider the product $\exp(B_1)\exp(B_2)$. It can be written in the form

$$e^{B_1} e^{B_2} = e^{C(B_1,B_2)}, \tag{15.30a}$$

with $C(B_1, B_2) \in$ Ł. According to CBH, $C(B_1, B_2)$ is given by

$$C(B_1, B_2) = \sum_{n=1}^{\infty} C_n(B_1, B_2), \tag{15.30b}$$

* The tensor d_{ABC} is completely symmetric in the indices. Its components are given by $d_{118} = d_{228} = d_{338} = -d_{888} = 1/\sqrt{3}$; $d_{156} = d_{157} = -d_{247} = d_{256} = d_{344} = d_{355} = -d_{366} = -d_{377} = 1/2$; $d_{448} = d_{558} = d_{668} = d_{778} = -\frac{1}{2\sqrt{3}}$.

where the contributions $C_n(B_1, B_2)$ are determined by the following recursion relations:

$$C_1(B_1, B_2) = B_1 + B_2,$$
$$(n+1)C_{n+1}(B_1, B_2) = \frac{1}{2}[B_1 - B_2, C_n(B_1, B_2)]$$
$$+ \sum_{\substack{p\geq 1\\ 2p\leq n}} k_{2p} \sum_{\substack{m_1,\ldots,m_{2p}>0\\ m_1+\ldots+m_{2p}=n}} \left[C_{m_1}(B_1, B_2), \left[\ldots, \left[C_{m_{2p}}(B_1, B_2), B_1 + B_2\right]\ldots\right]\right].$$

(15.30c)

Here k_{2p} are rational, and $k_{2p}(2p)!$ are the Bernoulli numbers. We now make repeated use of this formula to calculate the product

$$e^{B_1}e^{B_2}e^{B_3}e^{B_4} = e^{M(B)}$$

where B stands collectively for $B_1, \ldots, B_4$. Clearly

$$M = \sum_{i=1}^{4} B_i + O(B^2).$$

Consider the case where $\exp(B_i)$ are elements of SU(N) in some matrix representation. Since $M(B)$ is at least of $O(B)$, and since $Tr\ M = 0$, it follows that if we want to calculate $Tr\,(\mathbb{1} - \exp(M))$ up to fourth order in the B_i's,* we only need to know $M(B)$ up to $O(B^3)$. Hence we will also only need to know (15.30b) up to this order. From (15.30c) one obtains

$$C(B_i, B_j) = B_i + B_j + \frac{1}{2}[B_i, B_j]$$
$$+ \frac{1}{12}\left([B_i, [B_i, B_j]] + [B_j, [B_j, B_i]]\right) + \ldots .$$

Making repeated use of this expression and of the Jacobi identity, one finds after some algebra that $M(B)$ can be written in the form

$$M(B) = \sum_i B_i + \frac{1}{2}\sum_{i<j}[B_i, B_j] + \frac{1}{12}\sum_{i,j}[B_i, [B_i, B_j]]$$
$$+ \frac{1}{6}\sum_{i<j<k}\{[B_i, [B_j, B_k]] + [B_k, [B_j, B_i]]\} + O(B^4). \quad (15.31)$$

* This will suffice for our purposes.

We now apply this formula to the matrix product (15.24), and obtain an expression for (15.27), correct up to third order in the fields ϕ_μ:

$$\begin{aligned}\phi_{\mu\nu} = & \sum_i \theta_i + \frac{i}{2}\sum_{i<j}[\theta_i, \theta_j] - \frac{1}{12}\sum_{i,j}[\theta_i, [\theta_i, \theta_j]] \\ & - \frac{1}{6}\sum_{i<j<k}\{[\theta_i, [\theta_j, \theta_k]] + [\theta_k, [\theta_j, \theta_i]]\} + \dots .\end{aligned} \tag{15.32a}$$

Here the Lie-algebra valued fields θ_i are given by

$$\begin{aligned}\theta_1(n) &= \phi_\mu(n) \\ \theta_2(n) &= \phi_\nu(n+\hat\mu) = \phi_\nu(n) + \hat\partial^R_\mu \phi_\nu(n) \\ \theta_3(n) &= -\phi_\mu(n+\hat\nu) = -\phi_\mu(n) - \hat\partial^R_\nu \phi_\mu(n) \\ \theta_4(n) &= -\phi_\nu(n).\end{aligned} \tag{15.32b}$$

For simplicity we have suppressed the dependence of $\phi_{\mu\nu}$ and θ_i in (15.32a) on the lattice sites.

Consider for example the contribution of the first two terms appearing on the right–hand side of (15.32a). Inserting the definitions (15.32b) one finds that

$$\begin{aligned}\phi_{\mu\nu} = \hat\partial^R_\mu \phi_\nu - \hat\partial^R_\nu \phi_\mu + i[\phi_\mu, \phi_\nu] + i([\phi_\mu, \hat\partial^R_\mu \phi_\nu] - [\phi_\nu, \hat\partial^R_\nu \phi_\mu]) \\ + \frac{i}{2}([\phi_\nu, \hat\partial^R_\mu \phi_\nu] - [\phi_\mu, \hat\partial^R_\nu \phi_\mu]) \\ - \frac{i}{2}[\hat\partial^R_\mu \phi_\nu, \hat\partial^R_\nu \phi_\mu] + O(\phi^3).\end{aligned} \tag{15.33}$$

Introducing the dimensioned (matrix valued) gauge potentials $A_\mu(x)$, and the field strength tensor $F_{\mu\nu}$ in the familiar way, i.e.,

$$\begin{aligned}\phi_\mu(n) &\to g_0 a A_\mu(x), \\ \phi_{\mu\nu}(n) &\to g_0 a^2 F_{\mu\nu}(x), \qquad (x = na)\end{aligned}$$

we see that only the first two terms in (15.33) contribute to $F_{\mu\nu}$ in the continuum limit. Thus

$$F_{\mu\nu}(x) = \partial^R_\mu A_\nu(x) - \partial^R_\nu A_\mu(x) + ig_0\,[A_\mu(x), A_\nu(x)] + \dots,$$

where ∂^R_μ is the dimensioned right lattice derivative, and where the dots stand for terms which vanish for $a \to 0$. Hence in the continuum limit $F_{\mu\nu}$ coincides with the field strength tensor in QCD.

We now return to eq.(15.32a). Decomposing the matrices θ_i as follows

$$\theta_i = \sum_A \theta_i^A T^A,$$

and making use of the relations (15.2), one finds that the colour components of the field $\phi_{\mu\nu}$, defined in (15.27), are given by

$$\begin{aligned} \phi_{\mu\nu}^A(n) = &\sum_i \theta_i^A(n) - \frac{1}{2}\sum_{i<j} \theta_{ij}^A(n) + \frac{1}{12}\sum_{ij} \theta_{iij}^A(n) \\ &+ \frac{1}{6}\sum_{i<j<k} \left[\theta_{ijk}^A(n) + \theta_{kji}^A(n)\right] + \dots , \end{aligned} \tag{15.34a}$$

where

$$\begin{aligned} \theta_{ij}^A(n) &= \sum_{B,C} f_{ABC}\theta_i^B(n)\theta_j^C(n) \\ \theta_{ijk}^A(n) &= \sum_{B,C,D,E} f_{ABE}f_{CDE}\theta_i^B(n)\theta_j^C(n)\theta_k^D(n), \end{aligned} \tag{15.34b}$$

and where $\theta_i^A, \dots, \theta_4^A$ are related to the fields ϕ_μ^A, ϕ_ν^A and their derivatives by expressions analogous to (15.32b):

$$\begin{aligned} \theta_1^A(n) &= \phi_\mu^A(n) \\ \theta_2^A(n) &= \phi_\nu^A(n) + \hat{\partial}_\mu^R \phi_\nu^A(n) \\ \theta_3^A(n) &= -\phi_\mu^A(n) - \hat{\partial}_\nu^R \phi_\mu^A(n) \\ \theta_4^A(n) &= -\phi_\nu^A(n). \end{aligned} \tag{15.34c}$$

Inserting (15.34a) into (15.29) one arrives at the following expression for the action, correct up to fourth order in the fields $\{\theta_i^A\}$:*

$$\begin{aligned} S_G[\phi] = \frac{1}{g_0^2}\sum_{\substack{\mu,\nu \\ n}} \Big\{ &\frac{1}{4}\sum_{i,j} \theta_i^A\theta_j^A - \frac{1}{4}\sum_{\substack{i,j,k \\ j<k}} \theta_i^A\theta_{jk}^A \\ &+ \frac{1}{16}\sum_{\substack{i<j \\ k<l}} \theta_{ij}^A\theta_{kl}^A + \frac{1}{24}\sum_{i,j,k} \theta_i^A\theta_{jjk}^A \\ &+ \frac{1}{12}\sum_{\substack{i,j,k,l \\ j<k<l}} \theta_i^A\left(\theta_{jkl}^A + \theta_{lkj}^A\right) \\ &- \frac{1}{192} d_{ABE}d_{CDE}\sum_{i,j,k,l} \theta_i^A\theta_j^B\theta_k^C\theta_l^D \\ &- \frac{1}{288}\sum_{i,j,k,l} \theta_i^A\theta_j^A\theta_k^B\theta_l^B \Big\} + \dots . \end{aligned} \tag{15.35}$$

* From now on it will be understood that also repeated colour indices are summed.

Here a sum over repeated colour indices is understood. The dependence on n, μ, and ν of the quantity appearing within curly brackets is given implicitely by the relations (15.32b).

Clearly, (15.35) is a complicated function of the fields $\{\phi_\mu^A\}$ and their derivatives. Most of the contributions are however lattice artefacts and do not survive in the continuum limit.* This is easlily demonstrated. Introducing the definitions (15.34b,c) into (15.35), and the dimensioned colour components of the vector potentials according to

$$\phi_\mu^B(n) \to g_0 a A_\mu^B(x), \quad x = na, \tag{15.36}$$

one has that

$$\sum_i \theta_i^B(n) \to g_0 a^2 f_{\mu\nu}^B(x),$$

where

$$f_{\mu\nu}^B(x) = \partial_\mu^R A_\nu^B(x) - \partial_\nu^R A_\mu^B(x).$$

Hence the sum $\sum_i \theta_i^B$ actually vanishes with the second power of the lattice spacing. It is therefore evident, that in the naive continuum limit only the first three terms appearing on the right–hand side of (15.35) survive. For $a \to 0$ their contribution can be easily evaluated and yields the usual expression for the continuum action

$$S_G \underset{a\to 0}{\longrightarrow} \frac{1}{4}\int d^4x F_{\mu\nu}^B(x) F_{\mu\nu}^B(x),$$

where $F_{\mu\nu}^B(x)$ is the non–abelian field tensor defined in (6.16). On the other hand, for finite lattice spacing, only the contributions quadratic and cubic in θ_i^A have a simple form. We shall treat them in detail below.

15.4 Propagators and Vertices

i) *The gluon propagator*

Consider first the contributions to (15.22b) arising from S_G and S_{GF} which are quadratic in the fields ϕ_μ^A. Expressed in terms of the dimensioned gauge potentials one readily finds that**

$$S_G^{(0)}[A] = \frac{1}{2}\sum_{x,y} A_\mu^B(x) \Omega_{\mu\nu}^{BC}(x,y) A_\nu^C(y), \tag{15.37a}$$

---

* This is true for the classical action. But on the quantum level, we cannot simply ignore all these contributions, as we have pointed out repeatedly.

** We use the notation $\sum_x = \sum_n a^4$, introduced in chapter 14. The procedure for casting S_G into the form (15.37) is the same as that described in this chapter.

where

$$\Omega^{BC}_{\mu\nu}(x,y) = \delta_{BC}\Omega_{\mu\nu}(x,y), \tag{15.37b}$$

and $\Omega_{\mu\nu}$ is defined in (14.13b). Following the procedure discussed in chapter 14, we can immediately write down the propagator for the gauge potential in momentum space. To avoid the appearance of any superfluous phases, which will eventually cancel in the Feynman rules,* we Fourier decompose the fields $A^B_\mu(x)$ as follows

$$A^B_\mu(x) = \int_{BZ} \frac{d^4k}{(2\pi)^4} \tilde{A}^B_\nu(k) e^{ik\cdot x + ik_\mu a/2}. \tag{15.38}$$

Then the gluon propagator in k–space is given by

$$\underset{B,\mu \qquad C,\nu}{\overset{k}{\sim\!\sim\!\sim}} = \frac{1}{\tilde{k}^2}\left(\delta_{\mu\nu} - (1-\alpha_0)\frac{\tilde{k}_\mu \tilde{k}_\nu}{\tilde{k}^2}\right)\delta_{BC},$$

where $\tilde{k}_\mu$ is the lattice momentum defined by (13.7a).

ii) *The two–gluon vertex*

Consider next the contribution of $S_{meas}[\phi]$ quadratic in the gauge potential. From (15.15) one finds, using (15.9), that

$$S^{(2)}_{meas}[A] = \frac{1}{2!}\frac{g_0^2}{4a^2}\sum_x A^B_\mu(x)A^B_\mu(x). \tag{15.39}$$

This contribution is proportional to g_0^2 and hence should be considered as part of the interaction. What is striking about this contribution is, that it diverges like $1/a^2$ in the continuum limit!** In fact it has the typical structure of a mass–counter term. This is indeed the role it plays in lattice perturbation theory where it serves to eliminate quadratic "ultraviolet" ($a \to 0$) divergences in Feynman integrals contributing in $O(g_0^2)$ to the gluon self–energy. In this way the lattice provides its own counterterms to ensure the renormalizability of the theory. This demonstrates in a particularily drastic way, how important it is to include the effects of the lattice cut–off in the integration measure.

The two gluon vertex arising from (15.39) is given in momentum space by

$$\underset{\mu,B \quad \nu,C}{\overset{k \quad k'}{\sim\!\times\!\sim}} = -(2\pi)^4\delta^{(4)}_P(k+k')\frac{g_0^2}{4a^2}\delta_{\mu\nu}\delta_{BC},$$

* See the discussion in chapter 14.

** Since $\sum_x \equiv \sum_n a^4$, the contribution to S_{meas} arising from the Haar measure associated with an *individual* link variable actually vanishes for $a \to 0$.

where $\delta_P^{(4)}(k+k')$ is the periodic δ–function. Notice that we did not include the factor 1/2! appearing in (15.39) in the definition of the vertex, conforming to usual conventions.

iii) *The three–gluon vertex*

The contribution to S_G involving three gauge fields is obtained from the second term appearing on the right–hand side of (15.35).* Making the substitutions (15.34c) and (15.36), and using the antisymmetry of the structure constants f_{ABC} under exchange of any pair of indices, one finds that

$$S_G^{(3)} = g_0 \sum_x f_{ABC} \left(A_\mu^A(x) + \frac{a}{2} \partial_\nu^R A_\mu^A(x) \right) \left(\partial_\mu^R A_\nu^B(x) \right) A_\lambda^C(x) \delta_{\lambda\nu}.$$

Notice that this expression includes lattice artefacts vanishing linearly with a. It can be written in a more symmetric form by making use of the antisymmetry and cyclic symmetry of the structure constants f_{ABC} in A, B, C. Using the first mentioned property we have that

$$S_G^{(3)} = -\frac{g_0}{2} \sum_x f_{ABC} \left(\left(1 + \frac{a}{2} \partial_\nu^R \right) A_\mu^A(x) \right) \left(A_\nu^B(x) \overleftrightarrow{\partial}_\mu^R A_\lambda^C(x) \right) \delta_{\lambda\nu},$$

where the action of $\overleftrightarrow{\partial}_\mu^R$ is defined by

$$g(x) \overleftrightarrow{\partial}_\mu^R f(x) = g(x) \partial_\mu^R f(x) - \left(\partial_\mu^R g(x) \right) f(x).$$

Next we cyclically permute the pair of indices (A, μ), (B, ν) and (C, λ) and obtain

$$S_G^{(3)} = -\frac{g_0}{3!} \sum_x f_{ABC} \left\{ \left((1 + \frac{a}{2} \partial_\nu^R) A_\mu^A(x) \right) \left(A_\nu^B(x) \overleftrightarrow{\partial}_\mu^R A_\lambda^C(x) \right) \delta_{\lambda\nu} + cycl.perm. \right\}. \tag{15.40}$$

In this way, we have made manifest the Bose symmetry under the exchange of colour and Lorentz indices. Notice that by having cyclically permuted simultaneously the Lorentz and colour indices, the gauge potentials appearing in all three terms within the curly bracket in (15.40) are labeled by the pairs (A, μ), (B, ν) and (C, λ).

---

* Notice that $S_{meas}[\phi]$ does not contribute in this order.

Let us calculate explicitly the 3–point gluon vertex, $\Gamma^{ABC}_{\mu\nu\lambda}$, in momentum space. This vertex function is defined as follows

$$S_G^{(3)}[A] = -\frac{1}{3!}\sum_x \int \frac{d^4k}{(2\pi)^4}\frac{d^4k'}{(2\pi)^4}\frac{d^4k''}{(2\pi)^4}\tilde{A}^A_\mu(k)\tilde{A}^B_\nu(k')\tilde{A}^C_\lambda(k'')\Gamma^{ABC}_{\mu\nu\lambda}(k,k',k'') \times e^{i(k+k'+k'')\cdot x}. \tag{15.41}$$

We represent $\Gamma^{ABC}_{\mu\nu\lambda}(k,k',k'')$ by the following diagram

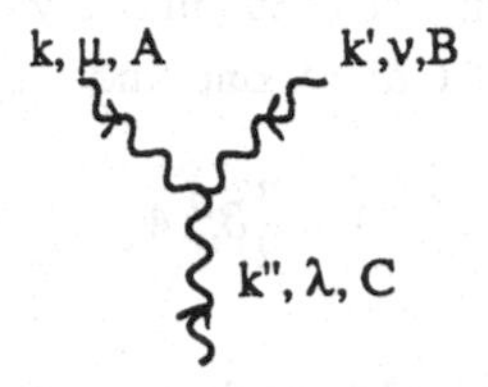

Consider the explicit expression displayed on the right–hand side of (15.40). In momentum space the operators $1+\frac{a}{2}\partial^R_\nu$ and $\overleftrightarrow{\partial}^R_\mu$ act multiplicatively as follows

$$\begin{aligned}(1+\frac{a}{2}\partial^R_\mu)e^{ik\cdot x} &\to (e^{ik_\mu\frac{a}{2}}\cos\frac{k_\mu a}{2})e^{ik\cdot x}\\ e^{ik\cdot x}\overleftrightarrow{\partial}^R_\mu e^{iq\cdot x} &\to i(\widetilde{q-k})_\mu e^{i(k+q)\cdot x}e^{i(k+q)_\mu\frac{a}{2}}\ ,\end{aligned} \tag{15.42}$$

where $\tilde{k}'_\mu$ and $\tilde{k}''_\mu$ are defined by an expression analogous to (13.7a). When the phase factors appearing in (15.42) are combined with the overall phase $\exp[i(k_\mu + k'_\nu + k''_\lambda)a/2]$, arising from the definition of the Fourier transform (15.38), and use is made of the fact that – because of the summation over x in (15.41) – the sum of the momenta flowing into the vertex vanishes, one finds after some trivial trigonometric algebra that

$$\begin{aligned}\Gamma^{ABC}_{\mu\nu\lambda}(k,k',k'') =& ig_0(2\pi)^4\delta^{(4)}_P(k+k'+k'')f_{ABC}\left[(\widetilde{k''-k'})_\mu\cos\frac{k_\nu a}{2}\delta_{\nu\lambda}\right.\\ &\left.+(\widetilde{k-k''})_\nu\cos\frac{k'_\mu a}{2}\,\delta_{\mu\lambda}+(\widetilde{k'-k})_\lambda\cos\frac{k''_\mu a}{2}\delta_{\mu\nu}\right],\end{aligned} \tag{15.43}$$

where $(\widetilde{p-q})_\mu$ stands for

$$\begin{aligned}(\widetilde{p-q})_\mu &= \frac{2}{a}\sin\left[\frac{(p-q)_\mu}{2}a\right].\\ &\xrightarrow[a\to 0]{} (p-q)_\mu.\end{aligned}$$

Hence in the continuum limit, (15.43) reduces to the familiar expression of the continuum formulation

$$\Gamma^{ABC}_{\mu\nu\lambda}(k,k',k'')\underset{a\to 0}{\longrightarrow} ig_0(2\pi)^4\delta^{(4)}(k+k'+k'')f_{ABC}\left[(k''-k')_\mu\delta_{\nu\lambda}\right.$$
$$\left.+(k-k'')_\nu\delta_{\mu\lambda}+(k'-k)_\lambda\delta_{\mu\nu}\right].$$

iv) The four-gluon vertex

The computation of the four-gluon vertex (see fig. on page 245) from the fourth-order contribution in $\{\theta_i^a\}$ to the action (given by the last five terms in (15.35)) is quite tedious.* Making use of the definition (15.34b) and of $\sum_i \theta_i^A = \hat{\partial}_\mu\phi_\nu^A - \hat{\partial}_\nu\phi_\mu^A$, following from (15.34c), as well as of the antisymmetry of f_{CDE} under $C \leftrightarrow D$, this contribution takes the form

$$(S_G)_{A^4} = S_G^{(f)} + S_G^{(d)}\ , \tag{15.44a}$$

where

$$\begin{aligned} S_G^{(f)} =& \frac{1}{4!}\frac{1}{g_0^2}\sum_n\sum_{\mu,\nu}\sum_{ABCDE} f_{ABE}f_{CDE}\Big\{(\hat{\partial}_\mu^R\phi_\nu^A - \hat{\partial}_\nu^R\phi_\mu^A)[\sum_{j\neq k}\theta_j^B\theta_j^C\theta_k^D \\ &+2\sum_{j<k<l}\theta_k^C(\theta_j^B\theta_l^D+\theta_l^B\theta_j^D)] \\ &+\frac{3}{8}\sum_{\substack{i<j\\ k<l}}(\theta_i^A\theta_j^B-\theta_j^A\theta_i^B)(\theta_k^C\theta_l^D-\theta_l^C\theta_k^D)\Big\} \end{aligned} \tag{15.44b}$$

and

$$\begin{aligned} S_G^{(d)} =& -\frac{1}{(4!)^2}\frac{1}{g_0^2}\sum_n\sum_{\mu,\nu}\sum_{ABCDE}\Big\{\frac{2}{3}(\delta_{AB}\delta_{CD}+\delta_{AC}\delta_{DB}+\delta_{AD}\delta_{BC})+ \\ &+(d_{ABE}d_{CDE}+d_{ACE}d_{DBE}+d_{ADE}d_{BCE})\Big\} \\ &\times(\hat{\partial}_\mu^R\phi_\nu^A-\hat{\partial}_\nu^R\phi_\mu^A)(\hat{\partial}_\mu^R\phi_\nu^B-\hat{\partial}_\nu^R\phi_\mu^B)(\hat{\partial}_\mu^R\phi_\nu^C-\hat{\partial}_\nu^R\phi_\mu^C)(\hat{\partial}_\mu^R\phi_\nu^D-\hat{\partial}_\nu^R\phi_\mu^D) \end{aligned} \tag{15.44c}$$

where $\{\theta_i^A\}$ have been defined in terms of the gauge potentials in (15.34c). We have suppressed the dependence of the fields on the lattice site n. The expression (15.44c) has been written in a manifestly symmetric form under the exchange of any pair of colour indices.

---

* The following derivation has also been carried out independently by P. Kaste.

Consider first the expression (15.44b). We need to express the r.h.s. in terms of the fields $\{\theta^A_\mu\}$ and their derivatives according to (15.34c). This clearly generates an anormous amount of terms which can be readily obtained by using a program like "Mathematica". The next step consists in classifying these terms according to the number of identical Lorentz indices of the gauge potentials. A large number of these terms can be combined by making use of the antisymmetry of the structure constants under the exchange of any pair of indices, and of the invariance of the individual contributions to the action under the exchange of μ and ν. Let us denote the contributions to $S^{(f)}_G$ involving two, three, and four gauge potentials with identical Lorentz indices by $S^{(f_1)}_G, S^{(f_2)}_G$ and $S^{(f_3)}_G$, respectively.

Consider for example the contribution $S^{(f_2)}$. After exploiting the above-mentioned symmetries one finds at the end of the day that the (hundreds!) of terms combine to

$$\begin{aligned} S^{(f_2)}_G =& \frac{1}{g_0^2}\sum_{n\mu,\nu}\sum_{ABCDE} f_{ABE}f_{CDE}\Big\{\frac{1}{96}(\phi^A_\mu \overleftrightarrow{\partial}^R_\nu \phi^B_\mu)(\phi^C_\nu \overleftrightarrow{\partial}^R_\mu \phi^D_\nu) \\ &+ \frac{1}{4}\phi^A_\mu\phi^B_\nu\phi^C_\mu\phi^D_\nu + \frac{1}{2}\phi^A_\mu(\hat\partial^R_\mu\phi^B_\nu)\phi^C_\mu\phi^D_\nu \\ &+ \frac{1}{12}(\hat\partial^R_\nu\phi^A_\mu)(\hat\partial^R_\mu\phi^B_\nu)\phi^C_\mu\phi^D_\nu + \frac{1}{6}(\hat\partial^R_\nu\phi^A_\mu)\phi^B_\nu\phi^C_\mu(\hat\partial^R_\mu\phi^D_\nu) \\ &- \frac{1}{48}(\hat\partial^R_\nu\phi^A_\mu)(\hat\partial^R_\mu\phi^B_\nu)(\hat\partial^R_\nu\phi^C_\mu)(\hat\partial^R_\mu\phi^D_\nu)\Big\}. \end{aligned}$$

This expression can be written in an even more convenient form:

$$\begin{aligned} S^{(f_2)}_G =& \frac{1}{4g_0^2}\sum_{n,\mu,\nu}\sum_{ABCDE} f_{ABE}f_{CDE}\Big\{\Big[[(1+\frac{1}{2}\hat\partial^R_\nu)\phi^A_\mu][(1+\frac{1}{2}\hat\partial^R_\nu)\phi^C_\mu] \\ &- \frac{1}{4}(\hat\partial^R_\nu\phi^A_\mu)(\hat\partial^R_\nu\phi^C_\mu)\Big]\Big[[(1+\frac{1}{2}\hat\partial^R_\mu)\phi^B_\nu][(1+\frac{1}{2}\hat\partial^R_\mu)\phi^D_\nu] - \frac{1}{4}(\hat\partial^R_\mu\phi^B_\nu)(\hat\partial^R_\mu\phi^D_\nu)\Big] \\ &- \frac{1}{12}(\hat\partial^R_\nu\phi^A_\mu)(\hat\partial^R_\mu\phi^B_\nu)(\hat\partial^R_\nu\phi^C_\mu)(\hat\partial^R_\mu\phi^D_\nu) - \frac{1}{12}(\phi^A_\mu\overleftrightarrow{\hat\partial}^R_\nu\phi^C_\mu)(\phi^B_\nu\overleftrightarrow{\hat\partial}^R_\mu\phi^D_\nu) \\ &+ \frac{1}{24}(\phi^A_\mu\overleftrightarrow{\hat\partial}^R_\nu\phi^B_\mu)(\phi^C_\nu\overleftrightarrow{\hat\partial}^R_\mu\phi^D_\nu)\Big\}. \end{aligned} \tag{15.45}$$

Next we introduce the dimensional gauge potentials through the identification $\phi^B_\mu = g_0 a A^B_\mu$, and Fourier decompose the fields A^B_μ according to (15.38). Let k, q, r and s denote the *incoming* momenta associated with the gauge potentials carrying colour indices A, B, C and D, respectively. Making use of (15.42), one readily finds, after carrying out the sum over n (which yields an energy momentum conserving δ-function), that

$$S_G^{(f_2)} = \frac{g_0^2}{4} \sum_{\mu\nu\lambda\rho} \sum_{ABCDE} \int_{BZ} \frac{d^4k}{(2\pi)^4} \frac{d^4q}{(2\pi)^4} \frac{d^4r}{(2\pi)^4} \frac{d^4s}{(2\pi)^4} f_{ABE} f_{CDE}$$
$$\times \Big\{ \delta_{\mu\lambda}\delta_{\nu\rho} \Big[\cos\frac{1}{2}a(k-r)_\nu \cos\frac{1}{2}a(q-s)_\mu + \frac{1}{12}a^2 (\widetilde{k-r})_\nu (\widetilde{q-s})_\mu - \frac{1}{12}a^4 \tilde{k}_\nu \tilde{q}_\mu \tilde{r}_\nu \tilde{s}_\mu \Big]$$
$$- \frac{1}{24}\delta_{\mu\nu}\delta_{\lambda\rho} a^2 (\widetilde{q-k})_\lambda (\widetilde{s-r})_\mu \Big\} \tilde{A}_\mu^A(k) \tilde{A}_\nu^B(q) \tilde{A}_\lambda^C(r) \tilde{A}_\rho^D(s)$$

,

$$(15.46a)$$

where, generically,

$$\tilde{k}_\mu = \frac{2}{a} \sin\frac{1}{2}ak_\mu. \qquad (15.46b)$$

The term proportional to $\delta_{\mu\lambda}\delta_{\nu\rho}$ can be written in a more symmetric form by making use of the antisymmetry of f_{CDE} under the exchange of C and D. The contribution proportional to $\delta_{\mu\nu}\delta_{\lambda\rho}$ is already invariant under the relabeling $(C, \lambda, r) \leftrightarrow (D, \rho, s)$. Adding to (15.46a) the expression with (C, λ, r) and (D, ρ, s) interchanged, and dividing the result by 2, we obtain

$$S_G^{(f_2)} = -\frac{1}{4\cdot 2} \sum_{\mu,\nu,\lambda,\rho} \sum_{ABCD} \int_{BZ} \frac{d^4k}{(2\pi)^4} \frac{d^4q}{(2\pi)^4} \frac{d^4r}{(2\pi)^4} \frac{d^4s}{(2\pi)^4} \qquad (15.47a)$$
$$\times \Gamma_{\mu\nu\lambda\rho}^{(f_2)ABCD}(k,q,r,s) \tilde{A}_\mu^A(k) \tilde{A}_\nu^B(q) \tilde{A}_\lambda^C(r) \tilde{A}_\rho^D(s)$$

where

$$\Gamma_{\mu\nu\lambda\rho}^{(f_2)ABCD}(k,q,r,s) =$$
$$-g_0^2 \sum_E f_{ABE} f_{CDE} \Big\{ \delta_{\mu\lambda}\delta_{\nu\rho} \Big[\cos\frac{1}{2}a(k-r)_\nu \cos\frac{1}{2}a(q-s)_\mu - \frac{1}{12}a^4 \tilde{k}_\nu \tilde{q}_\mu \tilde{r}_\nu \tilde{s}_\mu \Big]$$
$$-\delta_{\mu\rho}\delta_{\nu\lambda} \Big[\cos\frac{1}{2}a(k-s)_\nu \cos\frac{1}{2}a(q-r)_\mu - \frac{1}{12}a^4 \tilde{k}_\nu \tilde{q}_\mu \tilde{r}_\mu \tilde{s}_\nu \Big] \Big\}$$
$$+T_{\mu\nu\lambda\rho}^{ABCD}(k,q,r,s). \qquad (15.47b)$$

and

$$T_{\mu\nu\lambda\rho}^{ABCD}(k,q,r,s) = -g_0^2 \sum_E f_{ABE} f_{CDE} \Big\{ \frac{1}{12}\delta_{\mu\lambda}\delta_{\nu\rho} a^2 (\widetilde{k-r})_\nu (\widetilde{q-s})_\mu$$
$$-\frac{1}{12}\delta_{\mu\rho}\delta_{\nu\lambda} a^2 (\widetilde{k-s})_\nu (\widetilde{q-r})_\mu \qquad (15.47c)$$
$$-\frac{1}{12}\delta_{\mu\nu}\delta_{\lambda\rho} a^2 (\widetilde{q-k})_\lambda (\widetilde{s-r})_\mu \Big\}.$$

Let α, β, γ denote collectively the sets

$$\alpha = (A, \mu, k); \; \beta = (B, \nu, q); \; \gamma = (C, \lambda, r); \; \delta = (D, \rho, s). \tag{15.48}$$

Then the expression (15.47b) is symmetric under the relabelings i) $\alpha \leftrightarrow \beta$, ii) $\gamma \leftrightarrow \delta$, iii) $\alpha \leftrightarrow \gamma, \beta \leftrightarrow \delta$ and iv) $\alpha \leftrightarrow \delta, \beta \leftrightarrow \gamma$. The full Bose symmetry of the contribution $S_G^{(f_2)}$ to the 4-gluon vertex function can now be incorporated if we add to (15.47b) the expressions obtained by carrying out the permutations $\beta \leftrightarrow \gamma$, and $\beta \leftrightarrow \delta$, and dividing the result by 3. Then

$$S_G^{(f_2)} = -\frac{1}{4!} \sum_{\mu\nu\lambda\rho} \sum_{ABCD} \int_{BZ} \frac{d^4k}{(2\pi)^4} \frac{d^4q}{(2\pi)^4} \frac{d^4r}{(2\pi)^4} \frac{d^4s}{(2\pi)^4} \Big\{ \Gamma^{(f_2)ABCD}_{\mu\nu\lambda\rho}(k,q,r,s) + \begin{pmatrix} B & q & \nu \\ C & r & \lambda \end{pmatrix} + \begin{pmatrix} B & q & \nu \\ D & s & \rho \end{pmatrix} \Big\} \tilde{A}^A_\mu(k) \tilde{A}^B_\nu(q) \tilde{A}^C_\lambda(r) \tilde{A}^D_\rho(s). \tag{15.49}$$

Having written $S_G^{(f_2)}$ in this symmetrized form, one finds, upon making use of

$$\sum_E [f_{ABE} f_{CDE} + f_{ACE} f_{DBE} + f_{ADE} f_{BCE}] = 0 \,,$$

following from the Jacobi identity for double commutators of the generators of the group, that the last term in (15.47b), together with the permutations in (15.49) does not contribute to (15.49)

Consider next the contributions $S_G^{(f_3)}$ and $S_G^{(f_4)}$, with three and four Lorentz indices of the gauge potentials identified. Making again use of the symmetries of $f_{ABE} f_{CDE}$ one finds after a fair amount of work that they can be reduced to the simple forms

$$S_G^{(f_3)} = -\frac{1}{12 g_0^2} \sum_n \sum_{\mu,\nu} \sum_{ABCDE} f_{ABE} f_{CDE} (\phi^C_\nu \overset{\leftrightarrow}{\hat{\partial}}{}^R_\mu \phi^D_\nu)(\hat{\partial}^R_\nu \phi^A_\mu)[(1 + \frac{1}{2}\hat{\partial}^R_\mu)\phi^B_\nu] \tag{15.50a}$$

and

$$S_G^{(f_4)} = -\frac{1}{96 g_0^2} \sum_n \sum_{\mu,\nu} \sum_{ABCDE} f_{ABE} f_{CDE} (\phi^A_\mu \overset{\leftrightarrow}{\hat{\partial}}{}^R_\nu \phi^B_\mu)(\phi^C_\mu \overset{\leftrightarrow}{\hat{\partial}}{}^R_\nu \phi^D_\mu). \tag{15.50b}$$

Going over to momentum space, one finds, making again use of (15.42) that

$$S_G^{(f_3)} + S_G^{(f_4)} = -\frac{1}{96} \sum_{\mu\nu\lambda\rho} \sum_{ABCD} \int_{BZ} \frac{d^4k}{(2\pi)^4} \frac{d^4q}{(2\pi)^4} \frac{d^4r}{(2\pi)^4} \frac{d^4s}{(2\pi)^4} \times \Gamma^{(f_3+f_4)ABCD}_{\mu\nu\lambda\rho}(k,q,r,s) A^A_\mu(k) A^B_\nu(q) A^C_\lambda(r) A^D_\rho(s), \tag{15.51a}$$

where

$$\Gamma^{(f_3+f_4)ABCD}_{\mu\nu\lambda\rho}(k,q,r,s) = -g_0^2\sum_E f_{ABE}f_{CDE}[8\delta_{\nu\lambda}\delta_{\nu\rho}a^2(\widetilde{s-r})_\mu\tilde{k}_\nu\cos\frac{1}{2}aq_\mu + \delta_{\mu\nu}\delta_{\mu\lambda}\delta_{\nu\rho}a^2\sum_\sigma(\widetilde{q-k})_\sigma(\widetilde{s-r})_\sigma]. \tag{15.51b}$$

While the second term in this expression is symmetric under the permutations: i) $\gamma \leftrightarrow \delta$, ii) $\alpha \leftrightarrow \beta$, iii) $\alpha \leftrightarrow \gamma, \beta \leftrightarrow \delta$ and iv) $\alpha \leftrightarrow \delta$, $\beta \leftrightarrow \gamma$, where $\alpha, \beta, \gamma, \delta$ stand for the collection of indices (15.48), the first term is only symmetric under the exchange $\gamma \leftrightarrow \delta$. We therefore add to it the corresponding expressions obtained by implementing the permutations ii) - iv), making use of the symmetries of the structure constants, and divide the result by 4.

$$\begin{aligned}S_G^{(f_3)} + S_G^{(f_4)} =& \frac{g_0^2}{8}\sum_{\mu\nu\lambda\rho}\sum_{ABCD}\int_{BZ}\frac{d^4k}{(2\pi)^4}\frac{d^4q}{(2\pi)^4}\frac{d^4r}{(2\pi)^4}\frac{d^4s}{(2\pi)^4}\sum_E f_{ABE}f_{CDE}\\ &\times\Big\{\frac{1}{6}\delta_{\nu\lambda}\delta_{\nu\rho}a^2(\widetilde{s-r})_\mu\tilde{k}_\nu\cos\frac{1}{2}aq_\mu\\ &-\frac{1}{6}\delta_{\mu\lambda}\delta_{\mu\rho}a^2(\widetilde{s-r})_\nu\tilde{q}_\mu\cos\frac{1}{2}ak_\nu\\ &+\frac{1}{6}\delta_{\mu\nu}\delta_{\mu\rho}a^2(\widetilde{q-k})_\lambda\tilde{r}_\rho\cos\frac{1}{2}as_\lambda\\ &-\frac{1}{6}\delta_{\mu\nu}\delta_{\lambda\mu}a^2(\widetilde{q-k})_\rho\tilde{s}_\lambda\cos\frac{1}{2}ar_\rho\\ &+\frac{1}{12}\delta_{\mu\nu}\delta_{\mu\lambda}\delta_{\nu\rho}a^2\sum_\sigma(\widetilde{q-k})_\sigma(\widetilde{s-r})_\sigma\Big\}\tilde{A}^A_\mu(k)\tilde{A}^B_\nu(q)A^C_\lambda(r)A^D_\rho(s).\end{aligned} \tag{15.52}$$

The expression appearing within curly brackets is still not symmetric under all permutations of the collective variables (15.48). To exhibit the full Bose symmetry we add to it the corresponding expression obtained by the permutations appearing in (15.49), and divide the result by a factor 3.

The remaining contribution (15.44c) to the action, expressed in Fourier space, is obtained in a straightforward way. Combining it with that obtained above one is then led to the following expression for the four-gluon contribution to the action

$$\begin{aligned}(S_G)_{A^4} =& -\frac{1}{4!}\int_{BZ}\frac{d^4k}{(2\pi)^4}\frac{d^4q}{(2\pi)^4}\frac{d^4r}{(2\pi)^4}\frac{d^4s}{(2\pi)^4}\\ &\times\sum_{\mu\nu\lambda\rho}\sum_{ABCD}\Gamma^{ABCD}_{\mu\nu\lambda\rho}(k,q,r,s)\tilde{A}^A_\mu(k)\tilde{A}^B_\nu(q)\tilde{A}^C_\lambda(r)\tilde{A}^D_\rho(s)\,.\end{aligned} \tag{15.53a}$$

where*

$$
\begin{aligned}
&\Gamma^{ABCD}_{\mu\nu\lambda\rho}(k,q,r,s) = \\
&-g_0^2\Big[\sum_E f_{ABE}f_{CDE}\Big\{\delta_{\mu\lambda}\delta_{\nu\rho}[\cos\frac{1}{2}a(q-s)_\mu\cos\frac{1}{2}a(k-r)_\nu - \frac{a^4}{12}\tilde{k}_\nu\tilde{q}_\mu\tilde{r}_\nu\tilde{s}_\mu] \\
&\qquad - \delta_{\mu\rho}\delta_{\nu\lambda}[\cos\frac{1}{2}a(q-r)_\mu\cos\frac{1}{2}a(k-s)_\nu - \frac{a^4}{12}\tilde{k}_\nu\tilde{q}_\mu\tilde{r}_\mu\tilde{s}_\nu] \\
&\qquad + \frac{1}{6}\delta_{\nu\lambda}\delta_{\nu\rho}a^2(\widetilde{s-r})_\mu\tilde{k}_\nu\cos(\frac{1}{2}aq_\mu) \\
&\qquad - \frac{1}{6}\delta_{\mu\lambda}\delta_{\mu\rho}a^2(\widetilde{s-r})_\nu\tilde{q}_\mu\cos(\frac{1}{2}ak_\nu) \\
&\qquad + \frac{1}{6}\delta_{\mu\nu}\delta_{\mu\rho}a^2(\widetilde{q-k})_\lambda\tilde{r}_\rho\cos(\frac{1}{2}as_\lambda) \\
&\qquad - \frac{1}{6}\delta_{\mu\nu}\delta_{\mu\lambda}a^2(\widetilde{q-k})_\rho\tilde{s}_\lambda\cos(\frac{1}{2}ar_\rho) \\
&\qquad + \frac{1}{12}\delta_{\mu\nu}\delta_{\mu\lambda}\delta_{\mu\rho}a^2\sum_\sigma(\widetilde{q-k})_\sigma(\widetilde{s-r})_\sigma\Big\} \\
&\qquad + (B\leftrightarrow C, \nu\leftrightarrow\lambda, q\leftrightarrow r) + (B\leftrightarrow D, \nu\leftrightarrow\rho, q\leftrightarrow s)\Big] \\
&\qquad + \frac{g_0^2}{12}a^4\Big\{\frac{2}{3}(\delta_{AB}\delta_{CD} + \delta_{AC}\delta_{BD} + \delta_{AD}\delta_{BC}) \\
&\qquad + \sum_E(d_{ABE}d_{CDE} + d_{ACE}d_{BDE} + d_{ADE}d_{BCE})\Big\} \\
&\qquad \times\Big\{\delta_{\mu\nu}\delta_{\mu\lambda}\delta_{\mu\rho}\sum_\sigma\tilde{k}_\sigma\,\tilde{q}_\sigma\tilde{r}_\sigma\tilde{s}_\sigma - \delta_{\mu\nu}\delta_{\mu\lambda}\tilde{k}_\rho\tilde{q}_\rho\tilde{r}_\rho\tilde{s}_\mu \\
&\qquad - \delta_{\mu\nu}\delta_{\mu\rho}\tilde{k}_\lambda\tilde{q}_\lambda\tilde{s}_\lambda\tilde{r}_\mu - \delta_{\mu\lambda}\delta_{\mu\rho}\tilde{k}_\nu\tilde{r}_\nu\tilde{s}_\nu\tilde{q}_\mu - \delta_{\nu\lambda}\delta_{\nu\rho}\tilde{q}_\mu\tilde{r}_\mu\tilde{s}_\mu\tilde{k}_\nu \\
&\qquad + \delta_{\mu\nu}\delta_{\lambda\rho}\tilde{k}_\lambda\tilde{q}_\lambda\tilde{r}_\mu\tilde{s}_\mu + \delta_{\mu\lambda}\delta_{\nu\rho}\tilde{k}_\nu\tilde{r}_\nu\tilde{q}_\mu\tilde{s}_\mu + \delta_{\mu\rho}\delta_{\nu\lambda}\tilde{k}_\nu\tilde{s}_\nu\tilde{q}_\mu\tilde{r}_\mu\Big\}
\end{aligned}
\tag{15.53b}
$$

This concludes our discussion of the pure gluonic sector. We now proceed to the analysis of the fermionic and ghost sectors.

iv) *Fermionic and ghost contributions*

Consider first the contribution to (15.22b) arising from $S_F^{(W)}[A,\hat{\psi},\hat{\bar{\psi}}]$. For Wilson fermions it is given by (6.4), where the link variables are replaced by (15.1a). Expanding these in powers of ϕ_μ, and introducing the dimensional fermion fields and gauge potentials according to (4.3a) and (15.36), one obtains up to

* This is a corrected version of the expression given by Kawai et al. (1981).

second order in the gauge potentials $A_\mu^B(x)$:

$$S_F[\psi,\bar\psi,A] = S_F^{(0)}[\psi,\bar\psi] + S_F^{(1)}[A,\psi,\bar\psi] + S_F^{(2)}[A,\psi,\bar\psi], \tag{15.54a}$$

where

$$\begin{aligned} S_F^{(0)}[\psi,\bar\psi] = &\left(M + \frac{4r}{a}\right)\sum_x \bar\psi^a(x)\psi^a(x) \\ &- \frac{1}{2a}\sum_{x,\mu}\left[\bar\psi^a(x)(r-\gamma_\mu)\psi^a(x+a\hat\mu) + \bar\psi^a(x+a\hat\mu)(r+\gamma_\mu)\psi^a(x)\right], \end{aligned} \tag{15.54b}$$

$$\begin{aligned} S_F^{(1)}[A,\psi,\bar\psi] = &-\frac{ig_0}{2}\sum_{x,\mu} T_{ab}^B\left[\bar\psi^a(x)(r-\gamma_\mu)\psi^b(x+a\hat\mu)\right. \\ &\left.-\bar\psi^a(x+a\hat\mu)(r+\gamma_\mu)\psi^b(x)\right] A_\mu^B(x), \end{aligned} \tag{15.54c}$$

$$\begin{aligned} S_F^{(2)}[A,\psi,\bar\psi] = &\frac{g_0^2}{2!}\frac{a}{4}\sum_{x,\mu,\nu}\{T^B,T^C\}_{ab}\delta_{\mu\nu} \\ &\times\left[\bar\psi^a(x)(r-\gamma_\mu)\psi^b(x+a\hat\mu) + \bar\psi^a(x+a\hat\mu)(r+\gamma_\mu)\psi^b(x)\right] A_\mu^B(x)A_\nu^C(x). \end{aligned} \tag{15.54d}$$

A summation over the "quark" and gluon colour indices is understood*. Apart from some group theoretical factors, and the fact that the fields now carry colour indices, the structure of $S_F[A,\psi,\bar\psi]$ is quite similar to that discussed in the abelian U(1) case. Hence except for some obvious modifications, the structure of the fermion propagator and gluon–fermion vertices will also be the same as those obtained in the abelian case.

Finally, consider the contribution of the ghost fields to the action. It is given by (15.21b), where $\hat D_\mu[\phi]$ has been defined in (15.19b). By expanding the matrix $M^{-1}(\phi_\mu)$ up to terms linear in ϕ_μ (cf. eq. (15.19c)), we include lattice artefacts vanishing linearly with a in the naive continuum limit. Next we introduce the dimensioned Faddeev–Popov ghost fields $c^A(x)$ and $\bar c^A(x)$ according to **

$$\begin{aligned} \hat c^A(n) &\to ac^A(x), \\ \bar{\hat c}^A(n) &\to a\bar c^A(x). \end{aligned}$$

---

* Recall that small latin letters run from one to three, while capital letters run from one to eight.

** They carry the same dimension as the scalar field discussed in chapter 3.

Then (15.21b) becomes

$$\begin{aligned}S_{FP}[A,c,\bar{c}] = &-\sum_x \bar{c}^A(x)\delta_{AB}\Box c^B(x)\\ &-g_0\sum_x f_{ABC}\bar{c}^A(x)\partial^L_\mu[A^C_\mu(x)(1+\frac{a}{2}\partial^R_\mu)c^B(x)]\\ &-\frac{1}{2!}\frac{g_0^2a^2}{12}\sum_x \delta_{\mu\nu}\{t^C,t^D\}_{AB}(\partial^R_\mu\bar{c}^A(x))(\partial^R_\mu c^B(x))A^C_\mu(x)A^D_\nu(x)+\ldots,\end{aligned} \tag{15.55}$$

where we have made use of (15.8). By Fourier decomposing the quark and ghost fields in a way analogous to (14.17), one readily derives from (15.54b-d) the propagators and interaction vertices in momentum space. Except for some colour matrices, the quark propagator and the gluon–quark interaction vertices have the same structure as the corresponding expressions in the U(1)–gauge theory. Furthermore, from the quadratic contribution to (15.55) we see immediately that the ghost propagator is given by $\delta_{AB}/\tilde{k}^2$. The ghost–gluon interaction vertices in momentum space can also be read off immediately from (15.55) by making use of the property (15.42). Below we summarize the lattice propagators and vertices.

i) *Propagators and vertices which possess a non–vanishing continuum limit**

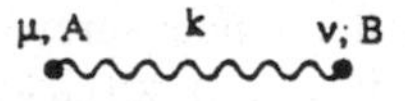

μ, A — k — ν, B

$$\frac{1}{\tilde{k}^2}\left(\delta_{\mu\nu}-(1-\alpha_0)\frac{\tilde{k}_\mu\tilde{k}_\nu}{\tilde{k}^2}\right)\delta_{AB}$$

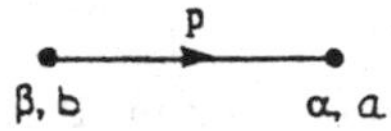

β, b — p — α, a

$$\left(\frac{1}{i\sum_\mu \frac{1}{a}\gamma_\mu \sin p_\mu a + M(p)}\right)_{\alpha\beta}\delta_{ab}$$

A — k — B

$$\frac{1}{\tilde{k}^2}\delta_{AB}$$

* For the definition of $M(p)$ confer eq. (4.29b).

p, β, b　　p', α, a
k, μ, A

$$-ig_0(2\pi)^4\delta_P^{(4)}(k+p-p')\left[\gamma_\mu\cos\left(\frac{(p+p')_\mu a}{2}\right)-ir\sin\left(\frac{(p+p')_\mu a}{2}\right)\right]_{\alpha\beta}T^A_{ab}$$

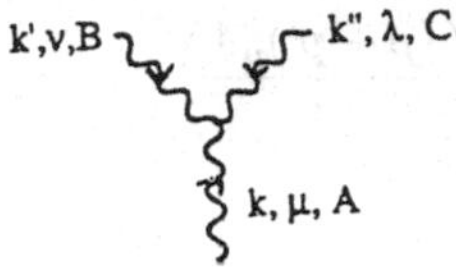

k',ν,B　　k'', λ, C
k, μ, A

$$ig_0(2\pi)^4\delta_P^{(4)}(k+k'+k'')f_{ABC}[\delta_{\nu\lambda}(\widetilde{k''-k'})_\mu\cos\frac{1}{2}k_\nu a$$
$$+\delta_{\mu\lambda}(\widetilde{k-k''})_\nu\cos\frac{1}{2}k'_\lambda a+\delta_{\mu\nu}(\widetilde{k'-k})_\lambda\cos\frac{1}{2}k''_\mu a]$$

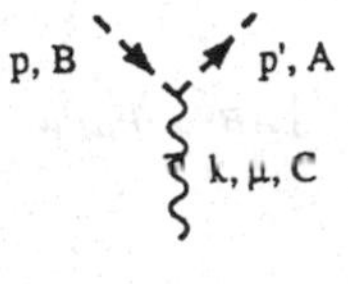

p, B　　p', A
k, μ, C

$$ig_0(2\pi)^4\delta_P^{(4)}(k+p-p')f_{ABC}\tilde{p'}_\mu\cos(p_\mu a/2)$$

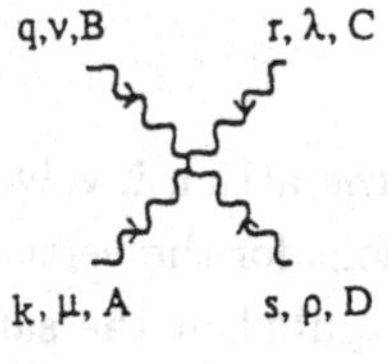

q,ν,B　　r, λ, C
k, μ, A　　s, ρ, D

(see eq. (15.53b))

ii) *Vertices which have no continuum analog*

μ, B ν, C

$$-(2\pi)^4\delta_P^{(4)}(k+k')\frac{g_0^2}{4a^2}\delta_{\mu\nu}\delta_{BC}$$

k′,ν,B p′, α, a

k, μ, A p, β, b

$$-\frac{a}{2}g_0^2(2\pi)^4\delta_P^{(4)}(k+k'+p-p')\delta_{\mu\nu}\{T^A,T^B\}_{ab}$$
$$\cdot\left[r\cos\left(\frac{(p+p')_\mu a}{2}\right)-i\gamma_\mu\sin\left(\frac{(p+p')_\mu a}{2}\right)\right]_{\alpha\beta}$$

p, B p′, A

k, μ, C k′,ν, D

$$\frac{1}{12}g_0^2a^2(2\pi)^4\delta_P^{(4)}(k+k'+p-p')\{t^C,t^D\}_{AB}\delta_{\mu\nu}\tilde{p'_\mu}\tilde{p_\mu}$$

Note that the factor $\frac{1}{n!}$ multiplying the contributions to the action involving couplings with n-gluons have not been included in the expressions for the vertices. The symmetry factor multiplying a Feynman integral is computed in the same way familiar from continuum perturbation theory.

In contradistinction to continuum perturbation theory, there are, in general, more diagrams to be considered when calculating a vertex function in a given order of the coupling. Thus consider for example the diagrams contributing to the gluon self energy in one–loop order. They are depicted in fig. (15-1). Their contributions have been calculated by Kawai, Nakayama and Seo (1981).

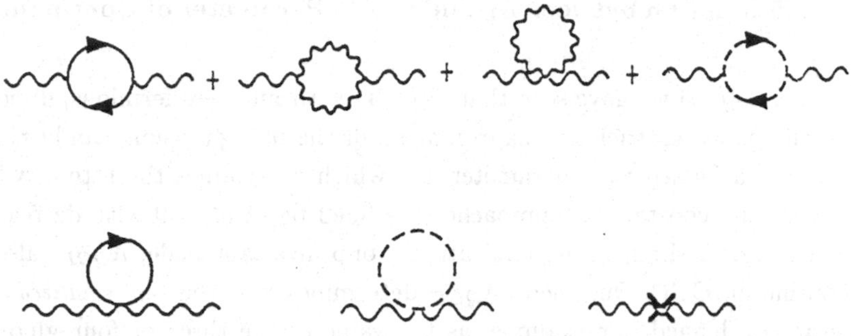

Fig. (15-1) Diagrams contributing to the gluon self- energy in one loop order

Each diagram contributes a quadratically divergent mass term. But when the graphs are summed these divergencies are found to cancel! There is another remarkable cancellation that occurs. When performing the calculation one encounters non–covariant terms of the type $p_\mu^2 \delta_{\mu\nu}$. But after summing all the contributions, these terms cancel out and one is left with a transverse expression for the gluon self energy, reflecting the gauge invariance of the theory!

Because of the complexity of lattice QCD Feynman rules (note that we have only expanded the action up to $\mathcal{O}(g_0^2)$), and the periodic structure of the integrands of Feynman integrals, perturbative calculations of more than one loop contributions become prohibitively difficult. In this connection the power counting theorem of Reisz discussed in chapter 13 is of great help, for it allows one at least to take the naive continuum limit in those cases where Feynman integrals satisfy the conditions for which the theorem applies.

In the following two sections we will apply the knowledge we have acquired so far to the computation of two important quantites: the ratio of two renormalization group invariant scales, and the ABJ anomaly of the axial vector current within the framework of QCD. As you shall see, the structure of this anomaly is, under very general conditions on the action, independent of the lattice regularization. What concerns the first mentioned quantity, we will not dwell on any technical details, but merely discuss the problem on a qualitative level.

15.5 Relation between Λ_L and the Λ–Parameter of Continuum QCD

In chapter 9 we have seen that in QCD with massless fermions, dimensioned physical quantities, such as a hadron mass, or the string tension, can be calculated in units of a lattice scale parameter Λ_L, which determines the rate at which the bare coupling constant g_0 approaches the fixed point $g_0^* = 0$ with decreasing lattice spacing. A similar renormalization group invariant scale, Λ_{QCD}, also occurs in continuum QCD. But there, Λ_{QCD} determines how the *renormalized* coupling constant g, defined, for example, as the value of the three or four–gluon vertex function at some momentum scale μ, changes with μ. The connection between g and μ, which ensures that physics does not depend on the choice of the renormalization point μ, can be obtained by studying the response of g to an infinitesimal change in μ. This response is measured by the following β–function

$$\beta(g) = \mu \frac{\partial g}{\partial \mu}. \tag{15.56}$$

In two–loop order this β–function is given by

$$\beta = -\beta_0 g^3 - \beta_1 g^5 + \ldots, \tag{15.57}$$

where β_0 and β_1 have the *same* values as those appearing in the two–loop expansion of the β–function considered in chapter 9 (cf. equation (9.21b)). Because $\beta_0 > 0$ the renormalized coupling constant is driven to $g \to g^* = 0$ as $\mu \to \infty$. This is the statement of asymptotic freedom. In the one loop approximation to the β–function, given by the first term on the right–hand side of (15.57), integration of (15.56) leads to the following relation between g and μ:

$$\frac{1}{\mu} = \frac{1}{\Lambda_{QCD}} e^{-\frac{1}{2\beta_0 g^2}}. \tag{15.58}$$

The value of the integration constant, Λ_{QCD}, depends on the definition of the renormalized coupling constant. Because of the asymptotic freedom property of QCD this scale can be measured in deep inelastic scattering processes, where the short distance dynamics can be described by renormalization group improved perturbation theory. Its value is found to be of the order of 200 MeV. On the other hand we have seen in chapter 9 that the connection between the bare coupling g_0 and the lattice spacing a, which ensures that physical observables remain unchanged as we remove the lattice structure, is given in the one–loop approximation to the β–function by

$$a = \frac{1}{\Lambda_L} e^{-\frac{1}{2\beta_0 g_0^2}}. \tag{15.59}$$

The lattice scale Λ_L, determined, for example, from a Monte Carlo calculation of the string tension is of the order of a few MeV. To confirm that QCD involves only a single scale which describes the large distance physics at quark separations of the order of 1 fm, as well as the short distance physics taking place at separations of 0.1 fm or less, one must check whether Λ_L, as determined from a measurement of, say, the string tension, corresponds to the value of Λ_{QCD} as obtained from deep inelastic scattering data.

The first calculation relating the two scales has been performed by Hasenfratz and Hasenfratz (1980) in the pure gauge theory. This calculation has been subsequently extended to the case of QCD with massless quarks by Kawai, Nakayama and Seo (1981). The basic idea underlying these computations is the following.

Suppose we calculate an observable O in QCD with massless fermions. In the lattice regulated theory the value of this observable depends on the bare coupling constant g_0 and the lattice spacing a:*

$$O = O(g_0, a, ...). \tag{15.60}$$

Now for sufficiently small lattice spacing, g_0 can be tuned to a in such a way that O remains fixed as we remove the lattice structure. In the two–loop approximation to the β–function this dependence is given by (9.21c,d). Hence g_0 is a function of the product $a\Lambda_L$: $g_0 = g_0(a\Lambda_L)$. On the other hand, we can also eliminate the cut–off dependence by introducing a renormalized coupling constant g in the way familiar from continuum perturbation theory. This coupling constant depends on g_0, the renormalization scale μ, and the cut–off a. Since g is dimensionless, it will depend on μ and a only through the product μa:

$$g = g(g_0, \mu a). \tag{15.61}$$

Solving this equation for g_0, one obtains the bare coupling constant as a function of g and μa:

$$g_0 = g_0(g, \mu a). \tag{15.62}$$

Upon inserting (15.62) into (15.60) one arrives at an expression in which the dependence on the lattice spacing has again been eliminated

$$O(g_0(g, \mu a), a; ...) \simeq O(g, \mu; ...). \tag{15.63}$$

---

* The "dots" stand for possible dependences on kinematical variables such as momenta; for example, O could be a scattering matrix element.

Now g can also be tuned to μ in such a way that the right–hand side of (15.63) remains fixed as we change the renormalization point μ. In the one–loop approximation to the β–function (15.56) the dependence of g on μ is given by (15.58); hence $g = g(\mu/\Lambda_{QCD})$. But both, the dependence of g_0 on $a\Lambda_L$, and of the renormalized coupling constant on μ/Λ_{QCD}, are determined from a single equation, namely from (15.61), or its inverse (15.62). Thus holding g_0 and a fixed, we determine the μ dependence of g. Alternatively, holding g and μ fixed we determine the a–dependence of g_0. This shows that Λ_L and Λ_{QCD} must be related. To obtain this relation one calculates the right–hand side of (15.61) in perturbation theory. In the one–loop approximation one obtains an expression of the form

$$g^2 = g_0^2 - 2\beta_0 g_0^4 \ln(\mu a/c) + O(g_0^6), \tag{15.64}$$

where c is a constant. From here one readily verifies the one–loop expressions for the β–functions (9.6b) and (15.56). Now to this order Λ_{QCD}/Λ_L, as determined from (15.58) and (15.59) is given by

$$\frac{\Lambda_{QCD}}{\Lambda_L} = \mu a \, e^{\frac{-1}{2\beta_0}\left(\frac{1}{g^2} - \frac{1}{g_0^2}\right)} .$$

But from (15.64) it follows that

$$\frac{1}{g^2} - \frac{1}{g_0^2} = 2\beta_0 \ln(\mu a/c) + O(g_0^2).$$

We therefore conclude that

$$\frac{\Lambda_{QCD}}{\Lambda_L} = c$$

Hence the purpose of a perturbative computation consists in calculating the constant c in (15.64). In the momentum subtraction scheme (MOM), Hasenfratz and Hasenfratz (1980) have calculated the ratio Λ_{MOM}/Λ_L for the pure SU(3) gauge theory. After a lengthy calculation they found that

$$\Lambda_{MOM}/\Lambda_L = 83.5 \qquad \text{(pure SU(3))}.$$

In full QCD with massless quarks this value was found to change as follows (Kawai et al., 1981)

$$\Lambda_{MOM}/\Lambda_L = 105.7 \qquad \text{(3 flavours)}$$

$$\Lambda_{MOM}/\Lambda_L = 117.0 \qquad \text{(4 flavours)}$$

These ratios are consistent with those determined from non–perturbative lattice calculations (Λ_L) and from the deep inelastic scattering data (Λ_{MOM}).

15.6 Universality of the Axial Anomaly in Lattice QCD

In this section we study the ABJ anomaly in the divergence of the colour singlet axial vector current in QCD, and show that not only for Wilson or Ginsparg-Wilson fermions the well known anomaly is reproduced in the continuum limit, but that the same result is obtained for any discretization of the action satisfying some very general conditions. In the case of a $U(1)$ gauge theory, this has been first shown to be the case by Reisz and Rothe (1999). Here we discuss the case of $SU(3)$ which has been studied subsequently by Frewer and Rothe (2001). From the analysis it will be evident that the same proof goes through for any $SU(N)$ gauge theory. The main steps we will follow are the following: we first discuss the general form of the lattice axial vector Ward identity. As you will see, its precise structure, which depends on the particular discretization of the lattice action, need not be known to compute the anomaly. Only very general properties thereof are required. In fact, the entire ambiguity in the lattice Ward identity, arising from different discretizations, will reside in a contribution which vanishes in the naive continuum limit. Although its structure depends on the way one has discretized the action, we will only make use of quite general properties thereof to generate the anomaly.

General Structure of the Ward Identity

In the following all expressions will be written in terms of dimensioned variables. We are interested in computing the anomalous contribution to the divergence of the colour singlet axial vector current $j_{\mu 5}(x) = \bar{\psi}(x)\gamma_5\gamma_\mu\psi(x)$ in an external gauge field, where $\psi(x)$ are 3-component fields in colour space.

Consider the fermionic contribution to the lattice action for QCD. It is of the form (4.61), i.e.

$$S_{ferm} = \sum_{x,y} \bar{\psi}(x)(D_U(x,y) + m\delta_{xy})\psi(y) \ , \tag{15.65}$$

where $D_U(x,y)$ is the Dirac operator (a matrix in Dirac-spin and colour space) depending on the matrix valued link variables which we denote collectively by U. As always $\sum_x = \sum_n a^4$, where n labels the lattice sites. $SU(3)$ Colour indices will be denoted in the following by small latin letters. The action is assumed to be gauge invariant, and to possess the discrete symmetries of the continuum theory. * The Dirac operator, which is a function of the link variables, can be expanded

* The Dirac operator can always be decomposed into a chirally symmetric, and

in the gauge potentials in the form

$$D_U(x,y) = \sum_{n,\mu_i,a_i,x_i} \frac{1}{n!} D^{(n)a_1\cdots a_n}_{\mu_1\cdots\mu_n}(x,y|x_1\cdots x_n) A^{a_1}_{\mu_1}(x_1)\cdots A^{a_n}_{\mu_n}(x_n) \ . \quad (15.66)$$

The next step consists in deriving a lattice Ward identity for the divergence of the singlet axial vector current. This is achieved by performing in the partition function the infinitessimal colour blind local axial transformation of the fermion fields analogous to (14.38). Note that, since we do not specify the Dirac operator, we have no other alternative to define a sensible axial transformation. Since the measure is invariant under this transformation one is led again to the statement (14.88). In the case of Wilson fermions the variation δS_{ferm} is given by (14.81), where $\bar{\Delta}(x)$ is an irrelevant operator vanishing in the continuum limit. For different lattice discretizations, $j_{5\mu}(x)$ and Δ will differ from (14.41) and (14.46) by terms which vanish in the naive continuum limit. What concerns the axial anomaly, however, the precise form of the various terms in (14.81) need not be known. This is quite remarkable. In fact, as we shall see, any lattice discretization of the action S with the following properties:

i) S has the correct continuum limit
ii) S is gauge invariant
iii) The Dirac operator is local
iv) Absence of species doubling

reproduces the axial anomaly in the continuum limit. This anomaly arises from an "irrelevant" (Δ) term in the Ward identity, which, in view of what has been said above, will necessarily have the form

$$< \partial^*_\mu j_{5\mu}(x) >_U = 2m < j_5(x) >_U + < \Delta >_U \ , \quad (15.67)$$

where $j_{5\mu}(x)$ and $j_5(x)$ possess the correct continuum limit. Thus for Wilson fermions, the Ward identity is of the above form (Rothe, 1998). Any other discretization of the action will differ only by lattice artefacts which can, in principle, be absorbed into the Δ-term. The reader may ask: is there no way of avoiding such a term? The answer is "no" as will be clear from our analysis.

---

a chiral symmetry breaking (sb) part as follows, $D(x,y) = D(x,y)_{sym} + D(x,y)_{sb}$, where $D(x,y)_{sym} = \frac{1}{2}[D,\gamma_5]\gamma_5$, and $D(x,y)_{sb} = \frac{1}{2}\{D,\gamma_5\}\gamma_5$. Any candidate for a lattice action should possess the correct continuum limit. It therefore follows that for $a \to 0$, $D(x,y)_{sym} \to \gamma_\mu D_\mu[A]$, where $D_\mu[A]$ is the covariant derivative, while $D(x,y)_{sb}$ vanishes in the continuum limit. It therefore also follows that for $a \to 0$ $D(x,y) \to \gamma_\mu D[A]$, where $D_\mu[A]$ is the covariant derivative.

Before proving the above assertion it is convenient to rewrite the Ward identity (15.67) in terms of correlators in momentum space. Let $\mathcal{O}(x)$ stand for any of the operators appearing in (15.67). Then $< \mathcal{O}(x) >_U$ has the following formal expansion in the gauge potentials

$$< \mathcal{O}(x) >_U = \sum_{n\geq 2} \frac{1}{n!} \sum_{\{x_i\},\{\mu_i\},\{a_i\}} \Gamma^{(\mathcal{O})a_1\cdots a_n}_{\mu_1\cdots\mu_n}(x|x_1,x_2,\cdots,x_n) A^{a_1}_{\mu_1}(x_1)\cdots A^{a_n}_{\mu_n}(x_n) \,, \tag{15.68}$$

where, because of the assumed symmetries of the action, the sum over n starts with $n = 2$. The correlation functions $\Gamma^{(\mathcal{O})a_1\cdots a_n}_{\mu_1\cdots\mu_n}(x|x_1,x_2,\cdots,x_n)$ are symmetric under the exchange of any pair of collective labels (x_i,μ_i,a_i). Defining the Fourier transform of $\Gamma^{(\mathcal{O})a_1\cdots a_n}_{\mu_1\cdots\mu_n}(x|x_1,x_2,\cdots,x_n)$ by

$$\Gamma^{(\mathcal{O})a_1\cdots a_n}_{\mu_1\cdots\mu_n}(x|x_1,\cdots,x_n) = \int_{-\frac{\pi}{a}}^{\frac{\pi}{a}} \frac{d^4q}{(2\pi)^4} e^{-iq\cdot x} \prod_{i=1}^{n} \frac{d^4k_i}{(2\pi)^4} e^{ik_i\cdot x_i} \hat{\Gamma}^{(\mathcal{O})a_1\cdots a_n}_{\mu_1\cdots\mu_n}(q|k_1,\cdots,k_n) \,, \tag{15.69a}$$

where, by translational invariance,

$$\hat{\Gamma}^{(\mathcal{O})a_1\cdots a_n}_{\mu_1\cdots\mu_n}(q|k_1,\cdots,k_n) = \delta(q - \sum_{i=1}^{n} k_i) \tilde{\Gamma}^{(\mathcal{O})a_1\cdots a_n}_{\mu_1\cdots\mu_n}(k_1,\cdots,k_n) \,, \tag{15.69b}$$

the Ward identity (15.67) translates as follows to momentum space,

$$\begin{aligned} -i\tilde{q}_\mu \tilde{\Gamma}^{a_1\cdots a_n}_{5\mu;\mu_1\cdots\mu_n}(k_1,\cdots,k_n) &= 2m\tilde{\Gamma}^{a_1\cdots a_n}_{5;\mu_1\cdots\mu_n}(k_1,\cdots,k_n) \\ &+ \tilde{\Gamma}^{(\Delta)a_1\cdots a_n}_{\mu_1\cdots\mu_n}(k_1,\cdots,k_n) \,, \end{aligned} \tag{15.70a}$$

where

$$\tilde{q}_\mu = e^{i\frac{q_\mu a}{2}} \frac{2}{a} \sin\frac{q_\mu a}{2} \,. \tag{15.70b}$$

As we shall see further below, gauge invariance implies that every term in (15.67) possesses a finite continuum limit. If in this limit the Δ-contribution is different from zero, we are faced with an anomaly, since the divergence of the axial vector current would not vanish in the chiral limit $m \to 0$ (as it would, if the axial symmetry would be implemented on quantum level). We now show that under the conditions i)-iv) this is indeed the case, and moreover, that this limit is universal.

Let us first study the implications of the assumptions i)-iv). Clearly the first assumption is a "must" and needs no further elaboration. Consider next assumption ii):

ii) Gauge invariance

Gauge invariance has strong implications. In fact the renormalizability of QED or QCD relies heavily on gauge invariance. Gauge invariance tells us that if $\mathcal{O}(A,\psi,\bar{\psi})$ is a gauge invariant operator, then its external field expectation value satisfies

$$F_{\mathcal{O}}[A^\omega] = F_{\mathcal{O}}[A]\ , \tag{15.71a}$$

where

$$F_{\mathcal{O}}[A] = \frac{\int D\psi D\bar{\psi}\ \mathcal{O}[A,\psi,\bar{\psi}]\ e^{-S[A,\psi,\bar{\psi}]}}{\int D\psi D\bar{\psi}\ e^{-S[A,\psi,\bar{\psi}]}}\ , \tag{15.71b}$$

with A^ω the gauge transformed potential. This can be readily verified by making use of the gauge invariance of the fermionic measure. On the lattice the variation of the gauge potentials induced by an infinitessimal gauge transformation is given by (15.19) with $\phi^b_\mu = gaA^b_\mu$, i.e., *

$$\delta A^a_\mu(x) = [gf_{abc}A^b_\mu(x) - M^{-1}_{ac}(gaA_\mu(x))\partial^R_\mu]\epsilon^c(x)\ , \tag{15.72}$$

where f_{abc} are the structure constants of $SU(3)$, ∂^R_μ is the dimensioned right lattice derivative, and the matrix M is given by (15.12b), or the expansion (15.19c). Because of the structure of the rhs of (15.68), the Γ's are symmetric functions under permutations of the labels $1,\cdots,n$. Hence the variation can be written in the form

$$\delta_\omega F_{\mathcal{O}}[A] = \sum_{n\geq 2}\frac{n}{n!}\sum_{\{x_i\},\{\mu_i\},\{a_i\}}\Gamma^{(\mathcal{O})a_1\cdots a_n}_{\mu_1\cdots\mu_n}(x|x_1,x_2,\cdots,x_n)\delta A^{a_1}_{\mu_1}(x_1)\cdots A^{a_n}_{\mu_n}(x_n)\ .$$

Inserting for $\delta A^{a_1}_{\mu_1}$ the expression (15.72), and considering in turn the coefficients of $\mathcal{O}(A^2)$ and $\mathcal{O}(A^3)$, one finds that (15.71a) implies that

$$\partial^L_{\mu_1}\Gamma^{(\mathcal{O})a_1a_2}_{\mu_1\mu_2}(x|x_1,x_2) = 0\ , \tag{15.73a}$$

and furthermore

$$\begin{aligned}\partial^L_{\mu_1}\Gamma^{(\mathcal{O})a_1a_2a_2}_{\mu_1\mu_2\mu_3}(x|x_1,x_2,x_3) &= g_0 f_{a_1a_2b}\Gamma^{(\mathcal{O})ba_3}_{\mu_2\mu_3}(x|x_2,x_3)\delta_{x_1x_2}\\ &\quad - \frac{1}{2!}g_0 a f_{a_1a_2b}\partial^L_{\mu_2}\left(\delta_{x_1x_2}\Gamma^{(\mathcal{O})ba_3}_{\mu_2\mu_3}(x|x_2,x_3)\right) + (2\leftrightarrow 3)\ .\end{aligned} \tag{15.73b}$$

These relations can be readily translated to momentum space. Defining the Fourier transform by (15.69), one finds that (15.73a) takes the form

$$(\tilde{k}^*_1)_{\mu_1}\tilde{\Gamma}^{(\mathcal{O})a_1a_2}_{\mu_1\mu_2}(k_1,k_2) = 0\ , \tag{15.74}$$

---

* In the following we prefer to use small letters for the colour indices.

where $*$ denotes complex conjugation, and where $\tilde{k}_\mu$ is defined in an analogous way to (15.70b). Because of Bose symmetry a corresponding statement holds for $(\tilde{k}_1)_{\mu_1}$ replaced by $(\tilde{k}_2)_{\mu_2}$. The second identity (15.73b) reads as follows in momentum space,

$$\begin{aligned} i(\tilde{k}_1^*)_{\mu_1}\tilde{\Gamma}^{(\mathcal{O})a_1a_2a_3}_{\mu_1\mu_2\mu_3}(k_1,k_2,k_3) &= gf_{a_1a_2c}\tilde{\Gamma}^{(\mathcal{O})ca_3}_{\mu_2\mu_3}(k_1+k_2,k_3) \\ &\quad - ia\frac{g}{2!}f_{a_1a_2c}(\tilde{k}_1^*)_{\mu_2}\tilde{\Gamma}^{(\mathcal{O})ca_3}_{\mu_2\mu_3}(k_1+k_2,k_3) \\ &\quad + (2\leftrightarrow 3)\ . \end{aligned} \tag{15.75}$$

Further relations connecting higher and lower order correlation functions follow from the requirement of gauge invariance. We will however not require them to calculate the anomaly.

iii) Locality

Gauge invariance alone will of course not allow us to compute the axial anomaly. But when combined with iii) and iv) (see p. 276) this will be possible. The point is that if iii) and iv) hold, then the correlation functions will be analytic in the momenta around vanishing momenta, and hence possess a Taylor expansion. This has important consequences. Thus consider (15.74). Analyticity around $\{k_i = 0\}$ tells us that

$$\tilde{\Gamma}^{(\mathcal{O})a_1a_2}_{\mu_1\mu_2}(k_1,k_2) = \tilde{\Gamma}^{(\mathcal{O})a_1a_2}_{\mu_1\mu_2}(0,0) + C^{(\mathcal{O})a_1a_2}_{\mu_1\mu_2\sigma}k_1^\sigma + \tilde{C}^{(\mathcal{O})a_1a_2}_{\mu_1\mu_2\sigma}k_2^\sigma + \cdots$$

Because of Bose symmetry under the exchange $k_1 \leftrightarrow k_2, \mu_1 \leftrightarrow \mu_2, a_1 \leftrightarrow a_2$ we must have that

$$C^{(\mathcal{O})a_1a_2}_{\mu_1\mu_2\sigma} = \tilde{C}^{(\mathcal{O})a_2a_1}_{\mu_2\mu_1\sigma}\ .$$

But since k_1 and k_2 are independent variables, it then follows from (15.74) that

$$\tilde{\Gamma}^{(\mathcal{O})a_1a_2}_{\mu_1\mu_2}(0,0) = 0\ ,$$

$$\tilde{C}^{(\mathcal{O})a_2a_1}_{\mu_2\mu_1\sigma} = 0\ ,$$

and therefore also

$$C^{(\mathcal{O})a_2a_1}_{\mu_2\mu_1\sigma} = 0\ .$$

Summarizing we therefore conclude that

$$T_1\tilde{\Gamma}^{(\mathcal{O})a_1a_2}_{\mu_1\mu_2}(k_1,k_2) = 0\ , \tag{17.76}$$

where T_1 denotes the Taylor expansion around zero momenta up to first order.

Consider next eq. (15.75) evaluated for $k_2 = k_3 = 0$, and small k_1. Because of (15.76) the rhs is of $\mathcal{O}(k_1^2)$. It therefore follows that

$$\tilde{\Gamma}^{(\mathcal{O})a_1a_2a_3}_{\mu_1\mu_2\mu_3}(0,0,0) = 0\ . \tag{15.77}$$

Relations (15.76) and (15.77) are weaker than those obtained from gauge invariance in the case of QED (Reisz. 1999), where an analogous statement to (15.76) holds for vertex functions involving an arbitrary number of gauge fields.

The reason that (15.76) and (15.77) play an important role for the computation of the anomaly, is that we shall derive an expression for the renormalized Ward identity in the continuum limit using the momentum-subtraction scheme with finite lattice spacing. We then have to show that the continuum limit of the potentially anomalous contribution to the divergence of the axial vector current is universal and given by the well known continuum expression. For this we will need to make use of the power counting theorem of Reisz (see section (13.3)), which requires not only that the lattice degree of divergence of all Zimmermann spaces of a Feynman integral is negative but also that the free fermion propagator is free of doublers. This is just our condition iv).

Universality of the ABJ Anomaly

We are now ready to prove our assertion. Our attention will be focused on the "irrelevant" contribution of the Δ-term in (15.67), which should generate the anomaly in the continuum limit. Because the expressions considered in the following are lattice regularized, all operations are well defined. Furthermore, since we are studying the Ward identity in an external background colour field, the only Feynman graphs which contribute are diagrams involving an arbitrary number of external gauge potentials attached to a fermion loop.

Consider the axial vector Ward identity (15.67). By power counting the ultraviolet lattice degree of divergence (LDD) of $\tilde{\Gamma}_{5\mu}$ and $\tilde{\Gamma}_5$ is given by $3-n$, where n is the number of external gluon fields, while the LDD of $\tilde{\Gamma}^{(\Delta)}$ is $4-n$. Hence the LDD of Feynman integrals contributing to $\Gamma^{(\Delta)}$ is negative for graphs involving more than 4 external gauge fields. Their contributions thus vanish by the Reisz theorem (see sec. 13.3) in the continuum limit, since Δ is an irrelevant operator. We therefore only need to consider the correlation functions $\tilde{\Gamma}^{(\Delta)a_1a_2}_{\mu_1\mu_2}(k_1,k_2)$,

$\tilde{\Gamma}^{(\Delta)a_1a_2a_3}_{\mu_1\mu_2\mu_3}(k_1,k_2,k_3)$, and $\tilde{\Gamma}^{(\Delta)a_1a_2a_3a_4}_{\mu_1\mu_2\mu_3\mu_4}(k_1,k_2,k_3,k_4)$ with $LDD = 2, 1$ and 0, respectively. Let us decompose these vertex functions as follows

$$\begin{aligned}\tilde{\Gamma}^{(\Delta)a_1a_2}_{\mu_1\mu_2}(k_1,k_2) &= (1-T_2)\tilde{\Gamma}^{(\Delta)a_1a_2}_{\mu_1\mu_2}(k_1,k_2)\\ &\quad + T_1\tilde{\Gamma}^{(\Delta)a_1a_2}_{\mu_1\mu_2}(k_1,k_2) + (T_2-T_1)\tilde{\Gamma}^{(\Delta)a_1a_2}_{\mu_1\mu_2}(k_1,k_2)\,,\end{aligned} \tag{15.78}$$

$$\begin{aligned}\tilde{\Gamma}^{(\Delta)a_1a_2a_3}_{\mu_1\mu_2\mu_3}(k_1,k_2,k_3) &= (1-T_1)\tilde{\Gamma}^{(\Delta)a_1a_2a_3}_{\mu_1\mu_2\mu_3}(k_1,k_2,k_3) + T_0\tilde{\Gamma}^{(\Delta)a_1a_2a_3}_{\mu_1\mu_2\mu_3}(k_1,k_2,k_3)\\ &\quad + (T_1-T_0)\tilde{\Gamma}^{(\Delta)a_1a_2a_3}_{\mu_1\mu_2\mu_3}(k_1,k_2,k_3)\,,\end{aligned} \tag{15.79}$$

and

$$\begin{aligned}\tilde{\Gamma}^{(\Delta)a_1a_2a_3a_4}_{\mu_1\mu_2\mu_3\mu_4}(k_1,k_2,k_3,k_4) &= (1-T_0)\tilde{\Gamma}^{(\Delta)a_1a_2a_3a_4}_{\mu_1\mu_2\mu_3\mu_4}(k_1,k_2,k_3,k_4)\\ &\quad + T_0\tilde{\Gamma}^{(\Delta)a_1a_2a_3a_4}_{\mu_1\mu_2\mu_3\mu_4}(k_1,k_2,k_3,k_4)\,,\end{aligned} \tag{15.80}$$

where T_n denotes the Taylor expansion around vanishing momenta up to n'th order. By the Reisz theorem the first term appearing on the rhs of (15.78) to (15.80), vanishes in the continuum limit, since it has negative LDD and Δ is an irrelevant operator. The respective second terms in (15.78) and (15.79) vanish by gauge invariance (cf. eqs (15.76) and (15.77)). Hence we conclude that

$$\lim_{a\to 0}\tilde{\Gamma}^{(\Delta)a_1a_2}_{\mu_1\mu_2}(k_1,k_2) = \lim_{a\to 0}(T_2-T_1)\tilde{\Gamma}^{(\Delta)a_1a_2}_{\mu_1\mu_2}(k_1,k_2)\,, \tag{15.81a}$$

$$\lim_{a\to 0}\tilde{\Gamma}^{(\Delta)a_1a_2a_3}_{\mu_1\mu_2\mu_3}(k_1,k_2,k_3) = \lim_{a\to 0}(T_1-T_0)\tilde{\Gamma}^{(\Delta)a_1a_2a_3}_{\mu_1\mu_2\mu_3}(k_1,k_2,k_3)\,, \tag{15.81b}$$

$$\lim_{a\to 0}\tilde{\Gamma}^{(\Delta)a_1a_2a_3a_4}_{\mu_1\mu_2\mu_3\mu_4}(k_1,k_2,k_3,k_4) = \lim_{a\to 0}T_0\tilde{\Gamma}^{(\Delta)a_1a_2a_3a_4}_{\mu_1\mu_2\mu_3\mu_4}(k_1,k_2,k_3,k_4)\,. \tag{15.81c}$$

Hence the contributions (15.81a) to (15.81c) are of second, first and zeroth order in the gluon momenta, respectively. As we now show, the Ward identity allows us to calculate the limits from the corresponding Taylor terms of continuum one-loop-fermion diagrams involving two and one external gluon fields, with a γ_5 insertion. It then follows that the anomalous contribution is necessarily universal, i.e., it does not depend on the particular form of the irrelevant term in the lattice Ward identity.

Consider first the rhs of (15.78). Applying the operation $(T_2 - T_1)$ to the (lattice regularized) axial vector Ward identity (15.70) with $n = 2$, and making use of (15.74), following from gauge invariance, we conclude that

$$(T_2-T_1)\tilde{\Gamma}^{(\Delta)a_1a_2}_{\mu_1\mu_2}(k_1,k_2) = -2m(T_2-T_1)\tilde{\Gamma}^{a_1a_2}_{5;\mu_1\mu_2}(k_1,k_2)\,. \tag{15.82}$$

The expression appearing on the rhs has negative LDD; hence its continuum limit is given, according to the Reisz theorem, by applying the $(T_2 - T_1)$ operation to the integrand of the corresponding continuum Feynman integral, i.e. the triangle diagram. One then readily finds that (15.81a) is given by

$$\lim_{a\to 0}\tilde{\Gamma}^{(\Delta)a_1a_2}_{\mu_1\mu_2}(k_1,k_2) = -\frac{g^2}{4\pi^2}\delta_{a_1a_2}\epsilon_{\mu_1\mu_2\sigma\rho}(k_1)_\sigma(k_2)_\rho\ . \tag{15.83}$$

Consider next the rhs of (15.81b). Applying the $(T_1 - T_0)$ operation to the Ward identity (15.70) with $n = 3$, and making use of (15.77) (following again from gauge invariance) one obtains

$$(T_1 - T_0)\tilde{\Gamma}^{(\Delta)a_1a_2a_3}_{\mu_1\mu_2\mu_3}(k_1,k_2,k_3) = -2m(T_1 - T_0)\tilde{\Gamma}^{a_1a_2a_3}_{5;\mu_1\mu_2\mu_3}(k_1,k_2,k_3)\ . \tag{15.84}$$

Since the LDD of the rhs is negative, its continuum limit is given by applying the $(T_1 - T_0)$ operation to the integrand of the continuum box Feynman diagram, involving three vector vertices and a γ_5 insertion. After some lengthy but straight forward calculation one finds that

$$\lim_{a\to 0}\tilde{\Gamma}^{(\Delta)a_1a_2a_3}_{\mu_1\mu_2\mu_3}(k_1,k_2,k_3) = i\frac{g^3}{4\pi^2}\epsilon_{\mu_1\mu_2\mu_3\sigma}f_{a_1a_2a_3}q_\sigma\ , \tag{15.85}$$

where $q = k_1 + k_2 + k_3$. Finally, consider (15.81c). Applying T_0 to the Ward identity (15.70) we obtain

$$\lim_{a\to 0} T_0\tilde{\Gamma}^{(\Delta)a_1a_2a_3a_4}_{\mu_1\mu_2\mu_3\mu_4}(k_1,k_2,k_3,k_4) = -2m\lim_{a\to 0} T_0\tilde{\Gamma}^{a_1a_2a_3a_4}_{5;\mu_1\mu_2\mu_3\mu_4}(k_1,k_2,k_3,k_4)\ . \tag{15.86}$$

Since the LDD of the vertex function on the rhs is again negative, the limit $a \to 0$ can be calculated by applying T_0 to the continuum pentagon graph involving a γ_5 insertion. A simple calculation shows that (15.81c) is given by

$$\lim_{a\to 0}\tilde{\Gamma}^{(\Delta)a_1a_2a_3a_4}_{\mu_1\mu_2\mu_3\mu_4}(k_1,k_2,k_3,k_4) = \frac{g^4}{4\pi^2}[\epsilon_{\mu_1\mu_2\mu_3\mu_4}Tr(T^{a_1}T^{a_2}T^{a_3}T^{a_4}) + perm.]\ , \tag{15.87}$$

where T^a are the $SU(3)$ generators in the fundamental representation. Notice that we have made constant use of the Ward Identity (15.70) which we know is true for arbitrary finite lattice spacing.

The (continuum) correlation functions in coordinate space, corresponding to (15.83), (15.85) and (15.87), are obtained by performing the Fourier transformation (15.69), where the limits of integration now extend to infinity. The anomalous contribution is then given by (15.68) with $\mathcal{O} \to \Delta$, where the sum over the coordinates are replaced by integrals. Only the two and three gluon vertex functions actually contribute to the anomaly, as can be shown by making use of the Jacobi identity. One then easily verifies that the anomalous contribution to the (euclidean) axial vector Ward identity takes the well known form

$$\mathcal{A}(x) = -\frac{g^2}{32\pi^2}\epsilon_{\mu\nu\lambda\rho}F^a_{\mu\nu}(x)F^a_{\lambda\rho}(x) , \tag{15.88a}$$

where

$$F^a_{\mu\nu}(x) = \partial_\mu A^a_\nu(x) - \partial_\nu A^a_\mu(x) - g f_{abc}A^b_\mu(x)A^c_\nu(x) \tag{15.88b}$$

is the non-abelian field strength tensor.

We have seen above that the Δ-term in (15.67), although superficially linearly divergent in the continuum limit, is actually finite in this limit, and a 4'th order polynomial in gluon field. The other two terms in this gauge identity are also finite. The reason is, that the latter have LDD $= 1$ and 0 for $n = 2$ and $n = 3$, respectively, and negative LDD for $n \geq 4$. Gauge invariance and locality of the Dirac operator, as embodied in the statements (15.76) and (15.77), allows us to replace the $n = 2$ and $n = 3$ vertex functions by the corresponding Taylor subtracted forms with negative LDD. These possess a finite continuum limit.

This completes our discussion of lattice perturbation theory. All analytic methods discussed in the last five chapters have their own merits, but they cannot be used to calculate non–perturbative observables in continuum QCD. One is therefore forced to compute these observables numerically. In the following chapter we now introduce the reader to various numerical methods that have been proposed in the literature.

CHAPTER 16

MONTE CARLO METHODS

16.1 Introduction

When one computes the expectation value of an observable in lattice QCD, one is faced with the problem of having to perform a tremendously large number of integrations. Thus suppose we want to compute the following ensemble average

$$\langle O \rangle = \frac{\int DUO[U]e^{-S[U]}}{\int DUe^{-S[U]}} \tag{16.1}$$

where $S[U]$ is a real functional of the link variables bounded from below. Using a 10^4 space–time lattice, the number of link variables is approximately 4×10^4. For the case of $SU(3)$, each of these link variables is a function of 8 real parameters; hence there are 320000 integrations to be done. Using a mesh of only 10 points per integration, it follows that the multiple integral will be approximated by a sum of 10^{320000} terms. This example shows that we must use statistical methods to evaluate the ensemble average (16.1). In fact, since most of the link configurations will have an action which is very large, only a small fraction of them will make a significant contribution to the integral (16.1). Hence an efficient way of computing the ensemble average would consist in generating a sequence of link variable configurations with a probability distribution given by the Boltzmann factor $\exp(-S[U])$. This is the technique of "importance sampling". If the sequence generated constitutes a representative set of configurations, then the ensemble average $\langle O \rangle$ will be approximately given by the following sum

$$\langle O \rangle \approx \frac{1}{N} \sum_{i=1}^{N} O(\{U\}_i) \tag{16.2}$$

where $\{U\}_i$ $(i = 1, \ldots, N)$ denote the link configurations generated.

For a few systems there is a simple method of generating field configurations on the lattice with the desired probability. This method is called the heat–bath algorithm. Let us demonstrate the basic idea of this algorithm by considering a one–dimensional integral

$$\langle F \rangle = \int_a^b dx F(x) P(x) \tag{16.3a}$$

where $P(x)$ is a normalized probability distribution

$$\int_a^b dx P(x) = 1 \ . \tag{16.3b}$$

Making the change of variable

$$y = \int_a^x dz P(z) \ , \tag{16.4}$$

(16.3) can be written in the form

$$\langle F \rangle = \int_0^1 dy \hat{F}(y) \ ,$$

where

$$\hat{F}(y) = F(x(y)) \ .$$

Hence the variable (16.4) is now distributed uniformly in the interval [0,1], and we can generate for y a representative set of values with the help of a random number generator. The crux of the whole matter is, of course, that one must be able to evaluate analytically the integral (16.4) and to obtain $x(y)$.

In praxis the right–hand side of (16.3a) is replaced by a multiple integral over the degrees of freedom of the physical system of interest. In this case one usually applies the above method to each variable in turn, holding all the others fixed; i.e., each variable is brought to equilibrium with a “heat bath reservoir”. For this reason it is called the “heat bath algorithm”. Unfortunately, the method can be applied only to a few systems of physical interest (for example, to the pure SU(2) gauge theory). We shall therefore not dwell on it any further, and proceed to a discussion of other algorithms which can be applied to general systems. But before we go into details, let us first explain what the general strategy will be to calculate the expectation value (16.1). We first need to construct an algorithm, that will generate a sequence of configurations which will eventually be distributed with the desired probability. Once the system has reached equilibrium we may use the *thermalized* configurations to measure our observable. In praxis the number of such configurations will be finite. But if the sequence generated by the algorithm constitutes a representative set, then the ensemble average (16.1) will be given approximately by the sum (16.2). Furthermore, if the N measurements are statistically independent, then the error in the mean will be of order $1/\sqrt{N}$. In praxis, however, the configurations we generate sequentially are not all statistically independent. Hence when computing the statistical error, one must make sure to select only configurations which are uncorrelated.

Summarizing, the procedure for calculating the ensemble average (16.1) is the following: given a rule for generating a sequence of configurations, one first *updates* the configurations a sufficiently large number of times until thermalization is achieved. The number of steps required for thermalization will depend on 1) the algorithm used, 2) the observable one is studying, and 3) on the values of the bare parameters. Thus it becomes increasingly more difficult to generate a thermalized ensemble as one approaches the continuum limit, where, as we have seen in chapter 9, the theory must exhibit critical behaviour. Indeed, the number of steps required increases with a power of the correlation length. This is known as *critical slowing down.* A first check of whether the system has reached thermal equilibrium, consists in choosing different starting configurations for generating the sequence. In particular, we can start from a disordered (hot start) or an ordered (cold start) configuration. Once the system is thermalized, it must have lost all memory of the configuration we have started with, and expectation values of observables must approach the same values independent of the starting configuration. Again for a finite set of configurations this will of course not be exactly true. But if the measurements of various observables for different runs agree within the estimated statistical errors, one has a good chance that the effects of the starting configuration have been eliminated.

16.2 Construction Principles for Algorithms. Markov Chains

In this section we will discuss some general principles for constructing algorithms which ensure that the sequence of configurations generated will constitute a representative ensemble which can be used to measure the observables. In particular, we shall be interested in the case where the above mentioned configurations are the elements of a Markov chain generated by a Markov process. So let us first discuss some elementary notions about Markov chains.* For simplicity we shall assume that the configurations can be labeled by a discrete index. If they are labeled by continuous variables then the summations will be replaced by integrals, with appropriately chosen measures, and the probability of occurrence of a given configuration must be replaced by a probability density.

---

* For a more detailed discussion, the reader may confer the books by Hammersley and Handscomb (1975), Clarke and Disney (1985).

Markov chains

Let $C_1, C_2, \ldots$ denote a countable set of states of the system. For example, in a pure lattice gauge theory, $C_i (i = 1, 2, \ldots)$ would correspond to different link variable configurations on the lattice. Consider now a stochastic process in which a finite set of configurations is generated one after the other according to some transition probability $P(C_i \to C_j) \equiv P_{ij}$ for going from configuration C_i to C_j. Let $C_{\tau_1}, C_{\tau_2} \ldots$ be the configurations generated sequentially in this way. Because of the probabilistic element built into this procedure, the state of the system at any given (simulation) time will be a random variable, whose distribution will depend only on the state preceding it, if the transition probability is independent of all states except for its immediate predecessor (i.e., if it does not involve any past history). This defines a Markov chain whose elements are the random variables mentioned above. Now given a set of configurations $\{C_{\tau_i}\}$ generated by the Markov process and an observable O evaluated on this set of states, we can define a "time" average of O taken over N elements of the Markov chain by

$$\langle O \rangle_N = \frac{1}{N} \sum_{i=1}^{N} O(C_{\tau_i}). \tag{16.5}$$

It is this quantity which we compute in praxis, and which we want to equal the ensemble average corresponding to a given Boltzmann distribution. With this in mind we shall therefore restrict the following discussion to so–called irreducible, aperiodic Markov chains whose states are positive, since there exist important theorems about such chains which are relevant for our problem. Let us first explain the terminology.

i) A chain is called irreducible if, starting from an arbitrary configuration C_i, there exists a finite probability of reaching any other configuration C_j after a finite number of Markov steps. In other words there exists a finite N such that

$$P_{ij}^{(N)} = \sum_{\{i_k\}} P_{ii_1} P_{i_1 i_2} \ldots P_{i_{N-1} j} \neq 0. \tag{16.6}$$

ii) A Markov chain is called aperiodic if $P_{ii}^{(N)} \neq 0$ for any N.

iii) A state is called positive if its mean recurrence time is finite. If $p_{ii}^{(n)}$ is the probability to get from C_i to C_i in n–steps of the Markov chain, without reaching this configuration at any intermediate step, then the mean recurrence time of C_i is given by

$$\tau_i = \sum_{n=1}^{\infty} n p_{ii}^{(n)}.$$

We now state some important results which hold for Markov chains satisfying i) to iii):

Theorem 1

If the chain is irreducible and the states are positive and aperiodic, then the limit $N \to \infty$ of (16.6) exists, and is unique; in particular one can show that *

$$\lim_{N\to\infty} P_{ij}^{(N)} = \pi_j \tag{16.7}$$

where $\{\pi_j\}$ are numbers which satisfy the following equations:

$$\pi_j > 0\ ,\ \sum_j \pi_j = 1, \tag{16.8a}$$

$$\pi_j = \sum_i \pi_i P_{ij}. \tag{16.8b}$$

Let us interprete these results. First of all eq. (16.7) states that the limit $N \to \infty$ of $P_{ij}^{(N)}$ is independent of the configuration used to start the Markov process. Furthermore, eq. (16.8) states that the set $\{\pi_i\}$ is left unchanged when we *update* these numbers with the transition probabilities $\{P_{ij}\}$. But this, together with (16.8a), is just the condition that the system is in equilibrium, with π_i the probability of finding the configuration C_i.

Theorem 2

If the chain is irreducible and its states are positive, and if

$$\tau_i^{(2)} \equiv \sum_{n=1}^{\infty} n^2 p_{ii}^{(n)} < \infty$$

then the time average (16.5) approaches the ensemble average

$$\langle O \rangle = \sum_i \pi_i O(C_i) \tag{16.9}$$

with a statistical uncertainty of order $O(\frac{1}{\sqrt{N}})$.

The above theorems provide the basis for using a Markov process to calculate the ensemble average (16.1). We must now determine the transition probabilities that generate a Markov chain of configurations, which will eventually be distributed with a given probability.

---

* See e.g. Hammersley and Handscomb (1975)

Since we shall be interested in the case where the configurations are labeled by a set of continuous variables we will drop from now on the discrete subscript on C_i and use capital letters to denote an arbitrary configuration. Thus, for example (16.8b) and (16.9) will be replaced by*

$$P_{eq}(C) = \sum_{C'} P_{eq}(C')P(C' \to C), \tag{16.10a}$$

$$\langle O \rangle = \sum_{C} P_{eq}(C)O(C), \tag{16.10b}$$

where $P_{eq}(C)$ now denotes the probability density for finding the configuration C at equilibrium, and $P(C \to C')$ is the transition probability density for going from C to C'. We will now show that for a Markov process to sample the distribution $\exp(-S(C))$, it is sufficient to require that the transition probability (density) satisfies detailed balance:

$$e^{-S(C)}P(C \to C') = e^{-S(C')}P(C' \to C) \tag{16.11}$$

for every pair C and C'. Notice that the transition probability $P(C \to C')$ is not a function of the variables labeling the configurations C and C', but stands for a rule which tells us how to select the next configuration in a Markov chain given the configuration immediately preceeding it.

Consider the following probability density which we want to generate:

$$P_{eq}(C) = \frac{e^{-S(C)}}{\sum_C e^{-S(C)}}. \tag{16.12}$$

Making use of the normalization condition

$$\sum_{C'} P(C \to C') = 1, \tag{16.13}$$

and assuming that $P(C \to C')$ satisfies (16.11), we see that (16.10a) is also satisfied. Hence if the conditions of Theorem 1 are satisfied, then (16.12) is the unique equilibrium distribution generated by the Markov process. In fact, it may be easily shown that the deviation from equilibrium decreases with each Markov step. The following proof is taken from Creutz (1983a).

---

* Actually, the sums must also be replaced by integrals with an appropriate measure. But for the following discussion it is convenient to keep the summation sign.

Let $P_N(C)$ be the probability of finding the configuration C at the N'th step of a Markov chain. If C_0 is the configuration with which we start the Markov process, then $P_N(C)$ is given by*

$$P_N(C) = \sum_{\{C_i\}} P(C_0 \to C_1)P(C_1 \to C_2)\ldots P(C_{N-1} \to C), \tag{16.14}$$

where each sum now runs over all possible configurations of the system. A useful quantity which measures the deviation of $P_N(C)$ from equilibrium is

$$\sigma_N = \sum_C |P_N(C) - P_{eq}(C)|. \tag{16.15}$$

Consider the deviation σ_{N+1}. Since according to (16.14) the probability distribution at the $(N+1)$'th step of the Markov chain is given by

$$P_{N+1}(C) = \sum_{C'} P_N(C')P(C' \to C) \tag{16.16}$$

we find, upon substituting (16.16) into the expression for σ_{N+1}, and making use of the normalization condition (16.13), as well as detailed balance, eq. (16.11), that

$$\begin{aligned} \sigma_{N+1} &= \sum_C \Big| \sum_{C'} (P_N(C')P(C' \to C) - P_{eq}(C)) \Big| \\ &= \sum_C \Big| \sum_{C'} (P_N(C')P(C' \to C) - P_{eq}(C)P(C \to C')) \Big| \\ &= \sum_C \Big| \sum_{C'} (P_N(C') - P_{eq}(C'))\, P(C' \to C) \Big| \\ &\leq \sum_C \sum_{C'} |P_N(C') - P_{eq}(C')| P(C' \to C), \end{aligned} \tag{16.17a}$$

or

$$\sigma_{N+1} \leq \sigma_N. \tag{16.17b}$$

Consider the equality sign. Then it follows from the last two lines in in eq. (16.17a) that

$$\sum_C \Big| \sum_{C'} (P_N(C') - P_{eq}(C'))P(C' \to C) \Big| = \sum_{C,C'} |P_N(C') - P_{eq}(C')| P(C' \to C).$$

* For finite N, $P_N(C)$ will depend on the initial configuration C_0. We have suppressed this dependence here.

But if for all C and C' $P(C \to C') \neq 0$ (i.e. the Markov chain is aperiodic), then this equation can only be satisfied if $P_N(C) = P_{eq}(C)$ for all C, since both $P_N(C)$ and $P_{eq}(C)$ are normalized distributions. We therefore conclude that the deviation σ_N decreases with each Markov step until we reach the desired equilibrium distribution (16.12). This does not mean, however, that for every configuration C, $P_{N+1}(C)$ is closer to $P_{eq}(C)$ than it was $P_N(C)$, as is evident from the definition (16.15).

The requirement of detailed balance does not determine the transition probability uniquely, and one can use this freedom to invent "efficient" algorithms adapted to the problem one is studying. In the following section we will give an example of an algorithm which is quite effective for studying systems whose action is a local function of the configuration space variables.

16.3 The Metropolis Method

This method was originally proposed by Metropolis et al. (1953) and is in principle applicable to any system. Let us first state the rule for generating the configurations in a sequence, and then prove that it satisfies detailed balance.

The rule: Let C be any configuration which is to be updated. We then suggest a new configuration C' with a transition probability $P_0(C \to C')$ which only satisfies the following microreversibility requirement:

$$P_0(C \to C') = P_0(C' \to C). \tag{16.18}$$

As an example consider the pure $U(1)$ gauge theory. A particular configuration C is then specified by the values of the link variables $U_\mu(n) = exp(i\theta_\mu(n))$ for $\mu = 1, \cdots, 4$ and for all lattice sites n. We can then suggest a new configuration by just choosing one of these link variables, and multiplying it by $exp(i\chi)$, where χ is a random number between $-\pi$ and π.* Then (16.18) is clearly satisfied. Having suggested a configuration C', we now must decide whether it should be accepted. Clearly the answer must depend on the actions $S(C)$ and $S(C')$, if the transition probability $P(C \to C')$ is to satisfy (16.11). The decision is made as follows: If $exp(-S(C')) > exp(-S(C))$, i.e. if the action has been lowered, then the configuration C' is accepted. On the other hand, if the action has increased, one

---

* Of course we can also suggest a new configuration which differs from the previous one in several link variables. But the usual procedure consists in updating only one variable at a time, sweeping through the lattice systematically, or choosing the variables in a more or less random way.

accepts the trial configuration only with probability $exp(-S(C'))/exp(-S(C))$. To this effect one generates a random number R in the interval $[0,1]$ and takes C' as the new configuration if

$$R \leq \frac{e^{-S(C')}}{e^{-S(C)}}.$$

Otherwise C' is rejected, and we keep the old configuration. Notice that it is this conditional acceptance which allows the system to increase its action. Thus while the classical configurations correspond to minima of the action, the quantum system is allowed to move away from the classical configuration. In this way the algorithm builds in the quantum fluctuations.

We now show that this algorithm satisfies detailed balance. Making use of the fact that the probability for the transition $C \to C'$ is the product of the probability $P_0(C \to C')$ for suggesting C' as the new configuration and the probability of accepting it, we see that the Metropolis algorithm for $P(C \to C')$ implies the following statements:

If $\exp(-S(C')) > \exp(-S(C))$, then

$$P(C \to C') = P_0(C \to C'),$$

and

$$P(C' \to C) = P_0(C' \to C)\frac{e^{-S(C)}}{e^{-S(C')}}.$$

Since $P_0(C' \to C) = P_0(C \to C')$, we see that (16.11) is satisfied. On the other hand:

If $\exp(-S(C')) < \exp(-S(C))$, then

$$P(C \to C') = P_0(C \to C')\frac{e^{-S(C')}}{e^{-S(C)}},$$

and

$$P(C' \to C) = P_0(C' \to C),$$

and detailed balance holds again.

As we have already mentioned, this algorithm is used in general to update a single variable at a time. The reader may ask, why not change all lattice variables at once? The reason is that such an updating procedure would involve in general large changes in the action. Hence the acceptance rate for those configurations where the action has increased, will be very small, and the system will move only slowly through configuration space. But the algorithm also becomes very slow for single variable updating, if the action depends non-locally on the coordinates, since in this case the ratio $exp(-S(C'))/exp(-S(C))$ will no longer be determined by the nearest neighbour interactions. This is the problem we shall be confronted with when taking fermions into account. Hence we shall need more efficient algorithms to handle such situations, and which allow one to update the entire lattice at once. In the following sections we will discuss some of the new algorithms which have been proposed in the literature. They are arranged in chronological order, so that the reader can see how these algorithms have improved in the past years, leading up to the so-called hybrid Monte Carlo algorithm which is free of systematic errors, and suited for handling systems described by a non-local action.

16.4 The Langevin Algorithm

The Langevin algorithm has been originally introduced by Parisi and Wu (1981), and was proposed as an updating method in numerical simulations in full QCD by Fukugita and Ukawa (1985) and Batrouni et al. (1985). In the following we discuss this algorithm for a bosonic system consisting of a finite number of degrees of freedom. Our presentation is based on the lectures of Toussaint (1988) and Negele (1988).

Let q_i $(i = 1, \cdots N)$ denote the coordinates of a system with action $S[q]$. In a pure gauge theory these are the link variables on the space-time lattice. We want to obtain a rule for updating these coordinates in a numerical simulation which satisfies the requirements of ergodicity and detailed balance. So let us think of these coordinates as depending on a new time variable τ which, when discretized, labels the elements of a Markov chain. Consider the following difference equation relating the variables at "time" $\tau_{n+1} = (n + 1)\epsilon_L$ to those at $\tau_n = n\epsilon_L$.*

$$q_i(\tau_{n+1}) = q_i(\tau_n) + \epsilon_L \left(-\frac{\partial S[q]}{\partial q_i(\tau_n)} + \eta_i(\tau_n) \right). \tag{16.19}$$

Here $\{\eta_i(\tau_n)\}$ is a set of independent Gaussian distributed random variables which probability distribution given by

$$P(\{\eta_i(\tau_n)\}) = \prod_i \sqrt{\frac{\epsilon_l}{4\pi}} exp[-\frac{\epsilon_L}{4}\eta_i(\tau_n)^2]. \tag{16.20}$$

In the limit $\epsilon_L \to 0$, eq. (16.19) becomes the Langevin equation:

$$\frac{dq_i}{d\tau} = -\frac{\partial S[q]}{\partial q_i} + \eta_i(\tau). \tag{16.21}$$

---

* ϵ_L is called the Langevin time step for reasons that will become clear below.

Since the width of the distribution (16.20) is of order $1/\sqrt{\epsilon_L}$, the Gaussian noise term in (16.19) is actually of order $\sqrt{\epsilon_L}$. We can make this explicit by introducing the new random variable $\tilde{\eta}_i(\tau_n) = \left(\sqrt{\epsilon_L/2}\right)\eta_i(\tau_n)$. Then (16.19) take the form

$$q_i(\tau_{n+1}) = q_i(\tau_n) - (2\epsilon_L)\frac{1}{2}\frac{\partial S[q]}{\partial q_i(\tau_n)} + \sqrt{2\epsilon_L}\tilde{\eta}_i(\tau_n), \tag{16.22}$$

where $\{\tilde{\eta}_i(\tau_n)\}$ are now random variables distributed according to

$$\tilde{P}(\tilde{\eta}(\tau_n)) = \left(\prod_i \frac{1}{\sqrt{2\pi}}\right)e^{-\sum_i \frac{1}{2}\tilde{\eta}_i(\tau_n)^2} \tag{16.23}$$

with variance

$$<\tilde{\eta}_i(\tau_n)\tilde{\eta}_j(\tau_m)>=\delta_{ij}\delta_{nm}.$$

This form of the difference equation will be useful when discussing the hybrid molecular dynamics algorithm later on.

Consider eq. (16.22).*It states that the probability for making the transition from q_i at "time" τ_n to q_i' at "time" τ_{n+1} is the probability for $\tilde{\eta}_i$ $(i = 1, 2, ...)$ to equal $\sqrt{2\epsilon_L}\frac{1}{2}\frac{\partial S[q]}{\partial q_i} + (q_i' - q_i)/\sqrt{2\epsilon_L}$. Hence we conclude that the transition probability $P(q \to q')$ is given by

$$P(q \to q') = N_0 \exp\left\{-\frac{1}{2}\sum_i \left[\frac{q_i' - q_i}{\sqrt{2\epsilon_L}} + \frac{1}{2}\sqrt{2\epsilon_L}\frac{\partial S}{\partial q_i}\right]^2\right\}, \tag{16.24}$$

where N_0 is a normalization constant. We now claim that in the limit $\epsilon_L \to 0$ this transition probability satisfies detailed balance, i.e. that

$$e^{-S(q)}P(q \to q') = e^{-S(q')}P(q' \to q). \tag{16.25}$$

Indeed, since $q' - q$ is of $O(\sqrt{\epsilon}_L)$, it follows from (16.24) that

$$\frac{P(q \to q')}{P(q' \to q)} \underset{\epsilon_L \to 0}{\longrightarrow} e^{-\sum_i (q_i' - q_i)\partial S/\partial q_i}$$
$$\underset{\epsilon_L \to 0}{\longrightarrow} e^{-[S(q') - S(q)]}\ .$$

We therefore conclude that in the limit $\epsilon_L \to 0$ the transition amplitude (16.24) satisfies (16.25). For finite Langevin time steps, however, this algorithm leads to systematic errors which need to be controled. In praxis this means that the

* The following argument is based on the lectures by Negele (1988).

data obtained using this algorithm must be extrapolated to $\epsilon_L \to 0$. Hence one is forced to generate configurations for different Langevin time steps. This is very time-consuming. On the other hand it is a simple algorithm which can be used to update all variables at once without having to worry about acceptance rates, since it does not involve any Metropolis acceptance test. The Langevin algorithm is therefore useful for treating systems with a non-local action, like QCD with dynamical fermions.

This concludes our discussion of the Langevin algorithm which, like the Metropolis method, is applicable to arbitrary systems. Its main drawback is the systematic error which is introduced by the finite Langevin-time step, ϵ_L, which in praxis one would like to choose as large as possible to ensure that the system covers much of the configuration space in a reasonable number of updating steps. A lot of effort has been invested in the past few years to invent new algorithms which are more effective and less subject to systematic errors. Most of these algorithms are modifications of the so-called molecular dynamics (or microcanonical) method of Callaway and Rahman (1982 and 1983). So let us first discuss this method.*

16.5 The Molecular Dynamics Method

The basic idea of the molecular dynamics method is that the euclidean path integral associated with a quantum theory can be written in the form of a partition function for a classical statistical mechanical system in four *spatial* dimensions with a canonical Hamiltonian that governs the dynamics in a new "time" variable (the simulation time). Invoking ergodicity, and making use of the fact that in the thermodynamic limit the canonical ensemble average can be obtained from the microcanonical ensemble average, computed at an energy determined by the parameters of the system**, one can calculate the expectation values of observables as time averages over "classical" trajectories. Hence, in the molecular dynamics approach, configurations are generated in a deterministic way, and the complexity of the motion in four space dimensions gives rise to the quantum fluctuations in the original theory of interest.

We now give the details. To keep the discussion as elementary as possible, we will consider a scalar field theory, with an action $S[\phi;\beta]$ depending on the scalar field ϕ and on some parameter β (e.g. a coupling constant).*** Consider the

---

* The microcanonical ensemble approach was also proposed by Creutz (1983b)

** In conventional statistical mechanics this energy would be fixed by the temperature of the system.

*** See Callaway and Rahman (1982) and (1983).

expectation value of an observable

$$< O >= \frac{1}{Z} \int D\phi O[\phi] e^{-S[\phi;\beta]}, \tag{16.26a}$$

where

$$Z = \int D\phi e^{-S[\phi;\beta]}. \tag{16.26b}$$

We assume that we have introduced a space-time lattice to define the integral (16.26). Hence the degrees of freedom of the system will be labeled by the coordinates of the lattice sites which we denote collectively by "i". Although the integral (16.26) resembles that encountered in statistical mechanics, Z is not the partition function of a classical Hamiltonian system. But one can cast it into such a form by introducing a set of canonically conjugate momenta π_i. Thus (16.26) can also be written in the form

$$< O >= \frac{1}{\bar{Z}} \int D\phi D\pi O[\phi] e^{-H[\phi,\pi;\beta]}, \tag{16.27a}$$

where

$$H[\phi,\pi;\beta] = \sum_i \frac{1}{2}\pi_i^2 + S[\phi;\beta], \tag{16.27b}$$

$$\bar{Z} = \int D\phi D\pi e^{-H[\phi,\pi;\beta]}. \tag{16.27c}$$

and the integration measure is defined as usual:

$$D\phi D\pi = \prod_i d\phi_i d\pi_i.$$

Clearly (16.27) coincides with the expression (16.26), since O does not depend on the momenta. Hence, as advertised above, the quantum theory in four-dimensional euclidean space-time resembles to a classical canonical ensemble in four spatial dimensions in contact with a heat reservoir. Notice that (16.27) is not to be confused with the usual phase space representation of the path integral (16.26), whose structure is completely different, and involves the Hamiltonian of the system in three space dimensions.

Having formulated our problem in terms of a classical statistical mechanical system, we can now make use of known results to calculate the expectation value (16.26a). In particular it is well known in statistical mechanics that in the thermodynamic limit (where the number of degrees of freedom become infinite), the canonical ensemble average can be replaced by the microcanonical ensemble average evaluated on an "energy" surface, which is determined by the parameters of the system. We have put "energy" into quotation marks to emphasize that, in the present case, it is the energy of a statistical mechanical system in four spatial dimensions. Hence, according to (16.27b), the action of the quantum mechanical system is allowed to fluctuate! This is the way how quantum fluctuations are taken into account. Let us recall how one arrives at this result. At the same time, this will give us the connection between the parameter β and the "energy" at which the microcanonical ensemble average is to be calculated.

Consider the expectation value (16.27). It can be written in the form

$$< O >_{can} (\beta) = \frac{\int D\phi D\pi O[\phi] \int dE \delta(H[\phi,\pi;\beta] - E)e^{-E}}{\int D\phi D\pi \int dE\delta(H[\phi,\pi;\beta] - E)e^{-E}}, \tag{16.28}$$

where we have introduced the subscript "can" to emphasize that we are calculating a canonical ensemble average. This quantity depends on β.

On the other hand, the microcanonical ensemble average $< O >_{mic}$, evaluated for a given β on the energy surface $H = E$, is given by

$$< O >_{mic} (E,\beta) = \frac{1}{Z_{mic}(E,\beta)} \int D\phi D\pi O[\phi]\delta(H[\phi,\pi;\beta] - E), \tag{16.29a}$$

where

$$Z_{mic}(E,\beta) = \int D\phi D\pi \delta(H[\phi,\pi;\beta] - E) \tag{16.29b}$$

is the phase space volume at energy E. We now want to establish a connection between the two ensemble averages in the thermodynamic limit. Define the "entropy" of the system by

$$s(E,\beta) = ln Z_{mic}(E,\beta). \tag{16.30}$$

Introducing the definitions (16.29) and (16.30) into (16.28), one finds that this expression can be written in the form

$$< O >_{can} (\beta) = \frac{\int dE < O >_{mic} (E,\beta)e^{-(E-s(E,\beta))}}{\int dE e^{-(E-s(E,\beta))}}. \tag{16.31}$$

Now comes the standard argument: when the number of degrees of freedom of the system becomes very large, the exponentials appearing in the integrands of (16.31) are strongly peaked about an energy $\bar{E}$ given implicitly by the equation

$$\left(\frac{\partial s(E,\beta)}{\partial E}\right)_{E=\bar{E}} = 1. \tag{16.32}$$

Hence in the thermodynamic limit we are led to the connection

$$\langle O\rangle_{can}(\beta) = [\langle O\rangle_{mic}]_{E=\bar{E}(\beta)}, \tag{16.33}$$

where β and $\bar{E}$ are related by (16.32). This relation is, however, not a useful one for doing numerical calculations. A more adequate expression can be obtained by making use of the equipartition theorem, which tells that the following equation holds for each momentum π_i:

$$< \pi_i \frac{\partial H}{\partial \pi_i} >_{mic} = \frac{1}{\partial s/\partial E} \quad .$$

Inserting for H the expression (16.27b) and making use of (16.32), one finds that

$$[< T_{kin} >_{mic}]_{E=\bar{E}(\beta)} = \frac{N}{2}, \tag{16.34}$$

where $T_{kin} = \sum_i \pi_i^2/2$ is the kinetic energy, and N is the number of degrees of freedom of the system. We have hence succeeded in rewriting (16.32) in terms of a microcanonical ensemble average. But we are still left with the problem that given β, $\bar{E}$ is only implicitly determined by (16.34). Fortunately this does not turn out to be a major stumbling block; $\bar{E}$ can be adjusted to any prescribed value of β by the heating/cooling procedure described by Callaway and Rhaman (1983).

A more familiar form of eq. (16.34) is obtained when the dependence of the action on the parameter β is multiplicative, i.e. if*

$$S[\phi, \beta] = \beta V[\phi].$$

In this case the expectation value (16.26) can be written as follows

$$< O > = \frac{1}{\tilde{Z}} \int D\phi D\pi O[\phi] e^{-\beta H[\phi,\pi]},$$

where

$$H[\phi, \pi] = \frac{1}{2}\sum_i \pi_i^2 + V[\phi],$$

and

$$\tilde{Z} = \int D\phi D\pi e^{-\beta H[\phi,\pi]}.$$

---

* This is the case in gauge theories within the pure gauge field sector, where the action is of the form $S = \beta V[U]$.

By following the arguments presented above, one arrives at the following connection between the canonical and microcanonical ensemble averages

$$< O >_{can} (\beta) =< O >_{mic} (\bar{E}),$$

where

$$< O >_{mic} (E) = \frac{1}{Z_{mic}(E)} \int D\phi D\pi O[\phi] \delta(H[\phi, \pi] - E),$$

$$Z_{mic}(E) = \int D\phi D\pi \delta(H[\phi, \pi] - E),$$

and β is related to $\bar{E}$ by

$$\frac{N}{2\beta} =< T_{kin} >_{mic} (\bar{E}).$$

This is the familiar form of the equipartition theorem with β playing the role of the inverse temperature.

With these remarks let us return to expression (16.33), and rewrite the right-hand side in a convenient form for numerical calculations. Assuming that the motion generated by the Hamiltonian (16.27b) is ergodic, we can use the Hamilton equations of motion

$$\dot{\phi}_i = \frac{\partial H[\phi, \pi]}{\partial \pi_i}, \tag{16.35a}$$

$$\dot{\pi}_i = -\frac{\partial H[\phi, \pi]}{\partial \phi_i}, \tag{16.35b}$$

to generate a representative ensemble of phase-space configurations with constant energy. Since the observables, whose expectation value we want to calculate, only depend on the coordinates, the microcanonical ensemble average can then be replaced by a time average over $\phi(\tau)$. We are therefore led to the following chain of equations valid for ergodic systems in the thermodynamic limit,

$$< O >_{can.} (\beta) \underset{therm.\ lim.}{\longrightarrow} [< O >_{mic.}]_{E=\bar{E}(\beta)} \underset{ergod.}{\longrightarrow} \lim \frac{1}{T} \int_0^T d\tau O(\{\phi_i(\tau)\}), \tag{16.36}$$

where the trajectory $\{\phi_i(\tau)\}$ is a solution to

$$\frac{d^2\phi_i}{d\tau^2} = -\frac{\partial S[\phi]}{\partial \phi_i}, \tag{16.37}$$

subject to the initial conditions that it carries energy $\bar{E}$.*

* Because the motion is assumed to be ergodic, one can use any starting configuration with this energy to integrate (16.37).

Notice the difference in the structure of (16.37) and the Langevin equation (16.21). While (16.37) is a deterministic set of coupled second order differential equations, the Langevin equation is of first order an involves a stochastic variable. It is instructive to compare the two equations when the times are discretized.* In the Langevin scheme the corresponding difference equations are given by (16.22). Consider the naive discretization of (16.37). Using the symmetric version for the discretized second derivative, $\ddot{\phi}_i(\tau_n) = [\phi_i(\tau_{n+1}) + \phi_i(\tau_{n-1}) - 2\phi_i(\tau_n)]/(\Delta\tau)^2$, we find that

$$\phi_i(\tau_{n+1}) = \phi_i(\tau_n) - \epsilon^2 \frac{1}{2} \frac{\partial S[\phi]}{\partial \phi_i(\tau_n)} + \epsilon \pi_i(\tau_n), \tag{16.38a}$$

where $\epsilon = \Delta\tau = \tau_{n+1} - \tau_n$ is the microcanonical time step, and

$$\pi_i(\tau_n) = \frac{1}{2\epsilon}(\phi_i(\tau_{n+1}) - \phi_i(\tau_{n-1})). \tag{16.38b}$$

Notice the striking similiarity between (16.38a) and (16.22)! The discretized version of the Langevin equation is obtained from (16.38a) by promoting the momenta π_i to random variables distributed according to (16.23), and identifying ϵ_L with $\epsilon^2/2$.**

In practical calculations, one integrates the Hamilton equations of motion (16.35) using the so called "leapfrog" method (see section 6). Let us compare the "distances" travelled in configuration space for the above-mentioned algorithms for $\epsilon_L = \epsilon^2/2$. In the molecular dynamics case the distance covered after n steps is of $O(n\epsilon)$, while in the Langevin approach it is only $O(\sqrt{n}\epsilon)$, because of its stochastic nature. Hence the classical algorithm moves the system faster through configuration space. But for this algorithm to be valid, a number of conditions had to be satisfied. For one thing, eq. (16.33) only holds for systems with an infinite number of degrees of freedom. This is clearly not the case in practical calculations where one is dealing with rather small lattices. Furthermore, the last step in the chain (16.36) requires the motion to be ergodic. But, in contrast to the Langevin approach, ergodicity is not explicitly built into the molecular dynamics algorithm. This suggests that one should look for a new algorithm which combines the good features of the two methods.

---

* see e.g., Toussaint (1988).

** The fact that the Langevin time step is the square of the microcanonical time step is not surprising, since the Langevin equation is of first order, while (16.37) involves the second time derivative.

16.6 The Hybrid Algorithm

A possible way of combining the stochastic Langevin approach and the microcanonical simulation into a new algorithm which takes into account ergodicity was proposed by Duane (1985). This so–called hybrid algorithm is suggested by considering the molecular dynamics algorithm in the form (16.38a), which - as we have pointed out before - becomes the discretized Langevin equation with time step $\epsilon_L = \epsilon^2/2$, if the momenta $\{\pi_i\}$ are replaced by random variables with probability density

$$P(\{\pi_i\}) = \left(\prod_i \frac{1}{\sqrt{2\pi}}\right) e^{-\sum_i \frac{1}{2}\pi_i^2} . \tag{16.39}$$

Notice that this is precisely the distribution of the momenta appearing in (16.27a)! In fact, if the motion generated by the Hamiltonian (16.27b) is ergodic, then the momenta would be eventually distributed according to (16.39).

This suggests that instead of integrating the Hamilton equations of motion (16.35) along a single trajectory with given energy $\bar{E}$, we interrupt this integration process once in a while and choose a new set of momenta with a probability density given by (16.39). This corresponds to performing a Langevin step, which if repeated many times builds in the ergodicity we want. At the same time one takes advantage of the fact that because of the molecular dynamics steps the system will move faster through configuration space than in the Langevin approach. Hence the hybrid algorithm contains a new parameter: the frequency with which momenta are refreshed. This parameter can be tuned to give the best results. In fact Duane and Kogut (1986) have suggested that at each step of the updating procedure a random choice is made between a molecular dynamics update with probability α (which corresponds to relating the momenta in (16.38a) to ϕ according to (16.38b)) and a Langevin update with probability $1-\alpha$ (which corresponds to replacing $\{\pi_i\}$ by Gaussian distributed random numbers). If α is close to one, then the system will move almost as fast through configuration space as in the molecular dynamics case, but ergodicity may not be fully realized. On the other hand if α is close to zero, ergodicity is ensured, but the system will move slower through configuration space. The optimal choice for α will lie between these two extremes, and has to be determined numerically.

After these qualitative remarks, we now demonstrate how this algorithm is implemented in praxis. For simplicity we shall again study the case of a scalar field theory.

Consider the equations of motion in their Hamiltonian form (16.35). We want to discretize them in the time with time step ϵ. To this effect let us expand $\phi_i(\tau+\epsilon)$ and $\pi_i(\tau+\epsilon)$ in a Taylor series up to order ϵ^2:

$$\begin{aligned}\phi_i(\tau+\epsilon) &= \phi_i(\tau) + \epsilon\dot{\phi}_i(\tau) + \frac{\epsilon^2}{2}\ddot{\phi}_i(\tau) + O(\epsilon^3), \\ \pi_i(\tau+\epsilon) &= \pi_i(\tau) + \epsilon\dot{\pi}_i(\tau) + \frac{\epsilon^2}{2}\ddot{\pi}_i(\tau) + O(\epsilon^3).\end{aligned} \tag{16.40}$$

But from the Hamilton equations of motion we have that $\dot{\phi}_i = \pi_i(\tau)$, and $\ddot{\phi}_i(\tau) = \dot{\pi}_i(\tau) = -\partial S/\partial\phi_i(\tau)$, so that

$$\ddot{\pi}_i(\tau) = -\sum_j \frac{\partial^2 S}{\partial\phi_i(\tau)\partial\phi_j(\tau)}\pi_j(\tau). \tag{16.41}$$

In leading order the right hand–side of (16.41) is given by

$$\sum_j \frac{\partial^2 S}{\partial\phi_i(\tau)\partial\phi_j(\tau)}\pi_j(\tau) = \frac{1}{\epsilon}\left(\frac{\partial S}{\partial\phi_i(\tau+\epsilon)} - \frac{\partial S}{\partial\phi_i(\tau)}\right) + O(\epsilon).$$

Introducing the above expressions into (16.40) we find after some rearrangements that

$$\begin{aligned}\phi_i(\tau+\epsilon) &= \phi_i(\tau) + \epsilon\left(\pi_i(\tau) - \frac{\epsilon}{2}\frac{\partial S}{\partial\phi_i(\tau)}\right) + O(\epsilon^3), \\ \left(\pi_i(\tau+\epsilon) - \frac{\epsilon}{2}\frac{\partial S}{\partial\phi_i(\tau+\epsilon)}\right) &= \left(\pi_i(\tau) - \frac{\epsilon}{2}\frac{\partial S}{\partial\phi_i(\tau)}\right) - \epsilon\frac{\partial S}{\partial\phi_i(\tau+\epsilon)} + O(\epsilon^3).\end{aligned} \tag{16.42}$$

But the quantities appearing within brackets are the momenta evaluated at the midpoint of the time intervals; hence up to $O(\epsilon^3)$, eqs. (16.42) are equivalent to the following pair

$$\phi_i(\tau+\epsilon) = \phi_i(\tau) + \epsilon\pi_i(\tau+\epsilon/2), \tag{16.43a}$$

$$\pi_i(\tau+\frac{3}{2}\epsilon) = \pi_i(\tau+\frac{\epsilon}{2}) - \epsilon\frac{\partial S}{\partial\phi_i(\tau+\epsilon)}, \tag{16.43b}$$

which, when iterated, amounts to integrating the Hamilton equations of motion (16.35) according to the so–called *leapfrog* scheme: while the coordinates are evaluated at times $\tau_n = n\epsilon$, momenta (i.e. derivatives of ϕ_i) are calculated at the midpoint of the time intervals.

In the hybrid algorithm the momenta are refreshed at the beginning of every molecular dynamics chain. Suppose we want to start the leapfrog integration at time τ. Let $\{\phi_i(\tau)\}$ be the coordinates at this time; then the momenta $\{\pi_i(\tau)\}$ are chosen from a Gaussian ensemble as described before. But to begin the iteration process (16.43), we need to know $\pi_i(\tau + \epsilon/2)$. We calculate this quantity by performing a half–time step as follows

$$\pi_i(\tau + \epsilon/2) = \pi_i(\tau) - \frac{\epsilon}{2}\frac{\partial S[\phi]}{\partial \phi_i(\tau)} + O(\epsilon^2), \tag{16.44}$$

which means that we are making an error of $O(\epsilon^2)$. But this error is made only once for each molecular dynamics trajectory generated, whose accumulated error after $1/\epsilon$ steps is also of order ϵ^2. Hence measurements of the observables will be correct up to this order.

Before we summarize the steps required for updating the configurations $\{\phi_i\}$ according to the hybrid algorithm, let us rewrite eqs. (16.43) and (16.44) in a way appropriate for numerical calculations. Denoting by $\{\phi_i(n)\}$ and $\{\pi_i(n)\}$ the momenta at time step $n\epsilon$, and by $\{\tilde{\pi}_i(n)\}$ the momenta at time $(n + 1/2)\epsilon$, eqs. (16.43) and (16.44) take the following form:

$$\phi_i(n+1) = \phi_i(n) + \epsilon\tilde{\pi}_i(n), \tag{16.45a}$$

$$\tilde{\pi}_i(n+1) = \tilde{\pi}_i(n) - \epsilon\frac{\partial S[\phi]}{\partial \phi_i(n+1)}, \tag{16.45b}$$

where

$$\tilde{\pi}_i(n) = \pi_i(n) - \frac{\epsilon}{2}\frac{\partial S[\phi]}{\partial \phi_i(n)}. \tag{16.45c}$$

In a lattice formulation, all quantities appearing in (16.45) will be dimensionless. The hybrid algorithm is then implemented by the following steps:

i) Choose the coordinate $\{\phi_i\}$ in some arbitrary way.
ii) Choose a set of momenta $\{\pi_i\}$ from the Gaussian ensemble (16.39).
iii) Perform the half step (16.45c) to start the integration process.
iv) Iterate eqs. (16.45a,b) for several time steps, and store the configurations generated in this way.
v) Go back to ii) and repeat the steps ii) to iv).

As has been shown by Duane and Kogut (1986), this algorithm generates a sequence of configurations $\{\phi_i\}$ which are eventually distributed according to $\exp(-S[\phi])$.

Suppose now that instead of refreshing the momenta only once in a while we refresh them at each time step. In this case eq. (16.45b) plays no longer any role in generating the coordinates $\{\phi_i\}$, which are now updated according to

$$\phi_i(n+1) = \phi_i(n) - \frac{\epsilon^2}{2}\frac{\partial S[\phi]}{\partial \phi_i(n)} + \epsilon\eta_i(n),$$

where $\{\eta_i(n)\}$are independent Gaussian distributed random numbers. But this is nothing but the Langevin equation with time step $\epsilon^2/2$. Hence the hybrid method differs only from the Langevin approach if the number of molecular dynamics steps is larger than one.

This concludes our general discussion of the hybrid algorithm. For more details the reader may confer to the lectures by Toussaint (1988).

16.7 The Hybrid Monte Carlo Algorithm

Although the hybrid algorithm is certainly an improvement over the previous two we have discussed, there remains the problem of having to control the systematic error introduced by the finite time step. In fact, of all the algorithms discussed so far, only the Metropolis method was free of systematic errors. But as we have already mentioned, this algorithm becomes very slow when updating configurations with an action depending non-locally on the fields. On the other hand, the hybrid algorithm allows one to update all variables at once, and hence is better suited for studying systems with a non-local action (as is the case for QCD when fermions are included). This suggests that one should try to eliminate the systematic error in the hybrid algorithm by combining it with a Metropolis acceptance test.

Originally, this idea was proposed by Scalettar, Scalapino, and Sugar (1986) in connection with the Langevin approach, and later applied to full QCD by Gottlieb et al. (1987a). Subsequently Duane, Kennedy, Pendleton, and Roweth (1987) suggested a similar modification of the hybrid algorithm. This modification consists in introducing a Metropolis acceptance test between step iv) and v) of the hybrid method discussed in the previous section. In other words, the phase space configurations generated at the end of each molecular dynamics chain, are used as trial configurations in a Metropolis test with the Hamiltonian (16.27b), which plays the role of the action in our discussion in section 16.3. For this reason this algorithm is called the hybrid Monte Carlo algorithm. We now state the rules for updating the configurations $\{\phi_i\}$ and then show that they generate a Markov chain satisfying detailed balance:

i) Choose the coordinates $\{\phi_i\}$ in some arbitrary way.

ii) Choose the momenta $\{\pi_i\}$ from the Gaussian ensemble (16.39).
iii) Calculate $\tilde{\pi}$ according to (16.45c).
iv) Iterate eqs. (16.45a,b) for some number of time steps. Let $\{\phi'_i, \pi'_i\}$ be the last configuration generated in the molecular dynamics chain, where π'_i is obtained from $\tilde{\pi}'_i$ according to (16.45c).
v) Accept $\{\phi'_i, \pi'_i\}$ as the new configuration with probability

$$p = min\{1, e^{-H[\phi',\pi']}/e^{-H[\phi,\pi]}\},$$

where H is the Hamiltonian (16.27b).*
vi) If the configuration $\{\phi'_i, \pi'_i\}$ is not accepted, keep the old configuration $\{\phi_i, \pi_i\}$, and repeat the steps starting from (ii). Otherwise use the coordinates $\{\phi'_i\}$ to generate a new configuration beginning with the second step.

Notice that if the Hamiltonian equations of motion could be integrated exactly, then H would be constant along a molecular dynamics trajectory and the configuration would always be accepted by the Metropolis test. But for finite $\epsilon, \delta H \neq 0$ and the Metropolis test eliminates the systematic error introduced by the finite time step. Indeed, as has been shown by Duane et al. (1987), this algorithm leads to a transition probability $P(\phi \to \phi')$ which satisfies the detailed balance equation (16.11) for an arbitrary time step ϵ. Since the proof is rather simple, we will present it here.

Let $\{\phi_i\}$ be the coordinates at the beginning of a molecular dynamics chain. Consider in turn the various steps required for updating this configuration according to the hybrid algorithm.

<u>Step 1</u>

Choose a set of momenta $\{\pi_i\}$ from a Gaussian ensemble with mean zero and unit variance. The corresponding probability density is given by

$$P_G(\pi) = \mathcal{N}_0 e^{-\frac{1}{2}\sum \pi_i^2}, \tag{16.46}$$

where $\mathcal{N}_0$ is a normalization constant.

<u>Step 2</u>

Let the phase space configuration** (ϕ, π) evolve deterministically for N time steps according to (16.45), including the half-time steps at the beginning and the

---

* We have suppressed here the dependence of H on the paramter β.
** For simplicity we suppress the subscript "i" on ϕ_i and π_i.

end of the molecular dynamics chain. Denote the configuration generated in this way by $(\phi^{(N)}, \pi^{(N)})$. Here $\phi^{(N)}$ and $\pi^{(N)}$ are functions of the starting configuration (ϕ, π). But because (16.37) involves the second time derivative, the motion from $\phi \rightarrow \phi^{(N)}$ can be reversed by changing the sign of all momenta in the initial and final phase–space configurations. Hence the probability for the transition $(\phi, \pi) \rightarrow (\phi', \pi')$ is the same as that for going from $(\phi', -\pi')$ to $(\phi, -\pi)$: i.e.

$$P_M((\phi, \pi) \rightarrow (\phi', \pi')) = P_M((\phi', -\pi') \rightarrow (\phi, -\pi)). \tag{16.47}$$

The subscript "M" stands for "molecular dynamics". Actually, since the system evolves deterministically, the state (ϕ', π') generated by the molecular dynamics chain will occur with unit probability. Hence, P_M is a δ -function, ensuring that (ϕ', π') evolved from (ϕ, π).

Step 3

Accept the configuration (ϕ', π') with probability

$$P_A((\phi, \pi) \rightarrow (\phi', \pi')) = min(1, e^{-H[\phi', \pi']}/e^{-H[\phi, \pi]}). \tag{16.48}$$

By taking the product of the probability densities for these three steps, and integrating over the momenta, we obtain the transition probability density $P(\phi \rightarrow \phi')$:

$$P(\phi \rightarrow \phi') = \int D\pi D\pi' P_G(\pi) P_M((\phi, \pi) \rightarrow (\phi', \pi')) P_A((\phi, \pi) \rightarrow (\phi', \pi')). \tag{16.49}$$

Consider next the product $e^{-S(\phi)} P(\phi \rightarrow \phi')$ corresponding to the left-hand side of eq. (16.11). Introducing the expression (16.46) into (16.49), this product can be written in the form

$$e^{-S(\phi)} P(\phi \rightarrow \phi') = \int D\pi D\pi' e^{-H[\phi, \pi]} P_M((\phi, \pi) \rightarrow (\phi', \pi')) P_A((\phi, \pi) \rightarrow (\phi', \pi')). \tag{16.50}$$

But from the definition (16.48) it follows that

$$e^{-H[\phi, \pi]} P_A((\phi, \pi) \rightarrow (\phi', \pi')) = e^{-H[\phi', \pi']} P_A((\phi', \pi') \rightarrow (\phi, \pi)), \tag{16.51}$$

which is the statement of detailed balance in phase space for the Metropolis acceptance test. Inserting this expression into (16.50), and making use of (16.47) and of the fact that $H[\phi, \pi] = H[\phi, -\pi]$, one finds that

$$\begin{aligned} e^{-S(\phi)} P(\phi \rightarrow \phi') =& e^{-S(\phi')} \int D\pi D\pi' [P_G(\pi') P_M((\phi', \pi') \rightarrow (\phi, \pi)) \\ & \times P_A((\phi', \pi') \rightarrow (\phi, \pi))] = e^{-S(\phi')} P(\phi' \rightarrow \phi). \end{aligned}$$

This is the relation we wanted to prove. The hybrid MC algorithm therefore generates a Markov chain of configurations $\{\phi_i\}$ distributed eventually according to $exp(-S[\phi])$.

Since according to this algorithm entire field configurations are updated at once, one may wonder whether one is not running into the same "acceptance" problems mentioned in section 4. This could indeed be the case if ϵ is not chosen sufficiently small and/or the number of molecular dynamics steps is too large. Then the acceptance probability (16.48) will in general be small, and the algorithm will move the system only slowly through configuration space. Fortunately, numerical "experiments" show that even for small ϵ, of the order of the time step needed in the hybrid algorithm for extrapolating the data to $\epsilon \to 0$, the system moves at least as fast through configuration space as with the hybrid algorithm. But the great advantage of the hybrid Monte Carlo method is that it is free of systematic errors arising from the finite time step. Hence no extrapolation of the data is required, which for one thing would be time-consuming, and furthermore a source of errors.

16.8 The Pseudofermion Method

The algorithms we have discussed in the previous sections can only be used to calculate ensemble averages of functionals depending on c-number variables. They can therefore not be applied directly to systems involving Grassmann fields. For this reason, one must first integrate out the fermionic degrees of freedom before performing a Monte Carlo calculation in QED or QCD. As we have pointed out in chapter 12, this can always be done for any correlation function involving the product of an arbitrary number of fermion fields, since the action is quadratic in the Grassmann variables. In the following we will restrict our discussion to the case of Wilson fermions. Then the ensemble average of $O[U,\psi,\bar{\psi}]$ is given by

$$\langle O\rangle = \frac{\int DU \langle O\rangle_{S_F} e^{-S_{eff}[U]}}{\int DU e^{-S_{eff}[U]}}, \tag{16.52a}$$

where

$$\langle O\rangle_{S_F} = \frac{\int D\bar{\psi}D\psi O[U,\psi,\bar{\psi}]e^{-S_F^{(W)}[U,\psi,\bar{\psi}]}}{\int D\bar{\psi}D\psi e^{-S_F^{(W)}[U,\psi,\bar{\psi}]}}, \tag{16.52b}$$

can be expressed in terms of Green functions calculated in a background field $U = \{U_\mu(n)\}$, and where

$$S_{eff}[U] = S_G[U] - \ln\det\tilde{K}[U]. \tag{16.53}$$

Here S_G is the action of the pure gauge theory, and $\tilde{K}[U]$ is the matrix appearing in the fermionic contribution to the action. For the case of QCD with N_f flavours of mass-degenerate quarks, the matrix elements of $\tilde{K}[U]$ are given by

$$\tilde{K}_{n\alpha af',m\beta bf}[U] = K_{n\alpha a,m\beta b}[U]\delta_{ff'}\,, \tag{16.54a}$$

where $K_{n\alpha a,m\beta b}[U]$ is defined implicitly by (6.25c), i.e.,*

$$\begin{aligned} K_{n\alpha a,m\beta b}[U] =& (\hat{M}_0 + 4r)\delta_{nm}\delta_{\alpha\beta}\delta_{ab} \\ & -\frac{1}{2}\sum_\mu [(r-\gamma_\mu)_{\alpha\beta}(U_\mu(n))_{ab}\delta_{n+\hat{\mu},m} \\ & + (r+\gamma_\mu)_{\alpha\beta}(U^\dagger_\mu(m))_{ab}\delta_{n-\hat{\mu},m}]. \end{aligned} \tag{16.54b}$$

The index $f(f')$ in (16.54a) labels the flavour degrees of freedom, and n, α, a (m, β, b) stand for the lattice site, and Dirac and colour degrees of freedom, respectively. Let us denote the latter triple by a single collective index i. Then the fermionic contribution to the action reads

$$S_F^{(W)} = \sum_{i,j} \bar{\psi}_i^{f'} K_{ij}[U]\delta_{f'f}\psi_j^f,$$

where $K_{ij}[U]$ is flavour-independent. It therefore follows that

$$det\tilde{K}[U] = (detK[U])^{N_f}. \tag{16.55}$$

An important property of the matrix $K[U]$ is that its determinant is real , i.e., $\det K[U] = \det K^\dagger[U]$, and that $\det K[U] > 0$ for values of the hopping parameter $\kappa = 1/(8r + 2\hat{M}_0)$ less than 1/8 (Seiler, 1982). We can therefore also write $\det K$ in the form

$$\det K[U] = \sqrt{\det Q[U]}, \tag{16.56a}$$

where

$$Q[U] = K^\dagger[U]K[U] \tag{16.56b}$$

is now a positive, hermitean matrix. This will be important for the following discussion. It follows from (16.55) and (16.56a) that (16.52a) is given by the ensemble average of $\langle O\rangle_{S_F}$, defined in (16.52b), calculated with the Boltzmann factor $\exp(-S_{eff})$, where

$$S_{eff}[U] = S_G[U] - \frac{N_f}{2}\ln\det Q[U]. \tag{16.57}$$

* In chapter 6 we only considered the case of a single flavour.

Suppose now that we use the Metropolis algorithm to update the link–variable configurations. Then the Metropolis acceptance/rejection test requires us to calculate the difference $S_{eff}[U'] - S_{eff}[U]$, where U denotes the original configuration to be updated, and U' is the new suggested configuration. Hence we must calculate the ratio

$$\rho(U, U') = \frac{\det Q[U']}{\det Q[U]}. \qquad (16.58)$$

Although $Q[U]$ is a sparse (but large) matrix, since it is constructed from $K[U]$ which only couples nearest neighbours, the exact computation of $\rho(U, U')$ is too slow to be useful for updating link configurations on large lattices, and one must look for a faster method for computing (16.58). Clever suggestions in this direction were made in 1981 by Fucito, Marinary, Parisi and Rebbi (1981) and by Petcher and Weingarten (1981).

The proposal of Fucito et al. was the following. Let δU denote the difference between the suggested link–variable configuration and the old configuration, i.e., $\delta U \equiv \{\delta U_\mu(n)\}$, where $\delta U_\mu = U'_\mu(n) - U_\mu(n)$. The corresponding change in the matrix $Q[U]$ we denote by δQ: $\delta Q = Q[U'] - Q[U]$. Then

$$\rho(U, U') = \frac{\det(Q + \delta Q)}{\det Q} = \det(1 + Q^{-1}\delta Q). \qquad (16.59)$$

For small matrices $Q^{-1}\delta Q$, this expression can be linearized in δQ:

$$\rho(U, U') \approx 1 + TrQ^{-1}\delta Q. \qquad (16.60)$$

This is a reasonable approximation if the link–variables are updated one by one, and if only small changes δU are allowed. Now the computation of the trace is easy. But Q^{-1} must be evaluated at each updating step. This is very time consuming. The usual procedure is therefore to use the same matrix $Q^{-1}[U]$ for an entire sweep through the lattice. This is consistent with the linearized approximation (16.60), but may lead to large systematic errors, since the errors introduced at each updating step can accumulate during a sweep. Even with this approximation the exact evaluation of Q^{-1} on the link–variable configurations generated after each sweep through the lattice is not feasable. But one may determine Q^{-1} approximately by noting that its matrix elements are given by the following ensemble average

$$Q^{-1}_{ij} = \langle \phi_i \phi^*_j \rangle = \frac{\int D\phi^* D\phi \; \phi_i \phi^*_j e^{-\sum_{i,j} \phi^*_i Q_{ij} \phi_j}}{\int D\phi^* D\phi e^{-\sum_{i,j} \phi^*_i Q_{ij} \phi_j}}. \qquad (16.61)$$

Here ϕ and ϕ^* are the so-called pseudofermionic (bosonic!) variables.* Because Q is a positive hermitean matrix, the exponential appearing in (16.61) can be interpreted as a Boltzmann factor. This is the reason why we have expressed $\det K[U]$ in the form (16.56a).

According to (16.61), one can calculate the inverse matrix Q^{-1} by a) generating a complex set of pseudofermionic variables with probability density $\exp(-\phi^* Q\phi)$, b) measuring the products $\phi_i\phi_j^*$ on each configuration generated, and c) taking the ensemble average. Hence the main work consists in generating the configurations.

At the same time when Fucito et al. made their proposal, Petcher and Weingarten (1981) made an alternative suggestion based on similar ideas. These authors made use of the fact that the determinant of Q is given by the following path integral expression over pseudofermionic variables

$$\det Q = \frac{1}{\det Q^{-1}} = \int D\phi^* D\phi e^{-\sum_{i,j}\phi_i^* Q_{ij}^{-1}\phi_j}.$$

As seen from (16.57), this is the relevant determinant for the case of *two* quark flavours. The corresponding partition function can therefore be written in the form

$$\begin{aligned} Z &= \int DU \det Q[U] e^{-S_G[U]} \\ &= \int DU D\phi^* D\phi e^{-S_G[U]-\sum_{i,j}\phi_i^* Q_{ij}^{-1}[U]\phi_j}. \end{aligned} \tag{16.62}$$

This shows that to compute ensemble averages, one must generate configurations with a *non-local* action. Hence the effectiveness of this method depends on the efficiency of the algorithm for calculating the inverse of $Q[U]$.

Following the work of Fucito et al., Scalapino and Sugar (1981) pointed out that in a local updating procedure, the computation of (16.58) is reduced to the evaluation of the determinant of a small matrix. This can be easily seen. Consider the logarithm of the ratio $\det K'/\det K \equiv \det K[U']/\det K[U]$:

$$\begin{aligned} \ln\frac{\det K'}{\det K} &= Tr\ln(1 + K^{-1}\delta K) \\ &= \sum_{N=1}^{\infty}\frac{(-1)^{N+1}}{N} Tr(K^{-1}\delta K)^N. \end{aligned}$$

* To every fermion degree of freedom there corresponds a pair of complex (bosonic) variables.

Now

$$Tr(K^{-1}\delta K)^N = Tr\{K^{-1}\delta K K^{-1}\delta K \ldots K^{-1}\delta K\}.$$

If only a single link–variable is changed in the updating procedure, then δK has only non–vanishing matrix elements between two fixed nearest neighbour sites. Because of the trace we only need to know K^{-1} in the subspace defined by $\delta K \neq 0$. Let us denote this matrix by $\widehat{K^{-1}}$. Then

$$\begin{aligned}
\ln \frac{\det(K+\delta K)}{\det K} &= \sum_{N=1}^{\infty} \frac{(-1)}{N}^{N+1} Tr(\widehat{K^{-1}}\delta K)^N \\
&= Tr \ln(1+\widehat{K^{-1}}\delta K) \\
&= \det(1+\widehat{K^{-1}}\delta K).
\end{aligned}$$

If the suggested configuration $U+\delta U$ is accepted by the Metropolis test, then the next updating step requires the knowledge of $[K(U+\delta U)]^{-1} = (K+\delta K)^{-1}$ in the appropriate new subspace. The inverse of the matrix $K+\delta K$ can be calculated as follows:

$$\begin{aligned}
(K+\delta K)^{-1} &= \frac{1}{1+K^{-1}\delta K} K^{-1} \\
&= \sum_{N=0}^{\infty} (-1)^N (K^{-1}\delta K)^N K^{-1} \\
&= K^{-1} - K^{-1}\delta K K^{-1} + K^{-1}\delta K K^{-1}\delta K K^{-1} + \ldots,
\end{aligned}$$

or

$$(K+\delta K)^{-1} = K^{-1} - K^{-1}[\delta K - \delta K K^{-1}\delta K + \ldots]K^{-1}.$$

The matrix appearing within square brackets has only non–vanishing support in the subspace mentioned before. Define

$$\delta M = \delta K - \delta K K^{-1}\delta K + \delta K K^{-1}\delta K K^{-1}\delta K + \ldots$$

Then

$$(K+\delta K)^{-1} = K^{-1} - K^{-1}\delta M K^{-1}, \tag{16.63}$$

and

$$\begin{aligned}
\delta M &= \delta K(1 - \widehat{K^{-1}}\delta K + \widehat{K^{-1}}\delta K \widehat{K^{-1}}\delta K + \ldots) \\
&= \delta K(1+\widehat{K^{-1}}\delta K)^{-1}.
\end{aligned}$$

Hence one never has to calculate directly determinants or inverses of large matrices. However, because after every change in a link variable the whole large inverse matrix must be updated according to (16.63), this algorithm is too slow to be useful in Monte Carlo calculations, except when studying field theories in two space-time dimensions, or four-dimensional field theories on very small lattices. But it can serve to check the efficiency of other approximate algorithms, such as those mentioned above, and for checking the results obtained by other approximate methods.

Finally, we want to mention an interesting modification of the algorithm of Fucito et al., proposed by Bhanot, Heller and Stamatescu (1983), which is not burdened with the systematic errors arising from the neglect of higher order corrections in δU. These authors suggested that one should compute the ratio (16.58) directly using the pseudofermionic method. Since

$$\frac{1}{\det Q[U]} = \int D\phi^* D\phi e^{-\sum_{i,j} \phi_i^* Q_{ij}[U]\phi_j}$$

this ratio is given by the following expression

$$\rho(U, U') = \frac{\int D\phi^* D\phi e^{-\phi^\dagger Q\phi}}{\int D\phi^* D\phi e^{-\phi^\dagger \delta Q\phi} e^{-\phi^\dagger Q\phi}}, \tag{16.64}$$

where $\delta Q = Q' - Q$, with $Q' = Q[U']$, $Q = Q[U]$. To simplify the expression we have used matrix notation. According to (16.64), ρ is given by the inverse of the ensemble average of $\exp(-\phi^* \delta Q\phi)$, calculated with the Boltzmann factor $\exp(-S_Q)$, where $S_Q = \phi^\dagger Q\phi$ is a local expression of the link variables:

$$\rho = \frac{1}{\langle \exp(-\phi^\dagger \delta Q\phi)\rangle_{S_Q}}. \tag{16.65a}$$

Alternatively, ρ is also given by

$$\rho = \langle \exp(\phi^\dagger \delta Q\phi)\rangle_{S_{Q'}}, \tag{16.65b}$$

where $S_{Q'}$ is the pseudofermion action for the *suggested* link–variable configuration U'. How does one recover the result of Fucito et al.? Consider the expression (16.65b). For small arguments of the exponential we can replace ρ by

$$\rho \approx 1 + \sum_{i,j} \langle \phi_i^* \phi_j\rangle_{S_Q} \delta Q_{ij},$$

or making use of (16.61)

$$\rho \approx 1 + TrQ^{-1}\delta Q.$$

This coincides with the approximation (16.60).

The advantages of the method of Bhanot et al. are that i) one is not limited to small changes in the link–variables; ii) the only errors introduced are statistical and iii) the accuracy of the method can be estimated by computing the ratio of the determinants either from (16.65a) or (16.65b).

This is all we want to say about the pseudofermion method. For details of how this method is implemented in praxis the reader should confer the published literature.

16.9 Application of the Hybrid Monte Carlo Algorithm to Systems with Fermions

We conclude this chapter by giving an example of how the hybrid Monte Carlo (HMC) algorithm of Duane et al. (1987) is implemented in gauge theories with fermions. To keep the discussion simple, we will restrict ourselves to lattice QED with two flavours of mass-degenerate Wilson fermions.

Before the HMC algorithm can applied to calculate a correlation function one must first write the path integral expressions in the form of an ensemble average over bosonic variables. This is done by proceeding in the way described in the previous section. The relevant partition function is given by (16.62), where Q is the positive hermitean matrix defined in (16.56b).

The next step consists in rewriting the partition function as a path integral in phase space. To this effect one introduces the momenta π_i^*, π_i and P_l canonically conjugate to ϕ_i, ϕ_i^* and A_l, where A_l is the angular variable parametrizing a link labeled by the collective index l.* The corresponding link variable we denote by U_l. Defining the "Hamiltonian"

$$H = \frac{1}{2}\sum_l P_l^2 + \sum_i \pi_i^* \pi_i + S_G[U] + S_{PF}[U, \phi, \phi^*], \tag{16.66}$$

where S_{FP} is the pseudofermionic action

$$S_{PF} = \sum_{i,j} \phi_i^* Q_{ij}^{-1}[U] \phi_j, \tag{16.67}$$

---

* l stands for the coordinates of the lattice site n and the direction μ labeling a particular link variable, and U_ℓ and A_ℓ are related by $U_\ell = exp(iA_\ell)$.

and S_G is the action of the pure $U(1)$ gauge theory defined in (5.21), the partition function (16.62) can be written in phase space as follows

$$Z = \int DUD\phi D\phi^* DPD\pi D\pi^* e^{-H}.$$

Recall that the Hamilton equation of motion derived from (16.66) describe the dynamics of a system in a new time τ, which is identified with the simulation time in a Monte Carlo calculation. These equations are given by

$$\dot{\phi}_i(\tau) = \pi_i(\tau), \tag{16.68a}$$

$$\dot{\pi}_i(\tau) = -\sum_j Q_{ij}^{-1}[U]\phi_j(\tau), \tag{16.68b}$$

$$\dot{A}_l(\tau) = P_l(\tau), \tag{16.68c}$$

$$\dot{P}_l(\tau) = -\frac{\partial S_G}{\partial A_l(\tau)} - \sum_{i,j} \phi_i^*(\tau)\frac{\partial Q_{ij}^{-1}[U]}{\partial A_l(\tau)}\phi_j(\tau). \tag{16.68d}$$

Expressed in terms of the link variables, (16.68c) becomes

$$\dot{U}(\tau) = iP_l(\tau)U(\tau).$$

The derivative of Q^{-1} appearing in (16-68d) can be written in a more convenient form by making use of the fact that Q is given by the product (16.56b), and that $\partial(K^{-1}K)/\partial A_l = \partial((K^{\dagger -1}K^{\dagger})/\partial A_l = 0$. One readily verifies that

$$\frac{\partial Q^{-1}}{\partial A_l} = Q^{-1}\frac{\partial Q}{\partial A_l}Q^{-1}.$$

Introducing this expression into (16.68d), we see that the equations (16.68) are equivalent to the following set:

$$\eta_i = \sum_j Q_{ij}^{-1}[U]\phi_j, \tag{16.69a}$$

$$\dot{\phi}_i = \pi_i, \tag{16.69b}$$

$$\dot{\pi}_i = -\eta_i, \tag{16.69c}$$

$$\dot{U}_l = iP_l U_l, \tag{16.69d}$$

$$\dot{P}_l = -\frac{\partial S_G}{\partial A_l} - \sum_{i,j} \eta_i^* \frac{\partial Q_{ij}}{\partial A_l}\eta_j, \tag{16.69e}$$

In principle these equations are to be used in the molecular dynamics part of the HMC algorithm (see section 7). But this is not what one does in practice, for there exists a faster way of generating a set of configurations in the pseudofermionic and link variables. Thus consider the pseudofermionic action (16.67). The inverse of the matrix Q is given by the product $K^{-1}(K^{\dagger})^{-1}$. Hence this action can be written in the form $S_{PF} = \sum_i \xi_i^* \xi_i$, where the vector $\xi = (\xi_1, \xi_2, ...)$ is constructed from $\phi = (\phi_1, \phi_2, ...)$ as follows

$$\xi = (K^{\dagger}[U]^{-1})\phi.$$

Now one can easily generate configurations in these new variables which are distributed according to $\exp(\xi^{\dagger}\xi)$. By calculating $\phi = K^{\dagger}[U]\xi$, one then obtains, for *fixed* $U = \{U_l\}$, an ensemble of configurations in the pseudofermionic variables distributed according to $\exp(-\phi^{\dagger}Q^{-1}\phi)$. Each of these configurations can be used as an background field in a molecular dynamics chain for the link variables. This is what one does in practice. Hence the implementation of the HMC algorithm goes as follows:

i) Choose a starting link variable configuration.

ii) Choose P_l from a Gaussian ensemble with Boltzmann factor $\exp(-\frac{1}{2}\sum_l P_l^2)$.

iii) Choose ξ to be a field of Gaussian noise.

iv) Calculate

$$\phi = K^{\dagger}[U]\xi$$

v) Allow the link variables and the canonical momenta P_l to evolve deterministically according to (16.69d,e), where $\eta = Q^{-1}[U]\phi$, with ϕ held fixed.

vi) Accept the new configuation (A', P') generated by the molecular dynamics chain with probability

$$p = min\left(1, \frac{e^{-\tilde{H}[A',P']}}{e^{-\tilde{H}[A,P]}}\right),$$

where $\tilde{H}$ is the Hamiltonian governing the dynamics of the link variables in the presence of a background field ϕ:

$$\tilde{H}[A, P_l] = \frac{1}{2}\sum_l P_l^2 + S_G[U] + \sum_{i,j} \phi_i^* Q_{ij}^{-1}[U]\phi_j$$

vii) Store the new configuration generated, or the old configuration, as dictated by the Metropolis test.

viii) Return to ii).

The step v) is realized using the leapfrog method, described in section 6. Thus (16.69e) is written in the form

$$P_l(\tau + \frac{3\epsilon}{2}) = P_l(\tau + \frac{\epsilon}{2}) - \epsilon \left[\frac{\partial S_G}{\partial A_l(\tau+\epsilon)} + \sum_{i,j} \eta_i^*(\tau+\epsilon) \frac{\partial Q_{ij}}{\partial A_l(\tau+\epsilon)} \eta_j(\tau+\epsilon) \right],$$

where the first half step is calculated as described in section 6. On the other hand the naive discretization of (16.69d) reads:

$$U_l(\tau+\epsilon) = U_l(\tau) + i\epsilon P_l(\tau + \frac{\epsilon}{2}) U_l(\tau).$$

Link variables updated in this way would, however, not be elements of the $U(1)$ group. For this reason one discretizes (16.69d) as follows:

$$U_l(\tau+\epsilon) = e^{i\epsilon P_l(\tau+\frac{\epsilon}{2})} U_l(\tau),$$

which for infinitesimal time steps is equivalent to the naive discretization.

The Hybrid Monte Carlo algorithm is the best algorithm available at present for simulating dynamical fermions. It allows one to choose large time steps without introducing any systematic errors. This is guaranteed by the Metropolis test. Since in step v) the system is allowed to evolve according to the Hamilton equations of motion, the only change in H will be due to the finite time-step approximation. Hence most of the configurations generated will be accepted by the Metropolis test, if the time step is not too large. This means that the system is moving fast through configuration space. The algorithm is however rather slow, since it requires the computation of the vector $\eta = Q^{-1}[U]\phi$ in step v). Furthermore, it is not clear how to implement it for an odd number of quark flavours.

CHAPTER 17

SOME RESULTS OF MONTE CARLO CALCULATIONS

We begin this chapter with an apology. As the reader can imagine, there has been an enormous amount of activity within the computational sector of lattice gauge theories since their invention in 1974. Most of the pioneering work has been performed on small lattices. Since then many calculations have been improved by working on larger and finer lattices, and by developing new computational techniques. Clearly we cannot possibly do justice to all the different groups of physicists who have made important contributions to this field. What we intend to do, is to acquaint the reader with some of the questions that physicists have tried to get an answer to, and to present some lattice calculations which illustrate the results one has obtained, placing special emphasis on pioneering work. Thus any comments should always be seen in the light of the Monte Carlo data obtained at that time. For a critical analysis of the data we present the reader should consult the original articles we cite. References to other contributions can be found in the proceedings to the numerous lattice conferences.

17.1 The String Tension and the $q\bar{q}$-Potential in the $SU(3)$ Gauge Theory

The string tension was defined in chapter 8 as the coefficient σ of the linearly rising part of the potential for large separations of a quark-antiquark pair. It is the force between a static quark and antiquark at infinite separation in the absence of pair production processes. In a numerical calculation this force can only be measured for relatively small $q\bar{q}$-separations (in lattice units). Hence if the scale on which this force is seen in nature is determined by the size of hadronic matter then one must choose the lattice to be sufficiently coarse (i.e., the coupling g_0 sufficiently large) for the hadron, represented by the $q\bar{q}$ system, to fit on it. But if the lattice is made too coarse, one cannot expect to extract continuum physics, and the dimensionless string tension σa^2 will not exhibit the behaviour predicted by the renormalization group. Thus one is faced in practice with a precarious situation: if one chooses too small a lattice spacing (or coupling constant), then one cannot expect to see the asymptotic form of the force on a lattice of small extent. On the other hand, if the lattice spacing is too large, then one cannot expect to see continuum physics. Thus at best one can hope to measure the physical string tension in a narrow range of the coupling constant. This is the so-called "scaling window".

Let us now see how things look in practice. To this effect recall first of all that the static $q\bar{q}$-potential in lattice units can in principle be determined by calculating the following limit:

$$\hat{V}(\hat{R}) = -\lim_{\hat{T}\to\infty}\left[\frac{1}{\hat{T}}\ln W(\hat{R},\hat{T})\right] \quad ,$$

where $W(\hat{R},\hat{T})$ is the expectation value of the Wilson loop with spatial and temporal extension $\hat{R}$ and $\hat{T}$, respectively. Since this expectation value is determined in general from a MC calculation where neither $\hat{R}$ nor $\hat{T}$ are very large, $\hat{V}(\hat{R})$ is not expected to be of the form (8.17a). In fact, already the self-energy contributions to $\ln W(\hat{R},\hat{T})$ will be proportional to the perimeter $\hat{R}+\hat{T}$, and will compete with the area term $\sigma\hat{R}\hat{T}$, for finite $\hat{T}$ and $\hat{R}$. Effects arising from terms proportional to the perimeter can, however, be easily eliminated. Assuming that $W(\hat{R},\hat{T})$ has the form

$$W(\hat{R},\hat{T}) = e^{-\hat{\sigma}\hat{R}\hat{T}-\hat{\alpha}(\hat{R}+\hat{T})+\hat{\gamma}}, \tag{17.1}$$

we can isolate the string tension $\hat{\sigma}$ by studying the Creutz ratios

$$\chi(\hat{R},\hat{T}) = -\ln\left(\frac{W(\hat{R},\hat{T})W(\hat{R}-1,\hat{T}-1)}{W(\hat{R},\hat{T}-1)W(\hat{R}-1,\hat{T})}\right) \quad .$$

If the Wilson loop depends on $\hat{R}$ and $\hat{T}$ in the way given by (17.1), then $\chi(\hat{R},\hat{T})$ will be independent of these variables, and will coincide with the string tension. In fig. (17-1) we show the first MC data for the SU(2) string tension (black dots) obtained by Creutz (1980) together with the results of the strong coupling expansion carried out by Münster (1981) up to twelvth order in $\beta = 4/g_0^2$. For comparison the lowest order result for $\hat{\sigma}$ in strong coupling has also been included. Notice that in the narrow "window" $2.2 < \beta < 2.5$ the MC data follow an exponential curve predicted by the renormalization group; indeed, for the case of SU(2), the relation between the lattice spacing a and the coupling g_0 analogous to (9.20) is found to read as follows:

$$a = \frac{1}{\Lambda_L}R(g_0)$$

where

$$R(g_0) \approx e^{-(3\pi^2/11)\beta}$$

and*

$$\beta = 4/g_0^2.$$

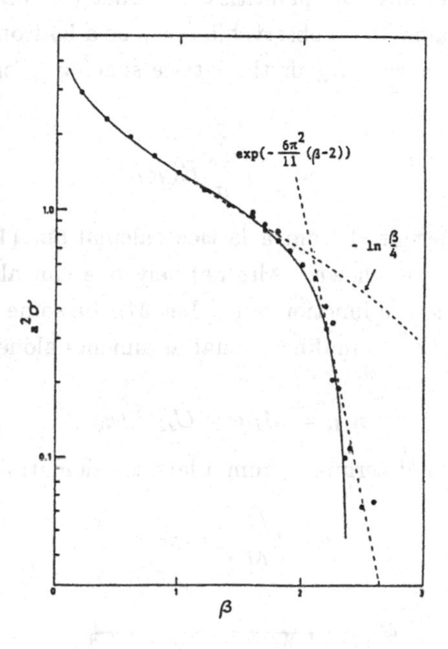

Fig. 17-1 The SU(2) string tension in lattice units as a function of $4/g_0^2$. The dots are MC results of Creutz (1980). The solid curve is the result of the cluster expansion carried out by Münster (1981).

Since the physical string tension σ has the dimension of $(\text{mass})^2$, $\hat{\sigma}(g_0)$ must behave as follows in the scaling region

$$a^2\sigma = \hat{\sigma}(g_0) \approx \hat{C}_\sigma[R(g_0)]^2. \tag{17.2}$$

This gives $\sqrt{\sigma}$ in units of the (dimensioned) lattice parameter Λ_L:

$$\sqrt{\sigma} = \sqrt{\hat{C}_\sigma}\Lambda_L. \tag{17.3}$$

The constant $\sqrt{\hat{C}_\sigma}$ can be determined from the weak coupling fit to the Monte Carlo data.

* Recall that for SU(N), β was defined as $\beta = 2N/g_0^2$. Λ_L is of course not to be identified with the corresponding lattice parameter in the $SU(3)$ gauge theory.

We want to point out that a lattice calculation can only determine a physical observable in units of the lattice parameter Λ_L; hence one can only calculate dimensionless ratios of physical quantities. Alternatively one may use the experimentally measured value for an observable (such as a hadron mass, string tension, etc.) to determine the scale Λ_L or the lattice spacing. For example solving eq. (17.2) for a we get

$$a = \sqrt{\frac{\hat{C}_\sigma}{\sigma}} R(g_0). \tag{17.4}$$

Since $\hat{C}_\sigma$ can be determined from a lattice calculation, (17.4) determines a in physical units, once σ is known. Alternatively one can also use the mass of a hadron to determine a as a function of g_0. Let M_H be some physical hadron mass (proton, pion, etc.); then from dimensional arguments alone we know that in the scaling region

$$\hat{M}_H = M_H a = \hat{C}_H \; R(g_0),$$

where $\hat{C}_H$ can again be determined from a lattice calculation; solving for a we get

$$a = (\frac{\hat{C}_H}{M_H}) R(g_0) \quad .$$

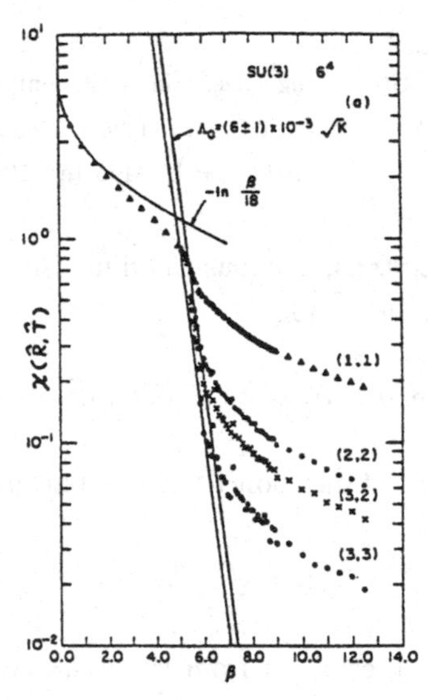

Fig. 17-2 Creutz ratios as a function of $\beta = 6/g_0^2$ for the SU(3) gauge theory. The figure is taken from Creutz and Moriarty (1982). The solid curve is the string tension in the leading strong coupling approximation.

Once the lattice spacing has been determined, one knows the physical extension of the lattice used in a MC calculation performed at a fixed value of the coupling. This gives us a handle of accessing the relevance of the lattice employed in studying a particular phenomenon.

Similar calculations to the one just described have been performed subsequently by Creutz and other authors for the case of SU(3). In fig. (17-2) we show the MC results of Creutz and Moriarty (1982) for the Creutz ratios constructed from different Wilson loops at various values of β, together with the strong coupling result in leading order. This calculation was carried out on a 6^4 lattice. From the figure it appears that asymptotic scaling sets in for values of β slightly below 6.0. Subsequent studies performed on larger lattices, however, indicated that scaling probably sets in for $\beta > 6.2$.

So far we have only considered the string tension, i.e. the linearly rising piece of the potential. For a proper determination of the string tension one should, however, also include in the fit the behaviour of the potential at short distances. For small separations of the $q\bar{q}$–pair, the potential is expected to exhibit a Coulomb type behaviour with a "running" coupling constant α depending logarithmically on the distance scale. This is predicted by asymptotic freedom.

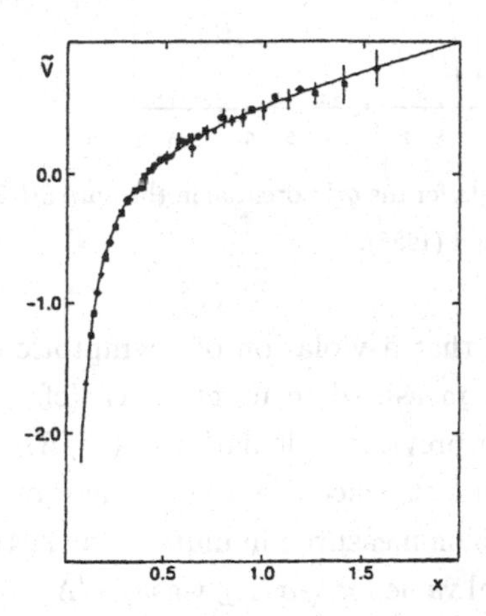

Fig. 17-3 MC data of Stack (1984) for the static $q\bar{q}$-potential at various $q\bar{q}$ separations. V and R are measured essentially in units of $\sqrt{\sigma}$ and $1/\sqrt{\sigma}$. The solid curve is a fit to the data based on the ansatz (17.5).

In fig. (17-3) we show the results of a MC calculation performed by Stack (1984) for the case of $SU(3)$ on a $8^3 \times 12$ lattice, with β values ($\beta = 6/g_0^2$) lying within the observed scaling window ($6.0 < \beta < 7.0$). The potential and the separation of the $q\bar{q}$ pair are measured essentially in units of $\sqrt{\sigma}$ and $1/\sqrt{\sigma}$. The solid line is a fit to the data based on a linear combination of a Coulomb and linear potential of the form

$$V(R) = -\frac{\alpha}{R} + \sigma R\,. \tag{17.5}$$

A large scale simulation was carried out later on by de Forcrand (1986) on a $24^3 \times 38$ lattice at $\beta = 6.3$. Fig. (17-4) shows the results of the MC calculation, together with a fit based on the form (17.5) for the potential. Similar results have been obtained by the same author at $\beta = 6.0$ on a 17^4 lattice.

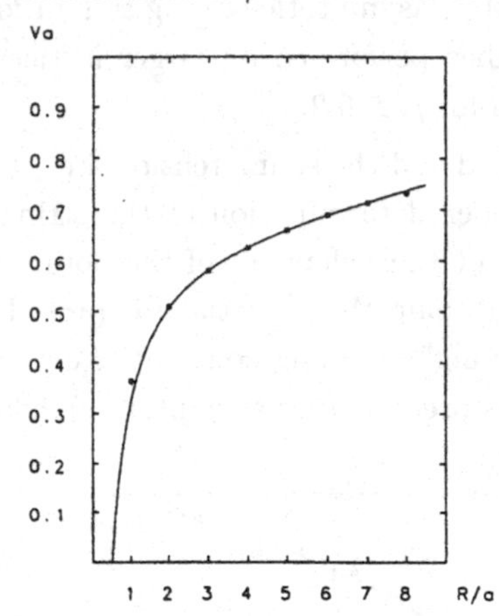

Fig. 17-4 MC data for the $q\bar{q}$-potential in the pure SU(3) gauge theory obtained by de Forcrand (1986).

What is noteworthy is that a violation of asymptotic scaling was observed, and that the string tension measured in units of Λ_L (cf. eq. (17.3)) was much lower than that obtained in previous calculations ($\sqrt{\sigma}/\Lambda_L \approx 92$ at $\beta = 6.0$, and $\sqrt{\sigma}/\Lambda_L \approx 79$ at $\beta = 6.3$). In fact, since the original determination of σ by Creutz and Moriarty, the string tension measured in units of the lattice parameter Λ_L, has kept decreasing. The original value for $\sqrt{\sigma}/\Lambda_L$ was $\sqrt{\sigma}/\Lambda_L = 167$. At the Berkeley conference (see Hasenfratz, 1986) the value had decreased to about 90. This value was extracted from measurements performed at rather small $q\bar{q}$–separations (up to 8 lattice sites). In a more recent calculation of the static $q\bar{q}$–potential performed with high statistics (Ding, Baillie and Fox, 1990) and for large separations of the quark and antiquark (up to $R = 12a$) an even lower value was obtained: $\sqrt{\sigma}/\Lambda_L \approx 77$. Also the value of the coupling α in eq. (17.5) was found to be quite different from that obtained in previous simulations ($\alpha = 0.58$ as compared to $\alpha \approx 0.3$). As Ding et al. point out, their values compare favourably with those of Eichten et al. (1980) obtained by fitting the (heavy quark) charmonium data with a Coulomb plus linear potential ansatz. Since charmonium is built from heavy quarks, it is reasonable to describe this system with a potential model. The values for the string tension $\sqrt{\sigma}$ and coupling α estimated by Eichten et al. were 420 MeV and 0.52, respectively. In a further paper Ding (1990) also finds that the dimensionless ratio $V/\sqrt{\sigma}$, which according to to (17.5) should be a function of $\sqrt{\sigma}R = \sqrt{\hat{\sigma}}\hat{R}$, is independent of the couplings used ($\beta = 6.0, 6.1$ and 6.2), as must be the case if one is in the scaling region. This is shown in Fig. (17-5).

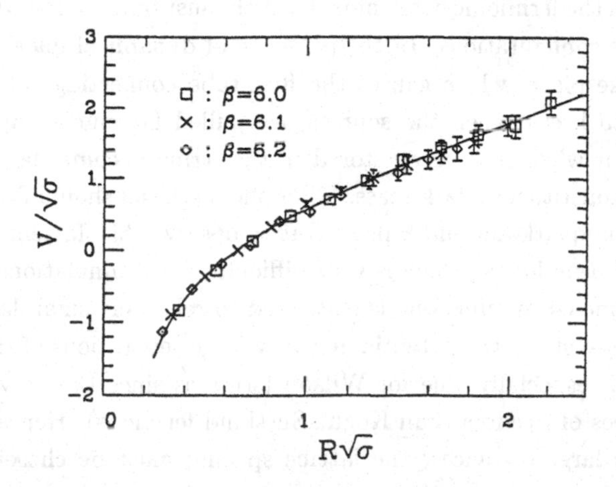

Fig. 17-5 MC data of Ding et al. for the $q\bar{q}$-potential obtained at various couplings $\beta = 6/g_0^2$. All the points are seen to lie on a single curve, indicating that one is in the scaling region. The figure is taken from Ding (1990).

A potential which rises linearly with the separation of the quarks is what is expected from a flux-tube picture of confinement. As we have mentioned in the introduction of chapter 7, it is generally believed that the non-abelian nature of the $SU(3)$ gauge theory causes the flux, linking the quark and antiquark, to be squeezed into a narrow tube. It is therefore of great interest to study the spatial distribution of the energy density between two static sources, immersed into a pure gluonic medium, in a Monte Carlo "experiment". Early work in this direction has been carried out by Fukugita and Niuya (1983) and by Flower and Otto (1985).

More complete studies of this problem have been carried out by Sommer (1987 and 1988) for $q\bar{q}$-separations up to four lattice units. For the case of the SU(2) gauge theory, detailed Monte Carlo studies of the flux distribution for a quark-antiquark pair have been reported by Haymaker et al. (1987-1992) and Bali et al. (1995). The authors find that the energy and action density are indeed confined to a very narrow region around the $q\bar{q}$-axis. For more details the reader may consult section 17.7.

17.2 The $q\bar{q}$-potential in full QCD

So far we have neglected the effects arising from pair production. These effects arise from the fermionic determinant which must be included when updating the link variable configurations. In the presence of dynamical quarks, quark-pair creation can take place, which causes the flux tube connecting the static quark and antiquark to break when the sources are pulled far enough apart. This is expected to occur when the energy stored in the string becomes larger than $2M$, where M is the constituent quark mass. Thus the potential should flatten for large separations of the quark-antiquark pair. But to observe this flattening, one needs to study large Wilson loops, which is very difficult. Since simulations in full QCD are extremely time-consuming, one is restricted to relatively small lattices, which do not allow one to study the potential for very large separations of the quark and antiquark (this is especially true for Wilson fermions since they involve a larger number of degrees of freedom than Kogut-Susskind fermions). Hence to calculate the potential at large distances, the lattice spacing must be chosen sufficiently large.

To identify screening effects due to quark-pair production processes, the full QCD data should, however, not be compared with the quenched data at the same value of the coupling. The reason is that part of the effects seen in numerical calculations can be interpreted in terms of a shift in the gauge coupling. Such a shift is expected, since when quarks are coupled to the gauge potential, vacuum polarization effects will lower the effective coupling. The amount by which the coupling is decreased will depend on the number of quark flavours. Also the relation between gauge coupling and lattice spacing is changed when dynamical fermions are included. To compare the full QCD results with the quenched data one should therefore first get an estimate of how the coupling is renormalized by vacuum polarization effects. The way that this is usually done in praxis is to determine the coupling in the quenched theory for which the expectation value of a Wilson loop or a plaquette variable agrees with that obtained in the full theory for a fixed choice of coupling.

The first calculation of the interquark potential with Wilson fermions which showed an impressive flattening of the potential for larger separations of the quark-antiquark pair was carried out by de Forcrand and Stamatescu (1986).

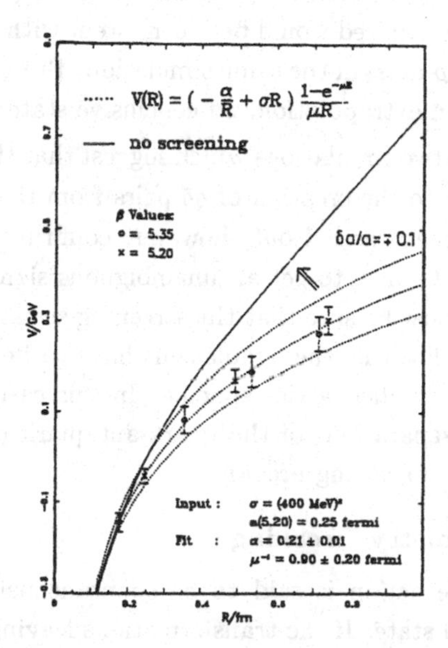

Fig. 17-6 MC data of Born et al. (1989) for the qq-potential in full QCD with Kogut-Susskind fermions. The data has been fitted with the screened potential (17.6). The neighbouring dotted lines show the shift in the potential if the lattice spacing is changed by 10%.

The lattice used was however very small, and it could not be ruled out that part of the screening observed was due to finite size (or temperature) effects. Kogut-Susskind fermions allow one to use larger lattices. Fig. (17-6) shows the potential in physical units obtained by Born et al. (1989) using Kogut-Susskind fermions. The calculation was performed on a $12^3 \times 24$ lattice for two couplings, $6/g_0^2 = 5.35$ and 5.20. The lattice spacing was determined by measuring the ρ mass in the same simulation. Hence $a = \hat{M}_\rho / M_\rho$, where $\hat{M}_\rho$ is the ρ mass in lattice units and $M_\rho = 770$ MeV is the physical ρ mass. These authors have parametrized the data according to

$$V(R) = \left(-\frac{\alpha}{R} + \sigma R \right) \frac{1 - e^{-\mu R}}{\mu R}, \tag{17.6}$$

where the string tension was fixed to be $\sqrt{\sigma} = 400$ MeV. The coupling and the screening mass μ were then determined from the fit to be $\alpha = 0.21 \pm 0.01$ and $\mu^{-1} = 0.9 \pm 0.2$ fermi. Although the data appears to indicate that the $q\bar{q}$ potential has been screened, this need not be the case. The conclusion is very sensitive to the lattice spacing. Thus by chosing a different lattice spacing, the data could also be fitted with an unscreened potential. The authors point out, however, that the change in lattice spacing required would be inconsistent with that determined from the measurement of the ρ-mass in the same simulation. But since the measurement of the ρ mass requires an extrapolation, no conclusive statement can be made.

We have mentioned two simulations which suggest that the potential is screened at "large" distances due to the creation of $q\bar{q}$-pairs from the vacuum. Not all the simulations that have been carried out, however, confirm this screening picture (see e.g. Gupta, 1989). Clearly to get an unambiguous signal for string-breaking at large distances, one must be sure that the screening seen is not a short distance or finite volume effect. For this the simulations have to be carried out on much larger lattices, and at a smaller lattice spacing. In this case the observation of a flat potential for large separations of the quark-antiquark pair would provide us with conclusive evidence for string breaking.

17.3 Chiral Symmetry Breaking

A symmetry of the action is said to be spontaneously broken if it is not respected by the ground state. If the transformations leaving the action invariant form a continuous group, then this spontaneous breakdown is accompanied by the appearance of massless particles, the so-called Goldstone bosons, which can propagate over large distances, giving rise to long-range correlations. The number of such Goldstone bosons equals the number of generators associated with the broken part of the symmetry group.*

In the limit of vanishing quark masses the pions observed in nature are believed to be the Goldstone bosons associated with a spontaneous breakdown of chiral symmetry. Let us illustrate what we mean by chiral symmetry for the case of an abelian $U(1)$ gauge theory. The continuum action for vanishing fermion mass has the form

$$S = \frac{1}{4}\int d^4x F_{\mu\nu}F_{\mu\nu} + \int d^4x \bar{\psi}\gamma_\mu(\partial_\mu + ieA_\mu)\psi \ .$$

---

* In the case of a gauge theory, there exist important cases where these Goldstone bosons do not appear, but are absorbed into the longitudinal degrees of freedom of the gauge field (Higgs mechanism).

Now the field ψ can always be decomposed into "left"- and "right"-handed parts as follows:

$$\psi = \psi_R + \psi_L \; ,$$

where

$$\psi_R = P_+\psi, \qquad \psi_L = P_-\psi \; ,$$
$$P_\pm = \frac{1}{2}(1 \pm \gamma_5); \qquad P_\pm^2 = P_\pm \; .$$

With

$$\bar{\psi}_R = \bar{\psi}P_-, \qquad \bar{\psi}_L = \bar{\psi}P_+ \; ,$$

the fermionic part of the action becomes

$$S_F = \int d^4x \bar{\psi}_L \gamma_\mu (\partial_\mu + igA_\mu)\psi_L + \int d^4x \bar{\psi}_R \gamma_\mu (\partial_\mu + igA_\mu)\psi_R \; .$$

This action is invariant under the following transformations

$$\begin{aligned} \psi_L &\longrightarrow e^{i\alpha}\psi_L, \qquad \bar{\psi}_L \longrightarrow e^{-i\alpha}\psi_L \; , \\ \psi_R &\longrightarrow e^{i\lambda}\psi_R, \qquad \bar{\psi}_R \longrightarrow e^{-i\lambda}\psi_R \; , \end{aligned} \tag{17.7}$$

where α and λ are arbitrary parameters. These transformations are elements of a $U_R(1)\times U_L(1)$ group: the chiral group. Consider now the ground state expectation value of $\bar{\psi}_R\psi_L$, i.e., $< \bar{\psi}_R\psi_L >$. $\bar{\psi}_R\psi_L$ transforms as follows under (17.7),

$$\bar{\psi}_R\psi_L \longrightarrow e^{i\Lambda}\bar{\psi}_R\psi_L \; ,$$

where $\Lambda = \alpha - \lambda$. If the transformations (17.7) are implemented by a unitary operator, and if the ground state is left invariant under the action of this operator, then $< \bar{\psi}_R\psi_L >$ must vanish. The same is true for $\langle\bar{\psi}_L\psi_R\rangle$. This need however not be the case in a quantum theory involving an infinite number of degrees of freedom, where the ground state may not be chirally invariant. Hence $\langle\bar{\psi}_R\psi_L\rangle$ may in fact be different from zero, implying that chiral symmetry has been broken spontaneously.

As we have just demonstrated, a good quantity for testing the spontaneous breakdown of chiral symmetry is $\langle\bar{\psi}_R\psi_L\rangle$. Alternatively, $\langle\bar{\psi}\psi\rangle = \langle\bar{\psi}_R\psi_L\rangle + \langle\bar{\psi}_L\psi_R\rangle$ will also do the job, and is the quantity usually studied in the literature. Since $\bar{\psi}\psi$ has the dimension of $(\text{length})^{-3}$, its expectation value in lattice units must have the following dependence on the bare coupling constant in the scaling region

$$\langle\hat{\bar{\psi}}\hat{\psi}\rangle = \hat{C}_{\bar{\psi}\psi}[R(g_0)]^3, \tag{17.8}$$

where $R(g_0)$ is given by eq. (9.21d). The constant $\hat{C}_{\bar{\psi}\psi}$ can be determined from a MC calculation. In physical units (17.8) then reads

$$\langle\bar{\psi}\psi\rangle = \hat{C}_{\psi\bar{\psi}}\Lambda_L^3$$

where Λ_L is the lattice scale appearing in (9.21c).

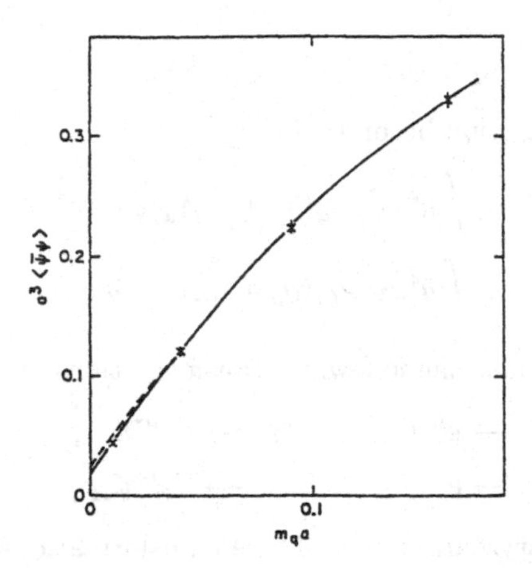

Fig. 17-7 MC data of Barkai, Moriarty and Rebbi (1985) for the chiral condensate $\langle\bar{\psi}\psi\rangle$ measured in lattice units, for various values of the dimensionless bare quark mass. The solid curve is a fit to the data, and has been extrapolated to the zero quark mass. The dashed line is a fit obtained by leaving out the point denoted by a cross which is rather sensitive to finite volume effects.

In fig. (17-7) we show the quenched data for $\langle\bar{\psi}\psi\rangle$ in lattice units as a function of the bare quark mass $\hat{m}_q$, obtained by Barkai, Moriarty and Rebbi (1985) on a $16^3 \times 32$ lattice at $\beta = 6.0$. The non-vanishing extrapolated value of $\langle\bar{\psi}\psi\rangle$ at $\hat{m}_q = 0$ is a sign of spontaneous symmetry breaking. If in the limit of vanishing quark mass the pion becomes the Goldstone boson associated with the breakdown of chiral symmetry, then its mass should vanish linearly with $\sqrt{\hat{m}_q}$. In fig. (17-8) we show the MC data of the above mentioned authors for $\hat{M}_\pi$ as a function of $\sqrt{\hat{m}_q}$. The linear fit extrapolates indeed to $\hat{M}_\pi = 0$ for vanishing quark mass. The calculation was performed using staggered fermions. Staggered fermions have been used also in most of the other computations of $\langle\bar{\psi}\psi\rangle$ in the literature. The reason is that, as we have seen in chapter 4, the action for Wilson fermions, (6.25c), breaks chiral symmetry explicitly also for $\hat{M}_0 \to 0$. Hence we cannot identify $\hat{M}_0$ with a bare quark mass. A possible definition of the quark mass is however suggested by the above made observation that M_π should vanish for $m_q \to 0$. This will occur for some critical value κ_c of the hopping parameter, corresponding to some $\hat{M}_0 = \hat{M}_c$:

$$\kappa_c = \frac{1}{8r + 2\hat{M}_c}.$$

This suggests the definition $\hat{m}_q = \hat{M}_0 - \hat{M}_c$, or equivalently

$$\hat{m}_q = \frac{1}{2}\left(\frac{1}{\kappa} - \frac{1}{\kappa_c}\right). \tag{17.9}$$

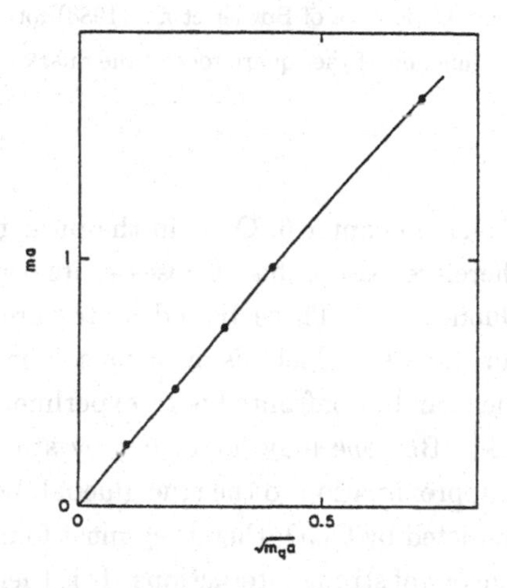

Fig. 17-8 MC data of Barkai, Moriarty and Rebbi (1985) for the pion mass as a function of the square root of the bare quark mass. Both masses are measured in lattice units.

Hence in a formulation using Wilson fermions, one must first determine the critical value of the hopping parameter where the pion mass vanishes, and then study the "chiral condensate" $\langle\bar{\psi}\psi\rangle$ in the limit $\kappa \to \kappa_c$. In fig. (17-9) we show a somewhat later calculation of the pion mass (in lattice units) as a function of $\sqrt{\hat{m}_q}$, performed by Bowler et al. (1988) on a $16^3 \times 24$ lattice, at $\beta = 6.15$. For a quark mass of less than 0.01 the data deviate from the linear behaviour. The above mentioned authors argue that this is probably due to finite-time effects which becomes increasingly important as the quark mass is lowered.

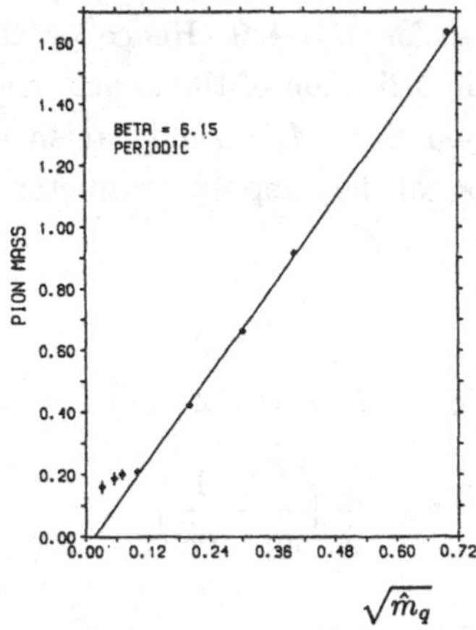

Fig. 17-9 Monte Carlo data of Bowler et al. (1988) for the pion mass in lattice units as a function of the square root of the quark mass $\hat{m}_q$.

17.4 Glueballs

As we have pointed out in chapter 6, QCD in the pure gauge sector is a very non trivial theory. We therefore expect that it posesses its own spectrum of bound states built from the gluon fields. These bound states are called glueballs. If the theory confines colour, then the glueballs must be colour singlets. Of course a realisitic calculation, which can be confronted with experiment, should include the effects arising from quarks. But one may hope that by studying the pure gauge sector one obtains a first approximation to the true glueball bound state spectrum. If glueballs are indeed predicted by QCD, then they must found experimentally, or QCD is ruled out as the theory of strong interactions. It is therefore very important to calculate the mass spectrum of the glueball states. The lattice formulation of QCD provides us with this possibility.

The basic idea that goes into the computation of the glueball mass spectrum is very simple. One first constructs a gauge invariant functional of the link variables located on a fixed time slice of the four-dimensional lattice, carrying the quantum numbers of the state one wishes to investigate. This functional is build from the trace of the product of link matrices along closed loops on the lattice. From these functionals one constructs zero momentum operators by summing the contributions obtained by translating the loops over the entire spatial lattice. Furthermore, since on the lattice we have only a cubic symmetry, but no rotational symmetry, the zero momentum operator should transform under an irreducible representation of the cubic group. These representations couple only to certain angular momenta in the continuum limit. *. The lowest glueball state is expected to be that carrying the quantum numbers of the vacuum ($J^{PC} = 0^{++}$), where P and C stands for parity and charge conjugation. An example of a zero momentum operator coupling to the spin zero state in the continuum limit is given by

$$O(\tau) = \sum_{\vec{x},orient.} \square \quad ,$$

where the square denotes an elementary plaquette variable with base at $(\vec{x}, \tau)$, and where the sum is carried out over all the positions and space orientations of the plaquette located in a fixed time slice. This is the simplest operator one can use to study a zero angular momentum glueball. In most calculations, however, other more complicated loops have been considered as well. Let us assume that we have chosen an operator $O(\tau)$ to study a particular glueball state. The lowest mass of a glueball carrying the quantum numbers of O can then, in principle, be determined from the behaviour of the correlation function $< O(\tau)O(0) >$ for large euclidean times. Indeed, recalling that $O(\tau) = \exp(H\tau)O(0)\exp(-H\tau)$, where H is the Hamiltonian, one has that

$$C(\tau) = \sum_n |\langle n|O\rangle|^2 e^{-E_n\tau} \ , \qquad (17.10)$$

where $|O>$ denotes the state created by the operator $O = O(0)$ from the vacuum, and E_n is the energy of the n'th eigenstate of H (measured relative to the vacuum.) In the large euclidean time limit, $C(\tau)$ is dominated by the lowest energy state carrying the quantum numbers of O . If these quantum numbers happen to coincide with that of the vacuum, then one is interested in the next higher energy state. Hence one must first subtract the vacuum contribution in (17.10) before taking the large euclidean time limit. This is accomplished by replacing $O(\tau)$ by $O(\tau)- < O(\tau) >$, or equivalently, subtracting $< O(\tau) >< O >$ from the correlation function. This yields the connected correlation function. Then

$$C(\tau)_{conn.} \underset{\tau\to\infty}{\longrightarrow} |\langle G|O\rangle|^2 e^{-E_G\tau} \ , \qquad (17.11)$$

---

* For a detailed discussion of this problem the reader may confer the lectures of Berg (Cargese 1983)

where $|G\rangle$ denotes the lowest glueball state above the vacuum with the quantum numbers of O. By studying the exponential decay of the correlation function for large euclidean times, one then extracts the glueball mass of interest. This sounds very simple indeed. But in praxis one is confronted with a number of serious problems. In fact, measuring the glueball masses is one of the most difficult tasks in numerical simulations of the SU(3) gauge theory. The ideal situation would be that the measured values of $C(\tau)$ would fall on a simple exponential curve, and that the lattice is fine enough for the gluon mass to exhibit the scaling behaviour predicted by the renormalization group. In particular, close to the continuum limit, the ratios of different glueball masses should become independent of the coupling. To see a single exponential would however require that we either are able to measure the correlation function for very large times, or that $|O\rangle$ has only a projection on the particular state of interest. This is certainly not the case in praxis. In the early days of glueball computations the operator O was built from simple plaquettes or from a linear combination of more complicated loops of different shapes and orientations. * The difficulty one encountered was that the signal was drowned within the statistical noise for times beyond two lattice spacings, even for the lightest glueball. One of the problems was that the overlap of the glueball states with $|O\rangle$ was too small. This overlap gets worse and worse as the lattice spacing is reduced. This is intuitively obvious since the physical extension of the glueball remains fixed, while the local operator O , constructed from small loops , probes an ever smaller region of the glue ball wave function as one decreases the lattice spacing. For this reason it is only possible to obtain a reasonable signal on coarse lattices, where the state $|O\rangle$ has a reasonable overlap with the glueball wave function. But even if the overlap is enhanced for larger lattice spacings, the glueball mass measured in lattice units is also increased, and hence the exponential in (17.11) leads to a stronger supression of the correlation function. Clearly what one needs are operators which have a strong overlap with the glueball state of interest for small lattice spacings. This would solve both problems mentioned above, and at the same time allow one to probe the scaling region. How important this is, is made evident by the observation that for local operators the overlap decreases with the fifth power of the lattice spacing.** This is clearly a disaster and makes it very hard to study continuum physics with

* for a review of early work see Berg (1983). For the earliest glueball calculations see Berg (1980); Bhanot and Rebbi (1981); Engels et al. (1981a).

** See the talk by Schierholz at Lattice 87 (Schierholz,1988), and the review talk by Baal and Kronfeld at Lattice 88 (1989).

local operators. After 1986 there have been several proposals for constructing non local operators for which $C(\tau)/C(0) \approx a$ (Berg and Billoire, 1986; Teper, 1986a; Albanese et. al., 1987; Kronfeld, Moriarty and Schierholz, 1988). The use of such (non-local, smeared, or fuzzy) operators improves the signal to noise ratio dramatically, and allows one to measure the correlation function of the 0^{++} and 2^{++} states for temporal separations much larger than than in previous calculations. As an example we show in fig. (17-10) the 0^{++} and 2^{++} correlation functions obtained by Brandstaeter et al. (Schierholz, 1989) at $6/g_0^2 = 6.0$ on a 16^4 lattice. A clear signal is obtained up to $t = 6$ for the 0^{++} state, and up to $t = 5$ for the 2^{++} state.*

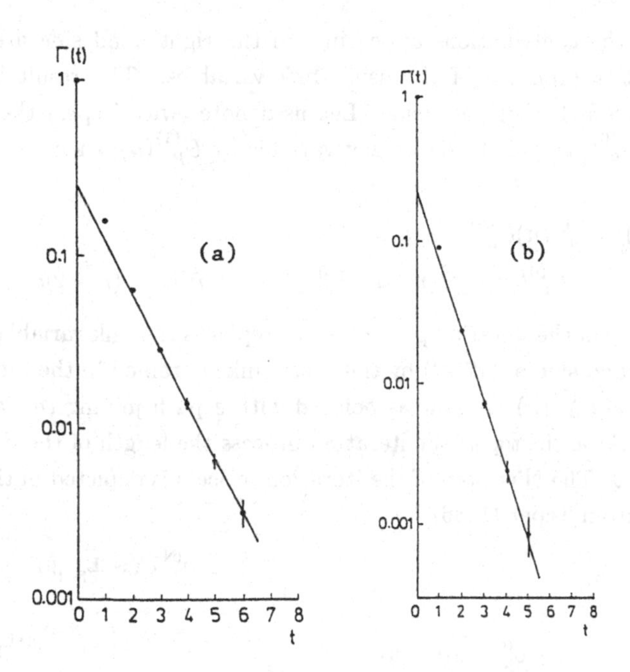

Fig. 17-10 MC data of Brandstaeter et al. for the (a) 0^{++} and (b) 2^{++} correlation function. The figure is taken from Schierholz (1989).

A particularily efficient prescription for constructing non local operators is that proposed by Teper (1986a). Since this prescription is very simple and intuitive, we describe it here. Teper constructs by an iterative method very complex non local operators O consisting of an enormous number of elementary paths. The

* For details regarding the operator used, see Schierholz (1988).

iterative scheme involves only the link variables located in a fixed time slice, and goes as follows. First one constructs so called *fuzzy* link variables associated with a path of length 2 along an arbitrary μ direction and with base at an arbitrary point n, by adding the contributions of the direct path and the spatial "staples" (as Teper calls them), as shown below:

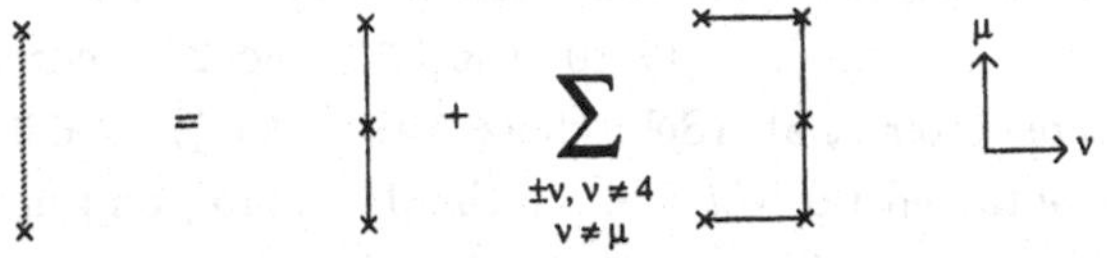

$$= \quad + \sum_{\substack{\pm\nu,\, \nu\neq 4 \\ \nu\neq\mu}} \qquad \mu \uparrow \; \rightarrow \nu \tag{17.12}$$

At this level the contributions appearing on the right hand side are calculated from the matrix products of the usual link variables. The result is the fuzzy "path" shown on the left hand side. Let us denote (with Teper) the usual links variables by $U_\mu^{(0)}(n)$ and the new fuzzy variable by $U_\mu^{(1)}(n)$; then

$$\begin{aligned} U_\mu^{(1)}(n) =& U_\mu^{(0)}(n)U_\mu^{(0)}(n+\hat{\mu}) \\ &+ U_\nu^{(0)}(n)U_\mu^{(0)}(n+\hat{\nu})U_\mu^{(0)}(n+\hat{\nu}+\hat{\mu})U_\nu^{(0)\dagger}(n+2\hat{\mu}). \end{aligned} \tag{17.13}$$

In the next step of the iteration processes one replaces the link variables appearing on the right hand side of (17.12) by the fuzzy links obtained in the first step. The new fuzzy links $U_\mu^{(2)}(n)$ are now associated with a path joining the lattice site n with $n+4\hat{\mu}$. At each step of the iteration process the length of the link increases by a factor of 2. The N'th step of the iteration process is depicted in the following figure, taken from Teper (1986):

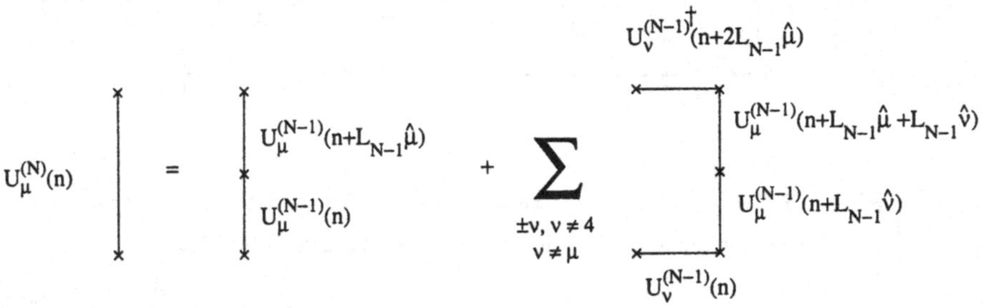

$U_\nu^{(N-1)\dagger}(n+2L_{N-1}\hat{\mu})$

$U_\mu^{(N)}(n)$ = $U_\mu^{(N-1)}(n+L_{N-1}\hat{\mu})$ $U_\mu^{(N-1)}(n)$ + $\sum_{\substack{\pm\nu,\, \nu\neq 4 \\ \nu\neq\mu}}$ $U_\mu^{(N-1)}(n+L_{N-1}\hat{\mu}+L_{N-1}\hat{\nu})$ $U_\mu^{(N-1)}(n+L_{N-1}\hat{\nu})$

$U_\nu^{(N-1)}(n)$

Here $l_N = 2^N$. Having generated at the N'th step the fuzzy "links" $U_\mu^{(N)}(n)$, one can e.g. construct from these superplaquettes in a manner completely analogous to the usual elementary plaquette:

$$= U_{ij}^{(N)}(n) = U_i^{(N)}(n)U_j^{(N)}(n+l_N\hat{\imath})U_i^{(N)\dagger}(n+l_N\hat{\jmath})U_j^{(N)\dagger}(n)$$

Hence loops of fuzzy links are now complicated objects when expressed in terms of elementary paths. Note that the matrizes $U_\mu^{(N)}(n)$ are not group elements. Only in the case of $SU(2)$ these are proportional to a unitary matrix, with the proportionality constant given by the determinant. But this is not of relevance here, since the aim is to merely construct sufficiently complex operators, to ensure that they create states having a large overlap with the glueball candidate of interest.

Because of the simplicity of the iterative scheme, one is now in the position to increase the number of iteration steps with decreasing coupling in such a way, that the size of the superplaquettes (or more complicated versions thereof) can be adjusted to the increasing volume (in lattice units) occupied by the glueball on the lattice. In this way one can achieve a large overlap of the operator with the glueball wave function even for small lattice spacing. Michael and Teper (1989) have applied the fuzzying prescription to the SU(3) gauge theory, and have calculated the glueball masses for all the JPC states, using couplings $6/g_0^2$ ranging between 5.9 and 6.2, and spatial lattices with volumes in the range from 10^3 to 20^3. They find, for example, that their fuzzy 0^{++} glueball operators have about 90% overlap with the corresponding glueball ground state, and this at a coupling $6/g_0^2$ as large as 6.2.

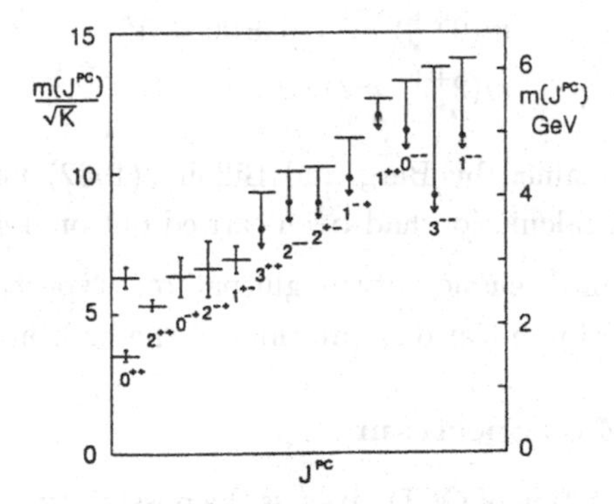

Fig. 17-11 Glueball masses in units of the square root of the string tension calculated by Michael and Teper (1989). The masses in physical units are given on the right hand scale and have been obtained using the value of 440 MeV for the string tension.

Fig. (17-11) shows the mass spectrum obtained by these authors. The masses in physical units are shown on the right hand scale. They have been obtained by using the value $\sqrt{\sigma} = 440MeV$ for the string tension.

The 0^{++} and 2^{++} glueball masses have so far been measured with greatest confidence. They can be extracted by studying the correlation function of the following operators: (see e.g., Ishikawa et al., 1983)

$$\begin{aligned}
\phi^{0^{++}}(\tau) =& ReTr \sum_{\vec{x}} (U_{12}(\vec{x},\tau) + U_{23}(\vec{x},\tau) + U_{31}(\vec{x},\tau)), \\
\phi^{2^{++}}(\tau) =& ReTr \sum_{\vec{x}} (U_{12}(\vec{x},\tau) - U_{13}(\vec{x},\tau)).
\end{aligned}$$

Here the "plaquette" variables U_{ij} can be ordinary plaquette variables or fuzzy superplaquette variables. Monte Carlo calculations performed with large lattice volumes by several groups (see the review of Michael, 1990) agree that the mass ratio $m(2^{++})/m(0^{++})$ is about 1.5, while early calculations based on the study of correlation functions of local operators, and $6/g_0^2 < 6.0$, suggested that this ratio is of the order of one or less. The physical values of the glueball masses can be expressed in terms of the string tension. Assuming a value of 420 MeV for the square root of the string tension, taken from potential models for heavy quark bound states (Eichten et al., 1980) the 0^{++} and 2^{++} masses reported by Baal and Kronfeld at Lattice 88 (Baal and Kronfeld, 1989) were

$$\begin{aligned}
m(0^{++}) =& 1370 \pm 90 MeV \\
m(2^{++}) =& 2115 \pm 125 MeV
\end{aligned}$$

The original value obtained by Berg and Billoire (1982) for the 0^{++} mass was $920 \pm 310 MeV$. This calculation had been carried out on a $4^3 \times 8$ lattice.

This concludes our discussion of the glueball mass spectrum. We now turn to the discussion of the Monte Carlo simulations of the hadron mass spectrum.

17.5 Hadron Mass Spectrum

The lattice formulation of QCD gives us the possibility of answering one of the most fundamental questions in elementary particle physics: what is the origin of the masses of the strongly interacting particles observed in nature? QCD should be able to predict the bound state spectrum of hadrons build from quarks and gluons, which are permanently confined within the hadrons. The determination of the masses of the lowest bound states with a given set of quantum numbers is based on the same ideas as discussed in the previous section; i.e., the masses are extracted from the large (euclidean) time behaviour of correlation functions for zero momentum operators carrying the appropriate quantum numbers to create the hadron of interest. For local operators constructed from Wilson fermions the generic form is given by

$$O_m(\tau) = \sum_{\vec{x}} \Gamma^{(m)}_{AB} \bar{\psi}_A(\vec{x},\tau)\psi_B(\vec{x},\tau), \qquad \text{(mesons)}$$

$$O_b(\tau) = \sum_{\vec{x}} S^{(b)}_{ABC} \psi_A(\vec{x},\tau)\psi_B(\vec{x},\tau)\psi_C(\vec{x},\tau), \qquad \text{(baryons)}$$

where we have used a continuum notation for convenience, and a summation over the collective indices (Dirac, colour and flavour) is understood. The coefficients $\Gamma^{(m)}_{AB}$ and $S^{(b)}_{ABC}$ are chosen in such a way that the operators are colour neutral* ($\Gamma^{(m)}_{AB} \propto \delta_{ab}$ for mesons, and $S^{(b)}_{ABC} \propto \epsilon_{abc}$ for baryons), and carry the quantum numbers (spin, parity, etc.) of the hadron of interest. Thus, for example, operators that can be used for studying the π^+, ρ^+, and proton are given by

$$O_{\pi^+} = \sum_{\vec{x}} \bar{d}^a \gamma_5 u^a; \quad \vec{O}_{\rho^+} = \sum_{\vec{x}} \bar{d}^a \vec{\gamma} u^a; \quad (O_p)_\alpha = \sum_{\vec{x}} \epsilon^{abc} u^a_\alpha [u^b_\beta (C\gamma_5)_{\beta\delta} d^c_\delta]$$

where u and d are the "up" and "down" quark fields, C is the charge conjugation matrix. The construction of the corresponding operators for Kogut-Susskind fermions is more complicated and we will not consider them here. **

For example, meson masses can be extracted by studying the behaviour of the correlation functions

$$\Gamma_m(\tau) = \langle O^\dagger_m(\tau) O_m(0)\rangle$$

for large euclidean times. This correlation function can be written as a sum of expectation values of products of two external field quark propagators calculated with a Boltzmann distribution $exp(-S_{eff}[U])$, where $S_{eff}[U]$ has been defined in (12.14b). In particular the pion correlation function for degenerate quark masses, describing the propagation of a quark-antiquark from $\vec{x}$ to $\vec{y}$, is given by

$$C_\pi(\tau) = \sum_{\vec{x},\vec{y}} \langle Tr(\gamma_5 \underset{\sim}{K}^{-1}_{\vec{x},0;\vec{y},\tau} \gamma_5 \underset{\sim}{K}^{-1}_{\vec{y},\tau;\vec{x}o})\rangle_{S_{eff}} \tag{17.14}$$

where K^{-1} is the quark propagator, i.e., the inverse of the matrix (12.2c). For baryons the correlation function analogous to (17.14) will involve the product of three quark propagators.

---

* Here a, b, c are the colour indices of the Dirac fields.

** See e.g., Morel and Rodrigues (1984); Golterman and Smit (1985); Golterman (1986b).

The procedure for calculating a correlation function is the following. Let us assume that one has decided on a set of values for the parameters appearing in the action, i.e., gauge coupling and quark masses in lattice units. The lattice size should in principle be chosen large enough to easily accommodate the hadron whose mass one wants to calculate. In particular, the extension of the lattice in the euclidean time direction must be large enough to allow one to study correlation functions for large euclidean times. This is important, since one doesn't want the measurements to be contaminated by contributions coming from higher excited states carrying the same quantum numbers of the hadron of interest. A priori, we do not know how large the lattice must (at least) be chosen, since we do not know the lattice spacing. In principle, this lattice spacing, which is controled by the gauge coupling, should be small enough to allow one to extract continuum physics. The values for the coupling constant and quark masses, and the linear extensions of the lattice used in numerical simulations, will be limited by the available computer facilities. Having fixed all these quantities, the next step consists in generating a set of link configurations which, in the quenched approximation, are distributed according to $\exp(-S_G[U])$, and in full QCD are generated with the probability density $\det K(U) \exp(-S_G[U])$. For each configuration one then computes the propagator $K^{-1}(U)$. It is this step which makes the computations of fermionic correlation functions – already in the quenched approximation – time-consuming. The algorithm most widely used to calculate $K^{-1}(U)$ is the conjugate gradient method. Once one has calculated the external field quark propagators, one constructs from these the hadron propagators, and averages the expression over the ensemble of field configurations. For mesons the ensemble average is carried out over products of two quark propagators, while for baryons the average is taken over products of three quark propagators. Only uncorrelated field configurations should be used for calculating the ensemble averages and the statistical errors. Thus for example in a local updating procedure, successive configurations generated by a Markov chain will be highly correlated, since one is changing only one, or a few links, at each updating step. Hence one must generate many more configurations than are actually used to calculate the ensemble average. In practice, only configurations separated by a fair number of sweeps through the lattice are used for measuring the observable. Finally the lowest hadron mass is extracted by studying the behaviour of the correlation function for large euclidean times. This yields the mass in lattice units. If one is working close to the continuum limit, then the ratios of different particle masses measured in lattice units should be independent of the coupling (or equivalently the lattice spacing) and can be identified with the corresponding ratios of the masses measured in physical units. In an actual Monte Carlo simulation these ratios will in general differ substantially from the experimentally measured values. The reason is that most calculations are performed at unrealistically high quark masses, of the order of the strange quark mass. Statistical fluctuations in the hadron propagators, which increase with decreasing quark masses, do not allow one to simulate such hadrons as the π, ρ and nucleon, for realistic dynamical bare (up and down) quark masses of the order of a few MeV.

Most earlier Monte Carlo simulations have been performed with degenerate quark masses. Hence the output of the calculation depends on the values used for the bare coupling constant, and on a single dimensionless quark mass. The quark mass in physical units at which the MC calculation has been performed can then be determined as follows. One calculates the masses of some hadrons such as the pion and the ρ for several values of the bare quark mass $\hat{m}$. One then fits the data for $\hat{M}_\pi$ and $\hat{M}_\rho$ using the following ansatz which is motivated from current algebra:

$$\hat{M}_\pi^2 = \lambda_\pi \hat{m}, \qquad (17.15a)$$

$$\hat{M}_\rho = \lambda_\rho \hat{m} + \beta_\rho. \qquad (17.15b)$$

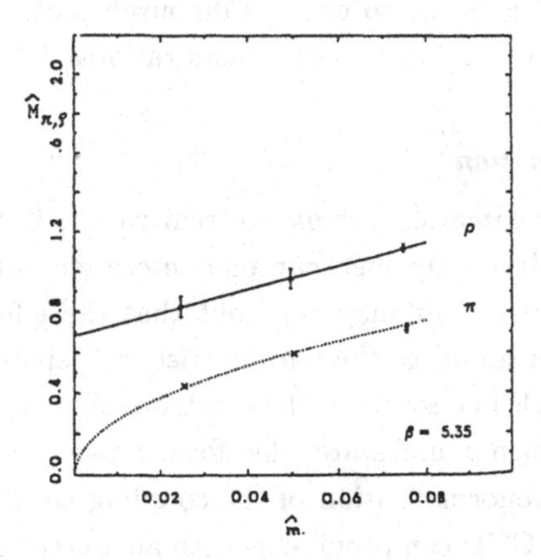

Fig. 17-12 Example of a fit to the MC data for the pion and rho mass with the ansatz (17.15). The figure is taken from Born et al. (1989).

In the case of Wilson fermions $\hat{m}$ is related to the hopping parameter by (17.9). Fig. (17-12) , taken from Born et al. (1989), gives an example of a fit to the MC data based on (17.15). From the fit one determines the coefficients λ_π, λ_ρ

and β_ρ. The lattice spacing can then be determined by recalling that $\hat{M}_\pi = M_\pi a$, and $\hat{M}_\rho = M_\rho a$. Thus by inserting for M_π and M_ρ their physical values, 140 MeV and 770 MeV, one calculates from (17.15) the lattice spacing and the quark mass. This information can be used to extract other hadron masses from Monte Carlo calculations. Assuming that the lattice spacing is only a function of the bare coupling (which was held fixed in the simulations performed at different quark masses) one then also knows the quark masses in physical units at which the MC calculations have been performed. The above way of proceeding involves an uncertainty in the extrapolation procedure. Furthermore it is only justified in simulations of full QCD, and not in the quenched approximation, since one doesn't know what the pion-rho mass ratio should be when the effects of dynamical fermions are ignored. Although the coefficients $\lambda_\pi, \lambda_\rho$ and β_ρ in (17.15) are dependent on the value of the bare coupling constant used, the ratio of hadrons masses should be independent of the coupling if one is working in the scaling region. In order to be able to compare directly the results obtained by various groups for Wilson and Kogut-Susskind fermions, and for different couplings and quark masses, the Edinburgh group (Bowler et al. (1985)) have suggested that one should exhibit, for example, the nucleon, pion, rho mass data as a plot of the nucleon-rho mass ratio versus the pion-rho mass ratio. In the scaling region all the data should then fall on a universal curve. This is the so called Edinburgh plot. In this way one can study the general trend of the nucleon-rho mass ratio as M_π/M_ρ approaches the physical value.

The Quenched Approximation

The quenched approximation is not the real world, but it is important to obtain reliable results, since only then can one determine how dynamical quarks influence the mass spectrum. It may turn out that the effects of pair creation (and annihilation) do not influence the hadron spectrum significantly. In fact, the success of potential models in describing the spectrum of heavy quark bound states suggest that at least for such bound states the effects arising from dynamical quarks can be absorbed into a renormalization of the coupling constant. Unfortunately, only a calculation in full QCD can provide us with an answer of how the quenched spectrum is modified by the dynamical quarks.

The earliest MC calculation of the quenched hadron spectrum in the SU(3) gauge theory date back to 1981 (Hamber and Parisi, 1981). Since then many groups have performed such calculations on larger lattices and with smaller lattice spacings using Wilson and Kogut-Susskind fermions. Up to 1988 the most ambitious calculation with Wilson fermions had been carried out by de Forcrand et al. (1988) on a $24^3 \times 48$ lattice. These authors used special blocking techniques in order to reduce the large number of degrees of freedom. Recall that for Wilson fermions every lattice site can accommodate all internal degrees of freedom. This makes these computations more demanding than those with staggered fermions. Fig. (17-13) taken from the above mentioned reference, illustrates the general form of meson correlation functions extracted from a Monte Carlo calculation performed on a periodic lattice in the time direction. The propagators displayed in Fig. (17-13) have been fitted in the interval $15 < t < 24$ with single mass exponentials symmetrized about $t = 24$.

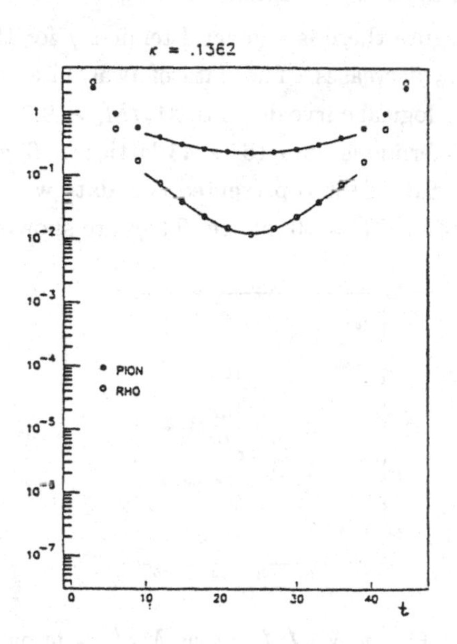

Fig. 17-13 MC data for the pion and rho propagators, obtained by Forcrand et al. (1988). The data has been symmetrized about $t = 24$.

The use of staggered fermions for computing the hadron mass spectrum poses a problem, since for finite lattice spacing the staggered fermion action brakes flavour symmetry (while Wilson fermions do not; see chapter 4). Already for this reason one does not expect that results obtained with Wilson and Kogut-Susskind fermions will agree, unless one is close to the continuum limit. And in fact, they did not agree for a long time. This situation has improved substantially and the calculations using Wilson and staggered fermions have given comparable results for $6/g_0^2 > 6.0$.*

Figure (17-14), taken from Yoshie, Iwasaki and Sakai (1990), shows the quenched results for the nucleon-rho mass ratio versus M_π/M_ρ reported prior to, or at the Capri lattice conference in 1989. The solid (open) points correspond to calculations performed with Wilson (Kogut-Susskind) fermions. The solid curve is obtained from a phenomenological mass formula (Ono, 1978). The open point at the left lower end of the curve corresponds to the experimental mass ratios. These data are from Barkai et al. (1985), Gupta et al. (1987), Hahn et al. (1987), Bowler et al. (1988), Bacilieri et al. (1988) and Iwasaki et al. (1989). The quark mass in all these calculations is, however, unrealistically large.

As seen from the figure there is a general tendency for the nucleon-rho mass ratio to drop as M_π/M_ρ decreases. The data of Iwasaki et al. agrees extremely well with the phenomenological curve down to $M_\pi/M_\rho \approx 0.7$. This calculation was performed with Wilson fermions on a $16^3 \times 48$ lattice at $6/g_0^2 = 5.85$. At lattice 89, Yoshie, Iwasaki and Sakai (1990) presented new data with increased statistics, including a calculation on a $24^3 \times 60$ lattice. They are shown in Fig. (17-15).

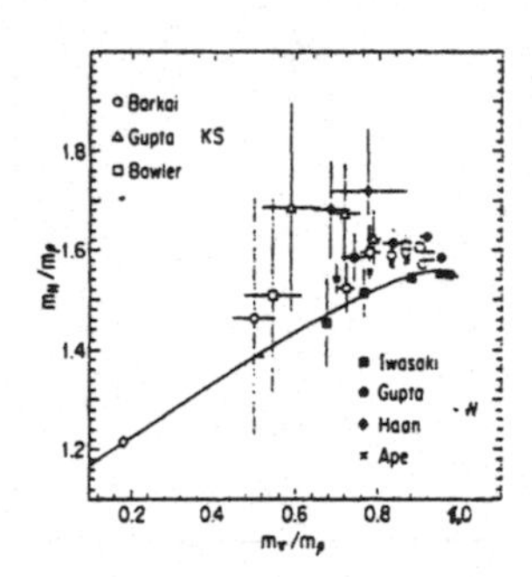

Fig. 17-14 Plot of M_N/M_ρ versus M_π/M_ρ in quenched QCD. The solid (open) points correspond to calculations performed with Wilson (Kogut-Süsskind) fermions. The figure, taken from Yoshie, Iwasaki and Sakai (1990), summarizes the results reported prior to, or at the Capri lattice conference.

The data are seen to be consistent with the phenomenological curve down to $M_\pi/M_\rho = 0.52$. One does however not expect that the quenched data (once it has settled down) follows this curve all the way down to the physical mass ratios. At some point the effects of light dynamical quarks must start to show up.

* see the plenary talk by Gupta at lattice 89 (Gupta, 1990).

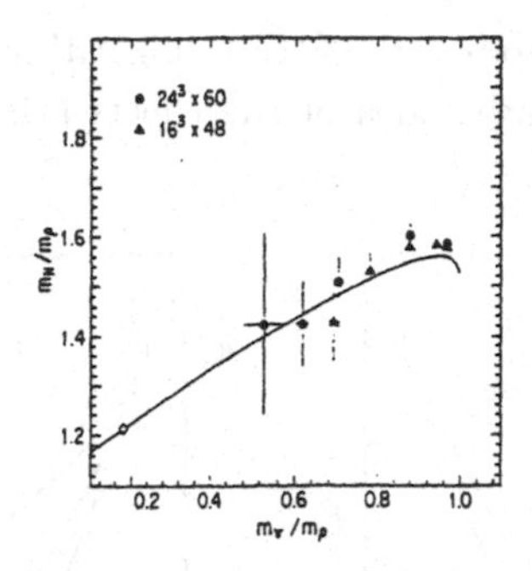

Fig. 17-15 MC data of Yoshie, Iwasaki and Sakai (1990) for M_N/M_ρ versus M_π/M_ρ.

We have only briefly discussed a few earlier measurements in quenched hadron spectroscopy. As we have pointed out, the calculation of the hadron spectrum is a difficult task, even in the quenched approximation. But even with precise numerical data available, one has no way of comparing it with experimental values, since we don't know what the result of a quenched calculation should be. Only when the effects of dynamical fermion are included, will we be able to answer the question whether QCD has survived this important test for being the correct theory of strong interactions. So let us take a look at some data that has been obtained including dynamical quarks. We will again restrict ourselves to earlier pioneering work, and leave it to the reader to confer the numerous proceedings for more recent data.

Hadrons in Full QCD

The quenched calculation of the hadron spectrum is already very time consuming since one needs to evaluate the quark propagators on the ensemble of link configurations distributed according to $\exp(-S_G[U])$. In full QCD, however, one must generate these configurations with the probability density $\det K[U] \exp(-S_G[U])$. Because of the appearance of the fermionic determinant, the times required for numerical simulations of the hadron spectrum in full QCD are very (!) much larger than in the quenched theory. For this reason the lattice volumes used are much smaller than in the quenched case. To avoid strong finite size effects one is therefore forced to work with lattice spacings which are much larger than those used in quenched simulations. By 1990 one was still far away from being able to compute numbers which can be confronted with experiment. Most of the calculations performed have served to test the various algorithms for handling dynamical fermions and to get a qualitative idea of the effects arising from the presence of quark loops.

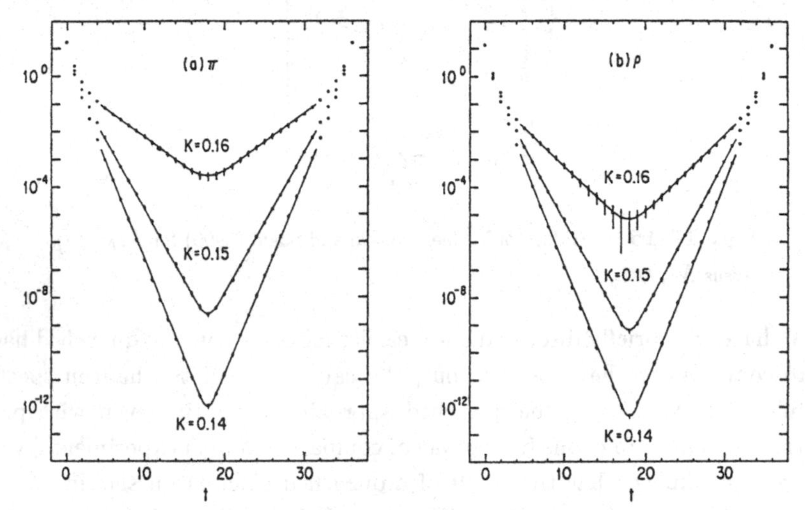

Fig. 17-16 (a) The pion (b) and ρ correlation functions obtained by Fukugita et al. (1988) in full QCD with Wilson fermions, for various values of the hopping parameter.

Calculations of the hadron mass spectrum in full QCD have been performed with Wilson and Kogut Susskind fermions. The most widely used algorithms for generating the link configurations have been the pseudofermion, Langevin, hybrid molecular dynamics and hybrid Monte Carlo (see chapter 16). Of these algorithms only the hybrid Monte Carlo is free of any systematic step size errors. As we have seen in chapter 16, this is achieved by subjecting a suggested configuration generated in a microcanonical step to a Metropolis acceptance/rejection test which eliminates the systematic errors introduced by the finite time step. There are many technical details that need to be discussed when simulating full QCD. We shall not discuss them here, and refer the reader to the proceedings of lattice conferences, and the literature cited there. Fig. (17-16), taken from Fukugita et. al. (1988), gives a nice example of the form of the correlation functions extracted in a numerical simulation with Wilson fermions. The correlation functions were computed on a $9^3 \times 36$ lattice using the Langevin algorithm and show a clear exponential decay at larger euclidean times. What concerns the Edinburgh plot, Fig. (17-17), taken from Laermann et al. (1990), exhibits the results obtained in some recent calculations. The dark points are those of Learmann et. al. The triangles, squares and inverted triangles are the results of Gottlieb et al. (1988), Hamber (1989), and Gupta et al. (1989), respectively.

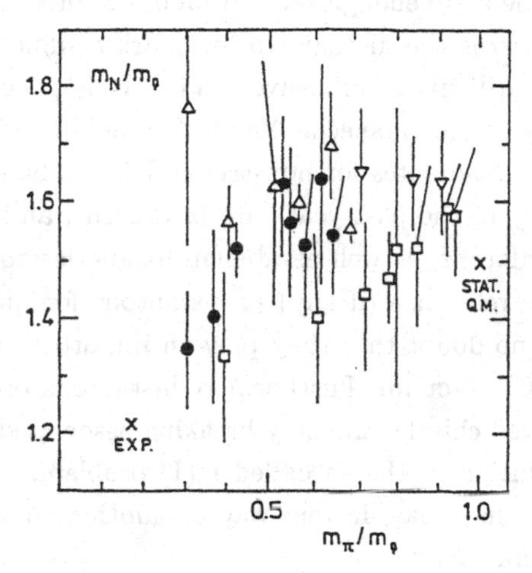

Fig. 17-17 Edinburgh plot showing the data of Gottlieb et al. (1988) (triangles), Hamber (1989) (squares), Gupta et al. (1989) (inverted triangles), and of Learmann et al. (1990) (dark points). The figure has been taken from Learmann et al. (1990)

Notice that this plot includes several measurements of the hadron ratios which are much closer to the experimental values (denoted by the cross) than in the quenched case. Clearly the data is still much too inacurate to allow one to estimate the effects of the dynamical quarks. So far there had been no definite signs that quark loops modify the quenched results in a significant way (see also Bitar et al., 1990; Campostrini et al., 1990). Even today the problem of determining the hadron mass spectrum from lattice calculations in full QCD remains a very challenging problem. Nevertheless, there has been substantial progress made, especially because of the available computer power. The effects of dynamical fermions appear to be small. But the quark masses used in the simulations are still too large.

17.6 Instantons

There is a general consensus among physicists that QCD accounts for quark confinement. Although there exists no analytic proof of this assertion, lattice simulations have demonstrated in a quite convincing manner that QCD confines quarks, and that the relevant degrees are to be identified within the pure Yang-Mills sector. The role played by dynamical quarks seems to be mainly that of allowing for string breaking, once the creation of a quark-antiquark pair is energetically favored at some large enough separation of the quarks, where hadronization sets in. What concerns the mechanism for quark confinement, however, this is a problem which is still under intensive study. It is generally believed that the mechanism is to be found in special Yang-Mills field configurations populating the QCD vacuum. Candidates for non-trivial field configurations which have been studied intensively in the past years are instantons, and finite temperature versions thereof, the calorons, as well as abelian magnetic monopoles and center vortices. Although the relevance (if any) of instantons for quark confinement is quite unclear, there is no doubt that they play an important role in determining the structure of the QCD vacuum. Furthermore instantons provide a natural expanation for the observed chiral symmetry breaking associated with confinement, and also provide a solution of the so called $U(1)$-problem.* Because of this it appears plausible that they may, in one way or another, play some role in the dynamics of quark confinement.

The instanton is a non-perturbative solution to the *classical euclidean* $SU(2)$ Yang-Mills equations of motion, carrying one unit of topological charge, and thus corresponds to a special field configuration associated with a subgroup of $SU(3)$. In fact this subgroup plays a distinguished role, as emphasized e.g. by Shifman (1999). The instanton solution was constructed by Belavin, Polyakov, Schwarz and Tyupkin (Belavin 1975), and is refered to in the literature as the BPST instanton Consider the $SU(2)$ euclidean Yang-Mills action,**

$$\begin{aligned} S_E &= \frac{1}{4}\int d^4x\; F^a_{\mu\nu}(x)F^a_{\mu\nu}(x) \\ &= \frac{1}{2}Tr\int d^4x\; F_{\mu\nu}F_{\mu\nu} \geq 0\;, \end{aligned} \tag{17.16}$$

where $F_{\mu\nu}$ is the matrix valued $SU(2)$ field strength tensor defined as in (6.13). Consider further the following integral,

$$I_\pm \equiv \frac{1}{8}\int d^4x\; (F^a_{\mu\nu} \pm \tilde{F}^a_{\mu\nu})(F^a_{\mu\nu} \pm \tilde{F}^a_{\mu\nu})\;, \tag{17.17}$$

* For more detailed discussions of instantons we refer the reader to Coleman (1985), Polyakov (1987) and Shifman (1999).

** Sums over repeated indices are always understood. For $SU(2)$ the "color" indices a run from 1 to 3.

where $\tilde{F}^a_{\mu\nu}$ is the dual field strength tensor, $\tilde{F}^a_{\mu\nu} = \frac{1}{2}\epsilon_{\mu\nu\lambda\rho}F^a_{\mu\nu}$. Clearly

$$I_\pm \geq 0 \, . \tag{17.18}$$

Since $\tilde{F}^a_{\mu\nu}\tilde{F}^a_{\mu\nu} = F^a_{\mu\nu}F^a_{\mu\nu}$, it follows that

$$S_E = \mp\frac{8\pi^2}{g^2}Q + I_\pm \, , \tag{17.19a}$$

where

$$Q = \frac{g^2}{32\pi^2}\int d^4x \; F^a_{\mu\nu}\tilde{F}^a_{\mu\nu} = \frac{g^2}{16\pi^2}\int d^4x \; Tr(F_{\mu\nu}\tilde{F}_{\mu\nu}) \, . \tag{17.19b}$$

Since $I_\pm \geq 0$ and $S_E \geq 0$, we have that

$$S_E \geq \frac{8\pi^2}{g^2}|Q| \, . \tag{17.20}$$

What makes this inequality interesting is that Q is a topologial invariant taking integer values $0, \pm 1, \pm 2, \cdots$ for all gauge fields with finite action. Finiteness of (17.16) requires that the integrand vanishes faster than $1/|x|^4$ for $|x| \to \infty$, or that the leading contribution to $\underset{\sim}{A}_\mu(x)$ is pure gauge (cf. (6.20)),

$$\underset{\sim}{A}_\mu(x) \underset{|x|\to\infty}{\longrightarrow} -\frac{i}{g}G(x)\partial_\mu G^{-1}(x) \, . \tag{17.21}$$

In the case of $SU(2)$, the unitary unimodular group elements $G(x)$ have the form

$$G(x) = a_4(x) + i\vec{a}(x)\cdot\vec{\sigma} \;\; ; \;\; a_4^2 + \vec{a}^2 = 1 \, , \tag{17.22}$$

where σ_i are the Pauli matrices. Field configurations satisfying (17.21) can be classified by a topological invariant taking positive and negative integer values, as we will comment in more detail below. Configurations characterized by different values of Q are not homotopic to each other, i.e. they cannot be continuously deformed into each other without violating the finiteness of the action. In each sector characterized by a given integer $Q = n$, the action satisfies

$$S_E \geq \frac{8\pi^2}{g^2}|n| \, . \tag{17.23}$$

As follows from (17.18) the lower bound in (17.23) corresponds to self dual, or anti-selfdual field configurations,

$$F^a_{\mu\nu}(x) = \tilde{F}^a_{\mu\nu}(x), \;\; (selfdual) \, ,$$

$$F^a_{\mu\nu}(x) = -\tilde{F}^a_{\mu\nu}(x). \ (anti-selfdual) \ . \tag{17.24}$$

Since the action (17.16) is non-negative, it follows from (17.19a) that for self-dual solutions (i.e. $I_- = 0$) Q is positive. For $Q = 1$ the solution is referred to in the literature as an *instanton.* On the other hand for $I_+ = 0$, Q must be negative. The corresponding $Q = -1$ solution is referred to as an anti-instanton.

Before constructing the so called BPST instanton let us first demonstrate why (17.19b) is an integer. There are different ways of showing this. The way we shall proceed, will at the same time shed light on the relevance of instanton configurations for the vacuum structure of the Yang-Mills theory.

Consider Q defined in (17.19b). One first shows that the integrand can be written as a divergence:*

$$Tr(F_{\mu\nu}\tilde{F}_{\mu\nu}) = \partial_\mu K_\mu(x) \ , \tag{17.25a}$$

where

$$K_\mu = 2\epsilon_{\mu\nu\lambda\rho} Tr\Big[A_\nu \partial_\lambda A_\rho + i\frac{2g}{3} A_\nu A_\lambda A_\rho\Big] \ . \tag{17.25b}$$

From here it follows that, if K_μ is regular within the domain of integration, then

$$Q = \frac{g^2}{16\pi^2}\int_{S^3_\infty} d^3\sigma \ n_\mu K_\mu(x) \ , \tag{17.26}$$

where $d^3\sigma$ is the 3-dimensional surface element of a sphere in four dimensions with radius $R \to \infty$ and normal $n_\mu(x)$. Now (17.25a) is gauge invariant. A particularily useful gauge for discussing the vacuum structure of the Yang-Mills theory is the "temporal" gauge $A^b_4(x) = 0$. The reason is, that in this gauge the Yang-Mills Hamiltonian takes a form quite similar to that familiar from Quantum Mechanics:

$$H = \int d^3x \Big[\frac{1}{2}\pi^a_i \pi^a_i + \frac{1}{4} F^a_{ij} F^a_{ij}\Big] \ , \tag{17.27}$$

---

* This can be shown by introducing the definition of $F_{\mu\nu}$ given in (6.13) into (17.19b), and making use of the antisymmetry of the ϵ-tensor, and of the invariance of the trace under cyclic permutations. Making further use of

$$\begin{aligned}\epsilon_{\mu\nu\lambda\rho} Tr\Big[(\partial_\mu A_\nu) A_\lambda A_\rho\Big] &= \epsilon_{\mu\nu\lambda\rho} Tr\Big[A_\nu (\partial_\mu A_\lambda) A_\rho\Big] = \epsilon_{\mu\nu\lambda\rho} Tr\Big[A_\nu A_\lambda (\partial_\mu A_\rho)\Big] \\ &= \frac{1}{3}\partial_\mu (A_\nu A_\lambda A_\rho)\end{aligned}$$

one then arrives at (17.25a)

where the canonical momenta $\pi_i^a(x) = F_{4i}^a(x)$ are subject to the Gauss law constraint,

$$\vec{\mathcal{D}}_{ab} \cdot \vec{\pi}^b = 0 . \tag{17.28}$$

Here $\mathcal{D}$ is the covariant derivative for $SU(2)$,

$$(\mathcal{D}_i)_{ab} = \delta_{ab}\partial_i + ig\epsilon_{abc}A_i^c . \tag{17.29}$$

Zero energy configurations (i.e., vacuum configurations) correspond to time independent gauge transformations of the trivial vacuum configuration $A_\mu^b = 0$, or in matrix form

$$A_i^{(vac)}(\vec{x}) = -\frac{i}{g}G(\vec{x})\partial_i G^{-1}(\vec{x}) . \tag{17.30}$$

In the temporal gauge these are the type of configurations which are approached for infinite times, ensuring the finiteness of the energy.

Consider now once more the topological charge (17.26), where S_∞^3 is now replaced by the three dimensional surface of a box in four dimensions. In the "temporal" gauge $A_4^b = 0$, the integrals over the surfaces with normal along the spacial directions will not contribute because of the ϵ- tensor in (17.25b) and the time independence of the asymptotic (vacuum) fields. Only the surface integrals with positive and negative oriented normals along the euclidean time direction will contribute, so that

$$Q = q_+ - q_- = \frac{g^2}{16\pi^2}\Big\{\int_{x_4=\infty} d^3x\ K_4(x) - \int_{x_4=-\infty} d^3x\ K_4(x)\Big\} . \tag{17.31}$$

For field configurations approaching vacuuum configurations (17.30) for $x_4 \to \pm\infty$ one has that

$$\begin{aligned} q_\pm &= \frac{g^2}{16\pi^2}\int_{x_4\to\pm\infty} d^3\sigma K_4(x) \\ &= \frac{1}{24\pi^2}\int d^3x\ \epsilon_{ijk}Tr\Big[(G_\pm\partial_i G_\pm^{-1})(G_\pm\partial_j G_\pm^{-1})(G_\pm\partial_k G_\pm^{-1})\Big] . \end{aligned} \tag{17.32}$$

This can be readily shown. Thus consider, more generally, (17.25b) evaluated on a vacuum configuration (17.21):

$$g^2K_\mu = -2\epsilon_{\mu\nu\lambda\rho}Tr\Big[(G\partial_\nu G^{-1})\partial_\lambda(G\partial_\rho G^{-1}) + \frac{2}{3}(G\partial_\nu G^{-1})(G\partial_\lambda G^{-1})(G\partial_\rho G^{-1})\Big] . \tag{17.33}$$

Now

$$\partial_\lambda(G\partial_\rho G^{-1}) = (\partial_\lambda G)(\partial_\rho G^{-1}) + G\partial_\lambda\partial_\rho G^{-1} .$$

Because of the ϵ-tensor in (17.33) the second term does not contribute. Writing

$$(\partial_\lambda G)(\partial_\rho G^{-1}) = (\partial_\lambda G)G^{-1}G(\partial_\rho G^{-1})$$

and making use of the identity

$$0 = \partial_\lambda(GG^{-1}) = (\partial_\lambda G)G^{-1} + G(\partial_\lambda G^{-1})$$

one obtains

$$g^2 K_\mu = \frac{2}{3}\epsilon_{\mu\nu\lambda\rho} Tr\Big[(G\partial_\nu G^{-1})(G\partial_\lambda G^{-1})(G\partial_\rho G^{-1})\Big] \tag{17.34}$$

Setting $\mu = 4$ we thus arrive at (17.32). If the group elements $G(\vec{x})$ are identified at spacial infinity, then (17.32) is an integer (the so called *winding number*"). The reason is that in this case, the three dimensional space becomes topologically equivalent to the three dimensional surface of a sphere, i.e. to S^3. On the other hand, the group elements $G(\vec{x})$ are parametrized by (17.22). Hence $G(\vec{x})$ defines a map $S^3 \to S^3$. Homotopy theory tells us that such maps can be classified by a topological invariant taking on integer values $n = 0, \pm 1, \pm 2, \cdots$, where $|n|$ gives the number of times S^3 in group space is covered as we sweep over S^3 in coordinate space. The corresponding expression is given by (17.32).* A $Q = 1$ instanton will therefore connect two vacuum configurations at $x_4 = \pm\infty$ carrying different winding numbers, which cannot be deformed continuously into each another. For this reason the instanton is interpreted as describing tunneling between classical inequivalent vacua. In a semiclassical approximation, the ground state of the quantum theory is then expected to be a linear combination of the infinite number of possible vacua carrying arbitrary winding numbers.**

Let us now briefly discuss the construction of the BPST instanton. *** The temporal gauge is not the most convenient one for constructing the self dual solution. Thus it is easier not to fix the gauge and to return to the expression (17.26) for the topological charge, which is written as a surface integral over S^3. Inserting the asymptotic behaviour for the potentials (17.21), and making use of (17.34), Q takes the form

$$Q = \frac{1}{24\pi^2}\int_{S^3} d^3\sigma \; n_\mu \epsilon_{\mu\nu\lambda\rho} Tr\Big\{(G\partial_\nu G^{-1})(G\partial_\lambda G^{-1})(G\partial_\rho G^{-1})\Big\}\,. \tag{17.35}$$

---

* We refer the reader to the literature quoted earlier for more details.

** For a discussion of the connection between instantons and tunneling in Minkowsky space the reader may confer the work of Bitar and Chang (Bitar 1978).

*** For further details see Belavin et. al (1975), Shifman (1999), Actor (1979).

As we have just learned this quantity is an integer. The BPST instanton corresponds to a group element $G(x)$ which maps compactified R_4 to the group element (17.22) in a one-to-one way:

$$G(x) = \frac{x_4 + i\vec{x}\cdot\vec{\sigma}}{|x|} \tag{17.36}$$

where $|x| = \sqrt{x_\mu x_\mu}$. Thus a_μ in (17.22) is identified with x_μ. From (17.21) and the definition $A_\mu = \sum_b A_\mu^b \frac{\sigma_b}{2}$ one then finds that for $|x| \to \infty$

$$A_\mu^a(x) \approx \frac{2}{g}\eta_{a\mu\nu}\frac{x_\nu}{x^2}\ , \tag{17.37}$$

where the $\eta_{a\mu\nu}$ are the so called 't Hooft symbols,

$$\eta_{a\mu\nu} = \begin{cases} \epsilon_{a\mu\nu} & ;\ \mu=\nu=1,2,3 \\ -\delta_{a\nu} & ;\ \mu = 4 \\ \delta_{a\mu} & ;\ \nu = 4 \\ 0 & ;\ \mu=\nu=4 \end{cases} \tag{17.38}$$

Having constructed the asymptotic form for the gauge potentials, we now make the following Ansatz for the $Q = 1$ instanton:

$$A_\mu^a(x) = \frac{2}{g}f(x^2)\eta_{a\mu\nu}\frac{x_\nu}{x^2} \tag{17.39}$$

with the condition that $f(x^2) \to 1$ for $|x| \to \infty$, and $f(x^2) \approx x^2$ for $|x| \to 0$. The latter condition insures that the field is non singular at the origen. The self dual (BPST) solution is found to be

$$A_\mu^a(x) = \frac{2}{g}\eta_{a\mu\nu}\frac{(x-x_0)_\nu}{(x-x_0)^2+\rho^2}\ . \tag{17.40}$$

The corresponding field strength reads

$$F_{\mu\nu}^a(x) = -\frac{4}{g}\eta_{a\mu\nu}\frac{\rho^2}{((x-x_0)^2+\rho^2)^2}\ , \tag{17.41}$$

and the topological density is given by

$$Q(x) = \frac{6}{\pi^2}\frac{\rho^4}{((x-x_0)^2+\rho^2)^4}\ . \tag{17.42}$$

Note that the solution (17.40) couples space-time and $SU(2)$ colour indizes. To actually see that the instanton solution connects different vacuum configurations of the Hamiltonian (17.27) carrying different winding numbers, one must of course make a gauge transformation to the temporal gauge.

Multi-instanton configurations in the $A_4^a = 0$ gauge with topological number larger than one connect vacua at $x_4 \to \pm\infty$ whose winding number differ by more than one unit. These vacuum configurations can be easily constructed (see e.g. Coleman, 1985). Approximate solutions corresponding to widely separated instantons, or instanton-antiinstantons, can of course be written down immediately as a linear superposition.

This completes our brief summary of the main concepts related to instantons in the continuum formulation. One obvious question now arises: can we "see" these instanton configurations on the lattice, and if so: what is their size distribution, and their possible relevance for confinement. To see the instantons is not so easy, since in a Monte Carlo simulation their structure will be blurred by quantum fluctuations. Having generated a set of link-variable configurations we must therefore strip off these fluctuations to see the underlying classical structure. This is called *Cooling* (Berg, 1981; Teper, 1986a; Ingelfritz, 1986). The idea is the following: with a given configuration $\{U_\mu(x)\}$ generated by a MC algorithm there is an associated euclidean action. In the continuum this field configuration would belong to a given topological sector, i.e, it would be characterized by a given integer, e.g. $Q = 1$. It would however not in general correspond to a minimum of the action, but rather be a field configuration which is a deformation of an underlying instanton configuration. To actually see this underlying instanton one must smoothen the short range quantum fluctuations, while keeping the long range physics unchanged. A way to proceed is to lower the action of a configuration in a systematic way until it takes the value $\frac{8\pi^2}{g^2}$. Once the action is minimized in the $Q = 1$ sector we are left in principle with a stable pure instanton configuration. In praxis this is of course not true, since we have discretized the action, and the above continuum arguments do not really apply. Because of lattice artefacts, lattice instantons will never be true instantons. In particular lattice instantons break in general scale invariance, i.e., the action depends on the size ρ of the instanton (see eq. (17.41)). The lattice manifests itself in that by cooling the configurations too long, the system will eventually end up in the trivial vacuum, i.e., the instanton has disappeared. One therefore needs to cool the system the right amount to see the instanton. The situation can be improved by working with so called "improved actions" where lattice artefacts have been eliminated up to higher orders in the lattice spacing. An important criterium for the identification of a $Q = 1$ instanton is that the topological charge of the cooled configuration should be unity up to lattice artefacts, and that the charge density and action density should be proportional with a proportionality factor given by $\frac{g^2}{8\pi^2}$. Multiinstanton configurations are identified by fitting the topological charge and action density to a superposition of instantons and anti-instantons.

How is the cooling performed in praxis? Suppose we have generated a particular configuration $\{U_\mu(x)\}$. Let $S_E[U]$ be the corresponding euclidean action. Consider the variation of a single link $\delta U_\mu(x)$ such that the action is lowered. In $SU(2)$ a link variable U is parametrized as in (17.22). The action S_E is given by a sum over plaquettes. A particular link variable U appears in all plaquettes having this link in comon. Hence the action (6.18a) is of the form

$$S_E = -\beta Tr(UW) + \cdots , \tag{17.43a}$$

where $\beta = \frac{2N}{g^2}$ for $SU(N)$,

$$U = a_4 + i\vec{\sigma} \cdot \vec{a} , \tag{17.43b}$$

and where the "dots" stand for contributions not containing U. W is a 2×2 matrix consisting of a sum of unitary matrices (staples). Hence W is not unitary. For $SU(2)$ we can however define a unitary matrix $\tilde{W}$ by

$$\tilde{W} = \frac{1}{\sqrt{det\ W}} W . \tag{17.44}$$

Then the action takes the form

$$S_E = -\beta\sqrt{det\ W}\ Tr(U\tilde{W}) + \cdots . \tag{17.45}$$

Keeping all link variables other than U fixed, and hence also $\tilde{W}$, we can minimize the action by choosing

$$U = \tilde{W}^{-1} . \tag{17.46}$$

since $U\tilde{W}$ is an element of $SU(2)$, whose trace is bounded from above by the trace of the unit matrix. One now repeats the procedure by starting from this new configuration and changes the value of another link variable such as to lower the action even further. Sweeping in this way through the lattice one thus generates a new configuration. The system can be further cooled by sweeping the lattice a number of times. Each sweep correspond to a complete cooling step.

An expression for the topological charge density can be obtained e.g. by transfering the expression $Tr(F_{\mu\nu}\tilde{F}_{\mu\nu}) = \frac{1}{2}\epsilon_{\mu\nu\rho\lambda}Tr(F_{\mu\nu}F_{\mu\nu})$, in (17.19b), to the lattice. A very natural definition was given by Peskin (1978). The most naive expression which reproduces (17.19b) in the continuum limit is given by

$$Q = -\frac{1}{32\pi^2} \sum_n \sum_{\mu\nu\sigma\rho} \epsilon_{\mu\nu\sigma\rho} Tr(U_{\mu\nu}(n) U_{\sigma\rho}(n)) \; .$$

With this definition, however, the charge density does not possess a definite parity. This deficiency can be easily corrected for, by symmetrizing the above expression with respect to $\mu \to -\mu, \nu \to -\nu,$, etc. (Di Vecchia, 1981)

$$Q = -\frac{1}{2^4 \cdot 32\pi^2} \sum_n \sum_{\{\mu\nu\sigma\rho\}=\pm 1}^{\pm 4} \epsilon_{\mu\nu\sigma\rho} \; Tr(U_{\mu\nu}(n) U_{\sigma\rho}(n)) \; . \tag{17.47}$$

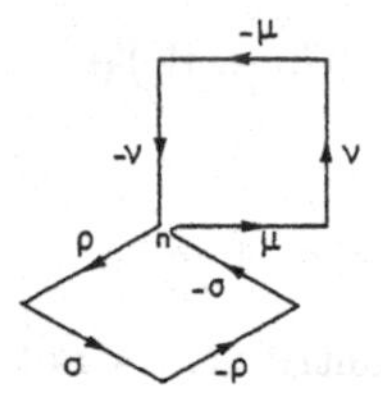

Fig. 17-18 Wilson loop involving two orthogonal planes, as introduced by Peskin (1978)

Fig. (17-18) shows the corresponding Wilson loop in Peskins definition of the topological charge. In praxis one takes for both, the action density and topological charge density, improved expressions whose lattice artefacts start only in $\mathcal{O}(a^6)$. This can be accomplished by constructing the lattice $F_{\mu\nu}$ operator from linear combinations of (gauge invariant!) 1×1, 1×2 and higher Wilson loop operators (de Forcrand, 1997; Garcia Perez, 1999). In fig. (17-19) we show an example of a $Q = 1$ configuration as generated by a Monte Carlo algorithm with an improved action, and the corresponding cooled configuation. For an instanton the action density, after rescaling, should equal the topological charge density. The particular shape of the cooled configuration is a consequence of the periodic structure. *

---

* Actually there exists no self-dual configurations on a hypertorus. To allow for a stable $Q = 1$ configuration one must impose twisted boundary condtions ('t Hooft, 1981a). The reason that the $Q = 1$ instanton configuration nevertheless appears, is that its size is fairly small compared to the extension of the lattice, and hence is not very sensitive to the choice of boundary conditions.

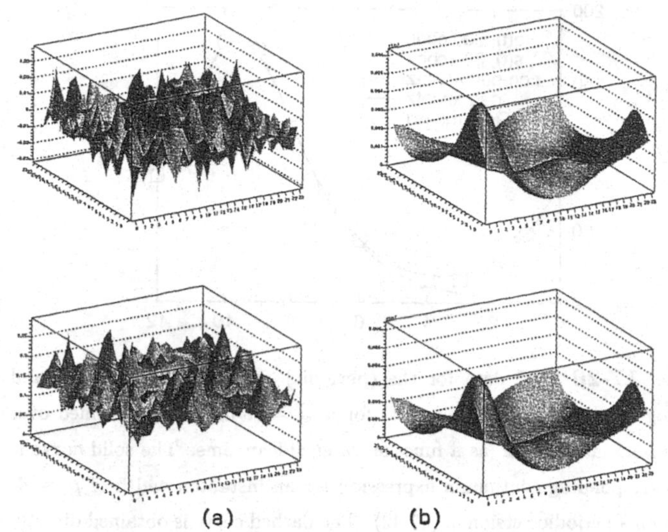

Fig. 17-19 Two-dimensional slice of the action density (top) and topological charge density (bottom) on a 24^4 lattice for a $Q = 1$ (a) uncooled and (b) cooled configuration after 500 improved cooling steps. The figure is taken from Wantz (2003).

In fig. (17-20) we show the MC data for the profile of the action density (circles) and topological charge density (squares) of an instanton configuration as a function of euclidean time, obtained by summing the action and charge density over the spacial lattice sites,

$$\hat{S}(\tau) = \sum_{\vec{x}} S(\vec{x}, \tau) , \tag{17.48a}$$

$$\hat{Q}(\tau) = \sum_{\vec{x}} Q(\vec{x}, \tau) , \tag{17.48b}$$

Notice that the action density (17.48a), after rescaling, coincides with the topological charge density (17.48b), as expected. The solid curve in fig. (a) is the (periodic) action and charge density for a $Q = 1$ instanton in the continuum. *

A similar plot for a two Instanton configuration is shown in fig. (17-21a). The plot (b) in the same figure exhibits the integrated topological charge density, and rescaled action as a function of the number of cooling steps. Special problems are

* We are very grateful to I.O. Stamatescu for providing us with this and also the following unpublished plots, extracted from the MC data of de Forcrand, Garcia-Perez and Stamatescu (de Forcrand 1997).

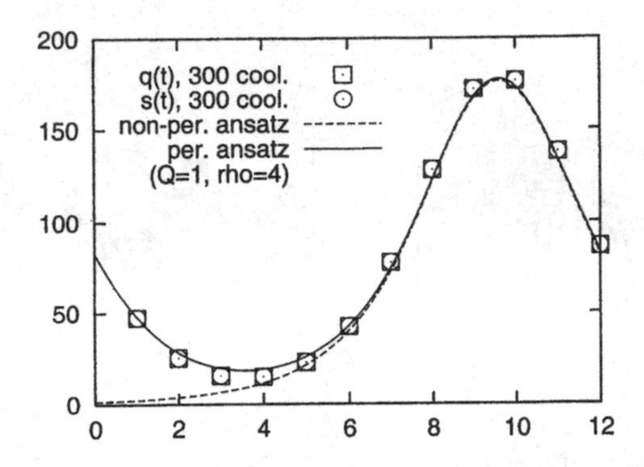

Fig. 17-20 MC data for the (periodic) action density (circles) and topological charge density (squares) for a $Q = 1$ instanton, summed over the spacial lattice sites, as a function of euclidean time. The solid curve is the corresponding continuum expression for an instanton width of $\rho = 4$, based on a periodic version of (17.42). The dashed curve is obtained directly from (17.42).

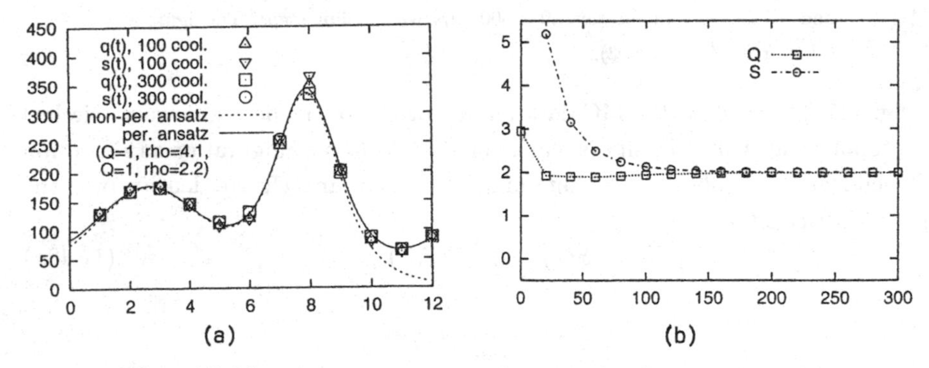

Fig. 17-21 (a) Similar plot as in fig. 17-20 but for a two instanton configuration. Fig. (b) gives the action and topological charge as a function of the number of cooling steps. Action and charge are seen to agree after 150 steps. The figure was provided to us by I.O. Stamatescu.

encountered when studying instanton-antiinstanton configurations on the lattice. Performing the same number of cooling steps as in the one instanton case destroys these configurations. This is not surprising since they carry the topological charge of the trivial vacuum configuration, whose action is lower than that corresponding to a superposition of an instanton and antiinstanton. The latter is a metastable state which upon cooling the system too long will decay into the trivial vacuum.

This is exhibited beautifully in fig. (17-22). The plots have been extracted from the data of Garcia-Perez et. al. (1999).

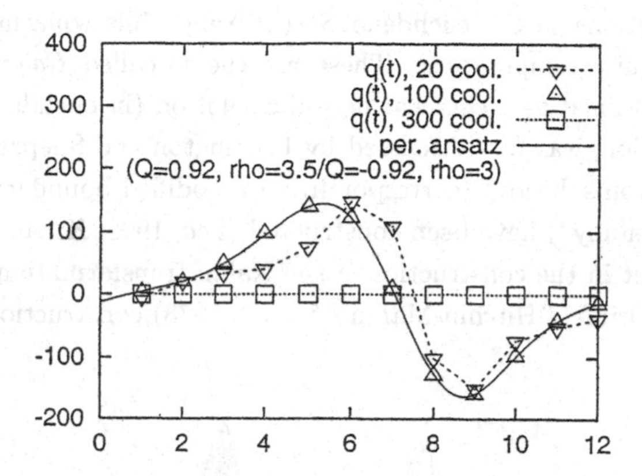

Fig. 17-22 Euclidean time profile of the topological charge density (17.48b) of an instanton-antiinstanton configuration for different number of cooling steps. The dashed and solid lines are fits with a periodic Ansatz consisting of a superposition of an instanton and antiinstanton. After 300 cooling steps the instanton-antiinstanton pair has dissappeared (boxes).

While the τ-profile of the charge density is fully consistent with an instanton-antiinstanton configuration up to 100 cooling steps, the topological charge density $Q(\tau)$ is seen to vanish for 300 cooling steps, indicating that one is left with a trivial vacuum configuration. To keep the configuration alive for a larger number of cooling steps, Garcia-Perez et. al. (1999) have introduced a physical criterion which essentially defines a cut off for the number of cooling steps, depending on the scale of fluctuations to be smoothened by cooling.

As we have already emphasized, instantons are an important ingredient for the structure of the Yang-Mills vacuum, and are believed to be the driving mechanism for chiral symmetry breaking*. But do instantons have anything to do with confinement? The instanton (17.40) is a field configuration at zero temperature where also confinement holds. In fact, as we shall see in chapter 20, Monte Carlo simulations show that quarks are actually confined up to high temperatures of the order of 10^{12} degrees! As we shall also see in chapter 18, field theory at finite temperature T is implemented in the path integral by integrating over the fields

* See e.g. the review by Schafer and Shuryak (Schafer, 1998)

on a space time manifold compactified in the euclidean time direction, with the bosonic (fermionic) fields satisfying periodic (anti-periodic) boundary conditions. Field configurations carrying a topological charge, and corresponding to self or anti-selfdual solutions to the euclidean $SU(2)$ Yang-Mills equations can also be constructed at finite temperature. These are the so called *Calorons*. The simplest caloron, obtained as a classical periodic solution (in euclidean time) to the equations of motion, was first discussed by Harrington and Shepard (Harrington, 1978). New caloron solutions (corresponding to modified boundary conditions $\rightarrow$ "non-trivial holonomy") have been constructed (Lee, 1998; Kraan, 1998). An important ingredient in the construction is the Nahm transformation (Nahm, 1984) and the Atiyah-Drinfeld-Hitchin-Manin (Atiyah, 1978) construction. **

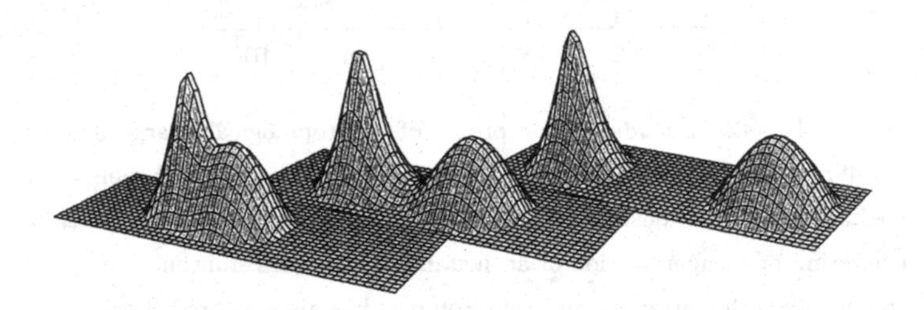

Fig. 17-23 Two dimensional profile of the action density of a $SU(2)$ caloron at three different temperatures, increasing from left to right. The figure is taken from Bruckmann et. al. (2003).

The discovery of these new calorons has given renewed impetus to the idea that instantons may, after all, play some role in the understanding of confinement. The new calorons have been the subject of intensive numerical investigations in the past years. Close to the phase transition, they are found to be much more abundant in the confined than in the deconfined phase (Grattinger, 2004), and to exhibit a composite structure (Lee, 1998; Kraan, 1998). $SU(N)$ calorons are made up of N so called BPS-monopoles (Prasad, 1975), carrying fractional topological charge. Conversely, monopoles can be looked upon as a periodic array of calorons (van Baal, 1998). In fig. (17-23) we show Monte Carlo data of Bruckmann et. al. (2003) for the action density of an $SU(2)$ caloron at three different temperatures $T = 1/\beta$. Thus the QCD vacuum appears to have a very rich structure, being populated by various types of topological excitations: instantons, monopoles, and

** See van Baal (1999) for a summary.

- as we will see in section 17.9 - vortices. The question whether instantons play at all a role for confinement is still waiting for a definite answer. This is all we will say about calorons. The details are fairly complex. The interest reader should confer the cited literature, and conference proceedings.

17.7 Flux Tubes in $q\bar{q}$ and qqq-Systems

In section 17.1 we have seen that Monte Carlo simulations of the pure SU(2) and SU(3) gauge theories confirm the long standing expectation that the $q\bar{q}$-potential rises linearly with the separation of the quarks for large separations. For small separations, asymptotic freedom predicts that the colour electric and magnetic fields should spread out in space in a similar way as for a dipole field in QED, leading to a Coulomb-like potential with a logarithmic dependence of the coupling constant on the $q\bar{q}$ separation (see chapter 9). The absence of free quarks, and the fact that meson resonances lie approximately on Regge trajectories, i.e., that there exists a linear relation between the angular momentum and the mass- squared, can be explained if one assumes that a quark-antiquark pair is connnected by a string (Goddard, 1973), with a constant string tension σ that is related to the slope of the Regge trajectory α' by $1/\alpha' = 2\pi\sigma$. This suggests that the non-perturbative dynamics at large distances squeezes the chromoelectric and magnetic fields into narrow flux tubes connecting the quark-antiquark pair. Whether this is indeed the case can in principle be checked by studying the distribution of the field energy in a Monte Carlo simulation. Some early analytical work on the flux tube behaviour has been carried out by Lüscher, Münster and Weisz (Lüscher, 1981), and Adler (1983).

On the lattice the flux tube problem has been first studied by Fukugita and Niuya (1983), followed by Flower and Otto (1985), Sommer (1987) and by Wosiek and Haymaker (1987). Since then the numerical data has improved substantially and there are now good indications that a flux tube is indeed formed for large separations of the quark-antiquark pair. To study the evolution of the flux tube as the $q\bar{q}$ separation is increased is a major challenge, since it requires large lattices, and special techniques for reducing fluctuations, and enhancing the projection onto the ground state. Because of limited computer power the earlier calculations have been restricted to rather small lattices. The energy (and action) density profiles for $SU(2)$ that were obtained e.g. by Haymaker et al. (1991, 1992) are consistent with those obtained more recently by Bali et al. (1995), who have studied, in particular, the distribution of the action density for quark-antiquark separations up to about 1.9 fm. The electric and magnetic contributions to the action and energy densities are determined by studying correlators of electric (space-time) plaquettes, and magnetic (space-space) plaquettes with a Wilson loop. The correlators have the generic form (10.24b). All authors find that the action density along the axis connecting the quark-antiquark pair is much larger than the energy density, and that they fall off fairly rapidly in the transverse direction. Fig. (17-24) shows this fall-off in a plane perpendicular to the axis of the $q\bar{q}$-pair at the midpoint, as measured by Haymaker et al. (1992).*

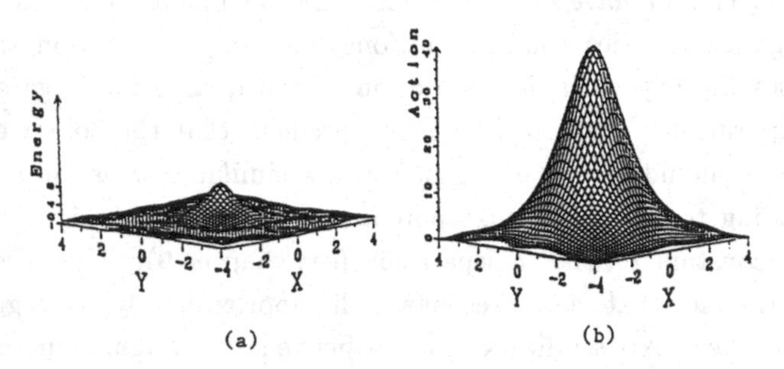

Fig. 17-24 (a) Energy density and (b) action density distribution in a plane midway between the $q\bar{q}$-pair (Haymaker et al, 1992)

In fig. (17-25) we show, as an example, the distribution of the energy and action density between a $q\bar{q}$-pair, obtained respectively by Haymaker et al (1992) and Bali et al. (1995). The peaks are the self energy contributions of the quark and antiquark.

In contrast to the $q\bar{q}$-potential, there have been only a few investigations of the three quark potential before 1999 (Sommer, 1984; Flower, 1986; Thacker, 1888). Possible flux-tube configurations that have been envisaged for the three quark system are the Δ and Y-type flux tube configurations. The Y-tube configuration is illustrated in diagram (b) of fig. (7-1). Within the string picture the classical ground state of the three quarks is envisaged to be given by three strings emanating from the quarks and meeting at a point whose position corresponds to the minmimal length of the three strings. Thus one may expect to "see" a Y-type flux tube configuration in a Monte Carlo simulation.

As in the case of the $q\bar{q}$-potential, the $3q$-potential can be extracted by studying the propagation in euclidean time of the corresponding gauge invariant 3-quark state

$$|qqq>=\epsilon_{a_1a_2a_3}U_{\Gamma_1}^{a_1b_1}U_{\Gamma_2}^{a_2b_2}U_{\Gamma_3}^{a_3b_3}q_{b_1}^{(1)}(\vec{x}_1,0)q_{b_2}^{(2)}(\vec{x}_2,0)q_{b_3}^{(3)}(\vec{x}_3,0)|0> , \qquad (17.49)$$

* See also Haymaker et al. (1996).

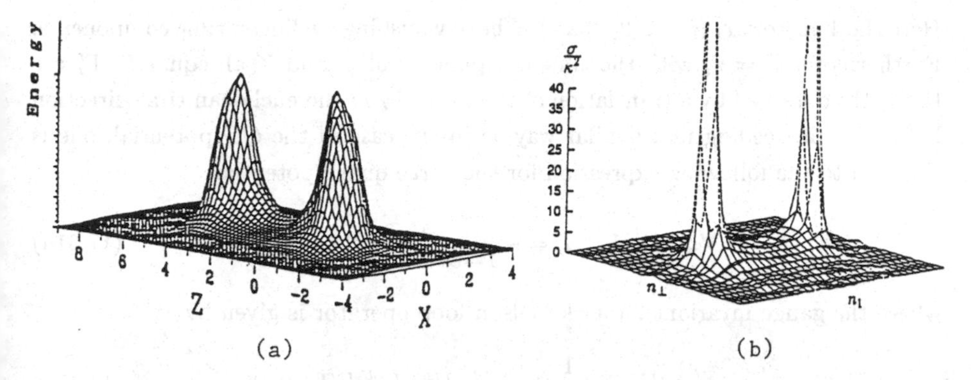

Fig. 17-25 (a) Energy density distribution between a $q\bar{q}$ pair, obtained by Haymaker et al. (1992); (b) Action density distribution, measured in units of the string tension, for a $q\bar{q}$ pair separated by about 1.35fm, at $\beta = 2.5$ (Bali et al. (1995))

where the unitary matrices U_{Γ_ℓ} are the path ordered product of link variables along the paths Γ_ℓ, $\ell = 1, 2, 3$ shown in fig. (17-26), connecting the quarks to the "center" ξ of the triangle with corners $\vec{x}_1$, $\vec{x}_2$ and $\vec{x}_3$, for which the sum of the respective distances is a minimum.

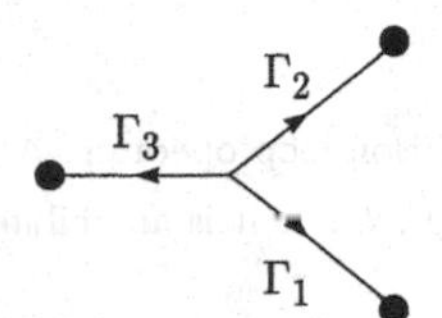

Fig. 17-26 The 3-quark state at $\tau = 0$. The heavy dots denote the quarks, and the lines stand for the ordered product of link variables which are tied together at the "center" by the ϵ-tensor.

Specifically, one considers the following correlator*

$$\begin{aligned} G(x_1,x_2,x_3;y_1,y_2,y_3) &= \epsilon_{a_1a_2a_3}\epsilon_{b_1b_2b_3} \\ &< 0|\bar{q}^{(1)}_{d_1}(y_1)\bar{q}^{(2)}_{d_2}(y_2)\bar{q}^{(3)}_{d_3}(y_3)U^{d_1b_1}_{\Gamma'_1}(y_1,Y)U^{d_2b_2}_{\Gamma'_2}(y_2,Y)U^{d_3b_3}_{\Gamma'_3}(y_3,Y) \\ &\times U^{a_3c_3}_{\Gamma_3}(X,x_3)U^{a_2c_2}_{\Gamma_2}(X,x_2)U^{a_1c_1}_{\Gamma_1}(X,x_1)q^{(3)}_{c_3}(x_3)q^{(2)}_{c_2}(x_2)q^{(1)}_{c_1}(x_1)|0> . \end{aligned} \quad (17.50)$$

* This the lattice analog of the continuum Green function studied by Brambilla et. al. (1995)

Here the 4-vectors x_i $(i = 1, 2, 3)$ and X have vanishing euclidean time components. Furthermore, $\vec{y}_i = \vec{x}_i$ with the time components of y_i and Y all equal T. Γ'_ℓ are the paths obtained by a translation of the paths Γ_ℓ in the euclidean time direction by T. By proceeding in a similar way as in the case of the $Q\bar{Q}$-potential, one is then led to the following expression for the three quark potential

$$V_{3Q}(\vec{x}_1, \vec{x}_2, \vec{x}_3) = -\lim_{T\to\infty} \frac{1}{T} \ln < W_{3Q} > , \tag{17.51a}$$

where the gauge invariant 3-quark Wilson loop operator is given by

$$W_{3Q} = \frac{1}{3!}\epsilon_{abc}\epsilon_{a'b'c'} U^{aa'}_{\Gamma_a} U^{bb'}_{\Gamma_b} U^{cc'}_{\Gamma_c} \tag{17.51a}$$

Here U_{Γ_a}, U_{Γ_b} and U_{Γ_c} are the path ordered product of the link variables along the paths shown in fig. (17-27).

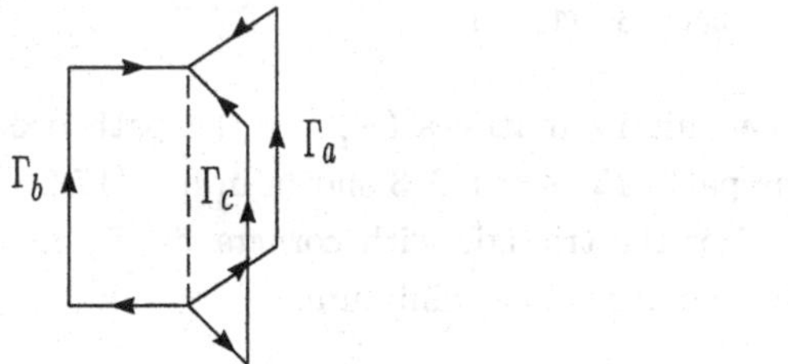

Fig. 17-27 The 3Q Wilson loop operator. A 3Q state created at time $\tau = 0$ propagates to $\tau = T$, where it is annihilated.

The extraction of the potential via (17.51a) requires the evaluation of $< W_{3Q} >$ for large euclidean times T. This poses of course the usual problems. Since the signal is suppressed exponentially with T, it is important to enhance the projection onto the ground state using a smearing technique, as described in sec. 17.4. Although the question whether the flux tube structure is of the Δ or Y-type is not yet settled, newer data obtained by Takahashi et. al. (2001-2003), and by Ichie et. al. (2003) support the Y-type flux tube picture.

In fig. (17-28) we show the action density in the presence of three quarks, obtained by Takahashi et. al. in a MC simulation performed at $\beta = 6.0$ in quenched QCD. The potential was fitted to the following conjectured Y-ansatz with a deviation of only 1%:

$$V_{3Q}(\vec{r}_1, \vec{r}_2, \vec{r}_3) = -A_{3Q} \sum_{i<j} \frac{1}{|\vec{r}_i - \vec{r}_j|} + \sigma_{3Q} L_{min} + C_{3Q} \ , \tag{17.52}$$

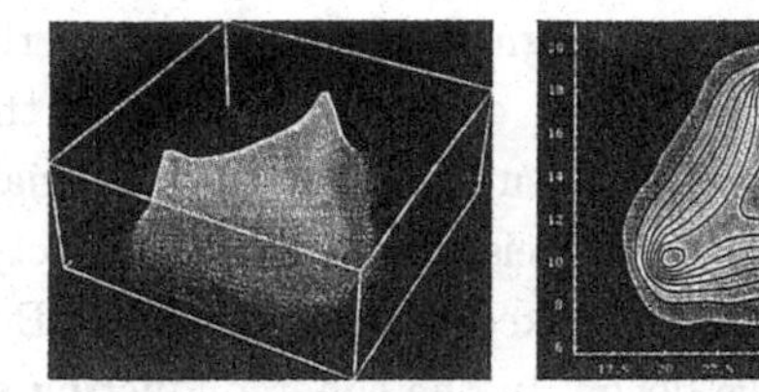

Fig. 17-28 Action density in the presence of 3 quarks, measured in a MC simulation on a $16^3 \times 32$ lattice at $\beta = 6.0$ for $SU(3)$. The figure is taken from Takahashi et al. (2004)

where L_{min} is the minimum length of the 3 strings. The authors also find that $\sigma_{3Q} \approx \sigma$, where σ is the two-body string tension, and that $A_{3Q} \approx \frac{1}{2}A_{Q\bar{Q}}$. A recent computation of the three quark potential in full QCD has been performed by Ichie et. al. (2003) in the "maximal abelian gauge" (see next section), and a similar flux tube profile was obtained.

17.8 The Dual Superconductor Picture of Confinement

Having obtained good indications that a flux tube is formed as the $q\bar{q}$-separation is increased, the next question one would like to have an answer to, concerns the dynamics responsible for the formation of the flux tube. It has been suggested a long time ago by Nielsen and Olesen (1973), and by Kogut and Susskind (1974) that confinement could be explained in a natural way if the QCD vacuum reacted to the application of a colour electric field, due to a quark-antiquark pair, in much the same way as a superconductor reacts to the application of a magnetic field. This could be achieved by adding to the gauge field an elementary charged scalar (Higgs) field, which has however not been detected so far.* The dual superconductor mechanism of 't Hooft (1976) and Mandelstam (1976) does not require the introduction of such a field, but assumes that dynamically generated topological excitations provide the persistent screening currents. Consider first a type I superconductor. Its ground state corresponds to a condensation of Bose particles

* The Higgs theory is the four-dimensional generalization of the Ginzburg-Landau theory.

(Cooper pairs). When an external magnetic field is applied, these Cooper pairs organize themselves to form persistent currents that expell the magnetic field, which is only allowed to penetrate the superconducting material a distance given by the London penetration depth (→ Meissner-Ochsenfeld effect). This is the case if the applied magnetic field does not exceed a critical value. Beyond this critical value superconductivity breaks down. In the superconducting state the currents only exist in a thin surface layer determined by the London penetration depth. A quantitative description of the relation between the surface currents and magnetic field is given by the London equations, * which in a stationary state, can be summarized in the Coulomb gauge by the following relation between the vector potential and the current density,

$$\vec{A} = -\frac{\lambda^2}{c}\vec{J}\,, \tag{17.53}$$

where $\lambda = m^*/n^*e^{*2}$, with m^*, n^* and e^* the mass, density and charge of the Cooper pairs. By taking the curl of (17.53), assuming λ to be constant, one obtains **

$$\vec{\nabla}\times\vec{J} + \frac{c}{\lambda^2}\vec{B} = 0\,. \tag{17.54}$$

Combining this equation with Ampere's law, $\vec{\nabla}\times\vec{B} = \vec{J}/c$ leads to the equation $\nabla^2\vec{B} = \frac{1}{\lambda^2}\vec{B}$, which implies that the magnetic field decays in the interior of the superconductor with a skin depth λ (London penetration depth). Since the Cooper-pair density is a function of the temperature, the same applies to λ. As one approaches the critical temperature λ increases strongly, and the magnetic field begins to penetrate more and more the superconductor. Clearly a superconductor of type I can have no analog in the ground state of a Yang-Mills theory with the colour electric field of a quark-antiquark pair squeezed into a narrow tube. In fact,

---

* The London equations relating the electric, magnetic fields and the current in a type I superconductor have the form

$$\vec{B} = -\frac{\lambda^2}{c}\vec{\nabla}\times\vec{j}\,,$$

$$\vec{E} = \frac{\lambda^2}{c^2}\partial_t\vec{j}\,.$$

** Our presentation is only qualitative. For a comprehensive discussion see e.g. the book by Tinkham (1975).

such property of the ground state (if true) suggests that it might actually be the dual analog of a superconductor of type II. A type II superconductor also exhibits a Meissner phase below a critical field strength B_{c_1}. The superconductivity is however not destroyed as the applied field is increased beyond this critical value. Instead a new phase appears (Shubnikov phase), in which the material is divided into normal and superconducting regions. Magnetic flux can now penetrate the material (mixed state), but only in narrow flux tubes ($\rightarrow$ Abrikosov flux tubes) whose separation decreases with increasing external field. The condition for this to happen (which implies that the new state is favored energetically) is determined by the ratio of the London penetration length λ to a new scale, the coherence length ξ. The coherence length is a measure of how strongly the Cooper-pair density varies in space, and has been introduced as a new parameter by Ginsburg and Landau, whose theory is an extension of the London theory. Abrikosov (1957) found a solution to these equations in which the flux tubes are arranged in a regular array, with each tube carrying one unit of flux $\Phi_0 = hc/e^*$. The energetically most favorable configuration turns out to be a triangular array. The relevant quantity which characterizes a superconductor of type I or II is

$$\kappa = \frac{\lambda}{\xi} \, .$$

We have the following simple criterion:

$$\kappa < \frac{1}{\sqrt{2}} \quad ; \ (type\ I)$$

$$\kappa > \frac{1}{\sqrt{2}} \quad ; \ (Type\ II)$$

For κ close to $1/\sqrt{2}$ the system is in a mixed state consisting of the Meissner and Shubnikov phase. In the Shubnikov phase, the flux tubes are trapped by circulating persistent currents. This phase persists up to a critical magnetic field $B_{c_2} > B_{c_1}$. Away from the superconducting-normal boundary, between the flux tubes, equation (17.54) again holds. This equation can be modified to take into account the presence of the core (Tinkham, 1975). If C is a closed curve encycling the flux tube, then

$$\int_S d\vec{S} \cdot (\vec{B} + \frac{\lambda^2}{c} \vec{\nabla} \times \vec{J}) = n\Phi_0 \, , \tag{17.55}$$

Note the appearance of Planck's constant. Indeed, the magnetic flux within the vortices, i.e., $\Phi = \oint d\vec{s} \cdot \vec{A}$ must be a multiple of Φ_0 in order that the wave function of the Cooper pairs be single valued. For an extreme type II superconductor the region of normal material consists of very thin filaments. The solution of the Ginzburg-Landau equations found be Abrikosov corresponds to one unit of flux concentrated in each of the vortices. Hence in the limit where the radius of the flux tube (taken along the z-direction) goes to zero one has that

$$\vec{B} + \frac{\lambda^2}{c}\vec{\nabla} \times \vec{J} = \Phi_0 \delta(\vec{x}_T)\hat{e}_z \ . \tag{17.56}$$

By analogy, if the QCD vacuum behaved like the dual version of a type II superconductor, one expects that the ground state consists of colour magnetic charges which in the presence of a quark-antiquark pair would organize themselves into persisting circulating magnetic currents which confine the colour-*electric* flux into narrow (dual Abrikosov)-flux tubes. 't Hooft (1981) has conjectured that the relevant degrees of freedom responsible for the confinement are actually U(1) degrees of freedom defined in the so called "maximal abelian gauge", and that the condensate in non-abelian gauge theories consists of U(1) Dirac monopoles. Since the U(1) lattice gauge theory is also known to confine charges for strong coupling, it is of interest to first check the above picture for the confinement mechanism in this theory. If the dual superconductor picture is correct, then for strong coupling the ground state should be strongly populated by Dirac magnetic monopoles. As the coupling is reduced this theory is known to undergo a transition to the Coulomb phase. In this phase the ground state should no longer show a condensation of magnetic charges. This has indeed been confirmed in numerical simulations by DeGrand and Toussaint (1980). These authors showed that on the lattice there are objects that can be naturally identified with Dirac monopoles. If a monopole is located inside an elementary spatial cube on the lattice, then the enclosed magnetic charge can be determined by measuring the total magnetic flux through the surface of this cube. The magnetic flux through the surface of a plaquette lying in the ij-plane is directly related to the phase of the plaquette variable

$$U_{P_{ij}}(n) = e^{iea^2 F_{ij}(n)} = e^{i\phi_{ij}(n)} \ ,$$

where F_{ij} is the component of the magnetic field in the direction perpendicular to the ij-plane. The measurement of this flux however involves a subtle point, resulting from the fact that $U_{P_{ij}}(n)$ is a periodic function of $\phi_{ij}(n)$. As we have seen in chapter 5 the plaquette variable is given in the U(1) gauge theory by the product of the oriented link variables around the boundary of the plaquette. A link variable associated with a link pointing in the μ-direction, with base located at the lattice site n, is given by $U_\mu(n) = e^{i\phi_\mu(n)}$, where $-\pi < \phi_\mu(n) < \pi$. Hence the product of such variables around the boundary of a plaquette lying in the $\mu\nu$-plane is

$$U_{P_{\mu\nu}}(n) = e^{i\phi_{\mu\nu}} , \tag{17.57a}$$

where

$$-4\pi < \phi_{\mu\nu} < 4\pi . \tag{17.57b}$$

For small angles $\phi_{\mu\nu}$ we can identify the space-space components of this quantity with $e\times$ (the magnetic flux through the surface of the plaquette). On the other hand, for large angles, the physical flux should be identified with $\phi_{\mu\nu}$ (mod $2n\pi$), since the plaquette value remains unchanged by shifting $\phi_{\mu\nu}$ by a multiple of 2π. The electric flux in the i'th direction through an elementary plaquette can be determined in a similar way from the phase of the plaquette variable $U_{P_{i4}}$.

Let us denote in the following the angle associated with a given plaquette P simply by ϕ_P. We then decompose this angle as follows (DeGrand, 1980)

$$\phi_P = \bar{\phi}_P + 2\pi n_P , \tag{17.58a}$$

where

$$-\pi < \bar{\phi}_P < \pi , \tag{17.58b}$$

and n_P are integers. Because of (17.57b), we then have that $n_P = 0, \pm 1, \pm 2$. Now it is clear that if we add up the plaquette angles ϕ_P of the six plaquettes bounding an elementary cube we will get a vanishing result, since each link is common to two plaquettes, and gives rise to the sum of two phases, equal in magnitude, but of opposite sign. It therefore follows that the magnetic flux $\frac{1}{e}\sum_P \bar{\phi}_P$ through the closed surface S bounding the elementary cube is given by

$$M = \sum_{P\in S} \frac{1}{e}\bar{\phi}_P = -\frac{2\pi}{e}\sum_{P\in S} n_P , \tag{17.59}$$

where M is the magnetic charge enclosed by the surface. This charge is therefore a multiple of $\frac{2\pi}{e}$.* If a magnetic monopole is located in an elementary volume, then at least one of the plaquette angles must be larger in magnitude than π, so that there is a Dirac string crossing the corresponding surface. By making a "large" gauge transformation (e.g., a particular link variable is mapped out of the principle domain $[-\pi, \pi]$) a Dirac string can be moved around. But the net

* We set $\hbar = c = 1$, as is appropriate to the lattice formulation.

number of such strings leaving the elementary volume will not be affected by the gauge transformation. Furthermore, it is clear, that the number of monopoles contained in a volume V is given by the sum of the monopole numbers of the elementary boxes making up the volume. The lattice simulations of DeGrand and Toussaint (DeGrand, 1980) have shown that the majority of the configurations in the confined phase of the $U(1)$ theory correspond to pairs of monopoles and antimonopoles located in adjacent boxes. Such a pair corresponds to the situation where only the plaquette angle associated with the common face exceeds π (in absolute value).

The observation that monopoles are abundant in the confined phase of the U(1) lattice gauge theory, and scarce in the deconfined phase, fulfills the first prerequisite for a test of the dual type II superconductor picture of confinement. * But if this picture is correct then one would also expect to "see" that when an oppositely charged pair is introduced into the vacuum these monopoles organize themselves to form persistent currents which squeeze the electric field into a narrow Abrikosov flux tube connecting the pair. Let the two charges be located on the z-axis. Then the magnetic supercurrents are expected to satisfy the *dual* version of the London equation for an Abrikosov vortex (we have set $\hbar = c = 1$),

$$E_z - \bar{\lambda}^2(\vec{\nabla} \times \vec{j}_M)_z = n\tilde{\Phi}_0\delta(\vec{x}_T) \tag{17.60}$$

where $\tilde{\Phi}_0$ is obtained from Φ_0 by replacing the electric charge of the Cooper pair by the monopole charge $q_m = \frac{2\pi}{e}$. The fluxoid density given by the left hand side vanishes everywhere except at the vortex. Consider a surface perpendicular to the z-axis on which the two charges are located. If this axis passes through the surface, then the electric flux through this surface is related to the circulation of the magnetic current around the boundary of the surface by

$$\int_S d\vec{S} \cdot \vec{E} - \bar{\lambda}^2 \int_{\partial S} d\vec{\ell} \cdot \vec{j}_M = n\tilde{\Phi}_0 \, . \tag{17.61}$$

If the surface is an elementary plaquette, then the line integral is roughly given by the z-component of the curl of the magnetic current multiplied by the area of the surface.

For the $U(1)$ lattice gauge theory this equation has been first studied in detail by Singh, Haymaker and Browne (Singh, 1993a), where $\bar{\lambda}$ has been considered to

* Di Giacomo et al. (2000) proposed a monopole creation operator, which serves as an order parameter for studying the deconfinement phase transition.

be a free parameter. These authors have chosen to define the i'th component of the electric field by $a^2E_i = \frac{1}{e}ImU_{P_{4i}}$, which takes account of the fact that the physical flux is determined from the phase of the plaquette variable, modulo $2n\pi$. To check (17.61) one also needs a lattice expression for the components of the magnetic current. They are defined (DeGrand, 1980) in terms of the dual field tensor $\tilde{F}_{\mu\nu} = \frac{1}{2}\epsilon_{\mu\nu\lambda\rho}F_{\lambda\rho}$ by

$$j_i^{(mag)} = \partial_\nu \tilde{F}_{\nu i} .$$

Consider for example the first component of the current. It can be written in the form

$$j_1^{(mag)} = \vec{\nabla} \cdot \vec{\mathcal{M}}_1 , \tag{17.62}$$

where $\vec{\mathcal{M}}_1 = (F_{43}, F_{24}, F_{32})$, and $\vec{\nabla} = (\partial_2, \partial_3, \partial_4)$. Integrating $j_1^{(mag)}$ over the volume of an elementary cube with edges along the 2, 3 and 4 directions, is equivalent to computing the flux of $\vec{\mathcal{M}}_1$ through the surface of this cube. Hence to compute the three components of $\vec{j}_M$ one needs to calculate the flux through the surfaces of 3 cubes, having one edge directed along the time-axis. This computation is carried out in a completely analogous way as described before for the magnetic flux. Hence by construction the three components of the current, when measured in lattice units, will be multiples of $\frac{2\pi}{e}$. To obtain $\vec{\nabla} \times \vec{j}^{(mag)}$ one performs the discrete line integral around the elementary square shown by the dotted lines in fig. (17-29), where the time direction of the 3-volumes determining the components of the currents has been suppressed.

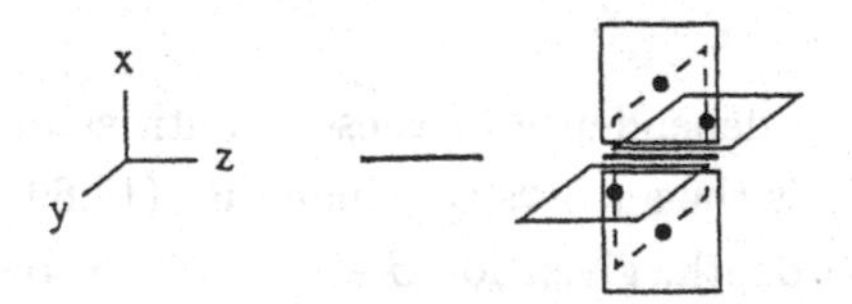

Fig. 17-29 The four elementary plaquettes shown in solid lines represent the four 3-volumes from which the components of the magnetic current are determined. The-time direction has been suppressed. The discrete line integral is performed along the dashed lines bounding the elementary square shown by the dotted lines. The fig. is taken from Singh et al. (1993b).

Singh et. al. (1993a) have measured the z-component of the electric field, and of the curl of the magnetic current, in the presence of a $q\bar{q}$ pair, by correlating these quantities with a Wilson loop located in the $z - \tau$ plane. The measurements were carried out in a plane perpendicular to the z-axis located at the midpoint between the two charges.

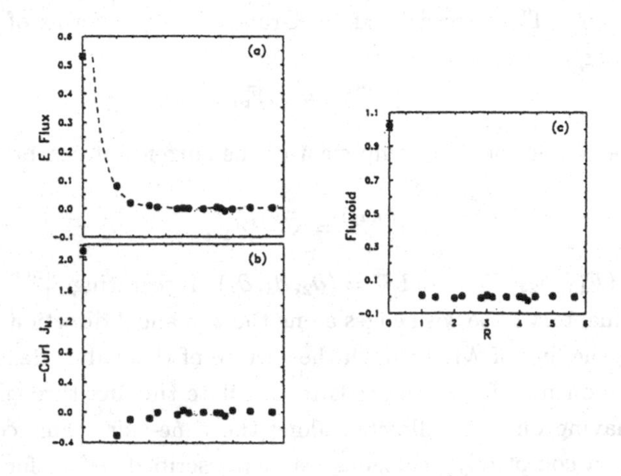

Fig. 17-30 Dependence of (a) the electric flux, (b) the curl of the monopole current, and (c) of the fluxoid in the U(1) gauge theory at $\beta = 0.95$ as a function of the perpendicular distance from the axis of the $q\bar{q}$ pair. The dashed line is the expected electrical flux obtained from the continuum equation (17.28) and the dual version of Ampere's law: $c\vec{\nabla} \times \vec{E} = -\vec{j}_M$. The fig. is taken from Singh et al. (1993a)

Fig. (17-30) shows the dependence of these quantities on the distance from the axis. From their analysis the authors conclude that (17.60) is indeed satisfied with a London penetration depth, given for $\beta \equiv \frac{1}{e^2} = 0.95$, by $\hat{\lambda} = 0.482 \pm 0.008$ (in lattice units), and one unit of electric flux.

The above result is encouraging. But what one really wants to study is the corresponding problem for the case of a non-abelian gauge theory. In the non-abelian case the situation is far less clear. As we have already mentioned at the beginning of this section, it is believed that the relevant degrees of freedom are actually U(1) degrees of freedom. It has been suggested by 't Hooft (1981) that these U(1) degrees of freedom can be isolated by going to the so-called Maximal Abelian Gauge (MAG). On the lattice the MAG has been first discussed by Kronfeld et. al (1987). Consider for example the $SU(2)$ gauge theory. An SU(2) link variable can be written in the form

$$U_\mu(n) = \begin{pmatrix} \sqrt{1-\eta_\mu^2(n)}e^{i\alpha_\mu(n)} & \eta_\mu(n)e^{i\psi_\mu(n)} \\ -\eta_\mu(n)e^{-i\psi_\mu(n)} & \sqrt{1-\eta_\mu^2(n)}e^{-i\alpha_\mu(n)} \end{pmatrix} \tag{17.63}$$

with $\eta_\mu(n) \geq 0$. Let $\{U_\mu(n)\}$ be some given link-variable configuration. We want to make a gauge transformation such that for all link variables the matrix is as diagonal as possible in the mean. Let us denote the transformed parameters with a "tilde". The transformed variables are given by $\tilde{U}_\mu(n) = G(n)U_\mu(n)G^{-1}(n + \hat{\mu})$. These variables will of course again be of the form (17.63). Then the above condition is equivalent to maximizing the quantity

$$R = \sum_{n,\mu} Tr[\sigma_3\tilde{U}_\mu(n)\sigma_3\tilde{U}_\mu^\dagger(n)] \; . \tag{17.64}$$

The gauge transformation which maximizes this expression is not unique, since no statement is made about the phases. Thus the value of (17.64) is left unchanged by gauge transformations generated by the group elements

$$\hat{G}(n) = \begin{pmatrix} e^{i\gamma(n)} & 0 \\ 0 & e^{-i\gamma(n)} \end{pmatrix} . \tag{17.65}$$

We can therefore restrict ourselves to gauge transformations of the type

$$U_\mu(n) \to G(n)U_\mu(n)G^{-1}(n + \hat{\mu}) \; ,$$

where

$$G(n) = \begin{pmatrix} \sqrt{1-\kappa^2(n)} & \kappa(n)e^{i\delta(n)} \\ -\kappa(n)e^{-i\delta(n)} & \sqrt{1-\kappa^2(n)} \end{pmatrix} ,$$

which is of the form $G(n) = g_0(n)1 + ig_1(n)\sigma_1 + ig_2(n)\sigma_2$, with σ_i the Pauli matrices. In the maximal abelian gauge the link variables take again the form (17.63), or equivalently

$$\tilde{U}_\mu(n) = \begin{pmatrix} \sqrt{1-|\xi_\mu(n)|^2} & \xi_\mu(n) \\ -\xi_\mu^*(n) & \sqrt{1-|\xi_\mu(n)|^2} \end{pmatrix} \begin{pmatrix} u_\mu(n) & 0 \\ 0 & u_\mu^*(n) \end{pmatrix} ,$$

where $u_\mu(n) = \exp[i\phi_\mu(n)]$. Under the residual gauge transformations, induced by (17.65), $\xi_\mu(n)$ and $u_\mu(n)$ transform as follows

$$\xi_\mu(n) \to e^{2i\gamma(n)}\xi_\mu(n) \; , \tag{17.66a}$$

$$u_\mu(n) \to e^{i\gamma(n)} u_\mu(n) e^{-i\gamma(n+\hat{\mu})} \,. \tag{17.66b}$$

Notice that the variables $u_\mu(n)$ transform like U(1) link variables under this gauge transformations, while the transformation of $\xi_\mu(n)$ is local. Having obtained the new link-variables from a given configuration in the maximal abelian gauge, the relevant U(1) gauge degrees of freedom are identified with $u_\mu(n)$, which, as we have just seen, transform in the desired way under the remaining U(1) gauge group. Monopoles are now identified by calculating the magnetic flux through the surface of a 3-volume bounded by plaquette variables constructed from these variables. A similar statement holds for the components of the magnetic current. These quantities are then correlated with a Wilson loop constructed from the abelian link variables.

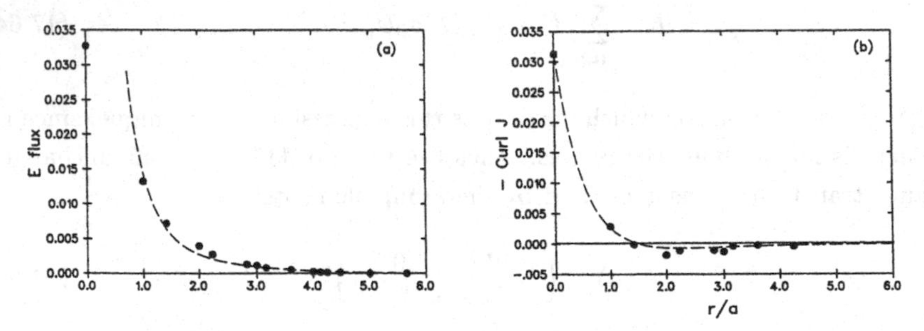

Fig. 17-31 Profile of (a) the electric flux, (b) the curl of the monopole current as a function of the perpendicular distance from the axis of the $q\bar{q}$ pair at $\beta = 2.4$. The fig. is taken from Singh et al. (1993b)

Singh, Haymaker and Browne (Singh, 1993b) have measured the monopole current and the electric field midway between an "quark-antiquark" pair and have shown that the results are consistent with the dual Ginzburg-Landau model, which is a generalization of the London theory that allows the magnitude of the condensate density to vary in space. The results for the flux and curl of the current are shown in fig. (17-31). These results have been confirmed by Matsubara et. al. (1994) with better statistics. These authors have also studied the same problem in the SU(3) gauge theory.

If monopoles defined in the MAG are the relevant degrees of freedom for confinement, then they must also account for the string tension in the full non-abelian theory. For $SU(2)$ this question has been first studied by Suzuki et. al. (1990, 1994).* A high precision MC measurement of the projected abelian string

* A more detailed analysis of isolating the monopole content in the light of the

tension in $SU(2)$ was carried out by Bali et.al. (1996). The authors find that the abelian monopoles reproduce the full string tension within 5%. This is referred to in the literature as *abelian dominance.*

17.9 Center Vortices and Confinement

As we have just seen, Monte Carlo simulations suggest that the mechanism of confinement is the condensation of $U(1)$ magnetic monopoles defined in the Maximal Abelian Gauge. The abelian degrees of freedom in the MAG seem to contain almost all the information regarding the long distance properties of the Yang-Mills theory relevant to confinement. It seems, however, that the relevant degrees of freedom for confinement can be reduced even further. When studying the deconfinement phase transition at high temperatures (see chapter 20) one finds that this transition goes along with a breakdown of the *center symmetry.** Not only that: chiral symmetry (see sec. 17.3) is restored at a critical temperature which in Monte Carlo simulations is found to coincide with the deconfinement transition. Hence the relevant degrees of freedom may be just be the *center* elements of the group. In the perturbative regime the link variables are close to the unit element. The relevant group is therefore $SU(3)/Z(3)$. On the other hand the center elements correspond to large fluctuations of the gauge fields of the order of $1/ga$.

A possible mechanism of confinement, which has been conjectured already a long time ago ('t Hooft, 1979; Mack, 1979; Ambjorn, 1980), and which has been discussed intensively in the past years, is the condensation of center vortices.** An example of a vortex in three spacial dimensions is a closed thin tube of magnetic flux. Such a vortex can be thought of as the magnetic field generated by a toroidal solenoid, in the limit of vanishing cross section. What is characteristic of such an arrangement, is that the vector potential is pure gauge outside the torus and has the form

$$\vec{A}(\vec{x}) = \frac{F}{4\pi}\vec{\nabla}\tilde{\Omega}(C;\vec{x}) \,, \tag{17.67}$$

De Grand-Toussaint prescription, discussed earlier, has been carried out by Shiba and Suzuki (1994), and by Stack et. al. (1994).

* The action is invariant under local $SU(3)$ transformations, and also under the multiplication of the link variables on a fixed time slice by a center element of $SU(3)$. As we will see in chapter 20 this center symmetry is broken in the confined phase. For $SU(2)$ and $SU(3)$ the center elements are given by $(\mathbf{1}, -\mathbf{1})$ and $(\exp(-2\pi i/3)\mathbf{1}, \mathbf{1}, \exp(2\pi i/3)\mathbf{1})$, respectively, where $\mathbf{1}$ is the unit matrix.

** For a recent review see e.g. Greensite, 2003.

where F is the magnetic flux in the torus, and $\tilde{\Omega}(C;\vec{x})$ is the solid angle subtended by the contour C of the infinitely thin torus at the observation point $\vec{x}$. This solid angle is a multivalued function, which changes by 4π along a closed contour piercing the surface Σ bounded by C. Thus the vortex (17.67) is introduced by making an a-periodic gauge transformation on the trivial vacuum configuration $\vec{A}(\vec{x}) = 0$:

$$\vec{A}(\vec{x}) = iG^{[C]}(\vec{x})\vec{\nabla}G^{[C]}(\vec{x})^{-1} , \tag{17.68a}$$

where

$$G^{[C]}(\vec{x}) = z^{\frac{\tilde{\Omega}(C;\vec{x})}{4\pi}} , \tag{17.68b}$$

with

$$z = e^{iF} . \tag{17.68c}$$

If a loop C', parametrized by $\vec{x}(s)$ $(0 \le s \le 1)$, winds through the hole of the torus, then

$$G^{[C]}(\vec{x}(1)) = zG^{[C]}(\vec{x}(0)) . \tag{17.69}$$

In the above discussion $\tilde{\Omega}$ is a multivalued function of $\vec{x}$. Alternatively one can define a function $\Omega(\vec{x})$ which is regular and single valued everywhere except on a surface Σ whose boundary is the loop C. As $\vec{x}$ crosses Σ, Ω jumps by $\pm 4\pi$. This function is given by

$$\begin{aligned} \Omega(\Sigma;\vec{x}) &= \int_\Sigma df'\vec{n}(\vec{x}')\cdot\frac{\vec{x}-\vec{x}'}{|\vec{x}-\vec{x}'|^3} \\ &= \int_\Sigma df'\vec{n}(\vec{x}')\cdot\vec{\nabla}'\frac{1}{|\vec{x}-\vec{x}'|} , \end{aligned} \tag{17.70}$$

where df' is the differential surface element, and $\vec{n}(\vec{x}')$ is the unit normal to the surface Σ at the point $\vec{x}'$. When $\vec{n}(\vec{x}')$ and $\vec{x}-\vec{x}'$ form an acute angle, this is nothing but the standard integral representation of the solid angle subtended by C at $\vec{x}$. This integral is discontinuous across the surface Σ, where it jumps by $\pm 4\pi$. * Now comes a subtle point. If one computes the rhs of (17.67) replacing $\tilde{\Omega}$ by Ω, this will not yield a gauge potential whose curl is the toroidal magnetic field. In fact, in this case, the corresponding expression is gauge equivalent to the trivial

---

* As a simple example the reader can consider the case of a circular loop in the x-y plane, and an observation point lying on the z-axis. Writing the integrand of (17.70) in cylindrical coordinates one readily finds the following dependence of the solid angle on z: $\Omega(z) = \frac{2\pi z}{\sqrt{R^2+z^2}} - 2\pi\epsilon(z)$, where $\epsilon(z) = 1$ for $z > 0$, and $\epsilon(z) = -1$ for $z < 0$.

configuration $\vec{A} = 0$. Indeed, the discontinuity of $\Omega(\vec{x})$ can be smoothened across the surface Σ. In this case the rhs of (17.67) with $\tilde{\Omega} \to \Omega$ is a bonafide pure gauge configuration, corresponding to a vanishing magnetic field everywhere. In order to obtain the toroidal magnetic field the gradient in (17.67) or (17.68a) should not operate on the discontinuity at the surface Σ. The following discussion is based on work of Engelhardt and Reinhardt (1999), who considered vortices within the continuum formulation of QCD.

The main goal is to isolate the contribution to $\vec{\nabla}\Omega$ arising from the above mentioned discontinuity. Consider

$$\partial_i \Omega = -\int_\Sigma df' n'_j \partial'_j \partial'_i \frac{1}{|\vec{x} - \vec{x}'|} \ . \tag{17.71}$$

where $\vec{n}' \equiv \vec{n}(\vec{x}')$, and, as always, sums over repeated indices are undertood. Making the decomposition

$$\partial_i \partial_j = \delta_{ij} \nabla^2 - (\delta_{ij} \nabla^2 - \partial_i \partial_j) \ ,$$

and use of the identity

$$\delta_{ij} \nabla^2 - \partial_i \partial_j = \epsilon_{ikm} \epsilon_{j\ell m} \partial_k \partial_\ell \ ,$$

as well as of

$$\nabla^2 \frac{1}{|\vec{x} - \vec{x}'|} = -4\pi \delta(\vec{x} - \vec{x}') \ ,$$

one finds that

$$\partial_i \Omega = 4\pi \int_\Sigma df' n'_i \delta(\vec{x} - \vec{x}') + \int_\Sigma df' \vec{n}' \cdot (\vec{\nabla}' \times \vec{F}_i) \tag{17.72a}$$

where

$$(\vec{F}_i)_m = -\sum_k \epsilon_{imk} \partial'_k \frac{1}{|\vec{x} - \vec{x}'|} \ . \tag{17.72b}$$

Application of Stokes theorem leads to*

$$\frac{1}{4\pi} \partial_i \Omega = \mathcal{A}_i(\Sigma; \vec{x}) - a_i(\partial\Sigma, \vec{x}) \tag{17.73a}$$

where

$$\mathcal{A}_i(\Sigma; \vec{x}) = \int_\Sigma df' n'_i \delta(\vec{x} - \vec{x}') \tag{17.73b}$$

---

* This is the 3-dimensional analog of the decomposition carried out by Engelhardt and Reinhard (1999) for the case of $SU(N)$ gauge theories.

and

$$a_i(\partial\Sigma, \vec{x}) = \int_{\partial\Sigma} dx'_k \; \epsilon_{ik\ell}\partial'_\ell D(\vec{x} - \vec{x}') \; , \tag{17.73c}$$

with

$$D(\vec{x} - \vec{x}') = \frac{1}{4\pi|\vec{x} - \vec{x}'|} \tag{17.73d}$$

satisfying

$$\nabla^2 D(\vec{x} - \vec{x}') = -\delta(\vec{x} - \vec{x}') \; . \tag{17.73e}$$

Note that $\mathcal{A}_i(\vec{x})$ has only a support on Σ. It is the contribution arising from the discontinuity mentioned above. Accordingly, ignoring this contribution to (17.73a), the relevant vortex vector potential is given by

$$\vec{A}_{vortex}(\vec{x}) = -F\vec{a}(\partial\Sigma, \vec{x}) \; , \tag{17.74}$$

which is now determined from the boundary of Σ, i.e. the location of the vortex. It is referred to in the literature at the *thin vortex*. Let us compute the line integral of the vortex field along a closed path C' piercing the surface Σ bounded by C n-times. A simple calculation yields

$$\int_{C'} d\vec{x} \cdot \vec{A}_{vortex}(\vec{x}) = -FL(C, C') \; , \tag{17.75a}$$

where

$$L(C, C') = \frac{1}{4\pi} \int_C dx_i \int_{C'} dy_j \; \epsilon_{ijk} \frac{x_k - y_k}{|\vec{x} - \vec{y}|^3} \tag{17.75b}$$

is the Gaussian linking number. This is, of course, the expected result for a toroidal field configuration. As one can also verify, the magnetic field computed from the curl of (17.74) is localized on C.

The configuration (17.74) is actually gauge equivalent to the so called *ideal vortex* field (17.73b), which only has non-vanishing support on the surface of discontinuity (Engelhardt, 1999). This can be seen as follows. Consider a gauge transformation generated by the inverse of the group element (17.68) with a single valued, but discontinuous, angular variable $\Omega(\vec{x})$. As already mentioned, one can imagine the discontinuity across Σ to be smeared out over an small interval, so that Ω is single valued and differentiable everywhere. After having performed the differentiation (17.68a) this interval is taken to vanish. A (bonafide!) gauge transformation on (17.74) carried out with this group element transforms the vortex field into the gauge equivalent potential $\vec{A}'_{vortex} = -F\vec{\mathcal{A}}(\Sigma, \vec{x})$, which now only has support on the surface of discontinuity. In this way one has eliminated the potential everywhere in space, except at points located on this surface.

Let us now leave the continuum and turn to the lattice. As we have already pointed out, it has been speculated for some time, that the relevant degrees of freedom for confinement in QCD may be $Z(3)$ degrees of freedom, and that confinement is intimately connected with the condensation of $Z(3)$ vortices. To expose the relevant $Z(3)$ content on the lattice one needs to carry out a "*center projection*". The idea is the same as for the abelian projection. One fixes a gauge, called the *maximal center gauge*, or MCG, where the link variables take a form as close as possible to $Z(3)$ elements, and then replaces these variables by the respective center elements. This is called "direct maximal center projection" or DMCP (Del Debbio, 1998). In terms of a concrete prescription this two step process reads as follows for $SU(2)$, where the link variables can be parametrized in the form (17.22). One first makes a gauge transformation on a given link variable configuration $U \to U^g$ which maximizes the quantity

$$F[U] = \sum_{n,\mu}(TrU_\mu(n))^2 = 2\sum_{n,\mu} a_4{}^2(n,\mu) ,$$

where $|a_4(n,\mu)| \leq 1$; i.e., one computes

$$max_{\{g\}} \sum_{n,\mu}(TrU^g_\mu(n))^2 , \tag{17.76}$$

where $U^g_\mu(n)$ is the gauge transform of $U_\mu(n)$. This leaves one with configurations which (on the average) are as close as possible to center elements. So far physics (which resides in gauge invariant quantities) has not been changed. The approximation comes with the second step: the gauge fixed link configurations $\{\tilde{U}_\mu\}$ are now projected to $Z(2)$ elements according to

$$\tilde{U}_\mu(n) \to z_\mu(n) = sgn[Tr\tilde{U}_\mu(n)] . \tag{17.77}$$

Del Debbio et al. (1997) have also proposed another way of performing the center projection, where the maximal center gauge itself is reached in a two step process. The first step consist in going into the maximal abelian gauge (MAG), and performing the abelian projection. The remaining $U(1)$ gauge symmetry is then further reduced by performing an additional gauge transformation which brings the link variables as close as possible to center elements. Center projection is then carried out by replacing the link variables by the respective center elements. For $SU(2)$, for example, the $U(1)$ matrix valued link variables, after abelian projection in the MAG, take the diagonal form $diag(e^{i\theta_\mu(n)}, e^{-i\theta_\mu(n)})$. Under the residual gauge transformation (17.66b) we have that $\theta_\mu(n) \to \theta_\mu(n) + \alpha_\mu(n)$. By a gauge transformation, the configurations are now brought as close as possible to center elements by maximizing the lattice average of $\sum_{n,\mu} \cos^2 \theta_\mu(n)$. This yields a set of $U(1)$-link variables $\tilde{U}_\mu(n)$ parametrized by $\{\tilde{\theta}_\mu(n)\}$. Finally one carries out the *center projection* by making the replacement

$$\tilde{U}_\mu(n) \to sgn(\cos \tilde{\theta}_\mu(n)) \tag{17.78}$$

This procedure is called *indirect* maximal center projection (IMCP) The IMCP has the merit that one can identify the $U(1)$ monopoles in the MAG, and study their possible correlations with the center projected link variable configurations.

In either way of carrying out the center projection one is finally left with a set of pure $Z(2)$ link variable configurations, i.e. one has stripped off all quantum fluctuations around the center elements.

How does one identify a so called P-vortex on the lattice? Consider again $SU(2)$ for simplicity. Consider also for simplicity a three dimensional lattice with link configurations consisting only of $Z(2)$ elements. Then a plaquette can only take the values ± 1. On a three dimensional spacial lattice this would determine the $Z(2)$-magnetic flux (or *center flux*) through the surface of the plaquette. Consider now the dual lattice consisting of the sites located at the center of the elementary cubes. Then a non-vanishing flux through a plaquette on the original lattice (corresponding to a center element -1), can be associated with a non-trivial $Z(2)$ link on the dual lattice, with the base located at the center of a cube and piercing this plaquette. In the $Z(2)$ scenario there is no arrow that can be attached to this link. Let Z_P denote the value of a plaquette P on the original lattice, and let V be the volume of an elementary hypercube bounded by 6 elementary plaquettes. It then follows trivially that

$$\prod_{P \in \partial V} Z_P = 1 \, .$$

This is just a special case of the *Bianchi identity*, $\prod_{P \in \partial V} \sigma_P = 1$, valid for $U(1)$. Here σ_P are the values of the plaquette variables bounding the volume V, computed from the product of $U(1)$ link variables with a sense of circulation determined, e.g., by the outward normal to ∂V. The links on the dual lattice, taking non-trivial values in the center of $SU(2)$ must therefore form closed loops in 3 dimensions. *.

* In four dimensions the P-vortices are closed two dimensional surfaces on the dual lattice. The existence of such $Z(2)$-vortices are believed to signal also the existence of "thick" vortices. Thick vortices are closed extended structures. They

To elucidate some of the above ideas let us consider again $SU(2)$ and a lattice in two space time dimensions. In two dimensions vortices are pointlike, and associated with sites on the dual lattice, located at the center of a plaquette with $Z_P = -1$.

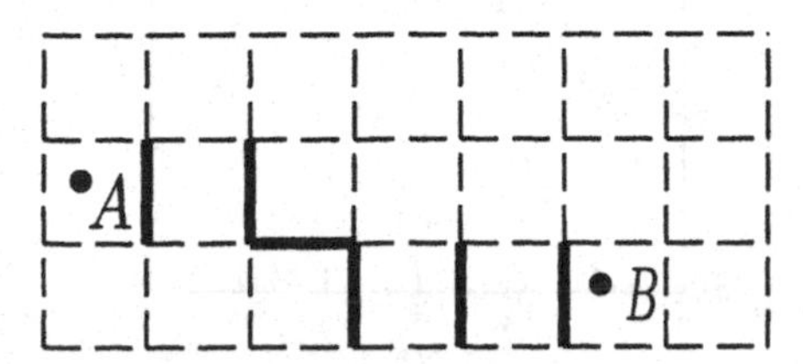

Fig. 17-32 Two vortices (heavy dots) on a two- dimensional lattice. The figure shows a possible link variable $Z(2)$ configuration. Links denoted by heavy lines take the value -1. Correspondingly, shaded plaquettes take the value -1. All other plaquettes take the trivial value +1.

In fig. (17-32) we show two vortices located at A and B. Dotted (solid) lines correspond to link variables taking the value +1 (-1). The plaquettes at A and B take both the value -1, while all other plaquettes take the value +1. The non-trivial links between the two vortices can be shifted around by performing an appropriate center gauge transformation. Their location have therefore no physical significance. Any Wilson loop which links with one of these vortices clearly yields a non trivial center element -1. If both vortices pierce the W-loop, then its value will be +1. * The value of the Wilson loop functional (7.24) on a particular $Z(2)$ configuration can only take the values $(-1)^n$, where n is the number of vortices piercing the area of the Wilson loop bounded by C'. Hence the potential can be calculated from the expectation value of this sign fluctuation.

Simulations of the $SU(2)$ string tension using the $Z(2)$ projected field configurations for the Wilson loop show that the full string tension is well reproduced. In fig. (17-33) we show the $q\bar{q}$-potential calculated by Kovacs and Tomboulis (1998) from the unprojected configurations, and from the sign average of the Wilson loop.

---

are expected to be the relevant configurations that scale in the continuum limit, and to lead to an area law behaviour for the Wilson loop.

* This is the analog of the observation made earlier in conjuction with the toroidal Aharonov-Bohm arrangement, except that there the vortex flux has a direction, and vortices piercing the W-loop in opposite directions will not affect the Wilson loop functional. Furthermore the center element is replaced by the $U(1)$ flux dependent element $\exp(iF)$.

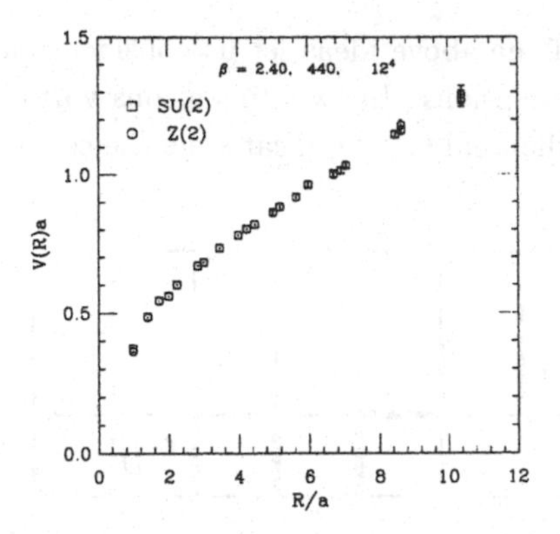

Fig. 17-33 The static $q\bar{q}$-potential for $SU(2)$ computed from unprojected configurations at $\beta = 2.4$ (squares), and from the sign average of the Wilson loop. The figure is taken from Kovacs (1998)

If center vortices are indeed the relevant degrees of freedom for confinement, then confinement should be lost when one removes configurations from the original ensemble, which give rise to vortices after center projection. This may not be an easy task to implement. In praxis it is easier to proceed as follows. After choosing the maximal center gauge the field configuration is written in the form

$$U_\mu(n) = Z_\mu(n)U'_\mu(n) ,$$

where $Z_\mu(n)$ is the center element after projection. By calculating the W-loop with the set of link variables $U'_\mu(n)$ one effectively throws out the center vortices. The result of following such a procedure for $SU(2)$ is shown in fig. (17-34), taken from Langfeld (2004).

As seen from the figure, the string tension vanishes upon removal of the center vortices. On the other hand the author finds that only 62% of the full $SU(3)$ string tension is recovered. Although many of the results obtained so far seem to point in the right direction, looking at the detailed numerical simulations show that there are many subtle problems such as e.g., the Gribov copy problem, the Casimir scaling problem, and the dependence of the results on the gauge in which the configurations are center projected. In this connection we only mention here, that center projected vortices in the so called "Laplacian gauge" reproduce not only the $SU(3)$ string tension, but are also free of Gribov copies (Alexandrou, 2000; de Forcrand, 2001). The question now arises: what is the quantum mechanical creation operator for a vortex? Such an operator was introduced by 't Hooft. To understand 't Hooft's formulation it is useful to go back to our *Aharonov –*

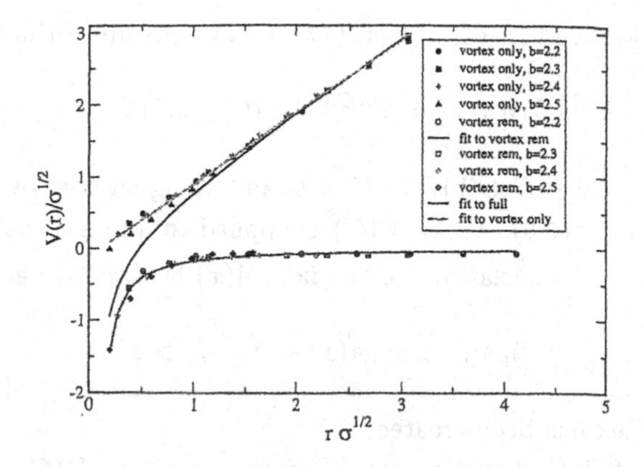

Fig. 17-34 The static $q\bar{q}$-potential for $SU(2)$ computed from unprojected, and center projected configurations, as well as from ensembles with the vortices removed. The figure is taken from Langefeld (2004)

Bohm arrangement of an infinitely thin toroidal solenoid, and construct a unitary operator which generates the gauge transformation

$$\vec{A}(\vec{x}) \to \vec{A}^{\Omega}(\vec{x}) = \vec{A}(\vec{x}) + \vec{A}_{vortex}(\vec{x}) \ . \tag{17.79}$$

where $\vec{A}_{vortex}(\vec{x})$ is given by (17.74). For this we must treat A_i as an operator. We shall denote it in the following with a "hat". Let $\hat{\pi}_i(\vec{x})$ be the momentum canonically conjugate to $\hat{A}_i(\vec{x})$. Making use of the operator relations

$$e^{-C}Be^{C} = B + [B, C] \, ,$$

and

$$e^{B}e^{C} = e^{[B,C]}e^{C}e^{B} \ ,$$

which are valid if $[B, C]$ is a c-number, one readily verifies that

$$\hat{V}(C)\hat{A}_i(\vec{x})\hat{V}^{-1}(C) = \vec{\hat{A}}(\vec{x}) + \vec{A}_{vortex}(\vec{x}) \ , \tag{17.80a}$$

where

$$\hat{V}(C) = e^{i\int d^3x \, \vec{A}_{vortex}(\vec{x})\cdot\vec{\hat{\pi}}(\vec{x})} \ . \tag{17.80b}$$

Consider now a Wilson loop. The W-loop functional (7.22b) evaluated on the gauge transformed field configuration (17.80a) is given by

$$\hat{V}(C)\hat{W}(C')\hat{V}^{-1}(C) = e^{-iF\int_{C'} d\vec{z}\cdot a(\vec{z})}\hat{W}(C') \ . \tag{17.81}$$

or, inserting for $\vec{a}$ the explicit expression (17.73c) one readily verifies that

$$\hat{V}(C)\hat{W}(C') = z^{L(C,C')}\hat{W}(C')\hat{V}(C) , \tag{17.82}$$

where $z = \exp(iF)$, and $L(C,C')$ is the Gausssian linking number of the two loops $C \equiv \partial\Sigma$ and C', i.e. (17.70). When $\hat{V}(C)$ is applied to an eigenstate $|\vec{A}(\vec{x}) >$ of $\vec{\hat{A}}(\vec{x})$ then it induces a "translation" of the field $\vec{A}(\vec{x})$ by (17.74) i.e.

$$\hat{V}(C)|\vec{A}(\vec{x}) >= |\vec{A}(\vec{x}) + \vec{A}_{vortex} > , \tag{17.83}$$

showing that a vortex has been created.

In 1978 't Hooft introduced a vortex creation operator $V(C)$ defined in an analogous way to (17.82), except that z is an element of the center of $SU(N)$ ('t Hooft, 1978). As was shown by 't Hooft, the expectation value of this operator serves as an order parameter for confinement. The area (perimeter) law for confinement (deconfinement) of the *temporal* Wilson loop is replaced by a perimeter (area) law for the 't Hooft loop operator, respectively. The corresponding vortex creation operator has been constructed by Reinhardt (2002) along the lines described above.

Let me close with some remarks. Monte Carlo calculations have shown that both, $U(1)$ monopoles, and center vortices can account for most part of the string tension in an appropriate center gauge. Independent of the choice of gauge one finds that if either the abelian monopoles or vortices are removed from the ensemble of configurations, confinement is lost. Not only that: chiral symmetry breaking is also lost. Although the various proposed mechanisms discussed in this and the previous section seem to account for quark confinement, a clear picture unifying all these observations is still lacking. Of all the proposed mechanisms, the dual superconductor picture of confinement is probably the closest to a dynamical picture leading to flux tube formation. The vortex and monopole pictures may just be alternative views of one and the same fundamental dynamics. Thus numerical simulations show that vortices and $U(1)$ monopole currents extracted in the indirect maximal center gauge are correlated (Kovalenko, 2004). With these remarks we conclude our discussion of this fascinating subject.

CHAPTER 18

PATH-INTEGRAL REPRESENTATION OF THE THERMODYNAMICAL PARTITION FUNCTION FOR SOME SOLVABLE BOSONIC AND FERMIONIC SYSTEMS

18.1 Introduction

So far we have studied the properties of hadronic matter at zero temperature. As we have seen, numerical calculations have given strong support for quark confinement. Thus QCD accounts for the fact that isolated quarks have never been seen in experiments performed in a normal environment. A natural question then arises whether quark confinement persists when one is dealing with hadronic matter under extreme conditions, such as at very high temperature or density.

It has been speculated already in the late seventies (Polyakov, 1978; Susskind, 1979) that there exists a phase transition from the low temperature regime, where quarks and gluons are confined and chiral symmetry is broken, to a chirally symmetric phase, consisting of a quark-gluon plasma in which the colour charge of quarks and gluons are Debye-screened. If such a plasma phase exists, then one should be able to detect it in high energy ion collisions. Such laboratory experiments could provide further tests of QCD as being the correct theory describing the strong interactions. The study of QCD at high temperatures is also of cosmological interest, for hadronic matter at high temperatures and densities was surely present in the early stages of the universe. Hence the study of QCD at very high temperatures and/or densities is important for constructing models of the universe.

Using renormalization group arguments, Collins and Perry (1975) have pointed out that at high densities the effective coupling of quarks and gluons should be small, and a perturbative description should be possible. The reason is that, when hadrons overlap, the separation between quarks will be small, and their dynamics will be determined by the asymptotic freedom property of QCD. But renormalization group arguments also suggest that the quark-gluon coupling constant decreases with increasing temperature. Hence at sufficiently large temperatures one might expect that thermodynamical observables can be computed in perturbation theory.

Actually, physics at high temperatures turned out to be more complex than originally expected. It was pointed out by Linde (1980) that the thermodynamics of massless Yang-Mills fields involves a severe infrared problem which leads to a breakdown of perturbation theory beyond a given order, which depends on the observable considered. For the thermodynamical potential the breakdown is expected to occur in $0(g^6)$, where g is the QCD coupling constant. This failure of perturbation theory is closely related to the fact that, in the presence of a heat bath, a screening mass of $0(g^2)$ is generated in the magnetic sector of QCD. Inspite of this "infrared problem", low-order perturbative calculations may still provide an adequate description of the thermodynamics of QCD at sufficiently large temperatures. At present, however, this can only be checked by calculating thermodynamical observables numerically, within the lattice formulation of QCD, and comparing the results with the perturbative predictions. One must therefore learn how to compute Feynman diagrams in finite temperature QCD.

Finite temperature continuum field theory has a long history. The pioneering work was carried out by Matsubara (1955) within a non-relativistic context. The development of perturbative methods for relativistic gauge field theories is, however, fairly new (Bernard, 1974; Weinberg, 1974; Dolan and Jackiw, 1974). We shall discuss these methods in the following chapter. The purpose of this chapter is to introduce the reader to a finite temperature formalism which will allow one to study the thermodynamics of a relativistic field theory like QCD by Monte Carlo methods.

The central object which is of interest when studying the thermodynamics of a system is the partition function. For simple systems, such as discussed in standard lectures on statistical mechanics, this partition function can be computed exactly. In the case of interacting field theories this is no longer possible and one has to recur to perturbation theory. The starting point for such computations is usually the path integral representation of the partition function. For non-interacting systems we can write down such a path integral representation which can be calculated exactly, and hence be checked against the expression obtained by other well established methods. In this chapter we shall study such models involving bosonic as well as fermionic degrees of freedom in quite some detail. This will give the reader some confidence in the path integral approach to studying the thermodynamics of relativistic quantum-mechanical systems.

18.2 Path-Integral Representation of the Partition Function in Quantum Mechanics

In chapter 2 we had derived a path integral representation for the imaginary time Green function in quantum mechanics. This allows us to write down immediately a corresponding path-integral representation for the thermodynamical partition function,

$$Z = Tr\ e^{-\beta H} \ . \tag{18.1}$$

Here H is the Hamiltonian, $\beta = \frac{1}{k_B T}$, with T the temperature, and k_B the Boltzmann constant. For convenience we shall set k_B equal to one in the following. Let n be the number of degrees of freedom of the system, and $|q\rangle = |q_1, q_2, \cdots q_n\rangle$ denote the simultaneous eigenstates of the coordinate operators Q_i with eigenvalues q_i. Then (18.1) is given by

$$Z = \int \prod_{\alpha=1}^{n} dq_\alpha \langle q|e^{-\beta H}|q\rangle \ . \tag{18.2}$$

The integrand has the phase-space path- integral representation (2.9) with $\tau' - \tau$ replaced by β, and with the coordinates at "time" $\tau = 0$ and $\tau = \beta$ identified. The partition function is obtained by integrating the expression over q (which we denote by $q^{(0)}$ in the following)

$$Z = \lim_{\substack{N\to\infty \\ \epsilon\to 0 \\ N\epsilon=\beta}} \int DqDp\ e^{i\phi[q,p]} e^{-\sum_{\ell=0}^{N-1} \epsilon H(q^{(\ell)},p^{(\ell)})}\Big|_{q^{(N)}=q^{(0)}} \ , \tag{18.3a}$$

where

$$DqDp = \prod_{\ell=0}^{N-1} \prod_{\alpha} \frac{dq_\alpha^{(\ell)} dp_\alpha^{(\ell)}}{2\pi} \ , \tag{18.3b}$$

and

$$\phi[q,p] = \sum_{\ell=0}^{N-1} \sum_{\alpha} p_\alpha^{(\ell)} (q_\alpha^{(\ell+1)} - q_\alpha^{(\ell)}) \ . \tag{18.3c}$$

The corresponding (formal) continuum form of this expression is given by

$$Z = \int_{per} Dq \int Dp\ e^{-\int_0^\beta d\tau [\sum_\alpha ip_\alpha(\tau)\dot{q}_\alpha(\tau) - H(q(\tau),p(\tau))} \ , \tag{18.4}$$

where the subscript "per" ($\to$ periodic) is to remind the reader that the coordinates at "time" $\tau = 0$ and $\tau = \beta$ are to be identified. Notice that only the coordinate degrees of freedom are required to satisfy periodic boundary conditions.

There are three features which distinguish the above expression from the classical partition function:

i) The phase space measure involves coordinate and momentum variables associated with every euclidean time support on the discretized interval $[0, \beta]$.

ii) The usual Boltzman factor is replaced by $\exp(-\beta\bar{H})$, where $\bar{H}$ is the following "time" average of the Hamiltonians defined at the discrete time supports: $\bar{H} = \frac{1}{N}\sum_{\ell=0}^{N-1} H(q^{(\ell)}, p^{(\ell)})$.

iii) The phase-space measure is multiplied by a phase factor which depends on the coordinates and momenta, and couples the coordinates at neighbouring lattice sites on the time (temperature) axis..

Since the integrand in (18.3a) involves the phase $e^{i\phi[q,p]}$, it cannot be interpreted as a probability distibution in phase space. It is for this reason that the phase-space path integral representation of thermodynamical observables (which can be obtained from the derivatives of Z in the usual way) does not lend itself to Monte Carlo simulations. But if the Hamiltonian is of the form (2.8), then by performing the Gaussian integration over the momenta we are led to a configuration space-path integral expression which is suited for such simulations:

$$Z = \int_{per} [dq] e^{-\int_0^\beta d\tau L_E(q,\dot{q})}, \tag{18.5a}$$

where

$$[dq] = \prod_{\ell=0}^{N-1}\prod_{\alpha=1}^{n} \frac{dq_\alpha^{(\ell)}}{\sqrt{2\pi\epsilon}}, \tag{18.5b}$$

and L_E is given by (2.10b), with $q = q^{(0)}$. Hence within the path-integral framework, temperature is introduced by merely restricting the euclidean time to the finite interval $[0, \beta]$, and imposing periodic boundary conditions on the coordinate degrees of freedom. This is truly a remarkable simple result. When (18.5) is generalized to field theories, this representation will provide the basis for computing thermodynamical observables using well known numerical methods in statistical mechanics.

18.3 Sum Rule for the Mean Energy

We now make use of the path integral representation of the partition function to obtain an expression for the mean energy which is suitable for Monte Carlo simulations. Such simulations are carried out on finite lattices. In quantum mechanics this lattice is one dimensional, with the lattice sites labeled by the discrete euclidean time supports on the finite compactified time interval $[0, \beta]$. Let N be the number of time steps of length ϵ, with $N\epsilon = \beta$. To compute the mean energy we have to differentiate the partition function with respect to temperature. We

Consider the sum over n of the logarithm *. Consider the following (also divergent) sum

$$\sum_{n=-\infty}^{\infty} \ln(n^2 + x^2) = \ln x^2 + 2g(x) \ , \tag{18.39a}$$

where

$$g(x) = \sum_{n=1}^{\infty} ln(n^2 + x^2), \tag{18.39b}$$

and

$$x = \beta E/2\pi. \tag{18.39c}$$

Whereas the right-hand side of (18.39b) diverges, its derivative

$$\frac{dg}{dx} = \sum_{n=1}^{\infty} \frac{2x}{n^2 + x^2} \tag{18.40}$$

is finite. The sum can be evaluated in closed form:**

$$\sum_{n=1}^{\infty} \frac{2x}{n^2 + x^2} = -\pi - \frac{1}{x} + \frac{2\pi}{1 - e^{-2\pi x}} \ .$$

By integrating (18.40) we can determine $g(x)$ up to a (divergent) constant. Inserting the result into (18.39a) one then finds that

$$\sum_{n=-\infty}^{\infty} ln\left(n^2 + (\frac{\beta E}{2\pi})^2\right) = \beta E + 2ln(1 - e^{-\beta E}) + C$$

where C is an integration constant. For (18.38) we therefore obtain (recall that we dropped the last term in (18.36a))

$$lnZ_0 = -V\int \frac{d^3k}{(2\pi)^3} ln(1 - e^{-\beta E(\vec{k})}) - \frac{\beta V}{2}\int \frac{d^3k}{(2\pi)^3} E(\vec{k}) - CV\int \frac{d^3k}{(2\pi)^3} \ .$$

The second and third term merely gives rise to a constant shift in the mean energy density and pressure. Dropping these (irrelevant) terms we therefore find that the

* We follow here the method of Dolan and Jackiw (1974). For an alternative procedure see Kapusta (1989).

** See e.g. "Table of Integrals, Series, and Products", by I. S. Gradshteyn and I. M. Ryzhik, Academic Press, New York and London (1965).

thermodynamical potential $\Omega = -\frac{1}{\beta V} \ln Z$ of an ideal gas of scalar neutral particles is given by

$$\Omega = -\frac{1}{\beta V} \ln Z = \frac{1}{\beta} \int \frac{d^3k}{(2\pi)^3} \ln(1 - e^{-\beta E}). \tag{18.41}$$

In the zero mass limit this expression reduces to

$$\Omega = -\frac{\pi^2}{90} T^4 \ , \quad (M = 0) \tag{18.42}$$

so that the energy density $\epsilon = \partial(\beta\Omega)/\partial\beta$ and pressure p of an ideal gas of massless scalar particles are given by the Stefan-Boltzmann law:

$$\epsilon = \frac{\pi^2}{30} T^4$$
$$p = \frac{1}{3}\epsilon \ .$$

18.6 The Photon Gas in the Path Integral Approach

Let us now consider the thermodynamics of a free photon gas within the path-integral framework. The euclidean finite temperature action, replacing (5.8b), is given by

$$S = \frac{1}{4} \int_0^\beta d\tau \int d^3x F_{\mu\nu}(x) F_{\mu\nu}(x) \ , \tag{18.43}$$

where $F_{\mu\nu} = \partial_\mu A_\nu - \partial_\nu A_\mu$ is the euclidean field strength tensor expressed in terms of the gauge potentials. Naively one would write down the following path integral expression for the partition function

$$Z = \mathcal{N} \int_{per} DA e^{-\frac{1}{4}\int_0^\beta d\tau \int d^3x F_{\mu\nu}(x) F_{\mu\nu}(x)} \ . \tag{18.44a}$$

where the gauge potentials are subject to the periodic boundary conditions

$$A_\mu(\vec{x}, 0) = A_\mu(\vec{x}, \beta) \ . \tag{18.44b}$$

Written in this form the integral (18.44a) diverges, since gauge field configurations which are related by a gauge transformation are weighted with the same exponential factor. The divergence can be isolated following the well known Fadeeev-Popov procedure, by which a gauge condition is introduced into the path integral. The following arguments are very formal, and we refer the reader to text books on field theory for more details.

Let us first write the partition function (18.44a) in the form

$$Z = \mathcal{N} \int_{per} DA e^{\frac{1}{2}\int_0^\beta d\tau \int d^3x A_\mu(x)(\delta_{\mu\nu}\Box - \partial_\mu\partial_\nu)A_\nu(x)} , \tag{18.45}$$

where we have made a partial integration in the action (18.43) expressed in terms of the gauge potentials. Next define the following gauge invariant functional $\Delta[A]$

$$\frac{1}{\Delta[A]} = \int D\Lambda \prod_x \delta[\partial_\mu(A_\mu(x) - \partial_\mu\Lambda(x)) - f(x)] , \tag{18.46}$$

where $f(x)$ is some arbitrary periodic function. Note that $A_\mu(x) - \partial_\mu\Lambda(x)$ is just a gauge transform of the potential $A_\mu(x)$. We now rewrite the δ-function in the form*

$$\begin{aligned}\prod_x \delta[\partial_\mu(A_\mu - \partial_\mu\Lambda) - f] &= \prod_x \delta[(-\Box)(\Lambda - \frac{1}{\Box}(\partial_\mu A_\mu - f))] \\ &= \frac{1}{\det(-\Box)} \prod_x \delta[\Lambda - \frac{1}{\Box}(\partial_\mu A_\mu - f)]\end{aligned}$$

Performing the functional integral over Λ in (18.46) we therefore have that

$$\Delta[A] = \det(-\Box) .$$

$\Delta[A]$ is the so called Faddeev-Popov determinant, which, although in the present case is independent of the gauge potentials, cannot be ignored. In fact at finite temperature, however, the determinant, which is given by the product of the eigenvalues of $-\Box$, with the eigenfunctions satisfying periodic boundary conditions analogous to (18.44b), is temperature dependent and hence is relevant for the thermodynamics of the system. This is the main message we wanted to convey.

The next step consists in introducing the identity

$$1 = \Delta[A] \int D\Lambda \prod_x \delta[\partial_\mu(A_\mu(x) - \partial_\mu\Lambda(x)) - f(x)]$$

into the path integral (18.45). After making a change of variables $A'_\mu = A_\mu - \partial_\mu\Lambda$, and using the fact that the action as well as integration measure are invariant under this transformation, one finds that

$$\begin{aligned}Z = \mathcal{N}[\int D\Lambda] \int_{per} DA \; \det(-\Box) \prod_x \delta[\partial_\mu(A_\mu(x) - f(x)] \\ \times e^{\frac{1}{2}\int_0^\beta d\tau \int d^3x A_\mu(x)(\delta_{\mu\nu}\Box - \partial_\mu\partial_\nu)A_\nu(x)} ,\end{aligned}$$

* We have chosen to factor out a $-\Box$ since the eigenvalues of this operator are non negative.

Here we have dropped the "prime" on A'_μ. The remaining steps leading to a covariant gauge fixed expression for the partition function are standard. Since the integral does not depend on the choice of $f(x)$ we can multiply (18.46) by $\exp[-\frac{1}{2\alpha}\int d^4x f^2(x)]$ and then functionally integrate the expression over $f(x)$ with the measure $Df = \prod_x df(x)$. After dropping the factor $\mathcal{N}[\int D\Lambda]$, since it should not affect the thermodynamics, we are left with the expression

$$Z = \int_{per} DA \; \det(-\Box) e^{\frac{1}{2}\int_0^\beta d\tau \int d^3x A_\mu(x)(\delta_{\mu\nu}\Box - (1-\frac{1}{\alpha})\partial_\mu\partial_\nu)A_\nu(x)} \; .$$

Choosing the Feynman gauge ($\alpha = 1$), we therefore have that*

$$Z = \det(-\Box)\left[\tilde{\mathcal{N}}\int_{per} DA \; e^{-\frac{1}{2}\int_0^\beta d\tau \int d^3x A_\mu(x)(-\delta_{\mu\nu}\Box)A_\nu(x)}\right]$$

The path integral factorizes into four path integrals having the form characteristic of a zero mass scalar neutral field. Hence we immediately conclude that

$$\begin{aligned}\ln Z &= \ln\left[\frac{1}{\sqrt{\det(-\Box)}}\right]^4 + \ln\det(-\Box) \\ &= 2\ln[\det(-\Box)]^{-\frac{1}{2}}\end{aligned} \; .$$

Hence $\ln Z$ is just twice the corresponding expression for a massless scalar neutral field. The factor two accounts for the fact that the photon has two transverse polarizations. We therefore conclude that the partition function for the photon gas is given by

$$\ln Z_{photon} = 2\ln Z_{\substack{scalar \\ m=0}} = -2V\int \frac{d^3k}{(2\pi)^3}\ln(1 - e^{-\beta|\vec{k}|}) \; , \tag{18.47}$$

---

* The determinant could in principle be incorporated into an effective action by introducing a set of Grassman fields (ghosts) $c(x)$ and $\bar{c}(x)$. Then $\det(-\Box) = \int DcD\bar{c}\; e^{\int_0^\beta d^4x \bar{c}(x)\Box c(x)}$. Since the determinant is the product of the eigenvalues of the operator $-\Box$ with the eigenfunctions satisfying periodic boundary conditions in euclidean time, this integral is to be evaluated with the ghost fields subject to the same boundary conditions. The relevance of the Faddeev-Popov determinant in obtaining the correct form for the partition function has been discussed in detail by Bernard (1974). The reader should consult this reference for a more general discussion of the gauge-fixing problem for the thermodynamical partition function.

and that the thermodynamical potential is just twice the expression (18.42), i.e.

$$\Omega = -\frac{\pi^2}{45}T^4 \ .$$

Summarizing, we have seen that the contribution of the Faddeev-Popov determinant was essential in deriving this result. Omission of the ghost contribution would have yielded a result which is twice as large. Furthermore it was important that the ghost fields obeyed the same boundary conditions as the photon fields which allowed for the cancellation of the unphysical degrees of freedom of the gauge potentials.

18.7 Functional Methods for Fermions. Basics

Having discussed in detail the path integral representation of the thermodynamical partition function in some simple bosonic theories, we now want to extend our discussion to the case of fermionic systems. A functional formalism which allows one to derive a path integral representation of the partition function for fermions is that of Berezin (1966).* In the following section we shall apply this formalism to a simple non-relativistic fermionic system, which allows us to derive a path integral representation for the partition function which is exact for an arbitrary choice of time step. This example will illustrate some important points which will be relevant when studying the thermodydnamical properties of relativistic field theories involving fermions on a lattice. In this section we will present the basic formulae that we shall need, and check them in a simple model. For a detailed discussion of the general functional formalism the reader may consult the book by Berezin (1966).

Consider first a fermionic system whose Hilbert space only consists of the vacuum state, $|0\rangle$, and the "one particle" state $|1\rangle = \hat{a}^\dagger|0\rangle$, where $|0\rangle$ is annihilated by $\hat{a}$. The operators $\hat{a}^\dagger$, and $\hat{a}$ satisfy the anticommutation relation $\{\hat{a}, \hat{a}^\dagger\} = 1$. All other anticommutators vanish. An general operator $\hat{A}$, acting on this space, then has the form

$$\hat{A} = K_{00} + K_{10}\hat{a}^\dagger + K_{01}\hat{a} + K_{11}\hat{a}^\dagger\hat{a} \ . \tag{18.48}$$

Note that the coefficients K_{ij} are not the matrix elements of the operator in the above mentioned basis. We have therefore denoted them with the symbol K. What we will need is an expression for the trace of an operator, and of a product

* See also Soper (1978)

of operators, in terms of an integral over Grassmann variables. This can be easily achieved in our example. We first associate with the operator (18.48) the so called *normal-form*

$$\tilde{A}(a^*, a) = K_{00} + K_{10}a^* + K_{01}a + K_{11}a^*a \ , \tag{18.49}$$

where a and a^* are the generators of a Grassmann algebra, i.e., $\{a, a\} = \{a^*, a^*\} = \{a, a^*\} = 0$. From the normal form, we construct the so called *matrix-form*

$$A(a^*, a) = e^{a^*a}\tilde{A}(a^*, a) \ . \tag{18.50}$$

Since $e^{a^*a} = 1 + a^*a$, one readily verifies that

$$A(a^*, a) = A_{00} + A_{10}a^* + A_{01}a + A_{11}a^*a \ ,$$

where A_{ij} are now the matrix elements of the operator $\hat{A}$ in the basis $|0\rangle$ and $|1\rangle = \hat{a}^\dagger|0\rangle$. Making use of the Grassmann integration rules

$$\int da^*a^* = \int daa = 1; \quad \int da = \int da^* = 0 \tag{18.51}$$

one finds that $Tr\hat{A} = A_{00} + A_{11}$ is given by the following Grassmann integral

$$Tr\hat{A} = \int dada^*e^{a^*a}A(a^*, a) \ . \tag{18.52}$$

To calculate the trace of a product of operators $\hat{C} = \hat{A}\hat{B}$ we need an expression for the matrix form of the operator $\hat{C}$, in terms of the matrix forms of the operators $\hat{A}$ and $\hat{B}$. One easily verfies for our simple example that

$$C(a^*, a) = \int d\alpha^*d\alpha e^{-\alpha^*\alpha}A(a^*, \alpha)B(\alpha^*, a) \ , \tag{18.53}$$

where a, a^*, α, α^*, $d\alpha$, $d\alpha^*$, are all anticommuting variables.

All the above expression have been written in a form which generalize to the case of a fermionic system with an arbitrary finite number of degrees of freedom. We summarize the main expressions we shall need in the following sections, and refer the reader to the book by Berezin (1966) for details.

Let $\hat{a}_i$ and $\hat{a}_i^\dagger$ $(i = 1, 2, \cdots, k)$ be operators satisfying the anticommuation relations

$$\{\hat{a}_i, \hat{a}_j^\dagger\} = \delta_{ij} \ ,$$

$$\{\hat{a}_i, \hat{a}_j\} = \{\hat{a}_i^\dagger, \hat{a}_j^\dagger\} = 0 \ . \tag{18.54}$$

A general function of these operators can then be written in *normal ordered* form:

$$\hat{A} = \sum_{n,m} \sum_{\{i_k\},\{j_k\}} K^{(nm)}_{i_1,i_2,\cdots,i_n;j_1,j_2,\cdots,j_m} \hat{a}^\dagger_{i_1} \hat{a}^\dagger_{i_2} \cdots \hat{a}^\dagger_{i_n} \hat{a}_{j_1} \hat{a}_{j_2} \cdots \hat{a}_{j_m} , \tag{18.55}$$

where the coefficients $K^{(nm)}_{i_1,i_2,\cdots,i_n;j_1,j_2,\cdots,j_m}$ are separately antisymmetric in the indices $i_1, i_2, \cdots i_n$ and $j_1, j_2, \cdots j_m$. By definition, $n = 0$, or $m = 0$ is understood to imply that the corresponding contribution to the sum (18.55) does not contain operators of the type $\hat{a}^\dagger_i$ or $\hat{a}_i$, respectively. A general state on which this operator acts is given by a linear combination of the vacuum state $|0\rangle$, which is annihilated by all the $\hat{a}_i$'s, and the states $\hat{a}^\dagger_{i_1} \cdots \hat{a}^\dagger_{i_n} |0\rangle$, where $1 \leq n \leq k$.

We now associate with (18.55) a *normal form*, obtained by replacing the operators $\hat{a}^\dagger_i$ and $\hat{a}_i$ by (anticommuting) Grassmann variables a^*_i and a_i, respectively:

$$\tilde{A}(a^*, a) = \sum_{n,m} \sum_{\{i_k\},\{j_k\}} K^{(nm)}_{i_1,i_2,\cdots,i_n;j_1,j_2,\cdots,j_m} a^*_{i_1} a^*_{i_2} \cdots a^*_{i_n} a_{j_1} a_{j_2} \cdots a_{j_m} . \tag{18.56}$$

The Grassmann variables a and a^* appearing in the argument of $\tilde{A}$ stand for the collection of variables $\{a^*_i\}$ and $\{a_i\}$, respectively. The coefficients $K^{(nm)}_{i_1,i_2,\cdots;j_1,j_2,\cdots}$ are not the matrix elements of the operator $\hat{A}$ in the above mentioned basis. From the normal form, we can construct the matrix form in a way analogous to (18.50):

$$A(a^*, a) = e^{\sum_i a^*_i a_i} \tilde{A}(a^*, a) . \tag{18.57}$$

Given the matrix forms of two operators $\hat{A}$ and $\hat{B}$, the matrix form associated with the product $\hat{A}\hat{B} = \hat{C}$ can be computed as follows:

$$C(a^*, a) = \int \prod_j d\alpha^*_j d\alpha_j e^{-\sum_i \alpha^*_i \alpha_i} A(a^*, \alpha) B(\alpha^*, a) , \tag{18.58}$$

where a^*_i, a_i, α^*_i, α_i are now generators of an extended Grassmann algebra.

Knowledge of the matrix form of an operator allows us to compute its trace according to

$$Tr\hat{A} = \int \prod_j da_j da^*_j e^{\sum_i a^*_i a_i} A(a^*, a) , \tag{18.59}$$

which is the generalization of (18.52). This expression can also be written in the form

$$Tr\hat{A} = \int \prod_j da^*_j da_j e^{-\sum_i a^*_i a_i} A(a^*, -a) . \tag{18.60}$$

Furthermore, using the Grassmann integration rules (18.51), one verifies that

$$\int \prod_j da_j \int \prod_j da_j^* e^{\sum_i a_i^*(a_i - b_i)} F(a,c) = F(b,c) \; ,$$

where again a, b, c stand for the collection of Grassmann variables $\{a_i\}$, $\{b_i\}$, $\{c_i\}$. Hence

$$\delta(a,b) = \int \prod_j da_j^* e^{\sum_i a_i^*(a_i - b_i)} \; . \tag{18.61}$$

acts as a δ-function.

As a simple application of the functional formalism, let us compute the grand canonical partition function for a system whose dynamics is governed by the Hamiltonian

$$\hat{H} = \sum_i E_i \hat{a}_i^\dagger \hat{a}_i \; , \tag{18.62a}$$

with the chemical potential μ coupled to the number operator

$$\hat{N} = \sum_i \hat{a}_i^\dagger \hat{a}_i \; . \tag{18.62b}$$

The partition function for this system is given by

$$\Xi(\beta, \mu) = Tr \hat{\Omega} \; , \tag{18.63a}$$

where

$$\hat{\Omega} = e^{-\beta(\hat{H} - \mu\hat{N})} = e^{-\beta \sum_i (E_i - \mu)\hat{a}_i^\dagger \hat{a}_i} \; . \tag{18.63b}$$

The exponential factorizes. Making use of the anticommutation relations (18.54) one finds after some algebra that

$$\hat{\Omega} = 1 + \sum_{\ell=1}^{k} \sum_{i_1 < i_2 < \ldots < i_\ell} \eta_{i_1} \cdots \eta_{i_\ell} \hat{a}_{i_1}^\dagger \cdots \hat{a}_{i_\ell}^\dagger \hat{a}_{i_\ell} \cdots \hat{a}_{i_1} \; ,$$

where

$$\eta_i = e^{-\beta(E_i - \mu)} - 1 \; ,$$

and where k is the number of degrees of freedom. The normal form associated with $\hat{\Omega}$ is obtained by replacing the operators $\hat{a}_i^\dagger$ and $\hat{a}_i$ by the corresponding Grassmann variables a_i^* and a_i. Making use of the fact that the square of a Grassmann variable vanishes, it can be written in the form

$$\tilde{\Omega}(a^*, a) = e^{\sum_i \eta_i a_i^* a_i} \; .$$

The corresponding *matrix* form is then obtained according to (18.57):

$$\Omega(a^*, a) = e^{\sum_i e^{-\beta(E_i-\mu)} a_i^* a_i} . \tag{18.64}$$

Finally, (18.63a) is calculated from (18.59). For $\ln \Xi$ one obtains

$$\ln \Xi(\beta, \mu) = \sum_i \ln[1 + e^{-\beta(E_i-\mu)}] , \tag{18.65}$$

which is the correct expression for the partition function.

18.8 Path Integral Represenation of the Partition Function for a Fermionic System valid for Arbitrary Time Step

We now use the functional formalism discussed in the previous section to derive a path integral expression for the partition function (18.63), which holds for an arbitrary finite "euclidean time"-step.

Consider the model whose Hamiltonian is given by (18.62a). To obtain a path integral representation of the grand canonical partition function we split the interval $[0, \beta]$ into N intervals of length $\epsilon = \beta/N$, and write the partition function in the form

$$\Xi = Tr(\hat{\Omega}_\epsilon)^N ,$$

where

$$\hat{\Omega}_\epsilon = e^{-\epsilon \sum_i (E_i-\mu)\hat{a}_i^\dagger \hat{a}_i} .$$

The matrix form of $\hat{\Omega}_\epsilon$ is given by (18.64) with β replaced by ϵ,

$$\Omega_\epsilon(\vec{a}^*, \vec{a}) = e^{\sum_i e^{-\epsilon(E_i-\mu)} a_i^* a_i} , \tag{18.66}$$

where, for notational reasons which will become clear below, we have denoted the sets of Grassman variables $\{a_i^*\}$ and $\{a_i\}$ by $\vec{a}^*$ and $\vec{a}$, respectively. We next calculate the matrix form of $\hat{\Omega}_\epsilon^N$ by making repeated use of the product formula (18.58). Let us label the integration variables with an extra index, which will be interpreted later as labeling different time slices. Thus in the following $\vec{a}_n$ denotes a set of k Grassmann variables $a_{i,n}$, $i = 1, 2, \cdots, k$, where k is the number of degrees of freedom of the system. A similar statement holds for $\vec{a}_n^*$. Then the matrix form of $(\hat{\Omega}_\epsilon)^N$ is given by

$$\Omega(\vec{a}_N^*, \vec{a}_N) = \int \prod_i \prod_{n=1}^{N-1} da_{i,n}^* da_{i,n} e^{-\vec{a}_n^* \cdot \vec{a}_n}$$
$$\times [\Omega_\epsilon(\vec{a}_N^*, \vec{a}_{N-1}) \Omega_\epsilon(\vec{a}_{N-1}^*, \vec{a}_{N-2}) ... \Omega_\epsilon(\vec{a}_1^*, \vec{a}_0)]_{\vec{a}_0 = \vec{a}_N} ,$$

The trace of $(\hat{\Omega}_\epsilon)^N$ is calculated according to (18.60):

$$Tr\hat{\Omega} = \int \prod_i da^*_{i,N} da_{i,N} e^{-\vec{a}^*_N \vec{a}_N} \Omega(\vec{a}^*_N, -\vec{a}_N) \ .$$

Writing out the components of $\vec{a}_n$ and $\vec{a}^*_n$ explicitely we are led to the following *exact* path integral representation for the partition function,

$$Tre^{-\beta(\hat{H}-\mu\hat{Q})} = \int \prod_{n=1}^{N} \prod_i da^*_{i,n} da_{i,n} e^{-S(\{a^*_{i,n}\},\{a_{i,n}\})}|_{a_{i,0}=-a_{i,N}} \ , \qquad (18.67a)$$

where

$$S(\{a^*_{i,n}\},\{a_{i,n}\}) = \sum_{n=1}^{N} \sum_i \{a^*_{i,n}[a_{i,n} - a_{i,n-1}] + (1 - e^{-\epsilon(E_i-\mu)}) a^*_{i,n} a_{i,n-1}\} \ . \qquad (18.67b)$$

Notice that the path integral (18.67a) is to be calculated subject to the antiperiodic boundary condition $a_{i,0} = -a_{i,N}$. While the index "i" labels the degrees of freedom of the system, the index "n" labels the different (euclidean) time slices.

The partition function can now be calculated by diagonalizing (18.67b). To this end we introduce a new set of variables $\tilde{a}_{i,n}$ and $\tilde{a}^*_{i,n}$ by the following unitary transformation (we take N to be even),

$$a_{i,n} = \sum_{\ell=-N/2}^{N/2-1} c_{n\ell} \tilde{a}_{i,\ell} \ ; \quad a^*_{i,n} = \sum_{\ell=-N/2}^{N/2-1} c^*_{n\ell} \tilde{a}^*_{i,\ell} \ , \qquad (18.68a)$$

where

$$c_{n\ell} = \frac{1}{\sqrt{N}} e^{i\hat{\omega}_\ell n} \ . \qquad (18.68b)$$

and $\hat{\omega}_\ell$ are the Matsubara frequencies,

$$\hat{\omega}_\ell = (\frac{2\ell+1}{N})\pi \ ; \quad (fermions) \qquad (18.69)$$

The coefficients (18.68b) satisfy the relations

$$\sum_{\ell=-N/2}^{N/2-1} c^*_{n'\ell} c_{n\ell} = \delta_{nn'}$$

$$\sum_{n=1}^{N} c^*_{n\ell} c_{n\ell'} = \delta_{\ell\ell'} \ . \qquad (18.70)$$

Notice that the antiperiodic boundary condition in (18.67a) has been incorporated in the expansion (18.68a). Upon inserting (16.68a) into (18.67b), and performing the sum over n by making use of (18.70) one finds that

$$Tr\hat{\Omega} = \int \prod_i \prod_{\ell=-\frac{N}{2}}^{\frac{N}{2}-1} d\tilde{a}^*_{i,\ell} d\tilde{a}_{i,\ell} e^{-\sum_{i,\ell} \tilde{a}^*_{i,\ell}\tilde{a}_{i,\ell}[1-e^{-\epsilon(E_i-\mu)}e^{-i\hat{\omega}_\ell}]}$$

$$= \prod_i \prod_{\ell=-\frac{N}{2}}^{\frac{N}{2}-1} \{1 - e^{-\epsilon(E_i-\mu)}e^{-i\hat{\omega}_\ell}\} .$$

We therefore see that the chemical potential is introduced into the partition function for $\mu = 0$ by the simple substitution rule

$$\hat{\omega}_\ell \to \hat{\omega}_\ell + i\hat{\mu} , \tag{18.71}$$

where $\hat{\mu} = \epsilon\mu$ is the chemical potential measured in lattice units. Combining the positive and negative frequency parts, $Tr\hat{\Omega} = \Xi$ can be written in the form

$$\Xi = \prod_i \prod_{\ell=0}^{\frac{N}{2}-1} \{1 + x_i^2 - 2x_i \cos\hat{\omega}_\ell\} = \prod_i (1 + x_i^N) , \tag{18.72a}$$

where

$$x_i = e^{-\epsilon(E_i-\mu)} . \tag{18.72b}$$

Upon setting $\epsilon N = \beta$, one recovers the result (18.65).

The path-integral representation (18.67) exhibits several interesting features which deserve a comment:

(i) As we have already mentioned, it is the *exact* path-integral expression for the partition function (18.63) for every choice of $\epsilon = \beta/N$.

(ii) An alternative expression for the path integral (which will turn out to be convenient when we compare it with the lattice actions employed in numerical simulations) is obtained by making the redefinitions $a_{i,n} \to a_{i,n+1}$ and $a^*_{i,n} \to e^{-\epsilon\mu} a^*_{i,n}$. This is always possible, since the a's and the a^*'s are independent Grassmann variables. One can then easily show that (18.67) can also be written in the form *

$$\Xi = [\prod_i e^{\beta\mu}] \int \prod_{n=1}^{N} \prod_i da^*_{i,n} da_{i,n} e^{-S(\{a^*_{i,n}\},\{a_{i,n}\})}|_{a_{i,N+1}=-a_{i,1}} , \tag{18.73a}$$

* Care must be taken of the fact that the integration measure for Grassmann variables does not transform in the way one is used to from integrations over c-number variables. Thus $\int da e^{\lambda a} = \lambda \int db e^b$

where

$$S(\{a^*_{i,n}\},\{a_{i,n}\}) = \sum_{n=1}^{N}\sum_{i}\{a^*_{i,n}[e^{-\epsilon\mu}a_{i,n+1}-a_{i,n}]+(1-e^{-\epsilon E_i})a^*_{i,n}a_{i,n}\}\,. \quad (18.73b)$$

The only subtle point in obtaining this result concerns the integration measure. The reäder can convince himself that it can be still written in the form given in (18.67a), by modifying the boundary conditions in the way exhibitėd in (18.73a).

iii) Still another form of the path-integral expression (18.73) can be obtained by introducing the variables

$$a'_{i,n} = e^{-\epsilon\mu n}a_{i,n}\,,$$

$$a'^*_{i,n} = e^{\epsilon\mu n}a^*_{i,n}\,. \quad (18.74)$$

which leaves the measure unchanged. In this way one can eliminate the μ-dependence in the action, and incorporate the chemical potential into the boundary condition. Dropping the "prime" one then obtains that

$$\Xi = \mathcal{N}\int\prod_{n=1}^{N}\prod_{i}da^*_{i,n}da_{i,n}[e^{-\sum_{i,n}\{a^*_{i,n}(a_{i,n+1}-a_{i,n})+(1-e^{-\epsilon E_i})a^*_{i,n}a_{i,n}\}}]_{a_{i,N+1}=-e^{-\beta\mu}a_{i,1}} \quad (18.75a)$$

where

$$\mathcal{N} = [\prod_i e^{\beta\mu}]\,. \quad (18.75b)$$

Note that the chemical potential now enters in the form of a boundary condition in the combination $\beta\mu$, which is independent of the chosen discretization. By expanding $e^{-\epsilon E_i}$ in (18.75) up to leading order in ϵ we can obtain an expression for the partition function, valid in the continuum limit, which can be generalized to the case where the Hamiltonian no longer has the simple form (18.62a),

$$\Xi = \mathcal{N}\lim_{\substack{\epsilon\to 0\\ N\to\infty\\ N\epsilon=\beta}}\int\prod_{n=1}^{N}\prod_{i}da^*_{i,n}da_{i,n}[e^{-\sum_{i,n}\{a^*_{i,n}(a_{i,n+1}-a_{i,n})+\epsilon\mathcal{H}(a^*_{i,n},a_{i,n})\}}]_{a_{i,N+1}=-e^{-\beta\mu}a_{i}} \quad (18.76a)$$

or equivalently

$$\Xi = \mathcal{N}\lim_{\substack{\epsilon\to 0\\ N\to\infty\\ N\epsilon=\beta}}\int\prod_{n=1}^{N}\prod_{i}da^*_{i,n}da_{i,n}[e^{-\sum_{i,n}\{a^*_{i,n}(e^{-\epsilon\mu}a_{i,n+1}-a_{i,n})+\epsilon\mathcal{H}(a^*_{i,n},a_{i,n})\}}]_{a_{i,N+1}=-a_{i}} \quad (18.76b)$$

where

$$\mathcal{H}(a^*_{i,n}, a_{i,n}) = E_i a^*_{i,n}, a_{i,n} \tag{18.76c}$$

is the Hamiltonian "density" defined on the n'th time slice. In obtaining (18.76b) we have again made a change of variables analogous to (18.74) in (18.76a). Expression (18.76b) has a form which most closely resembles that used in lattice calculation. Except for the overall factor multiplying the integral, the chemical potential is introduced in the bilinear terms coupling neighbouring lattice sites in the euclidean time direction in the way proposed by Kogut et.al. [Kogut (1983c)], and Hasenfratz and Karsch [Hasenfratz (1983)]:

$$a^*_{i,n} a_{i,n+1} \to a^*_{i,n} e^{-\hat{\mu}} a_{i,n+1} \,. \tag{18.77}$$

where $\hat{\mu}$ is the chemical potential measured in "lattice" units. As we have seen in this section, this prescription follows naturally from the functional formalism.

Let us verify that the partition function (18.76b) yields the correct answer for the mean energy, $< E >$ and mean particle number $< N >$. By diagonalizing the action as before one finds that

$$\Xi = \lim_{\substack{\epsilon\to 0\\ N\to\infty\\ N\epsilon=\beta}} \prod_i \prod_{\ell=-\frac{N}{2}}^{\frac{N}{2}-1} e^{i\hat{\omega}_\ell}[1-(1-\epsilon E_i)e^{-i\hat{\omega}_\ell+\epsilon\mu}]$$

Taking into account that the product includes as many positive as well as negative frequencies, this expression can again be written in the form (18.72a) with $x_i -(1-\epsilon E_i)e^{\epsilon\mu}$. Hence

$$\ln \Xi = \sum_i \ln[1+(1-\epsilon E_i)^N e^{\beta\mu}] \,.$$

Upon setting $\epsilon = \frac{\beta}{N}$, and taking the limit $N \to \infty$, we recover the result (18.65). For *fixed* N, the mean energy is given by (recall that $\beta\mu$ is to be held fixed),

$$< E >= - \lim_{\substack{\epsilon\to 0\\ N\to\infty\\ N\epsilon=\beta}} \frac{1}{N}\left[\frac{\partial}{\partial\epsilon}\ln\Xi\right]_{\epsilon\mu fixed} \tag{18.78}$$

Hence

$$\begin{aligned} < E > &= \lim_{\substack{\epsilon\to 0\\ N\to\infty\\ N\epsilon=\beta}} \sum_i \frac{E_i}{1-\frac{\beta}{N}E_i}\frac{1}{(1-\frac{\beta}{N}E_i)^{-N}e^{-\beta\mu}+1} \\ &= \sum_i \frac{E_i}{e^{\beta(E_i-\mu)}+1} \,. \end{aligned} \tag{18.79a}$$

A similar calculation for $< N >$ yields

$$< N >= \lim_{\substack{\epsilon\to 0 \\ N\to\infty \\ N\epsilon=\beta}} \frac{1}{N\epsilon}\frac{\partial}{\partial\mu}\ln\Xi = \sum_i \frac{1}{e^{\beta(E_i-\mu)}+1} \tag{18.79b}$$

As the reader will have noticed, the factor (18.75b) multiplying the integral in (18.76a) was important for obtaining the correct answer. Ignoring it would not influence the result for the mean energy, but would modify the expression (18.79b) by the potentially divergent sum $\sum_i = n_f$, where n_f is the number of degrees of freedom.

18.9 A Modified Fermion Action Leading to Fermion Doubling

We have seen in the previous section that the functional formalism for fermions tells us that the kinetic (time-derivative) contribution to the action is discretized using the right lattice derivative. The dependence on the chemical potential then appears in the form of a μ-dependent boundary condition. Alternatively, the chemical potential can be introduced into the action at $\mu = 0$ according to the rule (18.77). In this case the Grassmann variables satisfy antiperiodic boundary conditions. Although the action in the latter formulation resembles that used in lattice simulations, it actually differs from it in an essential way. When simulating relativistic field theories involving Dirac fermions on the lattice, one wants the action to exhibit a hypercubic symmetry, which is the lattice remnant of the 0(4) symmetry in the (euclidean) continuum formulation. The fermionic part of the action is therefore disretized using the symmetric form for the temporal (as well as spatial) lattice derivative. In chapter 3, where we discussed the fermion propagator, we have seen that using such a discretization leads to a serious problem: the fermion doubling problem. Of course we also expect to see the contribution of the doublers in the partition function. In fact, as we now show, the doubler contributions which arise from the use of a symmetric lattice *time*-derivative manifest themselves in a non-trivial way. This can be most clearly demonstrated for the non-relativistic fermionic system considered in the previous sections. The manifestation of the doubers in the relativistic case is more subtle and has been discussed by Bender et.al. [Bender (1993)].

The following partition function is a modification of (18.76b), where the kinetic term in the action (i.e., the "time"-derivative term) is modified in a way which resembles that used in actual Monte Carlo simulations of lattice gauge field theories:

$$\tilde{\Xi}(\beta,\mu) = \lim_{\substack{\epsilon\to 0\\ N\to\infty \\ N\beta\ fixed}} \left[\prod_i e^{\beta\mu}\right] \int \prod_i \prod_{n=1}^{N} da^*_{i,n} da_{i,n}$$
$$\times e^{-\sum_{n=1}^{N} \epsilon [\sum_i a^*_{i,n} \frac{(e^{-\hat{\mu}} a_{i,n+1} - e^{\hat{\mu}} a_{i,n-1})}{2\epsilon} + \mathcal{H}(a^*_{i,n}, a_{i,n})]} \Big|_{\substack{a_{i,0}=-a_{i,N}\\ a_{i,N+1}=-a_{i,1}}} . \tag{18.80}$$

Here we have introduced the chemical potential into the symmetric time-derivative at $\mu = 0$ according to the Kogut-Hasenfratz-Karsch prescription [Kogut (1983c); Hasenfratz (1983)]

$$a^*_{i,n} a_{i,n+1} \to a^*_{i,n} e^{-\hat{\mu}} a_{i,n+1} ,$$
$$a^*_{i,n} a_{i,n-1} \to a^*_{i,n} e^{\hat{\mu}} a_{i,n-1} , \tag{18.81}$$

where $\hat{\mu} = \epsilon\mu$ is the chemical potential in lattice units. Because of the modified kinetic term the anti-periodic boundary conditions in (18.76b) have now been replaced by $a_{i,0} = -a_{i,N}$, and $a_{i,N+1} = -a_{i,1}$. The intcgral can be performed by diagonalizing the action in the manner described earlier. Making the change of variables (18.68), which incorporate the above antiperiodic boundary conditions, one finds that

$$\tilde{\Xi} = \prod_i \prod_{\ell=-\frac{N}{2}}^{\frac{N}{2}-1} e^{\hat{\mu}} [i \sin(\hat{\omega}_\ell + i\hat{\mu}) + \epsilon E_i] .$$

From here we obtain for the mean energy

$$< E >= -\frac{1}{N} \left[\frac{\partial \ln \tilde{\Xi}}{\partial \epsilon}\right]_{N\hat{\mu} fixed} = -\frac{1}{N} \sum_i E_i \, [\sum_{\ell=-\frac{N}{2}}^{\frac{N}{2}-1} f(\hat{\omega}_\ell, E_i)] , \tag{18.82a}$$

where

$$f(\hat{\omega}_\ell, E_i) = \frac{1}{i \sin(\hat{\omega}_\ell + i\hat{\mu}) + \epsilon E_i} . \tag{18.82b}$$

Note that the frequency sum is performed over the finite interval

$$-\pi + \frac{\pi}{N} < \hat{\omega}_\ell < \pi - \frac{\pi}{N} , \tag{18.82c}$$

i.e., over frequencies lying within in the first Brillouin zone. This frequency sum can be written as a contour integral by making use of the formula

$$\frac{1}{N} \sum_{\ell=-\frac{N}{2}}^{\frac{N}{2}-1} F(\hat{\omega}_\ell) = -\frac{1}{2\pi} \int_C \frac{F(\hat{\omega})}{e^{iN\hat{\omega}} + 1} , \tag{18.83}$$

where the contour C encloses the poles of the integrand in the complex $\hat{\omega}$-plane, lying within the interval (18.82c), arising from the zeros of the denominator. The integration is carried out in the counterclockwise sense. These poles are located at $\hat{\omega} = \hat{\omega}_\ell$, where $\hat{\omega}_\ell$ has been defined in (18.69). The corresponding residues of the integrand are given by $R_\ell = \frac{i}{N} F(\hat{\omega}_\ell)$. If $F(\hat{\omega})$ does not possess any singularities on the real axis, then the above expression can also be written in the form

$$\frac{1}{N}\sum_{\ell=-\frac{N}{2}}^{\frac{N}{2}-1} F(\hat{\omega}_\ell) = -\frac{1}{2\pi}\int_{-\pi-i\epsilon}^{\pi-i\epsilon} d\hat{\omega}\frac{F(\hat{\omega})}{e^{iN\hat{\omega}}+1} + \frac{1}{2\pi}\int_{-\pi+i\epsilon}^{\pi+i\epsilon} d\hat{\omega}\frac{F(\hat{\omega})}{e^{iN\hat{\omega}}+1} . \tag{18.84}$$

A more convenient form can be obtained by introducing the integration variable $z = e^{i\hat{\omega}}$. Then

$$\frac{1}{N}\sum_{\ell=-\frac{N}{2}}^{\frac{N}{2}-1} F(\hat{\omega}_\ell) = -\frac{1}{2\pi i}\int_{|z|=1+\epsilon} dz\frac{\tilde{F}(z)}{z(z^N+1)} + \frac{1}{2\pi i}\int_{|z|=1-\epsilon} dz\frac{\tilde{F}(z)}{z(z^N+1)} , \tag{18.85}$$

where $\tilde{F}(e^{i\hat{\omega}}) = F(\hat{\omega})$, and where the integrations are performed on circles in the complex z-plane, with radii $|z| = 1\pm\epsilon$, in the counterclockwise direction. If $\tilde{F}(z)$ is a meromorphic function of z, satisfying $|z|^{-N}\tilde{F}(z) \to 0$ for $|z| \to \infty$, then we can distort the integration contour in the first integral to infinity, taking into account the contribution of the poles of $\tilde{F}(z)$ for $|z| > 1$. The second integral is just $2\pi i$ times the sum of the residues of $\frac{\tilde{F}(z)}{z}$ located inside the unit circle. Hence

$$\frac{1}{N}\sum_{\ell=-\frac{N}{2}}^{\frac{N}{2}-1} F(\hat{\omega}_\ell) = \sum_{|z_i|\neq 1}\frac{R(z_i)}{z_i^N+1} , \tag{18.86a}$$

where

$$R(z_i) = Res_{z_i}\left[\frac{\tilde{F}(z)}{z}\right] . \tag{18.86b}$$

We now use this expression to calculate the frequency sum in (18.82a). The position of the poles of (18.82b) in the z-plane are given by

$$z_\pm = [-\epsilon E_i \pm \sqrt{\epsilon^2 E_i^2 + 1}]e^{\epsilon\mu} = \pm e^{\mp arsinh\epsilon E_i + \epsilon\mu} .$$

Making use of (18.86) one then finds that

$$< E >= \sum_i \frac{E_i}{\sqrt{\epsilon^2 E_i^2 + 1}}\left[\frac{1}{e^{N(arsinh\epsilon E_i + \epsilon\mu)} + 1} - \frac{1}{e^{-N(arsinh\epsilon E_i - \epsilon\mu)} + 1}\right] .$$

Upon taking the limits $\epsilon \to 0, N \to \infty$, with $N\epsilon = \beta$ fixed, and making use of the identity $1/(e^x + 1) + 1/(e^{-x} + 1) = 1$, one obtains

$$< E >= \sum_i E_i [\frac{1}{e^{\beta(E_i - \mu)}} + \frac{1}{e^{\beta(E_i + \mu)}}] - \sum_i E_i \ . \tag{18.87}$$

This is an interesting result, for it deviates in a drastic way from (18.79a). Apart from a potentially diverent additive contribution, the remaining two contributions resemble those of a gas of particles and antiparticles! Hence (18.80) is not the path integral representation of the partition function (18.63b). This example clearly demonstrates that modifications in the action which are inconsistent with the general form dictated by the functional formalism, can give rise to spurious unphysical contributions. This is the case, although the actions appearing in (18.76b) and (18.80) differ *formally* only by terms of $\mathcal{O}(\epsilon)$. Nevertheless they give rise to very different expressions for the partition function. We had already been confronted with a similar situation when we discussed the propagator for Wilson fermions in chapter 3. There the action was modified by a so called irrelevant term, vanishing in the naive continuum limit. When introduced into the path integral, however, the fermion doubling problem disappears.

Whereas the doubler contributions to the partition function do possess a continuum limit, this is not true for Green functions, as we have already seen in chapter 3, and will be further demonstrated in section 9 of the following chapter, where we discuss in detail the Dirac propagator for naive (and Wilson) fermions at finite temperature and chemical potential.

18.10 The Free Dirac Gas. Continuum Approach

We have discussed in great detail the path integral representation of the partition function for a simple fermionic system, in order to point out some subtle points. In this and the following section we now extend our discussion to the relativistic case. In text books on finite temperature field theory the thermodynamical partition function for a free Dirac gas is usually derived from its path integral representation in the continuum, where such concepts as the right or left-derivative do not appear. Nor does the expression for the partition function reflect the fact that in the discretized version the chemical potential is introduced into the bilinear terms in the Grassmann variables coupling neighbouring sites along the euclidean time axis. In lattice regularized gauge field theories, however, the chemical potential should be introduced in the exponentiated form, discussed in the previous section, in order to ensure the renormalizabilty of the theory. For completeness sake, we will include the discussion of the free Dirac gas within the continuum formulation in this section. In the following section we then study the Dirac gas within the framework of the lattice regularization.

Consider the Hamiltonian for a free Dirac field

$$H = \int d^3x \psi^\dagger(x)\gamma_4[\gamma_j\partial_j + m]\psi(x) \ , \tag{18.88}$$

where γ_μ are the euclidean gamma matrices introduced in chapter 3, satisfying the anticommutation relations $\{\gamma_\mu, \gamma_\nu\} = 2\delta_{\mu\nu}$. The charge operator is

$$Q = \int d^3x \psi^\dagger(x)\psi(x) \ . \tag{18.89}$$

Within the continuum approach the (formal) path integral representation of the grand canonical partition function

$$\Xi = Tr\left\{e^{-\beta(H-\mu Q)}\right\} , \tag{18.90}$$

is given by

$$\Xi = \mathcal{N}\int_{antip.} D\psi^* D\psi \ e^{-\int_0^\beta d\tau \int d^3x[\psi^*(x)\gamma_4(\not{\partial}+m)\psi(x) - \mu\psi^*(x)\psi(x)]} \tag{18.91}$$

where $\mathcal{N}$ is a dimensioned normalization factor. The subscript "antip." is to remind the reader that the Grassmann integration is to be performed with the fields ψ satisfying antiperiodic boundary conditions in "time". Notice that this form for the partition function is formally obtained from (18.76b) by i) expanding $e^{-\epsilon\mu}$ to leading order in ϵ, ii) replacing the Grassmann variables $a_{i,n}$ and $a^*_{i,n}$ by* $\psi_\alpha(\vec{x},\tau)$ and $\psi^*_\alpha(\vec{x},\tau)$, and approximating $\epsilon\mu\psi^*(\vec{x},\tau)\psi(\vec{x},\tau+\epsilon)$ by the local form $\epsilon\mu\psi^*_\alpha(\vec{x},\tau)\psi_\alpha(\vec{x},\tau)$. This amounts to taking the *naive* continuum limit of the discretized action.

The definition (18.91) is the usual starting point for discussing the thermodynamics of a Dirac gas within the path integral framework (see e.g. Kapusta (1989)). We now proceed as in the case of the free scalar field, and rewrite the path integral in terms of dimensionless variables, by scaling the fields, space-time coordinates, and the mass m with an appropriate power of β according to their canonical dimensions. Thus introducing the dimensionless variables $\hat{\psi} = \beta^{3/2}\psi$,

---

* Recall that $(\vec{x}, \alpha)$ label the degrees of freedom.

$\hat{\psi}^* = \beta^{2/3}\psi^*$, $\hat{\tau} = \tau/\beta$, $\hat{x}_i = x_i/\beta$, $\hat{\mu} = \beta\mu$, $\hat{m} = \beta m$, the expression (18.91) takes the form

$$\Xi = \hat{\mathcal{N}} \int_{antip.} D\hat{\psi}^* D\hat{\psi} e^{-\int_0^1 d\hat{\tau} \int d^3\hat{x} \hat{\psi}^*(x)[\hat{\partial}_\tau - \hat{\mu} + \hat{\mathcal{H}}]\hat{\psi}(x)} , \tag{18.92a}$$

where

$$\hat{\mathcal{H}} = \vec{\alpha} \cdot \hat{\nabla} + \gamma_4 \hat{m} , \tag{18.92b}$$

with $\vec{\alpha} = \gamma_4 \vec{\gamma}$, is the Hamiltonian density measured in units of β^{-1}, and $\hat{\mathcal{N}}$ is now a dimensionless normalization factor. Since the integral (18.92a) is of the "Gaussian" type, it can be performed immediately. As we have seen in chapter 2, the result is just the determinant of the operator appearing within square brackets in (18.92a). Hence

$$\begin{aligned} \ln \Xi &= \ln \hat{\mathcal{N}} + \ln \det(\hat{\partial}_4 - \hat{\mu} + \hat{\mathcal{H}}) \\ &= \ln \hat{\mathcal{N}} + \ln \det[\beta(\partial_\tau - \mu + \mathcal{H})] , \end{aligned}$$

where $\mathcal{H} = \vec{\alpha} \cdot \vec{\nabla} + \gamma_4 m$ is the dimensioned Hamiltonian density. The determinant is given by the product of the eigenvalues of the operator acting in the space of functions satisfying antiperiodic boundary conditions on the compactified euclidean time interval $[0, \beta]$. The eigenfunctions of $\partial_\tau - \mu + \mathcal{H}$ have the form $\exp(i\omega_\ell \tau) \exp(i\vec{p} \cdot \vec{x}) W(\vec{p})$, where

$$\omega_\ell = \frac{(2\ell + 1)\pi}{\beta} \tag{18.93}$$

are the dimensionful Matsubara frequencies corresponding to (18.69), and where $W(\vec{p})$ are eigenvectors of the operator $i\vec{\alpha} \cdot \vec{p} + \gamma_4 m$. The eigenvalues are given by $\pm E(\vec{p})$, where

$$E(\vec{p}) = \sqrt{\vec{p}^2 + m^2} , \tag{18.94}$$

and are two-fold degenerate, corresponding to the two possible spin projections. Hence the eigenvalues of $\partial_\tau - \mu + \mathcal{H}$ are given by

$$\lambda^{(\pm)}_{\vec{p},\ell} = i\omega_\ell - \mu \pm E(\vec{p}) , \tag{18.95}$$

For the logarithm of the partition function we therefore obtain

$$\ln \Xi = \ln \hat{\mathcal{N}} + 2 \sum_{\vec{p}} \sum_{\ell} \{\ln[\beta(i\omega_\ell - \mu + E(\vec{r}))] + (E \to -E)\} ,$$

where ℓ runs over all possible integers. As in the case of the scalar field, we have enclosed the system into a box of finite volume to discretize the momenta. Collecting the positive and negative frequency contributions we can write this expression in an explicit real form:

$$\ln \Xi = \ln \hat{\mathcal{N}} + 2\sum_{\vec{p}}\sum_{\ell=0}^{\infty}(\ln\{[(2\ell+1)\pi]^2 + [\beta(E-\mu)]^2\} + (E \to -E)) .$$

The frequency sum diverges. To extract the finite temperature and chemical potential dependent part let us write the logarithm in the form*

$$\ln\{[(2\ell+1)\pi]^2+[\beta(E-\mu)]^2\} = \int_1^{[\beta(E-\mu)]^2} dy^2 \, \frac{1}{(2\ell+1)^2\pi^2+y^2} + \ln[(2\ell+1)^2\pi^2+1] .$$

Dropping the irrelevant constant contribution, and performing the sum over ℓ by making use of

$$\sum_{\ell=0}^{\infty} \frac{1}{(2\ell+1)^2\pi^2+y^2} = \frac{1}{4y}\tanh\frac{y}{2} ,$$

One then finds that

$$\begin{aligned} \ln \Xi = \ln \hat{\mathcal{N}} &+ 2V\beta \int \frac{d^3p}{(2\pi)^3} E(\vec{p}) \\ &+ 2V \int \frac{d^3p}{(2\pi)^3} \left\{ \ln(1+e^{-\beta(E-\mu)}) + \ln(1+e^{-\beta(E+\mu)}) \right\} , \end{aligned} \tag{18.96}$$

where we have replaced the formal sum over momenta by an integral according to (18.37). Apart from the first two terms, this is the expression familiar from statistical mechanics. The second term gives rise to a constant energy density shift and is just the contribution of the negative energy states in the Dirac sea. In this connection, recall that within the path integral approach, there is no normal ordering prescription. Of course we also would expect to see a corresponding shift in the mean charge density. In fact, comparison of the formal path integral expression (18.92a) with (18.76a), where the normalization factor is given by (18.75b), suggests that $\ln \hat{\mathcal{N}}$ is given by $\sum_i \beta\mu$, where i labels the (infinite) number of degrees of freedom. Since $\sum_i \approx 2V \int d^3p/(2\pi)^3$, one then easily verifies that the first term in (18.96) gives rise to the expected contribution to the mean charge $< Q >= \frac{1}{\beta}\frac{\partial \ln \Xi}{\partial \mu}$, arising from the Dirac sea.

---

* We follow here the method of Kapusta (1989)

18.11 Dirac Gas of Wilson Fermions on the Lattice

Let us now consider a lattice regularized version of the partition function for the Dirac field. Since we are interested in computing thermodynamical observables, we must be able to vary the temperature and volume in a continuous way. Hence we need an expression for the partition function formulated on an anisotropic space-time lattice, i.e., with different temporal and spatial lattice spacings, a_τ and a, respectively. The partition function will then be given by a path integral over Grassmann variables living on the space-time lattice sites $n = (\vec{n}, n_4)$, where $\vec{n}$ labels the coordinate degrees of freedom and takes all possible integer values, while n_4 runs over a finite number of lattice sites, $1 \leq n_4 \leq N_\tau$, where $N_\tau a_\tau$ is the inverse temperature β. Since N_τ is the inverse temperature measured in lattice units, we will use the more suggestive notation $\hat{\beta}$ in the following. The discretized version of the action having the correct *naive* continuum limit is not unique. We shall take it to be of a form which is consistent with the fermionic actions usually employed in Monte Carlo simulations of gauge theories. This means that, first of all, the kinetic term has a structure analogous to that appearing in the action of the expression (18.80). Secondly, the spatial derivative appearing in the Hamiltonian (18.88) will also be discretized using the symmetric lattice derivative, in order that the Hamiltonian be hermitean. With this prescription the action appearing in the exponential in (18.80) translates into

$$S_F = \sum_{n_4=1}^{\hat{\beta}} \sum_{\vec{n}} a^3 \{\psi^\dagger(n) \frac{1}{2} [e^{-\hat{\mu}} \psi(n + \hat{e}_4) - e^{\hat{\mu}} \psi(n - \hat{e}_4)] \tag{18.97a}$$

$$+ a_\tau \psi^\dagger(n) [\alpha_j \cdot \partial_j + \gamma_4 m] \psi(n)\} ,$$

where a summation over repeated indices is understood, and where

$$\partial_i \psi(n) = \frac{1}{2a} [\psi(n + \hat{e}_i) - \psi(n - \hat{e}_i)] \,. \tag{18.97b}$$

is the symmetric lattice derivative of $\psi(n)$, $\alpha_j = \gamma_4 \gamma_j$, with $\hat{e}_i$ a unit vector pointing in the i'th-direction. The action (18.97) can be written in an explicit dimensionless form by scaling the fields ψ and $\psi^\dagger$, as well as the spatial derivatives and fermion mass, with the *spatial* lattice spacing a according to their canonical dimension. The corresponding dimensionless quantities will be denoted as usual with a "hat". One is then led to the following path integral representation of the partition function

$$\Xi = \hat{\mathcal{N}} \int \prod_{\vec{n},\alpha} \prod_{n_4=1}^{\hat{\beta}} d\hat{\psi}^*_\alpha(n) d\psi_\alpha(n) \; e^{-S_F} \Big|_{\substack{\hat{\psi}(\vec{n},0) = -\hat{\psi}(\vec{n},\hat{\beta}) \\ \hat{\psi}(\vec{n},\hat{\beta}+1) = -\hat{\psi}(\vec{n},1)}} \,, \tag{18.98a}$$

where

$$S_F = \sum_{\vec{n}} \sum_{n_4=1}^{\beta} \{\hat{\psi}^\dagger(n)\frac{1}{2}[e^{-\hat{\mu}}\hat{\psi}(n+\hat{e}_4) - e^{\hat{\mu}}\hat{\psi}(n-\hat{e}_4)] + \frac{1}{\xi}\hat{\psi}^\dagger(n)[\alpha_j\cdot\hat{\partial}_j + \gamma_4\hat{m}]\hat{\psi}(n)\}\,, \tag{18.98b}$$

and $\xi = \frac{a}{a_\tau}$ is the anisotropy parameter. The (dimensionless) lattice derivative $\hat{\partial}_j$ is defined by (18.97b) with $a = 1$. For a fixed number of temporal lattice sites the temperature is now controled by the parameter ξ.

The action (18.98b) is expected to lead to fermion doubling. In particular, the kinetic term does not have the form dictated by the functional formalism. In section 9 we had demonstrated in a non-relativistic model that such a choice for the kinetic term leads to a non-trivial modification of the partition function. This is also the case for the relativistic Dirac gas, as has been discussed by Bender et al. [Bender (1993)].

The fermion doubling problem can be avoided by introducing a Wilson term. For vanishing chemical potential and zero temperature, where the extension of the lattice in the time direction is infinite, this Wilson term is given in (4.28), and exhibits a hypercubic symmetry. This symmetry is is the remnant of the $O(4)$ symmetry in the continuum, and is broken down to a spatial cubic symmetry at finite temperature due to the presence of the heat bath, as is evident from (18.98b). The Wilson parameters multiplying the contributions involving the second time-and space derivatives need therefore not be equal. The finite temperature parametrization of the Wilson term is ambiguous. The reason is that this (so called irrelevant) term vanishes linearly with the lattice spacing in the naive continuum limit. Expression (18.98b) however suggests a parametrization, where the ratio of the Wilson parameters associated with the temporal and spatial derivative contributions is given by the assymmetry parameter ξ. Thus for vanishing chemical potential we will take the Wilson term to be of the form

$$\delta S_F^{(W)} = -\frac{r}{2}\left[\sum_n \hat{\psi}^\dagger(n)\gamma_4\hat{\partial}_4^2\hat{\psi}(n) + \frac{1}{\xi}\sum_{n,j}\hat{\psi}^\dagger(n)\gamma_4\hat{\partial}_j^2\hat{\psi}(n)\right]\,, \tag{18.99a}$$

where

$$\hat{\partial}_\mu^2\hat{\psi}(n) = \hat{\psi}(n+\hat{e}_\mu) + \hat{\psi}(n-\hat{e}_\mu) - 2\hat{\psi}(n) \tag{18.99b}$$

is the discretized second derivative of $\hat{\psi}$, and where it is understood from now on that $\sum_n = \sum_{\vec{n}}\sum_{n_4=0}^{\beta}$. By expressing (18.99a) again in terms of the dimensioned fields and second derivatives, one easily verifies that such a parametrization implies that the temporal and spatial derivative contributions vanish in the naive continuum limit linearly with a_τ and a, respectively.

The chemical potential is now introduced into (18.99) according to the Hasenfratz Karsch-Kogut prescription [Hasenfratz (1983); Kogut (1983)]

$$\begin{aligned} \hat{\psi}^\dagger(n)\hat{\psi}(n+\hat{e}_4) &\to \hat{\psi}^\dagger(n)e^{-\hat{\mu}}\hat{\psi}(n+\hat{e}_4) , \\ \hat{\psi}^\dagger(n)\hat{\psi}(n-\hat{e}_4) &\to \hat{\psi}^\dagger(n)e^{\hat{\mu}}\hat{\psi}(n-\hat{e}_4) , \end{aligned} \tag{18.100}$$

which is the analog of (18.81). Then the fermion action (18.98b), supplemented by the Wilson term (18.99a), takes the form:

$$\begin{aligned} S_F^{(W)} &= [r + \frac{1}{\xi}(\hat{m}+3r)]\sum_n \hat{\psi}^\dagger(n)\gamma_4\hat{\psi}(n) \\ &- \frac{1}{2}\sum_n \hat{\psi}^\dagger(n)\gamma_4[(r-\gamma_4)e^{-\hat{\mu}}\hat{\psi}(n+\hat{e}_4) + (r+\gamma_4)e^{\hat{\mu}}\hat{\psi}(n-\hat{e}_4)] \\ &- \frac{1}{2\xi}\sum_{n,i} \hat{\psi}^\dagger(n)\gamma_4[(r-\gamma_i)\hat{\psi}(n+\hat{e}_i) + (r+\gamma_i)\hat{\psi}(n-\hat{e}_i)] . \end{aligned} \tag{18.101}$$

This is the generalization of (4.28) to finite temperature and chemical potential (recall that $\bar{\psi} = \psi^\dagger\gamma_4$). Notice that our finite temperature, finite chemical potential parametrization has preserved the $r \pm \gamma_\mu$ structure of the $T = \mu = 0$ Wilson action (4.28). We now diagonalize this action by Fourier decomposing the fields $\hat{\psi}(n)$ and $\hat{\psi}^\dagger(n)$ as follows (we take $\hat{\beta} = N_\tau$ to be even)

$$\begin{aligned} \hat{\psi}(n) &= \frac{1}{\hat{\beta}}\sum_{\ell=-\frac{\hat{\beta}}{2}}^{\frac{\hat{\beta}}{2}-1}\int_{-\pi}^{\pi}\frac{d^3\hat{p}}{(2\pi)^3}\tilde{\psi}(\hat{p},\hat{\omega}_\ell)e^{i(\vec{\hat{p}}\cdot\vec{n}+\hat{\omega}_\ell n_4)} , \\ \hat{\psi}^\dagger(n) &= \frac{1}{\hat{\beta}}\sum_{\ell=-\frac{\hat{\beta}}{2}}^{\frac{\hat{\beta}}{2}-1}\int_{-\pi}^{\pi}\frac{d^3\hat{p}}{(2\pi)^3}\tilde{\psi}^\dagger(\hat{p},\hat{\omega}_\ell)e^{-i(\vec{\hat{p}}\cdot\vec{n}+\hat{\omega}_\ell n_4)} , \end{aligned} \tag{18.102}$$

where $\hat{\omega}_\ell$ are the dimensionless Matsubara frequencies for fermions (18.69), with N identified with $\hat{\beta}$, i.e.,

$$\hat{\omega}_\ell = \frac{(2\ell+1)\pi}{\hat{\beta}} . \tag{18.103}$$

Notice that both, the frequencies and momenta, are restricted to the first Brillouin zone. Making use of the relation

$$\frac{1}{\hat{\beta}}\sum_{\vec{n}}\sum_{n_4=1}^{\hat{\beta}} e^{i(\vec{\hat{p}}-\vec{\hat{p}}')\cdot\vec{n}}e^{i(\hat{\omega}_\ell-\hat{\omega}_{\ell'})n_4} = \delta_{\ell\ell'}(2\pi)^3\delta_P^{(3)}(\vec{\hat{p}}-\vec{\hat{p}}') , \tag{18.104}$$

where $\delta_P^{(3)}(\vec{\hat{p}}-\vec{\hat{p}}')$ is the three dimensional version of the periodic δ-function (2.64), the action takes the following form for $r=1$,

$$S_F = \frac{1}{\hat{\beta}} \sum_{\ell=-\frac{\hat{\beta}}{2}}^{\frac{\hat{\beta}}{2}-1} \int_{-\pi}^{\pi} \frac{d^3\hat{p}}{(2\pi)^3} \tilde{\psi}^\dagger(\hat{p},\hat{\omega}_\ell) K(\vec{\hat{p}},\hat{\omega}_\ell,\xi,\hat{\mu}) \tilde{\psi}(\hat{p},\hat{\omega}_\ell) \ ,$$

where

$$K(\vec{\hat{p}},\hat{\omega}_\ell,\xi,\hat{\mu}) = i\sin(\omega_\ell + i\hat{\mu}) + \frac{1}{\xi} H(\vec{\hat{p}},\hat{\omega}_\ell,\xi,\hat{\mu}) \ ,$$

$$H(\vec{\hat{p}},\hat{\omega}_\ell,\xi,\hat{\mu}) = \sum_j i\alpha_j \sin\hat{p}_j + \gamma_4 \mathcal{M}(\vec{\hat{p}},\hat{\omega}_\ell,\xi,\hat{\mu}) \ ,$$

and

$$\mathcal{M}(\vec{\hat{p}},\hat{\omega}_\ell,\xi,\hat{\mu}) = \hat{M}(\vec{\hat{p}}) + 2\xi \sin^2\left(\frac{\hat{\omega}_\ell + i\hat{\mu}}{2}\right) \ ,$$

$$\hat{M}(\vec{\hat{p}}) = \hat{m} + 2\sum_j \sin^2 \frac{\hat{p}_j}{2} \ . \tag{18.105}$$

The matrix K can be further diagonalized in Dirac space. Then the partition function is given by the product of the eigenvalues (which are two-fold degenerate)

$$\lambda_\pm = i\sin(\omega_\ell + i\hat{\mu}) \pm \frac{1}{\xi}\hat{\mathcal{E}}(\vec{\hat{p}},\hat{\omega}_\ell,\xi,\hat{\mu}) \ ,$$

where

$$\hat{\mathcal{E}}(\vec{\hat{p}},\hat{\omega}_\ell,\xi,\hat{\mu}) = \sqrt{\sum_j \sin^2\hat{p}_j + \mathcal{M}^2(\vec{\hat{p}},\hat{\omega}_\ell,\xi,\hat{\mu})} \ . \tag{18.106}$$

Hence, formally, the logarithm of the partition function (ignoring an additive contribution arising from the normalization factor in (18.98a)), which should not influence the thermodynamics, is given by

$$\ln\Xi = 2\sum_{\vec{\hat{p}}=-\pi}^{\pi} \sum_{\ell=-\frac{\hat{\beta}}{2}}^{\frac{\hat{\beta}}{2}-1} \ln[\sin^2(\hat{\omega}_\ell + i\hat{\mu}) + \frac{1}{\xi^2}\hat{\mathcal{E}}^2(\vec{\hat{p}},\hat{\omega}_\ell,\xi,\hat{\mu})] \ , \tag{18.107}$$

where the factor 2 accounts for the two spin degrees of freedom. Making the replacement

$$\sum_{\vec{\hat{p}}=-\pi}^{\pi} \to \hat{V} \int_{-\pi}^{\pi} \frac{d^3\hat{p}}{(2\pi)^3} \ , \tag{18.108}$$

valid in the large volume limit ($\hat{V}$ is the volume measured in lattice units), we obtain

$$\ln \Xi = 2\hat{V} \sum_{\ell=-\frac{\hat{\beta}}{2}}^{\frac{\hat{\beta}}{2}-1} \int_{-\pi}^{\pi} \frac{d^3\hat{p}}{(2\pi)^3} \ln[\sin^2(\hat{\omega}_\ell + i\hat{\mu}) + \frac{1}{\xi^2}\hat{\mathcal{E}}^2(\vec{\hat{p}}, \hat{\omega}_\ell, \xi, \hat{\mu})] \,. \tag{18.109}$$

From here we compute the mean energy (in lattice units) and mean charge

$$< \hat{E} >= \frac{1}{\hat{\beta}} \left[\frac{\partial \ln \Xi}{\partial \xi}\right]_{\xi=1} , \tag{18.110a}$$

$$< Q >= \frac{1}{\hat{\beta}} \left[\frac{\partial \ln \Xi}{\partial \hat{\mu}}\right]_{\xi=1} . \tag{18.110b}$$

Consider for example the mean energy. Performing the indicated differentiation one finds after some algebra that

$$< \hat{E} >= 2\frac{\hat{V}}{\hat{\beta}} \sum_{\ell=-\frac{\hat{\beta}}{2}}^{\frac{\hat{\beta}}{2}-1} \int_{-\pi}^{\pi} \frac{d^3\hat{p}}{(2\pi)^3} F(\hat{\omega}_\ell + i\hat{\mu}, \vec{\hat{p}}) \,, \tag{18.111a}$$

where

$$F(\hat{\omega}_\ell + i\hat{\mu}, \vec{\hat{p}}) = -2\frac{\hat{E}^2(\vec{\hat{p}}) + \hat{M}(\vec{\hat{p}})[1 - \cos(\hat{\omega}_\ell + i\hat{\mu})]}{\hat{E}^2(\vec{\hat{p}}) + 2[1 + \hat{M}(\vec{\hat{p}})][1 - \cos(\hat{\omega}_\ell + i\hat{\mu})]} \,, \tag{18.111b}$$

with

$$\hat{E}(\vec{\hat{p}}) = \sqrt{\sum_j \sin^2 \hat{p}_j + \hat{M}^2(\vec{\hat{p}})} \,, \tag{18.111c}$$

and $\hat{M}(\vec{\hat{p}})$ given by (18.105). The frequency sum of $F(\hat{\omega} + i\hat{\mu}, \vec{\hat{p}})$ in (18.111a) can now be easily carried out by making use of either the summation formula (18.86) or, even more conveniently, with the help of the summation formula (D.3) derived in appendix D. The function analogous to $g(e^{i(\hat{\omega}+i\hat{\mu})}, \{\vec{\hat{p}}_i\})$ in (D.1) is now defined by $g(e^{i(\hat{\omega}+i\hat{\mu})}, \vec{\hat{p}}) = F(\hat{\omega} + i\hat{\mu}, \vec{\hat{p}})$. Correspondingly, $g(z)$ in (D.3) is now given by

$$g(z) = \frac{f(z)}{[z - z_+][z - z_-]} \,,$$

where

$$f(z) = -\frac{2}{(1 + \hat{M})}[\hat{M}z^2 - 2(\hat{E}^2 + \hat{M})z + \hat{M}] \,,$$

$$z_{\pm} = [\kappa(\vec{\hat{p}}) \pm \sqrt{\kappa^2(\vec{\hat{p}}) - 1}] ,$$

and

$$\kappa(\vec{\hat{p}}) = 1 + \frac{\hat{E}^2(\vec{\hat{p}})}{2(1 + \hat{M}(\vec{\hat{p}}))} . \tag{18.112}$$

The rest of the calculation is straight forward. After performing the frequency sum in (18.111a) one finds for the energy density measured in lattice units that

$$\frac{1}{\hat{V}} < \hat{E} >= 2 \int_{-\pi}^{\pi} \frac{d^3\hat{p}}{(2\pi)^3} \rho(\vec{\hat{p}}) \hat{E}(\vec{\hat{p}}) [\eta_F(\vec{\hat{p}}) + \bar{\eta}_F(\vec{\hat{p}})] + constant , \tag{18.113a}$$

where

$$\rho(\vec{\hat{p}}) = \frac{1 + \hat{M}(\vec{\hat{p}})/2}{(1 + \hat{M}(\vec{\hat{p}}))\sqrt{1 + \hat{M}(\vec{\hat{p}}) + (\hat{E}(\vec{\hat{p}})/2)^2}} \tag{18.113b}$$

is a function which approaches unity in the continuum limit. $\eta_F(\vec{\hat{p}})$ and $\bar{\eta}_F(\vec{\hat{p}})$ are lattice versions of the Dirac distribution functions for particles and antiparticles:

$$\eta_F(\vec{\hat{p}}) = \frac{1}{e^{\hat{\beta}[\ln(\kappa + \sqrt{\kappa^2 - 1}) - \hat{\mu}]} + 1} , \tag{18.114a}$$

$$\bar{\eta}_F(\vec{\hat{p}}) = \frac{1}{e^{\hat{\beta}[\ln(\kappa + \sqrt{\kappa^2 - 1}) + \hat{\mu}]} + 1} , \tag{18.114b}$$

where κ has been defined in (18.112). The continuum limit is realized by setting $\hat{m} = ma$, $\hat{\mu} = \mu a$, $\hat{\beta} = \frac{\beta}{a}$, and taking $a \to 0$. In this limit we can replace κ^2 by $1 + \frac{1}{2}(\vec{\hat{p}}^2 + \hat{m}^2)$,* One then finds that, apart from a temperature and chemical potential independent contribution, the mean energy density in physical units reduces to

$$\frac{1}{V} < E >= 2 \int_{-\infty}^{\infty} \frac{d^3p}{(2\pi)^3} E(\vec{p}) \left[\frac{1}{e^{\beta(E-\mu)} + 1} + \frac{1}{e^{\beta(E+\mu)} + 1} \right] . \tag{18.115}$$

A similar calculation of the mean charge density yields

$$\frac{1}{V} < Q >= 2 \int_{-\infty}^{\infty} \frac{d^3p}{(2\pi)^3} \left[\frac{1}{e^{\beta(E-\mu)} + 1} - \frac{1}{e^{\beta(E+\mu)} + 1} \right] . \tag{18.116}$$

We close this section with a remark. If we had chosen to work with naive fermions (i.e., no Wilson term) then the integrals (18.115) and (18.116) would have

---

* High momentum excitations, corresponding to finite $\vec{\hat{p}}$ do not contribute to (18.113a) in the continuum limit, as can be easily verfied.

been multiplied my an additional factor 2^4, arising from the doubler contributions. Thus by only looking at the partition function, the "doublers" manifest themselves in a rather trivial way. Our discussion in section 9 however suggests that this is only so, because the Dirac field excites particle *and* antiparticle states. In fact by studying separately the positive and negative energy contributions to the partition function, one finds that in both, the positive as well as negative energy sectors, the partition function resembles that of a gas of particles *and* antiparticles, except for an overall factor 2^3 arising from the "doublers" associated with high-three-momentum excitations at the corners of the Brillouine zone [Bender (1993)]. This agrees with our findings in section 9, where we discussed the doubling problem in a simple non-relativistic model.

CHAPTER 19

FINITE TEMPERATURE PERTURBATION THEORY OFF AND ON THE LATTICE

In the previous chapter we have studied in detail the path-integral representation of the partition function for some non-interacting bosonic, and fermionic systems. Because of the Gaussian nature of the integrations we were able to calculate the path integrals explicitely. In the presence of interactions this is in general no longer possible and one has to recur either to perturbation theory, or evaluate thermodynamical observables within the framework of a lattice regularized theory by Monte Carlo methods. In this chapter we shall show how the euclidean Feynman rules at zero temperature are modified when a relativistic system of interacting fields is placed in contact with a heat bath. We will first demonstrate this for the case of the $\lambda\phi^4$ theory in the continuum formulation. As we shall see, the prescription for making the transition from the zero-temperature, zero-chemical potential Feynman rules to the $T \neq 0$, $\mu \neq 0$ rules turns out to be very simple. The $\lambda\phi^4$-theory is considered in detail in the book by Kapusta (1989). We shall therefore only discuss some elementary aspects of this theory, since it provides a simple laboratory for studying the effects arising from the presence of a heat bath. The remaining part of this chapter will then be devoted to gauge field theories at finite temperature and chemical potential, both in the continuum as well as on the lattice. Although the prescription for making the transition from the zero-temperature, zero-chemical potential lattice Feynman rules to the $T \neq 0$, $\mu \neq 0$ rules turns out again to be very simple, the actual computation of lattice Feynman integrals at finite temperature and chemical potential is more involved than in the continuum, as we shall demonstrate in some sample calculations.

19.1 Feynman Rules for Thermal Green Functions in the $\lambda\phi^4$-Theory

Let $\phi(x)$ be a real scalar field whose dynamics is governed by a Hamiltonian H. At zero temperature all physical information about the system is contained in the ground state expectation value of time ordered products of the field operators $\Phi(x)$. When the system is placed in contact with a heat bath, all possible energy eigenstates of H are excited with a probabilty given by the Boltzmann factor. The thermal correlation functions are then defined by

$$< \phi(x_1)...\phi(x_n) >_\beta = \frac{Tr[e^{-\beta H} T(\Phi(x_1) \cdots \Phi(x_n))]}{Tr\ e^{-\beta H}}, \tag{19.1}$$

where $x_i = (\vec{x}_i, \tau_i)$, and $\Phi(x)$ are the field operators whose euclidean time dependence is given by (2.17). By introducing a complete set of eigenstates of the Hamiltonian, one readily verifies that in the limit of vanishing temperature (i.e., $\beta = \frac{1}{T} \to \infty$) the rhs of (19.1) reduces to the correlation function (2.14). It is now easy to derive a path-integral expression for (19.1) using the by now familiar techniques. We leave the details to the reader and only quote here the result, which is valid if the Hamiltonian density is given by the sum of a kinetic term, quadratic in the canonical momentum, and an interaction depending only on the fields $\phi(x)$:

$$< \phi(x_1)...\phi(x_n) >_\beta = \frac{\int_{per} D\phi \, \phi(x_1)...\phi(x_n) e^{-S^{(\beta)}[\phi]}}{\int_{per} D\phi e^{-S^{(\beta)}[\phi]}} . \tag{19.2a}$$

Here

$$S^{(\beta)}[\phi] = \int_0^\beta d\tau \int d^3x \mathcal{L}_E(\phi, \partial_\mu \phi) \tag{19.2b}$$

is the finite temperature action. The path integral is to be evaluated with the scalar field satisfying the periodic boundary condition (18.32). The correlation functions can be obtained from the generating functional

$$Z[J] = \int_{per} D\phi e^{-S^{(\beta)}[\phi] + \int_0^\beta d\tau \int d^3x J(\vec{x},\tau)\phi(\vec{x},\tau)} \tag{19.3}$$

in the usual way, by functionally differentiating this expression with respect to the sources $J(\vec{x}, \tau)$. If the interacting part of $S^{(\beta)}$, which we denote by $S_I^{(\beta)}[\phi]$, is a polynomial in the fields, we can also write (19.3) in the form

$$Z[J] = e^{-S_I^{(\beta)}[\frac{\delta}{\delta J}]} Z_0[J], \tag{19.4a}$$

where

$$Z_0[J] = \int D\phi e^{-S_0^{(\beta)}[\phi] + \int_0^\beta d\tau \int d^3x J(\vec{x},\tau)\phi(\vec{x},\tau)} \tag{19.4b}$$

is a path integral of the Gaussian type. Once $Z_0[J]$ is known, $Z[J]$ can be computed perturbatively by expanding the exponential in powers of the interaction.

For the ϕ^4-theory the finite temperature action is given by

$$S^{(\beta)}[\phi] = S_0^{(\beta)}[\phi] + S_I^{(\beta)}[\phi] , \tag{19.5a}$$

where

$$S_0^{(\beta)}[\phi] = \frac{1}{2} \int_\beta d^4x \phi(x)(-\Box + M^2)\phi(x) , \tag{19.5b}$$

$$S_I^{(\beta)}[\phi] = \frac{\lambda}{4!}\int_\beta d^4x(\phi(x))^4 \,, \tag{19.5c}$$

and $x = (\vec{x}, \tau)$. Here we have introduced the short-hand notation

$$\int_\beta d^4x = \int_0^\beta d\tau \int d^3x \,. \tag{19.5d}$$

The factor $1/4!$ in (19.5c) has been introduced for later convenience. Since $\phi(x)$ satisfies periodic boundary conditions, it has the following Fourier decomposition

$$\phi(\vec{x}, \tau) = \frac{1}{\beta}\sum_\ell \int \frac{d^3k}{(2\pi)^3}\tilde{\phi}(\omega_\ell, \vec{k})e^{i\vec{k}\cdot\vec{x}+i\omega_\ell\tau}, \tag{19.6a}$$

where ω_ℓ are the Matsubara frequencies

$$\omega_\ell = \frac{2\pi}{\beta}\ell \,. \qquad (bosons) \tag{19.6b}$$

Here ℓ takes all possible integers values. The factor $1/\beta$ has been inserted so that $\tilde{\phi}$ carries the same dimension as the corresponding field in the $T = 0$ formulation. Since

$$\int_0^\beta d\tau e^{i(\omega_\ell - \omega_{\ell'})\tau} = \beta\delta_{\ell\ell'} \,, \tag{19.7}$$

we can invert (19.6a):

$$\tilde{\phi}(\omega_\ell, \vec{k}) = \int_0^\beta d\tau \int d^3x\phi(\vec{x}, \tau)e^{-i\omega_\ell\tau - i\vec{k}\cdot\vec{x}}. \tag{19.8}$$

From (19.6a) and (19.8), we see that the finite temperature expressions follow from those at $T = 0$ by making the substitutions

$$k_4 \to \omega_\ell$$

$$\int \frac{dk_4}{2\pi} f(k_4) \to \frac{1}{\beta}\sum_\ell f(\omega_\ell) \,, \tag{19.9a}$$

and

$$\int d^4x \to \int_\beta d^4x \,. \tag{19.9b}$$

Consider now the generating functional of the free theory defined in (19.4b). Since the exponent in the integrand is quadratic in the fields, we can perform the integration and obtain

$$Z_0[J] = Z_0[0]e^{\frac{1}{2}\int_\beta d^4x \int_\beta d^4y \; J(x)\Delta^{(\beta)}(x-y)J(y)} \,, \tag{19.10}$$

where $\Delta^{(\beta)}(x-y)$ is the inverse of the operator $-\Box+M^2$ (i.e., Green function) on the space of functions satisfying the periodic boundary conditions (18.32). Hence $\Delta^{(\beta)}(z)$ is periodic in the euclidean time direction,

$$\Delta^{(\beta)}(\vec{z},0)=\Delta^{(\beta)}(\vec{z},\beta)$$

and therefore has the following Fourier expansion

$$\Delta^{(\beta)}(z)=\frac{1}{\beta}\sum_{\ell}\int\frac{d^3k}{(2\pi)^3}\tilde{\Delta}^{(\beta)}(\omega_\ell,\vec{k})e^{i\omega_\ell\tau+i\vec{k}\cdot\vec{z}}. \qquad (19.11)$$

The propagator in momentum space, $\tilde{\Delta}^{(\beta)}(\omega_\ell,\vec{k})$, can be determined by inserting this expression into the equation

$$(-\Box+M^2)\Delta^{(\beta)}(z)=\delta(\vec{z})\delta_P(\tau)\ ,$$

where $\delta_P(\tau)$ is the periodic δ-function

$$\delta_P(\tau)=\frac{1}{\beta}\sum_{\ell}e^{i\omega_\ell\tau}\ . \qquad (19.12)$$

One finds that

$$\tilde{\Delta}^{(\beta)}(\omega_\ell,\vec{k})=\frac{1}{\omega_\ell^2+\vec{k}^2+M^2}\ . \qquad (19.13)$$

The corresponding expression for $\Delta^{(\beta)}(z)$ reads

$$\Delta^{(\beta)}(z)=\frac{1}{\beta}\sum_{\ell}\int\frac{d^3k}{(2\pi)^3}\frac{e^{i\omega_\ell\tau+i\vec{k}\cdot\vec{z}}}{\omega_\ell^2+\vec{k}^2+M^2}\ . \qquad (19.14)$$

Comparing (19.13) and (19.14) with their zero temperature counterparts

$$\tilde{\Delta}(k)=\frac{1}{k^2+M^2}\ , \qquad (19.15a)$$

$$\Delta(z)=\int\frac{d^4k}{(2\pi)^4}\frac{e^{ik\cdot z}}{k^2+M^2}\ , \qquad (19.15b)$$

where $k=(\vec{k},k_4)$, and $k^2=k_4^2+\vec{k}^2$, we see that they are related by the rules (19.9a).

The temperature dependence of (19.14) is contained in the infinite sum over the temperature dependent Matsubara frequencies (19.6b). The frequency summation formula we shall derive below allows us to decompose $\Delta^{(\beta)}(z)$ into a zero and a finite temperature contribution arising from the presence of the heat bath. Suppose we want to calculate the sum

$$F(K) = \frac{1}{\beta}\sum_{\ell} f(\omega_\ell, K), \tag{19.16}$$

where K stands collectively for all remaining variables on which the function f may depend. Let us promote ω_ℓ to a continuous variable ω and define the function $f(\omega, K)$. Suppose that $f(\omega, K)$ has no singularities on the real axis. Consider the following function of the complex variable ω,

$$h(\omega) = \frac{i\beta}{e^{i\beta\omega} - 1}\,. \tag{19.17}$$

It has poles located at $\omega = \frac{2\pi}{\beta}\ell$ with unit residue. Then we can write the sum (19.16) as follows,

$$F(K) = \frac{1}{2\pi i\beta}\int_C d\omega\; h(\omega) f(\omega, K), \tag{19.18}$$

where C is the contour shown in fig. (19-1).

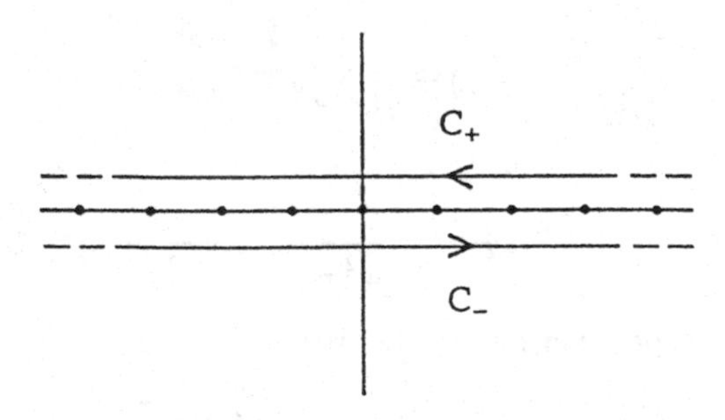

Fig. 19-1 Upper and lower branches of the contour C in eq. (19.18).

In the lower half plane (19.17) decreases exponentially for large imaginary parts of ω. In the upper half plane, on the other hand, $h(\omega)$ is finite for $Im(\omega) \to \infty$. Since we eventually want to close the contours at infinity, we shall make use of the following alternative form for (19.17) on the upper C_+-branch of C:

$$h(\omega) = -\frac{i\beta}{e^{-i\beta\omega} - 1} - i\beta.$$

Then (19.18) can also be written as follows:

$$\begin{aligned}\frac{1}{\beta}\sum_{\ell} f(\omega_\ell, K) =& \frac{1}{2\pi}\int_{-\infty}^{\infty} d\omega f(\omega, K) + \frac{1}{2\pi}\int_{-\infty-i\epsilon}^{+\infty-i\epsilon} d\omega \frac{f(\omega, K)}{e^{i\beta\omega} - 1} \\ &+ \frac{1}{2\pi}\int_{-\infty+i\epsilon}^{+\infty+i\epsilon} d\omega \frac{f(\omega, K)}{e^{-i\beta\omega} - 1}.\end{aligned} \tag{19.19}$$

If $f(\omega, K)$ is sufficiently well behaved for $|\omega| \to \infty$, so that we can close the integration contours in the second and third integrals in the lower and upper half planes at infinity, and if $f(\omega, K)$ is a meromorphic function of ω, then

$$\begin{aligned}\frac{1}{\beta}\sum_{\ell} f(\omega_\ell; K) = \frac{1}{2\pi}\int_{-\infty}^{\infty} d\omega f(\omega, K) + i \sum_{Im\bar{\omega}_i > 0} \frac{R_f(\bar{\omega}_i)}{e^{-i\beta\bar{\omega}_i} - 1} \\ - i \sum_{Im\bar{\omega}_i < 0} \frac{R_f(\bar{\omega}_i)}{e^{i\beta\bar{\omega}_i} - 1}\end{aligned} , \tag{19.20}$$

where $R_f(\bar{\omega}_i)$ are the residues of $f(\bar{\omega}_i, K)$ at the poles whose positions we have denoted by $\bar{\omega}_i$.

As an application of (19.19) we derive the expression for the propagator in *Minkowski* space, as the analytic continuation of the euclidean propagator (19.14) to real times. An alternative derivation, has been given by Dolan and Jackiw (1974).

Consider the finite temperature expression (19.14) for the euclidean propagator. The function $f(\omega, K)$ is now given by $\exp(i\omega\tau)/(\omega^2 + \vec{k}^2 + M^2)$. Since $|\tau|$ is restricted to the interval $[0, \beta]$, we can close the contours in the second and third integrals in (19.19) in the lower and upper half of the complex ω-plane, respectively, and obtain

$$\begin{aligned}\Delta^{(\beta)}(z) = &\int \frac{d^4k}{(2\pi)^4} \frac{e^{ik_4\tau + i\vec{k}\cdot\vec{z}}}{k^2 + M^2} \\ &+ \int \frac{d^3k}{(2\pi)^3} \frac{1}{E} \eta(E) \cosh(E\tau) e^{i\vec{k}\cdot\vec{z}},\end{aligned} \tag{19.21a}$$

where

$$E = \sqrt{\vec{k}^2 + M^2} \, ,$$

and

$$\eta(E) = \frac{1}{e^{\beta E} - 1} \tag{19.21b}$$

is the Bose-Einstein distribution function. In (19.21a) we have set $\omega = k_4$. Note that the first integral is just the propagator for vanishing temperature.

We next continue this expression to real time by setting $\tau = it$. Performing the usual Wick rotation ($k_4 \to -ik^0$) in the first integral one has that*

$$\begin{aligned}\Delta^{(\beta)}(z)\Big|_{\tau = it} = &i \int \frac{d^4k}{(2\pi)^4} \frac{e^{-ik\cdot z}}{k^2 - M^2 + i\epsilon} + \\ &+ 2\pi \int \frac{d^4k}{(2\pi)^4} \frac{1}{2E} \eta(E) (\delta(k^0 - E) + \delta(k^0 + E)) e^{-ik\cdot z},\end{aligned}$$

* The $i\epsilon$–prescription is that of the $T = 0$ field theory.

where $z = (\vec{z}, z^0)$, $k \cdot z = k^0 z^0 - \vec{k} \cdot \vec{z}$. Making use of the relation

$$\delta(k^2 - M^2) = \frac{1}{2E}[\delta(k^0 + E) + \delta(k^0 - E)] \tag{19.22}$$

we finally arrive at the following expression for the propagator in Minkowski space,

$$\Delta^{(\beta)}_{(Mink)}(z) = \int \frac{d^4k}{(2\pi)^4} \tilde{\Delta}^{(\beta)}_{(Mink)}(k) e^{-ik\cdot z}, \tag{19.23a}$$

where

$$\tilde{\Delta}^{(\beta)}_{(Mink)}(k) = \frac{i}{k^2 - M^2 + i\epsilon} + \frac{2\pi}{e^{\beta E} - 1}\delta(k^2 - \mu^2) . \tag{19.23b}$$

The second term can be interpreted as the contribution arising from on-mass shell particles in the heat bath.

The reader must be warned that (19.23) is not the propagator appearing in the Feynman rules for real-time Green functions in an interacting theory. The real-time Feynman rules are more complicated and require a doubling of degrees of freedom. *.

Having discussed in detail the generating functional of the free theory, we now obtain the Feynman rules for the thermal Green functions (19.1) in the standard way from the generating functional (19.4). This generating functional differs from that at zero temperature in that the zero temperature propagator (19.15b) is replaced by (19.14), and that the euclidean time integration is restricted to the interval $[0, \beta]$. It is therefore evident that the $T \neq 0$ Feynman rules in coordinate space are obtained from those at zero temperature by merely making the replacements

$$\Delta(z) \to \Delta^{(\beta)}(z) ,$$

$$\int_{-\infty}^{\infty} d\tau \to \int_0^\beta d\tau .$$

The corresponding rules in frequency-momentum space can also easily be obtained by making use of the Fourier decomposition (19.14) and of the orthogonality relation (19.7), which is the analog of $\int_{-\infty}^{\infty} d\tau \exp(ip_4\tau) = 2\pi\delta(p_4)$ at zero temperature. Thus the integration over the space-time coordinates of a vertex leads to the appearance of a factor β and a Kronecker-δ which enforces that the sum over Matsubara frequencies flowing into the vertex equals the sum over frequencies flowing

* See e.g. Niemi and Semenov (1984); Landsman and Weert (1987)

out of the vertex. A thermal correlation function then has the form

$$
\begin{aligned}
G(z_1, z_2, ...) = &\sum_{\ell_1} \frac{1}{\beta} \int \frac{d^3k_1}{(2\pi)^3} \sum_{\ell_2} \frac{1}{\beta} \int \frac{d^3k_2}{(2\pi)^3} \cdots \\
&\times \tilde{G}(\omega_{\ell_1}, \vec{k}_1; \omega_{\ell_2}, \vec{k}_2; \cdots) \prod_i e^{i\omega_{\ell_i}\tau_i + i\vec{k}_i \cdot \vec{z}_i} ,
\end{aligned}
\tag{19.24}
$$

where $z_i = (\vec{z}_i, \tau_i)$, and where ω_{ℓ_i} and k_i are the incoming frequencies and momenta associated with the external lines of a Feynman diagram. The Kernel $\tilde{G}(\omega_{\ell_1}, \vec{k}_1; \omega_{\ell_2}, \vec{k}_2; \cdots)$ has the following structure

$$
\tilde{G}(\omega_{\ell_1}, \vec{k}_1; \omega_{\ell_2}, \vec{k}_2; \cdots) = \beta(2\pi)^3 \delta_{\sum_i \omega_{\ell_i}, 0} \, \delta^{(3)}(\sum_i \vec{k}_i) \mathcal{G}(\omega_{\ell_1}, \vec{k}_1; \omega_{\ell_2}, \vec{k}_2; \cdots) \tag{19.25}
$$

and is computed with the following Feynman rules:

i) To each line of the diagram, which we label by an index i ($i = 1, 2, ...$), associate a propagator

$$
\tilde{\Delta}^{(\beta)}(\omega_{\ell_i}, \vec{k}_i) = \frac{1}{\omega_{\ell_i}^2 + \vec{k}_i^2 + M^2} , \tag{19.26}
$$

where $\vec{k}_i$ is the momentum carried by the i'th line, and ω_{ℓ_i} is the corresponding discrete energy, $2\pi\ell_i/\beta$.

ii) To each vertex assign the factor

$$
-\frac{\lambda}{4!}(2\pi)^3 \delta(\vec{K}) \beta \delta_{\omega,0} , \tag{19.27}
$$

where ω and $\vec{K}$ is the sum of the Matsubara frequencies and momenta flowing into the vertex.

iii) Sum over all discrete energies of the internal lines, and integrate over the momenta carried by these lines according to

$$
\sum_{\ell} \frac{1}{\beta} \int \frac{d^3k}{(2\pi)^3} . \tag{19.28}
$$

iv) Multiply the resulting expression with the factor, resulting from the number of distinct ways one can build the diagram from the given number n of vertices, and a factor $\frac{1}{n!}$ arising from the n'th order term in the expansion of $\exp(-S_I)$.

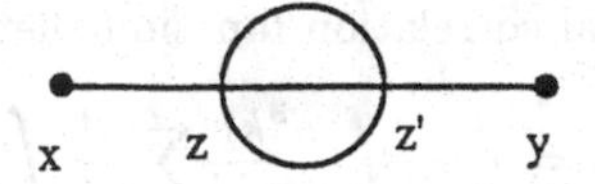

Fig. 19-2 Two loop contribution to the propagator.

As an example consider the second order contribution to the propagator shown in fig. (19-2).

$$D^{(\beta)}(x,y) = \frac{\lambda^2}{6}\int_\beta d^4z \int_\beta d^4z' \Delta^{(\beta)}(x-z)\left[\Delta^{(\beta)}(z-z')\right]^3 \Delta^{(\beta)}(z'-y). \quad (19.29)$$

The factor 6 arises as follows. There are eight possibilities to identify an external line emanating from the two ϕ^4-vertices with the space-time coordinates x. This leaves us with 4 possibilities for choosing y. The remaining lines can then be tied together in 3! ways. From the expansion of $\exp(-S_I^{(\beta)})$ to second order we get a factor 1/2!. This leaves us with a factor $(4!)^2/6$. Finally we must multiply the expression with $(\lambda/4!)^2$.

Applying the above Feynman rules, the corresponding momentum-space representation of (19.29) is given by

$$\tilde{D}^{(\beta)}(\omega_{\ell_1},\vec{k}_1;\omega_{\ell_2},\vec{k}_2) = \beta(2\pi)^3\delta_{\omega_{\ell_1}+\omega_{\ell_2},0}\delta^{(3)}(\vec{k}_1+\vec{k}_2)\bar{D}^{(\beta)}(\omega_{\ell_1},\vec{k}_1) \quad (19.30a)$$

where

$$\bar{D}^{(\beta)}(\omega_\ell,\vec{k}) = \frac{1}{\omega_{\ell_1}^2+\vec{k}_1^2+M^2}[\Pi^{(\beta)}(\omega_{\ell_1},\vec{k}_1)]\frac{1}{\omega_{\ell_1}^2+\vec{k}_1^2+M^2}, \quad (19.30b)$$

and where the self energy is given by

$$\Pi^{(\beta)}(\omega_{\ell_1},\vec{k}_1) = \frac{\lambda^2}{6}\frac{1}{\beta^2}\sum_{\ell',\ell''}\int\frac{d^3k'}{(2\pi)^3}\frac{d^3k''}{(2\pi)^3}\frac{1}{[\omega_{\ell'}^2+\vec{k'}^2+M^2][\omega_{\ell''}^2+\vec{k''}^2+M^2]}$$
$$\times\frac{1}{[(\omega_\ell-\omega_{\ell'}-\omega_{\ell''})^2+(\vec{k}-\vec{k'}-\vec{k''})^2+M^2]} \quad (19.30c)$$

Notice that this latter expression could have been immediately obtained from the corresponding zero temperature expression according to the rules (19.9a). The corresponding expression in coordinate space is then given according to (19.24).

19.2 Generation of a Dynamical Mass at $T \neq 0$

Having obtained the finite temperature Feynman rules, we are now want to study some features of the $\lambda\phi^4$-theory which arise from the presence of the heat bath.* The first phenomenon we will consider is the generation of a dynamical mass at $T \neq 0$. As we shall see below, a temperature-dependent mass is generated already on the one-loop level. To this order, only mass renormalization needs to be taken into account. The relevant euclidean action, including the mass counterterm, is therefore given by

$$S[\phi] = \frac{1}{2} \int_\beta d^4x \; \phi(x)(-\Box + M^2)\phi(x) + \lambda \int_\beta d^4x \; [\phi(x)]^4 + \frac{1}{2}\delta M^2 \int_\beta d^4x \; [\phi(x)]^2 \, . \tag{19.31}$$

We have not included the factor $\frac{1}{4!}$ in the interaction term, since the formulae we shall obtain then take a simpler form. The term proportional to δM^2 is the contribution of the mass counterterm. In lowest order perturbation theory the renormalized inverse propagator is given in momentum space by

$$[\Delta(\omega_\ell, \vec{k})]^{-1} = \omega_\ell^2 + \vec{k}^2 + M^2 + \pi_R^{(1)}(T) \, , \tag{19.32}$$

where $\pi_R^{(1)}(T)$ is the renormalized self-energy in the one loop approximation:

$$\pi_R^{(1)}(T) = -12 \; \underline{\bigcirc} \; + \delta M^2 . \tag{19.33}$$

We will determine δM^2 below in such a way that M is the mass of the ϕ field at zero temperature. The factor 12 multiplying the one–loop contribution arises as follows: there are 6 ways of contracting two lines emanating from a ϕ^4–vertex to form a loop, and 2 possibilities for connecting the remaining lines to two points x and y in a propagator. Hence the diagram is weighted with a factor 12. With the Feynman rules given in the previous section, we otain for (19.33)

$$\pi_R^{(1)}(T) = 12\lambda \sum_\ell \frac{1}{\beta} \int \frac{d^3k}{(2\pi)^3} \frac{1}{\omega_\ell^2 + \vec{k}^2 + M^2} + \delta M^2 . \tag{19.34}$$

The integral is infrared (IR) finite for $M \neq 0$, but ultraviolet (UV) divergent. This divergence is entirely contained in the zero-temperature contribution.** Indeed,

---

* Much of our discussion in this and the following section closely parallels that of Kapusta (1989). We have included it here, since the problems we address to are also of relevance in QCD.

** Since the presence of the heat bath should not affect the short distance behaviour of the correlation functions, we expect that thermal Green functions will be ultraviolet finite once the theory has been renormalized at zero temperature. See Norton and Cornwall (1975); Kislinger and Morley (1976, 1979).

carrying out the frequency sum, making use of the summation formula (19.20), we can write (19.34) in the form

$$\pi_R^{(1)}(T) = \pi_U^{(1)}(0) + 12\lambda \int \frac{d^3k}{(2\pi)^3} \frac{1}{E} \frac{1}{e^{\beta E} - 1} + \delta M^2,$$

where

$$\pi_U^{(1)}(0) = 12\lambda \int \frac{d^4k}{(2\pi)^4} \frac{1}{k^2 + M^2}$$

is the unrenormalized self energy in one-loop order for vanishing temperature, and $E = \sqrt{\vec{k}^2 + M^2}$. The second term appearing on the rhs is UV finite because of the appearance of the Bose-Einstein distribution function, and vanishes for $\beta \to \infty$, i.e., for $T = 0$. We therefore see that the UV-divergence is entirely contained in the $T = 0$ contribution, which is renormalized in the standard way. We first regulate the divergent integral by introducing a momentum cutoff Λ. By choosing

$$\delta M^2 = -12\lambda \int^{\Lambda} \frac{d^4k}{(2\pi)^4} \frac{1}{k^2 + M^2}, \tag{19.35}$$

we eliminate the above UV-divergence, and ensure that M is the physical mass of the ϕ-field in one loop order at zero temperature. With this renormalization prescription, we are thus led to the following expression for $\pi_R^{(1)}(T)$:

$$\pi_R^{(1)}(T) = 12\lambda \int \frac{d^3k}{(2\pi)^3} \frac{1}{E} \frac{1}{e^{\beta E} - 1} \; . \tag{19.36}$$

In the limit $M \to 0$, this expression reduces to

$$\pi_R^{(1)}(T) = \lambda T^2, \qquad (M = 0) \; . \tag{19.37}$$

Hence at $T \neq 0$ a dynamical mass is generated in one loop order.

19.3 Perturbative Expansion of the Thermodynamical Potential

Using the finite temperature Feynman rules derived in section 1 we can now calculate perturbative corrections to the thermodynamical potential

$$\Omega = -\frac{1}{\beta V} ln\ Z \tag{19.38}$$

of a free gas of neutral spinless bosons. The partition function is given by

$$Z = \mathcal{N} \int_{periodic} D\phi e^{-S_0[\phi] - S_I[\phi]} \tag{19.39}$$

where $S_0[\phi]$ is the finite temperature action of the free field, and $S_I[\phi]$ is the interaction term. Here we have dropped for simplicity the superscript β in the action $S^{(\beta)}[\phi]$. The normalization factor $\mathcal{N}$ carries the dimension of the inverse integration measure.

The perturbative espansion of Z is obtained by expanding the exponential $\exp(-S_I)$ in (19.39) in powers of S_I.* One easily verifies that

$$Z = Z_0 \left[1 + \sum_{l=1}^{\infty} \frac{(-1)^l}{l!} \langle (S_I)^l \rangle_{S_0} \right] , \tag{19.40a}$$

where

$$Z_0 = \mathcal{N} \int D\phi e^{-S_0[\phi]} , \tag{19.40b}$$

and

$$\langle S_I^l \rangle_{S_0} = \frac{\int D\phi (S_I[\phi])^l e^{-S_0[\phi]}}{\int D\phi e^{-S_0[\phi]}} \tag{19.40c}$$

is the expectation value of $S_I^l[\phi]$ calculated with the Boltzmann distribution of the free theory. From (19.40a) we have that ln Z is given by

$$\ln Z = \ln Z_0 + \sum_{n=1}^{\infty} \frac{(-1)^{n+1}}{n} \left[\sum_{l=1}^{\infty} \frac{(-1)^l}{l!} \langle S_I^l \rangle_{S_0} \right]^n . \tag{19.41}$$

The expectation value $\langle S_I^l \rangle_{S_0}$, which can be computed from the free generating functional (19.4b), receives contributions from connected and disconnected Feynman diagrams with no external lines. But only the former ones actually contribute to the sum. This is expected since the free energy is an extensive quantity. In fact, as we now show, (19.41) can also be written in the form

$$\ln Z = \ln Z_0 + \sum_{l=1}^{\infty} \frac{(-1)^l}{l!} \langle S_I^l \rangle_{S_0}^c , \tag{19.42}$$

where the superscript "c" stands for "connected". The proof of (19.42) goes as follows.

Consider the ensemble average of a power of S_I, i.e., $\langle S_I^l \rangle_{S_0}$. It is given by the sum over all possible pairwise contractions (i.e., propagators) of all the fields appearing in S_I^l. Hence $\langle S_I^l \rangle$ can be decomposed into a sum of products

* Our following presentation closely parallels that of Kapusta (1989). We nevertheless include it here for completeness sake.

of connected components, each of which consists of one or more vertices. The contribution of a connected component, consisting of n vertices, we denote by $\langle S_I^n \rangle_{S_0}^c$. Let a_n be the number of times that this connected component appears in a particular product. Since there are a total of l vertices, the products of connected components have the form

$$\langle S_I \rangle_c^{a_1} \langle S_I^2 \rangle_c^{a_2} ... \langle S_I^k \rangle_c^{a_k} , \tag{19.43a}$$

where

$$a_1 + 2a_2 + \cdots + ka_k = l , \tag{19.43b}$$

and where for simplicity of notation we have now written $\langle S_I^n \rangle_c$ instead of $\langle S_I^n \rangle_{S_0}^c$. There are however, in general, several terms appearing in the decomposition of $< S_I^l >_{S_0}$ yielding identical contributions (19.43a), since it does not matter which collection of n vertices are used to make up a particular product of connected components (19.43a). Hence $\langle S_I^l \rangle$ will be of the form

$$\langle S_I^l \rangle = \sum_k \sum_{a_1, a_2, \cdots, a_k} C_{a_1 a_2 \cdots a_k} \langle S_I \rangle_c^{a_1} \langle S_I^2 \rangle_c^{a_2} \cdots \langle S_I^k \rangle_c^{a_k} \delta_{a_1 + 2a_2 + \cdots + ka_k - l, 0} , \tag{19.44}$$

where the δ -function ensures that we pick up the contribution of order l. We now compute the combinatorial factor $C_{a_1 a_2 \cdots a_k}$. To this effect we first numerate the l vertices making up S_I^l from 1 to l. Consider a particular partition of the l vertices into connected components. By permuting the l vertices we generate $l!$ sets of connected components, all of which yield the same contribution to $\langle S_I^l \rangle$. These sets are however not all distinct, for the $l!$ permutations also include those which merely permute the a_n connected components of the type $< S_I^n >_{S_0}^c$. These must not be counted as distinct. We therefore must divide $l!$ by $a_1! a_2! ... a_k!$. But there is another factor which must be divided out. Recall, that the definition, $\langle S_I^n \rangle_{S_0}^c$ includes all possible pairwise contractions of the fields. Hence permutations of the vertices within a connected component of order n are already included in the definition of $\langle S_I^n \rangle_{S_0}^c$. We therefore must also divide by $(1!)^{a_1} (2!)^{a_2} ... (k!)^{a_k}$. Hence we conclude that the combinatorial factor in (19.44) is given by

$$C_{a_1 \cdots a_k} = \frac{l!}{\prod_{n=1}^k a_n! (n!)^{a_n}} .$$

Inserting this expression into (19.44), and carrying out the sum over l by making use of the δ-function, we conclude that

$$\sum_{l=0}^{\infty} \frac{(-1)^l}{l!} \langle S_I^l \rangle_{S_0} = \prod_{n=1}^{\infty} \left(\sum_{a_n=0}^{\infty} \frac{1}{a_n!} \left[\frac{(-1)^n}{n!} \langle S_I^n \rangle_{S_0}^c \right]^{a_n} \right)$$

$$= e^{\sum_{n=1}^{\infty} \frac{(-1)^n}{n!} \langle S_I^n \rangle_{S_0}^c} ,$$

where we have now again introduced the subscript S_0. Hence, according to (19.40a), $\ln Z$ is given by (19.42), which is the result we wanted to prove. The thermodynamical potential (19.38) is therefore given by

$$\Omega = \Omega_0 - \frac{1}{\beta V} \sum_{n=1}^{\infty} \frac{(-1)^n}{n!} \langle S_I^n \rangle_{S_0}^c . \tag{19.45}$$

Let us now calculate the lowest perturbative correction of $\mathcal{O}(\lambda)$ to the ideal gas formula (18.41), i.e.,

$$\Omega_{(1)} = \frac{1}{\beta V} \langle S_I \rangle_{S_0} ,$$

for the case of the $\lambda\phi^4$-theory, where $S_I[\phi]$ (including the mass counter term) is given by the last two terms in (19.31). The expectation value $\langle S_I \rangle_{S_0}$ is then given by

$$\langle S_I \rangle_{S_0} = \int_\beta d^4x \; \{3\lambda[\Delta^{(\beta)}(0)]^2 + \frac{1}{2}\delta M^2 \Delta^{(\beta)}(0)\} . \tag{19.46}$$

Here $\Delta^{(\beta)}(0)$ is the finite temperature propagator (19.14) evaluated for vanishing argument. The factor 3 takes into account the three distinct ways in which the four fields emanating from the vertex can be contracted to form a double loop. Because the integrand in (19.46) does not depend on x, $\langle S_I \rangle_{S_0}$ is proportional to βV:

$$\langle S_I \rangle_{S_0} = \beta V [3\lambda(\Delta^{(\beta)}(0))^2 + \frac{1}{2}\delta M^2 \Delta^{(\beta)}(0)] . \tag{19.47}$$

Let us next isolate the zero-temperature contribution to $\Delta^{(\beta)}(0)$. According to (19.21),

$$\Delta^{(\beta)}(0) = \Delta(0) + \int \frac{d^3k}{(2\pi)^3} \frac{1}{E} \frac{1}{e^{\beta E} - 1} , \tag{19.48}$$

where $\Delta(0)$ is given by (19.15b) with $z = 0$. Introducing the decomposition (19.48) into (19.47), and recalling that we had already determined δM^2 to be given by (19.35), one finds that, apart from an irrelevant additive constant, the $O(\lambda)$-correction to the thermodynamical potential is given by

$$\begin{aligned} \Omega_{(1)} &= \frac{1}{\beta V}\langle S_I\rangle = 3\lambda\left[\int \frac{d^3k}{(2\pi)^3}\frac{1}{E}\frac{1}{e^{\beta E}-1}\right]^2 \\ &\xrightarrow[M\to 0]{} \frac{\lambda}{48}T^4 \end{aligned} \tag{19.49}$$

Combining this result for $M = 0$ with (18.42) we are therefore left with the following expression valid up to $O(\lambda)$:

$$\Omega(\beta) = -\left(\frac{\pi^2}{90} - \frac{1}{48}\lambda\right)T^4 \; ; \quad (M = 0).$$

From $\ln Z = -\beta V\Omega$ we can calculate the mean energy $< E >$ and pressure according to

$$\begin{aligned} < E > &= -\frac{\partial}{\partial\beta}\ln Z \\ p &= \frac{1}{\beta}\frac{\partial}{\partial V}\ln Z \; . \end{aligned} \tag{19.50}$$

One then verifies that the mean energy density and pressure are lowered in $O(\lambda)$.

In higher orders of perturbation theory one is in general faced, in the zero mass limit, with infrared divergent integrals. By a partial resummation of higher order Feynman diagrams with self energy insertions, the propagators in a diagram of given order are replaced by massive propagators with a temperature dependent mass, which in lowest order is given by (19.37). This eliminates the infrared divergencies, but leads to a non-analytic behaviour in the coupling constant. The reader may consult the book by Kapusta (1989), where the self energy and thermodynamical potential is discussed in the so called "ring"-approximation. Here we shall illustrate the problems one encounters by considering the contribution to the thermodynamical potential arising from a class of Feynman graphs which are not of the ring type.

Consider the class of diagrams shown in fig. (19-3).

Suppose the diagram has N vertices. The number of propagators is 2N. Because of energy (frequency)-momentum conservation there are only $N + 1$ independent loop-momenta, and $N + 1$ frequency sums. In the limit $M \to 0$ the dominant infrared divergence arises from that term in the sum over Matsubara frequencies where all frequencies vanish. The corresponding contribution has the form

$$\Omega_N^{(\omega=0)} \sim \lambda^N T^{N+1}\int d^3k_1 \ldots d^3k_{N+1}\prod_{l=1}^{2N}\frac{1}{\vec{q}_l^{\,2}} \; , \tag{19.51}$$

Fig. 19-3 Generic diagram contributing to the thermodynamical potential, which is not of the "ring" type.

where "i" labels the the frequencies and 3-momenta of the internal lines, and where the q_l's are homogeneous linear combinations of the $N+1$ integration variables. The lowest order diagram corresponds to $N = 3$, and, by naive power counting, is infrared convergent for $M \to 0$. For $N > 3$ we are however faced with infrared divergent integrals. We can eliminate the IR-divergencies by summing diagrams with the same skeleton-structure as that depicted in fig. (19-3) , but with an arbitrary number of (renormalized) self energy insertions. This amounts to replacing the bare propagators by, i.e., by

$$\Delta(\omega_{n_i}, \vec{k}_i) = \frac{1}{\omega_{n_i}^2 + \vec{k}_i^2 + \pi_R} ,$$

where the self energy π_R depends in general not only on the temperature, but also on the frequency and momentum carried by the line. Inserting for π_R the expression (19.37), valid in lowest order perturbation theory, and setting $\omega_{n_i} = 0$, one therefore has that (19.51) is replaced by

$$\Omega_N^{(\omega=0)} \sim \lambda^N T^{N+1} \int d^3k_1 ... d^3k_{N+1} \prod_{l=1}^{2N} \frac{1}{\vec{q}_l^2 + m^2(T)} , \tag{19.52}$$

where $m(T) = \sqrt{\lambda} T$. This cures the infrared divergence. By scaling the momenta with $m(T)$ one is therefore led to the conclusion that Ω receives a contribution of the form

$$\sim \lambda^3 T^4 \left(\frac{\lambda T}{m(T)} \right)^{N-3} ; \quad (N \geq 3) \tag{19.53}$$

This is an interesting result, for it tells us that for $m(T) = \sqrt{\lambda} T$, the contribution to the thermodynamical potential is non-analytic in the coupling constant for

$N - 3 \neq 2k$ (k a positive integer), and that it is of lower order than λ^N. Notice that a naive perturbative expansion of the propagators in (19.52) would lead one to conclude that the leading order contribution is of $\mathcal{O}(\lambda^N)$. This conclusion is however incorrect, since each term in the expansion is IR-divergent for $N > 3$, with the divergence getting worse as one proceeds to higher orders. The resummation of these divergent contributions has thus led us to the above result.

Finally we remark that if the mass generated at $T \neq 0$ had been of $O(\lambda T)$ (rather than $O(\sqrt{\lambda}T)$), then we would be faced with the problem that there are an infinite number of non-trivial diagrams which contribute in $O(\lambda^3)$. As has been pointed out by Linde (1980) such a computational barrier arises in QCD, and is known there as the "infrared problem".

19.4 Feynman Rules for QED and QCD at non-vanishing Temperature and Chemical Potential in the Continuum

In section 1 we have shown for the ϕ^4-theory, that the transition from the $T = 0$ Feynman rules to the rules at finite temperature is effected in a very simple way. A temperature was introduced by merely compactifying the euclidean time direction, and imposing periodic boundary conditions on the fields. Hence the prescriptions given in section 1 apply to any bosonic theory. In the case of QED or QCD, where the gauge potentials are coupled to the fermion fields, these prescriptions must be supplemented by corresponding ones for the fermionic degrees of freedom. The ghost degrees of freedom are subject to the same rules as the gauge potentials, as we have already pointed out in chapter 18.

To keep our discussion general, we will allow also for a non-vanishing chemical potential μ coupled to the fermion charge density. The corresponding actions for QED and QCD have the form

$$\begin{aligned} S = S_G &+ \int_\beta d^4x \bar{\psi}(x)[\gamma_4(\partial_4 - \mu) + \gamma_i \partial_i + m]\psi(x) \\ &+ ie \int_\beta d^4x \bar{\psi}(x)\gamma_\mu A_\mu(x)\psi(x) + S_{GF} + S_{ghost}, \end{aligned} \tag{19.54}$$

where S_G is the action for the gauge fields, $\int_\beta$ has been defined in (19.5d), and S_{GF} and S_{ghost} are the gauge fixing term and ghost contributions arising from the Faddeev-Popov gauge fixing procedure. For QCD ψ and $\bar{\psi}$ are three-component colour vectors, and A_μ is the matrix valued field defined in (6.10). The fields $A_\mu(x)$ satisfy the boundary condition

$$A_\mu(\vec{x}, \beta) = A_\mu(\vec{x}, 0) \ . \tag{19.55}$$

The same is true for the ghost fields. The fermion fields, on the other hand, satisfy antiperiodic boundary conditions in euclidean time, as we had shown in chapter 18. This leads to a discretization of the fourth component of the fermion momentum, $p_4 \to \omega_\ell^- = \frac{1}{\beta}(2\ell+1)\pi$, where ω_ℓ^- are the Matsubara frequencies for fermions (18.93). We have denoted them here with a superscript "-" in order to distinguish them from the Matsubara frequencies (19.6b) for bosons, which will henceforth be denoted by ω_ℓ^+. From (19.54) we see that the chemical potential only appears in the kinetic term of the fermionic fields in the combination $\partial_4 - \mu$, which in frequency space takes the form $i(\omega_\ell^- + i\mu)$. It is thus evident that the chemical potential will only manifest itself in the fermion propagator.

From the above discussion it is evident that the euclidean continuum Feynman rules for QED and QCD at $T \neq 0$, $\mu \neq 0$ in frequency-momentum space follow immediately from those at $T = \mu = 0$ by the following simple prescriptions:

i) Consider the euclidean expressions for the propagators and vertices at $T = \mu = 0$ in momentum space. They are obtained by taking the naive continuum limit of the lattice expressions given in chapter 14 and 15. In particular for QCD they are given by*

Fermion propagator

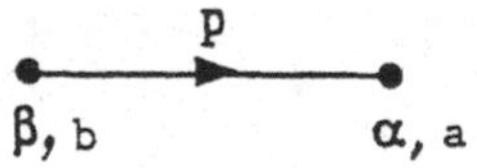

p

β, b α, a

$$[\frac{-i\gamma_\mu p_\mu + m}{p^2+m^2}]_{\alpha\beta}\delta_{ab}$$

Gluon propagator

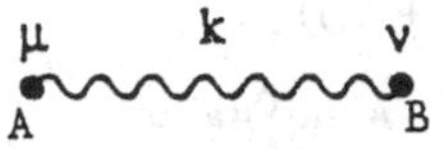

μ k ν

A B

$$\frac{1}{k^2}[\delta_{\mu\nu} - (1-\alpha_0)\frac{k_\mu k_\nu}{k^2}]\delta_{AB}$$

---

* In the case of QED in a linear gauge, the vertices coupling three and four gauge potentials, as well as the vertex involving the ghost fields are absent. The remaining expressions for the propagators and photon-fermion vertex are obtained by omitting the colour factors. The general prescription ii)-vi) given below therefore also hold for QED.

Ghost propagator

k

A B

$$\frac{1}{k^2}\delta_{AB}$$

Fermion-gluon vertex

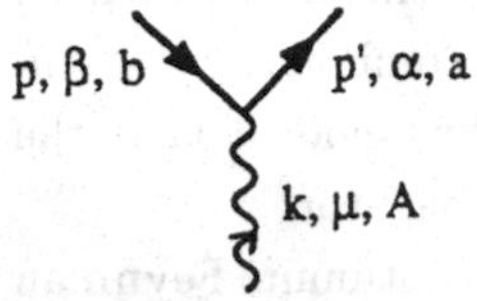

p, β, b

p', α, a

k, μ, A

$$-ig(2\pi)^4\delta^{(4)}(k+p-p')(\gamma_\mu)_{\alpha\beta}\frac{\lambda^A_{ab}}{2}$$

Ghost-gluon vertex

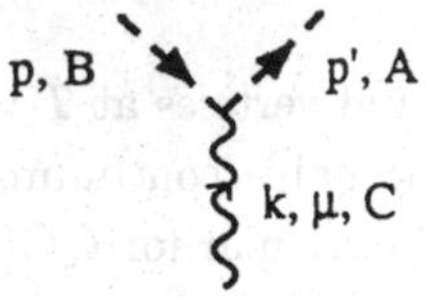

p, B

p', A

k, μ, C

$$ig(2\pi)^4\delta^{(4)}(k+p-p')f_{ABC}\ (p')_\mu$$

Three-gluon vertex

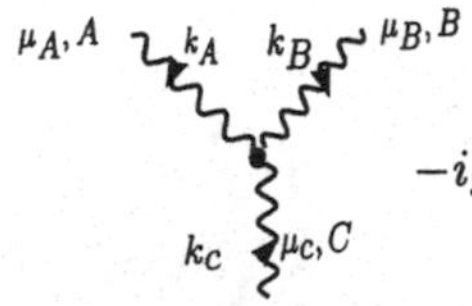

μ_A, A k_A k_B μ_B, B

k_C μ_C, C

$$-ig(2\pi)^4\delta^{(4)}(k_A+k_B+k_C)f_{ABC}\left[\delta_{\mu_A\mu_B}(k_A-k_B)_{\mu_C}+c.p.\right]$$

Four-gluon vertex

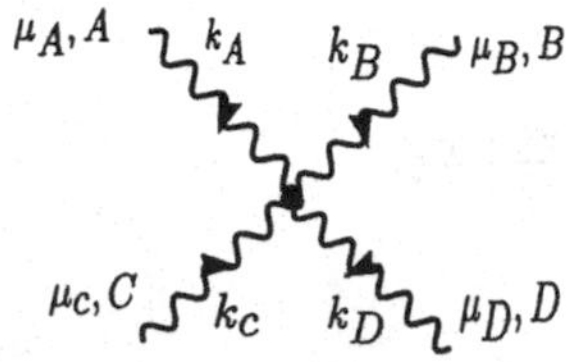

μ_A, A k_A k_B μ_B, B

μ_C, C k_C k_D μ_D, D

$$\begin{aligned}&-g^2(2\pi)^4\delta^{(4)}(k_A+k_B+k_C+k_D)\\&\times[f_{ABE}f_{CDE}(\delta_{\mu_A\mu_C}\delta_{\mu_B\mu_D}-\delta_{\mu_A\mu_D}\delta_{\mu_B\mu_C})\\&+f_{ACE}f_{BDE}(\delta_{\mu_A\mu_B}\delta_{\mu_C\mu_D}-\delta_{\mu_A\mu_D}\delta_{\mu_B\mu_C})\\&+f_{ADE}f_{CBE}(\delta_{\mu_A\mu_C}\delta_{\mu_B\mu_D}-\delta_{\mu_A\mu_B}\delta_{\mu_D\mu_C})]\end{aligned}$$

ii) Replace the fourth component of the momentum associated with a photon or ghost line as follows

$$k_4 \to \omega_\ell^+ \ ; \quad \omega_\ell^+ = \frac{2\pi\ell}{\beta}\ , \tag{19.56}$$

where ℓ is an integer.

iii) Replace the fourth component of the momentum associated with a fermion line as follows

$$p_4 \to \omega_\ell^- + i\mu \; ; \quad \omega_\ell^- = \frac{(2\ell+1)\pi}{\beta} \; . \tag{19.57}$$

iv) Integrations over the fourth component of momenta become *infinite* sums over Matsubara frequencies

$$\begin{aligned} &\int \frac{dk_4}{2\pi} f(k_4; \cdots) \to \frac{1}{\beta} \sum_{\omega_\ell^+} f(\omega_\ell^+; \cdots) \; ; \quad (bosons) \; , \\ &\int \frac{dp_4}{2\pi} f(p_4; \cdots) \to \frac{1}{\beta} \sum_{\omega_\ell^-} f(\omega_\ell^- + i\mu; \cdots) \; ; \quad (fermions) \; . \end{aligned} \tag{19.58}$$

v) At each vertex implement energy momentum conservation by *

$$\beta(2\pi)^3 \delta(\sum_i \omega_{\ell_i}^- + \sum_j \omega_{\ell_j}^+) \delta^{(3)}(\sum_i \vec{p}^{(i)} + \sum_j \vec{k}^{(j)}) \; , \tag{19.59}$$

where i and j label the different three-momenta and Matsubara frequencies flowing into the vertex.

vi) For every fermion or ghost loop include a minus sign.

The euclidean correlation functions are then given by

$$\begin{aligned} &< \psi_{\alpha_1}(x_1) \cdots \psi_{\alpha_n}(x_n) \bar{\psi}_{\beta_1}(y_1) \cdot\cdot \bar{\psi}_{\beta_n}(y_n) A_{\mu_1}(z_1) \cdot\cdot A_{\mu_m}(z_m) >_{\beta,\mu} = \int_{(-)} \prod_{i=1}^n d^4 p_i \\ &\cdot \int_{(-)} \prod_{i=1}^n d^4 p'_i \int_{(+)} \prod_{i=1}^m d^4 k_i G_{\alpha_1 \cdots \beta_1 \cdots \mu_1 \cdots}(\{p_j\}, \{p'_j\}, \{k_j\}) e^{i \sum_i p_i \cdot x_i + i \sum_i p'_i \cdot y_i + \sum_i i k_i \cdot z_i} \end{aligned} \tag{19.60}$$

---

* To emphasize the analogy with the $T = 0$ formulation, we have also written $\delta(\sum_i \omega_{\ell_i}^- + \sum_j \omega_{\ell_j}^+)$ instead of $\delta_{(\sum_i \omega_{\ell_i}^- + \sum_j \omega_{\ell_j}^+),0}$. Notice that since at each vertex two fermion lines are coupled to the gauge potential, the sum over fermionic Matsubara frequencies will add to a Matsubara frequency of the bosonic type.

where we have suppressed the colour indices on the fields and the momentum-space Green function (so that the expression also holds for QED), and have introduced the compact notations

$$p_i = (\vec{p}_i, \omega^-_{\ell_i}) \ ; \quad k_i = (\vec{k}_i, \omega^+_{\ell_i}) \ , \tag{19.61a}$$

$$\int_{(\pm)} d^4 q = \frac{1}{\beta} \sum_{\omega^\pm_\ell} \int \frac{d^3 q}{(2\pi)^3} \ . \tag{19.61b}$$

$G_{\alpha_1 \cdots \beta_1 \cdots \mu_1 \cdots}(\{p_j\}, \{p'_j\}, \{k_j\})$ is the correlation function in frequency-momentum space calculated with the above Feynman rules.

The Feynman integrals in QED or QCD at finite temperature will involve in general sums over bosonic as well as fermionic Matsubara frequencies. The evaluation of such sums for bosons has been discussed in section 1. We now derive the analog of (19.19) and (19.20) for the case of fermions.

Consider the function

$$h(\omega) = -\frac{i\beta}{e^{i\beta\omega} + 1} \ .$$

This function possesses simple poles with unit residue located at $\omega = \omega^-_\ell$. Assuming that $f(\omega)$ has no singularities on the real axis one finds, proceeding in an analogous way as in the bosonic case, that

$$\begin{aligned} \frac{1}{\beta} \sum_\ell f(\omega^-_\ell) = \int_{-\infty}^{\infty} \frac{d\omega}{2\pi} f(\omega) &- \frac{1}{2\pi} \int_{-\infty - i\epsilon}^{\infty - i\epsilon} d\omega \frac{f(\omega)}{e^{i\beta\omega} + 1} \\ &- \frac{1}{2\pi} \int_{-\infty + i\epsilon}^{\infty + i\epsilon} d\omega \frac{f(\omega)}{e^{-i\beta\omega} + 1} \ . \end{aligned} \tag{19.62}$$

The first integral is just the zero temperature ($\beta \to \infty$) limit of the lhs. If $f(\omega)$ is also a meromorphic function, and is sufficiently well behaved for $|\omega| \to \infty$, so that we can close the integration contours in the last two integrals in the lower and upper complex ω-planes, then (19.62) reduces to

$$\begin{aligned} \frac{1}{\beta} \sum_\ell f(\omega^-_\ell) = \int_{-\infty}^{\infty} \frac{d\omega}{2\pi} f(\omega) &+ i \sum_{Im\bar{\omega}_i < 0} \frac{R_f(\bar{\omega}_i)}{e^{i\beta\bar{\omega}_i} + 1} \\ &- i \sum_{Im\bar{\omega}_i > 0} \frac{R_f(\bar{\omega}_i)}{e^{-i\beta\bar{\omega}_i} + 1} \ , \end{aligned} \tag{19.63}$$

where $R_f(\bar{\omega}_i)$ are the residues of $f(\omega)$ at the poles, whose location we have denoted by $\bar{\omega}_i$. This expression allows us, for example, to decompose one-loop Feynman integrals into a zero-temperature contribution and a contribution arising from the presence of the heat bath.

19.5 Temporal Structure of the Fermion Propagator at $T \neq 0$ and $\mu \neq 0$ in the Continuum

Consider the fermion Feynman propagator for QED in Minkowski space at vanishing temperature and chemical potential,

$$S_F^{(Mink)}(z) = \int \frac{d^4p}{(2\pi)^4} \frac{\gamma^\mu p_\mu + m}{p^2 - m^2 + i\epsilon} e^{-ip\cdot z} , \tag{19.64}$$

where γ_μ are the usual Dirac gamma-matrices satisfying the anticommutation relations $\{\gamma_\mu, \gamma_\nu\} = 2g_{\mu\nu}$, and $p \cdot z = p^0 z^0 - \vec{p} \cdot \vec{z}$. $S_F(z)$ propagates positive (negative) energy states in the forward (backward) direction in time. In the Dirac hole theory the absence of a negative energy state is interpreted as an antiparticle with positive energy. The correspondence between positive (negative) energy states and forward (backward) propagation in time follows immediately from (19.64) by integrating this expression over p^0. The integrand possesses poles located at $p^0 = \pm E$ where $E = \sqrt{\vec{p}^2 + m^2}$. Depending on the sign of $t \equiv z^0$, we can close the integration contour at infinity in the upper or lower half of the complex p^0-plane. One then finds that

$$S_F^{(Mink)}(\vec{z}, t) = \int \frac{d^3p}{(2\pi)^3} \tilde{S}_F^{(Mink)}(\vec{p}, t) e^{i\vec{p}\cdot\vec{z}} , \tag{19.65a}$$

where

$$\tilde{S}_F^{(Mink)}(\vec{p}, t) = \Lambda_+(\vec{p})\Delta^{(+)}(\vec{p}, t) + \Lambda_-(\vec{p})\Delta^{(-)}(\vec{p}, t) , \tag{19.65b}$$

and

$$\Lambda_\pm(\vec{p}) = \frac{1}{2iE}[\pm\gamma^0 E + \gamma^j p_j + m] , \tag{19.65c}$$

$$\Delta^{(\pm)}(\vec{p}, t) = \theta(\pm t) e^{\mp iEt} . \tag{19.65d}$$

$$E = \sqrt{\vec{p}^2 + M^2} .$$

The $(\pm)$ superscripts denote the positive and negative energy contributions arising from the poles at $p^0 = \pm E$. A similar decomposition holds for the euclidean fermion propagator.* The purpose of this section is to study how the temporal

---

* Although this correlation function no longer describes propagation, we nevertheless shall refer to it as a propagator.

structure of this propagator is modified at finite temperature and chemical potential. Since the chemical potential controls the particle-antiparticle content, its effect on the temporal structure will be different for the positive and negative energy contributions. In section 9 we will repeat this exercise within the framework of the lattice regularization for naive as well as Wilson fermions. Comparison with the continuum results obtained below will shed further light on the nature of the "doubler" contributions which we discussed in chapter 4.

The euclidean fermion propagator at finite temperature and chemical potential can be immediately written down using the prescriptions given in the previous section:

$$S_F^{(\beta,\mu)}(\vec{z},\tau) = \int \frac{d^3p}{(2\pi)^3} \tilde{S}_F^{(\beta,\mu)}(\vec{p},\tau) e^{i\vec{p}\cdot\vec{z}} , \tag{19.66a}$$

where

$$\tilde{S}_F^{(\beta,\mu)}(\vec{p},\tau) = \frac{1}{\beta}\sum_{\ell} \hat{S}_F^{(\beta,\mu)}(\vec{p},\omega_\ell^-) e^{i\omega_\ell^-\tau} , \tag{19.66b}$$

and

$$\hat{S}_F^{(\beta,\mu)}(\vec{p},\omega_\ell^-) = \frac{-i\gamma_4^E(\omega_\ell^- + i\mu) - i\gamma_i^E p_i + m}{(\omega_\ell^- + i\mu)^2 + E^2(\vec{p})} . \tag{19.66c)}$$

The γ-matrices have been denoted here with a superscript 'E' ($\rightarrow$ euclidean) in order to distinguish them from the γ matrices appearing in (19.64). They are the γ matrices which we have been working with in the euclidean formulation.

The frequency sum (19.66b) can be carried out by making use of the summation formula (19.62). An even more convenient summation formula is given by (C.4) in Appendix C, where the first term on the rhs corresponds to the $T = \mu = 0$ contribution. * One then finds that

$$\tilde{S}_F^{(\beta,\mu)}(\vec{p},\tau) = \Gamma_+(\vec{p})\Delta^{(+)}_{(\beta,\mu)}(\vec{p},\tau) + \Gamma_-(\vec{p})\Delta^{(-)}_{(\beta,\mu)}(\vec{p},\tau) , \tag{19.67a}$$

where

$$\Gamma_\pm(\vec{p}) = \frac{1}{2E}[\pm\gamma_4^E E - i\gamma_j^E p_j + m] , \tag{19.67b}$$

is the euclidean analog of $\Lambda_\pm$ in (19.65c), and where $\Delta^{(\pm)}_{(\beta,\mu)}(\vec{p},\tau)$ is given by

$$\begin{aligned} \Delta^{(+)}_{(\beta,\mu)}(\vec{p},\tau) &= [\theta(\tau) - \eta_{FD}(E,\mu)]e^{-(E-\mu)\tau} , \\ \Delta^{(-)}_{(\beta,\mu)}(\vec{p},\tau) &= [\theta(-\tau) - \bar{\eta}_{FD}(E,\mu)]e^{(E+\mu)\tau} . \end{aligned} \tag{19.67c}$$

---

* The reader can easily convince himself that this formula can also be applied to the sum (19.66b) for $|\tau| \in [0,\beta]$.

Here

$$\eta_{FD}(E,\mu) = \frac{1}{e^{\beta(E-\mu)}+1} \quad , \; \bar{\eta}_{FD}(E,\mu) = \frac{1}{e^{\beta(E+\mu)}+1} \tag{19.67d}$$

are the Fermi-Dirac distribution functions for particles and antiparticles. For $\mu = 0$ and $\beta \to \infty$ the expressions (19.67) reduce to the euclidean analog of (19.65).

Consider the particular case $\mu = 0$, where particles and antiparticles contribute with the same weight to the propagator. Then (19.66a) can be written in the form

$$\begin{aligned} S_F^{(\beta,0)}(\vec{z},\tau) = & \int \frac{d^4p}{(2\pi)^4} \frac{(-ip_\mu\gamma_\mu^E + m)}{p^2+m^2} e^{ip_4\tau + i\vec{p}\cdot\vec{z}} \\ & + \int \frac{d^3p}{(2\pi)^3} \eta_{FD}(E) \frac{1}{E} \left\{ \gamma_4^E E sinh(E\tau) + (i\gamma_j^E p_j - m) cosh(E\tau) \right\} , \end{aligned} \tag{19.68a}$$

where

$$\eta_{FD}(E) = \frac{1}{e^{\beta E}+1} \; . \tag{19.68b}$$

The first integral is just the $T = 0$ contribution, and is the euclidean version of (19.64). Expression (19.68a) is the analog of (19.21a). Consider its analytic continuation to real times, $\tau \to it$. Performing a Wick rotation, $p_4 \to -ip_0$, one finds that*

$$\begin{aligned} S_F^{(\beta,0)}(z) \to & i \int \frac{d^4p}{(2\pi)^4} \frac{(-\gamma^\mu p_\mu + m)}{p^2 - m^2 + i\epsilon} e^{ip\cdot z} \\ & + \int \frac{d^3p}{(2\pi)^3} \eta_{FD}(E) \frac{1}{E} \left\{ i\gamma^0 E \sin(Et) + (\gamma^i p_i - m)\cos(Et) \right\} , \end{aligned}$$

where we have now introduced the γ-matrices in Minkowski space, $\gamma^0 = \gamma_4; \gamma^i = i\gamma_i$, satisfying $\{\gamma_\mu, \gamma_\nu\} = 2g_{\mu\nu}$. Expressing the trigonometric functions in terms of exponentials, and making use of (19.22) and of the relation

$$\delta(p_0 - E) - \delta(p_0 + E) = 2E(\vec{p})\epsilon(p_0)\delta(p^2 - m^2) \; ,$$

where $\epsilon(p^0) = \theta(p^0) - \theta(-p^0)$, and $p^2 = p^\mu p_\mu$, one finds, after making the change of variable $p^\mu \to -p^\mu$ (in order to conform to usual conventions), that

$$S_F^{(\beta,Mink)}(z) = \int \frac{d^4p}{(2\pi)^4} \tilde{S}_F^{(\beta,Mink)}(p) e^{-ip\cdot z} \; , \tag{19.69a}$$

---

* Note that, after making the change of variables $p_\mu \to -p_\mu$, the continuation to real times of the $T = \mu = 0$ contribution to the correlation function differs from (19.64) by a factor 'i'.

where

$$\tilde{S}_F^{(\beta,Mink)}(p) = \frac{i}{\not{p} - m + i\epsilon} - \frac{2\pi}{e^{\beta E}+1}(\not{p}+m)\delta(p^2 - m^2) \qquad (19.69b)$$

and $\not{p} = \sum_\mu \gamma^\mu p_\mu$. This is however not the propagator which appears in the real-time Feynman rules in an interacting theory. As in the bosonic case, the Feynman rules for computing finite temperature real-time correlation functions are far more complicated in the presence of interactions, and requires a doubling of degrees of freedom [Landsman and Weert (1987), Niemi and Semenoff (1984)]. The Dirac propagator in momentum space becomes a 2×2 matrix in the extended space, of which the first diagonal component is given by (19.69b).

19.6 The Electric Screening Mass in Continuum QED in One-Loop Order

As a less trivial application of the finite temperature, finite chemical potential formalism, we derive an expression for the electric screening mass to one loop order in QED. The result will be compared in section 10 with the screening mass computed in lattice perturbation theory. Let us first state what is meant by the electric screening mass.

When an external static charge Q is introduced into an electrically neutral QED plasma in thermal equillibrium, polarization of the medium will screen the Coulomb potential of the charge, leading to a short range potential with a Debye screening length given by the inverse screening mass. If the introduction of the external charge can be treated as a small perturbation, then one can use the theory of linear response (see e.g. Kapusta (1989)), to show that the screened Coulomb potential is given by

$$\Phi(\vec{r}) = Q \int \frac{d^3k}{(2\pi)^3} \frac{e^{i\vec{k}\cdot\vec{r}}}{\vec{k}^2 + \Pi_{44}^{(\beta,\mu)}(0,\vec{k})}\ , \qquad (19.70)$$

where $\Pi_{44}^{(\beta,\mu)}(0,\vec{k}) = \Pi_{44}^{(\beta,\mu)}(\omega = 0,\vec{k})$ is the 44-component of the vacuum polarization tensor at finite temperature and chemical potential, evaluated for vanishing Matsubara frequency. It is defined here as the negative of the one particle irreducible diagrams with two external photon lines. At large distances the behaviour of the integral is determined by contributions of momenta $|\vec{k}|$ of $\mathcal{O}(1/r)$. Defining the electric screening mass as the static infrared limit of $\Pi_{44}^{(\beta,\mu)}(0,\vec{k})$,

$$m_{el}^2 = \lim_{\vec{k}\to 0} \Pi_{44}(0,\vec{k})\ , \qquad (19.71)$$

we have that

$$\Phi(\vec{r}) \underset{r\to\infty}{\longrightarrow} Q\int \frac{d^3k}{(2\pi)^3}\frac{e^{i\vec{k}\cdot\vec{r}}}{\vec{k}^2+m_{el}^2} = \frac{Q}{4\pi}\frac{e^{-m_{el}r}}{r} . \tag{19.72}$$

Hence the static Coulomb potential of an external charge is screened exponentially in the presence of a plasma.

We only mention here that the electric screening mass can also be computed in a different way. As has been shown by Fradkin (1965) it can be related to the second derivative of the pressure with respect to the electric charge chemical potential:

$$m_{el}^2 = e^2\frac{\partial^2 p}{\partial\mu^2} .$$

This is an interesting relation, since it allows one to compute m_{el}^2 for a neutral QED plasma up to order e^5, since the pressure is known to $\mathcal{O}(e^3)$. For computations of the electric screening mass using this relation, see Kapusta (1992).

In the following we compute the electric screening mass according to (19.71) in one-loop order. Our presentation is adapted to be easily compared with the lattice calculation given in section 10.

In the euclidean continuum formulation of QED, the vacuum polarization tensor $\Pi_{\mu\nu}$ at finite temperature and chemical potential is given, using the rules discussed in section 4, in one-loop order by (see fig. (19-4))

$$\Pi_{\mu\nu}^{(\beta,\mu)}(\omega_{\ell_k}^+,\vec{k}) = (-ie)^2\frac{1}{\beta}\sum_{\omega_\ell^-}\int\frac{d^3p}{(2\pi)^3}Tr\{\frac{(-i\gamma_\rho(p+k)_\rho+m)\gamma_\mu(-i\gamma_\tau p_\tau+m)\gamma_\nu}{(p^2+m^2)[(p+k)^2+m^2]}\} , \tag{19.73a}$$

where

$$k_4 = \omega_{\ell_k}^+ \;\; ; \;\; p_4 = \omega_\ell^- + i\mu . \tag{19.73b}$$

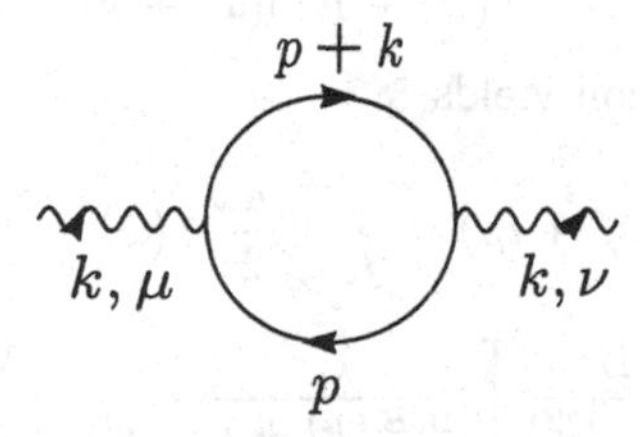

Fig.19-4 Diagram contributing to the vacuum polarization tensor in $\mathcal{O}(g^2)$.

Performing the trace by making use of the relations

$$\begin{aligned} Tr(\gamma_\mu\gamma_\nu) &= 4\delta_{\mu\nu} \\ Tr(\gamma_\mu\gamma_\nu\gamma_\rho\gamma_\lambda) &= 4(\delta_{\mu\nu}\delta_{\rho\lambda} - \delta_{\mu\rho}\delta_{\nu\lambda} + \delta_{\mu\lambda}\delta_{\nu\rho}) \end{aligned} , \tag{19.74}$$

expression (19.73a) takes the form

$$\Pi_{\mu\nu}^{(\beta,\mu)}(\vec{k}, \omega_{\ell_k}^+) = e^2 \frac{1}{\beta} \sum_{\omega_\ell^-} \int \frac{d^3p}{(2\pi)^3} \frac{V_{\mu\nu}(k,p)}{(p^2+m^2)[(p+k)^2+m^2]} \} , \tag{19.75a}$$

where

$$V_{\mu\nu}(k,p) = 8p_\mu p_\nu + 4(p_\mu k_\nu + k_\mu p_\nu) - 4(m^2 + p^2 + p \cdot k)\delta_{\mu\nu} . \tag{19.75b}$$

For vanishing photon frequency this expression reduces to

$$\Pi_{44}^{(\beta,\mu)}(0,\vec{k}) = -4e^2 \frac{1}{\beta} \sum_{\ell=-\infty}^{\infty} \int \frac{d^3p}{(2\pi)^3} \frac{G^2 - (\omega_\ell^- + i\mu)^2}{[(\omega_\ell^- + i\mu)^2 + E^2][(\omega_\ell^- + i\mu)^2 + F^2]} , \tag{19.76a}$$

where

$$E^2 = \vec{p}^2 + m^2, \quad F^2 = (\vec{p} + \vec{k})^2 + m^2, \quad G^2 = \vec{p} \cdot (\vec{p} + \vec{k}) + m^2 . \tag{19.76b}$$

To perform the frequency sum we can in principle make use of the summation formula (19.63), where the integral appearing on the rhs would however yield the contribution at finite chemical potential and vanishing temperature. It is therefore more convenient to use an alternative formula in which the $T = \mu = 0$ (vacuum) contribution is separated from the temperature and chemical potential dependent part. Only the $T = \mu = 0$ contribution will then need to be renormalized in the standard way. The relevant summation formula is given by (C.4) in Appendix C. The function $f(\omega)$ in (C.1) has the following form in the present case (we only exhibit the dependence on ω):

$$f(\omega) = \frac{G^2 - \omega^2}{(\omega^2 + E^2)(\omega^2 + F^2)} .$$

A straight forward calculation yields

$$\begin{aligned} \frac{1}{\beta} \sum_{\ell=-\infty}^{\infty} f(\omega_\ell^- + i\mu) &= \int_{-\infty}^{\infty} \frac{d\omega}{2\pi} f(\omega) \\ &- \frac{G^2 + E^2}{2E(F^2 - E^2)} \left[\frac{1}{e^{\beta(E+\mu)} + 1} + \frac{1}{e^{\beta(E-\mu)} + 1} \right] \\ &+ \frac{G^2 + F^2}{2F(F^2 - E^2)} \left[\frac{1}{e^{\beta(F+\mu)} + 1} + \frac{1}{e^{\beta(F-\mu)} + 1} \right] . \end{aligned} \tag{19.77}$$

The first term on the rhs is the $T = \mu = 0$ (vacuum) contribution to $\Pi_{44}(0,\vec{k})$. It is ultraviolet divergent and is renormalized in the standard way, yielding no contribution to the screening mass, as follows from Lorentz and gauge invariance. Hence we shall concentrate on the matter part, given by the last two terms. The dependence on the photon momentum $\vec{k}$ is contained in the functions G and F. To compute the screening mass we must take the limit $\vec{k} \to 0$. In this limit $F \to E$, so that the denominators in (19.77) vanish. Setting $F = E + \epsilon$, and taking the limit $\epsilon \to 0$, one finds that

$$m_{el}^2 = 2e^2\beta \int \frac{d^3p}{(2\pi)^3} \left\{ \frac{e^{\beta(E-\mu)}}{[e^{\beta(E-\mu)}+1]^2} + \frac{e^{\beta(E+\mu)}}{[e^{\beta(E+\mu)}+1]^2} \right\} . \qquad (19.78)$$

After carrying out the angular integration, this expression can be written in the form

$$m_{el}^2 = -\frac{e^2}{\pi^2}\int_0^\infty dp\, p\sqrt{p^2+m^2}\frac{\partial}{\partial p}[\eta_{FD}(E,\mu) + \bar{\eta}_{FD}(E,\mu)] \, .$$

where $\eta_{FD}(E,\mu)$ and $\bar{\eta}_{FD}(E,\mu)$ are the Fermi-Dirac distribution functions (19.67d). After a partial integration we finally arrive at

$$m_{el}^2 = \frac{e^2}{\pi^2}\int_0^\infty \frac{2p^2+m^2}{\sqrt{p^2+m^2}}[\eta_{FD}(E,\mu) + \bar{\eta}_{FD}(E,\mu)] \, . \qquad (19.79)$$

For vanishing temperature and finite chemical potential this expression reduces to

$$m_{el}^2 \underset{T\to 0}{\longrightarrow} \frac{e^2}{\pi^2}\int_0^\infty dp \frac{2p^2+m^2}{\sqrt{p^2+m^2}}\theta(|\mu| - \sqrt{p^2+m^2}) \, .$$

Carrying out the integral one obtains

$$\begin{aligned} m_{el}^2(T=0) &= \frac{e^2}{\pi^2}\mu\sqrt{\mu^2-m^2} \quad \text{for} \quad |\mu| > m \\ &= 0 \quad \text{for} \quad |\mu| < m \end{aligned} \, .$$

Hence for $|\mu| > m$ there is also an electric screening mass generated at zero temperature.

We close this section with a remark. In order to parallel the lattice calculation in section 10, we have taken the limit $\vec{k} \to 0$ before performing the angular integration. This yields the form (19.78) for the screening mass, which has a direct analog on the lattice. By performing first the angular integration (which cannot be carried out analytically on the lattice) one finds that

$$\begin{aligned} \Pi_{44}^{(\beta,\mu)}(0,\vec{k}) = \frac{e^2}{\pi^2}\int_{-0}^\infty dp \frac{p^2}{\sqrt{p^2+m^2}}[\eta_{FD}(E,\mu) + \bar{\eta}_{FD}(E,\mu)] \\ \times \left[1 + \frac{4(p^2+m^2) - \vec{k}^2}{4p|\vec{k}|} \ln\left(\frac{2p+|\vec{k}|}{2p-|\vec{k}|} \right) \right] \end{aligned} ,$$

where $p = |\vec{p}|$. For $|\vec{k}| \to 0$, the logarithm behaves like $|\vec{k}|/p$. Hence one is led to (19.79).

19.7 The Electric Screening Mass in Continuum QCD in One-Loop Order

The computation of the electric screening mass in QCD is more involved than in QED. In one loop order the vacuum polarization tensor receives contributions from the four diagrams shown in fig. (19-5). The quark loop contribution has the same form as in QED exept that there is an additional colour factor $Tr(\frac{\lambda^A}{2}\frac{\lambda^B}{2}) = \frac{1}{2}\delta_{AB}$, arising from the gluon-quark vertices. Hence we immediately conclude that the contribution of diagram (a) is related to that in QED (c.f. eq. (19.75)) by

$$\Pi^{AB}_{\mu\nu}(\omega_n^+, \vec{k})_{(a)} = \frac{1}{2}\delta_{AB}[\Pi_{\mu\nu}(\omega_n^+, \vec{k})]_{QED} \,. \tag{19.80}$$

where we have dropped the superscript (β, μ) for notational reasons. The remaining diagrams (b,c,d) involve summations over bosonic Matsubara frequencies and hence are expected to involve the Bose-Einstein distribution function. They do not depend on the chemical potential. Hence by dimensional arguments their contribution to the screening mass will be proportional to gT. Our objective is to compute the constant multiplying gT. In the following we discuss in turn the contributions of diagrams b,c, and d (Kaste, 1997).

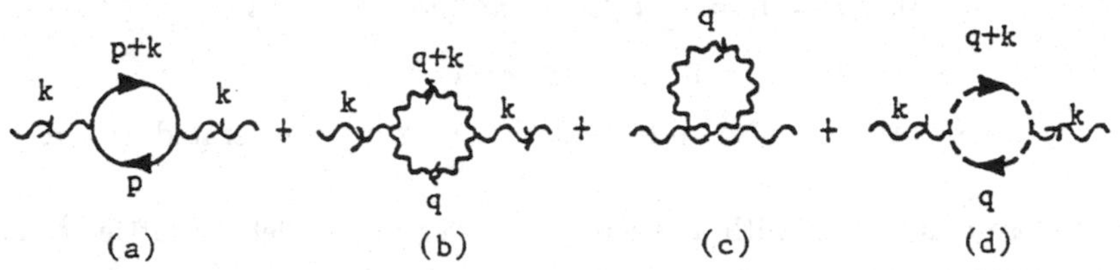

Fig. 19-5 Diagrams contributing to the vacuum polarization tensor.

Consider the diagram (b). The combinatorial factor which multiplies the finite temperature Feynman integral, obtained with the prescriptions stated in section 4, is calculated as follows: Consider the six lines emanating from the two vertices before they are contracted to form the loop. Label the lines from 1 ty 6. Then there are $6 \cdot 3$ possibilites to label the external lines of the diagram, and 2 possibilites to contract the remaining lines to form the loop. Thies yields a factor $(3!)^2$. Each vertex receives a factor $\frac{1}{3!}$ arising from the symmetrization of the 3-gluon interaction. Finally there is a factor $\frac{1}{2!}$ associated with the second order contribution, arising from the expansion of the exponential of the action. After some algebra one finds that in the Feynman gauge

$$\Pi^{AB}_{\mu\nu}(\omega_n^+, \vec{k})_{(b)} = \frac{g^2}{2!} f_{ACD} f_{BCD} \frac{1}{\beta} \sum_{\omega_\ell^+} \int \frac{d^3q}{(2\pi)^3} \frac{T^{(b)}_{\mu\nu}(q,k)}{q^2(q+k)^2} , \tag{19.81a}$$

where

$$\begin{aligned} T^{(b)}_{\mu\nu}(q,k) &= 2k_\mu k_\nu - 5(q_\mu k_\nu + k_\mu q_\nu) - 10 q_\mu q_\nu \\ &\quad - \delta_{\mu\nu}(2q^2 + 2q \cdot k + 5k^2) , \end{aligned} \tag{19.81b}$$

and a summation over repeated indices is understood. Here q^2, k^2 and $q \cdot k$ denote the four dimensional euclidean scalar products, and it is always understood from now on that

$$k_4 = \omega_n^+ \quad ; \quad q_4 = \omega_\ell^+ . \tag{19.82}$$

Next consider the contribution of the diagram (c). The factor multiplying the Feynman integral is given as follows: Take the 4 lines emanating from the vertex and label them from 1 to 4. Then there are $4 \cdot 3$ possibilities to label the external lines of the diagram. The remaining lines are then contracted to form the loop. Finally there is a factor $\frac{1}{4!}$ arising from the symmetrization of the four-gluon contribution to the action. One then readily verifies that in the Feynman gauge

$$\Pi^{AB}_{\mu\nu}(\omega_n^+, \vec{k})_{(c)} = 3g^2 \delta_{\mu\nu} f_{ACD} f_{BCD} \frac{1}{\beta} \sum_{\omega_\ell^+} \int \frac{d^3q}{(2\pi)^3} \frac{1}{q^2} . \tag{19.83}$$

Finally consider the contribution of the diagram (d). Since the ghost fields are Grassmann valued, there is a minus sign associated with the loop. There is only one possible way to contract the ghost-lines emanating from two vertices to form the loop. Hence the combinatorial factor multiplying the Feynman integral is determined by the two possibilities to label the external lines, and a factor $\frac{1}{2!}$ associated with the second order diagram. Application of the Feynman rules then yields

$$\Pi^{AB}_{\mu\nu}(\omega_n^+, \vec{k})_{(d)} = g^2 f_{ACD} f_{BCD} \frac{1}{\beta} \sum_{\omega_\ell^+} \int \frac{d^3q}{(2\pi)^3} \frac{(q+k)_\mu q_\nu}{q^2(q+k)^2} .$$

This expression can be cast into a symmetric form in μ and ν by noting that a change of variables $q \to -q-k$ interchanges μ and ν. Hence

$$\Pi^{AB}_{\mu\nu}(\omega_n^+, \vec{k})_{(d)} = \frac{1}{2} g^2 f_{ACD} f_{BCD} \frac{1}{\beta} \sum_{\omega_\ell^+} \int \frac{d^3q}{(2\pi)^3} \frac{(q+k)_\mu q_\nu + (q+k)_\nu q_\mu}{q^2(q+k)^2} . \tag{19.84}$$

Combining the contributions (19.81), (19.83) and (19.84), and making use of the relation

$$\sum_{C,D} f_{ACD} f_{BCD} = 3\delta_{AB}$$

one finds that

$$\Pi^{AB}_{\mu\nu}(\omega_n^+, \vec{k})_{(b+c+d)} = \delta_{AB} \Pi_{\mu\nu}(\omega_n^+, \vec{k})_{(b+c+d)} , \tag{19.85a}$$

where

$$\Pi_{\mu\nu}(\omega_n^+, \vec{k})_{(b+c+d)} = \frac{3}{2} g^2 \frac{1}{\beta} \sum_{\omega_\ell^+} \int \frac{d^3q}{(2\pi)^3} \frac{W_{\mu\nu}(q,k)}{q^2(q+k)^2} , \tag{19.85b}$$

and*

$$W_{\mu\nu}(q,k) = 2k_\mu k_\nu - 4(q_\mu k_\nu + q_\nu k_\mu) - 8q_\mu q_\nu + \delta_{\mu\nu}(4q^2 + k^2 + 10q \cdot k) \tag{19.85c}$$

As always, q_4 and k_4 are understood to be given by (19.82).

In the following we now restrict ourselves to the computation of the electric screening mass,

$$m_{el}^2 = \lim_{\vec{k} \to 0} \Pi_{44}(0, \vec{k}) ,$$

where $\Pi_{44}(\omega_n^+, \vec{k})$ is defined in (19.85a). For $\mu = \nu = 4$ and vanishing gluon-frequency, (19.85b) takes the form

$$\Pi_{44}(0, \vec{k})_{(b+c+d)} = \frac{3}{2} g^2 \frac{1}{\beta} \sum_{\omega_\ell^+} \int \frac{d^3q}{(2\pi)^3} \frac{4\vec{q}^2 + \vec{k}^2 + 10\vec{q}\cdot\vec{k} - 4\omega_\ell^{+2}}{[\omega_\ell^{+2} + \vec{q}^2][\omega_\ell^{+2} + (\vec{q}+\vec{k})^2]} .$$

The frequency sum can be calculated with the bosonic summation formula (19.20). For the same reason as discussed in the previous section, we only need to consider the contribution arising from the presence of the heat bath, i.e. the last two terms

* The expression for $W_{\mu\nu}$ differs from that obtained by Gross et al. [Gross (1981)] by a term which can be shown not to contribute to the integral (19.85b).

in (19.20). The $T = 0$ (vacuum) integral does not contribute to the screening mass. Performing the frequency sum in the above expression one then finds, after carrying out the angular integration, that this contribution, which we denote by $\Pi_{44}^{(\beta)}$, is given by

$$\Pi_{44}^{(\beta)}(0,\vec{k})_{(b+c+d)} = 3\frac{g^2}{2\pi^2}\int_0^\infty dq\; q\; \eta_{BE}(q)\Big\{2 + \frac{2q^2-\vec{k}^2}{2q|\vec{k}|}\ln\left[\frac{2q+|\vec{k}|}{2q-|\vec{k}|}\right]^2\Big\}\,, \tag{19.86a}$$

where $q = |\vec{q}|$, and where

$$\eta_{BE}(q) = \frac{1}{e^{\beta|\vec{q}|}-1} \tag{19.86b}$$

is the Bose-Einstein distribution function. The corresponding contribution to the electric screening mass is obtained by taking the limit $\vec{k} \to 0$:

$$(m_{el}^2)_{(b+c+d)} = \frac{6g^2}{\pi^2}\int_0^\infty dq\; q\; \eta_{BE}(q) = g^2T^2\,. \tag{19.87}$$

Combining this result with the contribution to the electric screening mass arising from the diagram (a) in fig. (19-4), which, according to (19.80), is given by one half of the screening mass (19.79) in QED, we conclude that

$$(m_{el}^2)^{QCD} = g^2T^2 + \frac{g^2}{2\pi^2}\int_0^\infty dp \frac{2p^2+m^2}{\sqrt{p^2+m^2}}[\eta_{FD}(E,\mu)+\bar{\eta}_{FD}(E,\mu)]\,, \tag{19.88}$$

where $\eta_{FD}(E,\mu)$ and $\bar{\eta}_{FD}(E,\mu)$ are the Fermi-Dirac distribution functions (19.67d). We therefore see that an electric screening mass is also generated in the pure gluonic sector, i.e. in the absence of quarks. For $SU(N)$, $\sum_{C,D} f_{ACD}f_{BCD} = N\delta_{AB}$, so that the contribution of the pure gauge part to the screening mass is given by

$$(m_{el}^2)_{gauge} = \frac{1}{3}Ng^2T^2 \qquad SU(N)$$

This concludes our discussion of QED and QCD at finite temperature and chemical potential in the continuum formulation. In the remaining sections of this chapter we now consider these theories within the framework of lattice regularized perturbation theory.

19.8 Lattice Feynman Rules for QED and QCD at $T \neq 0$ and $\mu \neq 0$

As we have seen, there is a simple prescription for making the transition from the continuum Feynman rules at vanishing temperature and chemical potential, to the $T \neq 0$, $\mu \neq 0$ rules. We now show that the finite temperature, finite chemical potential lattice Feynman rules can be obtained from those given in chapters 14 and 15 by the same prescription, except that momentum integrations and sums over Matsubara frequencies are restricted to the first Brillouin zone. This is not completely obvious. In fact the simplicity of the prescription relies heavily on the way the chemical potential is introduced in the lattice action. As will become clear from the following arguments, it suffices to consider QED. The same prescriptions then apply to QCD.

At zero temperature and chemical potential the lattice $U(1)$ action for Wilson fermions has the form (5.22), where the sum over plaquettes and lattice sites extend over a lattice of infinite extent. At finite temperature the lattice is compactified in the euclidean time direction, with the link variables satisfying the boundary condition

$$U_\mu(\vec{n}, \hat{\beta}) = U_\mu(\vec{n}, 0) , \tag{19.89}$$

where $\hat{\beta}$ is the inverse temperature measured in lattice units. The fermion fields satisfy antiperiodic boundary conditions. Introducing the chemical potential into the fermionic contribution to the action according to the Hasenfratz-Karsch-Kogut prescription (18.100), the U(1) action, written in dimensionless variables, takes the form*

$$\begin{aligned} S_{QED}^{(\beta,\mu)} &= \frac{1}{e^2}\sum_P [1 - \frac{1}{2}(U_P + U_P^\dagger)] + (\hat{m} + 4r)\sum_{n_4=1}^{\hat{\beta}}\sum_{\vec{n}} \bar{\hat{\psi}}(n)\hat{\psi}(n) \\ &- \frac{1}{2}\sum_{n_4=1}^{\hat{\beta}}\sum_{\vec{n}} [\bar{\hat{\psi}}(n)(r - \gamma_4)e^{-\hat{\mu}}U_4(n)\hat{\psi}(n + \hat{e}_4) + \bar{\hat{\psi}}(n + \hat{e}_4)(r + \gamma_4)e^{\hat{\mu}}U_4^\dagger(n)\hat{\psi}(n)] \\ &- \frac{1}{2}\sum_{n_4=1}^{\hat{\beta}}\sum_{\vec{n},i} [\bar{\hat{\psi}}(n)(r - \gamma_i)U_i(n)\hat{\psi}(n + \hat{e}_i) + \bar{\hat{\psi}}(n + \hat{e}_i)(r + \gamma_i)U_i^\dagger(n)\hat{\psi}(n)] . \end{aligned} \tag{19.90}$$

By introducing dimensioned parameters and fields according to $\hat{\mu} = \mu a$, $\hat{m} = ma$, $\hat{\beta} = \beta/a$, $\hat{\psi} = a^{3/2}\psi$, $\bar{\hat{\psi}} = a^{3/2}\bar{\psi}$, $\hat{A}_\mu = aA_\mu$, and taking the *naive* continuum limit, one recovers the continuum action, where the dependence on the chemical potential only appears in the kinetic term of the fermionic action in the form given in (19.54).

---

* Recall that this prescription was suggested by our analysis in section 8 of chapter 18

In weak coupling perturbation theory $U_\mu(n)$ is expanded as follows

$$U_\mu(n) = 1 + ie\hat{A}_\mu(n) - \frac{1}{2!}e^2\hat{A}_\mu^2(n) + \cdots , \qquad (19.91)$$

where $\hat{A}_\mu$ is the gauge potential in lattice units, i.e., $\hat{A}_\mu = aA_\mu$. From the structure of the lattice action (19.90) it is evident that the vertices generated by this expansion will depend, for finite lattice spacing, on the chemical potential. Our objective is to find out just how the chemical potential enters in the momentum-space Feynman rules. Although the answer is almost obvious, the following discussion will nevertheless be helpful.

To derive the Feynman rules in momentum space, the fermion fields are Fourier decomposed according to (18.102), while the Fourier expansion of the gauge potentials, replacing (14.22), now reads

$$\hat{A}_\mu(n) = \frac{1}{\hat{\beta}} \sum_{\ell=-\hat{\beta}}^{\frac{\hat{\beta}}{2}-1} \int_{-\pi}^{\pi} \frac{d^3\hat{k}}{(2\pi)^3} \tilde{A}_\mu(\hat{k}) e^{i\hat{k}\cdot(n+\hat{e}_\mu/2)} , \qquad (19.92a)$$

where

$$\hat{k}_4 := \hat{\omega}_\ell^+ = \frac{2\ell\pi}{\hat{\beta}} . \qquad (19.92b)$$

are the Matsubara frequencies (19.56) measured in units of the lattice spacing. For reasons which we discussed in chapter 14, the gauge potentials have been defined at the midpoints of the links connecting the lattice sites n and $n + \hat{e}_\mu$. Since the chemical potential only appears in the fermionic contribution to the action, we only need to consider the last three terms in (19.90).

Consider first the contribution of $\mathcal{O}(e^0)$. Making use of the completeness relation (18.104), it can be written as follows in frequency-momentum space,

$$[S_{ferm}^{(\beta,\mu)}]_{\mathcal{O}(e^0)} = \frac{1}{\hat{\beta}} \sum_{\ell=-\frac{\hat{\beta}}{2}}^{\frac{\hat{\beta}}{2}-1} \int_{-\pi}^{\pi} \frac{d^3\hat{p}}{(2\pi)^3} \bar{\tilde{\psi}}(\vec{\hat{p}}, \hat{\omega}_\ell^-) K^{(\beta,\mu)}(\vec{\hat{p}}, \hat{\omega}_\ell^-) \tilde{\psi}(\vec{\hat{p}}, \hat{\omega}_\ell^-) , \qquad (19.93a)$$

where $\hat{\omega}_\ell^-$ are the Matsubara frequencies for fermions (18.69) with $N = \hat{\beta}$, i.e.,

$$\hat{\omega}_\ell^- = \frac{(2\ell + 1)\pi}{\hat{\beta}} , \qquad (19.93b)$$

here denoted with the superscript "-", and where

$$K^{(\beta,\mu)}(\vec{\hat{p}}, \hat{\omega}_\ell^-) = i\gamma_4 \sin(\hat{\omega}_\ell^- + i\hat{\mu}) + i\gamma_j \sin\hat{p}_j + \hat{\mathcal{M}}(\vec{\hat{p}}, \hat{\omega}_\ell^- + i\hat{\mu}) , \qquad (19.93c)$$

$$\hat{\mathcal{M}}(\vec{\hat{p}}, \hat{\omega}_\ell^- + i\hat{\mu}) = \hat{M}(\vec{\hat{p}}) + 2r\sin^2(\frac{\hat{\omega}_\ell^- + i\hat{\mu}}{2}) \,, \tag{19.93d}$$

$$\hat{M}(\vec{\hat{p}}) = \hat{m} + 2r\sum_j \sin^2\frac{\hat{p}_j}{2} \,. \tag{19.93e}$$

A summation over repeated indices is always understood. The fermion propagator in frequency-momentum space is given by the inverse of the matrix $K^{(\beta,\mu)}(\vec{\hat{p}}, \hat{\omega}_\ell^-)$, i.e.,

$$\tilde{S}_F^{(\beta,\mu)}(\vec{\hat{p}}, \hat{\omega}_\ell^-) = \frac{-i\gamma_4\sin(\hat{\omega}_\ell^- + i\hat{\mu}) - i\gamma_j\sin\hat{p}_j + \hat{\mathcal{M}}(\vec{\hat{p}}, \hat{\omega}_\ell^- + i\hat{\mu})}{\sin^2(\hat{\omega}_\ell^- + i\hat{\mu}) + \sum_j \sin^2\hat{p}_j + \hat{\mathcal{M}}^2(\vec{\hat{p}}, \hat{\omega}_\ell^- + i\hat{\mu})} \,. \tag{19.94}$$

From here we see that a finite temperature and chemical potential is introduced into the $T = \hat{\mu} = 0$ propagator, discussed in chapter 4, according to the substitution rule $p_4 \to \hat{\omega}_\ell^- + i\hat{\mu}$.

Consider next the contribution of $\mathcal{O}(e^N)$. It involves N gauge potentials coupled to the fermion fields. By Fourier expanding the fields as before, one readily verifies that the contribution involving the fourth component of the gauge potential is given by*

$$[S_{ferm}^{(\beta,\mu)}]_{\mathcal{O}(e^N)}^{A_4} = \frac{(ie)^N}{N!}\sum_{n_4=1}^{\hat{\beta}}\sum_{\vec{n}}\int_{(-)} d^4\hat{p}' \int_{(-)} d^4\hat{p} \int_{(+)} \prod_{i=1}^{N} d^4\hat{k}_i \; \Lambda_{\alpha\beta}^{(\beta,\mu)}(\hat{p}_4', \hat{p}_4, \hat{K}_4) \,,$$
$$\times \, \bar{\tilde{\psi}}_\alpha(\hat{p}')\tilde{\psi}_\beta(\hat{p})\tilde{A}_4(\hat{k}_1)\cdots\tilde{A}_4(\hat{k}_N)e^{i(\hat{p}-\hat{p}'+\hat{K})\cdot n} \,, \tag{19.95a}$$

where

$$\Lambda_{\alpha\beta}^{(\beta,\mu)}(\hat{p}_4', \hat{p}_4, \hat{K}_4) = \frac{1}{2}\left[e^{i(\hat{p}_4 + \frac{\hat{K}_4}{2} + i\hat{\mu})} - (-1)^N e^{-i(\hat{p}_4' - \frac{\hat{K}_4}{2} + i\hat{\mu})}\right](\gamma_4)_{\alpha\beta}$$
$$- \frac{r}{2}\left[e^{i(\hat{p}_4 + \frac{\hat{K}_4}{2} + i\hat{\mu})} + (-1)^N e^{-i(\hat{p}_4' - \frac{\hat{K}_4}{2} + i\hat{\mu})}\right]\delta_{\alpha\beta} \,, \tag{19.95b}$$

and where we have used the compact notation

$$\int_{(\pm)} d^4\hat{q} f(\vec{q}, q_4; \cdots) = \sum_{\ell_q = \frac{\hat{\beta}}{2}}^{\frac{\hat{\beta}}{2}-1} \frac{1}{\hat{\beta}} \int_{-\pi}^{\pi} \frac{d^3\hat{q}}{(2\pi)^3} f(\vec{\hat{q}}, \hat{\omega}_{\ell_q}^{(\pm)}; \cdots) \,.) \,, \tag{19.95c}$$

---

* Vertices involving the spatial components of the gauge potentials are clearly not affected by the chemical potential

which is the lattice analog of (19.61b). The spatial and temporal components of the four-component vector $\hat{K}$ are given by the sum of the N photon momenta, and Matsubara frequencies, respectively. Carrying out the summation over the lattice sites, by making use of (18.104), we can write (19.95) as follows

$$[S^{(\beta,\mu)}_{ferm}]^{A_4}_{\mathcal{O}(e^N)} = \frac{1}{N!}\int_{(-)} d^4\hat{p}' \int_{(-)} d^4\hat{p} \int_{(+)} \prod_{i=1}^{N} d^4\hat{k}_i \, \Gamma^{(\beta,\mu)}_{\alpha\beta}(\hat{p}',\hat{p},\hat{K})$$
$$\bar{\tilde{\psi}}_\alpha(\hat{p}')\tilde{\psi}_\beta(\hat{p})\tilde{A}_4(\hat{k}_1)\cdots\tilde{A}_4(\hat{k}_N) \, ,$$

where

$$\Gamma^{(\beta,\mu)}_{\alpha\beta}(\hat{p}',\hat{p},\hat{K}) \equiv (ie)^N\hat{\beta}(2\pi)^3\delta(\vec{\hat{p}}-\vec{\hat{p}}'+\vec{\hat{K}})\delta_{\hat{\omega}^-_{\ell_{p'}}-\hat{\omega}^-_{\ell_p}-\hat{\omega}^+_{\ell_K},0}V^{(\beta,\mu)}_{\alpha\beta}(\frac{\hat{\omega}^-_{\ell_{p'}}+\hat{\omega}^-_{\ell_p}}{2}+i\hat{\mu}) \, .$$

and

$$V^{(\beta,\mu)}_{\alpha\beta}(q) = (\gamma_4)_{\alpha\beta}[(\bar{\xi}_N\cos q + i\xi_N\sin q) - r\delta_{\alpha\beta}(\xi_N\cos q + i\bar{\xi}_N\sin q)] \, ,$$

$$\xi_N = \frac{1}{2}[1+(-1)^N] \quad ; \; \bar{\xi}_N = \frac{1}{2}[1-(-1)^N] \, .$$

We therefore see that for finite lattice spacing the vertices are μ-dependent, and furthermore, that the chemical potential always appears in the combination $\hat{\omega}^-_{\ell_p} + i\hat{\mu}$, and $\hat{\omega}^-_{\ell_{p'}} + i\hat{\mu}$, as was the case in the continuum formulation. We emphasize that the simplicity of this result is a consequence of the fact that the chemical potential was introduced in the form $e^{\pm\hat{\mu}}$ into those terms of the fermionic action at $\mu = 0$, coupling neighbouring lattice sites along the time direction.

In our discussion we have chosen to work with a manifestly dimensionless action, where the lattice spacing does not appear, since this is the form of the action used in Monte Carlo simulations. The Feynman rules given in chapters 14 and 15 involve explicitely the lattice spacing. From the above analysis it is evident that the corresponding Feynman rules at finite temperature and chemical potential are obtained by the same prescriptions given in section 4, except that integrals over three-momenta, and sums of Matsubara frequencies are restricted to the first Brillouin zone. Since this result is a direct consequence of the periodic boundary conditions satisfied by the link variables and fermion fields, and the fact that the chemical potential was introduced in exponential form, it also applies to lattice QCD.

19.9 Particle-Antiparticle Spectrum of the Fermion Propagator at $T \neq 0$ and $\mu \neq 0$. Naive vs. Wilson Fermions

In section 5 we have studied in detail the fermion two-point correlation function at finite temperature and chemical potential in the continuum. We now carry out a similar analysis on the lattice for *naive* as well as *Wilson* fermions. In the case of Wilson fermions we of course expect to recover the results of section 5 in the limit of vanishing lattice spacing. For naive fermions, on the other hand, we will be confronted with the fermion doubling problem. In chapter 18 we had demonstrated in a simple model that the "doublers" modify the correct partition function in a quite non trivial way. Their unphysical nature is exposed most clearly by looking at the correlation function [Rothe (1995)], as we now show.

i) naive fermions

For naive fermions, (i.e., vanishing Wilson parameter r) the two-point fermion correlation function at finite temperature and chemical potential is given by (19.94) with $\hat{\mathcal{M}}$ replaced by $\hat{m}$, i.e.,

$$\hat{S}_F^{(\beta,\mu)}(\vec{\hat{p}},\hat{\omega}_\ell^-) = \frac{-i\gamma_4 \sin(\hat{\omega}_\ell^- + i\hat{\mu}) - i\gamma_j \sin\hat{p}_j + \hat{m}}{\sin^2(\hat{\omega}_\ell^- + i\hat{\mu}) + \hat{E}^2(\vec{\hat{p}})} , \tag{19.96a}$$

where

$$\hat{E}(\vec{\hat{p}}) = \sqrt{\sum_j \sin^2\hat{p}_j + \hat{m}^2} \tag{19.96b}$$

is the energy measured in lattice units. The (dimensionless) lattice analog of (19.66a) thus reads

$$\hat{S}_F^{(\beta,\mu)}(\vec{n},n_4) = \int_{-\pi}^{\pi} \frac{d^3\hat{p}}{(2\pi)^3} \tilde{S}_F^{(\beta,\mu)}(\vec{\hat{p}},n_4) e^{i\vec{\hat{p}}\cdot\vec{n}} , \tag{19.97a}$$

where

$$\tilde{S}_F^{(\beta,\mu)}(\vec{\hat{p}},n_4) = \frac{1}{\hat{\beta}} \sum_{\ell=-\frac{\hat{\beta}}{2}}^{\frac{\hat{\beta}}{2}-1} \hat{S}_F^{(\beta,\mu)}(\vec{\hat{p}},\hat{\omega}_\ell^-) e^{i\hat{\omega}_\ell^- n_4} . \tag{19.97b}$$

Note that the momentum integrals and sum over Matsubara frequencies are now restricted to the first Brillouin zone. The expression (19.97b) can be decomposed as follows

$$\tilde{S}_F^{(\beta,\mu)}(\vec{\hat{p}},n_4) = \hat{\Gamma}_+(\vec{\hat{p}})\mathcal{D}^{(+)}_{(\beta,\mu)}(\vec{\hat{p}},n_4) + \hat{\Gamma}_-(\vec{\hat{p}})\mathcal{D}^{(-)}_{(\beta,\mu)}(\vec{\hat{p}},n_4) , \tag{19.98a}$$

where

$$\hat{\Gamma}_{\pm}(\vec{\hat{p}}) = \frac{1}{2\hat{E}(\vec{\hat{p}})}[\pm\gamma_4\hat{E}(\vec{\hat{p}}) - i\gamma_j \sin\hat{p}_j + \hat{m}] \tag{19.98b}$$

is the lattice analog of (19.67b), and

$$\mathcal{D}^{(\pm)}_{(\beta,\mu)}(\vec{\hat{p}}, n_4) = \pm\frac{1}{\hat{\beta}} \sum_{\ell=-\frac{\beta}{2}}^{\frac{\beta}{2}-1} \frac{e^{i\hat{\omega}_\ell^- n_4}}{i\sin(\hat{\omega}_\ell^- + i\hat{\mu}) \pm \hat{E}(\vec{\hat{p}})} . \tag{19.98c}$$

The finite frequency sum can be evaluated by making use of the summation formula (18.85). As an example consider $\mathcal{D}^{(+)}_{(\beta,\mu)}(\vec{\hat{p}}, n_4)$. Defining $z = e^{i\omega}$, the function $\tilde{F}(z)$ in (18.85) now has the form

$$\tilde{F}(z) = \frac{2e^{\hat{\mu}} z^{n_4+1}}{z^2 + 2Ee^{\hat{\mu}}z - e^{2\hat{\mu}}} .$$

Hence the frequency sum in (19.98c) is given by

$$\begin{aligned} \mathcal{D}^{(+)}_{(\beta,\mu)}(\vec{\hat{p}}, n_4) = &-\frac{e^{\hat{\mu}}}{i\pi}\int_{|z|=1+\epsilon} dz \frac{z^{n_4}}{(z^{\hat{\beta}}+1)(z-z_+)(z-z_-)} \\ &+ \frac{e^{\hat{\mu}}}{i\pi}\int_{|z|=1-\epsilon} dz \frac{z^{n_4}}{(z^{\hat{\beta}}+1)(z-z_+)(z-z_-)} , \end{aligned} \tag{19.99a}$$

where

$$z_+ = e^{-(\hat{\mathcal{E}}-\hat{\mu})} ; \quad z_- = -e^{(\hat{\mathcal{E}}+\hat{\mu})} ,$$

$$\hat{\mathcal{E}} = ar\sinh \hat{E}(\vec{\hat{p}}) . \tag{19.99b}$$

The two circles $|z| = 1 \pm \epsilon$ enclose the zeros of $z^{\hat{\beta}} + 1$, located at the Matsubara frequencies. The integrations are carried out in the counterclockwise sense. For $|n_4| \leq \hat{\beta}$ (recall that euclidean time has been compactified), we can distort the contour in the first integral to infinity, taking into account the poles located outside of the unit circle. Hence the rhs of (19.99a) is twice the sum of the residues of the integrand at the poles lying inside and outside of the unit circle, multiplied by $e^{\hat{\mu}}$. For $n_4 < 0$ these poles also include a pole of order $|n_4|$. One then finds that $\mathcal{D}^{(+)}_{(\beta,\mu)}(\vec{\hat{p}}, n_4)$ can be written in the form

$$\mathcal{D}^{(+)}_{(\beta,\mu)}(\vec{\hat{p}}, n_4) = \hat{\Delta}^{(+)}_{(\beta,\mu)}(\vec{\hat{p}}, n_4) + (-1)^{n_4}\hat{\Delta}^{(-)}_{(\beta,\mu)}(\vec{\hat{p}}, n_4) , \tag{19.100a}$$

where

$$\hat{\Delta}^{(+)}_{(\beta,\mu)}(\hat{p}, \vec{n}_4) = [\theta(n_4) - \hat{\eta}_{FD}(\hat{\mathcal{E}}, \hat{\mu})]\frac{e^{-(\hat{\mathcal{E}}-\hat{\mu})n_4}}{\sqrt{1+\hat{E}^2}} , \tag{19.100b}$$

$$\hat{\Delta}^{(-)}_{(\beta,\mu)}(\vec{p},n_4) = [\theta(-n_4) - \bar{\hat{\eta}}_{FD}(\hat{\mathcal{E}},\hat{\mu})]\frac{e^{(\hat{\mathcal{E}}+\hat{\mu})n_4}}{\sqrt{1+\hat{E}^2}}\,, \qquad (19.100c)$$

and

$$\hat{\eta}_{FD}(\hat{\mathcal{E}},\hat{\mu}) = \frac{1}{e^{\hat{\beta}(\hat{\mathcal{E}}-\hat{\mu})}+1}\,,$$
$$\bar{\hat{\eta}}_{FD}(\hat{\mathcal{E}},\hat{\mu}) = \frac{1}{e^{\hat{\beta}(\hat{\mathcal{E}}+\hat{\mu})}+1}\,, \qquad (19.100d)$$

are the lattice analogs of the Fermi-Dirac distribution functions (19.67d). For $n_4 = 0$, $\theta(0) := \frac{1}{2}$. In a similar way one obtains

$$\mathcal{D}^{(-)}_{(\beta,\mu)}(\vec{p},n_4) = \hat{\Delta}^{(-)}_{(\beta,\mu)}(\vec{p},n_4) + (-1)^{n_4}\hat{\Delta}^{(+)}_{(\beta,\mu)}(\vec{p},n_4)\,. \qquad (19.101)$$

Introducing the dimensioned variables $\tau = an_4$, $p = \frac{\hat{p}}{a}$, $\beta = a\hat{\beta}$ and $\mu = \frac{\hat{\mu}}{a}$, one finds that (19.98b) and (19.100b-d) approach the continuum expressions (19.67b) and (19.67c,d). We therefore see that the terms proportional to $(-1)^{n_4}$ impede the propagator from having the correct continuum limit. These terms are the contributions of the doublers. They arise from the poles of the integrand in (19.99a) located on the negative z-axis, and are pure lattice artefacts having no continuum analog. The factor $(-1)^{n_4}$, which changes sign as one proceeds from one lattice site to the next in the euclidean time direction, is typical for the doubler contributions, as we have already seen in chapter 4. From (19.100b,c) we see that

$$\Delta^{(\pm)}_{(\beta,\mu)}(\vec{p},n_4) = \Delta^{(\mp)}_{(\beta,-\mu)}(\vec{p},-n_4)\,.$$

Making use of this relation, $\mathcal{D}^{(\pm)}_{(\beta,\mu)}(\vec{p},n_4)$ can also be written in the form

$$\mathcal{D}^{(\pm)}_{(\beta,\mu)}(\vec{p},n_4) = \Delta^{(\pm)}_{(\beta,\mu)}(\vec{p},n_4) + (-1)^{n_4}\Delta^{(\pm)}_{(\beta,-\mu)}(\vec{p},-n_4)\,,$$

which shows that, apart from the (non-trivial!) factor $(-1)^{n_4}$, the admixture of the doublers involves a reversal of the sign of the chemical potential as well as of euclidean time. It is thus not surprising that when one computes the logarithm of the partition function for a Dirac gas of naive fermions, one finds that it resembles that of a gas of particles and antiparticles in both, the positive as well as negative energy sectors [Bender et.al. (1993)].

ii) Wilson Fermions

Let us now see how the above results are modified for Wilson fermions. In the following we will choose r = 1 for the Wilson parameter. The propagator in frequency-momentum space is given by (19.94) with $r = 1$ in (19.93d,e). Hence (19.97b) is now replaced by

$$\tilde{S}_F^{(\beta,\mu)}(\vec{\hat{p}}, n_4) = \frac{1}{\hat{\beta}} \sum_{\ell=-\frac{\hat{\beta}}{2}}^{\frac{\hat{\beta}}{2}-1} F(\vec{\hat{p}}, \hat{\omega}_\ell^-; n_4) \,, \tag{19.102a}$$

where

$$F(\vec{\hat{p}}, \hat{\omega}_\ell^-; n_4) = \hat{S}^{(\beta,\mu)}(\vec{\hat{p}}, \hat{\omega}_\ell^-) e^{i\hat{\omega}_\ell^- n_4} \,, \tag{19.102b}$$

$$\hat{S}^{(\beta,\mu)}(\vec{\hat{p}}, \hat{\omega}_\ell^-) = \frac{-i\gamma_4 \sin(\hat{\omega}_\ell^- + i\hat{\mu}) - i\gamma_j \sin\hat{p}_j + \tilde{\mathcal{M}}(\vec{\hat{p}}, \hat{\omega}_\ell^- + i\hat{\mu})}{\sin^2(\hat{\omega}_\ell + i\hat{\mu}) + \sum_j \sin^2\hat{p}_j + \tilde{\mathcal{M}}^2(\vec{\hat{p}}, \hat{\omega}_\ell^- + i\hat{\mu})} \,, \tag{19.102c}$$

and

$$\begin{aligned} \tilde{\mathcal{M}}(\vec{\hat{p}}, \hat{\omega}_\ell^- + i\hat{\mu}) &= \hat{M}(\vec{\hat{p}}) + 2\sin^2\left(\frac{\omega_\ell^- + i\hat{\mu}}{2}\right) \,, \\ \hat{M}(\vec{\hat{p}}) &= \hat{m} + 2\sum_j \sin^2\frac{\hat{p}_j}{2} \,. \end{aligned} \tag{19.102d}$$

The frequency sum in (19.102a) can again be performed by making use of (18.85). Introducing the variable $z = e^{i\hat{\omega}}$, the function replacing $\tilde{F}(z)$ takes the form

$$\tilde{F}(z, \vec{\hat{p}}, n_4) = \frac{\frac{1}{2}\gamma_4(z^2 e^{-2\hat{\mu}} - 1) + \frac{1}{2}(ze^{-\hat{\mu}} - 1)^2 + (i\gamma_j \sin\hat{p}_j - \hat{M}(\vec{\hat{p}}))ze^{-\hat{\mu}}}{e^{-2\hat{\mu}}[1 + \hat{M}(\vec{\hat{p}})](z - z_+)(z - z_-)} z^{n_4} \,,$$

where

$$z_\pm = e^{\pm\tilde{\mathcal{E}} + \hat{\mu}} \,,$$

$$\tilde{\mathcal{E}} = \ln[K + \sqrt{K^2 - 1}] = ar\cosh K \tag{19.103a}$$

and

$$K(\vec{\hat{p}}) = 1 + \frac{\bar{E}^2(\vec{\hat{p}})}{2[1 + \hat{M}(\vec{\hat{p}})]} \,, \quad \bar{E}(\vec{\hat{p}}) = \sqrt{\sum_j \sin^2\hat{p}_j + \hat{M}^2(\vec{\hat{p}})} \,. \tag{19.103b}$$

Proceeding as before one then finds after some algebra that (19.102a) can be written as follows

$$\tilde{S}_F^{(\beta,\mu)}(\vec{\hat{p}}, n_4) = \tilde{\Gamma}_+(\vec{\hat{p}})\tilde{\mathcal{D}}^{(+)}_{(\beta,\mu)}(\vec{\hat{p}}, n_4) + \tilde{\Gamma}_-(\vec{\hat{p}})\tilde{\mathcal{D}}^{(-)}_{(\beta,\mu)}(\vec{\hat{p}}, n_4) \,, \tag{19.104a}$$

where

$$\tilde{\Gamma}_\pm(\vec{\hat{p}}) = \frac{1}{2\sinh\tilde{\mathcal{E}}}[\pm\gamma_4 \sinh\tilde{\mathcal{E}} - i\gamma_j \sin\hat{p}_j + \hat{m}(\vec{\hat{p}})] \, ,$$
$$\hat{m}(\vec{\hat{p}}) = \hat{m} + 2\sum_j \sin^2\frac{\hat{p}_j}{2} - 2\sinh^2\frac{\tilde{\mathcal{E}}}{2} \, , \tag{19.104b}$$

$$\tilde{\mathcal{D}}^{(+)}_{\beta,\mu}(\vec{\hat{p}}, n_4) = [\theta(n_4) - \hat{\eta}_{FD}(\tilde{\mathcal{E}}, \hat{\mu})]\frac{e^{-(\tilde{\mathcal{E}}-\hat{\mu})n_4}}{1 + \hat{M}(\vec{\hat{p}})}$$
$$\tilde{\mathcal{D}}^{(-)}_{\beta,\mu}(\vec{\hat{p}}, n_4) = [\theta(-n_4) - \bar{\hat{\eta}}_{FD}(\tilde{\mathcal{E}}, \hat{\mu})]\frac{e^{(\tilde{\mathcal{E}}+\hat{\mu})n_4}}{1 + \hat{M}(\vec{\hat{p}})} \, , \tag{19.104c}$$

and

$$\hat{\eta}_{FD}(\tilde{\mathcal{E}}, \hat{\mu}) = \frac{1}{e^{\hat{\beta}(\tilde{\mathcal{E}}-\hat{\mu})} + 1} \, ,$$
$$\bar{\hat{\eta}}_{FD}(\tilde{\mathcal{E}}, \hat{\mu}) = \frac{1}{e^{\hat{\beta}(\tilde{\mathcal{E}}+\hat{\mu})} + 1} \, . \tag{19.104d}$$

are the lattice Fermi-Dirac distribution function for Wilson fermions. Notice that in the case of Wilson fermions no terms proportional to $(-1)^{n_4}$ appear! In the continuum limit we have that $\hat{M}(\vec{\hat{p}}) \to \hat{m}$, $\sqrt{K^2 - 1} \to \hat{E}(\vec{\hat{p}})$, $\tilde{\mathcal{E}} \to \hat{E}(\vec{\hat{p}})$, and $\sin\hat{p}_j \to \hat{p}_j$ for finite physical momenta. For Wilson fermions these are in fact the only relevant momenta contributing to the integral (19.97a). This can be readily seen. In the continuum limit $\hat{\beta}$ goes to infinity. Hence the integral only receives contributions from momenta for which $\tilde{\mathcal{E}}$ goes to zero in this limit. Because of the momentum dependent Wilson mass $\hat{M}(\vec{\hat{p}})$, only (dimensionless) momenta $\vec{\hat{p}}$ in the immediate neighbourhood of $\vec{\hat{p}} = 0$ contribute to the integral. Hence in the continuum limit the expressions (19.104a-d) approach (19.67a-d), with the usual correspondence holding between particles (antiparticles) and forward (backward) propagation in time, translated into the euclidean language.

19.10 The Electric Screening Mass for Wilson Fermions in Lattice QED to One-Loop Order

In section 6 we had computed the electric screening mass in continuum QED to one-loop order. We now repeat this calculation within the framework of lattice regularized perturbation theory. The computation will be carried out for Wilson fermions, where we expect to recover the result of section 6 in the continuum limit. The calculations in this section are based on work carried out together with R. Pietig (Pietig, 1994). A similar computation for naive fermions, carried out by this author can be found in Appendix G.

The electric screening mass has been defined in (19.71) as the infrared limit of the 44-component of the vacuum polarization tensor evaluated for vanishing photon frequency. In lattice perturbation theory the vacuum polarization tensor in one-loop order receives contributions from the two diagrams shown in fig. (19-6). Diagram (b) has no continuum analog but is required by gauge invariance, which is implemented on the lattice for arbitrary lattice spacing. In the following we shall choose $r = 1$ for the Wilson parameter. This leads to a drastic simplification of the computations. For finite lattice spacing the result for the screening mass will of course depend on r, while in the continuum limit it is expected to be independent of this parameter. The following computation is carried out in the dimensionless formulation, i.e., all quantities are measured in lattice units.

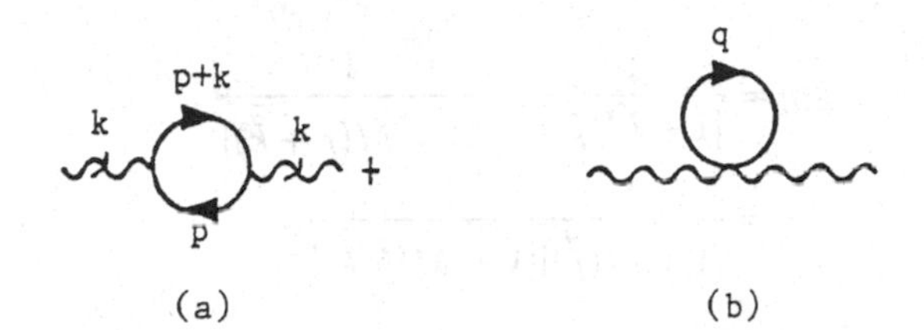

Fig.19-6 Diagrams contributing to the vacuum polarization tensor.

For $r = 1$ the fermion propagator in frequency-momentum space has the form (19.102c,d). For vanishing temperature and chemical potential the vertices are those given in chapter 14 with the lattice spacing a, and the Wilson parameter r replaced by unity. The transition to finite temperature and finite chemical potential is effected in the by now familiar way: the fourth component of the photon momentum is replaced by $\hat{\omega}_\ell^+$, defined in (19.92b), and the fourth component of the fermion momenta by $\hat{\omega}_\ell^- + i\hat{\mu}$, with $\hat{\omega}^-$ defined in (19.93b). Finally, integrals over the fourth component of momenta at $T = 0$ are replaced by *finite* frequency sums.

Consider first the contribution of the diagram (a) shown in fig. (19-6). For vanishing photon frequency it contributes as follows to $\hat{\Pi}_{44}^{(\beta,\mu)}(0, \vec{k})$,

$$\hat{\Pi}_{44}^{(\beta,\mu)}(0, \vec{\hat{k}})_{(a)} = (-ie)^2 \sum_{\ell=\frac{\hat{\beta}}{2}}^{\frac{\hat{\beta}}{2}-1} \frac{1}{\hat{\beta}} \int_{-\pi}^{\pi} \frac{d^3\hat{p}}{(2\pi)^3} Tr\Big\{[\gamma_4 \cos(\hat{\omega}_\ell^- + i\hat{\mu}) - i\sin(\hat{\omega}_\ell^- + i\hat{\mu})]$$
$$\times \hat{S}_F^{(\beta,\mu)}(\vec{\hat{p}}, \hat{\omega}_\ell^-)[\gamma_4 \cos(\hat{\omega}_\ell^- + i\hat{\mu}) - i\sin(\hat{\omega}_\ell^- + i\hat{\mu})]\hat{S}_F^{(\beta,\mu)}(\hat{\omega}_\ell^-, \vec{\hat{p}} + \vec{\hat{k}})\Big\} \quad (19.105)$$

where $\hat{S}_F^{(\beta,\mu)}(\vec{\hat{q}}, \hat{\omega}_\ell^-)$ is given by (19.102c,d). Upon carrying out the trace in Dirac space, making use of (19.74), and expressing the trigonometric functions in terms of exponentials, one finds after some algebra that the above expression can be written in the form

$$\hat{\Pi}_{44}^{(\beta,\mu)}(0,\vec{\hat{k}})_{(a)} = -e^2 \frac{1}{\hat{\beta}} \sum_{\ell=-\frac{\hat{\beta}}{2}}^{\frac{\hat{\beta}}{2}-1} \int_{-\pi}^{\pi} \frac{d^3\hat{p}}{(2\pi)^3} f^{(a)}(e^{i(\hat{\omega}_\ell^- + i\hat{\mu})}; \vec{\hat{p}}, \vec{\hat{k}}), \tag{19.106a}$$

where

$$f^{(a)}(z;\vec{\hat{p}},\vec{\hat{k}}) = \frac{2(z^4+1) - 2\eta(z^3+z) + 4\xi\mathcal{G}z^2}{\prod_{i=1}^4 (z-z_i)} . \tag{19.106b}$$

and

$$\begin{aligned} \eta &= \frac{1}{[1+\hat{M}(\hat{p})]} + \frac{1}{[1+\hat{M}(\vec{\hat{p}}+\vec{\hat{k}})]} , \\ \xi &= \frac{1}{[1+\hat{M}(\vec{\hat{p}})][1+\hat{M}(\vec{\hat{p}},\vec{\hat{k}})]} , \end{aligned} \tag{19.106c}$$

$$\mathcal{G} = 1 + \sum_j \sin\hat{p}_j \sin(\hat{p}+\hat{k})_j . \tag{19.106d}$$

The position of the poles of $f^{(a)}(z;\vec{\hat{p}},\vec{\hat{k}})$ are given by

$$\begin{aligned} z_1 &= e^{\phi}; \quad z_2 = e^{-\phi} \\ z_3 &= e^{\psi}; \quad z_4 = e^{-\psi} , \end{aligned} \tag{19.106e}$$

with

$$\begin{aligned} \phi &= \tilde{\mathcal{E}}(\vec{\hat{p}}) \\ \psi &= \tilde{\mathcal{E}}(\vec{\hat{p}}+\vec{\hat{k}}) , \end{aligned} \tag{19.106f}$$

with $\tilde{\mathcal{E}}(\vec{\hat{q}})$ defined in (19.103a,b). Note that

$$\phi \underset{\vec{\hat{p}}\to-\vec{\hat{p}}-\vec{\hat{k}}}{\longleftrightarrow} \psi , \tag{19.107}$$

while η, ξ and $\mathcal{G}$ are invariant under the transformation $\vec{p} \to -\vec{p} - \vec{k}$. This will be important further below.

The frequency sum in (19.106a) can in principle be again performed with the help of the summation formula (18.85). A more convenient formula can be derived however, which is based on the observation that $\hat{\omega}_\ell^-$ always appears in the combination $\hat{\omega}_\ell^- + i\hat{\mu}$. The following formula is derived in Appendix D.

$$\frac{1}{\hat{\beta}} \sum_{l=-\frac{\hat{\beta}}{2}}^{\frac{\hat{\beta}}{2}-1} g(e^{i(\hat{\omega}_l^- + i\hat{\mu})}) = \sum_i Res_{\bar{z}_i} \left(\frac{g(z)}{z} \right) \frac{1}{e^{\hat{\beta}\hat{\mu}} \bar{z}_i^{\hat{\beta}} + 1} , \tag{19.108}$$

where $Res_{\bar{z}_i} \frac{g(z)}{z}$ are the residues at the poles of $g(z)/z$, whose position we have denoted by $\bar{z}_i$. We now apply this formula to calculate the (finite) frequency sum in (19.106a). The function $g(z)$ in (19.108) is now replaced (19.106b). It has simple poles in z located at (19.106e) and a pole at $z = 0$. The evaluation of the residues of $f^{(a)}(z; \vec{\hat{p}}, \vec{\hat{k}})/z$ is straightforward. One finds

$$Res_{z_1} \left(\frac{f^{(a)}(z)}{z} \right) = -Res_{z_2} \left(\frac{f^{(a)}(z)}{z} \right) = h(\phi, \psi, \eta, \xi, \mathcal{G}) ,$$

$$Res_{z_3} \left(\frac{f^{(a)}(z)}{z} \right) = -Res_{z_4} \left(\frac{f^{(a)}(z)}{z} \right) = h(\psi, \phi, \eta, \xi, \mathcal{G}) ,$$

$$Res_0 \left(\frac{f^{(a)}(z)}{z} \right) = 2 ,$$

where

$$h(\phi, \psi, \eta, \xi, \mathcal{G}) = \frac{\cosh 2\phi - \eta \cosh \phi + \xi \mathcal{G}}{\sinh \phi (\cosh \phi - \cosh \psi)} . \tag{19.109}$$

For simplicity we have suppressed the dependence of $f^{(a)}(z; \vec{\hat{p}}, \vec{\hat{k}})$ on $\vec{\hat{p}}$ and $\vec{\hat{k}}$. Application of (19.108) then yields

$$\begin{aligned} \frac{1}{\hat{\beta}} \sum_{l=-\frac{\hat{\beta}}{2}}^{\frac{\hat{\beta}}{2}-1} f^{(a)}(e^{i(\hat{\omega}_l^- + i\hat{\mu})}; \vec{\hat{p}}, \vec{\hat{k}}) &= 2 + h(\phi, \psi, \eta, \xi, \mathcal{G}) \left[\frac{1}{e^{\hat{\beta}(\phi + \hat{\mu})} + 1} - \frac{1}{e^{-\hat{\beta}(\phi - \hat{\mu})} + 1} \right] \\ &+ h(\psi, \phi, \eta, \xi, \mathcal{G}) \left[\frac{1}{e^{\hat{\beta}(\psi + \hat{\mu})} + 1} - \frac{1}{e^{-\hat{\beta}(\psi - \hat{\mu})} + 1} \right] . \end{aligned} \tag{19.110}$$

To obtain $\hat{\Pi}_{44}^{(\beta,\mu)}(0; \hat{k})_{(a)}$, we must integrate this expression over $\vec{\hat{p}}$, with $\vec{\hat{p}} \in [-\pi, \pi]$. Noting that η, ξ and $\mathcal{G}$ are invariant under the transformation $\vec{\hat{p}} \to -\vec{\hat{p}} - \vec{\hat{k}}$, and making use of (19.107), as well as of the fact that the integrand in (19.106a) is a periodic function in $\hat{p}_i$ and $\hat{k}_i$, we can combine the last two contributions on the r.h.s. of (19.110) and obtain

$$\begin{aligned} \hat{\Pi}_{44}^{(a)}(0, \vec{\hat{k}}) &= 2e^2 \int_{-\pi}^{\pi} \frac{d^3\hat{p}}{(2\pi)^3} [h(\phi, \psi, \eta, \xi, \mathcal{G}) - 1] \\ &- 2e^2 \int_{-\pi}^{\pi} \frac{d^3\hat{p}}{(2\pi)^3} h(\phi, \psi, \eta, \xi, \mathcal{G}) [\hat{\eta}_{FD}(\phi) + \bar{\hat{\eta}}_{FD}(\phi)]. \end{aligned} \tag{19.111a}$$

where

$$\hat{\eta}_{FD}(\phi) = \frac{1}{e^{\hat{\beta}(\phi-\hat{\mu})}+1},$$
$$\bar{\hat{\eta}}_{FD}(\phi) = \frac{1}{e^{\hat{\beta}(\phi+\hat{\mu})}+1}, \tag{19.111b}$$

are the lattice Fermi-Dirac distribution functions for particles and antiparticles. Note that with the definition of ϕ, given in (19.106f), they coincide with those in (19.104d).

We next compute the contribution to $\hat{\Pi}_{44}^{(\beta,\mu)}(0,\vec{\hat{k}})$ of the Feynman diagram (b) depicted in Fig. (19-6). This diagram has no analog in the continuum. It is given by

$$\hat{\Pi}_{44}^{(\beta,\mu)}(0,\vec{\hat{k}})_{(b)} = -\frac{2e^2}{2!}\frac{1}{\hat{\beta}}\sum_{\ell=-\frac{\hat{\beta}}{2}}^{\frac{\hat{\beta}}{2}-1}\int_{-\pi}^{\pi}\frac{d^3\hat{p}}{(2\pi)^3}Tr\Big\{[\cos(\hat{\omega}_\ell^- + i\hat{\mu}) - i\gamma_4\sin(\hat{\omega}_\ell^- + i\hat{\mu})] \times \hat{S}_F^{(\beta,\mu)}(\vec{\hat{p}},\hat{\omega}_\ell^-)\Big\}. \tag{19.112}$$

The factor 2 arises from the two possibilities to label the two external photon lines emanating from the vertex, and the factor $\frac{1}{2!}$ from the expansion of the link variables to second order in the gauge potentials. The first factor in the argument of the trace is the finite temperature, finite chemical potential version of the vertex given in chapter 14, measured in lattice units. Performing the trace and expressing the trigonometric functions in terms of exponentials, one finds that

$$\hat{\Pi}_{44}^{(\beta,\mu)}(0,\vec{\hat{k}})_{(b)} = -2e^2\frac{1}{\hat{\beta}}\sum_{\ell=\frac{\hat{\beta}}{2}}^{\frac{\hat{\beta}}{2}-1}\int_{-\pi}^{\pi}\frac{d^3\hat{p}}{(2\pi)^3}f^{(b)}\left(e^{i(\hat{\omega}_\ell^- + i\hat{\mu})},\vec{\hat{p}}\right) \tag{19.113a}$$

where

$$f^{(b)}(z;\vec{\hat{p}}) = -\frac{z^2-2\rho z+1}{(z-z_1)(z-z_2)}, \tag{19.113b}$$

$$\rho = \frac{1}{1+\hat{M}(\vec{\hat{p}})}, \tag{19.113c}$$

and where z_1 and z_2 have been defined in (19.106e). Making again use of the frequency summation formula (19.108), one verifies that

$$\hat{\Pi}_{44}^{(\beta,\mu)}(0,\vec{\hat{k}})_{(b)} = -2e^2\int_{-\pi}^{\pi}\frac{d^3\hat{p}}{(2\pi)^3}\left(\coth\phi - \frac{\rho}{\sinh\phi} - 1\right) + 2e^2\int_{-\pi}^{\pi}\frac{d^3\hat{p}}{(2\pi)^3}\left(\coth\phi - \frac{\rho}{\sinh\phi}\right)[\hat{\eta}_{FD}(\phi) + \bar{\hat{\eta}}_{FD}(\phi)] \tag{19.114}$$

Combining this expression with (19.111a) one therefore finds that

$$\hat{\Pi}_{44}^{(\beta,\mu)}(0,\hat{\vec{k}}) = \hat{\Pi}_{44}^{(vac)}(0,\hat{\vec{k}}) + 2e^2 \int_{-\pi}^{\pi} \frac{d^3\hat{p}}{(2\pi)^3} H(\phi,\psi,\rho,\eta,\xi,\mathcal{G})[\hat{\eta}_{FD}(\phi) + \hat{\bar{\eta}}_{FD}(\phi)] \tag{19.115a}$$

where

$$H(\phi,\psi,\rho,\eta,\xi,\mathcal{G}) = \coth\phi - \frac{\rho}{\sinh\phi} - h(\phi,\psi,\eta,\xi,\mathcal{G}) \, . \tag{19.115b}$$

and

$$\hat{\Pi}_{44}^{(vac)}(0,\hat{\vec{k}}) = -2e^2 \int_{-\pi}^{\pi} \frac{d^3\hat{p}}{(2\pi)^3} H(\phi,\psi,\rho,\eta,\xi,\mathcal{G}) \tag{19.115c}$$

is the $T = \mu = 0$ contribution. As we now show, $\hat{\Pi}_{44}^{(vac)}(0,\hat{\vec{k}})$ vanishes in the limit $\hat{\vec{k}} \to 0$, and hence does not contribute to the screening mass.

Consider the function $h(\phi,\psi,\eta,\xi,\mathcal{G})$ defined in (19.109). It is singular for $\hat{\vec{k}} \to 0$, since in this limit $\psi \to \phi$. The singularity is however integrable. This can be seen as follows. Since according to (19.107), and the statement following it

$$h(\phi,\psi,\eta,\xi,\mathcal{G}) \underset{\vec{\hat{p}} \to -\vec{\hat{p}} - \vec{\hat{k}}}{\longrightarrow} h(\psi,\phi,\eta,\xi,\mathcal{G}) \tag{19.116}$$

we can also write (19.115c) in the form

$$\hat{\Pi}_{44}^{(vac)}(0,\hat{\vec{k}}) = -2e^2 \int_{-\pi}^{\pi} \frac{d^3\hat{p}}{(2\pi)^3} \left(\coth\phi - \frac{\rho}{\sinh\phi}\right) + e^2 \int_{-\pi}^{\pi} \frac{d^3\hat{p}}{(2\pi)^3} \tilde{h}(\phi,\psi,\eta,\xi,\mathcal{G}) \, , \tag{19.117a}$$

where

$$\tilde{h}(\phi,\psi,\eta,\xi,\mathcal{G}) = h(\phi,\psi,\eta,\xi,\mathcal{G}) + h(\psi,\phi,\eta,\xi,\mathcal{G}) \, . \tag{19.117b}$$

Although each term in this last expression is singular for $\hat{\vec{k}} \to 0$ ($\psi \to \phi$), the sum possesses a finite limit. Thus setting $\psi = \phi + \epsilon$ and taking the limit $\hat{\vec{k}} \to 0$ ($\epsilon \to 0$), one verifies that

$$\lim_{\hat{\vec{k}} \to 0} \tilde{h}(\phi,\psi,\eta,\xi,\mathcal{G}) = \frac{1}{\sinh^2\phi}\{ -\coth\phi(\cosh 2\phi - 2\rho\cosh\phi + \rho^2\mathcal{G}_0) + 2(\sinh 2\phi - \rho\sinh\phi)\} \, ,$$

where $\mathcal{G}_0 = \mathcal{G}(\vec{\hat{p}},0)$, with $\mathcal{G}(\vec{\hat{p}},\hat{\vec{k}})$ defined in (19.106d). From the definition of ϕ given in (19.106f), with $\tilde{\mathcal{E}}$ defined in (19.103a,b), one finds that

$$\mathcal{G}_0 = \mathcal{G}(\vec{\hat{p}},0) = \frac{1}{\rho^2}(2\rho\cosh\phi - 1) \, .$$

Hence

$$\lim_{\vec{k}\to 0} \tilde{h}(\phi,\psi,\eta,\xi,\mathcal{G}) = 2(\coth\phi - \frac{\rho}{\sinh\phi}) . \qquad (19.118)$$

From (19.117a) we therefore conclude that

$$\lim_{\vec{k}\to 0} \hat{\Pi}_{44}^{(vac)}(0,\hat{\vec{k}}) = 0 .$$

This result is not unexpected, since for vanishing temperature and chemical potential it is well known in the continuum formulation, that Lorentz and gauge invariance protects the photon from acquiring a mass. The screening mass is therefore determined by the finite temperature (f.T.), finite chemical potential contribution, given by the integral in (19.115a). By making again use of the fact that $\phi \leftrightarrow \psi$, when $\vec{\hat{p}} \to -\vec{\hat{p}} - \vec{\hat{k}}$, while η, ξ and $\mathcal{G}$ remain invariant under this change of variables, we can write this contribution in the form

$$\begin{aligned}\hat{\Pi}_{44}^{(\beta,\mu)}(0,\hat{\vec{k}})_{f.T.} = 2e^2 \int_{-\pi}^{\pi} \frac{d^3\hat{p}}{(2\pi)^3}(\coth\phi - \frac{\rho}{\sinh\phi})[\hat{\eta}_{FD}(\phi) + \hat{\bar{\eta}}_{FD}(\phi)] \\ -e^2 \int_{-\pi}^{\pi} \frac{d^3\hat{p}}{(2\pi)^3}\Big\{h(\phi,\psi,\eta,\xi,\mathcal{G})[\eta_{FD}(\phi) + \hat{\bar{\eta}}_{FD}(\phi)] \\ +h(\psi,\phi,\eta,\xi,\mathcal{G})[\eta_{FD}(\psi) + \hat{\bar{\eta}}_{FD}(\psi)]\Big\} .\end{aligned} \qquad (19.119)$$

Consider the second integral. It can be rewritten as follows

$$\int_{-\pi}^{\pi} \frac{d^3\hat{p}}{(2\pi)^3}\{\tilde{h}(\phi,\psi,\eta,\xi,\mathcal{G})[\eta_{FD}(\psi) + \bar{\eta}_{FD}(\psi)] + h(\phi,\psi,\eta,\xi,\mathcal{G})\Delta\hat{\eta}_{FD}(\phi,\psi)\} ,$$

where $\tilde{h}$ has been defined in (19.117b), and

$$\Delta\hat{\eta}_{FD} = [\hat{\eta}_{FD}(\phi) - \hat{\eta}_{FD}(\psi)] + [\hat{\bar{\eta}}_{FD}(\phi) - \hat{\bar{\eta}}_{FD}(\psi)] .$$

According to (19.118), $\tilde{h}$ approaches a finite limit for $\vec{\hat{k}} \to 0$. Hence the contribution proportional to $\tilde{h}$ is cancelled by the first integral in (10.119). We therefore conclude that

$$\lim_{\vec{k}\to 0} \hat{\Pi}_{44}^{(\beta,\mu)}(0,\hat{\vec{k}}) = -e^2 \lim_{\vec{k}\to 0} \int_{-\pi}^{\pi} \frac{d^3\hat{p}}{(2\pi)^3} h(\phi,\psi,\eta,\xi,\mathcal{G})\Delta\hat{\eta}_{FD}(\phi,\psi) . \qquad (19.120)$$

We have now dropped the subscript "f.T.", since in this limit only (19.119) contributes to the screening mass. To calculate this limit we proceed as before and set $\psi = \phi + \epsilon$. One then finds that

$$\Delta\hat{\eta}_{FD} = \epsilon\hat{\beta}\mathcal{N}_{FD}(\phi) + O(\epsilon^2) , \qquad (19.121a)$$

where

$$\mathcal{N}_{FD}(\phi) = \frac{e^{\hat{\beta}(\phi+\hat{\mu})}}{[e^{\hat{\beta}(\phi+\hat{\mu})}+1]^2} + \frac{e^{\hat{\beta}(\phi-\hat{\mu})}}{[e^{\hat{\beta}(\phi-\hat{\mu})}+1]^2} . \tag{19.121b}$$

From the definition of η and ξ given in (19.106c) it follows that for $\vec{\hat{k}} \to 0$, $\eta \to 2\rho$, $\xi \to \rho^2$. One then verifies that for small ϵ (or $\vec{\hat{k}} \to 0$)

$$h(\phi, \phi + \epsilon, \eta, \xi, \mathcal{G}) \underset{\vec{\hat{k}}\to 0}{\longrightarrow} -\frac{2}{\epsilon} + O(\epsilon^0) .$$

Inserting this expression and (19.121a) into (19.120) one therefore has that

$$\hat{m}^2_{el}(\hat{\beta}, \hat{\mu}, \hat{m}) = 2e^2\hat{\beta} \int_{-\pi}^{\pi} \frac{d^3p}{(2\pi)^3} \left\{ \frac{e^{\hat{\beta}(\phi+\hat{\mu})}}{[e^{\hat{\beta}(\phi+\hat{\mu})}+1]^2} + \frac{e^{\hat{\beta}(\phi-\hat{\mu})}}{[e^{\hat{\beta}(\phi-\hat{\mu})}+1]^2} \right\} . \tag{19.122}$$

Notice the similarity in structure of this expression with its continuum counterpart (19.78). For the same reason as discussed in section 9, only momenta $\vec{\hat{p}}$ in the immediate neighbourhood of $\vec{\hat{p}} = 0$ contribute to the integral in the continuum limit. But in this limit

$$\hat{\beta}\phi(\vec{\hat{p}}) = \frac{1}{a}\beta\tilde{\mathcal{E}}(\vec{p}a) \underset{a\to 0}{\longrightarrow} \beta\sqrt{\vec{p}^2 + m^2} ,$$

where $\tilde{\mathcal{E}}(\vec{\hat{p}})$ has been defined in (19.103a,b). Hence

$$\hat{\eta}_{FD}(\phi) \underset{a\to 0}{\longrightarrow} \frac{1}{e^{\beta(\sqrt{\vec{p}^2+m^2}+\mu)}+1} ,$$

$$\bar{\hat{\eta}}_{FD}(\phi) \underset{a\to 0}{\longrightarrow} \frac{1}{e^{\beta(\sqrt{\vec{p}^2+m^2}-\mu)}+1}$$

The screening mass in physical units is now obtained as the following limit

$$m^2_{el} = \lim_{a\to 0} \frac{1}{a^2} \hat{m}^2_{el}(\frac{\beta}{a}, \mu a, m a) . \tag{19.123}$$

Introducing in (19.122) the dimensioned momenta $\vec{p}$ as new integration variables one then verifies that one recovers the expression (19.78), or equivalently (19.79).

For naive fermions, on the other hand, one finds that (see Appendix E)

$$(\hat{m}^2_{el})_{naive} = 4e^2\hat{\beta} \int_{-\pi}^{\pi} \frac{d^3\hat{p}}{(2\pi)^3} \left\{ \frac{e^{\hat{\beta}(\hat{\phi}-\hat{\mu})}}{[e^{\hat{\beta}(\hat{\phi}-\hat{\mu})}+1]^2} + \frac{e^{\hat{\beta}(\hat{\phi}+\hat{\mu})}}{[e^{\hat{\beta}(\hat{\phi}+\hat{\mu})}+1]^2} \right\} ,$$

where

$$\hat{\phi} = ar\sinh \hat{E}(\vec{\hat{p}}) \ .$$

Notice that here $\hat{\phi}$ is just the function $\hat{\mathcal{E}}(\vec{\hat{p}})$ we encountered when studying the lattice fermion propagator for naive fermions (cf. eq. (19.99b)). The above expression for the screening mass differs in two important respects from that for Wilson fermions: First of all there is an extra factor 2 multiplying the integral. This factor arises from the doubler contributions associated with frequency excitations lying in the second half of the Brillouin zone, $\frac{\pi}{2} < \hat{\omega}_\ell^- < \pi$. Secondly, the function $\hat{\phi}$ now vanishes not only for $\vec{\hat{p}} = 0$, but also at the corners of the Brillouin zone, where one or more components of $\vec{\hat{p}}$ take the values $\pm\pi$. Hence the integral receives in the continuum limit 2^3 identical contributions having the form of the "normal" contribution arising from finite physical momenta, for which $\vec{\hat{p}} = pa$ is of $\mathcal{O}(a)$. Hence in the continuum limit

$$(m_{el}^2)_{naive} = 16 m_{el}^2$$

where m_{el}^2 is given by (19.79).

19.11 The Electric Screening Mass for Wilson Fermions in Lattice QCD to One-Loop Order

The computation of the electric screening mass in lattice QCD is more involved than in QED. The following discussion is based on work carried out together with P. Kaste (Kaste, 1997). In $\mathcal{O}(g^2)$ the vacuum polarization tensor receives contributions from the seven diagrams shown in fig. (19-7). The last three diagrams have no continuum analog but are required by gauge invariance. Diagram (g) is the contribution arising from the non-abelian compact integration measure. The Feynman integrals can be easily written down using the finite temperature, finite chemical potential Feynman rules discussed in section 8. As in the continuum case, the contribution of diagrams (a) and (e) to the vacuuum polarization tensor is related to that in QED by the relation (19.80). We therefore "only" need to consider the diagrams involving gluon and ghost lines only, which do not depend on the chemical potential. Furthermore, they only involve sums over Matsubara frequencies of the bosonic type. Since in the continuum limit the only dimensioned scale is the temperature, their contribution to the screening mass will be of the form *const* $\times\ gT$. For finite lattice spacing, however, the temperature dependence will of course be modified by lattice artefacts. In the following we first consider diagrams (b-d) which have an analog in the continuum. In all the expressions it will be understood that the fourth component of the momenta are Matsubara frequencies. The combinatorial factors associated with each diagram are of course the same as those in the continuum formulation.

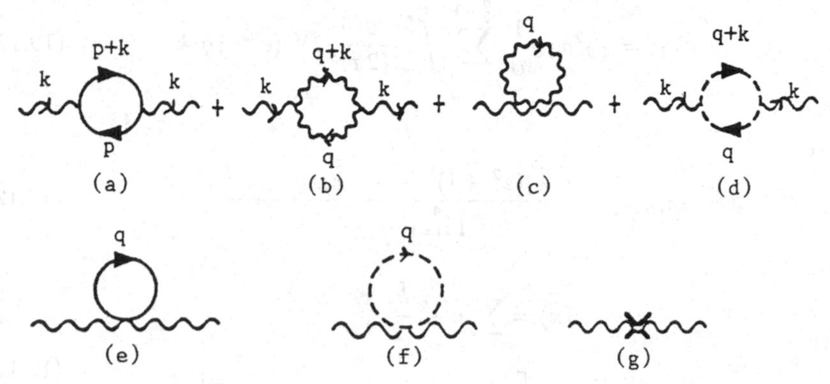

Fig.19-7 Diagrams contributing to the vacuum polarization tensor in lattice QCD in $\mathcal{O}(g^2)$.

Contribution of diagram (b)

Using the Feynman rules discussed in section 8 one finds, after making use of $\sum_{C,D} f_{ACD} f_{BCD} = 3\delta_{AB}$, that

$$\hat{\Pi}^{AB}_{\mu\nu}(\hat{\omega}^+_n, \vec{\hat{k}})_{(b)} = -\frac{3}{2} g^2 \delta_{AB} \int_+ d^4\hat{q} \frac{\hat{T}^{(b)}_{\mu\nu}(\hat{q}, \hat{k})}{\tilde{\hat{q}}^2 \widetilde{(\hat{q}+\hat{k})}^2}, \tag{19.124a}$$

where $\int_+ d^4\hat{q}$ has been defined in (19.95c); $\tilde{\hat{p}}_\mu$ and and $\tilde{\hat{p}}^2$ are defined generically by

$$\tilde{\hat{p}}_\mu = 2\sin\frac{\hat{p}_\mu}{2} \quad ; \quad \tilde{\hat{p}}^2 = \sum_\mu \tilde{\hat{p}}^2_\mu, \tag{19.124b}$$

and

$$\begin{aligned}
\hat{T}^{(b)}_{\mu\nu}(\hat{q}, \hat{k}) =& \Big\{ \widetilde{(\hat{q}-\hat{k})}_\mu \widetilde{(\hat{q}+2\hat{k})}_\nu \cos(\frac{1}{2}(\hat{q}+\hat{k})_\nu) \cos\frac{1}{2}\hat{q}_\mu \\
& - \widetilde{(\hat{q}-\hat{k})}_\mu \widetilde{(2\hat{q}+\hat{k})}_\nu \cos(\frac{1}{2}(\hat{q}+\hat{k})_\mu) \cos\frac{1}{2}\hat{k}_\mu \\
& - \widetilde{(2\hat{q}+\hat{k})}_\mu \widetilde{(\hat{q}+2\hat{k})}_\nu \cos\frac{1}{2}\hat{q}_\nu \cos\frac{1}{2}\hat{k}_\nu + (\mu \leftrightarrow \nu) \Big\} \\
& + \delta_{\mu\nu} \Big\{ \widetilde{(\hat{q}-\hat{k})}^2 \cos^2(\frac{1}{2}(\hat{q}+\hat{k})_\mu) + \widetilde{(\hat{q}+2\hat{k})}^2 \cos^2\frac{1}{2}\hat{q}_\mu \Big\} \\
& + \widetilde{(2\hat{q}+\hat{k})}_\mu \widetilde{(2\hat{q}+\hat{k})}_\nu \sum_\sigma \cos^2\frac{1}{2}\hat{k}_\sigma .
\end{aligned} \tag{19.124c}$$

To compute the corresponding contribution to the electric screening mass we only need to know $\hat{\Pi}_{44}(\hat{k})_{(b)}$ for vanishing gluon frequency. It is given by

$$\hat{\Pi}_{44}^{AB}(0,\vec{\hat{k}})_{(b)} = \frac{3}{2}g^2\delta_{AB}\frac{1}{\hat{\beta}}\sum_{\ell=-\frac{\hat{\beta}}{2}}^{\frac{\hat{\beta}}{2}-1}\int_{-\pi}^{\pi}\frac{d^3\hat{q}}{(2\pi)^3}f^{(b)}(e^{i\hat{\omega}_\ell^+};\vec{\hat{q}},\vec{\hat{k}}) , \tag{19.125a}$$

where

$$f^{(b)}(z;\vec{\hat{q}},\vec{\hat{k}}) = \frac{a(\vec{\hat{k}})(z^2-1)^2 - b(\vec{\hat{q}},\vec{\hat{k}})z(z+1)^2}{\prod_{i=1}^4[z-\bar{z}_i]} , \tag{19.125b}$$

and

$$a(\vec{\hat{k}}) = \sum_j \cos^2\frac{\hat{k}_j}{2} ,$$
$$b(\vec{\hat{q}},\vec{\hat{k}}) = \frac{1}{4}\left[\sum_j \widetilde{(\hat{q}-\hat{k})}_j^2 + \sum_j \widetilde{(\hat{q}+2\hat{k})}_j^2\right] . \tag{19.125c}$$

The zeros of the denominator in (19.125b) are located at

$$\bar{z}_1 = e^{\tilde{\phi}} \; ; \; \bar{z}_2 = e^{-\tilde{\phi}}$$
$$\bar{z}_3 = e^{\tilde{\psi}} \; ; \; \bar{z}_4 = e^{-\tilde{\psi}} , \tag{19.126a}$$

where

$$\tilde{\phi} = \text{arcosh } H(\vec{\hat{q}}) ,$$
$$\tilde{\psi} = \text{arcosh } H(\vec{\hat{q}}+\vec{\hat{k}}) ,$$
$$H(\vec{\hat{p}}) = 1 + 2\sum_j \sin^2\frac{\hat{p}_j}{2} . \tag{19.126b}$$

The frequency sum can be calculated by making use of (F.5) in appendix F. After some straight forward algebra one finds that

$$\begin{aligned}\hat{\Pi}_{44}^{AB}(0,\vec{\hat{k}})_{(b)} =& 6g^2\delta_{AB}\int_{-\pi}^{\pi}\frac{d^3\hat{q}}{(2\pi)^3}h(\tilde{\phi},\tilde{\psi},a,b)\hat{\eta}_{BE}(\tilde{\phi})\\ &+\frac{3}{2}g^2\delta_{AB}\left\{a(\vec{\hat{k}}) + \int_{-\pi}^{\pi}\frac{d^3\hat{q}}{(2\pi)^3}[h(\tilde{\phi},\tilde{\psi},a,b)+h(\tilde{\psi},\tilde{\phi},a,b)]\right\} ,\end{aligned} \tag{19.127a}$$

where

$$h(\tilde{\phi},\tilde{\psi},a,b) = \frac{-a\,\sinh^2\tilde{\phi} + \frac{1}{2}b[\cosh\tilde{\phi}+1]}{\sinh\tilde{\phi}[\cosh\tilde{\phi}-\cosh\tilde{\psi}]} , \tag{19.127b}$$

and

$$\hat{\eta}_{BE}(\tilde{\phi}) = \frac{1}{e^{\hat{\beta}\tilde{\phi}} - 1} \tag{19.127c}$$

is the lattice version of the Bose-Einstein distribution function. In obtaining this result we have made use of the fact that

$$\tilde{\phi} \underset{\vec{p}\to-\vec{p}-\vec{k}}{\longleftrightarrow} \tilde{\psi} \tag{19.128}$$

under the variable transformation $\vec{\hat{q}} \to -\vec{\hat{q}} - \vec{\hat{k}}$, while $a(\vec{\hat{k}})$ and $b(\vec{\hat{q}}, \vec{\hat{k}})$ are invariant under this transformation. Note that the function (19.127b) is singular for $\vec{\hat{k}} \to 0$, since in this limit $\tilde{\phi} \to \tilde{\psi}$. The singularity is however integrable as can be seen by making use of (19.128) to write (19.127a) in the form

$$\begin{aligned}\hat{\Pi}^{AB}_{44}(0, \vec{\hat{k}})_{(b)} &= 3g^2\delta_{AB}\int_{-\pi}^{\pi}\frac{d^3\hat{q}}{(2\pi)^3}h(\tilde{\phi}, \tilde{\psi}, a, b)\Delta\hat{\eta}_{BE}(\tilde{\phi}, \tilde{\psi}) \\ &+ \frac{3}{2}g^2\delta_{AB}\left\{a(\vec{\hat{k}}) + \int_{-\pi}^{\pi}\frac{d^3\hat{q}}{(2\pi)^3}[h(\tilde{\phi}, \tilde{\psi}, a, b) + h(\tilde{\psi}, \tilde{\phi}, a, b)][1 + 2\hat{\eta}_{BE}(\tilde{\phi})]\right\} ,\end{aligned} \tag{19.129a}$$

where

$$\Delta\hat{\eta}_{BE}(\tilde{\phi}, \tilde{\psi}) = \hat{\eta}_{BE}(\tilde{\phi}) - \hat{\eta}_{BE}(\tilde{\psi}) . \tag{19.129b}$$

The limit $\vec{\hat{k}} \to 0$ can now be easily be taken and one obtains the following contribution to the electric screening mass

$$\begin{aligned}(\hat{m}^2_{el})_{(b)} =& \frac{3}{2}g^2\left\{3 - \int_{-\pi}^{\pi}\frac{d^3\hat{q}}{(2\pi)^3}\left[3\ \mathrm{coth}\tilde{\phi} + \frac{1}{2\mathrm{sinh}\tilde{\phi}}\right][1 + 2\hat{\eta}_{BE}(\tilde{\phi})]\right\} \\ &+ \frac{15}{2}g^2\hat{\beta}\int_{-\pi}^{\pi}\frac{d^3\hat{q}}{(2\pi)^3}\mathcal{N}_{BE}(\tilde{\phi}) ,\end{aligned} \tag{19.130a}$$

where

$$\mathcal{N}_{BE}(\tilde{\phi}) = \frac{e^{\hat{\beta}\tilde{\phi}}}{[e^{\hat{\beta}\tilde{\phi}} - 1]^2} . \tag{19.130b}$$

Contribution of diagram (c)

This diagram involves the 4-gluon vertex (15.53b), which consists of types of terms differing in the colour structure: terms involving the structure constants f_{ABC}, and terms involving the completely symmetric colour couplings d_{ABC}. We denote the corresponding contributions to the vacuum polarization tensor by

$[\Pi^{AB}(\hat{k})_{(c)}]_{[f]}$ and $[\Pi^{AB}(\hat{k}))_{(c)}]_{[d]}$, respectively. Consider first $[\Pi^{AB}(\hat{k}))_{(c)}]_{[f]}$. It is given by

$$\left[\hat{\Pi}^{AB}_{\mu\nu}(\hat{k})_{(c)}\right]_{[f]} = \frac{3}{4}g^2\delta_{AB}\int_+ d^4\hat{q}\frac{[\hat{T}^{(c)}_{\mu\nu}(\hat{q},\hat{k})]_{[f]}}{\tilde{\hat{q}}^2}, \tag{19.131a}$$

where

$$\begin{aligned}\left[\hat{T}^{(c)}_{\mu\nu}(\hat{q},\hat{k})\right]_{[f]} =& \frac{1}{6}\delta_{\mu\nu}\Big\{24\cos\hat{q}_\mu\sum_\sigma\cos\hat{k}_\sigma \\ &- 12\cos^2(\frac{1}{2}(\hat{q}+\hat{k})_\mu) - 12\cos^2(\frac{1}{2}(\hat{q}-\hat{k})_\mu) \\ &+ 2\tilde{\hat{q}}^2_\mu[\tilde{\hat{k}}^2_\mu - \sum_\sigma\tilde{\hat{k}}^2_\sigma] + 4[\widetilde{(\hat{q}-\hat{k})}_\mu + \widetilde{(\hat{q}+\hat{k})}_\mu]\tilde{\hat{q}}_\mu\cos\frac{1}{2}\hat{k}_\mu\ . \\ &- \sum_\sigma[\widetilde{(\hat{q}+\hat{k})}^2_\sigma + \widetilde{(\hat{q}-\hat{k})}^2_\sigma]\Big\} \\ &+ \frac{1}{3}\Big\{[\widetilde{(\hat{q}+\hat{k})}_\mu - \widetilde{(\hat{q}-\hat{k})}_\mu]\tilde{\hat{k}}_\nu\cos\frac{1}{2}\hat{q}_\mu + (\mu\leftrightarrow\nu)\Big\}\end{aligned} \tag{19.131b}$$

For vanishing gluon frequency we have that

$$\left[\hat{\Pi}^{AB}_{44}(0,\vec{\hat{k}})_{(c)}\right]_{[f]} = \frac{3}{4}g^2\delta_{AB}\frac{1}{\hat{\beta}}\sum_{\ell=-\frac{\hat{\beta}}{2}}^{\frac{\hat{\beta}}{2}-1}\int_{-\pi}^{\pi}\frac{d^3\hat{q}}{(2\pi)^3}\left[f^{(c)}(e^{i\hat{\omega}^+_\ell};\hat{q},\hat{k})\right]_{[f]}, \tag{19.132a}$$

where

$$\left[f^{(c)}(z;\vec{\hat{q}},\vec{\hat{k}})\right]_{[f]} = \frac{-c(\vec{\hat{k}})(z^2+1) + d(\vec{\hat{k}})(z-1)^2 + P(\vec{\hat{q}},\vec{\hat{k}})z}{[z-\bar{z}_1][z-\bar{z}_2]} \tag{19.132b}$$

and

$$\begin{aligned} c(\vec{\hat{k}}) &= 1 + 2\sum_j\cos\hat{k}_j\ , \\ d(\vec{\hat{k}}) &= 1 - \frac{1}{3}\sum_j\tilde{\hat{k}}^2_j\ , \\ P(\vec{\hat{q}},\vec{\hat{k}}) &= 2 + \frac{1}{6}\sum_j[\widetilde{(\hat{q}+\hat{k})}^2_j + \widetilde{(\hat{q}-\hat{k})}^2_j]\ .\end{aligned} \tag{19.132c}$$

Performing the frequency sum in (19.132a) one finds that

$$\begin{aligned}[\hat{\Pi}^{AB}_{44}(0,\vec{\hat{k}})_{(c)}]_{[f]} =& \frac{3}{4}g^2\delta_{AB}\Big\{-c(\vec{\hat{k}}) + d(\vec{\hat{k}}) \\ &+ \int_{-\pi}^{\pi}\frac{d^3\hat{q}}{(2\pi)^3}[(c-d)\coth\tilde{\phi} + (d-\frac{1}{2}P)\frac{1}{\sinh\tilde{\phi}}][1+2\hat{\eta}_{BE}(\tilde{\phi})]\Big\}\ .\end{aligned} \tag{19.133}$$

Taking the limit $\vec{\hat{k}} \to 0$ one obtains the following expression for the screening mass

$$(\hat{m}^2_{el})^{(c)}_{[f]} = \frac{1}{2}g^2\Big[-9 + \frac{1}{2}\int_{-\pi}^{\pi}\frac{d^3\hat{q}}{(2\pi)^3}\Big[17\coth\tilde{\phi} + \frac{1}{\sinh\tilde{\phi}}\Big]\big[1 + 2\hat{\eta}_{BE}(\tilde{\phi})\big]\ . \quad (19.134)$$

Consider next the contribution $[\Pi^{AB}(\hat{k})_{(c)}]_{[d]}$. It is given by

$$\Big[\hat{\Pi}^{AB}_{\mu\nu}(\hat{k})_{(c)}\Big]_{[d]} = -\frac{1}{2}g^2\delta_{AB}\int_{+} d^4\hat{q}\frac{[\hat{T}^{(c)}_{\mu\nu}(\hat{q},\hat{k})]_{[d]}}{\tilde{\hat{q}}^2}\ , \quad (19.135a)$$

where

$$\Big[\hat{T}^{(c)}_{\mu\nu}(\hat{q},\hat{k})\Big]_{[d]} = \frac{1}{12}\left(\frac{20}{3} + d(A)\right)\Big\{\delta_{\mu\nu}\Big(\sum_\sigma \tilde{\hat{q}}^2_\sigma\tilde{\hat{k}}^2_\sigma + \tilde{\hat{q}}^2_\mu\sum_\sigma\tilde{\hat{k}}^2_\sigma\Big) - \tilde{\hat{q}}^2_\mu\tilde{\hat{k}}_\mu\tilde{\hat{k}}_\nu - \tilde{\hat{q}}^2_\nu\tilde{\hat{k}}_\mu\tilde{\hat{k}}_\nu\Big\}\ . \quad (19.135b)$$

Here $d(A)$ is defined by

$$\sum_{E,F=1}^{8}\left[d_{AFE}d_{BFE} + d_{BFE}d_{AFE} + d_{FFE}d_{ABE}\right] = d(A)\delta_{AB}\ .$$

For vanishing gluon frequency expression (19.135a) can be written in the form

$$\Big[\hat{\Pi}^{AB}_{44}(0,\vec{\hat{k}})_{(c)}\Big]_{[d]} = -\frac{1}{2}g^2\delta_{AB}\frac{1}{\hat{\beta}}\sum_{\ell=-\frac{\hat{\beta}}{2}}^{\frac{\hat{\beta}}{2}-1}\int_{-\pi}^{\pi}\frac{d^3\hat{q}}{(2\pi)^3}\Big[f^{(c)}(e^{i\hat{\omega}^+_\ell};\vec{\hat{q}},\vec{\hat{k}})\Big]_{[d]}\ , \quad (19.136a)$$

where

$$\Big[f^{(c)}(z;\vec{\hat{q}},\vec{\hat{k}})\Big]_{[d]} = \frac{1}{12}\left(\frac{20}{3} + d(A)\right)\frac{K(\vec{\hat{k}})(z-1)^2 - L(\vec{\hat{q}},\vec{\hat{k}})z}{[z-\bar{z}_1][z-\bar{z}_2]}\ , \quad (19.136b)$$

and

$$K(\vec{\hat{k}}) = \sum_j \tilde{\hat{k}}^2_j\ ,\qquad L(\vec{\hat{q}},\vec{\hat{k}}) = \sum_j \tilde{\hat{q}}^2_j\tilde{\hat{k}}^2_j\ . \quad (19.136c)$$

Performing the frequency sum one obtains

$$\Big[\hat{\pi}^{AB}_{44}(0,\vec{\hat{k}})_{(c)}\Big]_{[d]} = g^2\delta_{AB}\frac{1}{24}\left(\frac{20}{3} + d(A)\right) \times\Big\{-K + \int_{-\pi}^{\pi}\frac{d^3\hat{q}}{(2\pi)^3}\Big[K\coth\tilde{\phi} - \frac{K + \frac{1}{2}L}{\sinh\tilde{\phi}}\Big]\big[1 + \hat{\eta}_{BE}(\phi)\big]\Big\}\ .$$

Since $K(\vec{\hat{k}})$ and $L(\vec{\hat{q}},\vec{\hat{k}})$ vanish for $\vec{\hat{k}} \to 0$, it does not contribute to the screening mass, i.e.,

$$[(\hat{m}_{el})_c]_{[d]} = 0 \; . \tag{19.137}$$

Contribution of diagram (d)

The only other diagram possessing an analog in the continuum is the ghost loop shown in fig. (19-7d). Its contribution is given by

$$\hat{\Pi}^{AB}_{\mu\nu}(\hat{k})_{(d)} = \frac{3}{2} g^2 \delta_{AB} \int_+ d^4\hat{q}\, \frac{\hat{T}^{(d)}_{\mu\nu}(\hat{q},\hat{k})}{\tilde{\hat{q}}^2 \widetilde{(\hat{q}+\hat{k})}^2} \, , \tag{19.138a}$$

where

$$\hat{T}^{(d)}_{\mu\nu}(\hat{q},\hat{k}) = \tilde{\hat{q}}_\mu \widetilde{(\hat{q}+\hat{k})}_\nu \cos(\frac{1}{2}(\hat{q}+\hat{k})_\mu) \cos \frac{1}{2}\hat{q}_\nu + (\mu \leftrightarrow \nu) \; . \tag{19.138b}$$

For vanishing gluon frequency we have that

$$\hat{\Pi}^{AB}_{44}(0,\vec{\hat{k}})_{(d)} = \frac{3}{2} g^2 \delta_{AB} \frac{1}{\hat{\beta}} \sum_{\ell=-\frac{\hat{\beta}}{2}}^{\frac{\hat{\beta}}{2}-1} \int_{-\pi}^{\pi} \frac{d^3\hat{q}}{(2\pi)^3} f^{(d)}(e^{i\hat{\omega}^+_\ell};\vec{\hat{q}},\vec{\hat{k}}) \, , \tag{19.139a}$$

where

$$f^{(d)}(z;\vec{\hat{q}},\vec{\hat{k}}) = -\frac{1}{2} \frac{(z^2-1)^2}{\prod_{i=1}^4 [z - \bar{z}_i]} \; . \tag{19.139b}$$

Performing the frequency sum one finds that

$$\begin{aligned} \hat{\Pi}^{AB}_{44}(0,\vec{\hat{k}})_{(d)} = \frac{3}{4} g^2 \delta_{AB} \Big\{ -1 + & \int_{-\pi}^{\pi} \frac{d^3\hat{q}}{(2\pi)^3} [g(\tilde{\phi},\tilde{\psi}) + g(\tilde{\psi},\tilde{\phi})] \\ & + 4 \int_{-\pi}^{\pi} \frac{d^3\hat{q}}{(2\pi)^3} g(\tilde{\phi},\tilde{\psi}) \hat{\eta}_{BE}(\tilde{\phi}) \Big\} \, , \end{aligned} \tag{19.140a}$$

where

$$g(\tilde{\phi},\tilde{\psi}) = \frac{\sinh \tilde{\phi}}{\cosh \tilde{\phi} - \cosh \tilde{\psi}} \; . \tag{19.140b}$$

This function is again singular for $\vec{\hat{k}} \to 0$. To compute the limit we proceed as discussed earlier and write (19.140a) in the form

$$\begin{aligned} \hat{\Pi}^{AB}_{44}(0,\vec{\hat{k}})_{(d)} = \frac{3}{4} g^2 \delta_{AB} \Big\{ -1 + & \int_{-\pi}^{\pi} \frac{d^3\hat{q}}{(2\pi)^3} [g(\tilde{\phi},\tilde{\psi}) + g(\tilde{\psi},\tilde{\phi})][1 + 2\hat{\eta}_{BE}(\tilde{\phi})] \\ & + 2 \int_{-\pi}^{\pi} \frac{d^3\hat{q}}{(2\pi)^3} g(\tilde{\phi},\tilde{\psi}) \Delta\hat{\eta}_{BE}(\tilde{\phi},\tilde{\psi}) \, , \end{aligned}$$

where $\Delta\hat{\eta}_{BE}(\tilde{\phi})$ has been defined in (19.129b). Taking the infrared limit one obtains for the contribution of diagram (d) to the electric screening mass:

$$(\hat{m}_{el}^2)_{(d)} = \frac{3}{4}g^2\Big\{-1+\int_{-\pi}^{\pi}\frac{d^3\hat{q}}{(2\pi)^3}[1+2\hat{\eta}_{BE}(\tilde{\phi})]\coth\tilde{\phi} \\ -2\hat{\beta}\int_{-\pi}^{\pi}\frac{d^3\hat{q}}{(2\pi)^3}\mathcal{N}_{BE}(\tilde{\phi})\Big\}, \tag{19.141}$$

with $\mathcal{N}_{BE}(\tilde{\phi})$ defined in (19.130b). Combining the results (19.130a), (19.137), (19.134), and (19.141), we therefore find that those diagrams possessing a continuum analog yield the following contribution to the electric screening mass for finite lattice spacing

$$(\hat{m}_{el}^2)_{(b+c+d)} = g^2\Big\{-\frac{3}{4}+\frac{1}{2}\int\frac{d^3\hat{q}}{(2\pi)^3}\left(\coth\tilde{\phi}-\frac{1}{\sinh\tilde{\phi}}\right)[1+2\hat{\eta}_{BE}]\Big\} \\ +6g^2\hat{\beta}\int\frac{d^3\hat{q}}{(2\pi)^3}\mathcal{N}_{BE}(\tilde{\phi}). \tag{19.142}$$

As we now show, the remaining diagrams (f) and (g), which are a consequence of the lattice regularization, precisely cancel the first term in (19.142).

Contribution of Diagram (f)

This contribution is given by

$$[\hat{\Pi}_{\mu\nu}^{AB}]_{(f)} = \frac{1}{2}g^2\delta_{AB}\delta_{\mu\nu}\int_+ d^4\hat{q}\frac{\tilde{\hat{q}}_\mu^2}{\tilde{\hat{q}}^2}. \tag{19.143}$$

For vanishing gluon frequency and $\mu=\nu=4$ this expression reduces to

$$(\hat{\Pi}_{44}^{AB})_{(f)} = \frac{1}{2}g^2\delta_{AB}\frac{1}{\hat{\beta}}\sum_{\ell=-\frac{\hat{\beta}}{2}}^{\frac{\hat{\beta}}{2}-1}\int_{-\pi}^{\pi}\frac{d^3\hat{q}}{(2\pi)^3}f^{(f)}(e^{i\hat{\omega}_\ell^+}), \tag{19.144a}$$

where

$$f^{(f)}(z) = \frac{(z-1)^2}{[z-\bar{z}_1][z-\bar{z}_2]}. \tag{19.144b}$$

The corresponding contribution to the screening mass is found to be

$$(\hat{m}_{el}^2)_{(f)} = \frac{1}{2}g^2\Big\{1-\int_{-\pi}^{\pi}\frac{d^3\hat{q}}{(2\pi)^3}[\coth\tilde{\phi}-\frac{1}{\sinh\tilde{\phi}}][1+2\hat{\eta}_{BE}(\tilde{\phi})]\Big\}. \tag{19.145}$$

Contribution of diagram (g)

Finally, the contribution of diagram (g) to the screening mass (arising from the link-integration measure) is trivially given by

$$(\hat{m}^2_{el})_{(g)} = \frac{1}{4}g^2 \tag{19.146}$$

From (19.145) and (19.146) we see that their sum just cancels the first term appearing on the rhs of (19.142) leaving us with the following simple expression for the contribution to the screening mass arsing from diagrams containing only gluon and ghost lines:

$$(\hat{m}^2_{el})_G = 6g^2\hat{\beta}\int_{-\pi}^{\pi}\frac{d^3\hat{q}}{(2\pi)^3}\mathcal{N}_{BE}(\tilde{\phi}) \; . \tag{19.147}$$

The corresponding expression for the dimensioned screening mass squared is given in the continuum limit (19.123) by

$$\begin{aligned}(m^2_{el})_G &= \lim_{a\to 0} 6g^2\beta\int_{-\frac{\pi}{a}}^{\frac{\pi}{a}}\frac{d^3q}{(2\pi)^3}\frac{e^{\frac{1}{a}\beta\tilde{\phi}(\vec{q}a)}}{[e^{\frac{1}{a}\beta\tilde{\phi}(\vec{q}a)}-1]^2}\\ &= \frac{3}{\pi^2}g^2\beta\int_0^\infty dq q^2\frac{e^{\beta q}}{[e^{\beta q}-1]^2} \; .\end{aligned}$$

where $q = |\vec{q}|$. After a partial integration this expression takes the form

$$(m^2_{el})_G = \frac{6}{\pi^2}g^2T^2\int_0^\infty dx\frac{x}{e^x-1} \; .$$

Making use of the formula

$$\int_0^\infty dx\frac{x^{\alpha-1}}{e^x-1} = \Gamma(\alpha)\zeta(\alpha) \; ,$$

where $\Gamma(\alpha)$ is the Euler Gamma function, and $\zeta(\alpha)$ the Riemann Zeta-function, we finally obtain

$$(m^2_{el})_G = g^2T^2 \; . \tag{19.148}$$

Including the contribution of diagrams (a) and (e) to the screening mass (which are just one-half of the expressions obtained in the previous section for the screening mass in QED) we therefore have that

$$\begin{aligned}\hat{m}^2_{el}(\hat{\beta},\hat{\mu},\hat{m}) &= g^2\hat{\beta}\int_{-\pi}^{\pi}\frac{d^3p}{(2\pi)^3}\left\{\frac{e^{\hat{\beta}(\phi+\hat{\mu})}}{[e^{\hat{\beta}(\phi+\hat{\mu})}+1]^2}+\frac{e^{\hat{\beta}(\phi-\hat{\mu})}}{[e^{\hat{\beta}(\phi-\hat{\mu})}+1]^2}\right\}\\ &\quad + 6g^2\hat{\beta}\int_{-\pi}^{\pi}\frac{d^3\hat{q}}{(2\pi)^3}\frac{e^{\hat{\beta}\tilde{\phi}}}{[e^{\hat{\beta}\tilde{\phi}}-1]^2} \qquad (lattice)\end{aligned} \tag{19.149}$$

where ϕ and $\tilde{\phi}$ have been defined in (19.106f) and (19.126b). In the continuum limit this expression reduces to

$$m_{el}^2 = \frac{g^2}{2\pi^2} \int_0^\infty dp \, \frac{2p^2 + m^2}{\sqrt{p^2 + m^2}} \left[\eta_{FD}(E,\mu) + \bar{\eta}_{FD}(E,\mu)\right] + g^2 T^2 \qquad (continuum) \tag{19.150}$$

In lattice simulations of the pure gauge theory the electric screening mass is extracted from correlators of Polyakov loops or from the long distance behaviour of the gluon propagator [see e.g. Irbäck (1991), Heller (1995)]. In these simulations the number of lattice sites $N_\tau \equiv \hat{\beta}$ is fixed and the temperature is varied by varying the lattice spacing $a = \frac{1}{TN_\tau}$. If the lattice expression for the (dimensioned) screening mass is to approximate the continuum, then the lattice spacing must be small compared to all physical length scales in the problem. Hence we must have that $a << \frac{1}{T}$, $a << \frac{1}{m}$ and $a << \frac{1}{\mu}$. This implies, that $ma = (\frac{m}{T})\frac{1}{N_\tau} <<$ 1, or $N_\tau >> \frac{m}{T}$. Similarily we must have that $N_\tau >> \frac{\mu}{T}$. For N_τ fixed, the electric screening mass in physical units, divided by the temperature, is given by $[m_{el}]_{latt}/T = N_\tau \hat{m}_{el}(N_\tau, (\mu/T)N_\tau^{-1}, (m/T)N_\tau^{-1})$, while in the continuum limit ($N_\tau \to \infty$) this ratio is just a function of m/T and μ/T. In fig. (19-8) we have plotted the ratio $(m_{el})_{latt}/(m_{el})_{con}$ as a function of $\frac{m}{T}$ and $\frac{\mu}{T}$ for $N_\tau = 8, 16$. In the parameter range considered the deviation from unity is seen to be at most 1.7% for $N_\tau = 16$, and 10% for $N_\tau = 8$.

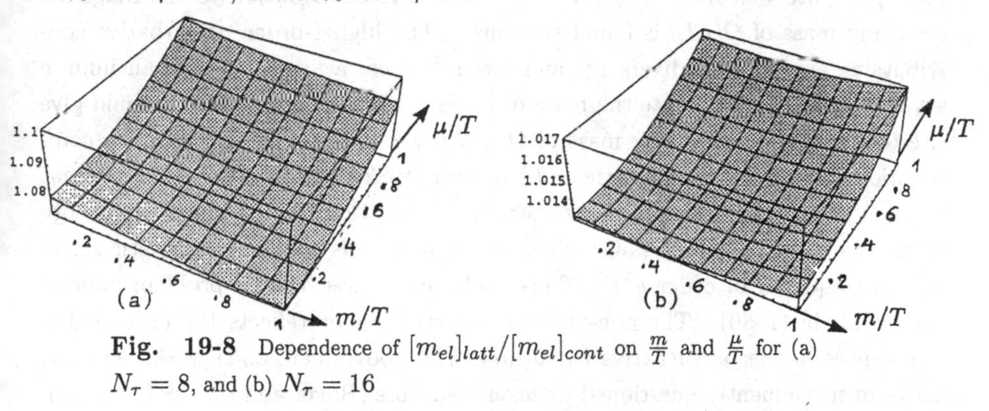

Fig. 19-8 Dependence of $[m_{el}]_{latt}/[m_{el}]_{cont}$ on $\frac{m}{T}$ and $\frac{\mu}{T}$ for (a) $N_\tau = 8$, and (b) $N_\tau = 16$

The deviations are mainly due to the fermion loop contribution as seen from fig. (19-9), where we have plotted $[m_{el}]_{latt}/[m_{el}]_{cont}$ for the pure SU(3) gauge theory for various values of N_τ. The solid line is drawn to guide the eye. This ratio only depends on the number of lattice sites N_τ. As seen from the figure the deviation from the continuum is very small for $N_\tau \geq 8$. For $N_\tau = 8$, and $N_\tau = 16$, it is about 2%, and 0.4%, respectively.

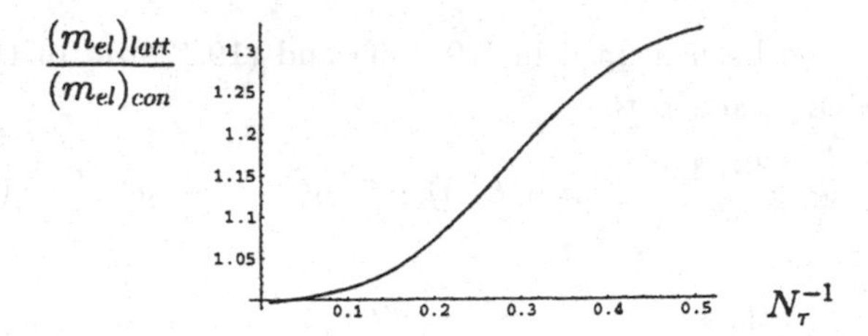

Fig. 19-9 Dependence of the pure gluonic contribution to $[m_{el}]_{latt}/[m_{el}]_{con}$ on the number of lattice sites N_τ. The solid line is drawn to guide the eye.

19.12 The Infrared Problem

When we discussed the $\lambda\phi^4$-theory we have seen that a mass of $O(\lambda)$ was generated in the presence of a heat bath. As we have just seen a similar phenomenon is encountered in QCD, except that there the situation is more complex. Because the heat bath singles out a preferred reference frame the vacuum polarization tensor is no longer Lorentz covariant, but only covariant under spatial rotations. Hence different screening masses can be generated in the electric and magnetic sectors.

The generation of a magnetic screening mass turns out to pose a serious problem, although the appearance of an infrared cut-off in the magnetic sector is in principle quite welcome. The problem is that the contribution to the magnetic screening mass of $O(gT)$ is found to vanish. The higher-order perturbative contributions are infrared divergent, and one is therefore forced to sum an infinite set of diagrams to calculate the mass in lowest order. In principle this could give rise to a magnetic screening mass of $O(g^\alpha T)$ where $1 < \alpha < 2$. In fact, a computation carried out by Kajantie and Kapusta (1982) based on the self-consistent solution of an approximated Schwinger-Dyson equation for the gluon self-energy in the temporal ($A_0^B = 0$) gauge, suggested that gluons acquire a magnetic screeing mass-squared of order g^3T^2. This could have cured the IR-problem pointed out by Linde (1980). The non-analytic structure in g^2 reflects the original IR-divergences in the perturbative expansion. The above mentioned work, however, has been subsequently questioned by several authors (Baker and Li, 1983; Toimela, 1985). If the magnetic screening mass is of $O(g^2T)$ then one is stuck with a serious computational impass. Consider, for example, the contribution to the free energy arising from diagrams involving only four-gluon couplings. Since the four-gluon interaction vertex is dimensionless the same general argument as given for the case of the ϕ^4 theory leads to an expression similar to (19.53), except for an important difference: the four-gluon coupling is of order g^2. Hence the contribution to the thermodynamical potential arising from a diagram containing N such vertices is now given by (19.53) with λ replaced by g^2:

$$\Omega_N(T) \sim g^6 T^4 \left(\frac{g^2 T}{m(T)} \right)^{N-3} .$$

If we substitute for $m(T)$ the magnetic screening mass $m_M(T)$, which we assume to be of $O(g^2T)$, then an infinite set of diagrams contribute to $O(g^6)$. Even if we could calculate the contributions from the individual diagrams (which we cannot) their sum could very well diverge. But the situation is even worse. For a similar reason the magnetic screening mass itself cannot be computed even in lowest order. (Gross, Pisarski and Yaffe, 1981).

What we have just described is the notorious infrared problem in QCD. Being unable to estimate higher order corrections to the free energy, one could take the pessimistic point of view that perturbation theory can tell us nothing about the thermodynamical properties of QCD, even at high temperatures, where renormalization group arguments suggest that the coupling becomes small (Collins and Perry, 1975). It may however turn out that MC calculations confirm the early expectations that at sufficiently high temperatures the non-perturbative corrections to thermodynamical observables are small. In section 8 of chapter 20 we shall present some numerical results which give us some insight into this problem.

We now briefly summarize the results obtained in the literature for the contributions of $O(g^2)$ and $O(g^3)$ to the thermodynamical potential Ω of an $SU(N)$ gauge theory with N_f flavours and at zero chemical potential.

The correction of $O(g^2)$ requires the calculation of a set of two-loop diagrams involving gluon, quark, and ghost fields (Kapusta, 1979). The zero temperature renormalization prescriptions are sufficient to eliminate all ultraviolet divergencies. After subtracting the vacuum contributions, these diagrams make a finite contribution to Ω:

$$\Omega_2 = \frac{g^2 T^4}{144} (N^2 - 1)(N + \frac{5}{4} N_f).$$

In fourth and higher orders of the coupling one encounters infrared divergencies. These become more and more severe with increasing number of loops. By summing the infrared divergent contributions one therefore expects to encounter corrections of lower order than g^4. Indeed, summing the ring diagrams of the type depicted in fig. (19-10), where the shaded blobs stand for the gluon self-energy in $O(g^2)$, Kapusta (1979) finds that there is an $O(g^3)$ correction to the thermodynamical potential:

$$\Omega_3 = -\frac{g^3T^4}{12\pi}(N^2-1)[\frac{1}{3}(N+\frac{N_f}{2})]^{3/2}.$$

This is the so-called plasmon term. It arises from those interactions which give the gluon an electric screening mass at one loop level.

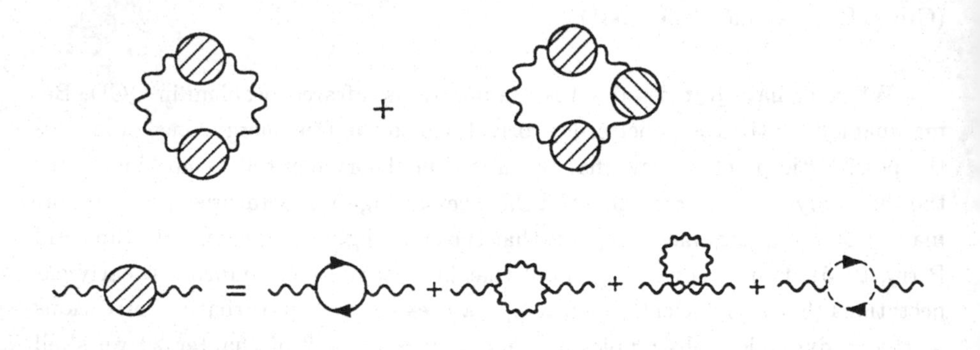

Fig. 19-10 Ring diagrams which when summed give rise to a contribution of $O(g^3)$ to the thermodynamical potential.

Concluding, we are left with the following somewhat disappointing situation in QCD: while the electric screening mass is of $O(gT)$, the magnetic screening mass is expected to be of $O(g^2T)$, but cannot be computed. As a consequence, the thermodynamical potential can at best be calculated up to fifth order. If one is lucky, non-perturbative effects do not play an important role, and the potential obtained by analytic means suffices to describe the thermodynamics of QCD at very high temperatures, where the coupling is expected to be small. The only way to test this is to compute thermodynamical observables numerically by starting from the lattice formulation of QCD.

CHAPTER 20

NON-PERTURBATIVE QCD AT FINITE TEMPERATURE

In the remaining part of the book we want to study the behaviour of hadronic matter at finite temperature as predicted by QCD. Here are some questions that one would like to have an aswer to: i) Does QCD predict a phase transition from a low temperature confining phase, where chiral symmetry is broken, to a high temperature phase where quarks and gluons are deconfined, and chiral symmetry is restored? ii) If so, what is the critical transition temperature and the nature of the phase transition? Is it of first or second order? iii) What is the nature of the high temperature phase? Is it that of a quark-gluon plasma where the interactions of quarks are Debye-screened? iv) Does perturbation theory provide an adequate description of the thermodynamical properties at very high temperatures? v) What is a possible signal for plasma formation in heavy ion collisions?

The non-perturbative framework provided by the lattice formulation of QCD allows us, at least in principle, to obtain an answer to the above questions. In the following four sections we will set up the theoretical framework which provides the basis for studying a possible deconfining phase transition. The theoretical ideas introduced will be exemplified in a simple lattice model in section 5. In sections 6-8 we then present some Monte Carlo results on this transition, and on the high temperature phase of QCD. As always, we shall restrict ourselves to early pioneering work, and we leave it to the reader to confer the numerous proceedings for more recent results. Finally in section 9, we discuss some possible signatures for plasma formation in heavy ion collisions.

20.1 Thermodynamics on the Lattice

For the computation of thermodynamical observables, like the mean energy, pressure and entropy, we need a non-perturbative expression for the partition function. From what we have learned in the previous chapters we immediately conjecture that the partition function is given by

$$Z_{QCD} = \int DUD\psi D\bar{\psi} e^{-S_{QCD}^{(\beta,\mu)}[U,\psi,\bar{\psi}]} . \tag{20.1}$$

For Wilson fermions, $S_{QCD}^{(\beta,\mu)}[U,\psi,\bar{\psi}]$ is the finite-temperature, finite chemical potential action having the form (19.90), where ψ and U_μ are vectors and matrices in colour space, and $\frac{1}{e^2}$ is replaced by $\frac{6}{g_0^2}$. The link variables and Dirac fields are subjected to periodic and antiperiodic boundary conditions, respectively.

Consider for example the implementation of the antiperiodic boundary conditions on the fermionic degrees of freedom, i.e.,

$$\begin{aligned} \hat{\psi}(\vec{n},1) &= -\hat{\psi}(\vec{n},\hat{\beta}+1) \ , \\ \hat{\bar{\psi}}(\vec{n},1) &= -\hat{\bar{\psi}}(\vec{n},\hat{\beta}+1) \ . \end{aligned} \tag{20.2}$$

The only contribution to the action which is sensitive to these boundary conditions, is the fermionic contribution involving the link variables pointing along the time direction. By implementing these conditions, this piece of the action becomes a function of the Grassmann variables located on the time slices $n_4 = 1, \cdots \hat{\beta}$, and can be written in matrix form as follows,

$$\sum_{n_4,m_4=1}^{\hat{\beta}} \sum_{\vec{n}} \hat{\bar{\psi}}(\vec{n},n_4)\mathcal{M}_{n_4 m_4}(\vec{n})\hat{\psi}(\vec{n},m_4) \ ,$$

where $\mathcal{M}(\vec{n})$ is a $\hat{\beta} \times \hat{\beta}$ matrix with the non-vanishing entries given by

$$\mathcal{M}(\vec{n}) = \begin{pmatrix} \zeta & y(\vec{n},1) & \cdot & \cdot & \cdot & \cdot & \cdot & -\bar{y}(\vec{n},\hat{\beta}) \\ \bar{y}(\vec{n},1) & \zeta & y(\vec{n},2) & & & & & \cdot \\ \cdot & \bar{y}(\vec{n},2) & \zeta & y(\vec{n},3) & & & & \cdot \\ \vdots & \vdots & \vdots & \vdots & \ddots & & & \vdots \\ -\bar{y}(\vec{n},\hat{\beta}) & \cdot & \cdot & \cdot & & & \bar{y}(\vec{n},\hat{\beta}-1) & \zeta \end{pmatrix}$$

Here

$$\zeta = \hat{m} + 4r \ ,$$

and

$$\begin{aligned} y(\vec{n},n_4) &= -\frac{1}{2}(r-\gamma_4)e^{-\hat{\mu}}U_4(n) \ , \\ \bar{y}(\vec{n},n_4) &= -\frac{1}{2}(r+\gamma_4)e^{\hat{\mu}}U_4^\dagger(n) \ . \end{aligned}$$

The fermionic contribution to the action can therefore be written in the form

$$\tilde{S}_F^{(\beta,\mu)} = \sum_{n_4,m_4=1}^{\hat{\beta}} \sum_{\vec{n},\vec{m}} \hat{\bar{\psi}}(\vec{n},n_4)K_{\vec{n},n_4;\vec{m},m_4}[U]\hat{\psi}(\vec{m},m_4) \ ,$$

and the Grassmann integration in (20.1) extends over the independent variables $\hat{\psi}(\vec{n},n_4)$ and $\hat{\bar{\psi}}(\vec{n},n_4)$ with $n_4 = 1,\cdots,\hat{\beta}$. The Grassmann integral just yields the determinant of the matrix K, so that

$$Z = \int_{per} DUe^{-S_G[U]+\ln detK[U]} \ . \tag{20.3}$$

Note that we did not bother to fix a gauge. The integral will therefore include field-configurations which are related by a gauge transformation. Since the group-volume for a compact group is finite, (20.3) is a well defined expression for any finite lattice as is used in Monte Carlo simulations.*

Given the partition function, one can proceed to study the behaviour of thermodynamical observables as a function of the temperature, and to determine the critical properties of the theory. For example, the existence of a discontinuity in the energy density (latent heat) or a jump in an order parameter would tell us that the transition is first order. Of course, in practice all calculations are carried out on lattices having finite spatial extensions. Hence one can only hope to see a rapid change in these quantities near the transition temperature. Looking for metastable states and hysteresis effects will further help one to determine the order of the transition.

In the following we will restrict ourselves to the pure $SU(N)$ Yang-Mills theory. Then the partition function has the form

$$Z = \int_{per} DU e^{-S_G[U]} . \tag{20.4}$$

Because of the non-abelian structure of the action, this theory is quite non-trivial. In fact, as we have seen in chapter 17, the pure $SU(3)$ Yang-Mills theory confines a static quark-antiquark pair at low temperatures. The question is whether confinement persists as the temperature is raised, or whether there exists a critical temperature where deconfinement sets in. To answer this question we need a non-perturbative expression for such observables as the mean energy density, and pressure.

The energy density and pressure are obtained from the partition function in the usual way:

$$\begin{aligned} \epsilon &= -\frac{1}{V}\frac{\partial}{\partial\beta}(\ln Z)_V \\ p &= \frac{1}{\beta}\frac{\partial}{\partial V}(\ln Z)_\beta . \end{aligned} \tag{20.5}$$

---

* Recall that in perturbation theory we were forced to fix the gauge. The gauge condition was implemented in the path integral by introducing a judiciously chosen factor 1. One was then led to a gauge fixed path-integral expression where the field configurations are weighted with an effective action involving the logarithm of the Faddeev-Popov determinant. Apart from an overall group-volume factor, which plays no role for the thermodynamics, the gauge fixed partition function was completely equivalent to the non-gauge fixed expression.

These expressions must be translated on the lattice into expectation values of gauge-invariant expressions constructed from the link variables. Since according to (20.5) we are to keep the physical volume or temperature fixed while varying, respectively, the temperature or volume, we must be able to vary independently the extension of the lattice in the time and space directions. For a given lattice, this can be done by choosing different lattice spacings a and a_τ along the space and time directions. In the continuum limit physics should of course be independent of the lattice regularization. Here we shall only use the anisotropic formulation to calculate the derivatives in (20.5). The resulting expressions are then evaluated for $a = a_\tau = a$.

The action on an anisotropic lattice has already been constructed in chapter 10. It is given by (10.16a,b), with $\mathcal{P}_s$ and $\mathcal{P}_\tau$ defined in (10.15b,c). At finite temperature the lattice has a finite physical extension in the euclidean time direction, and the link variables satisfy periodic boundary conditions. Note that the couplings associated with the time-like and space-like plaquettes depend on the spatial lattice spacing, as well as on the anisotropy parameter. This dependence is a consequence of quantum fluctuations, while the explicit dependence of the action on ξ is that dictated by naive arguments alone. If a_τ is the lattice spacing in the temporal direction, then the inverse temperature is given by $N_\tau a_\tau$, where N_τ is the number of lattice sites along the temporal direction. Of course in a numerical simulation the number of lattice sites in the spatial direction, N_s, will also be finite. The spatial extension of the lattice in physical units should however be large enough for finite volume effects to be negligible. In particular it should be large compared with the largest correlation length which is determined by the lowest glueball mass. A finite temperature then implies that the extension of the lattice in the temporal direction will be necessarily smaller than its linear spatial extensions. Given the action on an anisotropic lattice, we can now vary the temperature by keeping N_τ fixed, and varying the temporal lattice spacing. Hence the mean energy density and pressure can be calculated according to

$$\epsilon = \frac{< E >}{V} = -\frac{1}{N_\tau N_s^3 a^3} \left[\frac{\partial \ln Z(a, a_\tau)}{\partial a_\tau} \right]_{a_\tau = a},$$

$$p = \frac{1}{3 N_\tau N_s^3 a^3} \left[\frac{\partial \ln Z(a, a_\tau)}{\partial a} \right]_{a_\tau = a},$$

where we have returned to an isotropic lattice after having performed the differentiation. Here $\ln Z$ is considered to be a function of the spatial and temporal lattice spacings a and a_τ. These expressions can also be written in the form

$$\epsilon = \frac{1}{N_\tau N_s^3 a^4} \frac{\partial \ln Z(a,\xi)}{\partial \xi}|_{\xi=1} ,$$

$$p = \frac{1}{3a^4 N_\tau N_s^3} \left[a \frac{\partial \ln Z(a,\xi)}{\partial a} + \xi \frac{\partial \ln Z(a,\xi)}{\partial \xi} \right]_{\xi=1} ,$$

where the partition function is now considered to be a function of the spatial lattice spacing a and the asymmetry parameter ξ. From (20.4) it then follows that

$$\epsilon = \frac{1}{N_\tau N_s^3 a^4} < -\frac{\partial S_G}{\partial \xi} >_{\xi=1} , \tag{20.6a}$$

$$\epsilon - 3p = \frac{1}{N_\tau N_s^3 a^3} < \frac{\partial S_G}{\partial a} >_{\xi=1} . \tag{20.6b}$$

The derivatives can be readily calculated from the expression (10.16) for the action, where g_0 is the coupling defined on an isotropic lattice, whose dependence on the spatial lattice spacing is dictated by (9.21c,d). For $SU(3)$ one finds that [Engels et al. (1982)]

$$\epsilon = \frac{1}{N_\tau N_s^3 a^4} \frac{6}{g_0^2(a)} [< \mathcal{P}_s - \mathcal{P}_\tau > - g_0^2(a) < c_s \mathcal{P}_s + c_\tau \mathcal{P}_\tau >] , \tag{20.7a}$$

where $\mathcal{P}_s$ and $\mathcal{P}_\tau$ have been defined in (10.15a,b), and where

$$c_\sigma = \left(\frac{\partial g_\sigma^{-2}(a,\xi)}{\partial \xi} \right)_{\xi=1} . \tag{20.7b}$$

For the difference $\epsilon - 3p$ one obtains

$$\epsilon - 3p = -\frac{1}{N_\tau N_s^3 a^4} \frac{2}{g_0} \left(a \frac{\partial g_0(a)}{\partial a} \right) \langle S_G \rangle_{\xi=1} . \tag{20.8}$$

In obtaining the above expressions, we have made use of the relations

$$g_\sigma(a,1) = g_0(a),$$
$$\left(\frac{\partial g_\sigma(a,\xi)}{\partial a} \right)_{\xi=1} = \frac{\partial g_0(a)}{\partial a} . \tag{20.9}$$

Notice that (20.8) tells us that the difference $\epsilon - 3p$ is proportional to the derivative of the bare coupling with respect to the lattice spacing. For a noninteracting ideal gas of massless vector particles the right–hand side of (20.8) therefore vanishes, since in the absence of interactions g_0 does not depend on a.

Let us now return to expression (20.8). It may be used to obtain a formula for the latent heat produced in a first order deconfinement phase transition (which is expected to be seen in the pure SU(3) gauge theory). The reason is that the pressure should vary continuously in the transition region, while, in the limit of infinite spatial lattice volume, ϵ will exhibit a discontinuity at the critical temperature. On a finite lattice this discontinuity will of course be smoothed out, and therefore only manifest itself as a rapid change in ϵ. But if the pressure varies continuously in the transition region, then the discontinuity in ϵ (latent heat) is determined from (20.8) to be

$$\Delta\epsilon = \frac{1}{a^4}\frac{1}{N_s^3 N_\tau}\frac{12}{g_0^3}\beta(g_0)\Delta\langle\mathcal{P}\rangle, \tag{20.10}$$

where $\mathcal{P} = \mathcal{P}_s + \mathcal{P}_\tau$, and $\beta(g_0) = -a\partial g_0/\partial a$ is the β-function (9.6b). For small coupling $\beta(g_0)$ is given by (9.21a,b) with $N_F = 0$. Inserting this expression into (20.10) we are left with the following formula for $\Delta\epsilon$:

$$\Delta\epsilon = -\frac{1}{a^4}\frac{12}{N_s^3 N_\tau}\left(\frac{11}{(4\pi)^2} + \frac{102}{(4\pi)^4}g_0^2 + \ldots\right)\Delta\langle\mathcal{P}\rangle. \tag{20.11}$$

Hence $\Delta\epsilon$, measured in units of a^{-4} can be obtained directly in a MC–simulation. To obtain $\Delta\epsilon$ measured in physical units, we must eliminate the lattice spacing a in favour of the physical lattice scale Λ_L, which is a renormalization group invariant. The relevant relation is given by (9.21c,d). Inserting the expression for a into (20.11), we obtain

$$\frac{\Delta\epsilon}{\Lambda_L^4} = -\frac{12}{N_s^3 N_\tau}\left(\frac{11g_0^2}{16\pi^2}\right)^{\frac{204}{121}}\left(\frac{11}{(4\pi)^2} + \frac{102}{(4\pi)^4}g_0^2 + \ldots\right)e^{\frac{32\pi^2}{11g_0^2}}\Delta\langle\mathcal{P}\rangle. \tag{20.12}$$

Hence by measuring the discontinuity in $\langle\mathcal{P}\rangle$ at the (first order) phase transition, for a given value of the bare coupling, one can determine $\Delta\epsilon$ in units of the lattice parameter. The above relation holds of course only close to the continuum limit.

20.2 The Wilson Line or Polyakov Loop

At zero temperature, the potential of a static quark-antiquark pair can be determined by studying the ground state expectation value of the Wilson loop for large euclidean times. At finite temperatures the lattice has a finite extension in the time direction, and the Wilson loop no longer plays this role. The question therefore arises which is the corresponding object to be considered when studying QCD at finite temperature. To this effect we notice that, because of the periodic structure of the lattice, we can also construct gauge-invariant quantities by taking the trace of the product of link variables along topologically non-trivial loops winding around the time direction. In fig. (20-1) we show a two-dimensional picture of the simplest loop we can construct, located at some spatial lattice site $\vec{n}$. Consider the following expression, constructed from the link variables located on this loop,

$$L(\vec{n}) = Tr \prod_{n_4=1}^{\hat{\beta}} U_4(\vec{n}, n_4), \tag{20.13}$$

This expression is invariant under periodic gauge transformations:

$$\begin{aligned} U_4(\vec{n}, n_4) \rightarrow G(\vec{n}, n_4)&U_4(\vec{n}, n_4)G^{-1}(\vec{n}, \hat{n}_4 + 1) \\ G(\vec{n}, 1) =&G(\vec{n}, \hat{\beta} + 1). \end{aligned}$$

$L(\vec{n})$ is referred to in the literature as the *Wilson line*, or *Polyakov loop*. As we now show, its expectation value has a simple physical interpretation. The argument given below is only qualitative. For a more detailed discussion we refer the reader to the work of McLerran and Svetitsky (1981). To keep the presentation as simple as possible, we will consider the $U(1)$ gauge theory in the absence of dynamical fermions. Furthermore we shall use the continuum formulation.

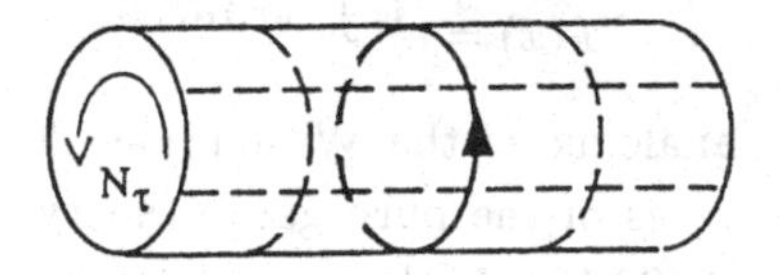

Fig. 20-1 Two dimensional picture of a loop winding around the time direction.

Consider the partition function of the system consisting of an infinitely heavy quark coupled to a fluctuating gauge potential.

$$Z = \sum_s \langle s| e^{-\beta H} |s\rangle. \tag{20.14}$$

The sum over s extends over all states of the form $\Psi^\dagger(\vec{x},0)|s'\rangle$, where $\Psi^\dagger(\vec{x},0)$ creates a quark located at $\vec{x}$ and at time $x_4 = 0$ when acting on the states $|s'\rangle$ which do not contain the heavy quark. Using the fact that $exp(-\beta H)$ generates euclidean time translations by β, we can write (20.14) in the form

$$\begin{aligned} Z =& N\sum_{s'}\langle s'|\Psi(\vec{x},0)e^{-\beta H}\Psi^\dagger(\vec{x},0)|s'\rangle \\ =& N\sum_{s'}\langle s'|e^{-\beta H}\Psi(\vec{x},\beta)\Psi^\dagger(\vec{x},0)|s'\rangle, \end{aligned} \tag{20.15}$$

where N takes into account the normalization of the quark state. But in the static limit the dependence of $\Psi(\vec{x},\beta)$ on β is determined by*

$$(\partial_\tau - ieA_4(\vec{x},\tau))\,\Psi(\vec{x},\tau) = 0.$$

Hence

$$\Psi(\vec{x},\beta) = e^{ie\int_0^\beta d\tau A_4(\vec{x},\tau)}\bar{\Psi}(\vec{x},0).$$

Inserting this result into (20.15), we are left with an expression involving the operator $\Psi(\vec{x},0)\Psi^\dagger(\vec{x},0)$. This operator describes the creation and destruction of a heavy quark at the same space time point. It merely gives rise to an (infinite) constant which is cancelled by the normalization factor N. A more careful treatment, similar to that presented in chapter 7, shows that (20.15) can be replaced by

$$Z = Tr(e^{-\beta H}L(\vec{x}))\ , \tag{20.16a}$$

where

$$L(\vec{x}) = e^{ie\int_0^\beta d\tau A_4(\vec{x},\tau)} \tag{20.16b}$$

is the $U(1)$ continuum analogue of the Wilson line (20.13), and where the trace is performed over the states of the pure gauge theory. The corresponding path integral representation of (20.16a) is therefore given by

$$Z = \int DA\ L(\vec{x})e^{-S_G^{(\beta)}[A]}, \tag{20.17}$$

where $S_G[A]$ is the finite temperature gauge field action, and where the integration extends over all fields $A_\mu(x)$ satisfying periodic boundary conditions. Notice that

* This corresponds to neglecting the contribution of the kinetic term and of the vector potential in the Dirac equation. We have dropped the mass term since it gives merely rise to an exponential factor.

$L(\vec{x})$ is invariant under periodic gauge transformations. In the non-abelian case the potential A_4 is a Lie algebra-valued matrix, and (20.16b) must be replaced by the trace in colour space of the corresponding time-ordered expression. From (20.17) we conclude that the free energy of the system with a single heavy quark, measured relative to that in the absence of the quark is given by

$$e^{-\beta F_q} = \langle L \rangle = \frac{1}{V} \sum_{\vec{x}} \langle L(\vec{x}) \rangle. \tag{20.18}$$

where V is the spatial volume of the lattice. In writing the last equality we have made use of the translational invariance of the vacuum. On the lattice, the left-hand side of (20.18) is replaced by $exp(-\hat{\beta}\hat{F}_q)$, where $\hat{\beta}$ and $\hat{F}_q$ are the inverse temperature and the free energy measured in lattice units.

The Polyakov loop resembles the world line of a static quark in a Wilson loop. This suggests that the free energy of a static quark and antiquark located at $\vec{x} = \vec{n}a$ and $\vec{y} = \vec{m}a$, with a the lattice spacing, can be obtained from the correlation function of two such loops with base at $\vec{n}$ and $\vec{m}$, and having opposite orientations (see fig. (20-2)):

$$\Gamma(\vec{n}, \vec{m}) = \langle L(\vec{n}) L^{\dagger}(\vec{m}) \rangle. \tag{20.19}$$

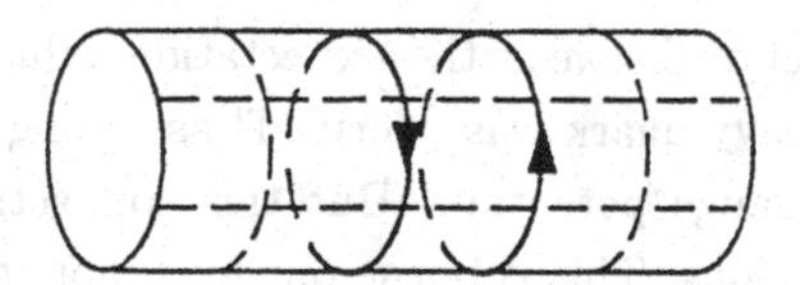

Fig. 20-2 Two Polyalov loops, winding around the time direction, used to measure the $q\bar{q}$-potential.

Indeed, by the same type of arguments which led to the connection between the Wilson loop and the static $q\bar{q}$-potential in chapter 7, but with T now replaced by β, one can show that (20.19) is related to the free energy $\hat{F}_{q\bar{q}}(\vec{n}, \vec{m})$ of a static quark-antiquark pair, measured relative to that in the absence of the $q\bar{q}$ pair, as follows:

$$\Gamma(\vec{n}, \vec{m}) = e^{-\hat{\beta}\hat{F}_{q\bar{q}}(\vec{n}, \vec{m})}. \tag{20.20}$$

Assuming that $\Gamma(\vec{n},\vec{m})$ satisfies clustering, we have that

$$\langle L(\vec{n})L^\dagger(\vec{m})\rangle \underset{|\vec{n}-\vec{m}|\to\infty}{\longrightarrow} |\langle L\rangle|^2. \qquad (20.21)$$

From here we conclude that if $\langle L\rangle = 0$, then the free energy increases for large $|\vec{n}-\vec{m}|$ with the separation of the quarks. We interpret this as signalizing confinement.

$$\langle L\rangle = 0 \quad \text{(confinement)}. \qquad (20.22)$$

On the other hand, if $\langle L\rangle \neq 0$, then the free energy of a static quark-antiquark pair approaches a constant for large separations. In the absence of vacuum polarization effects, arising from dynamical quark, we interpret this as signalizing deconfinement:

$$\langle L\rangle \neq 0, \quad \text{(deconfinement in the pure gauge theory)} \qquad (20.23)$$

The $q\bar{q}$ potential in the deconfined phase is obtained by dividing (20.20) by $|\langle L\rangle|^2$, which removes the self-energy contributions of the individual quarks, i.e.,

$$e^{-\hat{\beta}V_{q\bar{q}}(\hat{R})} = \frac{\langle L(\vec{n})L^\dagger(\vec{m})\rangle}{|\langle L\rangle|^2}, \qquad (20.24)$$

where $\hat{R} = |\vec{n}-\vec{m}|$.

The above connection between the expectation value of the Wilson line and the free energy of a heavy quark was "derived" assuming that no quarks of finite mass coupled to the gauge potential. But this connection also holds when we include dynamical fermions. This is borne out by a more careful treatment, where one first considers the partition function including "light" and heavy quarks, and then takes the infinite mass limit for the heavy quarks. In this way one finds that the free energy F_q is again related to $\langle L\rangle$ by (20.18), except that now the expectation value is calculated with the finite temperature action involving the dynamical fermions [McLerran and Svetitsky (1981)].

According to (20.22) and (20.23), the expectation value of the Wilson line evaluated in a pure gluonic medium serves as an order parameter for distinguishing a confined from a deconfined phase in the pure $SU(3)$ gauge theory. In statistical mechanics, phase transitions are usually associated with a breakdown of a global symmetry. We now show that this is also expected to be true in the $SU(3)$ gauge theory.

20.3 Spontaneous Breakdown of the Center Symmetry and the Deconfinement Phase Transition

The lattice action of the pure $SU(3)$ gauge theory is not only invariant under periodic gauge transformations, but also possesses a further symmetry which is not shared by the Wilson line. Consider the elements of $SU(3)$ belonging to the center $\mathcal{C}$ of the group.* They are given by $exp(2\pi il/3) \in Z(3)$ where $l = 0, 1, 2$. Clearly the action (10.16) is invariant if we multiply all time-like oriented link variables U_4 between two neighbouring spatial sections of the lattice by an element of the center

$$U_4(\vec{n}, n_4) \rightarrow zU_4(\vec{n}, n_4),$$

where $z \in \mathcal{C}$. But the Wilson line is not invariant since it contains one link variable which transforms non–trivially:

$$L(\vec{n}) \rightarrow zL(\vec{n}).$$

Now if the ground state of the quantum system respects the symmetry of the classical action, then link configurations related by the center symmetry will occur with the same probability. Thus the same number of configurations will yield the values $L_l = e^{2\pi il/3}L \quad (l = 0, 1, 2)$ for the Wilson line. Since $\sum_l exp(2\pi il/3) = 0$, it follows that the expectation value of the Wilson line must vanish. But this we had interpreted as signalizing confinement. Hence we expect that the center symmetry is realized in the low temperature, confining phase of the pure $SU(3)$ gauge theory. On the other hand, if $\langle L \rangle \neq 0$, then the center symmetry is necessarily broken. We hence expect that a deconfinement phase transition is accompanied by a breakdown of the center symmetry and that the phases of the Polyakov loops cluster around any one of the $Z(3)$ roots. As we shall see in section 6, this is borne out by MC calculations. That the $Z(3)$ symmetry plays a crucial role in the confinement problem has been suggested some time ago by Svetitsky and Yaffe (1982). These authors argued that the critical behaviour of the $SU(3)$ gauge theory is that of an $Z(3)$ spin model. Since this model exhibits a first-order phase transition, one also expects that a deconfinement phase transition in the pure $SU(3)$ gauge theory, is of first order.

The above considerations applied only to the pure gauge sector, or equivalently in the infinite quark mass limit. The case of infinitely heavy quarks is of

---

* The center $\mathcal{C}$ of a group G consists of all elements z for which $zgz^{-1} = g$, with $g \in G$.

course unphysical. Nevertheless the study of the pure $SU(3)$ gauge theory provides us with important information regarding the role played by the non-abelian gauge field for quark confinement. As we have pointed out before, one generally believes that quark confinement is a consequence of the non-abelian self-couplings of the gauge potential. At low temperatures, this non-abelian coupling is expected to lead to the formation of a flux tube ($\rightarrow$ string) connecting the quark and antiquark. The energy stored in the string is proportional to its length which is of the order of the separation of the quark and antiquark pair. As the temperature is increased, the string begins to fluctuate more and more, its length now being larger than the separation between the quarks. At the same time the number of possible string configurations with a given length will increase. That is, the entropy will start competing with the energy carried by the string. At sufficiently large temperature, the system may then prefer to form networks of flux tubes to which the quark and antiquark are attached to. Thus above some critical temperature a new phase may be formed, consisting of huge networks of flux tubes (Patel, 1984). The energy of such a configuration will be roughly the same, no matter where the quark and antiquark are hooked onto a network. This corresponds to deconfinement. When dynamical quarks are coupled to the gauge potentials, then flux tubes can break. The probability for this to occur increases with decreasing quark mass. This is also true for the single flux tube connecting the quark and antiquark at low temperatures. In both cases the forces between the quarks get screened. But whereas the screening at low temperatures is due to pair production processes along the single string connecting the two quarks, the breakdown of the network in the presence of light dynamical quarks resembles more the Debye screening in a plasma.

Finally we want to mention that a fermionic contribution to the action breaks the center symmetry explicitly. Hence the expectation value of the Wilson line need not vanish in the confining phase. In this case the free energy of a $q\bar{q}$ pair approaches a finite constant for infinite separations, which is consistent with the expectation that the force between two quarks is screened by vacuum polarization effects.

20.4 How to determine the Transition Temperature

When studying QCD at finite temperature in a Monte Carlo simulation one must first decide on the spatial and temporal extension of the lattice to be used. In principle, the spatial volume should be chosen large enough so that finite volume effects do not play an important role. For a given set of lattice sites the linear physical extension of the lattice will depend on the lattice spacing, and hence on the value of the bare coupling constant. Again, in principle, this coupling should be taken small enough to approximate continuum physics. Let N_s be the number of lattice sites along any one of the three spatial directions. It then follows from the euclidean space–time symmetry of the path integral formulation that for lattices with temporal extent $N_\tau > N_s$ physics should be insensitive to the periodic boundary conditions in the time direction - if it is not sensitive to the finite spatial volume. Hence such a choice of lattice is appropriate for studying QCD at "zero" temperatures.* On the other hand, to study QCD at high temperatures, the temporal extension of the lattice must be much smaller than in the space directions. This is the reason why Monte Carlo computations are carried out on lattices with $N_\tau < N_s$. For a given lattice spacing a the physical temperature is given by

$$T = \frac{1}{N_\tau a}.$$

Thus the temperature can be varied by either changing N_τ (then the temperature varies in a discontinuous way), or by varying the lattice spacing, which can be done by changing the coupling g_0. In principle the best way of proceeding would be to introduce two types of couplings which allow one to vary the lattice spacing in the temporal direction without changing the spatial volume of the lattice. But if this volume is large enough for finite volume effects to be negligible (which they are not in general) then a single coupling constant should suffice.

To determine the critical temperature at a phase transition one studies the temperature dependence of an order parameter, and of such thermodynamical quantities as the energy density, specific heat, etc.. If the QCD coupling $6/g_0^2$ is large enough, then the lattice spacing is related to the bare coupling constant by (9.21c,d) and the temperature is given by

$$T = \frac{\Lambda_L}{N_\tau R(g_0)}.$$

Let g_c be the critical coupling at which the phase transition takes place. Then the critical temperature is given by

$$T_c = \frac{\Lambda_L}{N_\tau R(g_c)}. \tag{20.25}$$

* Actually, the temperature is never really zero on a lattice with finite temporal extension.

To test whether one is really determining a physical transition temperature, one should repeat the simulation for lattices of different temporal extensions, determine g_c in each case (g_c will of course depend on the choice of N_τ) and check that

$$N_\tau R(g_c(N_\tau)) = const. \tag{20.26}$$

Another way of obtaining the transition temperature in physical units is to measure some other observable, such as a hadron mass. Since the critical temperature and hadron mass in lattice units are given by $\hat{T}_c = T_c a$ and $\hat{m}_H = m_H a$, where m_H is the physical hadron mass, it follows that $\hat{T}_c/\hat{m}_H = T_c/m_H$. Hence the temperature in physical units is given by

$$T_c = \left(\frac{\hat{T}_c}{\hat{m}_H}\right) m_H .$$

As we shall see in section 6 there exists strong evidence that QCD undergoes a deconfinement phase transition at high temperatures. If so, this would be an interesting prediction of this theory which, if confirmed by experiment, would constitute an important test of QCD as being the correct theory of strong interactions.

20.5 A Two-Dimensional Model. Test of Theoretical Concepts

In the previous sections we have introduced a number of important concepts which are relevant for studying the thermodynamics of a pure gauge theory. Since our discussion has been in part quite formal, it is instructive to check the theoretical ideas in a solvable model. The simplest lattice model one can consider is the $U(1)$ gauge theory in one time and one space dimension. This theory will confine a $q\bar{q}$-pair, since in one space dimension the field energy cannot spread out in space. In this section we will use the compact $U(1)$ gauge theory as a toy model to calculate the following quantities: i) the expectation value of the Polyakov loop, ii) the expectation value of two oppositely oriented Polyakov loops (which is related to the free energy of a static $q\bar{q}$-pair) , and iii) the ensemble average of the electric field energy density stored in the string connecting an oppositely charged pair. We will compute these quantities using the character expansion, in order to illustrate at the same time a technique which is relevant for strong coupling expansions in $SU(N)$ lattice gauge theories. * For the simple model we consider the calculations could actually be carried out without making use of this sophisticated method.

* See e.g. Migdal (1975); Drouffe and Zuber (1983); Munster (1981). The character expansion has been used by Rusakov (1990) to compute observables in $U(N)$ gauge theories on arbitrary two-dimensional manifolds.

The basic idea of the character expansion is the following. In a lattice gauge theory the pure gauge part of the action has the form

$$S_G[U] = \hat{\kappa} \sum_P F(U_P) ,$$

where U_P are the plaquette variables, and $F(U_P)$ is a real valued function of the plaquette variables which is invariant under gauge transformations, $F(g^{-1}U_P g) = F(U_P)$, with g an element of the unitary gauge group G. We have denoted the coupling by $\hat{\kappa}$, instead of $\hat{\beta}$, which we reserve for the inverse temperature measured in lattice units. Since $\exp(-\hat{\kappa}F(U_P))$ is a class function on G it can be expanded in terms of irreducible characters (traces of irreducible representations) of U_P as follows

$$\exp(-\hat{\kappa}F(U_P)) = \sum_\nu d_\nu \lambda_\nu(\hat{\kappa}) \chi_\nu(U_P) , \tag{20.27}$$

where ν labels the irreducible representations of G, χ_ν is the character of U_P in the ν'th irreducible representation, and d_ν is the dimension of the representation. For a real class function $F(U)$, conjugate representations χ_ν and $\chi_{\bar{\nu}}$ contribute with the same weight in (20.27). The plaquette variables U_P must be chosen with a conventional orientation. The convention chosen is immaterial, for changing the convention merely replaces U_P by $U_P^\dagger$, and hence the representation "ν" by its conjugate "$\bar{\nu}$". Below we list some important formulae which we will need for our discussion.

Let V, W, V_1, Ω_1, etc. be elements of a compact unitary group G, and dV the normalized Haar measure on G (i.e., $\int DV = 1$), satisfying

$$dV = dV^{-1} = d(VW) \ ; \ W \in G . \tag{20.28}$$

Then the following relations hold:*

$$\int dV \chi_\nu(V) \chi_{\nu'}(V^\dagger) = \delta_{\nu\nu'} , \tag{20.29a}$$

$$\int dV \chi_{\nu_1}(V_1 V) \chi_{\nu_2}(V^\dagger V_2) = \frac{1}{d_{\nu_1}} \delta_{\nu_1\nu_2} \chi_{\nu_1}(V_1 V_2) , \tag{20.29b}$$

$$\int dV \chi_\nu(V\Omega_1 V^\dagger \Omega_2) = \frac{1}{d_\nu} \chi_\nu(\Omega_1) \chi_\nu(\Omega_2) . \tag{20.29c}$$

---

* Making use of (20.28), these formulae are seen to hold also if V denotes a product of link variables and dV, the corresponding product of Haar measures.

From the character expansion (20.27), and the orthogonality relation (20.29a) it follows that

$$\lambda_\nu(\hat{\kappa}) = \frac{1}{d_\nu}\int dU \chi_\nu(U^\dagger)e^{-\hat{\kappa}F(U)} \,. \tag{20.30}$$

In the case where $G = U(1)$, the irreducible representations are one dimensional, and the characters are given by* $\chi_\nu(U) = e^{i\nu\theta}$, where ν is an integer. The $U(1)$ action has the form (see (8.10))

$$S_G[U] = \hat{\kappa}\sum_P[1 - \frac{1}{2}(U_P + U_P^\dagger)] = \hat{\kappa}\sum_P[1 - \cos\theta_P] \,, \tag{20.31}$$

where $\hat{\kappa} = \frac{1}{\hat{e}^2}$, with $\hat{e}$ the charge e measured in lattice units, i.e., $\hat{e} = ea$. θ_P is the sum of the phases of the directed link variables around a plaquette P. The Haar measure, normalized to unity, is given by $dU = d\theta/2\pi$. When calculating the expectation value of an observable we can replace the Boltzmann factor $\exp(-\hat{\kappa}S_G[U])$ by $\prod_P \exp(\hat{\kappa}\cos\theta_P)$. Then

$$< O >= \frac{\int DUO[U]\prod_P e^{\hat{\kappa}f(U_P)}}{\int DU\prod_P e^{\hat{\kappa}f(U_P)}} \,, \tag{20.32a}$$

where

$$f(U_P) = \cos\theta_P \,. \tag{20.32b}$$

Accordingly, eq. (20.27) is replaced in the $U(1)$ gauge theory by

$$e^{\hat{\kappa}f(U_P)} = \sum_\nu I_\nu(\hat{\kappa})\chi_\nu(U_P) \,, \tag{20.33}$$

where the coefficients have been calculated from (20.30) with $\hat{\kappa}F(U)$, dU and $\chi_\nu(U)$ replaced by $-\hat{\kappa}\cos\theta$, $d\theta/2\pi$ and $\exp(i\nu\theta)$, respectively. $I_\nu(\hat{\kappa})$ is the modified Bessel function of order ν.

Consider two plaquettes P_1 and P_2 having a link in common. With each plaquette there is a factor $\exp(\hat{\kappa}f(U_P))$ asssociated with it. The character expansion allows us to perform the integration over the common link-variable, which we denote by V. This is also the case for non-abelian gauge theories, where this method

---

* For SU(2) the link variables in the fundamental representation can be parame trized in form $U = \cos\frac{\theta}{2} + i\vec{\sigma}\cdot\vec{n}\sin\frac{\theta}{2}$, where σ_i are the Pauli matrices and $\vec{n}$ a unit vector. The Haar measure and characters are given in this case by $dU = \frac{1}{8\pi^2}\sin^2\frac{\theta}{2}d\theta d\Omega$ and $\chi_\nu(U) = \frac{\sin(j+1/2)\theta}{\sin\frac{\theta}{2}}$, where j takes integer or half integer values.

acquires its real power. For the U(1) gauge theory the integral of interest is given by

$$\int dV e^{\hat{\kappa} f(U_{P_1})} e^{\hat{\kappa} f(U_{P_2})} = \sum_{\nu_1 \nu_2} I_{\nu_1}(\hat{\kappa}) I_{\nu_2}(\hat{\kappa}) \int dV \, \chi_{\nu_1}(U_{P_1}) \chi_{\nu_2}(U_{P_2}) \, .$$

Let Ω_1 and Ω_2 denote the path ordered products* of link variables along the solid and dottet paths shown in fig (20-3). Then $U_{P_1} = \Omega_1 V$ and $U_{P_2} = V^\dagger \Omega_2$

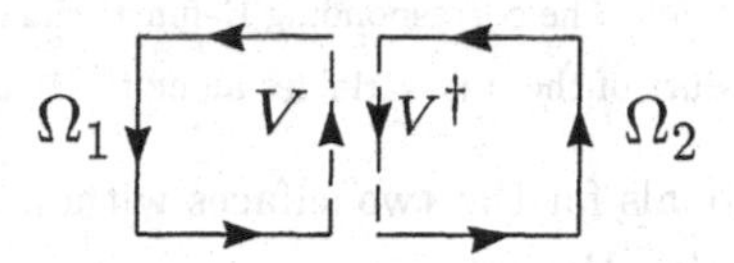

Fig. 20-3 Two plaquettes which are glued together at the common link.

Making use of (20.29b) with $d_\nu = 1$, one obtains

$$\int dV e^{\hat{\kappa} f(U_{P_1})} e^{\hat{\kappa} f(U_{P_2})} = \sum_\nu [I_\nu(\hat{\kappa})]^2 \chi_\nu(U_{\partial P_{12}}) \, ,$$

where $U_{\partial P_{12}}$ is the "path ordered" product of link variables around the boundary of the area covered by the two plaquettes. This area, measured in lattice units, equals the power of the coefficient $I_\nu(\hat{\kappa})$. Proceeding in this way, one can carry out successively the integration over all link variables lying inside of a simply connected two-dimensional lattice of area A, bounded by a simple closed contour ∂A. The result is the K-functional [Migdal (1975)],

$$K_A(U_{\partial A}) = \sum_\nu [I_\nu(\hat{\kappa})]^{\hat{A}} \chi_\nu(U_{\partial A}) \, , \tag{20.34}$$

where $U_{\partial A} = \Omega_0 \Omega' \Omega_\beta \Omega$ is the path ordered product of link variables around the boundary of the lattice shown in fig. (20-4), and $\hat{A}$ is the number of plaquettes in the domain bounded by ∂A. This expression holds for a lattice bounded by an arbitrarily shaped contour.

The only other K-functional we shall need is that associated with a lattice with a hole, depicted in fig. (20-5a). We shall denote this functional by $K_{A,o}$, where $\hat{A}$ is the number of plaquettes bounded by the two closed curves. It is obtained by

---

* Of course in the abelian case path ordering is irrelevant. We shall nevertheless use this language and always write the product of link variables in the path ordered form, as would be required in the non-abelian case.

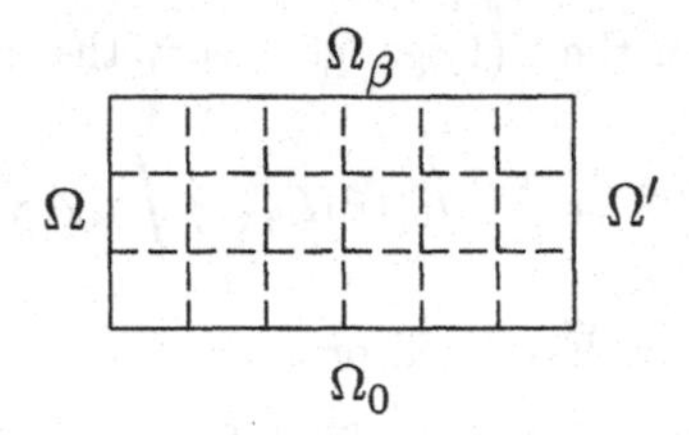

Fig. 20-4 All link variables except for those on the boundary of the lattice have been integrated out. The corresponding K-functional only depends on the path ordered product of the link-variables along the boundary

glueing together the K-functionals for the two sufaces without holes, shown in fig. (20-5b), along the common links. Hence

$$K_{A,o} = \int dV_1 dV_2 K_{A_1}(U_{\partial A_1}) K_{A_2}(U_{\partial A_2}) \; .$$

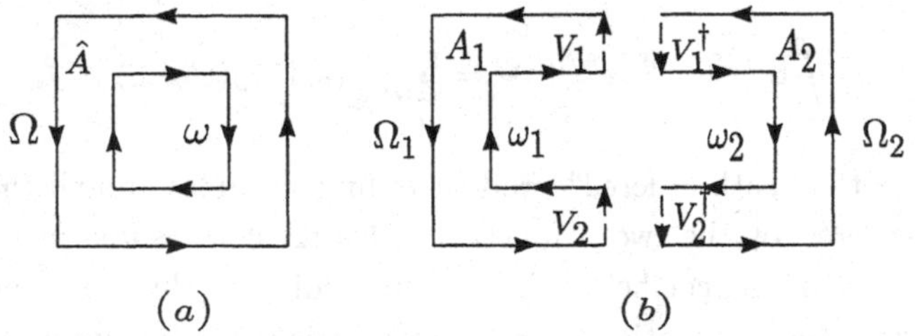

Fig. 20-5 Two surfaces without holes which are glued together along the common links

Written out explicitely we have that

$$K_{A,o} = \sum_{\nu_1 \nu_2} [I_{\nu_1}(\hat{\kappa})]^{\hat{A}_1} [I_{\nu_2}(\hat{\kappa})]^{\hat{A}_2} \int dV_1 dV_2 \chi_{\nu_1}(\Omega_1 V_2 \omega_1 V_1) \chi_{\nu_2}(V_1^\dagger \omega_2 V_2^\dagger \Omega_2) \; .$$

Making use of (20.29b) with $d_\nu = 1$, we can perform the integral over V_1 and obtain

$$K_{A,o} = \sum_\nu [I_\nu(\hat{\kappa})]^{\hat{A}} \int dV_2 \chi_\nu(\Omega_1 V_2 \omega_1 \omega_2 V_2^\dagger \Omega_2) \; .$$

This integral is of the form (20.29c).* Hence

$$K_{A,o}(\Omega,\omega) = \sum_\nu [I_\nu(\hat{\kappa})]^{\hat{A}} \chi_\nu(\Omega) \chi_\nu(\omega) \; , \tag{20.35}$$

---

* In the non-abelian case one makes use of the invariance of the character (trace) under cyclic permutations of the matrix valued group elements.

where Ω and ω are the products of the link variables along the closed contours shown in fig. (20-5a). As we shall see, all other K-functionals which we will need can be obtained from (20.34) and (20.35). These K-functionals will depend on i) the link variables living on the boundary of the lattice, and ii) the link variables on which the observable depends whose expectation value we want to calculate.

At finite temperature the lattice has a finite extension in the euclidean time direction, given by the inverse temperature β, and the link variables at times $\tau = 0$ and $\tau = \beta$ are identified. For the time-like oriented link variables on the spatial boundary of the lattice one can, for example, impose periodic, or free boundary conditions. Let Γ_L denote the set of unconstrained variables on the boundary of the lattice, and Γ the set of variables on which the observable O depends. The corresponding finite temperature K-functional we denote generically by $K^{(\beta)}(\Gamma_L, \Gamma)$. Then the expectation value (20.32a) will be given by

$$< O >= \frac{\int D\Gamma_L D\Gamma \; O[\Gamma] K^{(\beta)}(\Gamma_L, \Gamma)}{\int D\Gamma_L K^{(\beta)}(\Gamma_L)} , \qquad (20.36a)$$

where

$$K^{(\beta)}(\Gamma_L) = \int D\Gamma K^{(\beta)}(\Gamma_L, \Gamma) . \qquad (20.36b)$$

Having set up the general framework, we now proceed to some concrete calculations.

(i) *The Polyakov Loop*

Consider the expectation value of a Polyakov loop (or Wilson line) in the $\nu = 1$ representation, i.e.,

$$< L >=< \chi_1(U_\Gamma) > ,$$

where U_Γ denotes the (path ordered) product of link variables along a closed loop Γ winding around the compactified surface along the temperature axis. To calculate this expectation value we must evaluate the numerator and denominator in (20.32a).

Consider first the denominator. The relevant finite-temperature K-functional is obtained from (20.34) by identifying $\Omega_\beta^\dagger$ with Ω_0 in $U_{\partial A} = \Omega_0 \Omega' \Omega_\beta \Omega$ (see fig. (20-4)), and integrating the expression over Ω_0, making use of (20.29c) with $d_\nu = 1$. One then finds that

$$K_A^{(\beta)}(\Omega, \Omega') = \sum_\nu [I_\nu(\hat{\kappa})]^{\hat{A}} \chi_\nu(\Omega) \chi_\nu(\Omega') . \qquad (20.37)$$

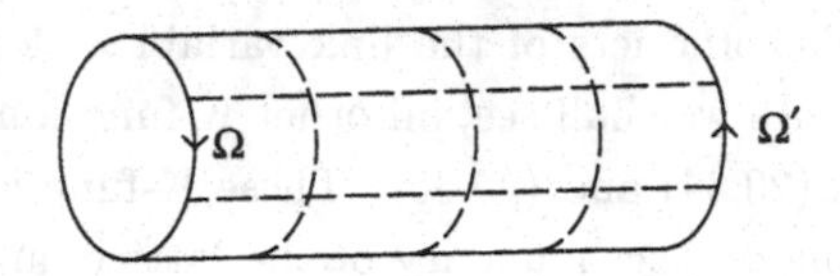

Fig. 20-6 Lattice which is compactified along the temperature axis. The coresponding finite temperature K-functional only depends on the path ordered products of link variables Ω and Ω'.

This is the K-functional associated with the surface shown in fig. (20-6).

We next must decide what type of boundary conditions we want to impose on the link variables at the ends of the cylinder. Imposing periodic boundary conditions, i.e., identifying Ω and $\Omega'^\dagger$, and integrating the expression over Ω, fig. (20-6) is glued together to the torus. Making use of (20.29a) the corresponding K-functional is given by

$$K_A^{(\beta,torus)} = Z_{(torus)} = \sum_\nu [I_\nu(\hat{\kappa})]^{\hat{A}} , \tag{20.38}$$

which is just the denominator of (20.32a).

If instead of toroidal boundary conditions we impose free spatial boundary conditions, then integrating (20.37) over Ω and Ω' projects out the trivial representation. Hence

$$K_A^{(\beta,free)} = Z_{free} = [I_0(\hat{\kappa})]^{\hat{A}} . \tag{20.39}$$

Z_{free} is just the denominator of (20.32a) for free spatial boundary conditions.

Consider next the numerator of (20.32a), or of (20.36a) with $O[\Gamma]$ replaced by $\chi_1(U_\Gamma)$. The relevant K-functional is that associated with fig. (20-7a). It is given by the product of the K-functionals associated with the two surfaces shown in fig. (20-7b):

$$K_{A_1A_2}^{(\beta)} = K_{A_1}^{(\beta)}(\Omega, U_\Gamma)K_{A_2}^{(\beta)}(U_\Gamma^\dagger, \Omega') . \tag{20.40}$$

Imposing the periodic boundary condition $\Omega'^\dagger = \Omega$, and integrating over Ω, we obtain

$$K_{A_1A_2}^{(\beta,torus)}(U_\Gamma, U_\Gamma^\dagger) = \sum [I_\nu(\hat{\kappa})]^{\hat{A}} \chi_\nu(U_\Gamma)\chi_\nu(U_\Gamma^\dagger) ,$$

where $\hat{A}$ is the total area of the lattice, measured in lattice units. The expectation value of the Polyakov loop $< L >=< \chi_1(U_\Gamma) >$ is now given by

$$< L >_{(torus)} = \frac{1}{Z_{(torus)}} \int DU_\Gamma \chi_1(U_\Gamma) K_{A_1,A_2}^{(\beta,torus)}(U_\Gamma, U_\Gamma^\dagger) .$$

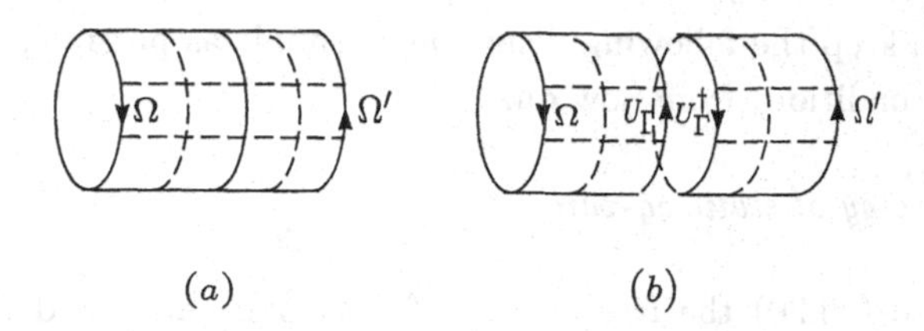

Fig. 20-7 (a) All link variables except for those living on the closed contours are integrated out. The corresponding K-functional is given by product of the K-functionals associated with the two surfaces shown in (b).

A pictorial representation of the numerator is given in fig. (20-8). Evaluation of the integral yields

$$< L >_{torus} = \frac{1}{Z_{(torus)}} \sum_{\nu} [I_\nu(\hat{\kappa})]^{\hat{A}} \mathcal{D}^{\nu}_{1\nu} ,$$

where the Wigner coefficient $\mathcal{D}^k_{rr'}$ is defined by

$$\begin{aligned} \mathcal{D}^k_{rr'} &= \int dU \chi_r(U) \chi_{r'}(U) \chi_k(U^\dagger) \\ &= \delta_{k,r+r'} \quad (for\ U(1)) \end{aligned} \quad . \tag{20.41}$$

Hence

$$< L >_{torus} = 0 ,$$

which according to the criterium (20.22) implies confinement. This is of course expected to be the case in one space dimension, as we have pointed out at the beginning of this section. As the reader can readily verify, the same result is obtained by imposing free boundary conditions.

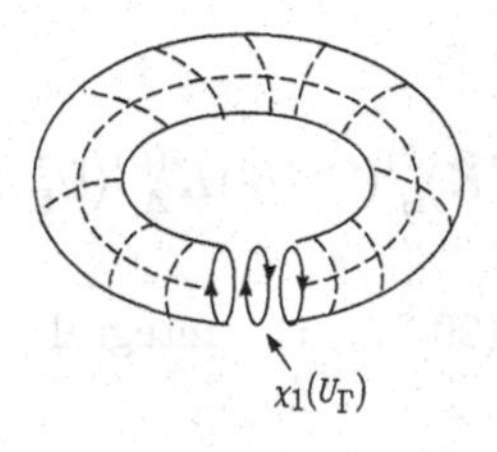

Fig. 20-8 Fig. (20-7b) glued together to a torus. Also shown is the insertion of a Polyakov loop.

In order to keep the following disussion as simple as possible, we will choose free boundary conditions from now on.

(ii) The free energy of static $q\bar{q}$-pair

Acording to (20.20) the free energy of a $q\bar{q}$ pair, measured relative to the vacuum, is related to the expectation value of two oppositely oriented Wilson lines by

$$\hat{F}_{q\bar{q}}(\hat{R},\hat{\beta}) = -\frac{1}{\hat{\beta}} \ln < L_n L^\dagger_{n'} > , \tag{20.42a}$$

where $\hat{R} = |n - n'|$, and

$$L_n = \chi_1(U_\Gamma) \;\; ; \;\; L^\dagger_{n'} = \chi_1(U^\dagger_{\Gamma'}) . \tag{20.42b}$$

Here U_Γ and $U^\dagger_{\Gamma'}$ denote the product of link variables along the two closed paths shown in fig. (20.9a), located at the spatial lattice sites n and n' of the two charges.

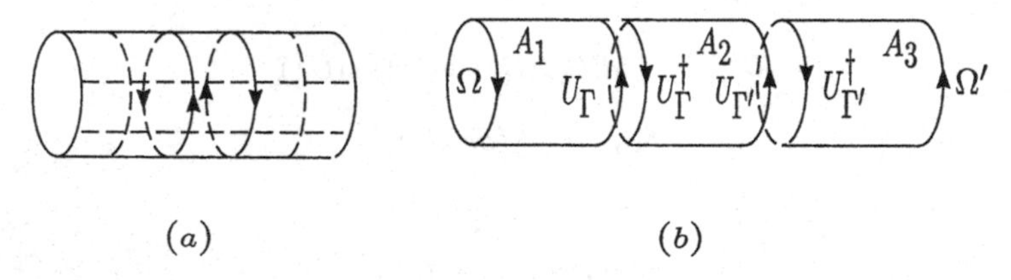

Fig. 20-9 (a) Two oppositely oriented Polyakov loops winding around the compactified lattice; (b) The relevant K-functional is given by the product of the K-functionals associated with the 3 surfaces.

For free spatial boundary conditions the relevant K-functional is given by the product of the K-functionals associated with the three surfaces in fig. (20-9b), integrated over Ω and Ω',

$$K^{(\beta,free)}_{A_1A_2A_3} = \int d\Omega d\Omega' K^{(\beta)}_{A_1}(\Omega, U_\Gamma) K^{(\beta)}_{A_2}(U^\dagger_\Gamma, U_{\Gamma'}) K^{(\beta)}_{A_3}(U^\dagger_{\Gamma'}, \Omega') .$$

Making use of the expression (20.37), the integral can be readily performed and one obtains

$$K^{(\beta,free)}_{A_1A_2A_3}(U^\dagger_\Gamma, U_{\Gamma'}) = [I_0(\hat{\kappa})]^{\hat{A}_1+\hat{A}_3} \sum_\nu [I_\nu(\hat{\kappa})]^{\hat{A}_2} \chi_\nu(U^\dagger_\Gamma)\chi_\nu(U_{\Gamma'}) .$$

The expectation value in (20.42a) is then given by

$$< L_n L_{n'}^{\dagger} >= \frac{1}{Z_{(free)}} \int dU_{\Gamma} dU_{\Gamma'} \chi_1(U_{\Gamma}) \chi_1(U_{\Gamma'}^{\dagger}) K_{A_1 A_2 A_3}^{(\beta, free)}(U_{\Gamma}^{\dagger}, U_{\Gamma'}) .$$

The integral can be evaluated immediately by making use of the orthogonality relation (20.29a) and of (20.39):

$$< L_n L_{n'}^{\dagger} >= \left(\frac{I_1(\hat{\kappa})}{I_0(\hat{\kappa})} \right)^{\hat{A}_2} . \tag{20.43}$$

where $\hat{A}_2$ is the number of plaquettes enclosed by the two Polyakov loops. According to (20.42a), the free energy of a static oppositely charge pair is therefore given in lattice units by

$$\hat{F}_{q\bar{q}}(\hat{R}, \hat{\kappa}) = \left(\ln \frac{I_0(\hat{\kappa})}{I_1(\hat{\kappa})} \right) \hat{R} . \tag{20.44}$$

The right hand side coincides with the expression we derived in chapter 8 for the $q\bar{q}$- potential at zero temperature (cf. eq. (8.17)) . This is not surprising, since in one space dimension the string connecting the charged pair cannot fluctuate. From our discussion in chapter 8 it therefore follows immediately that in the continuum limit ($\hat{\kappa} = \frac{1}{\hat{e}^2} = \frac{1}{e^2 a} \to \infty$) the free energy, measured in physical units, is given by

$$F_{q\bar{q}}(R) = \frac{1}{2} e^2 R . \tag{20.45}$$

Hence in this model the $q\bar{q}$-system is confined.

iii) *Energy Sum Rule*

In the following we first derive an energy sum rule which relates the mean energy of a static quark-antiquark pair to the field energy stored in the string. The sum rule is then evaluated using the character expansion.

Consider the partition function for a static $q\bar{q}$ pair:

$$Z_{q\bar{q}} = \int DU L_n L_{n'}^{\dagger} e^{-S_G} = Z_0 < L_n L_{n'}^{\dagger} >$$

where $Z_0 = \int DU e^{-S_G}$. The mean energy, measured relative to the vacuum, is then given by

$$< E >= - \frac{\partial}{\partial \beta} \ln < L_n L_{n'}^{\dagger} > .$$

To evaluate the rhs we must be able to vary the temperature in a continuous way. Since $\beta = \hat{\beta} a_\tau$, where a_τ is the lattice spacing in the euclidean time direction, we can vary the temperature by varying a_τ, keeping the number of lattice spacings $\hat{\beta}$ in the temporal direction fixed. To calculate the β-derivative we must therefore first evaluate the expectation value $< L_n L^\dagger_{n'} >$ on an anisotropic lattice. From our discussion in section 2 of chapter 10 it is evident that the finite temperature action on an anisotropic lattice is given for our $U(1)$ model by

$$S^{(\beta)}[U;\xi] = \hat{\kappa}\xi \sum_P [1 - ReU_P] \ , \tag{20.46}$$

where $\xi = \frac{a}{a_\tau}$, with a the spatial lattice spacing. Since the naive continuum limit of this action is that of a free theory, we do not expect that the coupling $\hat{\kappa}$ must be tuned with the lattice spacings a and a_τ when taking the continuum limit (as was the case for the non-abelian theory). The mean energy, measured in lattice units is therefore given by

$$< \hat{E} >= \frac{1}{\hat{\beta}} \left(\frac{\partial}{\partial \xi} \ln < L_n L^\dagger_{n'} > \right)_{\xi=1} \ . \tag{20.47a}$$

where

$$< L_n L^\dagger_{n'} >= \frac{\int DU L_n L^\dagger_{n'} e^{-\hat{\kappa}\xi \sum_P [1-ReU_P]}}{\int DU e^{-\hat{\kappa}\xi \sum_P [1-ReU_P]}} \ . \tag{20.47b}$$

In (20.47a) we have returned to an isotropic lattice after differentiation. Performing the differentiation in ξ we obtain

$$< \hat{E} >= \frac{\hat{\kappa}}{\hat{\beta}} < -\mathcal{P} >_{q\bar{q}-0} \ , \tag{20.48a}$$

where

$$\mathcal{P} = \sum_P [1 - ReU_P] \ , \tag{20.48b}$$

and

$$< \mathcal{P} >_{q\bar{q}-0}= \frac{< L_n L^\dagger_{n'} \mathcal{P} >}{< L_n L^\dagger_{n'} >} - < \mathcal{P} > \ . \tag{20.48c}$$

In the naive continuum limit $\mathcal{P} \to \frac{1}{2} \int d^2x F_{12} F_{12}$. In one space dimension the field tensor $F_{\mu\nu}$ has only one non-vanishing component, F_{12}, corresponding to the electric field. One of the objectives of this calculation will be to confirm that the euclidean formulation, with this minus sign in (20.48a), gives the expected answer for the mean energy. A similar expression had been obtained in chapter 10 for the contribution of the electric field to the energy sum rule in the pure $SU(3)$ gauge theory.

Because of translational invariance in the euclidean time direction, we can also write (20.48a) in the form

$$< \hat{E} >= \hat{\kappa} < -\mathcal{P}' >_{q\bar{q}-0} , \tag{20.49}$$

where $\mathcal{P}'$ denotes the contribution to (20.48b) arising from plaquettes located on a fixed time slice. From here we infer that the energy density, as probed by a plaquette, is given in lattice units by

$$< \hat{\mathcal{E}} >= \hat{\kappa}\Big[\frac{< L_n L_{n'}^\dagger ReU_P >}{< L_n L_{n'}^\dagger >} - < ReU_P >\Big] . \tag{20.50}$$

The expectation values are calculated on an isotropic lattice according to (20.32). For free spatial boundary conditions $< L_n L_{n'}^\dagger >$ is given by (20.43). The relevant K-functional for evaluating $< L_n L_{n'}^\dagger ReU_P >$ *before* imposing free spatial boundary conditions, is given by the product of i) the K-functionals associated with the surfaces with areas A_1 and A_3 in fig. (20-10), ii) the K-functional associated with the surface A_2, where the window has the size of a single plaquette, and iii) the K-functional of a single plaquette, whose Fourier Bessel expansion is given by (20.33). The *finite-temperature* K-functional associated with the surface A_2 is obtained from (20.35) in the by now familiar way. One readily finds that

$$K_{A_2,o}^{(\beta)} = \sum_\nu [I_\nu(\hat{\kappa})]^{\hat{A}_2} \chi_\nu(U_\Gamma^\dagger)\chi_\nu(U_{\Gamma'})\chi_\nu(U_P^\dagger) . \tag{20.51}$$

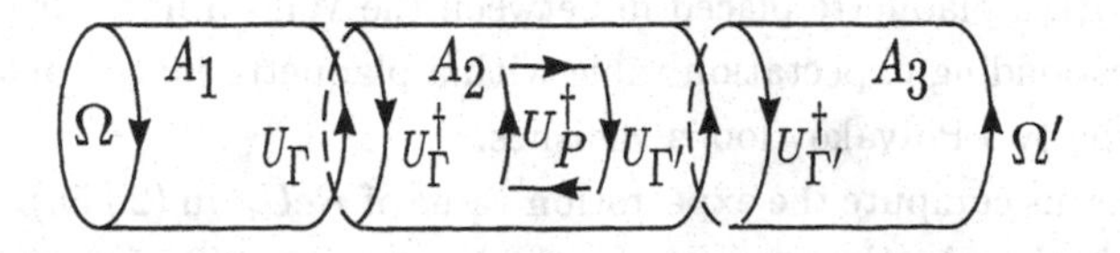

Fig. 20-10 The K-functional relevant for the computation of $< L_n L_{n'}^\dagger ReU_P >$ is given, before imposing spatial boundary conditions, by the product of the K-functionals associated with the 3 surfaces with areas A_1, A_2, and A_3, and the K-functional for a single plaquette

Hence the K-functional associated with fig. (20-10), integrated over Ω and Ω' ($\rightarrow$ free boundary conditions) is given by

$$K^{(\beta,free)}_{A_1A_2A_3A_4} = [I_0(\hat{\kappa})]^{\hat{A}_1+\hat{A}_3}[\sum_{\nu_2}[I_{\nu_2}(\hat{\kappa})]^{\hat{A}_2}\chi_{\nu_2}(U_P^\dagger)\chi_{\nu_2}(U_\Gamma^\dagger)\chi_{\nu_2}(U_{\Gamma'})]$$
$$\times \sum_{\nu_4} I_{\nu_4}(\hat{\kappa})\chi_{\nu_4}(U_P) .$$

This K-functional has to be folded with

$$L_n L^\dagger_{n'} ReU_P = \chi_1(U_\Gamma)\chi_1(U^\dagger_{\Gamma'})\frac{1}{2}[\chi_1(U_P) + \chi_1(U_P^\dagger)] .$$

One then finds that

$$< L_n L^\dagger_{n'} ReU_P >= \frac{1}{2}\left(\frac{I_1(\hat{\kappa})}{I_0(\hat{\kappa})}\right)^{\hat{A}_2} \sum_{\nu_4}\left(\frac{I_{\nu_4}(\hat{\kappa})}{I_0(\hat{\kappa})}\right)(\mathcal{D}^1_{1\nu_4} + \mathcal{D}^{\nu_4 *}_{11}) ,$$

where the Wigner coefficients have been defined in (20.41). Hence

$$< L_n L^\dagger_m \frac{1}{2}(U_P + U_P^\dagger) >= \frac{1}{2}\left(\frac{I_1(\hat{\kappa})}{I_0(\hat{\kappa})}\right)^{\hat{A}_2}\left(1 + \frac{I_2(\hat{\kappa})}{I_0(\hat{\kappa})}\right) .$$

This result is also valid if the plaquette touches a Wilson line. Dividing this expression by (20.43), with $\hat{A}_2$ replaced by $\hat{A}_2 + 1$, (i.e., the area enclosed by the two Polyakov loops, including the area of the window), we obtain

$$\frac{< L_n L^\dagger_m \frac{1}{2}(U_P + U_P^\dagger) >}{< L_n L^\dagger_m >} = \frac{I_0(\hat{\kappa}) + I_2(\hat{\kappa})}{2I_1(\hat{\kappa})} . \tag{20.52}$$

The reader will have noticed that in obtaining this expression we have probed the field energy with a plaquette placed in between the Wilson lines. It can be shown that the corresponding expectation value with a plaquette placed outside the area bounded by the two Polyakov loops vanishes.

Finally, let us compute the expectation value of ReU_P in (20.50). The relevant K- functional is given by the product of i) the K-functional (20.51) with $\hat{A}_2 = \hat{A}-1$, integrated over U_Γ and $U_{\Gamma'}$ ($\rightarrow$ free boundary conditions), and ii) the K-functional for a single plaquette is given by (20.33). One then finds that

$$< ReU_P >= \frac{1}{Z_{free}}[I_0(\hat{\kappa})]^{\hat{A}} \sum_\nu \frac{I_\nu(\hat{\kappa})}{I_0(\hat{\kappa})}\int DU_P \frac{1}{2}[\chi_1(U_P) + \chi_1(U_P^\dagger)]\chi_\nu(U_P)$$

or

$$< ReU_P >= \frac{I_1(\hat{\kappa})}{I_0(\hat{\kappa})} . \tag{20.53}$$

Here we have made use of the fact that $\chi_\nu(U^\dagger) = \chi_{-\nu}(U)$, and $I_{-\nu}(\hat{\kappa}) = I_\nu(\hat{\kappa})$. Taking the difference of (20.52) and (20.53), we obtain the mean energy density in lattice units

$$< \hat{\mathcal{E}} >= \hat{\kappa}\frac{I_0(\hat{\kappa}) + I_2(\hat{\kappa})}{2I_1(\hat{\kappa})} - \hat{\kappa}\frac{I_1(\hat{\kappa})}{I_0(\hat{\kappa})} . \tag{20.54}$$

In the continuum limit $\hat{\kappa} \to \infty$. Expanding the ratio of modified Bessel functions in powers of $\frac{1}{\hat{\kappa}}$ one finds that each of the two terms in the above expression is singular in this limit. The singular parts however cancel in the difference, and after setting $\hat{\kappa} = \frac{1}{e^2a^2}$, one finds that the mean energy per unit length, measured in physical units, i.e. $\mathcal{E} = \frac{\hat{\mathcal{E}}}{a^2}$, is given by $e^2/2$, or

$$< E >= \frac{1}{2}e^2 R .$$

Notice that the mean energy coincides with the free energy (20.45). This is particular to the simple model we have considered, where the free energy does not depend on the temperature.

The expression (20.54) could also have been derived directly from (20.48a). Thus on an anisotropic lattice $< L_n L^\dagger_{n'} >$ is given by (20.43) with $\hat{\kappa}$ replaced by $\xi\hat{\kappa}$ (as follows from the form of the action (20.46)). Performing the differentiation in (20.47a) one readily verifies that one recovers the result (20.54). But our work has not been in vain, for it served to test some subtle points we mentioned in chapter 10. Thus it not only served to test the energy sum rule, but in particular the minus sign in (20.49a), which is typical for the euclidean formulation, as we have already seen in chapter 10. There is another lesson we have learned. As we have mentioned above both terms in (20.54) diverge in the continuum limit, but their difference is finite. We had encountered a similar situation in chapter 18, when we discussed the energy sum rule for the harmonic oscillator. Hence the computation of the *electric* field energy density from correlators of plaquette variables with two Wilson lines, and averaged plaquette variables, involves, close to the continuum limit, taking the difference of two large numbers of the same order of magnitude. This may be a serious stumbling block for numerical simulations. These are the main messages we wanted to convey.

20.6 Monte Carlo Study of the Deconfinement Phase Transition in the Pure $SU(3)$ Gauge Theory

Numerical simulations of the pure SU(3) gauge theory at finite temperature have been carried out since the early eighties. * Since then much effort has been invested in studying the deconfinement phase transition, which is expected to be associated with a breakdown of the $Z(3)$ center symmetry, and which had been predicted by Polyakov (1978) and Susskind (1979). Svetitsky and Yaffe (1982) then conjectured that the phase transition should be first order. The order parameter which distinguishes the two phases is the expectation value of the Wilson line (or Polyakov loop). When quarks are coupled to the gauge fields the action is no longer $Z(3)$ symmetric and the Wilson line no longer plays the role of an order parameter. But at least for large quark masses, such as those of the "charmed", "bottom" and "top" quarks, it is still expected to show a rapid variation across the transition region. QCD with only heavy quarks is however not the theory one is ultimately interested in. In the real world we also have the light "up" and "down" quark, and a heavier "strange" quark. Of these, at least the "up" and "down" quarks, are expected to influence the phase transition in a decisive way. Thus in the limit of vanishing quark masses, the continuum action possesses a chiral symmetry. This symmetry is broken in the low temperature phase and is expected to be restored at sufficiently high temperatures (Pisarski and Wilczek, 1984). The order parameter which characterizes the two phases in the zero quark mass limit is the chiral condensate $<\bar{\psi}\psi>$. It now replaces the Wilson line which tests the center symmetry in the pure gauge theory.

In this section we present some numerical results of lattice calculations which strongly support the above mentioned expectations that the deconfinement phase transition is of first order. Clearly, it is impossible to discuss the numerous contributions made in the literature , and to present a critical analysis of the results. But this is also not the purpose of this section. Our objective is to stimulate the readers interest in this subject. To this end we shall select a few representative results of early Monte Carlo simulations. Hence the figures we present, do not represent the best numerical data available today, and we apologize to the many physicists that have made important contributions in this field, and whom we do not mention here explicitely. The reader should confer the proceedings for

---

* Of course many calculations have also been performed in the pure SU(2) gauge theory, which is much easier to simulate.

more recent results, and the original articles cited for a critical assessment of the numerical results presented here.

After these general remarks, let us now first take a look at what Monte Carlo "experiments" tell us about the phase transition in the pure SU(3) gauge theory, i.e., in the infinite quark mass limit. In particular we are interested in establishing the nature of the transition (if it exists), and the critical temperature at which the phase transition takes place.

In the infinite volume limit a first order deconfinement phase transition should show up as a discontinuity in the Wilson line and energy density. The latent heat associated with the transition tells us how much energy has to be pumped into the system to produce the new state of matter. On a finite lattice any discontinuous behaviour of an observable will be smoothed out, but a rapid variation across the transition region should still be seen. Such a variation would however not exclude the possibility that the transition is second order. There are several characteristic features of a first order phase transition: i) the coexistence of phases at the critical temperature. In a Monte Carlo calculation this should manifest itself in the more or less frequent flip of the system between the "ordered" and "disordered" phases. The frequency with which these flipps occur will depend on the lattice volume used, decreasing with increasing volume, since the system will tend to remain in either of the two metastable states for a longer simulation time; ii) On a finite lattice the critical coupling $6/g_0^2$ at which the phase transition occurs should show a specific finite size scaling behaviour. In particular, for a first order phase transition, the location of the transition is expected to be shifted by an amount proportional to 1/V , where V is the spatial volume of the lattice.; iii) additional information about the nature of the transition can be obtained from the finite size scaling analysis of the peak and width in the specific heat or the susceptibility of the Polyakov loop, which should show a specific dependence on the volume.* Thus, if the transition is first order, then the height and width in the susceptibility of the order parameter is expected to increase linearly with the volume and shrink like $1/V$, respectively.** Studying the response of thermodynamical observables to a change in lattice volume is probably the best method to establish the order of the phase transition. But it is very time consuming. In most simulations one has therefore looked for signs of metastable states. Such states are however

* The susceptibility of the Polyakov loop is defined by $\chi = V(< (ReL)^2 >- < ReL >^2)$.

** The scaling behaviour of first-order phase transitions has been discussed e.g. by Imry (1980); Fisher and Berker (1982); Binder and Landau (1984).

not easily detected, especially if one is working on small lattices, since the flips between different thermodynamical states cannot be clearly disentangled from the statistical fluctuations.

What concerns the measurement of the transition temperature T_c, there are different ways one can proceed. If the transition is first order, and the spatial lattice volume large enough, then localizing the (smoothed out) discontinuity in the energy density or order parameter should suffice to determine T_c. Alternatively, one may test the Z(3) symmetry directly by looking at the distribution of the real and imaginary parts of Polyakov loops, measured on a large number of link configurations, as a function of the temperature. In the Z(3) symmetric phase, configurations related by the Z(3) symmetry should occur with equal probability. On the other hand, in the Z(3) broken phase the system will spend substantial simulation time in one of the three vacua, before tunnelling between the vacua will restore the Z(3) symmetry.

After these general remarks, let us now take a look at some specific examples of early Monte Carlo calculations.

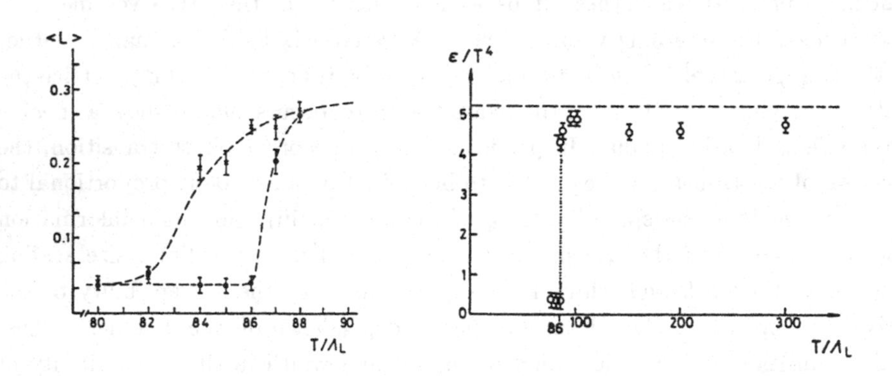

Fig. 20-11 MC data of Celic et al. (1983) showing (a) the hysteresis pattern for the order parameter, and (b) the temperature dependence of the energy density in the phase transition region.

First evidence for the existence of a first order phase transition in the pure SU(3) gauge theory came from computer simulations dating back to 1981 (Kajantie, Montonen and Pietarinen, 1981).* In the first few years most of the cal-

* For a review of early calculations see Cleymans et al. (1986).

culations were performed on lattices of small spatial volume. The observation of hysteresis effects, of coexisting states, and of rapid changes in the energy density and Polyakov loop suggested that the transition is of first order. In fig. (20-11) we show data obtained by Celic, Engels and Satz (1983) exhibiting the hysteresis pattern for the order parameter, and the energy density as a function of the temperature measured on a $8^3 \times 3$ lattice. These data are suggestive of a first order phase transition. The latent heat was found to be $\Delta\epsilon = (3.75 \pm 0.25)T_c^4$. Using Monte Carlo data for the string tension available at that time, the critical temperature in physical units was determined to be $T_c = 208 \pm 20MeV$.*

As another example we show in fig. (20-12) later data obtained by Kogut et al. (1985) for the averaged Wilson line and the energy density on a $6^3 \times 2$ lattice. Notice that both quantities exhibit a very steep variation in the transition region at the same value of the coupling (and therefore also temperature).

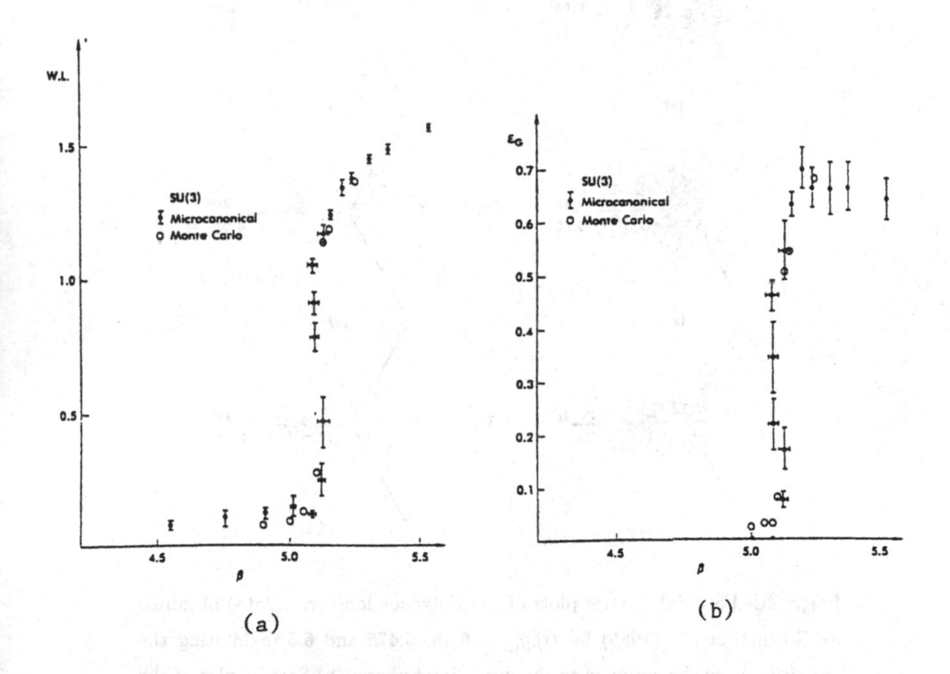

Fig. 20-12 (a) The Wilson line and (b) the gluon energy density as a function of the coupling $6/g_0^2$. The figure is taken from Kogut et al. (1985).

* For other early calculations of the latent heat see e.g. Kogut et al. (1983b); Svetitsky and Fucito (1983).

In both examples the lattices are however still too small to allow one to be sure that one is seeing continuum physics. Strong evidence that the SU(3) gauge theory does indeed exhibit a first order phase transition came from numerical calculations performed by Gottlieb et al. (1985) on lattices with varying temporal extent, ranging between $N_\tau = 8$ and $N_\tau = 16$. These authors performed long Monte Carlo runs at different couplings, and measured the real and imaginary parts of the Polyakov loop averaged over the spatial lattice for each configuration generated after every ten sweeps. The results for the largest lattice they have used ($19^3 \times 14$) are shown in fig. (20-13a).

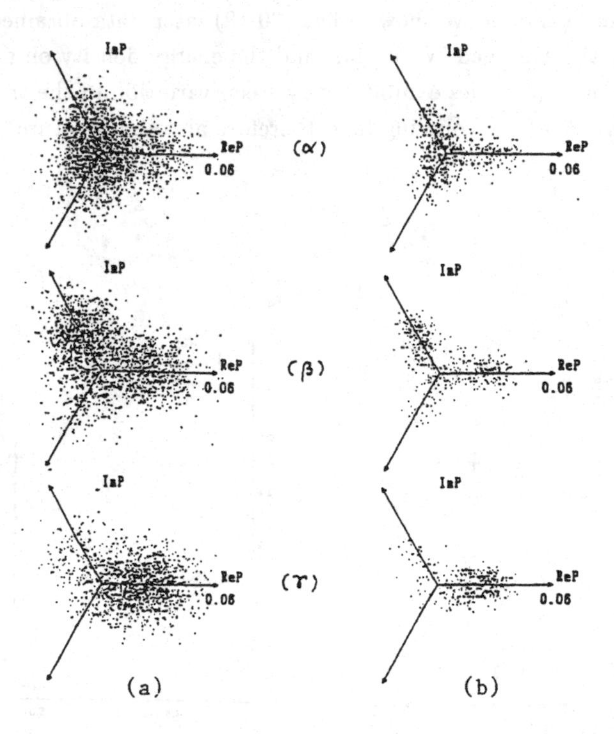

Fig. 20-13 (a) Scatter plots of the Polyakov loop (raw data) obtained by Gottlieb et al. (1985) for $6/g_0^2 = 6.45$, 6.475 and 6.5, exhibiting the transition from the confined to the deconfined phase; (b) Scatter plot of the same data as in (a), where always five successive measurements have been averaged.

The averaged Polyakov loop is denoted by P. Fig. (20-13b) shows the same data, where always five successive measurements have been averaged. In the figures labeled by (α) the system is in the deconfined phase, with the data distributed more or less uniformely around the origin. The figures labeled by (β) show the coexistence of the deconfined and confined phases, while the figures labeled by (γ) correspond to the $Z(3)$ broken phase, with the data clustering around a non-vanishing expectation value of the Polyakov loop. Notice that the transition from the $Z(3)$ symmetric to the $Z(3)$ broken phase occurs within the narrow interval $6.45 < 6/g_0^2 < 6.5$.* For another choice of lattice the critical coupling will of course be different. To test whether one is extracting continuum physics, one must check whether relation (20.26) holds. If so, the critical temperature in units of the renormalization group invariant scale, Λ_L, is given by (20.25).

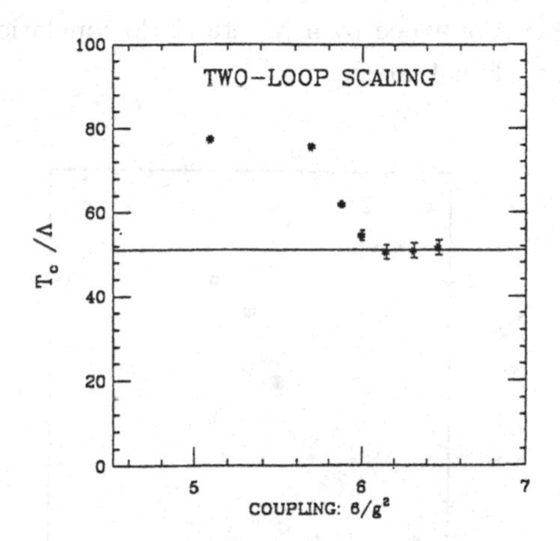

Fig. 20-14 The critical temperature measured in units of the lattice parameter calculated on lattices with temporal extensions N_τ =2,4,6 (Kennedy et al. 1985) and N_τ = 8,10, 12 and 14 (Gottlieb et al. 1985). Asymptotic scaling appears to set in for $6/g^2 > 6.15$.

Fig. (20-14) shows the results for T_c/Λ_L obtained by these authors for various choices of N_τ ranging between $N_\tau = 8$ and $N_\tau = 14$. The data at $N_\tau = 2, 4$, and 6 (first three points) are earlier results obtained by Kennedy et al. (1985). The plateau observed for $6/g_0^2 > 6.15$ strongly suggests that one is extracting continuum physics.** This need however not be so. Just like the first two points

* Recall that varying the coupling corresponds to varying the lattice spacing and hence the temperature.

** In fact, Gottlieb et al. used the measurement of T_c to determine the coupling in the figure, when considered by themselves, could have suggested that scaling sets in already for $6/g_0^2 = 5.1$, it could happen that the plateau observed in the range $6.15 < 6/g_0^2 < 6.5$ does not correspond to scaling but is followed by a scaling violating behaviour similar to that observed for $5.1 < 6/g_0^2 < 5.7$. To obtain the temperature in physical units one must eliminate the lattice scale parameter. This can be done by measuring another physical quantity such as the string tension. Then $T_c = (\hat{T}_c/\sqrt{\hat{\sigma}})\sqrt{\sigma}$, where $\hat{T}_c$ and $\hat{\sigma}$ are the critical temperature and string tension measured in lattice units. Using the value 400 MeV for the square root of the string tension one finds that $T_c = 250 MeV$.

Further evidence for a first order phase transition came from measurements of the latent heat on large spatial lattices, and from the observation of metastable states. A nice example is provided by a Monte Carlo simulation of Brown et al. (1988) performed on a $24^3 \times 4$ lattice.

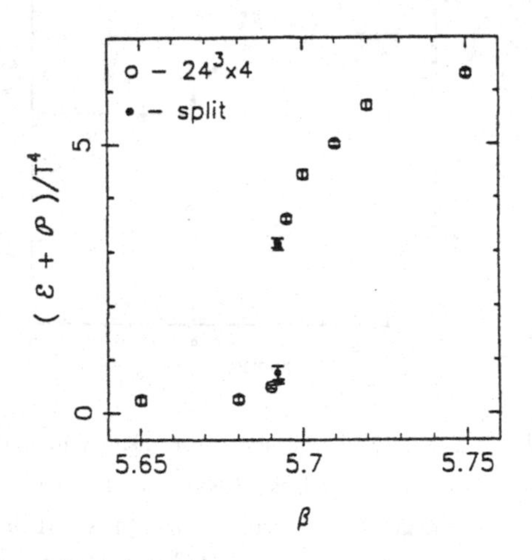

Fig. 20-15 MC data of Brown et al. (1988) for the sum of the energy density and pressure measured in units of T^4 plotted versus $6/g_0^2$ for a $24^3 \times 4$ lattice. The solid points have been obtained by dividing the events in a single MC run, showing flip-flop behaviour, by hand into confined and deconfined parts.

One of the quantities which has been measured by these authors is the sum of the energy density and pressure measured in units of T^4. Since the pressure should

---

at which scaling sets in.

vary continuously across the transition region, a measurement of the discontinuity in $(\epsilon + p)/T^4$ yields directly the latent heat in units of T^4.

Fig. (20-15) shows the behaviour of $(\epsilon + p)/T^4$ in the transition region, as obtained by Brown et al. on a $24^3 \times 4$ lattice. In the same Monte Carlo experiment the coexistence of the two phases at the critical temperature was also clearly seen.

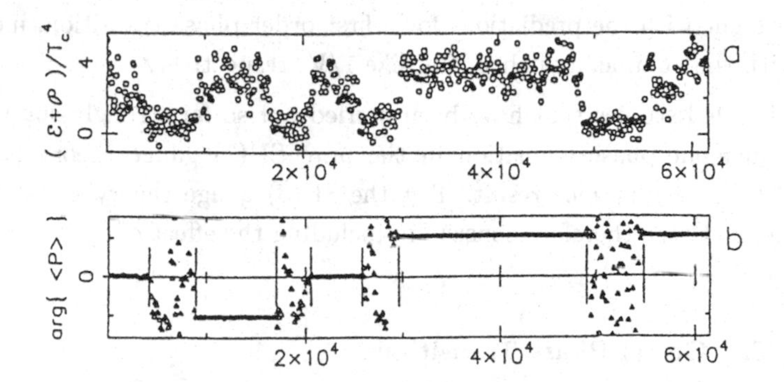

Fig. 20-16 Evolution of $(\epsilon + p)/T^4$ (a) and the argument of the Polyakov loop (b) on a $24^3 \times 4$ lattice as a function of the simulation time. The solid horizontal lines are MC data showing that the system resides in one of the three vacua with broken Z(3) symmetry. In the remaining time intervals the system is in the Z(3) symmetric confining phase. The figure is taken from Brown et al. (1988).

In Figs. (20-16a,b) taken from Brown et al. (1988), we show the evolution of $(\epsilon+p)/T^4$ and of the argument of the Polyakov loop as a function of the simulation time at the phase transition. The values for $(\epsilon + p)/T^4$ obtained by dividing the events in a single Monte Carlo run -showing flip-flop behaviour- by hand into confined and deconfined parts are displayed in fig (20-15) by the dark dots. As seen from fig. (20-16b), the "time" evolution of the argument of the Polyakov loop also displays the $Z(3)$ symmetric phase as well as the three vacua with broken $Z(3)$ symmetry in which the system remains for longer simulation times. The observation of coexisting phases allows one to determine the discontinuity indirectly from fig. (20-15). The authors find that $\Delta\epsilon/T^4 = 2.54 \pm 0.12$.

Fukugita, Okawa and Ukawa (1989) have analyzed in a very large-scale simulation the finite size dependence of the flip-flop behaviour at the phase transition, and of the peak and width in the susceptibility of the Polyakov loop on lattices for spatial volumes ranging between 8^3 and 36^3 , and fixed temporal extension, $N_\tau = 4$. They found, in particular, that on a $24^3 \times 4$ lattice and at a coupling $6/g_0^2 = 5.6925$, the system "appears to stay in one phase over $(1-3) \times 10^4$ sweeps before flipping to the other phase". The average duration between flip-flops was found to increase strongly with the lattice volume and to be consistent with that expected for a first order phase transition. Furthermore the volume dependence of the height and width of the susceptibility of the Polyakov loop also turned out to be consistent with the predictions for a first order phase transition, increasing linearly with the volume, and shrinking like $1/V$, respectively.

The detailed studies that have been carried out so far strongly support that the deconfinement phase transition in the pure SU(3) gauge theory is of first order.*. This is a very nice result. But the SU(3) gauge theory is not the real world. The next step therefore consists in including the effect of dynamical (finite mass) quarks.

20.7 The Chiral Phase Transition

Let us now ask what happens to the phase transition when we couple dynamical quarks to the gauge potentials. For sufficiently large quark masses it is reasonable to expect that the presence of quarks will not influence the results obtained in the pure gauge theory very much. A realistic simulation, however, should be performed with "up" and "down" quarks with a bare mass of the order of a few MeV, and a heavier "strange" quark with a bare mass about twenty times larger. The influence of the heavy "charmed", "top", and "bottom" quarks on the phase transition can very likely be ignored. This may also to be true for the "strange" quark. Hence as a first approximation one would like to study the case of QCD with two light, roughly degenerate, quarks. This case is interesting, since the chiral phase transition is expected to be driven by the light quarks. But also the case of four degenerate low mass quarks is interesting from the theoretical point of view, since there exist general arguments (Pisarski and Wilczek, 1984) predicting a first order chiral symmetry restoring phase transition for three or more flavours of massless quarks.

The majority of the Monte Carlo simulations have been carried out with Kogut-Susskind fermions. The reason for this is the following. As we have seen in chapter four, the staggered fermion action possesses a continuous axial flavour

* There had been some doubts regarding this raised by the APE collaboration (Bacilieri et al., 1989)

symmetry in the limit of vanishing quark masses. This allows one to study spontaneous chiral symmetry breaking and the associated Goldstone phenomenon without having to tune any parameters, as would be required in the case for Wilson fermions. Thus in the case of staggered fermions the chiral limit just corresponds to setting the quark masses to zero. On the other hand, for Wilson fermions chiral symmetry is broken explicitely by the fermionic action, and the chiral limit is realized at a critical value of the hopping parameter κ, corresponding to vanishing pion mass (see section 3 of chapter 17). This critical value can however only be determined by generating ensembles of gauge field configurations at different values of κ, and extrapolating the data to $\kappa_{crit.}$. This turns out to be a quite non-trivial and time consuming problem. In addition one is faced with the problem that the number of degrees of freedom for Wilson fermions is much larger than for Kogut-Susskind fermions. Consequently, simulations with Wilson fermions have been performed on much smaller lattices than those employed for staggered fermions. For these reasons, staggered fermions are more convenient for studying the chiral phase transition.

By 1991, numerous calculations with dynamical fermions had already been performed on lattices with a spatial extension up to 16 lattice sites, and temporal extention $N_\tau = 4, 6, 8$. In most simulations the quarks have been taken to have the same mass. The general scenario that emerged was the following. As the quark mass is decreased from $\hat{m} = \infty$, the first-order deconfinement phase transition weakens,* and even seems to disappear for intermediate quark masses. As the quark masses are decreased further, the transition gathers eventually again in strength and, in the case of three or four quark flavours, exhibits characteristics of a first order phase transition for sufficiently small quark masses. Clear signals of metastable states have been observed. ** In the case of two quark flavours the transition turned out to be much weaker.

Let us now look at some Monte Carlo simulations. As in the case of the pure SU(3) gauge theory we have selected only a few representative early examples which will hopefully stimulate the reader to further reading.

In the early days, before the advent of the supercomputers and the development of more refined algorithms for including the effects of the fermionic determi-

* See for example Hasenfratz, Karsch, and Stamatescu (1983); Fukugita and Ukawa (1986).

** See for example: Gavai, Potvin and Sanielevici (1987), and Gottlieb et al. (1987b).

nant in numerical simulations, physicists have studied the temperature dependence of the chiral condensate $< \bar{\psi}\psi >$ and of the quark and gluon internal energies in the quenched approximation. In fig. (20-17) we show early results obtained by Kogut et al. (1983a) for the chiral condensate and the Wilson line. The chiral condensate was computed from the fermion propagator by taking the limit of vanishing bare fermion mass.

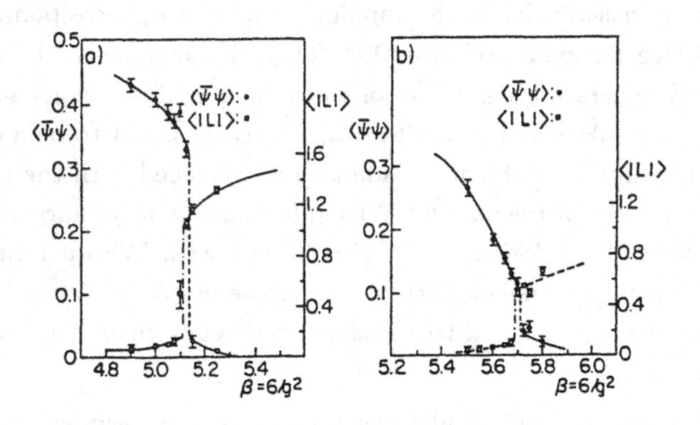

Fig. 20-17 MC data of Kogut et al. (1983a) for the chiral order parameter and the expectation value of the Wilson line as a function of $6/g_0^2$ in quenched QCD. The calculations were performed on a (a) $8^3 \times 2$ and (b) $8^3 \times 4$ lattice.

Notice that there exists only a single transition region across which both parameters show a very rapid variation. Since the fermionic determinant was neglected in this calculation, the sharp rise in the Wilson line corresponds to the deconfinement phase transition in a pure gluonic medium. A similar strong variation has been observed at the same transition temperature separately in the fermion and gluon internal energies (see e.g., Kogut et al., 1983b).

Until the mid-eighties, simulations including dynamical fermions were still in an exploratory stage. Early results for the Wilson line and chiral condensate for four light flavours showed again a strong variation of both quantities at the same critical temperature (Polonyi et al., 1984). Since then many simulations, using different algorithms, quark masses, and larger lattices have been performed. In Fig. (20-18) we show the data for the Wilson line and chiral condensate obtained by Karsch, Kogut, Sinclair and Wyld (1987) for four degenerate light quarks, using the hybrid algorithm. The results suggest a first order chiral phase transition. Notice that the Wilson line, although it is not an order parameter in full QCD, still rises steeply across the transition region. This suggests that the mechanism for deconfinement in the presence of dynamical fermions is that responsible for the restauration of chiral symmetry.

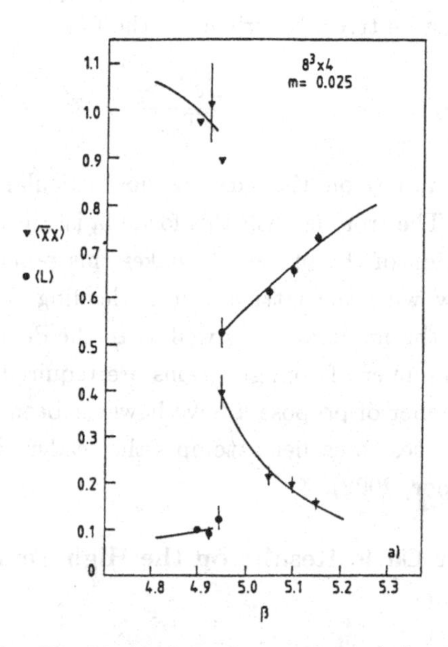

Fig. 20-18 The chiral order parameter (triangles) and the expectation value of the Wilson line (dots) calculated by Karsch et al. (1987) on an $8^3 \times 4$ lattice for QCD with staggered fermions.

Most of the simulations have been carried out with two or four flavours of mass-degenerate quarks. For some earlier simulations of the more realistic case of two equal light quarks and a heavier strange quark see e.g., Kogut and Sinclair (1988), Brown et al. (1990), and more recently Karsch and Laermann (1994), where again chiral symmetry restauration was found to take place at the same critical temperature where deconfinement sets in. This critical temperature was determined to be $T_C \approx 173 MeV$.

In the following we will take a look at some Monte Carlo studies of the high temperature phase, emphasizing, as always, the earlier pioneering work. We shall say nothing about the full phase diagram of QCD in the temperature-chemical potential plane, which is the subject of intensive current research. Monte Carlo simulations with a non-vanishing chemical potential are problematic, since the fermion determinant associated with the lattice action (19.90) is complex. Recall that the logarithm of this determinant contributes to the effective action. Hence a direct implementation of Monte Carlo techniques with a Boltzmann factor $e^{-S_{eff}}$ is not possible. In principle one could write $detK = |detK|e^{i\Gamma}$, and only include $|detK|$ into the effective action. Denote this action by $\tilde{S}_{eff}$. The ensemble average of an observable can then be trivially written in the form

$$< \mathcal{O} >= \frac{< \mathcal{O}e^{i\Gamma} >}{< e^{i\Gamma} >} ,$$

where the expectation values on the rhs are now calculated with the (phase quenched) action $\tilde{S}_{eff}$. The trouble with this formula is, that in actual numerical simulations the fluctuation of the phase $e^{i\Gamma}$ makes this procedure impracticable. These fluctuations grow with the lattice volume leading to large cancellations among contributions to the numerator, as well as to the denominator. As a consequence an enormous number of configurations are required to achieve any reasonable accuracy. A number of proposals have however been made to circumvent this problem. For references to earlier attempts the reader can confer the article by Fodor and Katz (Fodor, 2002). *

20.8 Some Monte Carlo Results on the High Temperature Phase of QCD

The confirmation that quarks are confined in hadronic matter at low temperatures, and the (strongly suggested) existence of a phase transition at temperatures of the order of 10^{12} Kelvin, are the two most spectacular predictions of lattice QCD. Such a transition would probably have occured about 10^{-6} seconds after the big bang. It has been speculated for some time that the high temperature phase is that of a quark-gluon plasma (QGP), where hadronic matter is disolved into its constituents. The possibility of implementing the QCD phase transition in the laboratory has become feasable within the past decade through heavy ion collisions carried out at ultrarelativistic energies. In these collisions the initial kinetic energy is deposited in a very short time interval in a small spatial region, creating matter with densities 10 to 100 times that of ordinary nuclear matter. Ion beams of ^{197}Ag, with 11.4 GeV per nucleon have been produced in 1992 by the AGS accelerator at Brookhaven National Laboratory. At the CERN SPS accelerator beams of ions as heavy as ^{32}S have been obtained with energies up to 160

---

* For more recent proposals see (Fodor, 2002; Allton, 2002; de Forcrand, 2002; Anagnostopoulos, 2002).

GeV per nucleon. The verification that a quark-gluon plasma is actually formed in such collisions is a very non trivial problem since it requires an unambiguous signature for its formation. For this a detailed understanding of the dynamics of a plasma is necessary. Lattice calculations can only provide us at present with bulk quantities like the mean energy density, entropy and pressure, but not with the expected particle yields, momentum spectra, etc., at freeze out.

If the high temperature phase is indeed that of a quark-gluon plasma consisting of a weakly interacting gas of quarks and gluons, as is suggested by renormalization group arguments (Collins and Perry, 1975), then one would expect that the energy density and pressure are approximately that of an ideal gas of quarks and gluons, and that the interquark potential is Debye screened. But because of the severe infrared divergencies encountered in perturbation theory, the behaviour of these observables in the deconfined phase could very well turn out to be non perturbative even at very high temperatures. The only way to estimate the non perturbative effects is to calculate the above quantities numerically. In the following we will take a look at some early Monte Carlo simulations of the high temperature phase of QCD. We will be very brief, and present here only a few results without going into details. Most of the material in this section has been taken from the review article by Karsch published by World Scientific (1990), where the reader can also find an extensive list of references to the work published on this subject.

Energy density and pressure above T_c

Early Monte Carlo calculations performed in the pure SU(3) gauge theory showed that above the critical temperature the energy density approached very quickly the Stefan-Boltzmann limit for a free gluon gas. An example is shown in fig. (20-11). This behaviour of the energy density has also been confirmed in later calculations performed on larger lattices.

On the other hand, it has been found that the pressure approaches the Stefan-Boltzman limit only slowly. Fig. (20-19) taken from the above mentioned review by Karsch, shows the Monte Carlo data for the energy density and pressure obtained by Attig et al. (unpublished), (dots and triangles), and Brown et al. (1988), (squares), on a $12^3 \times 4$ lattice and $24^3 \times 4$ lattice, respectively. The numerical results are compared with the perturbative prediction (dashed horizontal lines) of Heller and Karsch (1985) for a $12^3 \times 4$ lattice. As seen from this figure, the energy density agrees rather well with (lattice) perturbation theory already at temperatures slightly above T_c , while the pressure rises only slowly above the phase transition and approaches the perturbative prediction at best for temperatures above $3T_c$. As we will see below, however, the approach of the energy density to the Stefan-Boltzmann limit is probably much slower than it appears in the figure.

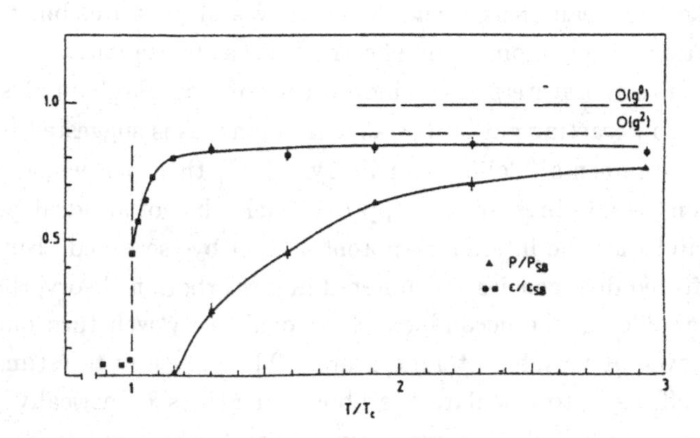

Fig. 20-19 Energy density and pressure in units of the ideal gas values as a function of T/T_c. The dots and triangles are data of Attig et al. (unpublished), and the squares are data of Brown et al. (1988). The lines are drawn to guide the eye. The figure is taken from Karsch (1990)

Note that the pressure seems to be negative slightly above T_c. In a high statistics simulation performed on a lattice with spatial volume 16^3, Deng (1989) has studied, in particular, the behaviour of the pressure just above the phase transition. The results are shown in fig. (20-20), where we also display, for comparison, the data obtained by this author for the energy density.

As seen from fig. (20-20b), the pressure is not only negative slightly above T_c but also appears to be discontinuous at the phase transition. Hence the extraction of the latent heat from a measurement of the discontinuity in $(\epsilon + p)$, or $(\epsilon - 3p)$, which assumes that the pressure is continuous at the phase transition, will lead to wrong conclusions. Engels et al. (1990) have subsequently discussed the origin of the problem. As we have seen in section 1, the energy density and pressure can be obtained from the expectation value of plaquette variables, if one knows the derivatives of the bare gauge coupling with respect to the spatial and temporal lattice spacings, a_s and a_τ , evaluated at $a_s = a_\tau$. In most simulations, the thermodynamical quantities have been extracted by assuming that the above mentioned derivatives can be approximated by their leading order weak coupling

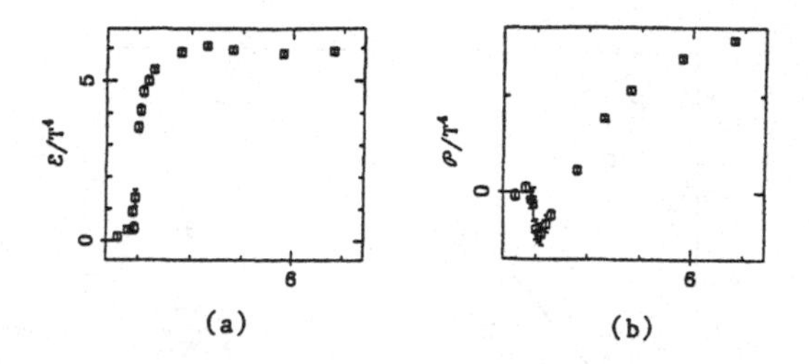

Fig. 20-20 MC data of Deng (1989) for (a) the energy density and (b) the pressure, measured in units of T^4, as a function of $6/g_0^2$. The calculation was performed on a $16^3 \times 4$ lattice.

expressions. But for the couplings at which the MC calculations are performed, these derivatives actually deviate substantially from the low order perturbative results (Burgers et al., 1988).

The above mentioned authors have therefore used an alternative, non perturbative, procedure for calculating the pressure. The basic idea is to calculate this quantity from the relation $p = -f$, where f is the free energy density: $f = -T(lnZ)/V$.* Of course the free energy cannot be computed directly in a Monte Carlo simulation (which only allows one to calculate expectation values). But its derivative with respect to the coupling $\lambda = 6/g_0^2$ can be calculated since $-\lambda \partial lnZ/\partial\lambda = < S_G >$, where S_G is the gauge action.**

By integrating this equation one then obtains, apart from an additive constant, the free energy, and hence the pressure, as a function of the coupling (and therefore also of the temperature). The free energy was normalized by subtracting the vacuum contribution at zero temperature (actually, the temperature is never really zero on a finite periodic lattice). By proceeding in this way one circumvents the problem of having to compute derivatives of the bare coupling constant. Fig. (20-21) shows the pressure as a function of $6/g_0^2$ obtained by these authors (solid line) together with the MC data of of Deng (1989) and Brown et al. (1988). The pressure is seen to be positive everywhere and continuous at the phase transition.

---

* This relation assumes that the system is homogenous. The authors discuss the validity of this assumption in their paper.

** Recall that we are still discussing the case of a pure SU(3) gauge theory. In order to avoid any confusion with the inverse temperature, $\beta = 1/T$, we have denoted the coupling $6/g_0^2$ by λ and not by β as is usually done in the literature.

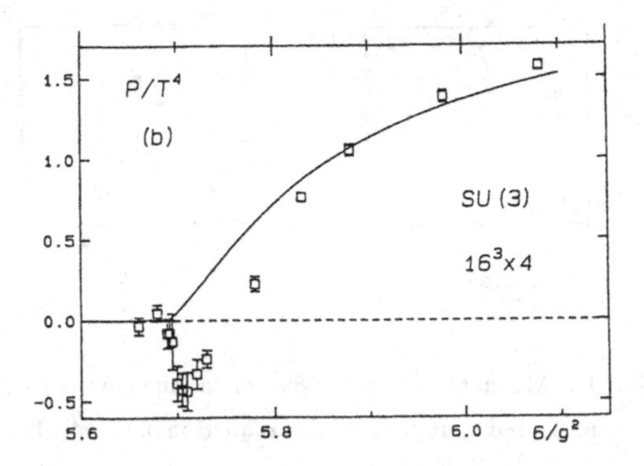

Fig. 20-21 The pressure in units of T^4 obtained by Engels et al. (1990) in the pure SU(3) gauge theory (solid line) together with the MC data of Deng (1989) and Brown et al. (1988) based on the perturbative expressions for the β-function.

The above authors have also calculated the energy density, which - as seen from eq. (20.8) - requires the knowledge of the β-function. For couplings $6/g_0^2 <$ 6.1 this β-function (which can be obtained from a Monte Carlo renormalization group analysis), deviates considerably from that given by the perturbative expression (9.21a,b). By using the non-perturbative β-function, and their results for the pressure, the authors find that the energy density approaches the Stefan-Boltzmann limit much slower that originally believed, and that the "discontinuity" at the phase transition is much weaker than that suggested by the data displayed in fig. (20-19). This is shown in fig. (20-22).

So far we have only considered the pure SU(3) gauge theory. For full QCD there exists no such detailed analysis. Simulations with dynamical fermion indicated that the behaviour of the energy density and pressure (obtained by using the leading order perturbative expressions for the derivatives of the bare coupling constant) is similar to that encountered in the pure gauge theory. Again the energy density was found to rise steeply in the transition region, approaching rapidly the ideal gas value (from above), while the pressure was found to rise only slowly above the critical temperature.

As an example we show in fig. (20-23), taken from the above mentioned review by Karsch, the energy density and pressure as a function of the coupling $6/g_0^2$ for QCD with two light quarks. The data shown is that of Gottlieb et al.,

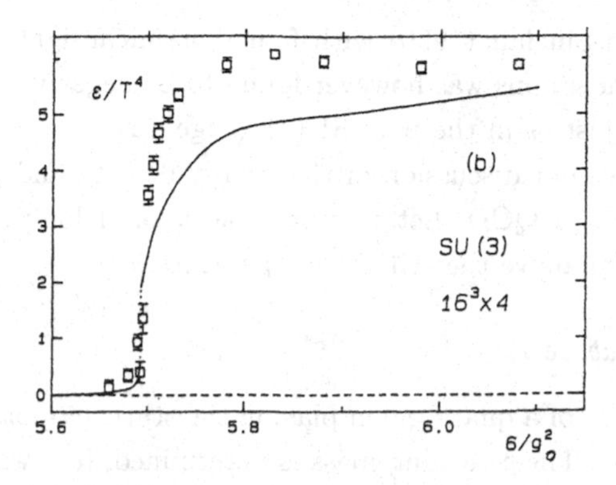

Fig. 20-22 The energy density in units of T^4 obtained by Engels et al. (1990) as a function of $6/g_0^2$ (solid line). The calculation is based on the results for the pressure shown in fig. (20.22) and on a non-perturbative expression for the β-function. The MC data are from Brown et al. (1988), and Deng (1989).

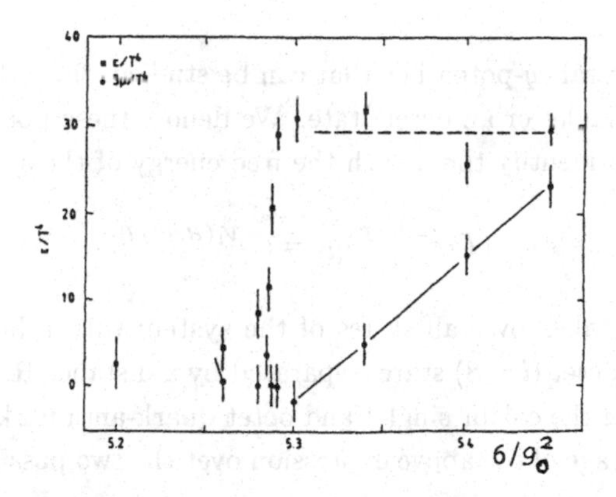

Fig. 20-23 Energy density and pressure in units of T^4 as a function of $6/g_0^2$ for QCD with two flavours. The data is from Gottlieb et al. (1987).

(1987). In contradistinction to the pure SU(3) gauge theory, the pressure, due to the quarks and gluons, appears to be positive everywhere and continuous at the phase transition. The same behaviour was also observed in a simulation carried out by Kogut and Sinclair (1990) with four dynamical light quarks. The partial pressure due to the gluons was however found to be negative in the vicinity of the phase transition, just as in the pure SU(3) gauge case.

This concludes our discussion of the energy density and pressure in the high temperature phase of QCD. Let us now take a brief look at the static quark-antiquark potential above the critical temperature.

The $q\bar{q}$-potential above T_c

In the presence of a quark gluon plasma the static $q\bar{q}$-potential is expected to be Debye screened. The screening mass is determined, in a way analogous to that for an ordinary plasma, from the zero momentum limit of the time component of the vacuum polarization tensor evaluated at zero frequency. In one loop order the electric screening mass is given by (19.88). In two loop order one is faced with infrared divergencies, and an infinite number of graphs need to be summed to yield an infrared finite result (Toimela, 1985). Hence one can only trust the lowest order perturbative result. The lattice formulation of QCD provides us with the possibility to study the $q\bar{q}$-potential non perturbatively. The results of the Monte Carlo simulations can then be compared with the low order perturbative calculations.

There are several $q\bar{q}$-potentials that can be studied, since the quark-antiquark pair can be in a singlet or an octet state. We denote these potentials by $V_1(R,T)$ and $V_8(R,T)$, and identify them with the free energy of the system:

$$(Tre^{-H/T})_{(l)} = e^{-V_l(R,T)/T}.$$

Here the trace is taken over all states of the system with a heavy $q\bar{q}$ pair in the singlet ($l=1$) or octet ($l=8$) state, separated by a distance R. We now define the thermal average of the colour singlet and octet quark-antiquark potential $V(R,T)$ by taking the average of the above expression over the two possible states weighted with their degeneracy:

$$e^{-V(R,T)/T} = \frac{1}{9}(e^{-V_1(R,T)/T} + 8e^{-V_8(R,T)/T}).$$

The colour averaged potential can be extracted from the following correlation function of two Polyakov loops (McLerran and Svetitsky, 1981),

$$e^{-V(R,T)/T} = \frac{\langle TrL(\vec{0})TrL^\dagger(\vec{R})\rangle}{\langle |L|\rangle^2},$$

where

$$L = \frac{1}{N_s^3} \sum_{\vec{x}} TrL(\vec{x})$$

is the Polyakov loop averaged over the N_s^3 spatial lattice sites. By normalizing the correlation function in this way, one eliminates divergent self-energy contributions to the potential. For a detailed discussion of the singlet and octet potentials we refer the reader to the paper by Nadkarni (1986), and to the review article by Karsch cited before. Here we only make some general remarks.

Perturbation theory predicts that the singlet and octet potentials are related by

$$\frac{V_1(R,T)}{V_8(R,T)} = -8 + O(g^4),$$

where

$$V_1(R,T) = -\frac{g^2(T)}{3\pi}\frac{1}{R}e^{-m_E(T)R}$$

in leading order. Hence the singlet potential is attractive, while the octet potential is repulsive. Their relative strength is such that the colour averaged potential behaves like $[\exp(-2m_E(T)R)]/R^2$.

The determination of the screening mass from Monte Carlo simulations is very difficult since one needs to study the behaviour of the correlation functions over a large range of separations of the quark and antiquark. But for large separations the signal tends to get drowned in the statistical noise. And this situation worsens with increasing temperature, since the screening mass is expected to increase with temperature. For this reason the region slightly above the phase transition, which is more accessible to numerical computations, has been studied in greatest detail. Knowledge of the potential in this region is in fact of great importance, since for temperatures just above T_c a quark gluon plasma may have been formed in the experiments performed at CERN, involving the high energy collisions of heavy nuclei (Abreu et al., 1988). A strongly screened potential would inhibit the formation of bound states with a large radius. The suppression of such bound states could (possibly) be used as a signal for plasma formation. What numerical simulations tell us is that close to the critical temperature the colour averaged potential can actually be well approximated by a simple screened Coulomb form with an effective screening mass. It is therefore non-perturbative in this region. There exists however some evidence that for temperatures well above the critical temperature the colour averaged potential, as well as the potentials in the singlet and octet channels, aquire the Debye screened form predicted by perturbation theory. This is supported, in particular, by a high statistics calculation of Gao (1990) performed on a $24^3 \times N_\tau$ lattice, with $N_\tau = 4, 6$, and 16 . This author found good indications that perturbative behaviour sets in at temperatures $T > 3.5T_c$.

20.9 Some Possible Signatures for Plasma Formation

In the previous sections we have seen that there are strong indications from lattice calculations that at high temperatures QCD undergoes a phase transition to a new state of matter, the quark gluon plasma. Such a plasma is expected to be formed in very high energy collisions of heavy nuclei. It is therefore important to look for a unambiguous experimental signal for plasma formation. Our following discussion of a possible signature will be more than incomplete and is mainly intended to stimulate the reader to confer the extensive literature on this fascinating subject. For earlier comprehensive reviews the reader may consult the book "*Quark-Gluon Plasma 2*" published by World Scientific (1995), and the review article by Meyer-Ortmanns (1996).

Of the signatures that have been considered in the literature we have chosen to discuss in greatest detail the "J/Ψ suppression", proposed by Matsui and Satz (1986), because of its simplicity, and because it has been the trigger for many subsequent investigations. This signature has been the subject of much dispute, since it is believed not to provide an unambiguous signal for plasma formation. Other signatures have been intensively discussed since then, and we shall only mention them briefly at the end.

Many years ago Matsui and Satz (1986) made an interesting proposal for a possible signal of plasma formation. These authors had emphasized that, because the $q\bar{q}$-potential is Debye screened in a deconfining medium, the formation of $q\bar{q}$-bound states will be strongly influenced by the presence of a quark-gluon plasma if the screening length (which can be estimated from lattice calculations) is sufficiently small. This led them to predict that the formation of the J/ψ (a $3S_1$ $\bar{c}c$-bound state with a mass of 3.1 GeV) should be strongly suppressed in high energy nucleus-nucleus collisions if a hot quark-gluon plasma is formed. Since the $\mu^+\mu^-$ decay of the J/ψ provides a very clear experimental signal for its formation, and since the mechanism for background muon pair production is fairly well understood, this resonance appears to be a good candidate for studying the properties of the deconfining medium.

Experiments involving high energy collisions of heavy nuclei performed at CERN (Abreu et al. (1988); Grossiord (1989); Baglin (1989)) have shown that J/ψ formation is substantially suppressed for small transverse momenta of the J/ψ, and for large values of the total transverse energy released in the collisions.

The plasma hypothesis is able to account for the observed suppression pattern. But if J/ψ suppression is to be an unambiguous signal for plasma formation, then one must rule out the possibility that other more conventional mechanisms can also explain the observed effects. In fact, following the pioneering work of Matsui and Satz, several authors have pointed out that inelastic scattering of the J/ψ in a dense nuclear medium can also lead to substantial suppression (see e.g., Gavin, Gyulassy and Jackson, 1988; Gerschel and Hüfner, 1988). Nuclear absorption alone, however, does not suffice to reproduce the experimental data. But by including also initial state interactions, one is able to obtain results compatible with the observed J/ψ suppression pattern. (Gavin and Gyulassy, 1988; Hüfner, Kurihara and Pirner, 1988; Blaizot and Ollitraut, 1989).

In the following we shall take the point of view that a plasma has been formed in the heavy ion experiments performed at CERN, and that the observed J/ψ suppression pattern is due to Debye screening in a plasma. For a review of other possible suppression mechanisms see Blaizot and Ollitraut (1990).

What is so attractive about the plasma hypothesis is that it gives a simple qualitative explanation of the effects observed in the NA38 experiments. That a plasma could have been formed is not out of this world. The above mentioned collaboration has studied, in particular, the J/ψ production in collisions of oxygen and sulfur at 200 GeV/nucleon incident on a uranium target. A rough (may be too optimistic) estimate of the energy density deposited in the collisions (Satz, 1990) yields the value $\epsilon = 2.8$ GeV/fm^3. From lattice calculations one estimates the energy density required for deconfinement to be 2-2.5 GeV/fm^3. From these estimates one can at least conclude that the formation of the plasma in the above collision process cannot be excluded. But, given the uncertainties in the estimates, one must accept the possibility that a plasma has not been formed.

In the plasma picture, J/ψ suppression is not a consequence of individual collisions between nucleons in the projectile and target nuclei, but is the result of a collective property of the medium, i.e. the existence of a Debye-screened $q\bar{q}$ potential. The details of the suppression pattern will, however, depend on the characteristics of the plasma formed in the collision, such as its spatial extension, temperature, hydrodynamical expansion etc.. But independent of such details one can make the following general statement: since the screening length decreases with increasing temperature, the influence of the plasma on bound state formation should become more pronounced with increasing temperature. Hence there should exist a Debye temperature T_D above which the potential is so short-ranged that it can no longer bind a given quark-antiquark pair. This temperature will depend on the particular pair considered.

For temperatures below the transition to the quark-gluon plasma the static $q\bar{q}$-potential is of the form

$$V(r) = -\frac{\alpha}{r} + \sigma r, \tag{20.55}$$

where σ is the temperature dependent string tension, and α the temperature dependent coupling constant. From renormalization group arguments α is expected to decrease with increasing temperature. At the critical temperatur T_c, the string tension vanishes, and above T_c the colour singlet potential is expected to be replaced by a Debye screened Coulomb potential of the form

$$V(r) = -\frac{\alpha}{r} e^{-r/r_D(T)}, \tag{20.56}$$

where $r_D(T)$ is the Debye screening length, which can be estimated from lattice calculations. In the WKB approximation one can determine the minimum screening radius $r_D^{(min)}$ for which the potential (20.6) can still accommodate a bound state of a quark and antiquark with zero angular momentum and radial excitation number n (Blaizot and Ollitraut, 1987):

$$r_D^{(min)}(T) = n^2 \pi \hbar^2 / 2m\alpha. \tag{20.57}$$

Here m is the mass of the quarks. The J/ψ resonance is a $^3S_1$ bound state of a c and $\bar{c}$ quark with $n = 1$. Inserting in (20.57) some typical values for the charmed quark mass, m_c, and coupling constant $\alpha(m_c \approx 1.37$ GeV, $\alpha \approx 0.5)$ determined from spectrum calculations (Quigg and Rosner, 1979; Eichten et al., 1980), one obtains that $r_D^{(min)} \approx 0.45 fm$. The actual value is however expected to be larger, since the effective coupling constant decreases with increasing temperature. Making use of the estimates for $r_D(T)$ obtained from lattice calculations, one is led to the expectation that the J/ψ cannot be formed already at temperatures slightly above the deconfinement phase transition. Other resonances like the ψ', which also decays into a $\mu^+\mu^-$ pair, but have a larger binding radius than the J/ψ should be suppressed already at a lower temperature.

Let us now follow the fate of a $c\bar{c}$ pair, which is expected to be produced within a very small space-time volume in the early state of the collision process before the plasma had the time to form. Because of the large mass of the charmed quarks, it is unlikely that the $c\bar{c}$-pair is produced within the plasma. Thus the creation of charmed quarks at temperature T should be suppressed by the Boltzmann factor

$exp(-m_cT)$.* In the absence of a quark-gluon plasma the c and $\bar{c}$ quarks would begin to separate and combine to form a J/ψ once their separation has reached the binding radius of the J/ψ (which is about $0.2 - 0.5 fm$). The time required for this binding process to take place in the rest frame of the quark pairs is called the formation time τ_0. It can be estimated from the radius of the J/ψ and the average radial momentum of the charmed quarks in the J/ψ bound state. In addition to the formation time there are two other important time scales which are relevant for discussing the fate of the $\bar{c}c$-system: the time required for a plasma to be formed in the collision process, and the time required for the plasma temperature to drop below T_D, where Debye-screening is no longer effective in inhibiting J/ψ formation. This cooling process is associated with an expansion of the plasma whereby the central hot region shrinks with time. Hence to study the fate of a $\bar{c}c$-pair we must follow their motion through the dense nuclear medium taking into account the hydrodynamical expansion of the system.** This is a complicated problem and one must resort to simple model calculations*** But the general qualitative features of J/ψ suppression can be deduced without performing any explicit calculation. The general scenario is the following.

Let us assume that the $\bar{c}c$-pair is produced before a plasma has been formed. If the c and $\bar{c}$ quarks find themselves in a deconfining medium by the time they could form a J/ψ, then the J/ψ will not be formed. Hence the c and $\bar{c}$ quarks will eventually leave the hot plasma region and combine with other non-charmed quarks (of which there are many around) to form charmed particles ($\rightarrow$ open charm). The number of $\bar{c}c$-pairs which are still trapped within the plasma by the time their separation is of the order of the bound state radius of the J/ψ will depend on various factors. First of all, if the $\bar{c}c$-system carries sufficiently large transverse momentum p_T, then it will be able to escape the hot plasma region before having reached the bound state radius of the J/ψ. Hence normal J/ψ formation will be possible. These J/ψ's decay subsequently into a $\mu^+\mu^-$ pair, which can be easily detected in the experiment. Hence for sufficiently large p_T, J/ψ formation should not be suppressed at all. But just how large p_T must be for this to be the case depends on the plasma life time and on the region occupied by the hot plasma in the course of its expansion. Thus if the plasma lifetime

---

* See e.g. the review of Satz in "Quark Gluon Plasma", edited by R.C. Hwa (World Scientific, 1990).

** Unless they have already escaped the critical region by the time the plasma has been formed. In this case the plasma will not affect J/ψ formation.

*** See the review by Karsch in the reference given in the previous footnote.

is sufficiently short, then even J/ψ's with low transverse momentum should not be suppressed. But the plasma lifetime is a function of the initial temperature. And the initial temperature depends on the energy deposited in the collision. The higher the energy density ϵ, the higher the initial temperature, and hence the plasma life time. If the initial temperature of the plasma is high enough, it will take a longer time to cool below the critical temperature T_D where Debye screening no longer is effective. Now a measure for the energy deposited in a collision is the total transverse energy E_T released in the event. This energy tells us how violent the collision was. Hence for a given initial spatial extension of the hot plasma, we expect that the screening mechanism will become more effective as E_T is increased.

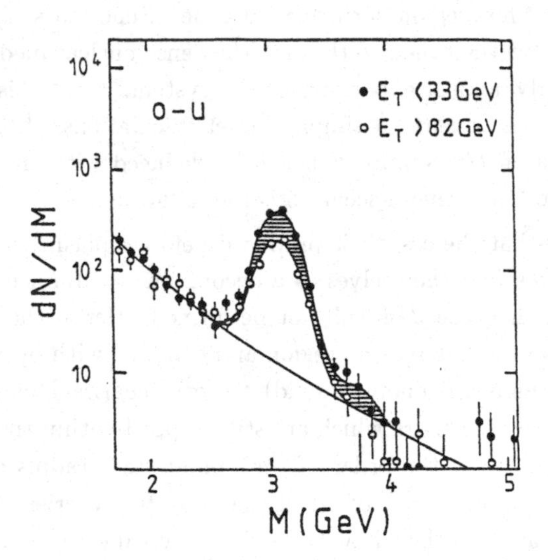

Fig. 20-24 Dilepton spectrum in oxygen-uranium collisions for two transverse energy cuts observed by the NA38 collaboration at CERN (Abreu et al., 1988). The figure is taken from the review by Satz (1990).

Let us now look at the experimental situation. Formation of the J/ψ should show up as a peak in the invariant mass M of the $\mu^+\mu^-$ pairs at the mass of the J/ψ. In fig. (20-24) we show the dilepton spectrum in oxygen-uranium collisions obtained by the NA38 collaboration at CERN (Abreu et al., 1988) for two transverse energy cuts. This figure is taken from the review by Satz cited earlier. To exhibit the suppression of the J/ψ at larger transverse energies, the fitted

$\mu^+\mu^-$ continuum* in the $E_T < 33$ GeV and $E_T > 81$ GeV data (arising from other processes than the decay of the J/ψ, ψ' etc.) have been matched.

Figure (20-25) shows the dependence of the ratio of the number of J/ψ events to the number of continuum events N_ψ/N_c as a function of the transverse energy. The figure is taken from the review of Kluberg (1988). The ratio is seen to decrease by about a factor of 2 between the lowest and highest transverse energy bins considered.

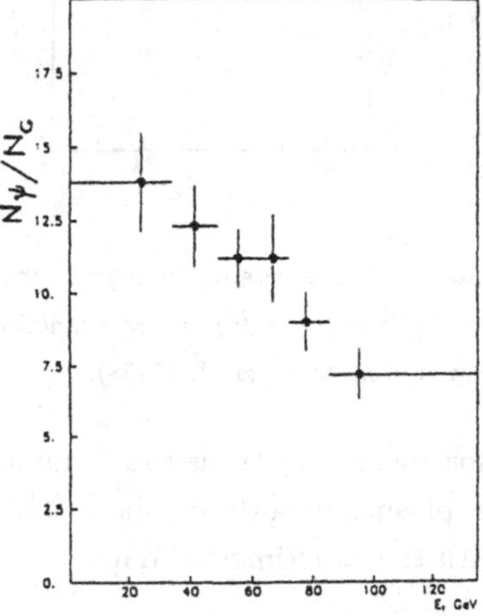

Fig. 20-25 Ratio of the number of J/ψ events to the number of continuum events, in oxygen-uranium collisions, as a function of the transverse energy. The figure is taken from the review by Kluberg (1988).

As we have pointed out earlier, the formation of a plasma should also lead to stronger J/ψ suppression for smaller transverse momenta of the J/ψ's. This effect has been clearly seen by the NA38 collaboration.

In fig. (20-26) we show the p_T dependence of the ratio of J/ψ events in the highest transverse energy bin, to the number of events in the lowest E_T bin. The figure is taken from Abreu et al., (1988), and shows the expected decrease

---

* There are several sources for background $\mu^+\mu^-$ production which need to be considered in detail in order to determine the actual strength of the observed effect. In particular one requires detailed information about the dependence of the production rate on the transverse energy and momentum of the background $\mu^+\mu^-$ pairs before one can acertain that the observed effect is not due to an enhancement of $\mu^+\mu^-$ production in the continuum. For a discussion of this problem see the reviews of Satz and of Blaizot and Ollitrault (1990).

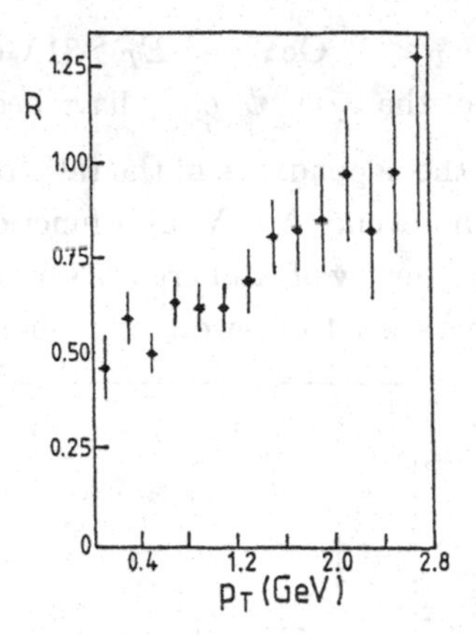

Fig. 20-26 Ratio of J/ψ events in the highest transverse energy bin to the number of events in the lowest E_T bin as a function of the transverse momentum. The data is from Abreu et al. (1988).

in the J/ψ suppression for increasing transverse momenta of the J/ψ. Model calculations based on the plasma hypothesis show that one can get reasonable quantitative agreement with the experimental data.

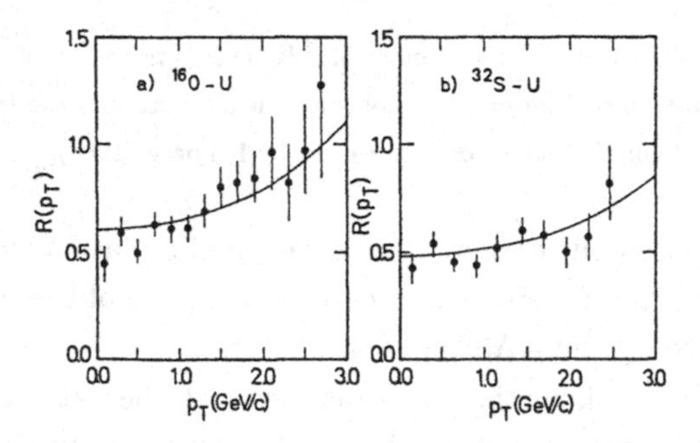

Fig. 20-27 Ratio $R(p_T)$ of the number of events in the highest transverse energy bin to the number of events in the lowest transverse energy bin as a function of the transverse momentum of the J/ψ. The solid lines are the results obtained by Hüfner, Kurihara and Pirner (1988). The p_T dependence is due to initial state interactions. The data is from Abreu et al. (1988).

But, as we have already mentioned, a large amount of J/ψ suppression can also be obtained assuming that the J/ψ disintegrates due to inelastic scattering processes in a dense nuclear medium. By including also the effects arising from initial state interactions, one can obtain a p_T-dependence of the J/ψ suppression compatible with the experimental data (Gavin and Gyulassy, 1988; Hüfner, Kurihara and Pirner, 1988; Blaizot and Ollitraut, 1989). As an example we show in fig. (20-27) the results obtained by Hüfner, Kurihara and Pirner (1988) for the ratio of the number of high E_T to low E_T events as a function of the transverse momentum in oxygen-uranium and sulfur-uranium collisions. The data are from Abreu et al. (1988). In their analysis gluon multiple scattering was the dominant mechanism for the p_T distribution of the J/ψ in nuclear collisions. As seen from the figure, the agreement with the experimental data is quite good. Hence one does not know which of the scenarios for J/ψ suppression (if any) is correct. Probably, J/ψ suppression is a result of the interplay of several different mechanisms.

We close this chapter with a very brief discussion of two other proposals made in the literature for a possible signature of the formation of a quark-gluon plasma. For details the reader should consult the book mentioned at the beginning of this section, and the review article by Meyer-Ortmanns (1996). The main objective is, as always, to look for features of particle spectra which are sensitive to their production in a deconfined medium.

Dilepton Production

It has been argued already many years ago that dilepton production could provide a clean signal for the formation of a QGP.* Since leptons only interact electromagnetically or weakly they can escape from the dense nuclear matter without further rescattering, and therefore carry the information about their production at all stages of the collision process, from the initial hadronic phase, through the evolution of the plasma until freeze-out. There are several mechanisms for dilepton production which are operative in different kinematical regions. At high invariant masses (larger than 2-2.5 GeV) we have Drell-Yan production due to parton-antiparton annihilation, and lepton pairs originating from the decay of different hadrons. These processes are well understood. Low mass dileptons (with an invariant mass less than 1 GeV) are mainly produced from the decay of neutral mesons. If they originate at the late stage of the collision process, where the system has cooled down, then they carry no information about the hot and dense matter in a plasma. On the other hand suppression lepton pairs in the low mass region could be interpreted as being due to the melting of the ρ and ϕ within the

* For a review see Ruuskanen (1992).

plasma. Candidates for a signal of plasma formation are the thermal leptons with an invariant mass in the intermediate mass region. They originate from parton collisions in a hot and dense medium and their spectrum is sensitive to the temperature of the plasma at the time of their formation, as well as to the nature of the phase transition, close to T_c (Cleymans et al. (1987)). A typical observable is the differential multiplicity per invariant mass-squared, transverse momentum and rapidity interval. The prediction of the dilepton spectrum is carried out within the hydrodynamical framework, and involves several assumptions and approximations (see e.g. the review by Meyer-Ortmanns (1996)).

Strangeness Production

It was predicted already some time ago that the production of strange hadrons should be enhanced in heavy ion collisions [Rafelsky (1981)]. Such an enhancement was observed for example for the ratio K^+/π^+ [Abbott et al. (1990)]. The mechanism for strangeness production is quite different in a quark-gluon plasma (where strange quark pairs are produced from a very dense medium of quarks and gluons) and in a gas of hadrons (where hadrons with opposite strangeness are produced in inelastic hadron-hadron collisions). Although the observed enhancement of strange particles is consistent with the idea that a QGP has been formed, there are numerous assumptions that go into predicting this enhancement, and the involved sources of possible systematic errors may be difficult to control [see e.g. the review by Meyer-Ortmanns (1996)]. The problem is that because strange particles interact strongly, the accumulation of strangeness proceeds throughout all stages of the collision process, from the very beginning to the very end. In fact the observed K^+/π^+ enhancement can also be be explained in a more conventional way. That strangeness could play a crucial role in the search for a signature of QGP formation had been advocated by many physicists. It has also been proposed that the formation of a quark-gluon plasma in heavy ion collisions could lead to the production of exotic droplets of stable or metastable strange quark matter, consisting of approximately equal numbers of strange, "up", and "down" quarks [Greiner et al. (1987)]. For a review we refer the reader to the book "Quark-Gluon Plasma 2" mentioned earlier.

But what does experiment tell us? The Relativistic Heavy Ion Collider (RHIC) at Brookhaven National Laboratory began its operation in 2000. Two ion beams are brought to collision with a center of mass energy of 200 GeV per Nucleon. Whatever the precise implications of the results may turn out to be, it appears that one statement can be made rather safely: the data strongly suggest that the phenomena observed in the nuclear collisions reflect collective behaviour. The energy density deposited in the early stage of the collision has been estimated to be 20 GeV/fm^3, which exceeds the theoretical estimate given earlier by far. May be we are in for many more surprises once sufficient precise data is available from RHIC.

Much can be said about the very important problem of finding an unambigous signal for the formation of quark-gluon plasma. We have only taken here a glimpse at a few of the proposals made in the literature so far. Several other signals, like direct photons, jet production, hadron mass modifications in hot dense media have also been discussed. But a definite signature for plasma formation is still waiting to be found. Clearly, the experimental verification of the existence of a quark-gluon plasma would be the most spectacular prediction of lattice QCD.

APPENDIX A

The energy sum rule (10.40) has been checked in lattice perturbation theory by Feuerbacher (2003a,b) up to $\mathcal{O}(g_0^4)$. In this appendix we give a few technical details which are useful for carrying out the very extensive computations. In the following all quantities are understood to be measured in lattice units, i.e., we suppress the "hat" on the dimensioned quantities.

The first step in verifying the energy sum rule consists in making use of the *exact* action sum rule (10.24a) to cast the energy sum rule in a form which minimizes the computational effort.

With the definition (10.39b) the energy sum rule (10.30) takes the form

$$\hat{V}(\hat{R},\hat{T}) = \lim_{\hat{T}\to\infty} \frac{1}{\hat{T}} \left[\eta_- < -\mathcal{P}_\tau + \mathcal{P}_s >_{q\bar{q}-0} + \frac{\beta_L(g_0)}{2g_0}\hat{\beta} < \mathcal{P}_\tau + \mathcal{P}_s >_{q\bar{q}-0}\right] , \tag{A.1}$$

where $< \mathcal{O} >_{q\bar{q}-0}$ has been defined in (10.24b), and where, in accordance with (10.25a), we have made the replacement

$$< \mathcal{P}'_\sigma >_{q\bar{q}-0} = \lim_{T\to\infty} \frac{1}{T} < \mathcal{P}_\sigma >_{q\bar{q}-0} \tag{A.2}$$

Next we make the decomposition $< -\mathcal{P}_\tau + \mathcal{P}_s > = - < \mathcal{P}_\tau + \mathcal{P}_s > +2 < \mathcal{P}_s >$. Then (A.1) takes the following form for $SU(N)$,

$$\hat{V}(\hat{R},\hat{T}) = \lim_{\hat{T}\to\infty} \frac{1}{\hat{T}} \left[-\frac{g_0^2}{2N}\eta_- < S >_{q\bar{q}-0} + \frac{g_0^2}{N}\eta_- < S_s >_{q\bar{q}-0} + \frac{\beta_L(g_0)}{2g_0} < S >_{q\bar{q}-0}\right] \tag{A.3}$$

where S is the action (10.22) on an isotropic lattice, and S_s is the contribution to the action arising from the spacial plaquettes only. This form of the energy sum rule is particularily convenient, since one can make use of the *exact* action sum rule (10.24a)* to express $\lim_{T\to\infty}(< S >_{q\bar{q}-0} /T)$ in terms of the potential and derivatives thereof. Thus the action sum rule can be rewritten in the form

$$\lim_{\hat{T}\to\infty} \frac{1}{\hat{T}} < S >_{q\bar{q}-0} = -g_0^2 \frac{\partial \hat{V}}{\partial g_0^2} . \tag{A.4}$$

The energy sum rule (for $SU(N)$) then becomes equivalent to the following statement

$$\hat{V} - \eta_- \frac{g_0^4}{2N}\frac{\partial \hat{V}}{\partial g_0^2} + \frac{\beta_L(g_0)}{2g_0} g_0^2 \frac{\partial \hat{V}}{\partial g_0^2} = \lim_{\hat{T}\to\infty} \frac{1}{\hat{T}} \eta_- \frac{g_0^2}{N} < S_s >_{q\bar{q}-0} . \tag{A.5}$$

---

* Recall that the action sum rule follows directly from the definition of the potential via the Wilson loop.

Since the leading contribution to the potential is of $\mathcal{O}(g_0^2)$, and since the same is true for $< S_\sigma >_{q\bar{q}-0}$ $(\sigma = \tau, s)$, as we shall see below, we only need to know η_- up to $\mathcal{O}(g_0^0)$. Now up to $\mathcal{O}(g_0^0)$ η_- is given by (Karsch 1982),

$$\eta_- = \frac{2N}{g_0^2} + cN \tag{A.6}$$

where c has been determined by this author as a function of the anisotropy (which here is set equal to 1).

The potential has been calculated in lattice perturbation theory up to $\mathcal{O}(g_0^4)$ by Kovacs (1982), and by Heller and Karsch (1985), and is given by (9.8). To compute the connected correlator

$$< S_s >_{q\bar{q}-0} = \frac{< S_s W >}{< W >} - < S_s > \tag{A.7}$$

up to this order, we must expand the action and the Wilson loop in powers of the coupling. Consider first the expansion of the Wilson loop, normalized for $SU(N)$ conveniently as follows,

$$W[U] = \frac{1}{N} Tr \prod_{\ell \in C} U_\ell \,, \tag{A.8}$$

where C is a rectangular loop, and U_ℓ denotes the matrix valued link variable associated with the link labeled by ℓ. Written in terms of the gauge potentials we have

$$U_\ell = e^{i g_0 A_\ell} \,. \tag{A.9}$$

The expansion of $W[U]$ has the form

$$W = 1 - g_0^2 \omega^{(2)} - g_0^3 \omega^{(3)} - g_0^4 \omega^{(4)} + \mathcal{O}(g^4) \,. \tag{A.10}$$

Note that because of the trace in (A.8) there is no $\mathcal{O}(g_0)$ term. * Since (A.10) starts with the unit element, the leading term in W which contributes to (A.7) is of $\mathcal{O}(g_0^2)$. We therefore only need to expand S_s up to this order:

$$S_s = S_s^{(0)} + g_0 S_s^{(1)} + g_0^2 S_s^{(2)} + \mathcal{O}(g_0^3) \,. \tag{A.11}$$

---

* Expanding the product of the link variables in (A.8) in powers of the coupling, the leading term is just given by the sum of gauge potentials associated with the links along the Wilson loop. For $SU(N)$ $(N \geq 2)$ these are elements of a Lie algebra with vanishing trace.

The perturbative computation of the (gauge invariant) rhs of (A.5) is done most conveniently in the Feynman gauge. Since we are computing expectation values of operators which are at least of $\mathcal{O}(g_0^2)$, the "Boltzmann" factor in the path integral expressions can replaced by $e^{-S_{eff}}$, where

$$S_{eff} = S + g_0^2 S_{FP}^{(2)} + g_0^2 S_{meas}^{(2)} + \mathcal{O}(g_0^3) \,. \tag{A.12}$$

Here $S_{FP}^{(2)}$ and $S_{meas}^{(2)}$ are the contributions arising from the Faddeev-Popov determinant (associated with the gauge fixing), and from the integration measure (see chapter 15). Up to $\mathcal{O}(g_0^2)$ the Boltzmann factor $e^{S_{eff}}$ can be expanded as follows

$$e^{S_{eff}} \approx e^{S^{(0)}} \left(1 + g_0 S^{(1)} + g_0^2 \left[\frac{1}{2}(S^{(1)})^2 + S^{(2)} + S_{FP}^{(2)} + S_{meas}^{(2)}\right]\right) .$$

Then up to $\mathcal{O}(g_0^4)$ $< S_s >_{q\bar{q}-0}$ is given by (Feuerbacher, 2003)

$$\begin{aligned}
< S_s >_{q\bar{q}-0} = &-g_0^2 < S_s^{(0)} \omega^{(2)} >_{con} + g_0^4 < S_s^{(0)} S^{(2)} \omega^{(2)} >_{con} \\
&- \frac{1}{2} < S_s^{(0)} (S^{(1)})^2 \omega^{(2)} >_{con} + g_0^4 < S_s^{(0)} S_{FP}^{(2)} \omega^{(2)} >_{con} \\
&+ g_0^4 < S_s^{(0)} S_{meas}^{(2)} \omega^{(2)} >_{con} + g_0^4 < S_s^{(1)} S^{(1)} \omega^{(2)} >_{con} \\
&- g_0^4 < S_s^{(2)} \omega^{(2)} >_{con} + g_0^4 < S_s^{(0)} S^{(1)} \omega^{(3)} >_{con} \\
&- g_0^4 < S_s^{(0)} \omega^{(4)} >_{con} - g_0^4 < S_s^{(1)} \omega^{(3)} >_{con} \\
&- g_0^4 < S_s^{(0)} \omega^{(2)} >_{con} < \omega^{(2)} > + \mathcal{O}(g_0^6),
\end{aligned} \tag{A.13a}$$

where, generically

$$< \mathcal{O} \omega^{(n)} >_{con} \equiv < \mathcal{O} \omega^{(n)} > - < \mathcal{O} >< \omega^{(n)} > \tag{A.13.b}$$

and $S^{(0)}$ is the free action. The subsript "con" stands for "connected". Note that all expectation values are now calculated with the Boltzmann factor $\exp(-S^{(0)})$. Note also that $g_0 S^{(1)}$ and $g_0^2 S^{(2)}$ are the contributions to the action of the 3 and 4-gluon vertex (see chap. 15). For a perturbative calculation one must express the above expectation values in terms of expectation values of products of gauge potentials living on the links of the lattice. Consider e.g. the Wilson loop whose expansion determines the $\omega^{(n)}$'s in (A.10),

$$W_C = \frac{1}{N} Tr \prod_{\ell \in C} e^{i g_0 A_\ell} = \frac{1}{N} Tr e^{B_C} \tag{A.14}$$

where B_C is again an element of the Lie-algebra of $SU(N)$, and where the product of the exponentials is ordered along the contour C in the counterclockwise sense.

The index $\ell = 1, 2, 3 \cdots$ labels the successive links along the contour. Making repeated use of the Campbell-Baker-Hausdorff formula, the exponent B_C will be given by the sum of $\sum_{l \in C} B_l$ and higher order commutators in the B_l's, where $B_\ell = ig_0 A_\ell$. Hence every term in this expansion will be an element of the Lie-algebra and therefore traceless. For this reason we only need to know B_C up to order g_0^3 in order to calculate the contribution to the Wilson loop up to order g_0^4. We had already made use of this when expanding the plaquette contributions to the action in section (15.3). In particular we had made use of (15.31) to calculate the argument of the exponential associated with an elementary plaquette. In fact, this formula remains valid for an arbitrary product of link variables. Thus one can easily convince oneself that if (15.31) holds for the product $e^{B_1} e^{B_2} \cdots e^{B_n} \equiv e^{M_n}$, it also holds for $e^{M_n} e^{B_{n+1}}$. An alternative form for (15.31) has be used by Feuerbacher (2003). To keep in the spirit of this author we will make use of it below. Up to $\mathcal{O}(g_0^3)$ the exponent B_C in (A.14), expressed in terms of the gauge potentials, is given by

$$\begin{aligned} B_C = ig_0 \sum_l A_l - \frac{1}{2} g_0^2 \sum_{l_1 < l_2} [A_{l_1}, A_{l_2}] - \frac{i}{4} g_0^3 \sum_{l_1 < l_2 < l_3} [[A_{l_1}, A_{l_2}], A_{l_3}] \\ - \frac{i}{12} g_0^3 \sum_{(l_1, l_2) < l_3} [A_{l_1}, [A_{l_2}, A_{l_3}]] - \frac{i}{12} g_0^3 \sum_{l_1 < l_2} [[A_{l_1}, A_{l_2}], A_{l_2}] + \mathcal{O}(g_0^4). \end{aligned} \tag{A.15}$$

Expanding the exponential (A.14) one finds that the coefficients ω_ℓ in (A.10) are given as follows for $SU(N)$ (Heller, 1985),

$$\omega^{(2)} = \frac{1}{4N} \left(\sum_l A_l^B \right)^2 ,$$

where $B = 1, \cdots N$, and

$$\omega^{(3)} = \frac{i}{6N} Tr \left(\sum_l A_l \right)^2 + \frac{i}{2N} Tr \left(\sum_l A_l \sum_{l_1 < l_2} [A_{l_1}, A_{l_2}] \right) ,$$

$$\begin{aligned}\omega^{(4)} = &-\frac{1}{24N}Tr\left(\sum_l A_l\right)^4 - \frac{1}{8N}Tr\left(\sum_{l_1<l_2}[A_{l_1}, A_{l_2}]\right)^2 \\ &-\frac{1}{4N}Tr\left(\sum_l A_l \sum_{l_1<l_2<l_3}[[A_{l_1}, A_{l_2}], A_{l_3}]\right) \\ &-\frac{1}{12N}Tr\left(\sum_l A_l \sum_{(l_1,l_2)<l_3}[A_{l_1}, [A_{l_2}, A_{l_3}]]\right) \\ &-\frac{1}{12N}Tr\left(\sum_l A_l \sum_{l_1<l_2}[[A_{l_1}, A_{l_2}], A_{l_2}]\right) \\ &-\frac{1}{4N}Tr\left(\left(\sum_l A_l\right)^2 \sum_{l_1<l_2}[A_{l_1}, A_{l_2}]\right).\end{aligned} \tag{A.16}$$

In perturbation theory the expectation values in (A.13a) involve propagators connecting sites located on the Wilson loop and on the boundary of *spacial* plaquettes, modified by 3 and 4-gluon interactions arising from S_{eff}. In fig. (A-1) we show relevant diagrams contributing to (A.13a). The figure is taken from Feuerbacher (2003).

The evaluation of the terms in (A.13a) is evidently quite non trivial, and requires substancial gymnastics. When computing the expectation values in (A.13a) one is confronted with various types of partial sums involving gauge potentials living on the square contour of the Wilson loop. In the following we give two simple examples demonstrating the technicalities involved.

Example 1

Consider e.g. the sum over potentials along the square contour of a Wilson loop. Let N_0 denote the base point of the Wilson loop with spacial and temporal extent R and T (measured in lattice units). The sum of interest has the form

$$\begin{aligned}\sum_{\ell\in C} A_\ell = &\sum_{n=0}^{R-1}(A_\mu(N_0+n\hat{\mu}) - A_\mu(N_0+T\hat{\nu}+n\hat{\mu}) \\ &+\sum_{n=0}^{T-1}(A_\nu(N_0+R\hat{\mu}+n\hat{\nu}) - A_\nu(N_0+n\hat{\nu}))\end{aligned} \quad . \tag{A.17}$$

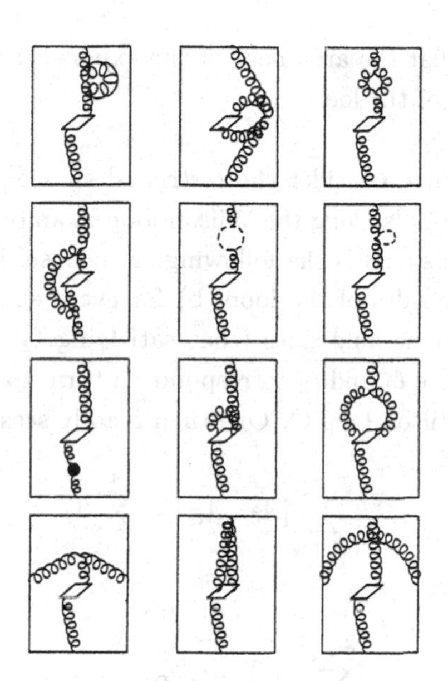

Fig. A-1 Connected diagrams contributing to $< S_s >_{q\bar{q}-0}$, with $< S_s >_{q\bar{q}-0}$ given by (A.13a), in next to leading order (Feuerbacher, 2003b).

Fourier decomposing the potentials as follows*

$$A_\mu(N) = \int_{BZ} \frac{d^4p}{(2\pi)^4} \tilde{A}(p) e^{ip\cdot(N+\frac{\hat{\mu}}{2})} \,, \tag{A.18}$$

and making use of the identity

$$\sum_{n=0}^{R-1} e^{i(np_\mu + \frac{p_\mu}{2})} = e^{iR\frac{p_\mu}{2}} \frac{\sin\left(R\frac{p_\mu}{2}\right)}{\sin\left(\frac{p_\mu}{2}\right)} \tag{A.19}$$

one readily finds that the above sum can be written in the form

$$\begin{aligned}\sum_{\ell\in C_W} A_\ell = -2i \int d^4p \; A_\alpha(p) e^{ip\cdot(n_0 + R\frac{\hat{\mu}}{2} + T\frac{\hat{\nu}}{2})} \sin(Tp_\nu/2)\sin(Rp_\mu/2) \\ \cdot\left(\delta_{\alpha\mu}\frac{1}{\sin(p_\mu/2)} - \delta_{\alpha\nu}\frac{1}{\sin(p_\nu/2)}\right),\end{aligned} \tag{A.20}$$

* Recall that the potentials are evaluated at the midpoints of the links (see chapter 14).

where $\mu \neq \nu$. Note that the argument of the exponential involves the lattice site located at the center of the loop.

Example 2

As another example consider the restricted sum $\sum_{\ell_1<\ell_2}[A_{\ell_1}, A_{\ell_2}]$, where ℓ_1 and ℓ_2 label the potentials along the Wilson loop in an ordered way. A convenient way for evaluating this sum is the following: a) choose, in turn, ℓ_1 to correspond to points on the four sides of the loop; b) for every such choice sum over all ℓ_2 associated with the remaining sides (thus satisfying $\ell_2 > \ell_1$); c) finally consider the contributions where ℓ_1 and ℓ_2 correspond, in turn, to points lying on the same line element of the Wilson loop C. One then readily sees that

$$\sum_{\ell_1<\ell_2}[A_{\ell_1}, A_{\ell_2}] = \sum_{k=1}^{4} T_k \tag{A.21}$$

where

$$\begin{aligned}
T_1 &= [\sum_{n=0}^{R-1} A_\mu(N_0+n\hat{\mu}), \sum_{n'=0}^{T-1}(A_\nu(N_0+R\hat{\mu}+n'\hat{\nu}) - A_\nu(N_0+n'\hat{\nu}))] \\
&\quad - [\sum_{n=0}^{R-1} A_\mu(n_0+n\hat{\mu}), \sum_{n'=0}^{R-1} A_\mu(N_0+T\hat{\nu}+n'\hat{\mu})] \\
&\quad + [\sum_{n=0}^{R-2} A_\mu(N_0+n\hat{\mu}), \sum_{n'=n+1}^{R-1} A_\mu(N_0+n'\hat{\mu})] \\
T_2 &= [\sum_{n=0}^{T-1} A_\nu(N_0+R\hat{\mu}+n\hat{\nu}), -\sum_{n'=0}^{R-1} A_\mu(N_0+T\hat{\nu}+n'\hat{\mu})] \\
&\quad + [\sum_{n=0}^{T-1} A_\nu(N_0+R\hat{\mu}+n\hat{\nu}), -\sum_{n'=0}^{T-1} A_\nu(N_0+n'\hat{\nu})] \\
&\quad + [\sum_{n=0}^{T-2} A_\nu(N_0+R\hat{\mu}+n\hat{\nu}), \sum_{n'=n+1}^{T-1} A_\nu(N_0+R\hat{\mu}+n'\hat{\nu})] \\
T_3 &= [-\sum_{n=0}^{R-1} A_\mu(N_0+T\hat{\nu}+n\hat{\mu}), -\sum_{n'=0}^{T-1} A_\nu(N_0+n'\hat{\nu})] \\
&\quad + [-\sum_{n=1}^{R-1} A_\mu(N_0+T\hat{\nu}+n\hat{\mu}), -\sum_{n'=0}^{n-1} A_\mu(N_0+T\hat{\nu}+n'\hat{\mu})] \\
T_4 &= [-\sum_{n=1}^{T-1} A_\nu(N_0+n\hat{\nu}), -\sum_{n'=0}^{n-1} A_\nu(N_0+n'\hat{\nu})]\ .
\end{aligned} \tag{A.22}$$

The corresponding Fourier integral expressions for T_k can now be obtained making use of (A.17). Thus for example one readily finds that

$$\sum_{n=0}^{R-1}\sum_{n'=0}^{T-1}[A_\mu(N_0+n\hat{\mu}), A_\nu(N_0+R\hat{\mu}+n'\hat{\nu})] = \int_{BZ}(dp)(dp')[A_\mu(p), A_\nu(p')]$$

$$\times\, e^{i(p+p')\cdot(N_0+R\frac{\hat{\mu}}{2}+T\frac{\hat{\nu}}{2})}e^{ip'_\mu\frac{R}{2}}e^{-ip_\nu\frac{T}{2}}\frac{\sin(p_\mu\frac{R}{2})\sin(p'_\nu\frac{T}{2})}{\sin(\frac{p_\mu}{2})\sin(\frac{p'_\nu}{2})}\,.$$

Clearly, restricted multiple sums of double commutators of the gauge potentials, lead to expressions which are far more complicated. Major simplifications however result when one considers the limit $T\to\infty$, which is of interest when checking the energy sum rule. Furthermore, a large number of lattice integrals can be calculated almost analytically. In evaluating the integrals extensive use is made of the cubic lattice symmetry, and of partial integration. This is demonstrated by the following examples taken from Feuerbacher (2003). All relations hold in d dimensions.

Consider the following integral which vanishes by construction.

$$\int_{BZ}\frac{d^dp}{(2\pi)^d}\frac{\partial}{\partial p_\mu}\frac{\sin(p_\mu)}{\tilde{p}^2}=0$$

where

$$\tilde{p}_\mu = 2\sin(\frac{1}{2}p_\mu)$$

and

$$\tilde{p}^2=\sum_{\mu=1}^{d}\tilde{p}_\mu^2$$

By performing the differentiation in the integrand one is readily led to

$$\int_{BZ}\frac{d^dp}{(2\pi)^d}\left(\frac{1-\frac{1}{2}\tilde{p}_\mu^2}{\tilde{p}^2}-2\frac{\tilde{p}_\mu^2-\frac{1}{4}\tilde{p}_\mu^4}{(\tilde{p}^2)^2}\right)=0\,.$$

Because of the cubic symmetry we can replace $\tilde{p}_\mu^2$ by $\frac{1}{d}\tilde{p}^2$. One therefore finds that

$$\int_{BZ}\frac{d^dp}{(2\pi)^d}\frac{\tilde{p}_\mu^4}{(\tilde{p}^2)^2}=\frac{4-2d}{d}\Delta_0+\frac{1}{d}\,, \qquad (A.23a)$$

where

$$\Delta_0=\int_{BZ}\frac{d^dp}{(2\pi)^d}\frac{1}{\tilde{p}^2}\,. \qquad (A.23b)$$

This integral can be reduced to a one-dimensional integral as follows: We first write (A.22b) in the form

$$\begin{aligned}
\Delta_0 &= \int_0^\infty dt \int_{BZ} \frac{d^d p}{(2\pi)^d} \exp\left(-4t \sum_{\mu=4}^{d} \sin^2(p_\mu/2)\right) \\
&= \int_0^\infty dt \left(\int_{-\pi}^{\pi} \frac{dp}{2\pi} \exp(-4t \sin^2(p/2))\right)^d \\
&= \int_0^\infty dt\ e^{-8t} \left(\int_{-\pi}^{\pi} \frac{dp}{2\pi} e^{-2t\cos(p)}\right)^d \\
&= \frac{1}{2}\int_0^\infty dt e^{-dt} I_0^d(t)\ ,
\end{aligned} \qquad (A.23c)$$

where

$$I_0(t) = \frac{1}{2\pi} \int_{-\pi}^{\pi} dx\ e^{-t\cos x} \qquad (A.23d)$$

is the Bessel function. This integral can be easily evaluated numerically.

Another trivial integral is given e.g. by

$$1 = \int_{BZ} \frac{d^d p}{(2\pi)^d} \frac{(\tilde{p}^2)^2}{(\tilde{p}^2)^2}\ .$$

Using the cubic symmetry of the lattice we can also write it in the form

$$1 = d \int_{BZ} \frac{d^d p}{(2\pi)^d} \frac{\tilde{p}_\mu^4}{(\tilde{p}^2)^2} + d(d-1) \int_{BZ} \frac{d^d p}{(2\pi)^d} \frac{\tilde{p}_\mu^2 \tilde{p}_\nu^2}{(\tilde{p}^2)^2}\ ,$$

where $\mu \neq \nu$. Making use of (A.21a) one therefore has that

$$\int_{BZ} \frac{d^d p}{(2\pi)^d} \frac{\tilde{p}_\mu^2 \tilde{p}_\nu^2}{(\tilde{p}^2)^2} = \frac{2d-4}{d(d-1)} \Delta_0 \quad ;\ \mu \neq \nu\ . \qquad (A.24)$$

There are many further lattice integrals that can be evaluated with similar techniques. For the readers convenience we present some further useful integrals taken from Feuerbacher (2003a):

$$\begin{aligned}
\int_{BZ} \frac{d^dp}{(2\pi)^d} \frac{\tilde{p}_\mu^2}{(\tilde{p}^2)^3} &= \frac{1}{d}\Delta_1 \\
\int_{BZ} \frac{d^dp}{(2\pi)^d} \frac{(\tilde{p}_\mu^2)^2}{(\tilde{p}^2)^3} &= \frac{1}{2d}\Delta_0 - \frac{1}{d}\Delta_1 \\
\int_{BZ} \frac{d^dp}{(2\pi)^d} \frac{\tilde{p}_\mu^2\tilde{p}_\nu^2}{(\tilde{p}^2)^3} &= \frac{1}{2d(d-1)}\Delta_0 + \frac{1}{d(d-1)}\Delta_1 \ ; \ \mu \neq \nu \\
\int_{BZ} \frac{d^dp}{(2\pi)^d} \frac{(\tilde{p}_\mu^2)^3}{(\tilde{p}^2)^3} &= \frac{3-2d}{d}\Delta_0 + \frac{1}{d} - \frac{4}{d}\Delta_1 \\
\int_{BZ} \frac{d^dp}{(2\pi)^d} \frac{\tilde{p}_\mu^2\tilde{p}_\nu^2}{(\tilde{p}^2)^3} &= \frac{1}{d(d-1)}\Delta_0 + \frac{4}{d(d-1)}\Delta_1 \ ; \ \mu \neq \nu \\
\int_{BZ} \frac{d^dp}{(2\pi)^d} \frac{\tilde{p}_\mu^2\tilde{p}_\nu^2\tilde{p}_\lambda^2}{(\tilde{p}^2)^3} &= \frac{2d-6}{d(d-1)(d-2)}\Delta_0 - \frac{8}{d(d-1)(d-2)}\Delta_1 \ ; \ \mu \neq \nu\lambda \, ,
\end{aligned} \tag{A.25a}$$

where Δ_1 is given by

$$\Delta_1 = (d-4)\int_{BZ} \frac{d^dp}{(2\pi)^d} \frac{1}{(\tilde{p}^2)^2} \, . \tag{A.25b}$$

In the limit $d \to 4$ we have that $\Delta_1 \to \frac{1}{2(2\pi)^2}$. There are many other type of lattice integrals which need to be evaluated in order to check the energy sum rule (A.1). The complete set of calculations, which go beyond checking the energy sum rule, can be found in (Feuerbacher, 2003).

APPENDIX B

Up to $\mathcal{O}(e^2)$ the vertices in momentum space associated with the axial vector and pseudoscalar currents (14.41) and (14.42), and with the operator Δ defined in (14.43), are easily obtained by Fourier transforming the field as follows,

$$\psi(x) = \int_{BZ} (dp)\ \tilde{\psi}(p)e^{ip\cdot x}$$

$$\bar{\psi}(x) = \int_{BZ} (dp)\ \tilde{\bar{\psi}}(p)e^{-ip\cdot x}$$

$$A_\mu(x) = \int_{BZ} (dk)\ \tilde{A}_\mu(k)e^{ik\cdot x}\ ,$$

where, generically,

$$(dq) \equiv \frac{d^4q}{(2\pi)^4}\ .$$

If $O(x)$ stands for any of the operators $j_{5\mu(x)}$, $j_5(x)$ or $\Delta(x)$, defined in (14.41), (14.42), and (14.43), then we define the Fourier transform $\tilde{O}$ by

$$O(x) = \int_{BZ} (dq)\ e^{-iq\cdot x}\tilde{O}(q)\ , \qquad (B.1a)$$

where

$$\tilde{O}(q) = \int (dp')(dp) \cdots\ V(q;p',p,\cdots)\ . \qquad (B.1b)$$

Here the dots stand for possible photon momenta k_i (see below), and

$$V(q;p',p,\cdots) = (2\pi)^4\delta^{(4)}(p'-p-q-\cdots)\tilde{V}((q;p',p,\cdots)\ , \qquad (B.1c)$$

with the vertices $\tilde{V}(q;p',p,\cdots)$ defined as follows

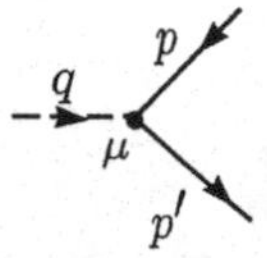

$$\to \qquad e^{-i\frac{q_\mu a}{2}}\cos((\frac{p+p'}{2})_\mu a)\gamma_\mu\gamma_5\ , \qquad (B.2a)$$

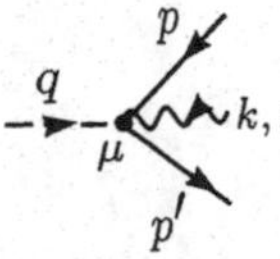

$$\to \qquad -ea\delta_{\mu\nu}e^{-i\frac{q_\mu a}{2}}\sin((\frac{p+p'}{2})_\mu a)\gamma_\mu\gamma_5\ , \qquad (B.2b)$$

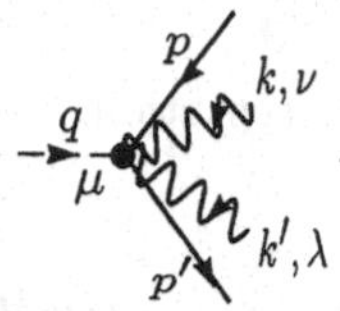

$$\rightarrow \quad -a^2e^2\delta_{\mu\nu}\delta_{\nu\lambda}e^{-i\frac{q_\mu a}{2}}\cos((\frac{p+p'}{2})_\mu a)\gamma_\mu\gamma_5 \,, \tag{B.2c}$$

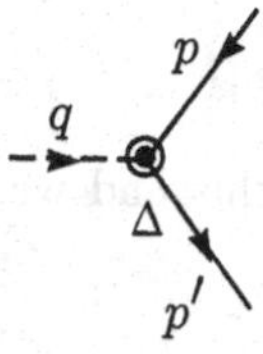

$$\rightarrow \quad [M_r(p,a)+M_r(p',a)]\gamma_5 \,, \tag{B.2d}$$

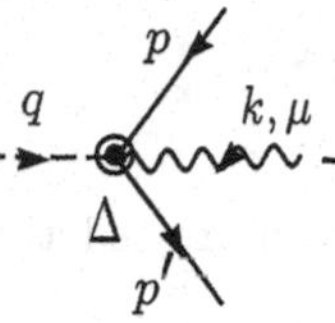

$$\rightarrow \quad er\Big\{\sin(p+\frac{k}{2})_\mu a+\sin(p'-\frac{k}{2})_\mu a\Big\}\gamma_5 \,, \tag{B.2e}$$

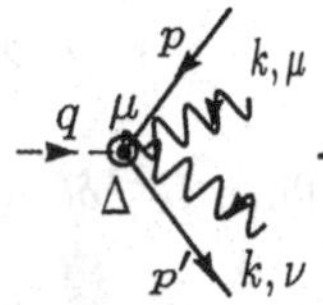

$$\rightarrow \quad are^2\delta_{\mu\nu}\Big\{\cos(p+\frac{k+k'}{2})_\mu a+\cos(p'-\frac{k+k'}{2})_\mu a\Big\}\gamma_5 \,. \tag{B.2f}$$

Taking the left derivative of $j_{5\mu}(x)$ amounts to contracting the vertices (B.2a-c) with $-2i\sin(q_\mu a/2)\exp(iq_\mu a/2)$. Hence the exponentials appearing in the vertices (B.2a-c) are eliminated by this contraction.

APPENDIX C

Consider a general group element of SU(3) in the fundamental representation:

$$U(\phi) = e^{i\phi},$$
$$\phi = \sum_{A=1}^{8} \phi^A T^A. \qquad (C.1)$$

We want to compute $U^{-1}\delta U$, where $\delta U = U(\phi + \delta\phi) - U(\phi)$. To this end we introduce the following matrices (Boulware, 1970)

$$W(\lambda) = U(\lambda\phi),$$
$$\delta W(\lambda) = U(\lambda(\phi + \delta\phi)) - U(\lambda\phi),$$

where λ is a real parameter, and where for simplicity, we have suppressed the dependence of W on ϕ. Next, we derive a differential equation for

$$\delta\chi(\lambda) = -iW^{-1}(\lambda)\delta W(\lambda), \qquad (C.2)$$

and obtain its solution, subject to the condition that $\delta\chi(0) = 0$. Then $U^{-1}\delta U$ is given by $i\delta\chi(1)$. We now give the details.

From (C.2) we obtain

$$\frac{\partial}{\partial\lambda}\delta\chi(\lambda) = -i\frac{\partial W^{-1}(\lambda)}{\partial\lambda}\delta W(\lambda) - iW^{-1}(\lambda)\frac{\partial}{\partial\lambda}\delta W(\lambda). \qquad (C.3)$$

Inserting the expressions

$$\frac{\partial W^{-1}(\lambda)}{\partial\lambda} = -i\phi W^{-1}(\lambda),$$
$$\frac{\partial}{\partial\lambda}\delta W(\lambda) = i\delta W(\lambda)\phi + i(W(\lambda) + \delta W(\lambda))\delta\phi,$$

into (C.3), one finds that

$$\frac{\partial}{\partial\lambda}\delta\chi(\lambda) = i[\delta\chi(\lambda), \phi] + \delta\phi + i\delta\chi(\lambda)\delta\phi.$$

Since $\delta\chi(\lambda)$ is itself of order $\delta\phi$, we have that up to $O(\delta\phi)$,

$$\frac{\partial}{\partial\lambda}\delta\chi(\lambda) = i[\delta\chi(\lambda), \phi] + \delta\phi. \qquad (C.4)$$

Furthermore, up to this order, $\delta\chi$ is an element of the Lie algebra of SU(3).* Hence we may write $\delta\chi(\lambda)$ in the form

$$\delta\chi(\lambda) = \sum_A T^A \delta\chi^A(\lambda).$$

An analogous decomposition holds of course for $\delta\phi$:

$$\delta\phi = \sum_A T^A \delta\phi^A.$$

Making use of the commutation relations (15.2a), one finds that (C.4) implies the following differential equation for the components $\delta\chi^A$:

$$\frac{\partial}{\partial\lambda}\delta\chi^A(\lambda) = -\sum_{B,C} f_{ABC}\delta\chi^B(\lambda)\phi^C + \delta\phi^A. \tag{C.5}$$

But according to (15.8)

$$\sum_C f_{ABC}\phi^C = i\sum_B t^C_{AB}\phi^C,$$

where t^C are the generators of SU(3) in the adjoint representation. Hence we may write (C.5) in the form

$$\frac{\partial}{\partial\lambda}\delta\vec{\chi}(\lambda) = -i\Phi\delta\vec{\chi}(\lambda) + \delta\vec{\phi}, \tag{C.6}$$

where $\delta\vec{\chi}$ and $\delta\vec{\phi}$ are vectors with components $\delta\chi^A$ and $\delta\phi^A$ $(A = 1, \ldots, 8)$, respectively, and where

$$\Phi = \sum_A \phi^A t^A$$

is an element of the Lie algebra of SU(3) in the adjoint representation. The corresponding generators t^A are normalized according to (15.9). The solution to (C.6), subject to the requirement that $\delta\chi^A(0) = 0$, is now immediately obtained:

$$\delta\vec{\chi}(\lambda) = \left(\frac{e^{-i\lambda\Phi} - 1}{-i\Phi}\right)\delta\vec{\phi}.$$

---

* This can be easily seen by applying the Campbell–Baker–Hausdorff formula to $i\delta\chi(\lambda) = [e^{-i\lambda\phi}e^{i\lambda(\phi+\delta\phi)} - 1]$.

Setting $\lambda = 1$, we therefore find that

$$U^{-1}\delta U = i\sum_B T^A M_{AB}(\phi)\delta\phi^B, \tag{C.7a}$$

where the matrix $M(\phi)$ is given by

$$M(\phi) = \frac{1 - e^{-i\Phi}}{i\Phi}. \tag{C.7b}$$

This is the result we have been looking for. A similar expression to (C.7a) can be derived for $\delta U U^{-1}$ by solving the differential equation for

$$\delta\tilde{\chi}(\lambda) = -i\delta W(\lambda)W^{-1}(\lambda),$$

with $W(\lambda)$ as defined above. One readily finds that

$$\delta U(\phi)U^{-1}(\phi) = i\sum_{A,B} T^A M_{AB}(-\phi)\delta\phi^B. \tag{C.8}$$

This expression will be useful in the following, where we study how $U(\phi)$ transforms under infinitesimal gauge transformations.

APPENDIX D

In this appendix we give a proof of formula (15.19).

Let $U(\phi)$ be the group element defined in (C.1). Consider the following infinitesimal transformation

$$e^{i\delta\omega}U(\phi)e^{-i\delta\omega'} = U(\phi+\delta\phi), \tag{D.1}$$

where $\delta\omega$ and $\delta\omega'$ are infinitesimal elements of the Lie algebra of SU(3) in the fundamental representation. They can be decomposed as follows

$$\delta\omega = \sum_A T^A \delta\omega^A,$$
$$\delta\omega' = \sum_A T^A \delta\omega'^A.$$

We want to calculate $\delta\phi$ up to terms linear in $\delta\omega$ and $\delta\omega'$. This can be easily accomplished by making use of the results obtained in appendix C. Consider first the product

$$U(\phi)e^{-i\delta\omega'} = U(\phi+\delta^R_{(\omega')}\phi), \tag{D.2}$$

where the superscript R on $\delta^R_{(\omega')}$ is to remind us that we are interested in the change of ϕ arising from the group multiplication of $U(\phi)$ with $e^{-i\delta\omega'}$ from the right. Clearly the leading contribution to $\delta^R_{(\omega')}\phi$ is of order $\delta\omega'$. To calculate this change we write the right-hand side of (D.2) in the form

$$U(\phi+\delta^R_{(\omega')}\phi) = U(\phi)\{1+U^{-1}(\phi)\delta U(\phi)\},$$

where $U^{-1}\delta U$ is given by (C.7) with $\delta\phi$ replaced by $\delta^R_{(\omega')}\phi$. But to this order we may replace $\exp(-i\delta\omega')$ in (D.2) by $1-i\delta\omega'$. Hence we conclude that

$$\delta\omega'^A = -\sum_B M_{AB}(\phi)\delta^R_{(\omega')}\phi^B, \tag{D.3a}$$

or

$$\delta^R_{(\omega')}\phi^A = -\sum M^{-1}_{AB}(\phi)\delta\omega'^B. \tag{D.3b}$$

Next we calculate

$$e^{i\delta\omega}U(\phi)e^{-i\delta\omega'} = e^{i\delta\omega}U(\phi'),$$

where $\phi' = \phi+\delta^R_{(\omega')}\phi$. Let us denote the change in ϕ' arising from the left multiplication of $U(\phi')$ with $\exp(i\delta\omega)$ by $\delta^L_{(\omega)}\phi'$; then

$$e^{i\delta\omega}U(\phi') = U(\phi'+\delta^L_{(\omega)}\phi'). \tag{D.4}$$

The right–hand side can be written in the form

$$U(\phi' + \delta^L_{(\omega)}\phi') = \{1 + \delta U(\phi')U^{-1}(\phi')\}U(\phi'),$$

where $\delta U(\phi') = U(\phi' + \delta^L_{(\omega)}(\phi')) - U(\phi')$. On the other hand the left–hand side of (D.4) can be approximated by $(1 + i\delta\omega)U(\phi')$. Now $\delta U(\phi')U^{-1}(\phi')$ is given by (C.8) with ϕ replaced by (ϕ'). Hence we are led to the following relation between $\delta\omega^A$ and $\delta^L_{(\omega)}\phi^B$ in leading order

$$\delta\omega^A = \sum_B M_{AB}(-\phi)\delta^L_{(\omega)}\phi^B.$$

Inversion of this equation gives

$$\delta^L_{(\omega)}\phi^A = \sum M^{-1}_{AB}(-\phi)\delta\omega^B. \tag{D.5}$$

The total change in ϕ induced by the infinitesimal transformation (D.1) is therefore given by the sum of (D.3b) and (D.5):

$$\delta\phi^A = \sum_B \left[M^{-1}_{AB}(-\phi)\delta\omega^B - M^{-1}_{AB}(\phi)\delta\omega'^B\right]. \tag{D.6}$$

Consider now an infinitesimal gauge transformation of the link variables

$$U_\mu(n) \to e^{i\delta\omega(n)}U_\mu(n)e^{-i\delta\omega(n+\hat\mu)}.$$

Let $\delta_{(\omega)}\phi^A_\mu(n)$ denote the change in the group–parameters, defined in (15.1a,b), induced by this transformation. By making the appropriate substitutions in (D.6), one obtains

$$\delta_{(\omega)}\phi^A_\mu(n) = \sum_B \left[M^{-1}_{AB}\left(-\phi_\mu(n)\right)\delta\omega^B(n) - M^{-1}_{AB}\left(\phi_\mu(n)\right)\delta\omega^B(n+\hat\mu)\right]. \tag{D.7}$$

This expression may also be written in the form (15.19). To this effect we set

$$\delta\omega^B(n+\hat\mu) = \delta\omega^B(n) + \hat\partial^R_\mu\delta\omega^B(n),$$

where $\hat\partial^R_\mu$ is the right lattice derivative. Then (D.7) becomes

$$\begin{aligned}\delta_{(\omega)}\phi^A_\mu(n) = \sum_B \Big\{&[M^{-1}(-\phi_\mu(n)) - M^{-1}(\phi_\mu(n))]_{AB}\delta\omega^B(n)\\ &- M^{-1}_{AB}(\phi_\mu(n))\hat\partial^R_\mu\delta\omega^B(n)\Big\}.\end{aligned} \tag{D.8}$$

Making use of the explicit form for $M(\phi)$ given in (C.7b), one can show after some simple algebra, that

$$M^{-1}\left(-\phi_{\mu}(n)\right) - M^{-1}\left(\phi_{\mu}(n)\right) = -i\Phi_{\mu}(n), \tag{D.9a}$$

where

$$\Phi_{\mu}(n) = \sum_{A} \phi_{\mu}^{A}(n) t^{A}. \tag{D.9b}$$

Substituting (D.9) into expression (D.8) we finally obtain

$$\delta_{(\omega)}\phi_{\mu}^{A}(n) = -\sum_{B}\left(i\Phi_{\mu}(n) + M^{-1}\left(\phi_{\mu}(n)\right)\hat{\partial}_{\mu}^{R}\right)_{AB}\delta\omega^{B}(n), \tag{D.10}$$

which is the result we wanted to prove.

APPENDIX E

In this appendix we derive a formula which allows us to carry out sums over fermionic Matsubara frequencies in the continuum formulation, where these frequencies are not restricted to a finite interval.

Consider the function

$$h(\omega) = \frac{-i\beta}{e^{i\beta(\omega - i\mu)} + 1} ,$$

where ω is a complex variable. It has simple poles with unit residue located at $\omega = \omega_\ell^- + i\mu$, where $\omega = \omega_\ell^-$ are the fermionic Matsubara frequencies (19.57). Let $f(\omega)$ be a function of the complex variable ω which is non-singular for $Im\ \omega \in [\mu - \epsilon, \mu + \epsilon]$, with ϵ infinitessimal. Then

$$\frac{1}{\beta} \sum_{\ell=-\infty}^{\infty} f(\omega_\ell^- + i\mu) = \frac{1}{2\pi i\beta} \int_C d\omega f(\omega) h(\omega) , \tag{E.1}$$

where, for $\mu > 0$, C is the closed contour depicted in Fig. E-1.

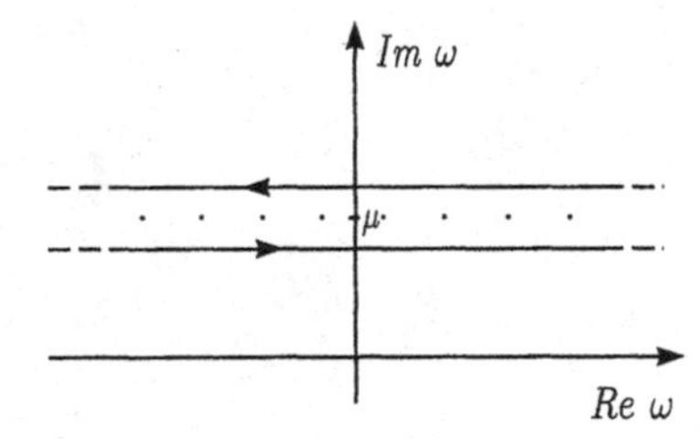

Fig. E-1 Contour of integration C in eq. (E.1)

By making use of the relation

$$\frac{1}{e^x + 1} = 1 - \frac{1}{e^{-x} + 1} \tag{E.2}$$

on the upper branch of the contour we can write (E.1) in the form

$$\frac{1}{\beta} \sum_{l=-\infty}^{\infty} f(\omega_\ell^- + i\mu) = \frac{1}{2\pi} \int_{-\infty+i\mu+i\epsilon}^{\infty+i\mu+i\epsilon} d\omega\ f(\omega) - \frac{1}{2\pi} \int_{-\infty+i\mu+i\epsilon}^{\infty+i\mu+i\epsilon} d\omega\ \frac{f(\omega)}{e^{-i\beta(\omega-i\mu)} + 1}$$
$$- \frac{1}{2\pi} \int_{-\infty+i\mu-i\epsilon}^{\infty+i\mu-i\epsilon} d\omega\ \frac{f(\omega)}{e^{i\beta(\omega-i\mu)} + 1}$$

Consider the case where $f(\omega)$ is of the form $f(\omega) = p(\omega)/q(\omega)$, where $p(\omega)$ and $q(\omega)$ are polynomials in ω. Then we can close the integration contours in the last two integrals in the lower and upper planes, respectively, and obtain

$$\frac{1}{\beta}\sum_{\ell=-\infty}^{\infty} f(\omega_\ell^- + i\mu) = \frac{1}{2\pi}\int_{-\infty+i\mu+i\epsilon}^{\infty+i\mu+i\epsilon} d\omega\ f(\omega) + i \sum_{Im\bar{\omega}_i<\mu+\epsilon} \frac{R_f(\bar{\omega}_i)}{e^{i\beta(\bar{\omega}_i-i\mu)}+1} - i\sum_{Im\bar{\omega}_i>\mu+\epsilon} \frac{R_f(\bar{\omega}_i)}{e^{-i\beta(\bar{\omega}_i-i\mu)}+1}, \tag{E.3}$$

where $R_f(\bar{\omega}_i)$ stand for the residues at the poles of $f(\omega)$ whose location we have denoted by $\hat{\omega}_i$.

Next consider the integral in (E.3). If $f(\omega)$ vanishes faster than $|\omega|^{-1}$ for $|\omega| \to \infty$, then the residue theorem tells us that

$$\frac{1}{2\pi}\int_{-\infty+i\mu+i\epsilon}^{\infty+i\mu+i\epsilon} d\omega\ f(\omega) = \frac{1}{2\pi}\int_{-\infty}^{\infty} d\omega\ f(\omega) - i \sum_{0<Im\bar{\omega}_i<\mu+\epsilon} R_f(\bar{\omega}_i).$$

Making use of the identity (E.2) we have

$$\sum_{0<Im\bar{\omega}_i<\mu+\epsilon} R_f(\bar{\omega}_i) = \sum_{0<Im\bar{\omega}_i<\mu+\epsilon} R_f(\bar{\omega}_i)\left(\frac{1}{e^{i\beta(\bar{\omega}_i-i\mu)}+1} + \frac{1}{e^{-i\beta(\bar{\omega}_i-i\mu)}+1}\right),$$

so that

$$\frac{1}{\beta}\sum_{\ell=-\infty}^{\infty} f(\omega_\ell^- + i\mu) = \int_{-\infty}^{\infty}\frac{d\omega}{2\pi} f(\omega) + i\sum_{Im\bar{\omega}_i<0}\frac{R(\bar{\omega}_i)}{e^{i\beta(\bar{\omega}_i-i\mu)}+1} - i\sum_{Im\bar{\omega}_i>0}\frac{R(\bar{\omega}_i)}{e^{-i\beta(\bar{\omega}_i-i\mu)}+1}. \tag{E.4}$$

where we have now set $\epsilon = 0$, since by assumption $f(\omega)$ is non-singular in the strip $Im\omega \in [\mu - \epsilon, \mu + \epsilon]$.

APPENDIX F

In this Appendix we derive expressions which are useful for performing sums over fermionic and bosonic Matsubara frequencies on the lattice. We first consider the fermionic case.

i) Fermionic Frequency Sums

As we have seen in section (19.11) the following type of sums are of interest:

$$I(\{\vec{\hat{p}}_i\}) = \frac{1}{\hat{\beta}} \sum_{l=-\frac{\hat{\beta}}{2}}^{\frac{\hat{\beta}}{2}-1} g(e^{i(\hat{\omega}_l^- + i\hat{\mu})}; \{\vec{\hat{p}}_i\}) ,$$

where $\hat{\omega}_\ell^- = \frac{(2\ell+1)\pi}{\hat{\beta}}$ are the fermionic Matsubara frequencies, with $\hat{\beta}$ the inverse temperature measured in units of the lattice spacing. The dependence of $g(e^{i(\hat{\omega}_l^- + i\hat{\mu})}; \{\vec{\hat{p}}_i\})$ on the momentum variables $\{\vec{\hat{p}}_i\}$ will be suppressed from now on.

Consider the following function of the complex variable $\hat{\omega}$:

$$h(\hat{\omega}) = \frac{-i\hat{\beta}}{e^{i\hat{\beta}(\hat{\omega} - i\hat{\mu})} + 1} . \tag{F.1}$$

It has simple poles located at $\hat{\omega} = \hat{\omega}_l^- + i\hat{\mu}$, with unit residue. Hence if $g(e^{i\hat{\omega}})$ has no singularities for $Im\hat{\omega} =\in [\hat{\mu} - \epsilon, \hat{\mu} + \epsilon]$, then

$$\frac{1}{\hat{\beta}} \sum_{l=-\frac{\hat{\beta}}{2}}^{\frac{\hat{\beta}}{2}-1} g(e^{i(\hat{\omega}_l^- + i\hat{\mu})}) = -\frac{1}{2\pi} \oint_C d\hat{\omega} \frac{g(e^{i\hat{\omega}})}{e^{i\hat{\beta}(\hat{\omega} - i\hat{\mu})} + 1} , \tag{F.2}$$

where C is the closed contour depicted in fig. (F-1).

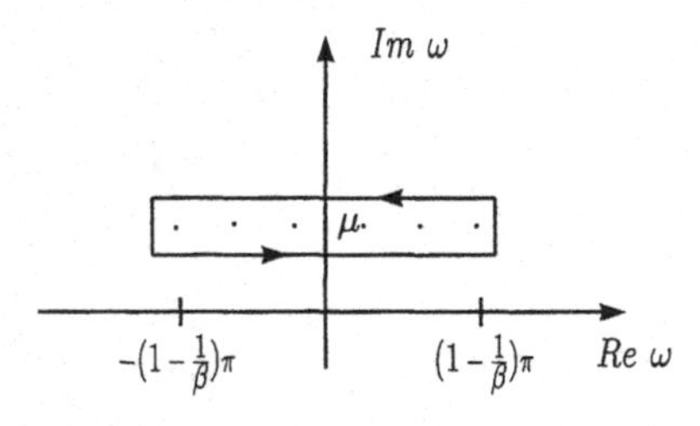

Fig. F-1 Contour of integration C.

Introducing the variable $z = e^{i\hat{\omega}}$, the poles of $h(\hat{\omega})$ are located on a circle with radius $e^{-\hat{\mu}}$ in the complex z-plane. The upper and lower branches of the contour in fig. (F-1) are mapped onto two circles of radius $|z| = \exp(-\hat{\mu} + \epsilon)$ and $|z| = \exp(-\hat{\mu} - \epsilon)$, respectively, traversed in a counterclockwise sense. Hence (F.2) takes the form

$$\begin{aligned} \frac{1}{\hat{\beta}} \sum_{l=-\frac{\hat{\beta}}{2}}^{\frac{\hat{\beta}}{2}-1} g(e^{i(\hat{\omega}_l^- + i\hat{\mu})}) = &- \frac{1}{2\pi i} \oint_{|z|=e^{-\hat{\mu}+\epsilon}} \frac{dz}{z} \frac{g(z)}{[e^{\hat{\beta}\hat{\mu}} z^{\hat{\beta}} + 1]} \\ &+ \frac{1}{2\pi i} \oint_{|z|=e^{-\hat{\mu}-\epsilon}} \frac{dz}{z} \frac{g(z)}{[e^{\hat{\beta}\hat{\mu}} z^{\hat{\beta}} + 1]} \,. \end{aligned} \tag{F.3}$$

If $|z|^{-\hat{\beta}} g(z) \to 0$ for $|z| \to \infty$, then we can distort the contour in the first integral to infinity, taking proper account of the singularities. For the case where $g(z)$ is a meromorphic function of z, the combined contributions of the two integrals in (F.3) yields

$$\frac{1}{\hat{\beta}} \sum_{l=-\frac{\hat{\beta}}{2}}^{\frac{\hat{\beta}}{2}-1} g(e^{i(\hat{\omega}_l^- + i\hat{\mu})}) = \sum_i \left(Res_{\bar{z}_i} \frac{g(z)}{z} \right) \frac{1}{e^{\hat{\beta}\hat{\mu}} \bar{z}_i^{\hat{\beta}} + 1} \,, \tag{F.4}$$

where $Res_{\bar{z}_i} \frac{g(z)}{z}$ are the residues of $g(z)/z$ at the poles, whose position we have denoted by $\bar{z}_i$.

ii) Bosonic Frequency Sums

Bosonic frequency sums on the lattice of the type required in section 11 of chapter 19, i.e.

$$K(\{\vec{\hat{p}}_i\}) = \frac{1}{\hat{\beta}} \sum_{l=-\frac{\hat{\beta}}{2}}^{\frac{\hat{\beta}}{2}-1} f(e^{i\hat{\omega}_l^+}; \{\vec{\hat{p}}_i\}) \,,$$

with $\hat{\omega}^+ = 2\ell\pi/\hat{\beta}$ can also be readily performed. By considering instead of $h(\hat{\omega})$ in (F.1) the function

$$\tilde{g}(\hat{\omega}) = \frac{i\hat{\beta}}{e^{i\hat{\omega}\hat{\beta}} - 1} \,,$$

which has simple poles with unit residue located at the bosonic Matsubara frequencies $\hat{\omega}_\ell^+$, and following the same line of arguments as in the fermionic case, one readily derives the following summation formula:

$$\frac{1}{\hat{\beta}} \sum_{\ell=-\frac{\hat{\beta}}{2}}^{\frac{\hat{\beta}}{2}-1} g(e^{i\hat{\omega}_\ell^+}) = -\sum_i \frac{Res_{\bar{z}_i}\left(\frac{g(z)}{z}\right)}{\bar{z}_i^{\hat{\beta}} - 1} \,. \tag{F.5}$$

APPENDIX G

The Electric Screening Mass for Naive Fermions in Lattice QED to One-Loop Order

The electric screening mass squared has been defined in section 6 of chapter 19 as the infrared limit of the 44-component of the vacuum polarization tensor evaluated for vanishing photon frequency (cf. eq. (19.71)). In this Appendix we compute this screening mass for naive fermions (i.e., for vanishing Wilson parameter), following closely the work of R. Pietig (1994).

In the following the vacuum polarization tensor is defined as the negative of the one-particle irreducible diagrams with two external photon lines. The Feynman diagrams contribution in $\mathcal{O}(g^2)$ are shown in fig. (19-6). Using the finite temperature lattice Feynman rules discussed in chapter 19, one finds, after carrying out the traces in Dirac space that

$$\hat{\Pi}_{44}^{(\beta,\mu)}(\hat{\omega}_n^+,\vec{\hat{k}}) = 4e^2\frac{1}{\hat{\beta}}\sum_{\ell=-\frac{\hat{\beta}}{2}}^{\frac{\hat{\beta}}{2}-1}\int_{-\pi}^{\pi}\frac{d^3\hat{p}}{(2\pi)^3}\frac{\sin^2(\hat{\omega}_\ell^- + i\hat{\mu})}{\sin^2(\hat{\omega}_\ell^- + i\hat{\mu}) + \hat{E}^2}$$

$$- 4e^2\frac{1}{\hat{\beta}}\sum_{\ell=-\frac{\hat{\beta}}{2}}^{\frac{\hat{\beta}}{2}-1}\int_{-\pi}^{\pi}\frac{d^3\hat{p}}{(2\pi)^3}\frac{\cos^2(\frac{1}{2}\hat{\omega}_n^+ + \hat{\omega}_\ell^- + i\hat{\mu})[\hat{G}^2 - \sin(\hat{\omega}_n^+ + \hat{\omega}_\ell^- + i\hat{\mu})\sin(\hat{\omega}_\ell^- + i\hat{\mu})]}{[\sin^2(\hat{\omega}_\ell^- + i\hat{\mu}) + \hat{E}^2][\sin^2(\hat{\omega}_n^+ + \hat{\omega}_\ell^- + i\hat{\mu}) + \hat{F}^2]} \qquad (G.1a)$$

where

$$\begin{aligned}\hat{E}^2 &= \sum_i \sin^2\hat{p}_i + \hat{m}^2,\\ \hat{F}^2 &= \sum_i \sin^2(\hat{p}+\hat{k})_i + \hat{m}^2,\\ \hat{G}^2 &= \sum_i \sin\hat{p}_i\sin(\hat{p}+\hat{k})_i + \hat{m}^2.\end{aligned} \qquad (G.1b)$$

and where $\hat{\omega}_\ell^+$ and $\hat{\omega}_\ell^-$ are the Matsubara frequencies for bosons and fermions defined in (19.92b) and (19.93b). All quantities are measured in lattice units. The first integral is the contribution of diagram (b) in fig. (19-6), which has no analog in the continuum. As always, the fermionic Matsubara frequencies appear in the combination $\omega_\ell^- + i\hat{\mu}$. The frequency sum can be performed by making use of the summation formula (F.4).

Expression (G.1a) can be written in the form (recall that we always take $\hat{\beta}$ to be even)

$$\hat{\Pi}_{44}^{(\beta,\mu)}(\hat{\omega}_n^+,\vec{\hat{k}}) = 4e^2\frac{1}{\hat{\beta}}\sum_{\ell=-\frac{\hat{\beta}}{2}}^{\frac{\hat{\beta}}{2}-1}\int_{-\pi}^{\pi}\frac{d^3\hat{p}}{(2\pi)^3}f(e^{i\hat{\omega}_\ell^-})\,, \tag{G.2a}$$

where

$$f(z) = -\frac{(z^2e^{-2\hat{\mu}}-1)^2}{(2\hat{E}e^{-\hat{\mu}}z)^2-(z^2e^{-2\hat{\mu}}-1)^2}$$
$$-\left[\frac{(z^2e^{-2\hat{\mu}}e^{i\frac{\hat{\omega}_n^+}{2}}+e^{-i\frac{\hat{\omega}_n^+}{2}})^2}{(2\hat{E}e^{-\hat{\mu}}z)^2-(z^2e^{-2\hat{\mu}}-1)^2}\right]\frac{(2ze^{-\hat{\mu}}\hat{G})^2+(z^2e^{-2\hat{\mu}}-1)(e^{i\hat{\omega}_n^+}e^{-2\hat{\mu}}z^2-e^{-i\hat{\omega}_n^+})}{(2\hat{F}e^{-\hat{\mu}}z)^2-(z^2e^{i\hat{\omega}_n^+}e^{-2\hat{\mu}}-e^{-i\hat{\omega}_n^+})^2} \tag{G.2b}$$

is a meromorphic function of z. We have suppressed for simplicity the dependence of $f(z)$ on the momentum variables. Let us rewrite this expression as follows,

$$f(z) = \frac{e^{4\hat{\mu}}(z^2e^{-2\hat{\mu}}-1)^2}{\prod_{i=1}^4(z-z_i)}$$
$$-\left[e^{8\hat{\mu}-2i\hat{\omega}_n^+}(z^2e^{-2\hat{\mu}+i\frac{\hat{\omega}_n^+}{2}}+e^{-i\frac{\hat{\omega}_n^+}{2}})^2\right]\frac{4z^2\hat{G}^2e^{-2\hat{\mu}}+(z^2e^{-2\hat{\mu}}-1)(z^2e^{-2\hat{\mu}}e^{i\hat{\omega}_n^+}-e^{-i\hat{\omega}_n^+})}{\prod_{i=1}^4(z-z_i)(z-z_i')}\,, \tag{G.3a}$$

where

$$\begin{aligned} z_1 &= -z_3 = -e^{\hat{\mu}-\hat{\phi}}\,,\\ z_2 &= -z_4 = e^{\hat{\mu}+\hat{\phi}}\,,\\ z_1' &= -z_3' = -e^{\hat{\mu}-\hat{\psi}}e^{-i\hat{\omega}_n^+}\,,\\ z_2' &= -z_4' = e^{\hat{\mu}+\hat{\psi}}e^{-i\hat{\omega}_n^+}\,,\\ \hat{\phi} &= ar\sinh\hat{E}\,,\\ \hat{\psi} &= ar\sinh\hat{F}\,. \end{aligned} \tag{G.3b}$$

Now for $|z|\to\infty$, $f(z)$ approaches a constant. Hence we can make direct use of the summation formula (F.4) to calculate the frequency sum (G.2a). Thus

$$\frac{1}{\hat{\beta}}\sum_{\ell=-\frac{\hat{\beta}}{2}}^{\frac{\hat{\beta}}{2}-1}f(z_\ell) = \sum_i\left[\frac{R(z_i)}{z_i^{\hat{\beta}}+1}+\frac{R(z_i')}{z_i'^{\hat{\beta}}+1}\right], \tag{G.4}$$

where

$$R(z_i) = Res_{z_i}\left(\frac{f(z)}{z}\right)$$

are the residues of $f(z)/z$ at $z = z_i$. A similar statement holds for $R(z_i')$. Notice that only the poles at $z = z_i$, $z = z_i'$ contribute, since the residue of the aparent pole at $z = 0$ vanishes. The computation of the residues is straightforward, although tedious. One finds that

$$\begin{aligned} R(z_1) &= R(z_3) = -R(z_2)^* = -R(z_4)^* = -\frac{1}{2}\tanh\hat{\phi} + \hat{H}(\hat{\phi},\hat{\psi},\hat{G}) \\ R(z_1') &= R(z_3') = -R(z_2')^* = -R(z_4')^* = \hat{H}^*(\hat{\psi},\hat{\phi},\hat{G}) \ , \end{aligned} \tag{G.5a}$$

where*

$$\hat{H}(\hat{\phi},\hat{\psi},\hat{G}) = \frac{\cosh^2(\hat{\phi} - i\frac{\hat{\omega}_n^+}{2})[\hat{G}^2 + \sinh\hat{\phi}\sinh(\hat{\phi} - i\hat{\omega}_n^+)]}{\sinh(2\hat{\phi})\sinh(\hat{\phi} - \hat{\psi} - i\hat{\omega}_n^+)\sinh(\hat{\phi} + \hat{\psi} - i\hat{\omega}_n^+)} \ . \tag{G.5b}$$

Hence (G.4) is given by

$$\begin{aligned} \frac{1}{\hat{\beta}}\sum_{\ell=-\frac{\hat{\beta}}{2}}^{\frac{\hat{\beta}}{2}} f(z_\ell) = &-\tanh\hat{\phi}\left[\frac{1}{e^{\hat{\beta}(\hat{\mu}-\hat{\phi})}+1} - \frac{1}{e^{\hat{\beta}(\hat{\mu}+\hat{\phi})}+1}\right] \\ &+ 2\left[\hat{H}(\hat{\phi},\hat{\psi},\hat{G})\frac{1}{e^{\hat{\beta}(\hat{\mu}-\hat{\phi})}+1} + \hat{H}^*(\hat{\psi},\hat{\phi},\hat{G})\frac{1}{e^{\hat{\beta}(\hat{\mu}-\hat{\psi})}+1}\right] \\ &- 2\left[\hat{H}(\hat{\psi},\hat{\phi},\hat{G})\frac{1}{e^{\hat{\beta}(\hat{\mu}+\hat{\psi})}+1} + \hat{H}^*(\hat{\phi},\hat{\psi},\hat{G})\frac{1}{e^{\hat{\beta}(\hat{\mu}+\hat{\phi})}+1}\right] \ . \end{aligned} \tag{G.6}$$

where we have made used of the fact that $e^{i\hat{\beta}\hat{\omega}_\ell^+} = 1$, since $\hat{\beta}$ is even. Now from the definitions of $\hat{\phi}$ and $\hat{\psi}$ given in (G.3b), and the definition (G.1b) we see that for any function $K(\hat{\phi},\hat{\psi},\hat{G})$,

$$K(\hat{\phi},\hat{\psi},\hat{G}) \underset{\vec{p}\to-\vec{p}-\vec{k}}{\longrightarrow} K(\hat{\psi},\hat{\phi},\hat{G}) \ .$$

Hence if $K(\hat{\phi},\hat{\psi},\hat{G})$ is a periodic function of $\vec{p}$ with period 2π, then

$$\int_{-\pi}^{\pi} d^3\hat{p}\ K(\hat{\phi},\hat{\psi},\hat{G}) = \int_{-\pi}^{\pi} d^3\hat{p}\ K(\hat{\psi},\hat{\phi},\hat{G}) \ . \tag{G.7}$$

Making use of this relation, with K given by the rhs of (G.6), we can write (G.1a) in the form

$$\hat{\Pi}_{44}^{(\beta,\mu)}(\hat{\omega}_n^+,\vec{\hat{k}}) = -4e^2\int_{-\pi}^{\pi}\frac{d^3p}{(2\pi)^3}[\tanh\hat{\phi} - 4Re\,\hat{H}(\hat{\phi},\hat{\psi},\hat{G})]\left[\frac{1}{e^{-\hat{\beta}(\hat{\phi}-\hat{\mu})}+1} - \frac{1}{e^{\hat{\beta}(\hat{\mu}+\hat{\phi})}+1}\right.$$

* We have supressed in $\hat{H}$ the dependence on the photon frequency

This expression can be decomposed as follows by making use of the identity $(e^x + 1)^{-1} + (e^{-x}+1)^{-1} = 1$,

$$\hat{\Pi}_{44}^{(\beta,\mu)}(\hat{\omega}_n^+, \vec{\hat{k}}) = \hat{\Pi}_{44}^{(vac)}(\hat{\omega}_n^+, \vec{\hat{k}}) + \hat{\Pi}_{44}(\hat{\omega}_n^+, \vec{\hat{k}})_{f.T.} \,, \tag{G.8}$$

where

$$\hat{\Pi}_{44}(\hat{\omega}_n^+, \vec{\hat{k}})_{f.T.} = 4e^2 \int_{-\pi}^{\pi} \frac{d^3\hat{p}}{(2\pi)^3} [\tanh\hat{\phi} - 4Re\, \hat{H}(\hat{\phi}, \hat{\psi}, \hat{G})] \left(\frac{1}{e^{\hat{\beta}(\hat{\phi}-\hat{\mu})}+1} + \frac{1}{e^{\hat{\beta}(\hat{\phi}+\hat{\mu})}+1} \right) \tag{G.9a}$$

$$\hat{\Pi}_{44}^{(vac)}(\hat{\omega}_n^+, \vec{\hat{k}}) = -4e^2 \int_{-\pi}^{\pi} \frac{d^3\hat{p}}{(2\pi)^3} [\tanh\hat{\phi} - 4Re\, \hat{H}(\hat{\phi}, \hat{\psi}, \hat{G})] \,. \tag{G.9b}$$

As we shall see below (G.9a) is the finite temperature ($\to$ f.T.) contribution arising from the presence of the heat bath, while $\hat{\Pi}_{44}^{(vac)}(\hat{\omega}_n^+, \vec{\hat{k}})$ is the lattice expression which one obtains by using the zero temperature, zero chemical potential lattice Feynman rules, with the fourth component of the photon momentum evaluated at $\hat{\omega}_\ell^+$. Indeed, for $\hat{\mu} = 0$ and $\hat{\beta} \to \infty$, the expression corresponding to (G.1a) is obtained by setting $\hat{\mu} = 0$, replacing $\hat{\omega}_\ell^-$ by the continuous variable p_4, and the frequency sum by an integral according to $\beta^{-1}\sum_\ell \to \int_{-\pi}^{\pi} dp_4/2\pi$. Upon introducing $z = e^{ip_4}$ as a new integration variables, we therefore have that

$$\hat{\Pi}_{44}^{(T=\mu=0)}(\hat{k}) = \frac{2e^2}{i\pi} \int_{-\pi}^{\pi} \frac{d^3\hat{p}}{(2\pi)^3} \oint dz \frac{f(z)}{z} \,,$$

where $f(z)$ is given by (G.3a) with $\hat{\mu} = 0$, and where the contour integration is carried out over a circle in the complex z-plane with unit radius. Since for $\hat{\mu} = 0$, the singularities of $f(z)$ inside the circle are located at $\pm e^{-\hat{\phi}}$ and $\pm e^{-\hat{\psi}-i\hat{\omega}_n^+}$, one is then led to the result (G.9b). As we now show, (G.9) does not contribute to the screening mass.

Consider (G.9b) for vanishing frequency. One readily verifies that

$$\hat{\Pi}_{44}^{(vac)}(0, \vec{\hat{k}}) = -4e^2 \int_{-\pi}^{\pi} \frac{d^3\hat{p}}{(2\pi)^3} [\tanh\hat{\phi} - 4h(\hat{\phi}, \hat{\psi}, \hat{G})] \,, \tag{G.10a}$$

where

$$h(\hat{\phi}, \hat{\psi}, \hat{G}) = \frac{\hat{G}^2 \coth\hat{\phi} + \frac{1}{2}\sinh 2\hat{\phi}}{2\sinh(\hat{\phi}-\hat{\psi})\sinh(\hat{\phi}+\hat{\psi})} \,. \tag{G.10b}$$

To compute the corresponding contribution to the screening mass we must take the limit $\vec{\hat{k}} \to 0$. In this limit $\hat{\psi} \to \hat{\phi}$, so that $h(\hat{\phi}, \hat{\psi}, \hat{G})$ becomes singular. This singularity is however integrable. To see this we make use of (G.7) to rewrite (G.10) in the form

$$\hat{\Pi}_{44}^{(vac)}(0,\vec{\hat{k}}) = -4e^2 \int_{-\pi}^{\pi} \frac{d^3\hat{p}}{(2\pi)^3} [\tanh\hat{\phi} - 2\hat{f}(\hat{\phi},\hat{\psi},\hat{G})] \,, \tag{G.11}$$

where

$$\hat{f}(\hat{\phi},\hat{\psi},\hat{G}) = \hat{h}(\hat{\phi},\hat{\psi},\hat{G}) + \hat{h}(\hat{\psi},\hat{\phi},\hat{G}) \,.$$

Making use of the relations for hyperbolic functions, $\hat{f}(\hat{\phi},\hat{\psi},\hat{G})$ can be written as follows

$$\hat{f}(\hat{\phi},\hat{\psi},\hat{G}) = \frac{1}{2\sinh(\hat{\phi}+\hat{\psi})}\Big[-\frac{\hat{G}^2}{\sinh\hat{\phi}\sinh\hat{\psi}} + \cosh(\hat{\phi}+\hat{\psi})\Big] \,.$$

Now for $\vec{\hat{k}} \to 0$, $G^2 \to \sinh^2\hat{\phi}$, and $\hat{\psi} \to \hat{\phi}$. Hence

$$\hat{f}(\hat{\phi},\hat{\psi},\hat{G}) \underset{\vec{\hat{k}}\to 0}{\longrightarrow} \frac{1}{2}\tanh\hat{\phi} \,. \tag{G.12}$$

Inserting this expression into (G.11) we therefore find that

$$\lim_{\vec{\hat{k}}\to 0} \hat{\Pi}_{44}^{(vac)}(0,\vec{\hat{k}}) = 0 \,. \tag{G.13}$$

Hence there is no contribution to the screening mass arising from $\hat{\Pi}_{44}^{(vac)}(\hat{\omega}^+,\vec{\hat{k}})$.

Consider now the second term on the rhs of (G.8), i.e., (G.9a). Let us replace the second term in the integrand by an expression symmetrized in ϕ and ψ. One then verifies that (G.9a) can be written in the form

$$\hat{\Pi}_{44}^{(\beta,\mu)}(0,\vec{\hat{k}}) = 4e^2 \int_{-\pi}^{\pi} \frac{d^3\hat{p}}{(2\pi)^3} \Big\{\tanh\hat{\phi} - 2\{\hat{f}(\hat{\phi},\hat{\psi},\hat{G})[\hat{\eta}_{FD}(\hat{\psi}) + \bar{\hat{\eta}}_{FD}] + h(\hat{\phi},\hat{\psi},G)\Delta\hat{\eta}_{FD}\Big\} \tag{G.14}$$

where

$$\Delta\hat{\eta}_{FD} = [\hat{\eta}_{FD}(\phi) - \hat{\eta}_{FD}(\psi)] + [\bar{\hat{\eta}}_{FD}(\phi) - \bar{\hat{\eta}}_{FD}(\psi)] \,,$$

and ($\hat{\rho} = \hat{\phi}$, $\hat{\psi}$)

$$\hat{\eta}_{FD}(\hat{\rho}) = \frac{1}{e^{\hat{\beta}(\hat{\rho}-\hat{\mu})}+1} \,,$$

$$\bar{\hat{\eta}}_{FD}(\hat{\rho}) = \frac{1}{e^{\hat{\beta}(\hat{\rho}+\hat{\mu})}+1} \,, \tag{G.15}$$

are the lattice Fermi-Dirac distribution functions for particles and antiparticles. While $h(\hat{\phi}, \hat{\psi}, G)$ is singular in the limit $\vec{\hat{k}} \to 0$, the product $h(\hat{\phi}, \hat{\psi}, G)\Delta\hat{\eta}_{FD}$ is finite, as can be seen by setting $\hat{\psi} = \hat{\phi} + \epsilon$ and taking the limit $\epsilon \to 0$. The remaining terms in (G.14) are finite in this limit, as follows from (G.12). One then finds that

$$\lim_{\vec{\hat{k}} \to 0} \hat{\Pi}_{44}(0, \vec{\hat{k}}) = \hat{m}_{el}^2 = 4e^2\hat{\beta} \int_{-\pi}^{\pi} \frac{d^3\hat{p}}{(2\pi)^3} \left\{ \frac{e^{\hat{\beta}(\hat{\phi}-\hat{\mu})}}{[e^{\hat{\beta}(\hat{\phi}-\hat{\mu})} + 1]^2} + \frac{e^{\hat{\beta}(\hat{\phi}+\hat{\mu})}}{[e^{\hat{\beta}(\hat{\phi}+\hat{\mu})} + 1]^2} \right\} . \tag{G.16}$$

It is now evident that this expression possesses a finite continuum limit, since for $\hat{\beta} \to \infty$, $\hat{\beta}\hat{\mu} = \beta\mu$ fixed, only momenta $\hat{p}_i$ contribute for which $\hat{\beta}\hat{\phi}$ is finite. This implies that $\hat{E} = \sin h\hat{\phi}$ is of $O(1/\hat{\beta})$. Nevertheless, the above expression is not the lattice analog of (19.78). First of all there is an extra factor two multiplying the integral. This factor arises from excitation in frequency space near the corner of the Brillouin zone. Such a factor was of course expected since we have carried out our computations with naive fermions. Of course we also expect to see the effects of the 2^3 douplers arising from excitations in 3-momenta at the edges of the Brillouin zone. This can be easily seen. To this effect we notice the momentum dependence of $\hat{\phi}$ appears in the form $\sin^2 \hat{p}_j$, which also vanishes at the edges of the Brillouin zone. Hence the integral (G.16) is just 2^3 times the integral extending over only half the Brillouin zone. This integral is now dominated for $\hat{\beta} \to \infty$ by momenta $\hat{p}_j$ of the order of $1/\hat{\beta}$ (for which $\hat{\beta}\hat{\phi}$ takes finite values). We are therefore now allowed to replace $\hat{\phi} = arsinh\hat{E}$ by $\sqrt{\hat{p}^2 + \hat{m}^2}$. Introducing the dimensionful variables p_j, m, μ and β by $\hat{p}_j = p_j a$, $\hat{m} = ma$, $\hat{\mu} = \mu a$ and $\hat{\beta} = \beta/a$, we therefore find that the physical screening mass, $m_{el} = \hat{m}_{el}/a$, is given by

$$\begin{aligned} m_{el}^2(\beta, \mu, m) &= \lim_{a \to 0} \frac{1}{a^2} \hat{m}_{el}^2(\frac{\beta}{a}, \mu a, ma) \\ &= 32e^2\beta \int_{-\infty}^{\infty} \frac{d^3p}{(2\pi)^3} \left\{ \frac{e^{\beta(\sqrt{\vec{p}^2+m^2}-\mu)}}{[e^{\beta(\sqrt{\vec{p}^2+m^2}-\mu)} + 1]^2} + \frac{e^{\beta(\sqrt{\vec{p}^2+m^2}+\mu)}}{[e^{\beta(\sqrt{\vec{p}^2+m^2}+\mu)} + 1]^2} \right\} . \end{aligned} \tag{G.17}$$

Peforming the angular integrations, and an additional partial integration in $|\vec{p}|$ we finally obtain

$$m_{el}^2 = 16\frac{e^2}{\pi^2} \int_0^{\infty} dp \frac{2p^2 + m^2}{\sqrt{p^2 + m^2}} \left[\frac{1}{e^{\beta(\sqrt{p^2+m^2}-\mu)} + 1} + \frac{1}{e^{\beta(\sqrt{p^2+m^2}+\mu)} + 1} \right] , \tag{G.18}$$

which is just sixteen times the result we obtained in continuum perturbation theory.

We want to call the readers attention to the fact that the first term in (G.1a), arising from the diagram (b) in fig. (19-6), which has no continuum analog, played an essential role in obtaining (except for the factor 16, of course) the correct continuum limit.

REFERENCES

Abbot T. et al., E802 Collaboration, Phys. Rev. Lett. **64**. 847 (1990); Phys. Rev. Lett. **66**, 1567 (1991)

Abers, E.S. and B.W. Lee, Phys. Rep. **9C**, 1 (1973)

Abreu M.C. et al. (NA 38 collab.), Z. Phys. C**38** 117 (1988)

Abrikosov A. A., Zh. Eksp. Teor. Fiz. **32**, 1442 (1057); Sov. Phys. JETP **5** 1174 (1957)

Actor A., Rev. Mod. Phys. **51**, 461 (1979)

Adler S.L., Phys. Rev. **177**, 2426 (1969)

Adler S.L., Nucl. Phys. B217. 381 (1983)

Albanese M. et al. (APE collab.), Phys. Lett. **192B**, 163 (1987); **197B**, 400 (1987);

Alexandrou C., M. D'Elia and P. de Forcrand, Nucl. Phys. (Proc. Suppl) **83**, 437 (2000)

Allton C.R., S. Ejiri, S.J. Hands, O. Kaczmarek, F. Karsch, E. Laermann, Ch. Schmidt and L. Scorzato, Phys. Rev. **D66**, 074507 (2002)

Ambjorn J. and P. Oleson, Nucl. Phys. **B170** [FS1], 265 (1980)

Anagnostopoulos K.N. and J. Nishimura, Phys. Rev. **D66**, 106008 (2002)

Atiyah M.F. and I.M. Singer, Ann. Math. **87**, 596 (1968); **93**, 139 (1971)

Atiyah M.F., N.J. Hitchin, V. Drinfeld and Yu. I. Manin, Phys. Lett. **65A** 185 (1978)

Attig N., F. Karsch, B. Petersson, H. Satz and U. Wolff, unpublished

Bacilieri P. et al., Phys. Lett. **214B**, 115 (1988)

Bacilieri P. et. al. (APE collab.), Nucl. Phys. B (Proc. Suppl.) **9**, 315 (1989). See also, Phys. Rev. Lett. **61**, 1545 (1988)

Baglin C. et al. (NA38 collab.), Phys. Lett. **220B**, 471 (1989)

Baker M. and S.P. Li, Phys. Lett **124B**, 397 (1983)

Baker M., J.S. Ball and F. Zachariasen, Phys. Rev. **D41**, 2612 (1990)

Bali G.S., K. Schilling and Ch. Schlichter, Phys. Rev. **D51**. 5165 (1995)

Bali G.S. et al., Phys. Rev. **D54**, 2863 (1996)

Banks T., D.R.T. Jones, J. Kogut, S. Raby , P.N. Scharbach, D.K. Sinclair, and L. Susskind D.K. Sinclair, Phys. Rev. **D15**, 1111 (1977).

Barcielli A., E. Montaldi and G.M. Prosperi, Nucl Phys. **B296**, 625 (1988); Erratum Nucl Phys. **B303**, 752 (1988).

Barkai D., K.J.M. Moriarty and C. Rebbi, Phys. Lett. **156B**, 385 (1985)

Batrouni G.G., G.R. Katz, A.S. Kronfeld, G.P. Lepage, B. Svetitsky and

K.G. Wilson, Phys. Rev. **D32**, 2736 (1985)
Belavin A.A. and A.A. Migdal, JETP Lett. **19**, 181 (1974)
Belavin A., A. Polyakov, A. Schwarz and Yu. Tyupkin, Phys. Lett. **59B** 85 (1975)
Bell J.S. and R. Jackiw, Nuovo Cimento **60A**, 47 (1969)
Bender I., D. Gromes, K.D. Rothe and H.J. Rothe, Nucl. Phys. **B136**, 259 (1978)
Bender I., H.J. Rothe, I.O. Stamatescu and W. Wetzel, Z. Physik **C58**, 333 (1993)
Bender C.M. and A. Rebhan, Phys. Rev. **D41**, 3269 (1990)
Berezin F.A., *The Method of Second Quantization*, Academic Press (1966)
Berg B., Phys. Lett. **97B**, 401 (1980)
Berg B., Phys. Lett. **B104**, 475 (1981)
Berg B and A. Billoire, Phys. Lett. **113B**, 65 (1982)
Berg B., *Progress in Gauge Field Theory*, Cargese (1983)
Berg B. and A. Billoire, Phys. Lett. **166B**, 203 (1986)
Berker A.N. and M.E. Fisher, Phys. Rev. **B26**, 2507 (1982)
Bernard C.W., Phys. Rev. **D9**, 3312 (1974)
Bernreuther W. and W. Wetzel, Phys. Lett. **132B**, 382 (1983)
Bhanot G. and C. Rebbi, Nucl. Phys. **B180** [$FS2$], 469 (1981)
Bhanot G., U. Heller, and I.O. Stamatescu, Phys. Lett. **129B**, 440 (1983)
Binder K. and D.P. Landau, Phys. Rev. **B30**, 1477 (1984)
Bitar K.M. and S.-J. Chang, Phys. Rev. **D17**, 486 (1978)
Bitar K.M. et. al., Phys. Rev. **D42**, 3794 (1990)
Blaizot J.P. and J.Y. Ollitrault, Phys. Lett. **199B**, 199 (1987)
Blaizot J.P.and J.Y. Ollitrault, Phys. Lett. **217B**, 392 (1989)
Blaizot J.P and J.Y. Ollitrault, in *Quark Gluon Plasma*, ed. by R.C. Hwa (World Scientific, 1990)
Born K.D., E. Learmann, N. Pirch, T.F. Walsh and P.M. Zerwas, Phys. Rev. **D40**, 1653 (1989); see also Nucl. Phys. B (Proc. Suppl.) **9**, 269 (1989)
Boulware G., Ann. of Phys. **56** (N.Y.), 140 (1970)
Bowler K.C., C.B. Chalmers, A. Kenway, R.D. Kenway, G.S. Pawley and D.J. Wallace, Phys. Lett. **162B**, 354 (1985)
Bowler K.C., C.B. Chalmers, R.D. Kenway, D. Roweth and D. Stephenson, Nucl Phys. **B296**, 732 (1988)
Boyko P. Yu, M. Polikarpov and V.I. Zakharov, Nucl Phys. Proc. Suppl.

119, 724 (2003)
Brambilla N., G.M. Prosperi and A. Vario, Phys. Lett. **B362**, 113 (1995)
Brown L.S., and W.I. Weisberger, Phys. Rev. **D20**, 3239 (1979)
Brown F.R., N.H. Christ, Y. Deng, M.Gao and T.J. Woch,
Phys. Rev Lett. **61**, 2058 (1988);
Brown F.R., F.P. Butler, H. Chen, N.H. Christ, Z. Dong, W. Schaffer,
L.I. Unger and A. Vaccarino, Phys. Rev. Lett. **65**, 2491 (1990).
Bruckmann F., D. Nogradi abd P. van Baal, Acta Phys. Polon. **B34**,
5717 (2003), hep-th/0309008
Burgers G., F. Karsch, A. Nakamura and I.O. Stamatescu, Nucl. Phys.
B304, 587 (1988)
Burkitt A., K.H. Mütter and P. Überholz; talk presented by K.H. Mütter
at the International Symposium *Field Theory on the Lattice*, Seillac,
France (1987)
Callan C.G., Phys. Rev. **D2**, 1541 (1970)
Callan C., R. Dashen and D. Gross, Phys. Rev. **D17**, 2717 (1978)
Callaway D.J.E. and A. Rahman, Phys. Rev. Lett. **49**, 613 (1982);
Callaway D.J.E. and A. Rahman, Phys. Rev. **D28**, 1506 (1983)
Campostrini et al, Phys. Rev. **D42**, 203 (1990)
Caracciolo S., P. Menotti and A. Pelisetto, Nuc. Phys. **B375**, 195 (1992)
Caswell W.E., Phys. Rev. Lett. **33**, 244 (1974)
Celic T., J. Engels and H. Satz, Phys. Lett. **125B**, 411 (1983); **129B**,
323 (1983)
Chiu T.W., Phys. Lett. **B445**, 371 (1999)
Clarke A.B. and R.L. Disney, *Probability and Random Processes: A first
Course with Applications*, John Wiley & Sons (1985)
Cleymans J, R.V. Gavai and E. Suhonen, Phys. Rep. **130**, 217 (1986)
Cleymans J., J. Fingberg and K. Redlich, Phys. Rev. **D35**, 2153 (1987)
Close F.E., *An Introduction to Quarks and Partons*, Academic Press (1979)
Coleman S., *Uses of Instantons* in *Aspects of Symmetry*, selected
Erice Lectures, Cambridge University press, (1985)
Collins J.C. and M.J. Perry, Phys. Rev. Lett. **34**, 1353 (1975)
Collins J.C., A. Duncan and S.D. Joglekar, Phys. Rev. **D16**, 438 (1977)
Creutz M., Phys. Rev. **D15**, 1128 (1977)
Creutz M., J. Math. Phys. **19**, 2043 (1978)
Creutz M., Phys. Rev. **D21**, 2308 (1980)
Creutz M. and K.J.M. Moriarty, Phys. Rev. **D26**, 2166 (1982)

Creutz M., *Quarks, Gluons and Lattices*, Cambridge University Press (1983a).
Creutz M., Phys. Rev. Lett. **50**, 1411 (1983b)
de Forcrand P. and I.O. Stamatescu, Proc. of the XXIII Int. Conf. in High Energy Physics, Berkeley 1986; see also P. de Forcrand, V. Linke and I.O. Stamatescu, Nucl. Phys. **B304**, 645 (1988); P. de Forcrand and I.O. Stamatescu, Nucl. Phys. B (proc. Suppl.) **9**, 276 (1989)
de Forcrand P., J. Stat. Phys. **43**, 1077 (1986)
de Forcrand P., R. Gupta, S. Güsken, K.H. Mütter, A. Patel, K. Schilling and R. Sommer, Phys. Lett. **200B**, 143 (1988)
de Forcrand P., M. Garcia Perez and I.O. Stamatescu, Nucl. Phys. **B499** 409 (1997)
de Forcrand P. and M. D'Elia, Phys. Rev. Lett. **82**, 4582 (1999)
de Forcrand P. and M. Pepe, Nucl. Phys. **B598**, 557 (2001)
de Forcrand P. and O. Philipsen, Nucl. Phys. **B642**, 290 (2002); M. D'Elia and M.P. Lombardo, Phys. Rev. **D67**, 014505 (2003)
Del Debbio L. et al., Phys. Rev. **D55**, 2298 (1997)
Del Debbio L., M. Faber, J. Giedt, J. Greensite and S. Olejnik, Phys. Rev. **D58**, 094501 (1998)
DeGrand T.A. and T. Toussaint, Phys. Rev. **D22**, 2478 (1980)
Deng Y., Nucl. Phys. B (Proc. Suppl.) **9**, 334 (1989)
Di Giacomo A., B. Lucini, L. Montesi and G. Paffuti, Phys. Rev. **D61** 034503 (2000); 034504 (2004)
Di Vecchia P., K. Fabricius, G.C. Rossi and G. Veneziano, Nucl. Phys. **B192**, 392 (1981)
Ding H.-Q, C.F. Baillie and G.C. Fox, Phys. Rev. **D41**, 2912 (1990)
Ding H.-Q., "Heavy quark potential in Lattice QCD. A review of recent progress", CALTECH preprint no. 32681 (1990). See also Phys. Rev. **D42**, 2350 (1990)
Dolan L. and R. Jackiw, Phys. Rev. **D9**, 3320 (1974)
Dosch H.G., O. Nachtmann and M. Rueter, hep-ph 9503386
Drouffe J.M. and J.B. Zuber, Phys. Rep. **102**, 1 (1983)
Duane S., Nucl. Phys. **B257** [FS14], 652 (1985)
Duane S. and J.B. Kogut, Nucl. Phys. **B275** [FS17], 398 (1986)
Duane S., A.D. Kennedy, B.J. Pendleton and D. Roweth, Phys. Lett. **195B**, 216 (1987)
Duru I.R. and H. Kleinert, Phys. Lett. **84B**, 185 (1979)

Eichten E., K. Gottfried, T. Kinoshita, K.D. Lane and T.M. Yan Phys. Rev. **D21**, 203 (1980)
Eichten E. and F. Feinberg, Phys. Rev. **D23**, 2724 (1981)
Engelhardt M. and H. Reinhard, Nucl. Phys. **B567**, 249 (2000)
Engels J., F. Karsch, I. Montvay and H. Satz, Phys. Lett. **102B**, 332 (1981a)
Engels J., F. Karsch, I. Montvay and H. Satz, Phys. Lett. **101B**, 89 (1981b)
Engels J., F. Karsch, I. Montvay and H. Satz, Nucl. Phys. **B205** [*FS*5,], 545 (1982)
Engels J., J. Fingberg, F. Karsch, D. Miller and M. Weber, Phys. Lett. **252B**, 625 (1990)
Feuerbacher J., (2003a) hep-lat/0309016
Feuerbacher J., Nucl. Phys. **B674**, 484 (2003b)
Feynman R.P., Rev. Mod. Phys. **20**, 367 (1948)
Feynman R.P., and A.R. Hibbs, *Quantum Mechanics and Path Integrals*, McGraw Hill (1965).
Flower J.W. and S.W. Otto, Phys. Lett. **160B**, 128 (1985)
Flower J., Caltech preprints, CALT-68-1369 (1986); CALT-68-1377 (1987); CALT-68-1378 (1987)
Fodor Z. and S.D. Katz, JHEP **0203**, 014 (2002); JHEP **0404** 050 (2004)
Fradkin E.S., Proc. Lebedev Physics Institute **29**, 6 (1965)
Frewer M. and H.J. Rothe, Phys. Rev. **D63**, 054506 (2001)
Fritzsch H. and M. Gell-Mann, in *Proc. XVI'th Intern. Conf. on High Energy Physics*, Chicago-Batavia, 1972
Fritzsch H., M. Gell-Mann, and H. Leutwyler, Phys. Lett. **47B**, 365 (1973)
Fucito F., E. Marinari, G. Parisi and C. Rebbi, Nucl Phys. **B180** [*FS*2], 369 (1981)
Fujikawa K., Phys. Rev. Lett. **42**, 1195 (1979); Phys. Rev. **D21**, 2848 (1980); ibid **D22**. 1499(E) (1980); **D29**, 285 (1984)
Fukugita M. and I. Niuya, Phys. Lett. **132B**, 374 (1983)
Fukugita M. and A. Ukawa, Phys. Rev. Lett. **55**, 1854 (1985)
Fukugita M. and A. Ukawa, Phys. Rev. Lett. **57**, 503 (1986)
Fukugita M., Y. Oyanagi and A. Ukawa, Phys. Lett. **203B**, 145 (1988)
Fukugita M., M. Okawa and A. Ukawa, Phys. Rev. Lett. **63**, 1768 (1989); see also Nucl. Phys. B (proc. Suppl.) **17**, 204 (1990)
Gao M., Phys. Rev. **D41**, 626 (1990)
Garcia Perez M., O. Phillipsen, I.O. Stamatescu, Nucl. Phys. **B551**,

293 (1999)
Gavai R.V., J. Potvin and S. Sanielevici, Phys. Rev. Lett. **58**, 2519 (1987)
Gavai R.V., J. Potvin and S. Sanielevici, Phys. Rev. **D37**, 1343 (1988)
Gavin S., M. Gyulassy and A. Jackson, Phys. Lett. **207B**, 257 (1988)
Gavin S. and M. Gyulassy, Phys. Lett. **214B**, 241 (1988)
Gerschel C. and J. Hüfner, Phys. Lett., **207B**, 253 (1988)
Ginsparg P.H. and K.G. Wilson, Phys. Rev. **D25**, 2649 (1982)
Goddard P., J. Goldstone, C. Rebbi and C.B. Thorn, Nucl. Phys. **B56** 109 (1973)
Golterman M.F.L. and J. Smit, Nucl. Phys. **B245**, 61 (1984)
Golterman M.F.L.and J.Smit, Nucl. Phys. **B255**, 328 (1985)
Golterman M.F.L., Nucl. Phys. **B278**, 417 (1986a)
Golterman M.F.L., Nucl. Phys. **B273**, 663 (1986b)
Gottlieb S.A., J. Kuti, A.D. Kennedy, S. Meyer, B.J. Pendleton, R.L. Sugar and D. Toussaint, Phys. Rev. Lett. **55**, 1958 (1985)
Gottlieb S., W. Liu, D. Toussaint, R.L.Renken and R.L. Sugar, Phys. Rev. **D35**, 2531 (1987a)
Gottlieb S., W. Liu, D. Toussaint, R.L.Renken and R.L. Sugar, Phys. Rev. **D35**, 3972 (1987b)
Gottlieb, S., W. Liu, R.L. Renken, R.L. Sugar and D. Toussaint, Phys. Rev. **D38**, 2245 (1988)
Grattinger C., Nucl. Phys. **B**(Proc. Suppl.) **129 & 130**, 653 (2004)
Greensite J., Prog. Part. Nucl. Phys. **51**, 1 (2003); hep-lat/0301023
Greiner C., P. Koch and H. Stöcker, Phys. Rev. Lett. **58**, 1825 (1987); C. Greiner, D.-H. Rischke, H. Stöcker and P. Koch, Z. Physik **C38**, 283 (1988); Phys. Rev. **D38**, 2797 (1988); C. Greiner and H. Stöcker, Phys. Rev. **D44** 3517 (1991)
Griffiths L.A., C. Michael and P.E.L. Rakow, Phys. Lett. **129B**, 351 (1983)
Gromes D., Zeitschrift f. Physik **C22**, 265 (1984)
Gromes D., see the review by this author in Phys. Rep. **200** (Part II), 186 (1991)
Gross, D.J. and F. Wilczek, Phys. Rev. **D8**, 3633 (1973); Phys. Rev. Lett. **30**, 1343 (1973)
Gross D.J., R.D. Pisarski and L.G.Yaffe, Rev. Mod. Phys. **53**, 43 (1981)
Grossiord J.Y. (NA38 coll.), Nucl. Phys. **A498**, 249c (1989)
Gruber C. and H. Kunz, Commun. Math. Phys. **22**, 133 (1971)

Gupta R., G. Guralnik, G. Kilcup, A. Patel, S.R. Sharpe and T. Warnock, Phys. Rev. **D36**, 2813 (1987)

Gupta R., Nucl. Phys. B (Proc. Suppl.) **9**, 473 (1989)

Gupta, R., A. Patel, C.F. Baillie, G. Guralnik, G.W. Kilcup and S.R Sharpe, Phys. Rev. **D40**, 2072 (1989).

Gupta R., Nucl. Phys. B (Proc. Suppl.) **17**, 70 (1990)

Guth A.H., Phys. Rev. **21**, 2291 (1980)

Hahn O., E. Schnepf, E. Learmann, K.H. Mütter, K. Schilling and R. Sommer, Phys. Lett. **190B**, 147 (1987)

Hamber H.W. and G. Parisi, Phys. Rev. Lett. **47**, 1792 (1981)

Hamber H.W., Phys. Rev. **D39**, 896 (1989)

Hammersley J.M. and D.C. Handscomb, *Monte Carlo Methods*, Methuen's Monographs, London (1975)

Hart A. and M. Teper, Phys. Rev. **D58**, 014504 (1998)

Harrington B.J. and H.K. Shepard, Phys. Rev. **D17**, 2122 (1978); Phys. Rev. **D18**, 2990 (1978)

Hasenfratz A. and Hasenfratz P., Phys. Lett. **93B**, 165 (1980);

Hasenfratz A. and Hasenfratz P., Phys. Lett. **104B**, 489 (1981a)

Hasenfratz A. and Hasenfratz P., Nucl. Phys. **B193**, 210 (1981b)

Hasenfratz P., F. Karsch and I.O. Stamatescu, Phys. Lett. **133B**, 221 (1983)

Hasenfratz P. and F. Karsch, Phys. Lett. **125B**, 308 (1983)

Hasenfratz P., Proc. of the XXIII Int. Conf. on High Energy Physics Berkeley (1986)

Hasenfratz P., V. Laliena and F. Niedermayer, Phys. Lett. **B427** 125 (1998); P. Hasenfratz, Nucl. Phys. **B 525**,401 (1988)

Haymaker R.W. and J. Wosiek, Phys. Rev. Rapid Comm. **D36**, 3297 (1987); Acta Physica Polonica **B21**, 403 (1990); R.W. Haymaker, Y. Peng, V. Singh and J. Wosiek, Nucl. Phys. B (proc. Suppl) **17**, 558 (1990); R.W. Haymaker; V. Singh and Wosiek, Nucl. Phys. B (Proc. Suppl) **20**, 207 (1991)

Haymaker R. and J. Wosiek, Phys. Rev. **43**, 2676 (1991); R. Haymaker, V. Singh, D. Browne and J. Wosiek, *Paris Conference on the QCD Vacuum (1992)*

Haymaker R., V. Singh and Y. Peng, Phys. Rev. **D53**, 389 (1996)

Heller U. and F. Karsch, Nucl. Phys. **B251** [FS13], 254 (1985) Nucl. Phys. **B258**, 29 (1985)

Heller U.M., F. Karsch and J. Rank, Phys. Lett. **B355**, 511 (1995)

Hernandez P., K. Jansen and M. Lüscher, Nucl. Phys. **B552**, 363 (1999)
Hüfner J., Y. Kurihara and H.J. Pirner, Phys. Lett. **215B**, 218 (1988)
Ichie H., V. Bornyakov, T. Streuer and G. Schierholz, Nucl. Phys. **A721**, 899c (2003)
Imry Y., Phys. Rev. **B21**, 2042 (1980)
Ingelfritz E.M., M.L. Laursen, M. Müller-Preußker, G. Schierholz and H. Schiller, Nucl. Phys. **B268**, 693 (1986)
Irbäck A., P. Lacock, D. Miller and T. Reisz, Nucl. Phys. **B363**, 34 (1991)
Ishikawa K., A. Sato, G. Schierholz and M. Teper, Z. Physik **C21**, 167 (1983)
Itzykson C. and J.-M. Drouffe, *Statistical Field Theory*, Cambridge Monographs on Mathematical Physics, Vol 1 (1989)
Iwasaki Y. and T. Yoshie, Phys. Lett. **216B**, 387 (1989); see also Y. Iwasaki, Nucl. Phys. B (Proc. Suppl.) **9**, 254 (1989)
Ji X., Phys. Rev. **D52**, 271 (1995)
Jolicoeur T., A. Morel and B. Petersson, Nucl. Phys. **B274**, 225 (1986).
Jones D.R.T., Nucl. Phys. **B75**, 531 (1974)
Joos H. and I. Montvay, Nucl. Phys. **225**, 565 (1983)
Kajantie K., C. Montonen and E. Pietarinen, Z. Physik **C9**, 253 (1981)
Kajantie K. and J. Kapusta, Phys. Lett. **110B**, 299 (1982)
Kapusta J.I., Nucl. Phys. **B148**, 461 (1979)
Kapusta J., *Finite-Temperature Field Theory*, Cambridge Monographs on Mathematical Physics, Cambridge University Press (1989)
Kapusta J., Phys. Rev. **D46**, 4749 (1992)
Karsch F. , Nucl. Phys. **B205** [$FS5$], 285 (1982)
Karsch F., J.B. Kogut, D.K. Sinclair and H.W. Wyld, Phys. Lett. **188B**, 353 (1987)
Karsch F., in *Quark Gluon Plasma* (World Scientific, 1990), ed. by R.C. Hwa
Karsten L.H. and J. Smit, Nucl. Phys. **B183**, 103 (1981)
Kaste P. and Rothe H.J., Phys. Rev. **D56**, 6804 (1997)
Kawai H., R. Nakayama, and K. Seo, Nucl. Phys. **B189**, 40 (1981)
Kennedy A.D., J. Kuti, S. Meyer and B.J. Pendleton, Phys. Rev. Lett. **54**, 87 (1985)
Kilcup G.W. and Sharpe, Nucl. Phys. **B283**, 497 (1987)
Kislinger M.B. and P.D. Morley, Phys. Rev. **D13**, 2771 (1976); Phys. Rep. **51**, 63 (1979)
Kluberg-Stern H., A. Morel, O. Napoly and B. Petersson, Nucl. Phys. **B220**

[*FS*8], 447 (1983)
Kluberg L., Nucl. Phys. **A488**, 613c (1988)
Kogut J.B. and L. Susskind, Phys. Rev. **D9**, 3501 (1974)
Kogut J.B. and L. Susskind, Phys. Rev, **D11**, 395 (1975)
Kogut J.B., Rev. Mod. Phys. **51**, 659 (1979)
Kogut J.B., R.P. Pearson, and J Shigemitsu, Phys. Lett. **98B**, 63 (1981); ibid, Phys. Rev. Lett. **43**, 484 (1979).
Kogut J.B., in *Recent Advances in Field Theory and Statistical Mechanics*, Les Houches (1982)
Kogut J.B., Rev. Mod. Phys. **55**, 775 (1983)
Kogut J.B., M. Stone, H.W. Wyld, W.R. Gibbs, J. Shigemitsu, S.H. Shenker and D.K. Sinclair, Phys. Rev. Lett. **50**, 393 (1983a)
Kogut J.B., H. Matsuoka, S. Shenker, J. Shigemitsu, D.K. Sinclair, M. Stone and H.W. Wyld, Phys. Rev. Lett. **51**, 869 (1983b)
Kogut J.B., H. Matsuoka, S. Shenker, J. Shigemitsu, D.K. Sinclair, M. Stone and H.W. Wyld, Nucl. Phys. **B225** [FS9] 93 (1983c)
Kogut J.B., *Proceedings of the International School of Physics Enrico Fermi* (1984).
Kogut J.B., J. Polonyi, H.W. Wyld, J. Shigemitsu, and D.K. Sinclair, Nucl. Phys. **B251** [*FS*13], 311 (1985)
Kogut J.B. and D.K. Sinclair, Phys. Rev. Lett. **60**, 1250 (1988)
Kogut J.B. and D.K. Sinclair, Nucl. Phys. **B344**, 238 (1990)
Kovacs E., Phys. Rev. **D25**, 871 (1982)
Kovacs T.G. and E.T. Tomboulis, Phys. Rev. **D57**, 4054 (1998)
Kovalenko A.V., M.I. Polikarpov, S.N. Syritsyn and V.I. Zakharov, Nucl. Phys. (Proc. Suppl.) **129 & 130**, 665 (2004)
Kraan T.C. and P. van Baal, Phys. Lett. **B435**, 389 (1998); Phys. Lett. **B428**, 268 (1998); Nucl. Phys. **533**, 627 (1998)
Kronfeld A.S., M.L. Laursen, G. Schierholz and U.-J. Wiese, Phys. Lett. **B198**, 516 (1987); A.S. Kronfeld, G. Schierholz and U.-J. Wiese, Nucl. Phys. **293**, 461 (1987)
Kronfeld A.S., J.K.M. Moriarty and G. Schierholz, Comp. Phys. Comm. **52**, 1 (1988)
Kuti J., J. Polonyi and K. Szlachanyi, Phys. Lett. **98B**, 199 (1981)
Landsman, N.P., and Ch. G. van Weert, Phys. Rep. **145**, 141 (1987)
Langfeld K., Phys. Rev. **D69**, 014503 (2004)
Lautrup B., and M. Nauenberg, Phys. Lett. **95B**, 63 (1980)

Learmann E., R. Altmeyer, K.D. Born, W. Ibes, R. Sommer, T.F. Walsh and P.M. Zerwas, Nucl. Phys. B (Proc. Suppl.) **17**, 436 (1990)

Lee K. and P. Yi, Phys. Rev. **D56**, 3711 (1997)

Lee K. and C. Lu, Phys. Rev. **D58**, 025011 (1998)

Linde A.D., Phys. Lett. **96B**, 289 (1980)

Lüscher M., G. Münster and P. Weiss, Nucl. Phys. **B180** [FS2] 1 (1981)

Lüscher M., Lectures given at the Summer School on *Critical Phenomena, Random Systems and Gauge Theories*, Les Houches (1984)

Lüscher M., Lectures given at the Summer School on *Fields, Strings* and Critical Phenomena, Les Houches (1988)

Lüscher M., Phys.Lett. **B428**, 342 (1999a)

Lüscher M., Nucl. Phys. **B538**, 515 (1999b)

Mack G., in "Recent Developments in Gauge Theories", edited by 't Hooft et al. (Plenum, New York, 1980)

Mandelstam S., Phys. Rep. **23C**, 245 (1976)

Matsubara T., Progr. Theor. Phys. **14**, 351 (1955)

Matsubara Y., S. Ejiri and T. Suzuki, Nucl. Phys. **B34** (Proc. Suppl.) 176 (1994)

Matsui T. and H. Satz, Phys. Lett. **178B**, 416 (1986). For a review see H. Satz, Z. Physik **C62**, 683 (1994)

McLerran L.D. and B. Svetitsky, Phys. Rev. **D24**, 450 (1981)

Metropolis N., A.W. Rosenbluth, M.N. Rosenbluth, A.H. Teller and E. Teller, J. Chem. Phys. **21**, 1087 (1953)

Meyer B. and C. Smith, Phys. Lett. **B123**, 62 (1883)

Meyer-Ortmanns H., Rev. Mod. Phys. **68**, 473 (1996)

Michael C., Nucl. Phys. **B280** [FS18], 13 (1987)

Michael C. and M. Teper, Nucl. Phys. **B314**, 347 (1989).

Michael C., Nucl. Phys. (Proc. Suppl.) **17**, 59 (1990)

Michael C., Phys. Rev. **D53**, 4102 (1996)

Migdal A.A., ZETF **69**, 810 (1975)

Montvay I. and G. Münster, *Quantum Fields on a Lattice* (Cambridge University Press, 1994)

Morel A. and J.P. Rodrigues, Nucl. Phys. **B247**, 44 (1984).

Münster G., Nucl. Phys. **B180** [*FS*2], 23 (1981)

Nahm W., "*Selfdual Monopoles and Calorons*", Lecture Notes in Physics **201**, 189 (1984)

Nadkarni S., Phys. Rev. **D33**, 3738 (1986); **D34**, 3904 (1986)

Negele J.W., *Monte Carlo Methods for Hadronic Physics*; lectures presented at the *International Spring School on Medium and High Energy, Nuclear Physics*, Taiwan (1988)

Neuberger H., Phys. Lett. **B417**, 141 (1998); *ibit* **B427**, 353 (1998)

Nielsen H.B. and P. Olesen, Nucl. Phys. **B160**, 45 (1973)

Nielsen H.B. and P. Olesen, Nucl. Phys. **B61**, 380 (1979)

Nielsen H.B. and M. Ninomiya, Nucl. Phys. **185**, 20 (1981)

Niemi A.J. and G.W. Semenov, Ann. Phys. **152**, 105 (1984)

Niemi A.J. and G.W. Semenov, Nucl. Phys. **B230** [FS10], 181 (1984)

Norton R.E. and J.M. Cornwall, Ann. Phys. **91**, 106 (1975)

Ono S., Phys. Rev. **D17**, 888 (1978)

Osterwalder K. and R. Schrader, Comm. Math. Phys. **31**, 83 (1973); **42**, 281 (1975); See also K. Osterwalder in *Constructive Quantum Field Theory*, ed. by G. Velo and A.S. Wightman, Lecture Notes in Physics **25** (Springer, Berlin)

Parisi G. and Y.-S. Wu, Sci. Sin.**24**, 483 (1981)

Patel A., Nucl. Phys. **B243**, 411 (1984)

Peskin M.A., SLAC-PUB-3273 (1983), 11'th SLAC Summer Institute

Peskin M., Cornell University preprint **CLNS 395** (1978), thesis

Petcher D.N. and D.H. Weingarten, Phys. Lett. **99B**, 333 (1981)

Pietig R., Master thesis (1994) unpublished

Pisarski R.D. and F. Wilczek, Phys. Rev. **D29**, 338 (1984)

Politzer H.D., Phys. Rev. Lett. **30**, 1346 (1973)

Politzer H.D., Phys. Rep. **14C**, 129 (1974)

Polonyi J., J.B. Kogut, J. Shigemitsu, D.K. Sinclair, and H.W. Wyld, Phys. Rev. Lett. **53**, 644 (1984)

Polyakov A.M., Phys. Lett. **72B**, 477 (1978)

Polyakov A.M,," *Gauge Fields and Strings*", Harwood Academic Publishers, (1987)

Prasad M.K. and C.M. Sommerfield, Phys. Rev. Lett. **35**. 760 (1975); E.B. Bogomol'nyi, Sov. J. Nucl. Phys. **24**, 449 (1976)

Quigg C. and J.L. Rosner, Phys. Rep. **56**, 167 (1979)

Rafelski J., in " *Workshop on Future Relativistic Heavy Ion Experiments*, GSI-Report 81-6 (1981); see also Phys. Rep. **88**, 331 (1982)

Reinhardt H., Phys. Lett. **B557**, 317 (2003)

Reisz T., Commun. Math. Phys. **117**, 79 (1988a); **117**, 639 (1988b); **116**, 81 (1988c)

Reisz T. and H.J. Rothe, Phys. Lett. **B455**, 246 (1999)
Reisz T. and H.J. Rothe, Phys. Rev. **D62**, 014504 (2000)
Roepstorff G., *Path Integral Approach to Quantum Physics. An introduction* Springer Verlag (1994)
Rossi P. and U. Wolff, Nucl. Phys. **B248**, 105 (1984)
Rothe H.J., *An introduction to lattice Gauge Theories*, Proceedings of the *Autumn College on Techniques in Many-Body Problems*, ed. by. A. Shaukat (World Scientific, 1989)
Rothe H.J., Phys. Lett. **B355**, 260 (1995a)
Rothe H.J., Phys. Lett. **B364**, 227 (1995b)
Rothe H.J., Phys. Rev. **D52**, 332 (1995)
Rothe H.J. and N. Sadooghi, Phys. Rev. **D58**, 074502 (1998).
Rusakov B. Ye, Mod. Phys. Lett. **A5**, 693 (1990)
Ruuskanen P.V., in Proc. of Quark Matter '91, Gatlinburg, edited by V. Plasil et al., Nucl. Phys. **A544** (1992)
Satz H., in "Quark Gluon Plasma", ed. by R.C. Hwa (World Scientific, 1990)
Satz H., in Proc. of Int. Lepton-Photon Symp., Geneve (1991)
Scalapino D.J. and R.L. Sugar, Phys. Rev. Lett. **46**, 519 (1981)
Scalettar R.T., D.J. Scalapino and R.L. Sugar, Phys. Rev. **B34**, 7911 (1986).
Schafer T. and E.V. Shuryak, Rev. Mod. Phys. **70**, 323 (1998)
Schierholz G., Nucl. Phys. B (Proc. Suppl.) **4**, 11 (1988)
Schierholz G., Nucl. Phys. B (proc. Suppl.) **9**, 244 (1989)
Seiler E., in *Gauge Theories as a Problem of Constructive Quantum Field Theory and Statistical Mechanics*, Springer Verlag, Berlin, Heidelberg, New York (1982), and references quoted there
Sharatchandra H.S., H.J. Thun, and P.Weisz, Nucl. Phys. **B192**, 205 (1981)
Shiba H. and T. Suzuki, Phys. Lett. **B 333**, 461 (1994)
Shifman M.A., *Particle Physics and Field Theory*, World Scientific Vol. 62, 1999
H. Shiba, T. Suzuki, PL **B333**, 461 (1994)
Singh V., R.W. Haymaker and D.A. Browne, Phys. Rev. **D47**, 1715 (1993a)
Singh V., D.A. Browne and R.W. Haymaker, Phys. Lett. **B306**, 115 (1993b)
Sommer R. and J. Wosiek, Phys. Lett. **B149**, 497 (1984); Nucl. Phys. **B267**, 531 (1986)
Sommer R., Nucl. Phys. **B291**, 673 (1987); **B306**, 180 (1988).
Soper D.E., Phys. Rev. **D18**, 4590 (1978)
Stack J.D., Phys. Rev. **D29**, 1213 (1984)

Stack, J.D., S.D. Neiman and R.J. Wensley, Phys. Rev. **D50**, 3399 (1994)
Stamatescu I.O., Phys. Rev. **D25**, 1130 (1982)
Susskind L., in *Weak and Electromagnetic Interactions at High Energy*, Les Houches (1976)
Susskind L., Phys. Rev. **D16**, 3031 (1977)
Susskind L., Phys. Rev. **D20**, 2610 (1979)
Suzuki T. and I. Yotsuyanagi,Phys. Rev. D 42, 4257 (1990)
Svetitsky B. and L.G. Yaffe, Nucl. Phys. **B210** [*FS*6], 423 (1982);
Svetitsky B. and F. Fucito, Phys. Lett. **131B**, 165 (1983)
Svetitsky B., Phys. Rep. **132**, 1 (1986)
Symanzik K., Commun. Math. Phys. **18**, 227 (1970)
Symanzik K., in *Mathematical Problems in Theoretical Physics*, eds. R. Schrader et al. (Springer, Berlin, 1982)
Symanzik K., Nucl. Phys. **B226**, 187 and 205 (1983)
Takahashi T.T., H. Matsufuru, Y. Nemoto and H. Suganuma, Phys. Rev. Lett. **86**, 18 (2001)
Takahashi T.T., H. Suganuma, Y. Nemoto and H. Matsufuru, Phys. Rev. **D65**, 114509 (2002)
Takahashi T.T., H. Suganuma, H. Ichie, H. Matsufuru and Y. Nemoto, Nucl. Phys. **A721**, 926c (2003)
Takahashi T.T., H. Suganuma and H. Ichie, Conf. Proc., *Color Confinement and Hadrons in Quantum Chromodynamics*, World Scientific (2004)
Teper M., Phys. Lett. **B171**, 81 (1986); ibit **183B**, 345 (1986)
Teper M., Phys. Lett. **183B**, 345 (1986a)
't Hooft G., unpublished comments at the conference in Marseille (1972)
't Hooft G., in "High Energy Physics", Ed. A. Zichichi (Editrice Compositori; Bologna (1976)
't Hooft G., Nucl. Phys. **B138**, 1 (1978)
't Hooft G., Nucl. Phys. **B190** [FS3], 455 (1981)
't Hooft G., Comm. Math. Phys. **81**, 267 (1981a)
Thacker H.B., E. Eichten and J.C. Sexton, Nucl. Phys. (Proc. Suppl.) **B4**, 234 (1988)
Tinkham M., *Introduction to Superconductivity* (McGraw-Hill (1996))
Toimela T., Z. Physik **C27**, 289 (1985)
Toussaint D., *Introduction to Algorithms for Monte Carlo Simulations*

and their Application to QCD; lectures presented at the *Symposium on New Developments in Hardware and Software for Computational Physics*, Buenos Aires (1988); Comp. Phys. Commun. **56**, 69 (1989)

van Baal P. and A.S. Kronfeld, Nucl. Phys. B (Proc. Suppl.) **9**, 227 (1989)

van Baal P., Nucl. Phys. B (Proc. Suppl.) **63**, 126 (1998)

van Baal P., hep-th/9912035 (1999)

van den Doel and J. Smit, Nucl. Phys. **B228**, 122 (1983)

Wantz O., Diploma thesis, University of Heidelberg (2003)

Wegner F.J., J. Math. Phys. **12**, 2259 (1971).

Weinberg S., Phys. Rev. **D9**, 3357 (1974)

Wetzel W., Nucl. Phys. **B255**, 659 (1985)

Wilson K.G., Phys. Rev. **D10**, 2445 (1974)

Wilson K.G., "New Phenomena in Subnuclear Physics" (Erice, 1975), ed. A. Zichichi (New York, Plenum, 1975)

Wolff U., Phys. Lett. **153B**, 92 (1985)

Wosiek J. and R. Haymaker, Phys. Rev. **D36**, 3297 (1987); J. Wosiek Nucl. Phys. **B**(Proc. Suppl.) **4**, 52 (1988)

Yang C.N. and R. Mills, Phys. Rev. **96**, 191 (1954)

Yoshie, T., Y. Iwasaki and S. Sakai, Nucl. Phys. B (proc. Suppl.) **17**, 413 (1990)

INDEX